国外优秀数学著作
原版丛书（第三十辑）

数学物理精选专题讲座
——李理论的进一步应用
（英文）

Lectures on Selected Topics in Mathematical Physics—Further Applications of Lie Theory

［美］威廉·A. 施瓦姆（William A. Schwalm）著

HITP
哈爾濱工業大學出版社
HARBIN INSTITUTE OF TECHNOLOGY PRESS

黑版贸登字 08-2021-049 号

Lectures on Selected Topics in Mathematical Physics: Further Application of Lie Theory

图书在版编目(CIP)数据

数学物理精选专题讲座:李理论的进一步应用=
Lectures on Selected Topics in Mathematical
Physics: Further Applications of Lie Theory:英文/
(美)威廉·A. 施瓦姆(William A. Schwalm)著. —
哈尔滨:哈尔滨工业大学出版社,2024. 10
(国外优秀数学著作原版丛书. 第三十辑)
ISBN 978-7-5767-1340-4

Ⅰ. ①数…　Ⅱ. ①威…　Ⅲ. ①数学物理方法-英文
Ⅳ. ①O411. 1

中国国家版本馆 CIP 数据核字(2024)第 073693 号

SHUXUE WULI JINGXUAN ZHUANTI JIANGZUO: LI LILUN DE JINYIBU YINGYONG

策划编辑　刘培杰　杜莹雪
责任编辑　刘家琳　张嘉芮
封面设计　孙茵艾
出版发行　哈尔滨工业大学出版社
社　　址　哈尔滨市南岗区复华四道街 10 号　邮编 150006
传　　真　0451-86414749
网　　址　http://hitpress. hit. edu. cn
印　　刷　哈尔滨圣铂印刷有限公司
开　　本　787 mm×1 092 mm　1/16　印张 45. 5　字数 641 千字
版　　次　2024 年 10 月第 1 版　2024 年 10 月第 1 次印刷
书　　号　ISBN 978-7-5767-1340-4
定　　价　252. 00 元(全 4 册)

(如因印装质量问题影响阅读,我社负责调换)

This volume is dedicated to three of our teachers, John Hermanson, William Kinnersley and Mark Peterson.

Contents

Preface

My intended audience includes beginning graduate students in physics taking a first-year graduate course in methods of theoretical physics and practicing physicists who find a need to become acquainted with Lie theory. Some basic ideas were presented in an earlier volume of this series, where the application was mostly to the analysis and solution of ordinary differential equations. As before, the presentation will be informal and is intended for physics students. It assumes some familiarity with certain physical ideas, such as quantum mechanics, and some common vocabulary. In addition to a knowledge of calculus and familiarity with basic differential equations and physics, the reader is assumed to be familiar with or have access to such material on Lie groups as is presented in the previous volume.

Acknowledgements

I want to acknowledge support and a great deal of discussion and input from my wife and colleague, Mizuho Schwalm. Where the pronouns we or us appear without another obvious antecedent, they refer to her and me. I am pleased to acknowledge significant conversations with her and with my other colleagues on some of the material in this volume, particularly Jerry Brooks, Anthony Bevelacqua, Jerry Metzger and Yen Lee Loh.

We are grateful to our teachers. In 1970 we both were enroled in an undergraduate math course on Lie groups given by Professor Philip Gold of the Mathematics Department at Portland State University. We are indebted to Dr Gold, no doubt, for many ideas. We should like to thank, as well, Dr William Kinnersley for his lectures on a variety of topics including generating functions. We thank Dr Mark Peterson and especially Dr John Hermanson of the Physics Department at Montana State University for stimulating courses on groups and other theoretical methods. Particularly, our dissertation advisor Dr Hermanson taught a very useful summer course in group representations and quantum mechanics.

Thanks to my students working through certain parts of the lectures and making suggestions occasionally.

Finally, we must thank the University of North Dakota for use of the facilities and for a challenging work environment.

Author Biography

William A Schwalm

Dr William A Schwalm has been in the Department of Physics and Astrophysics at the University of North Dakota since 1980. His research is in condensed matter theory and application of mathematical methods to physical problems. He has taught lots of different physics courses at all levels.

Current research involves application of Lie groups to finding generating functions for the stationary states of quantum systems, and also applying them to decoupling discrete dynamical systems. Another area of active interest is in finding Green functions for certain classes of lattice problems involving electron transport, vibrations and other collective excitations.

Dr Schwalm has received two outstanding teaching awards, the University of Utah Physics Outstanding Undergraduate Instructor (1979) and the McDermott Award for Excellence in Teaching, UND (1995). He is also the author of *Lectures on Selected Topics in Mathematical Physics: Elliptic Functions and Elliptic Integrals*, and *Lectures on Selected Topics in Mathematical Physics: Introdution to Lie Theory with Applications*.

Introduction

This volume is devoted mostly to Lie groups, Lie algebras and generating functions, both for standard special functions and for solution of certain types of physical problems. It is by no means complete or mathematically rigorous. It is an informal treatment of these topics intended for physics graduate students or others with a physics background wanting a brief and informal introduction to the subjects addressed in a style and vocabulary not completely unfamiliar. It is a sequel to a previous volume in this series [1] devoted to Lie groups and differential equations.

In the development presented below, the topic of Lie continuous groups, also treated in the previous volume, is reviewed in some detail, particularly the construction and use of Lie's infinitesimal generators for local groups. Examples include the Euclidean group, the special conformal subgroup in two dimensions, groups of Cremona transformations, quaternions, higher dimensional rotation groups, and several particular examples of continuous symmetries in specific problems. These are used, for example, to construct generating functions for problems such as the mechanical Green functions for vibrations of a standard chain of springs and weights. The symmetry groups allow one to navigate between symmetry-related problems and from one formal result to another, or sometimes from a special case to a general solution. Also treated briefly is the important problem complementary to that of finding group generators, namely the problem of, when given a set of generators, reconstructing the local Lie group.

An outstanding characteristic of group generators is that as operators they do not generally commute with one another. With respect to commutators the vector space of generators becomes a Lie algebra. The commutators of quantum operators or Poisson brackets of mechanical or thermodynamical quantities form systems with Lie structure. The topics of associative and non-associative algebras, commutators and the Lie bracket product, matrix representations and representations as differential operators, and classification of Lie algebras are dealt with in somewhat more detail than is common in physics books. Although the treatment is quite consistently an informal, conversational one, it is hoped that the discussion will be general enough to uncover some underlying structure and to be a useful introduction to a more abstract treatment.

Particularly, the notions of a Cartan subalgebra, of root vectors and roots, and some discussion of how to construct root vectors and their relationship to ladder operators in certain physical problems are all dealt with in some detail and illustrated with examples. Every Lie algebra has a Cartan subalgebra. In a semi-simple Lie algebra this is a set of mutually commuting vectors that is maximal in some sense. In the context of quantum theory, for instance, where the Lie brackets are commutators, one can think of a collection of mutually commuting operators that thus represent simultaneously measurable physical quantities. In the quantum system one may be able to find a set of ladder operators, or raising and lowering operators, that transform from one mutual eigenfunction of the commuting operators to another, as do the raising and lowering operator in the simple oscillator problem or in the case of angular momentum. Or in classical mechanics, the Cartan subalgebra may represent a set of mutually conserved quantities such that the Poisson brackets vanish in pairs. The following material generalizes these ideas. The root vectors to be introduced below are abstractions of the ladder operators in the context of a Lie algebra. A basic strategy for finding the root vectors, or roughly for finding ladder operators for elementary excitations of a system, is developed and illustrated in some detail. Examples treated include, for instance, the algebra $so(5)$ of five-dimensional rotations and the two-dimensional harmonic oscillator with applied magnetic field. What is not dealt with here, and is often addressed in other books, is the very important, useful topic of reducibility of matrix representations. It was decided to focus instead in this short tract on a few topics and ideas less apt to be treated at an elementary level elsewhere.

The concept of a generating function, both for manipulating special functions and as a means of solving physical problems, is a reoccurring theme. It is often easier to treat the general case of an infinite set of applications or to find all the solutions in an infinite class of problems, or all the matrix elements for a certain operator and so on, than it is to treat a few special cases. In the treatment below Lie groups are consistently used to deform a trivial solution for a given problem into a generating function for an infinite set of solutions. For instance, the Coulombic potential for an isolated point charge is deformed by continuous group transformations, first into a generating function for Legendre polynomials and then via the Lie group $SO(3)$ into generating functions for the whole set of spherical harmonics. For another example, the ground state wave function for the one-dimensional oscillator is transformed, in a way illustrating Weisner's group theoretic method, to a generating function for all the eigenfunctions. The generating functions in this case turn out to be the coherent states. Weisner's method of extending the action of a Lie group also produces a generating function for the angular momentum-resolved eigenstates of a charged harmonic oscillator in a magnetic field, which in turn gives a matrix element generating function with which other problems are addressed. As an illustrative example, these matrix elements are used in a sketch of a variational study of a charged particle moving in a sombrero potential under the influence of a magnetic field.

Reference

[1] Schwalm W 2017 *Lectures on Selected Topics in Mathematical Physics: Introduction to Lie Theory with Applications IOP Concise Physics* (San Rafael, CA: Morgan & Claypool)

William A Schwalm

Chapter 1

Generating functions

1.1 The basic idea

Before reviewing or applying some of the material on groups, it is useful to have on hand some rudimentary ideas about generating functions. These functions appear in several places. First, they appear in counting or enumeration problems. For instance, partition functions are generating functions that enumerate microstates in a thermostatical system. Green functions are generating functions that count or enumerate paths between points in a space, where the space may be somewhat abstract. On the other hand, a generating function is a convenient way to handle a set of special functions, or the set of eigensolutions to some boundary value problem, by stringing the discrete set of functions together like beads into a single function with one or more, extra 'counting variables.' The resulting single generating function is often much easier to work with than the original discrete set. It facilitates calculating matrix elements or sums of matrix elements or for extracting properties, often much more easily.

An elementary introduction to the generating function concept directed toward undergraduate physics students is our paper of some time ago in the *American Journal of Physics* [1] which treats both counting and special functions applications in basic physics problems. A useful introduction to group theoretic methods for finding generating functions is McBride [2], especially the chapters on Weisner's method and Truesdell's method. Weisner's paper [3] is readable and interesting also. A very complete, mathematically rigorous and detailed treatment of special functions, factorization and generating functions is [4].

There are two common uses for a power series. One use is as an asymptotic approximant to a function near some particular value of its argument, for instance near zero. But the other use is of more interest in the work presented here, and it is the following.

One can look at a power series as a one-to-one correspondence between an analytic function and the set of its Taylor coefficients. The Taylor coefficients can be

doi:10.1088/2053-2571/aae789ch1

numbers or can be functions of another variable. For instance, if $\{f_n(t)\}$ is a set of functions such that the series

$$F(t, s) = \sum_n A_n s^n f_n(t) \tag{1.1}$$

converges for some range of t with s inside some circle of convergence, where $\{A_n\}$ is a set of specially chosen, particular coefficients depending only on n, then this relation provides a correspondence between the family $\{f_n(t)\}$ and the generating function $F(t, s)$. One importance of this is that hidden properties of the family $\{f_n(t)\}$ can be exposed by manipulating $F(t, s)$. It turns out that, in the context of this use of the series for manipulating the coefficients, convergence is often not very important. Equation (1.1) thus defines a generating function for this family of functions.

1.2 Elementary examples of generating functions

As a simple example consider

$$F(s) = e^s = \sum_{n=0}^{\infty} \frac{1}{n!} s^n.$$

One may think of this as a generating function for $1/n!$. As such, it contains facts about $1/n!$ that may not be obvious. It is useful to observe that the range need not be delimited so carefully in this case, since $n!$ is infinite for a negative integer n. This is a general observation about factorial quantities appearing in denominators. So I write simply

$$F(s) = \sum_n \frac{1}{n!} s^n.$$

Then,

$$e^{2s} = F(s)^2 = \sum_m \sum_n s^{m+n} \frac{1}{m!n!} = \sum_{k=0}^{\infty} s^k \sum_n \frac{1}{n!(k-n)!}$$

and

$$e^{2s} = \sum_k \frac{2^k}{k!} s^k.$$

Comparing, one finds

$$\sum_n \frac{1}{n!(k-n)!} = \frac{2^k}{k!}.$$

Truly speaking, I would not have thought of this identity, although in retrospect it is an elementary property of binomial coefficients. The point is that a generating function can uncover such things easily.

Two fairly common examples of generating functions are

$$F(t, s) = \exp\left(\frac{x}{2}\left(s - \frac{1}{s}\right)\right) = \sum_{m=-\infty}^{\infty} s^m J_m(t), \tag{1.2}$$

for the Bessel functions of the first kind, $\{J_m(t)\}$ and

$$F(t, s) = \exp\left(2st - s^2\right) = \sum_{n=0}^{\infty} \frac{1}{n!} s^n H_n(t), \tag{1.3}$$

for the Hermite polynomials $\{H_n(t)\}$. Notice in the first case, the summation on m is from $-\infty$ to ∞ and that in the second case there is a non trivial $A_n = 1/n!$.

Putting aside for the moment the question of how these generating functions arise, I can deduce some interesting properties of the functions they generate. For instance, from the Bessel generator,

$$F(s, t) = \exp\left(\frac{t\, s}{2}\right) \exp\left(\frac{-t}{2\, s}\right) = \sum_{n,k} \frac{(-1)^k s^{n-k}}{n!k!} \left(\frac{t}{2}\right)^{n+k}.$$

Now let $n - k = m$ so that $n = m + k$,

$$F(s, t) = \sum_{m=-\infty}^{\infty} s^m J_m(t) = \sum_{m=-\infty}^{\infty} s^m \left(\frac{t}{2}\right)^m \sum_{k=0}^{\infty} \frac{(-1)^k}{k!(k + m)!} \left(\frac{t}{2}\right)^{2k}.$$

This gives the Frobenius expansion for the mth Bessel function of the first kind,

$$J_m(t) = \left(\frac{t}{2}\right)^m \sum_{k=0}^{\infty} \frac{(-1)^k}{k!(k + m)!} \left(\frac{t}{2}\right)^{2k}.$$

Another expression for $J_m(t)$ comes from the residue theorem applied to the essential non-branching singularity of $F(t, s)$ at $s = 0$. One has, by integrating right-handedly around 0 in the s plane,

$$\oint \frac{1}{s^{m+1}} F(t, s)\, ds = 2\pi i\, J_m(t) = \oint \frac{1}{s^{m+1}} \exp\left(\frac{t}{2}\left(s - \frac{1}{s}\right)\right) ds.$$

Now make the Fourier substitution $s = e^{i\theta}$, noting that $ds = i\, s\, d\theta$:

$$\begin{aligned} J_m(t) &= \frac{1}{2\pi} \int_{-\pi}^{\pi} e^{-i\, m\, \theta} \exp\left(\frac{t}{2}(e^{i\theta} - e^{-i\theta})\right) d\theta. \\ &= \frac{1}{2\pi} \int_{-\pi}^{\pi} \exp i\, (t \sin\theta - m\, \theta)\, d\theta. \end{aligned}$$

The exponent factor $(t \sin\theta - m\,\theta)$ is an odd function of θ. Using Euler's identity, the integrand has a real cosine and an imaginary sine part. The sine part is altogether an odd function of θ and the cosine part is even, so the integration becomes

$$J_m(t) = \frac{1}{\pi} \int_0^{\pi} \cos(t \sin\theta - m\,\theta)\, d\theta.$$

This expression leads to the asymptotic expansions of the Bessel functions at large t and so on.

It is easy to derive other properties from this Bessel function generator. For instance,

$$F(x, s)F(y, s) = \exp\left(\frac{x+y}{2}\left(s - \frac{1}{s}\right)\right) = \sum_m s^m J_m(x+y),$$

and on the other hand,

$$F(x, s)F(y, s) = \sum_{k,n} s^{k+n} J_k(x) J_n(y) = \sum_m s^m \sum_n J_{m-n}(x) J_n(y),$$

so that

$$J_m(x+y) = \sum_{n=-\infty}^{\infty} J_{m-n}(x) J_n(y).$$

Turning to the generating function for the Hermite polynomials, I find, for example, that

$$\int_{-\infty}^{\infty} F(t, s) \exp(-t^2) F(t, u)\, dt = \int_{-\infty}^{\infty} e^{-s^2+2st-t^2+2tu-u^2} dt = \sqrt{\pi}\, e^{2su},$$

where the integration is done easily by completing the square in the exponent. So,

$$\sum_{m,n} \frac{s^m u^n}{m!\, n!} \int_{-\infty}^{\infty} H_m(t)\, e^{-t^2} H_n(t)\, dt = \sum_k \frac{1}{k!} 2^k s^k u^k.$$

The resulting expression is a generating function for a set of inner product integrals involving pairs of polynomials. Evidently,

$$\int_{-\infty}^{\infty} H_m(t) e^{-t^2} H_n(t)\, dt = \delta_{m,n}(n!\, 2^n).$$

One sees that it is because the generating function integral results in a function of the product su, and not of each one separately, that the H inner product integrals vanish unless $m = n$. No such result appears for the Bessel functions, since the Bessel functions are not orthogonal in this way.

One more example will show how this generating function idea applies directly to a physical problem [1]. Figure 1.1 shows a portion of an infinite line of springs and weights. The masses are all the same m and the spring constants are all κ. For the nth one, the displacement from equilibrium is x_n. The motion is along the direction of the chain in each case.

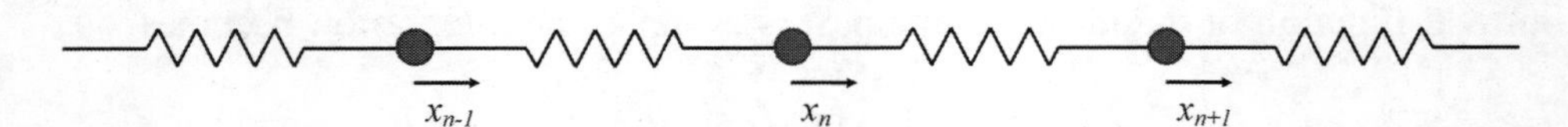

Figure 1.1. Section of an infinite line of springs and weights.

The normal way to begin analyzing a problem like this is to notice that the equilibrium state of the physical system has a discrete translation symmetry, and then to use this symmetry to find normal modes of vibration, and finally to use the symmetry and the normal modes to calculate other properties. This procedure is quite general, and it will be developed in later sections of the book. However, in order to cultivate another kind of insight, I choose to formulate a problem that is solvable directly by finding a generating function.

The motion of each weight is governed by Newton's second law, the force being due to the springs,

$$m\,\frac{d^2}{dt^2}x_n = -\kappa(x_n - x_{n+1}) - \kappa(x_n - x_{n-1}).$$

For simplicity, suppose all the weights are initially at equilibrium, except that $x_0 = a$ with all masses at rest. Notice that this initial condition with one weight displaced breaks the translation symmetry of the equilibrium state. The generating function

$$G(t,\,s) = \sum_{n=-\infty}^{\infty} s^m\,x_n(t),$$

will be convenient. For the initial condition chosen, $G(0,\,s) = a$, and the initial time derivative vanishes. Multiply the differential equation for $x_n(t)$ by s^n and sum on n:

$$m\sum_n s^n \frac{d^2}{dt^2}x_n = -2\,\kappa\sum_n s^n\,x_n + \kappa\sum_n s^n\,(x_{n-1} + x_{n+1}).$$

The two terms in the last summation can be simplified and combined, moving n first up by 1, and then down by 1,

$$\cdots\sum_n s^n\,(x_{n-1} + x_{n+1}) = \cdots\sum_n s^n\left(s\,x_n + \frac{1}{s}x_n\right),$$

so that the differential equation for the generating function $G(t,\,s)$ becomes

$$m\frac{\partial^2 G}{\partial t^2} = \kappa\left(s - 2 + \frac{1}{s}\right)G.$$

This is really an ordinary differential equation in the independent variable t where s appears as a parameter. Making use of the identity

$$\left(q - \frac{1}{q}\right)^2 = \left(q^2 - 2 + \frac{1}{q^2}\right),$$

its general solution is

$$G(t,\,s) = A(s)\cosh\left(\left(\sqrt{s} - \frac{1}{\sqrt{s}}\right)\sqrt{\frac{\kappa}{m}}\;t\right) + B(s)\sinh\left(\left(\sqrt{s} - \frac{1}{\sqrt{s}}\right)\sqrt{\frac{\kappa}{m}}\;t\right).$$

The initial conditions $G(0, s) = a$ and $(\partial G(t, s)/\partial t)_0 = 0$ require that $A(s) = a$ and $B(s) = 0$:

$$G(t, s) = a \cosh\left(\left(\sqrt{s} - \frac{1}{\sqrt{s}}\right)\sqrt{\frac{\kappa}{m}}\, t\right).$$

To compare with the Bessel function generator of equation (1.3), set

$$\left(\sqrt{s} - \frac{1}{\sqrt{s}}\right)\sqrt{\frac{\kappa}{m}}\, t = \frac{\tau}{2}\left(y - \frac{1}{y}\right).$$

Thus,

$$\tau = 2\sqrt{\frac{\kappa}{m}}\, t \qquad \text{and} \qquad y = \sqrt{s}.$$

The hyperbolic cosine separates into two exponential parts, and the two exponential parts each match a Bessel generator, thus

$$\begin{aligned} G(t, s) &= a \cosh\left(\left(\sqrt{s} - \frac{1}{\sqrt{s}}\right)\sqrt{\frac{\kappa}{m}}\, t\right) \\ &= \frac{a}{2} F\left(2\sqrt{\frac{\kappa}{m}}\, t, \sqrt{s}\right) + \frac{a}{2} F\left(2\sqrt{\frac{\kappa}{m}}\, t, -\sqrt{s}\right). \end{aligned}$$

Expanding,

$$\begin{aligned} G(t, s) = &\frac{a}{2}\left(\sum_n s^{n/2} J_n\left(2\sqrt{\frac{\kappa}{m}}\, t\right)\right) \\ &+ \frac{a}{2}\left(\sum_n (-1)^n s^{n/2} J_n\left(2\sqrt{\frac{\kappa}{m}}\, t\right)\right). \end{aligned}$$

The odd powers of $\sqrt{s}$ cancel:

$$G(t, s) = a \sum_n s^n J_{2n}\left(2\sqrt{\frac{\kappa}{m}}\, t\right).$$

The motion of the nth particle in the chain is given as a function of time by

$$x_n(t) = a\, J_{2n}\left(2\sqrt{\frac{\kappa}{m}}\, t\right).$$

These examples should give at least some indication that the generating functions can be useful, if they are known. How to find a generating is a topic of interest. The answer is not always the same, and the methods are not routine. Sometimes group theory can be helpful, although it is sometimes said that clever series manipulation may often be the most fruitful approach. Notwithstanding, the emphasis in the current monograph will be on applying Lie groups to find generating functions.

References

[1] Schwalm W and Schwalm M 1998 Generating functions in physics *Am. J. Phys.* **51** 230–5

[2] McBride E B 1971 *Obtaining Generating Functions* (Berlin: Springer)

[3] Weisner L 1955 Group-theoretic origin of certain generating functions *Pac J. Math.* **5** 1033–9

[4] Miller W 1968 *Lie Theory and Special Functions* (New York: Academic)

Chapter 2

Groups

2.1 Introduction

This section summarizes very briefly first the basic idea of a group and then of a one-parameter Lie group. The interested reader may find more material treated in somewhat greater depth but written in the same, informal style in a previous volume in this series [1].

Background material on groups in a general physics context can be found in the popular book by Hamermesh [5]. Application of continuous symmetry groups to differential equations was described originally by Sophus Lie, for instance in German in [7, 8]. My introduction to this subject was through a course on Lie theory applied to differential equations based on Abraham Cohen's very accessible undergraduate text [9]. This is a good reference especially for physics or engineering students. It may be read online via the Cornell University Library Historical Math Monographs [10]. The more recent book by Olver [11] is more rigorous, very readable, and contains a very complete and useful bibliography. General material on Lie groups in physics, especially semi-simple groups and their algebras, is found in [18] and a descriptive outline is given in [17].

2.2 Groups in general and finite groups in particular

A *group* is a mathematical system with one operation $\cdot$, which may or may not be commutative. A group then consists of a set S of elements and the operation $\cdot$ such that

1. (closure) If $a \in S$ and $b \in S$ then $a \cdot b \in S$ and also $b \cdot a \in S$,
2. (associativity) If a, b and c belong to S, then $a \cdot (b \cdot c) = (a \cdot b) \cdot c$,
3. (existence of identity) There is a particular element e in S such that, for each $a \in S$, $e \cdot a = a$ and $a \cdot e = a$,
4. (existence of inverse elements) For each $a \in S$ there is a unique element in S denoted by a^{-1} which has the property $a^{-1} \cdot a = e$ and also $a \cdot a^{-1} = e$.

doi:10.1088/2053-2571/aae789ch2

In general, a group need not be commutative or 'Abelean'. An Abelean group is one which, in addition to these four properties, has the property

1. (commutativity) For each a and b in S, $a \cdot b = b \cdot a$.

A group that is not Abelean is sometimes called non-Abelean to emphasize this fact.

Groups can be either finite or infinite, meaning there is either a finite or infinite number of elements in S. For finite groups, one can write a *Cayley table* to define the operation. This is a group multiplication table except that the redundant column and row for 'e times' and 'times e' are deleted to save space.

When studying finite groups in physics, it is usually in connection with either vibrations, quantum mechanics of a few particles, etc. Thus there are certain special objectives and one usually encounters topics in roughly the following order: subgroups, homomorphisms, representations, irreducible representations, the orthogonality theorem for representations, characters and orthogonality for characters, character tables, symmetry of eigenstates, selection rules etc. Some of these topics will be encountered in material to follow, sometimes in different order.

2.3 Continuous groups

The topics of Lie groups and their Lie algebras is introduced as brief sketches in the two general books on mathematical topics for physicists, [12, 13]. Cohen's book, already mentioned [9], is one of the best elementary introductions. The more complete, if less elementary, book by R Hermann [14] is also recommended.

In classical mechanics (other than small oscillations) in field theory, and often in quantum mechanics, one is apt to be interested in *continuous groups*, and especially *Lie groups*. A Lie group is an infinite group where the different group elements are indicated by different values of a continuous parameter, say α. Then, since clearly there can be no Cayley table, the result of 'multiplying' group elements together must be given by a formula relating values of the parameter. If one denotes group elements as g_α, where α denotes the parameter value, the parameter value specifies *which* group element, and one must have a formula,

$$\alpha_3 = \phi(\alpha_1, \alpha_2)$$

to define the product

$$g_{\alpha_3} = g_{\alpha_1} g_{\alpha_2} \equiv g_{\alpha_1} \cdot g_{\alpha_2},$$

where α_1, α_2 and α_3 in this notation are different values of the single parameter α. When we say a Lie group is a continuous group, it implies also that the function ϕ is continuous in each of its arguments. As a matter of fact, these functions are assumed to have convergent power series representations.

One can phrase this property explicitly in terms of mappings, and define a Lie group succinctly in the following way. Let $G = \{g_\alpha\}$ be the (uncountable) set of group elements. A formal definition is that a Lie group $(G, \cdot)$ is a group with operation $\cdot$ and G is a smooth manifold such that the maps

$$G \times G \to G \quad \text{taking} \quad (g_\alpha, g_{\alpha'}) \mapsto g_\alpha \cdot g_{\alpha'} \qquad \text{(product)}$$

and

$$G \to G \quad \text{taking} \quad g_\alpha \mapsto (g_\alpha)^{-1} \qquad \text{(inverse)}$$

are both smooth maps. A digression on some aspects of this definition, including a brief discussion of manifolds in physical language, is provided in the previous volume in this series [1].

A Lie group can depend on several parameters, rather than just one parameter α. In this case α becomes a set $\alpha = \{\alpha^1, \alpha^2, \ldots, \alpha^r\}$ for an r-parameter group. One should not really think of this as a vector, as it is not generally useful to think of componentwise addition. This extension to several parameters will be important, and it leads to the idea of a Lie algebra, which will be encountered in subsequent sections. But the focus will be on single-parameter groups for the moment.

The parameters $\{\alpha^1, \alpha^2, \ldots \alpha^m\}$ of the set α can be either real or complex. Given a group with real parameters, it may be of interest to extend it to one with complex parameters. On the other hand, a group with complex parameters may be restricted so that the parameters take on only real values. The dimension of the group is the number of independent parameters. So a group with complex parameters has a complex dimension equal to the number of independent complex parameters. The same group can sometimes be considered as a real-parameter group of higher dimension. The real dimension would be the number of real parameters. It is often easier to make general statements about the complex parameter case for several reasons. One reason is that the group manifold is apt to be more connected in the complex case. Often, though, in physical applications the groups that arise naturally are real-parameter Lie groups.

It is interesting that the requirement of associativity forces a restriction on the multiplication function ϕ in the case of a single real parameter such that one finds, on solving a partial differential equation implied by associativity, that such a one-parameter Lie group is always Abelean. This means that in fact one can parameterize it in such a way that

$$g_{\alpha_1} g_{\alpha_2} = g_{\alpha_1 + \alpha_2},$$

always. Notice here once again that α_1 and α_2 are two specific values of the same single parameter α. Hence, when parameterized in this way, one sees that

$$g_0 = e,$$

where e is the identity. The identity is normally taken as the origin in the parameter space, and the parametrization is usually assumed to be additive for one-parameter groups.

2.4 Group action and infinitesimal generators

In many instances, including the examples given below, groups are defined by the ways they act on points in space, transforming them into other points, or the way they act on functions of the coordinates in some coordinate system. Formally [12] a group action is a mapping from a group to a set of transformations on a set or a space, such as a function space. If $(G, \cdot)$ is a group and V is a set of points, then an action π of G on V is a mapping π from G to transformations on V. Thus it is a map

$$\pi : \qquad G \times V \to V \qquad (g, x) \mapsto \pi(g)x,$$

with the following two properties:

$$\pi(e)x = x,$$

where e is the identity in G, and

$$\pi(a \cdot b) = \pi(a)(\pi(b)x).$$

Notice that a group can be viewed as acting on itself. Usually, as long as no confusion is apt to arise, I will indicate the action of a group element g_α on $x \in V$ just by writing the group element to the left of x.

We will view the group elements usually as transformations that act on a space of functions, for instance functions of x, y and z, or in a general problem, functions of all the appropriate coordinates. In this section we encounter the infinitesimal group generators and the exponential maps that reconstruct from these generators the subgroup of a given Lie group that is continuously connected to the identity.

Each of the following is an example of a one-parameter real Lie group acting in three spatial dimensions. The action is defined as action on the Cartesian coordinates, but of course this also defines the way the same group will act in terms of other, equivalent coordinates.

Examples:

1. Translation in x direction: $g_\alpha f(x, y, z) = f(x + \alpha, y\ z)$.
2. Rotation about y axis: $g_\alpha f(x, y, z) = f(x \cos\alpha - z \sin\alpha, y, z \cos\alpha + x \sin\alpha)$.
3. Dilation: $g_\alpha f(x, y, z) = f(\alpha\, x, \alpha\, y, \alpha\, z)$.

The dilation group example above is not parameterized in the standard way. In order to have the parameters add upon group multiplication it should really be written as

$$g_\alpha f(x, y, z) = f(e^\alpha\, x, e^\alpha\, y, e^\alpha\, z),$$

but sometimes the other parametrization is convenient.

Infinitesimal generators: The Taylor series expansion of $f(x + \alpha)$ can be written in terms of the exponentiation of a derivative operator. Thus

$$\begin{aligned} f(x+\alpha) &= f(x) + \alpha f'(x) + \frac{1}{2!}\alpha^2 f''(x) + \cdots \\ &= \left(1 + \alpha\frac{d}{dx} + \frac{1}{2!}\alpha^2\left(\frac{d}{dx}\right)^2 + \cdots\right) f(x) = \exp\left(\alpha\frac{d}{dx}\right) f(x). \end{aligned}$$

More briefly,

$$\exp\left(\alpha \frac{d}{dx}\right) f(x) = f(x + \alpha).$$

An interesting way to describe this is to say that the displacement $x \to x + \alpha$ is a simple example of a Lie group action. In fact it is just a translation group acting on $f(x)$.

To do the same thing for a rotation in x, y one can use a method sometimes used to derive the Taylor expansion in several variables. Define

$$\begin{aligned} x \to \tilde{x}(x, y, \alpha) &= x + \xi(x, y)\, \alpha + \cdots, \\ &= x \cos\alpha - y \sin\alpha, \\ y \to \tilde{y}(x, y, \alpha) &= y + \eta(x, y)\, \alpha + \cdots \\ &= y \cos\alpha + x \sin\alpha, \end{aligned}$$

where $\xi(x,y)$ and $\eta(x,y)$ are coefficients of the linear terms in the Taylor series in α. The significance of these will be seen presently. Then to focus on the expansion in α, let

$$h(\alpha) = f(\tilde{x}, \tilde{y}).$$

Thus via the chain rule,

$$\frac{d}{d\alpha} h(\alpha) = \frac{\partial f(\tilde{x}, \tilde{y})}{\partial \tilde{x}} \frac{\partial \tilde{x}}{\partial \alpha} + \frac{\partial f(\tilde{x}, \tilde{y})}{\partial \tilde{y}} \frac{\partial \tilde{y}}{\partial \alpha}$$

and so

$$\left(\frac{dh(\alpha)}{d\alpha}\right)_{\alpha=0} = -y \frac{\partial f(x, y)}{\partial x} + x \frac{\partial f(x, y)}{\partial y} \equiv U f(x, y),$$

where

$$U = -y \frac{\partial}{\partial x} + x \frac{\partial}{\partial y}$$

is an example of a Lie derivative acting on scalar functions, sometimes called a tangent vector. For a general group action in two dimensions,

$$\xi(x, y) = \frac{\partial \tilde{x}}{\partial \alpha}\Big|_{\alpha\to 0}, \qquad \eta(x, y) = \frac{\partial \tilde{y}}{\partial \alpha}\Big|_{\alpha\to 0},$$

and the tangent vector at (x,y) is

$$U = \xi \frac{\partial}{\partial x} + \eta \frac{\partial}{\partial y}.$$

Lie called this an infinitesimal generator [7, 9] and I will usually refer to it this way also. In the case at hand where $\xi = -y$ and $\eta = x$, U is the infinitesimal generator of rotation about the z-axis. Repeating the same steps, one finds also that

$$\left(\frac{d^2h(\alpha)}{d\alpha^2}\right)_{\alpha=0} = U^2 f(x, y), \qquad \text{or generally} \qquad \left(\frac{d^nh(\alpha)}{d\alpha^n}\right)_{\alpha=0} = U^n f(x, y).$$

Thus the whole rotation group can be recovered by exponentiating the Lie derivative, i.e. by Taylor series,

$$\begin{aligned} f(\tilde{x}, \tilde{y}) &= u(0) + \alpha\, h'(0) + \frac{1}{2}\alpha^2\, h''(0) + \cdots \\ &= \left(1 + \alpha\, U + \frac{1}{2!}\alpha^2\, U^2 + \cdots\right) f(x, y) \end{aligned}$$

so that

$$f(x \cos\alpha - y \sin\alpha, y \cos\alpha + x \sin\alpha) = h(\alpha) = \exp(\alpha\, U) f(x, y).$$

It may surprise the reader that the entire group action derives from knowing only the generator U.

To generalize, when a one-parameter group acts on a space of functions of n variables $x = \{x^1, x^2, x^3, \ldots, x^n\}$,

$$x^k \to \tilde{x}^k(x, \alpha) = x^k + \xi^k(x)\,\alpha + \mathcal{O}(\alpha^2)$$

where

$$\xi^k(x) = \lim_{\alpha\to 0}\left(\frac{\partial \tilde{x}^k}{\partial \alpha}\right)_x.$$

The infinitesimal generator for the group action expressed in this coordinate set is

$$U = \xi^k(x)\,\frac{\partial}{\partial x^k},$$

where one is to sum over the repeated index k from 1 to n. Notice that the generator U acts as a derivative in the sense that it is linear with respect to constants,

$$U\,(a_1 f_1(x) + a_2 f_2(x)) = a_1\, Uf_1(x) + a_2\, Uf_2(x),$$

and satisfies a Leibniz rule when acting on a product of functions:

$$U\big(f_1(x)f_2(x)\big) = \big(Uf_1(x)\big)f_2(x) + f_1(x)\big(Uf_2(x)\big).$$

It is not quite true in the general case that one can recover the whole action of the group in this way starting from the identity, or in other words starting from the case $\alpha = 0$ and expanding in α. More generally, the group manifold where the parameters live may consist of disconnected parts. An example would be an extension of the Euclidean group, or simply the rotation group acting on the plane, to include reflections, which up to now have been excluded. One part of the group connects by smooth variation of the parameter to the identity $\alpha = 0$. This is the part without the reflections, so that each operation takes a right-handed frame of two vectors into another right-handed frame, preserving the angle between the vectors. This is the part that has been constructed by exponentiating the infinitesimal generator.

Both rotations and reflections preserve the lengths of vectors and the angles between them, however the reflections, including the product of a pure rotation with one reflection, flip a right-handed frame into a left-handed frame and vice versa. Often the latter operations are called improper rotations. In either case the action on coordinates is a linear transformation, and so it can be written in matrix form. The matrices in either case are orthogonal, comprising columns that are orthogonal to one another as vectors. But the determinant is +1 for the proper and −1 for the improper rotations. Since the determinant is a continuous function of the matrix entries, and hence also of the group parameter, it is clear that the improper rotations inhabit a region, or component of the group manifold, that cannot be connected to the identity. The part disconnected from the origin is not itself a group, since it has no identity, while the part with determinant +1 contains the identity and constitutes the subgroup of proper rotations.

Thus in general, the parameter manifold of a Lie group may consist of several parts. Varying the parameters continuously moves the point representing g_α around inside a given component, but there are also, in general, discrete operations of the group that shift from one component to another. The part connected to the identity is always a Lie subgroup and it can be reconstructed by exponentiating combinations of the infinitesimal generators. It is sometimes called the local Lie group defined by the generators.

A one-parameter group acting in two or three dimensions: Rather than the general notation $U = \xi^k \frac{\partial}{\partial x^k}$ it is often convenient to use a simplified, more explicit notation in two or three dimensions. Consider a Lie group of transformations of the three variables (x, y, z) into

$$\begin{aligned} \tilde{x}(x, y, z, \alpha) &= x + \xi(x, y, z)\, \alpha + \cdots, \\ \tilde{y}(x, y, z, \alpha) &= y + \eta(x, y, z)\, \alpha + \cdots \\ \tilde{z}(x, y, z, \alpha) &= z + \zeta(x, y, z)\, \alpha + \cdots \end{aligned} \tag{2.1}$$

where of course

$$\xi = \left(\frac{\partial \tilde{x}}{\partial \alpha}\right)_{\alpha=0}, \qquad \eta = \left(\frac{\partial \tilde{y}}{\partial \alpha}\right)_{\alpha=0}, \qquad \zeta = \left(\frac{\partial \tilde{z}}{\partial \alpha}\right)_{\alpha=0}. \tag{2.2}$$

The Lie derivative, or infinitesimal generator, is then written out in this notation as

$$U = \xi \frac{\partial}{\partial x} + \eta \frac{\partial}{\partial y} + \zeta \frac{\partial}{\partial z}, \tag{2.3}$$

and the group is recovered (locally) by

$$f(\tilde{x}, \tilde{y}, \tilde{z}) = \exp(\alpha\, U) f(x, y, z). \tag{2.4}$$

Some observations are in order. The first is that by writing the group operation as exponential of a generator, it becomes fairly clear why a one-parameter group is Abelean. For any $f(x,y)$,

$$g_\alpha\, g_\beta f(x, y) = e^{\alpha U} e^{\beta U} f(x, y) = e^{(\alpha+\beta)U} f(x, y).$$

The group infinitesimal generator U is the same for each, so the operations commute,

$$\beta + \alpha = \alpha + \beta.$$

The one-parameter Lie groups are like the cyclic finite groups in that each element is a power of the some generating element, which in this case can be thought of as e^U. Multi-parameter groups where the infinitesimal generators need not commute will not in general be Abelean. In the examples seen so far, the groups are all of the one-parameter type. In some ways they are in fact all the same group. In each case there is only one infinitesimal generator. If I think of the general group element $g_\alpha = e^{\alpha U}$ as e^U raised to the power α then each of these groups is indeed rather like a finite cyclic group, although there are subtle differences. Since α is continuous, the difference can have to do with where α takes on its values. It can be an angle, parameterizing a circle, or it can range over $(-\infty, \infty)$. This distinction is important, but the possibilities seem rather limited. However, even groups that are in this algebraic sense isomorphic, or essentially the same, can be different in the way they act. This distinction between different group actions is also quite important in physical problems.

For instance, the dilation group acting in two dimensions with infinitesimal generator

$$U_{dil} = x\frac{\partial}{\partial x} + y\frac{\partial}{\partial y},$$

and the rescaling transformation acting on the x coordinate only, with generator

$$U_{rex} = x\frac{\partial}{\partial x}$$

are both one-parameter groups of the form $g_\alpha = e^{\alpha U}$ with $\alpha \in (-\infty, \infty)$, but the effect on a general $f(x, y)$ is significantly different. So one must be concerned usually with the action of the group.

One other comment is a subtle one. Often one would like to think of the substitution rules that define the group action in terms of operators that operate on functions. The operators change or 'move' the functions in a way that is opposite to the way the coordinates change. For instance, the rule $x \to x + \alpha$ moves x to the right if $\alpha > 0$. But if I graph a function such as e^{-x^2} and then apply the substitution $x \to x + \alpha$ and consider the function still as a function of only x, I see the position of the peak moves to the left by an amount α. For now, just let g_α applied to any expression imply making a substitution for the coordinate variables as described above in this section. (See [6, 13].)

Recovering a one-parameter group action g_α from its generator U:

It is possible [9], generally, to recover the (local) group action from U by solving a system of differential equations. This topic will be addressed in more detail below in the case of multi-parameter groups, but here is the idea in simple form. Suppose

$$U = \xi\frac{\partial}{\partial x} + \eta\frac{\partial}{\partial y},$$

where $\xi(x, y)$ and $\eta(x, y)$ are given. Then, differentiating

$$\tilde{x}(x, y, \alpha) = e^{\alpha U} x, \qquad \tilde{y}(x, y, \alpha) = e^{\alpha U} y$$

with respect to α holding x and y fixed gives an autonomous system of differential equations,

$$\frac{d\tilde{x}}{d\alpha} = \xi(\tilde{x}, \tilde{y}) \qquad \text{with} \qquad \tilde{x}(0) = x,$$

and

$$\frac{d\tilde{y}}{d\alpha} = \eta(\tilde{x}, \tilde{y}) \qquad \text{with} \qquad \tilde{y}(0) = y,$$

where the initial conditions have, as is often done, been suppressed in the notation. Solving these equations gives the group action.

For example, consider the rotation generator

$$U_{rot} = -y \frac{\partial}{\partial x} + x \frac{\partial}{\partial y},$$

from which

$$\frac{d\tilde{x}}{d\alpha} = -\tilde{y} \qquad \frac{d\tilde{y}}{d\alpha} = \tilde{x}$$

with the general solution

$$\tilde{x} = c_1 \cos\alpha - c_2 \sin\alpha, \qquad \tilde{y} = c_1 \sin\alpha + c_2 \cos\alpha,$$

which becomes on evaluating the constants at the identity $\alpha = 0$,

$$\tilde{x} = x \cos\alpha - y \sin\alpha, \qquad \tilde{y} = x \sin\alpha + y \cos\alpha.$$

The topic of reconstructing the action will come up again shortly in the multi-parameter case.

2.5 Three examples of generating functions from one-parameter groups

Before moving on, it may be instructive to have some additional examples of generating functions that come, directly or indirectly, from the action of a one-parameter group. If one member of a sequence of functions is known, then it may be possible to find a group action that will take a trivial generating function, such as a single power of s times the single member of the sequence of functions, into a non-trivial generating function for the whole sequence. Here are some examples.

1. Laplace transforms: The Laplace transform is a common tool of operational calculus. If $f(t)$ is a function of real variable t, and is such that $f(t) = 0$ for $t < 0$ and the integral

$$F(s) = \int_0^\infty f(t)e^{-st}dt$$

converges, then $F(s)$ is the Laplace transform of $f(t)$. It is useful to have formulas for Laplace transforms of

$$t^n e^{-a\,t}\sin(b\,t) \qquad \text{and} \qquad t^n e^{-a\,t}\cos(b\,t).$$

Each of these really represents a function that is zero when $t < 0$. The significance of them is that in linear combination they make up the solution set for the linear homogeneous differential equations with constant coefficients. The transform integrals are easy to do and the results are familiar. However the point here is to use a generating function obtained by a group action. Start with the constant function

$$f(t) = 1 \qquad (\text{for t} > 0).$$

Then,

$$F(s) = \frac{1}{s}.$$

The group of simple translations in the s domain is applied:

$$g_\alpha\, F(s) = F(s+\alpha),$$

$$\frac{1}{s+\alpha} = e^{\alpha\frac{\partial}{\partial s}}\int_0^\infty e^{-st}dt = \int_0^\infty e^{-(s+\alpha)t}dt.$$

Expanding each side in a Taylor series in the group parameter,

$$\sum_{n=0}^{\infty}(-1)^n\frac{\alpha^n}{s^{n+1}} = \sum_{n=0}^{\infty}(-1)^n\frac{\alpha^n}{n!}\int_0^\infty t^n e^{-st}dt.$$

Comparing like powers of α gives the transforms of powers of t,

$$\int_0^\infty t^n e^{-st}dt = \frac{n!}{s^{n+1}}.$$

One could say that the group of translations of the s coordinate distorts the trivial result for the Laplace transform of 1 into the less trivial result for the transform of t^n. The latter result comes from comparing generating functions.

Starting from

$$\frac{1}{s+\alpha} = \int_0^\infty e^{-(s+\alpha)t}dt$$

and performing an additional distortion, $\alpha \to \alpha + a + ib$ where a and b are real positive,

$$\frac{1}{s+\alpha+a+ib} = \int_0^\infty e^{-(s+\alpha+a+ib)t}dt,$$

and expanding again in α, first on the left,

$$\frac{1}{s+\alpha+a+ib} = \frac{1}{s+a+ib}\left(1+\frac{\alpha}{s+a+ib}\right)^{-1} = \sum_{n=0}^{\infty}\frac{(-1)^n\alpha^n}{(s+a+ib)^{n+1}},$$

then the right, and equating coefficients,

$$\frac{(-1)^n}{(s+a+ib)^{n+1}} = \frac{(-1)^n}{n!}\int_0^\infty t^n e^{-(s+a+ib)t}dt.$$

The real and imaginary parts give

$$\int_0^\infty t^n e^{-at}\cos(bt)e^{-st}dt = \frac{n!}{\left((a+s)^2+b^2\right)^{\frac{n+1}{2}}}\cos\left((n+1)\tan^{-1}\left(\frac{b}{a+s}\right)\right),$$

$$\int_0^\infty t^n e^{-at}\sin(bt)e^{-st}dt = \frac{n!}{\left((a+s)^2+b^2\right)^{\frac{n+1}{2}}}\sin\left((n+1)\tan^{-1}\left(\frac{b}{a+s}\right)\right).$$

These formulas follow from applying translation to distort the result for the trivial case where $f(t) = 1$.

2. Another family of integrals: One more example of this type is the family of integral formulas generated by

$$\int_{-1}^{1}\frac{1}{x^2+a}\frac{dx}{\sqrt{1-x^2}} = \frac{\pi}{\sqrt{a(a+1)}},$$

under translation of a. Thus the left-hand side becomes

$$\exp\left(\alpha\frac{\partial}{\partial a}\right)\int_{-1}^{1}\frac{1}{x^2+a}\frac{dx}{\sqrt{1-x^2}} = \int_{-1}^{1}\frac{1}{x^2+a+\alpha}\frac{dx}{\sqrt{1-x^2}},$$

and the right-hand side

$$= \frac{\pi}{\sqrt{(a+\alpha)(a+\alpha+1)}} = \frac{\pi}{\sqrt{a(a+1)}}\left(1+\frac{\alpha}{a}\right)^{-1/2}\left(1+\frac{\alpha}{1+a}\right)^{-1/2},$$

so, after manipulating some factorial expressions, which in turn come from the binomial coefficients, one finds

$$\sum_{n=0}^{\infty} \alpha^n (-1)^n \int_{-1}^{1} \frac{1}{(x^2+a)^{n+1}} \frac{dx}{\sqrt{1-x^2}} =$$

$$\sum_{k=0}^{\infty}\sum_{m=0}^{\infty} \alpha^{k+m}(-1)^{k+m} \frac{\pi}{\sqrt{a(a+1)}} \frac{(2m)!\,(2k)!}{4^{m+k}(m!)^2(n!)^2 a^m (1+a)^k}$$

$$= \sum_{n=0}^{\infty}\sum_{k=0}^{n} \alpha^n(-1)^n \frac{\pi}{\sqrt{a(a+1)}} \frac{(2(n-k))!\,(2k)!}{4^n((n-k)!)^2(n!)^2 a^{n-k}(1+a)^k}.$$

Comparing coefficients of α,

$$\int_{-1}^{1} \frac{1}{(x^2+a)^{n+1}} \frac{dx}{\sqrt{1-x^2}}$$

$$= \frac{1}{4^n(n!)^2} \frac{\pi}{\sqrt{a(a+1)}} \sum_{k=0}^{n} \frac{(2(n-k))!\,(2k)!}{((n-k)!)^2 a^{n-k}(1+a)^k}.$$

3. Legendre polynomials: The electrostatic potential of a charge distribution $\rho(\vec{r})$ satisfies Poisson's differential equation,

$$\nabla^2 \varphi(\vec{r}) = \frac{1}{\epsilon_o} \rho(\vec{r}).$$

In the case where the source is a point charge of strength q at the origin, one knows from a basic physics course that

$$\varphi(\vec{r}) = \frac{1}{4\pi\epsilon_o} \frac{q}{r}.$$

The charge density in this case is

$$\rho(\vec{r}) = q\, \delta^3(\vec{r}),$$

so it is zero everywhere except at the origin. Except right on top of the point charge, the potential satisfies the Laplace homogeneous equation

$$\nabla^2 \varphi(\vec{r}) = 0.$$

Now again a translation group is appropriate. The group of translations along the z-axis moves the source up to $z = a$, or equivalently it moves the field point $\vec{r}$ down to $\vec{r} - a\hat{z}$. Applying this to the Laplace equation, it appears that

$$e^{-a\frac{\partial}{\partial z}} \nabla^2 \frac{1}{r} = \nabla^2 e^{-a\frac{\partial}{\partial z}} \frac{1}{r} = \nabla^2 \frac{1}{|\vec{r} - a\,\hat{z}|} = 0$$

in some neighborhood of every point except $\vec{r} = a\,\hat{z}$. This works since, from Clairaut's theorem, ∇^2 commutes with $\partial/\partial z$ and hence with $\exp(-\alpha\, \partial/\partial z)$. The plan now is to expand in powers of a and then the coefficient of each power of a will be an independent solution to the Laplace equation:

$$\frac{1}{|\vec{r} - a\,\hat{z}|} = \frac{1}{\sqrt{r^2 - 2a\,\hat{z}\cdot\vec{r} + a^2}},$$

and so there is an inner and an outer binomial expansion. When $r < a$,

$$\frac{1}{|\vec{r} - a\,\hat{z}|} = \frac{1}{a}\frac{1}{\sqrt{\left(\frac{r}{a}\right)^2 - 2\mu\,\frac{r}{a} + 1}},$$

where $\mu = \cos\theta$ with θ the angle from $\hat{z}$ to $\vec{r}$. When $r > a$, one can use

$$\frac{1}{|\vec{r} - a\,\hat{z}|} = \frac{1}{r}\frac{1}{\sqrt{\left(\frac{a}{r}\right)^2 - 2\mu\,\frac{a}{r} + 1}},$$

and so in either case the relevant expansion is

$$f(\mu, s) = \frac{1}{\sqrt{1 - 2\,\mu\,s + s^2}} = \sum_{\ell=0}^{\infty} s^\ell\, P_\ell(\mu). \tag{2.5}$$

The coefficients of the powers of s are polynomials $P_\ell(\mu)$. Equation (2.5) defines the Legendre polynomials in terms of their generating function $f(\mu, s)$. Applying this to the electrostatics problem, the inner and outer expansions

$$\frac{1}{|\vec{r} - a\,\hat{z}|} = \sum_{\ell=0}^{\infty} \frac{r^\ell}{a^{\ell+1}} P_\ell(\cos\theta) \qquad \text{for} \quad r < a,$$
$$\frac{1}{|\vec{r} - a\,\hat{z}|} = \sum_{\ell=0}^{\infty} \frac{a^\ell}{r^{\ell+1}} P_\ell(\cos\theta) \qquad \text{for} \quad r > a.$$

Because of the linearity of ∇^2, one arrives at a set of axially symmetric solutions to Laplace's differential equation $\nabla^2\varphi(\vec{r}) = 0$, namely the set

$$\left\{r^\ell\, P_\ell(\cos\theta),\ \frac{1}{r^{\ell+1}} P_\ell(\cos\theta)\right\}, \qquad \text{for} \quad \ell = 0, 1, 2, \dots, \infty.$$

Action of the translation group in the z direction has uncovered these solutions, starting from the original solution $\frac{1}{r}$. The reason this works is that the original Laplacian partial differential equation is unchanged by translation or is translation invariant; and when the source point is translated along z the problem still has axial symmetry about the z-axis. These ideas will be developed further below.

2.6 Quantum oscillator example

The mechanical oscillator consisting of a particle of mass m anchored to a fixed point by a spring of force constant κ is a model for general vibrating systems. A standard quantum mechanical problem is to solve the time-independent Schrödinger equation for the stationary states and quantized energies of such an oscillator with one degree of freedom. This is a boundary value problem,

$$H\,\psi(x) = E\,\psi(x).$$

The Hamiltonian operator

$$H = \frac{p^2}{2m} + \frac{1}{2}\kappa x^2,$$

where

$$p = -i\,\hbar\,\frac{d}{dx},$$

acts on a Hilbert space generated by sufficiently differentiable functions that go rapidly to zero as $x \to \pm\infty$. The spectrum of H is discrete, meaning that the energies are quantized. One would usually like to find the set $\{E_n\}$ of energy eigenvalues and the corresponding set $\{\psi_n\}$ of eigenfunctions. In this subsection the objective is to find a generating function for the eigenfunctions and illustrate how to use it to compute matrix elements.

Consider the resulting differential equation

$$\left(-\frac{\hbar^2}{2m}\frac{d^2}{dx^2} + \frac{1}{2}\,m\,\omega^2\,x^2\right)\psi = E\,\psi,$$

with $\kappa = m\,\omega^2$. There are several, superficially different ways to construct solutions by factoring the operator. In this subsection, the development will be similar to the one given originally by Schrödinger [15]. Later it will be interesting to revisit the problem from another direction. Readers already comfortable with the eigenfunction construction should skip to the discussion following equation (2.13) and refer back if desired.

To rewrite the differential equation in dimensionless form, since H has dimensions of energy, select the standard energy unit $\hbar\,\omega$,

$$\hbar\,\omega\,\frac{1}{2}\left(\frac{m\,\omega}{\hbar}\,x^2 - \frac{\hbar}{m\,\omega}\frac{d^2}{dx^2}\right)\psi = E\,\psi.$$

One finds that $b = \sqrt{\hbar/m\omega}$ must have length units. Let $\epsilon = E/(\hbar\omega)$. Introducing undetermined coefficients α, β,

$$\frac{1}{\sqrt{2}}\left(\frac{x}{b} + \alpha\, b\, \frac{d}{dx}\right)\frac{1}{\sqrt{2}}\left(\frac{x}{b} + \beta\, b\, \frac{d}{dx}\right)\psi$$
$$= \frac{1}{2}\left(\frac{x^2}{b^2} + (\alpha + \beta)x\, \frac{d}{dx} + \alpha + \alpha\, \beta\, b^2 \frac{d^2}{dx^2}\right)\psi.$$

To cancel the first derivative term and make the second derivative term agree, one has

$$\alpha + \beta = 0 \qquad \text{and} \qquad \alpha\, \beta = -1.$$

The two roots, $\alpha = +1$ and $\alpha = -1$, give two separate factorizations:

$$\frac{1}{\sqrt{2}}\left(\frac{x}{b} + b\, \frac{d}{dx}\right)\frac{1}{\sqrt{2}}\left(\frac{x}{b} - b\, \frac{d}{dx}\right)\psi = \left(\epsilon + \frac{1}{2}\right)\psi, \tag{2.6}$$

$$\frac{1}{\sqrt{2}}\left(\frac{x}{b} - b\, \frac{d}{dx}\right)\frac{1}{\sqrt{2}}\left(\frac{x}{b} + b\, \frac{d}{dx}\right)\psi = \left(\epsilon - \frac{1}{2}\right)\psi. \tag{2.7}$$

One generally introduces the operator

$$a = \frac{1}{\sqrt{2}}\left(\frac{x}{b} + b\, \frac{d}{dx}\right).$$

Integration by parts shows that with respect to the standard inner product, or in other words the inner product relevant to the quantum mechanical problem at hand,

$$\langle f, g\rangle = \int_{-\infty}^{\infty} f^*(x) g(x)\, dx,$$

the other operator

$$a^\dagger = \frac{1}{\sqrt{2}}\left(\frac{x}{b} - b\, \frac{d}{dx}\right),$$

is Hermitian conjugate to a, since $\langle f, a\, g\rangle = \langle a^\dagger f, g\rangle$. So equations (2.6) and (2.7) become

$$a\, a^\dagger\, \psi = \left(\epsilon + \frac{1}{2}\right)\psi, \tag{2.8}$$

$$a^\dagger\, a\, \psi = \left(\epsilon - \frac{1}{2}\right)\psi. \tag{2.9}$$

These two equations represent two ways to factor the operator H in the Schrödinger eigenvalue problem. If ψ is an eigenstate with dimensionless energy ϵ then either form of the equation is satisfied. Operate on both sides of equation (2.9) with a:

$$a\,(a^\dagger\, a)\,\psi = \left(\epsilon - \frac{1}{2}\right) a\,\psi, \tag{2.10}$$

or

$$(a\, a^\dagger)\,(a\,\psi) = \left(\epsilon - \frac{1}{2}\right)(a\,\psi). \tag{2.11}$$

Comparing this result with equation (2.8) shows that either $a\,\psi = 0$ or $a\,\psi$ is another eigenstate of the oscillator with lower dimensionless energy $\epsilon - 1$.

If $a\,\psi \neq 0$, one can apply a again multiple times and form a sequence of states $\{\psi,\, a\,\psi,\, a^2\,\psi,\, \ldots\}$ for which, as long as each entry in the sequence is non-zero, the corresponding sequence of energies is

$$S = \{\epsilon,\, \epsilon - 1,\, \epsilon - 2,\, \ldots\}.$$

These energy values continually decrease. However it can be shown that there must be a lowest energy state in the sequence of states for which, on applying a one more time, the result must be zero. To see this, consider again some eigenstate with energy ϵ:

$$a^\dagger\, a\,\psi = \left(\epsilon - \frac{1}{2}\right)\psi.$$

If a is applied to ψ, consider the norm of the result. That is, consider the integral $\langle a\,\psi,\, a\,\psi\rangle$. It is the integral of the square of some function so it is not negative. However,

$$\langle a\,\psi,\, a\,\psi\rangle = \langle \psi,\, a^\dagger\, a\,\psi\rangle = \left(\epsilon - \frac{1}{2}\right)\langle \psi,\, \psi\rangle \;\geqslant\; 0,$$

because $\langle \psi,\, \psi\rangle > 0$. This shows that the sequence S of energies must terminate before any of its entries goes below $\frac{1}{2}$. That means that there must be some eigenfunction ψ_0 such that $a\,\psi_0 = 0$ and thus $\epsilon_0 = \frac{1}{2}$. This eigenfunction must be unique, since it satisfies the differential equation

$$a\,\psi_0 = \frac{1}{\sqrt{2}}\left(\frac{x}{b} + b\,\frac{d}{dx}\right)\psi_0, =0.$$

This is separable, and the solution is

$$\psi_0(x) = \frac{e^{-\frac{x^2}{2b^2}}}{\sqrt[4]{\pi}\sqrt{b}}, \tag{2.12}$$

where the integration constant is determined by the probability interpretation of the wave function, which requires that

$$\langle \psi_0, \psi_0 \rangle = 1.$$

At this point it has turned out that any stationary state of the oscillator belongs to a sequence of states with successively lower dimensionless energies, differing by integer values, that must terminate at the unique ground state ψ_0. The energy of the ground state is $\epsilon_0 = \frac{1}{2}$, so each state has an energy of the form $\epsilon_n = n + \frac{1}{2}$ where $n \in \{0, 1, 2, \ldots\}$. The operator a is a lowering operator that annihilates the ground state and takes any other state with energy ϵ_n into one with lower energy $\epsilon_{n-1} = \epsilon_n - 1$.

In a similar way it can be seen that $a^\dagger$ is a raising operator that takes a state with energy ϵ_n into one with energy $\epsilon_{n+1} = \epsilon_n + 1$. Consider again some eigenfunction of the scaled Hamiltonian operator with eigenvalue ϵ_n. Start by operating from the left on equation (2.8) with $a^\dagger$. The result shows that either $a^\dagger \psi$ is indeed another eigenfunction with eigenvalue $\epsilon_n + 1$, or else $a^\dagger \psi = 0$. But solving the first-order differential equation

$$a^\dagger \psi = 0$$

yields

$$\psi(x) = A e^{+\frac{x^2}{2b^2}},$$

which is not normalizable. Thus there is no eigenfunction that is annihilated by $a^\dagger$.

Start with any eigenfunction ψ. The result of first lowering by a and then raising by $a^\dagger$ is proportional to the same state, ψ, as seen from equation (2.9),

$$\langle \psi_n, a^\dagger a \, \psi_n \rangle = \left(\epsilon_n - \frac{1}{2} \right) \langle \psi_n, \psi_n \rangle = n \, \langle \psi_n, \psi_n \rangle.$$

Thus if ψ_n is normalized, so is

$$\psi_{n-1} = \frac{1}{\sqrt{n}} \psi_n.$$

Similarly, one finds

$$\psi_{n+1} = \frac{1}{\sqrt{n+1}} a^\dagger \, \psi_n.$$

Starting from the ground state and applying $a^\dagger$ n times gives an important result that may be familiar:

$$\begin{aligned} \psi_n &= \frac{1}{\sqrt{n}} \, a^\dagger \psi_{n-1} = \frac{1}{\sqrt{n}} \frac{1}{\sqrt{n-1}} (a^\dagger)^2 \psi_{n-2} = \cdots \\ \psi_n(x) &= \frac{1}{\sqrt{n!}} (a^\dagger)^n \psi_0(x). \end{aligned} \tag{2.13}$$

Now one can construct a generating function for the eigenstates $\{\psi_n\}$ of the quantum oscillator with one degree of freedom, which in this case is the x coordinate. Taking into account equation (2.13), consider the sum

$$f(x, s) = \sum_{n=0}^{\infty} \frac{s^n}{\sqrt{n!}} \psi_n(x).$$

One can write

$$f(x, s) = \sum_{n=0}^{\infty} \frac{1}{n!}(s\, a^\dagger)^n \psi_0(x) = \exp(s\, a^\dagger)\, \psi_0(x).$$

If $a^\dagger$ were an infinitesimal generator, this would indicate the result of some one-parameter group acting on the ground state ψ_0. However, the operator

$$a^\dagger = \frac{1}{\sqrt{2}}\left(\frac{x}{b} - b\frac{d}{dx}\right),$$

is not in the form of a tangent vector, i.e. of an infinitesimal generator. Nevertheless $\exp(s\, a^\dagger)$ can be expressed in terms of a group action in the following way.

Extended action of a one-parameter group: Suppose in general U is the infinitesimal generator for some one-parameter group

$$G = \{g_\alpha\}.$$

The usual action of the group on some arbitrary function of the coordinates is

$$g_\alpha f(x) = \exp(\alpha\, U) f(x),$$

where $f(x)$ is an arbitrary test function. But a more general action [2–4] is

$$\begin{aligned} \pi(g_\alpha) f(x) &= \left(\phi(x)\, g_\alpha \frac{1}{\phi(x)}\right) f(x) \\ &\equiv \phi(x)\, g_\alpha\left(\frac{1}{\phi(x)} f(x)\right), \end{aligned}$$

where $\phi(x)$ is some particular, fixed function of the coordinates that is not zero or infinity anywhere in the domain. This is a group action [12] since it satisfies the two essential properties,

$$\pi(e) f(x) = f(x) \qquad \text{and} \qquad \pi(g_{\alpha_1} \cdot g_{\alpha_2}) f(x) = \pi(g_{\alpha_1})\big(\pi(g_{\alpha_2}) f(x)\big).$$

The second of these identities follows simply because ϕ and $1/\phi$ cancel in the calculation. Thus, for instance, the exponential of $\alpha\phi U\frac{1}{\phi}$ becomes

$$\exp\left(\alpha\,\phi\,U\,\frac{1}{\phi}\right)=\left(1+\alpha\,\phi\,U\,\frac{1}{\phi}+\frac{1}{2!}\alpha^2\,\phi\,U\,\frac{1}{\phi}\phi\,U\,\frac{1}{\phi}\right.$$
$$\left.+\frac{1}{3!}\alpha^3\,\phi\,U\,\frac{1}{\phi}\phi\,U\,\frac{1}{\phi}\phi\,U\,\frac{1}{\phi}+\cdots\right)$$
$$=\phi\left(1+\alpha U+\frac{1}{2!}\alpha^2U^2+\frac{1}{3!}\alpha^3\,U^3+\cdots\right)\frac{1}{\phi}.$$

Then making use of the Leibniz property, or in other words of the fact that U acts as a derivative,

$$\phi(x)\,U\left(\frac{1}{\phi(x)}f(x)\right)=-f(x)\left(\frac{U\phi(x)}{\phi(x)}\right)+Uf(x).$$

One can set the function

$$-\left(\frac{U\phi(x)}{\phi(x)}\right)=A(x).$$

That means that when given an operator of the form

$$B=A(x)+U,$$

where U generates a group, then B acting on a general $f(x)$ can be understood as

$$Bf(x)=\phi(x)\,U\left(\frac{1}{\phi(x)}f(x)\right),$$

or in simplified operator notation,

$$B=\phi\,U\,\frac{1}{\phi},$$

where ϕ is such that, if B is considered as an operator acting on the function ϕ,

$$B\,\phi=0.$$

Then

$$\exp\,(s\,B)f(x)=\left(1+s\,\phi\,U\,\frac{1}{\phi}+\frac{s^2}{2!}\,\phi\,U\,\frac{1}{\phi}\phi\,U\,\frac{1}{\phi}+\cdots\right)f(x)$$

or

$$\exp\,(s\,B)f(x)=\phi(x)\exp\,(sU)\left(\frac{1}{\phi(x)}f(x)\right)=\frac{\phi(x)}{\phi(\tilde{x})}f(\tilde{x}).$$

Now we may return to finding the oscillator generating function.

Evaluating the eigenstate-generating function: In order to evaluate the generating function

$$\exp\,(s\,a^{\dagger})\,\psi_0(x) = \sum_{n=0}^{\infty} \frac{s^n}{\sqrt{n!}}\,\psi_n(x),$$

the first step is to find $\phi(x)$ such that $a^{\dagger}\,\phi = 0$ and thus express the operation on the left-hand side as an extended action of some group. It was already noted above that such a function is

$$\phi(x) = e^{+\frac{x^2}{2b^2}}.$$

Starting from the raising operator,

$$a^{\dagger} = \frac{1}{\sqrt{2}}\left(\frac{x}{b} - b\,\frac{d}{dx}\right),$$

one sees that

$$U = -\frac{b}{\sqrt{2}}\,\frac{d}{dx}.$$

Applying the general pattern,

$$\begin{aligned}\exp\,(s\,a^{\dagger})\,\psi_0(x) &= e^{+\frac{x^2}{2b^2}}\exp\left(-\frac{b\,s}{\sqrt{2}}\frac{d}{dx}\right)e^{-\frac{x^2}{2b^2}}\psi_0(x)\\ &= \frac{1}{\pi^{1/2}\sqrt{b}}e^{+\frac{x^2}{2b^2}}\exp\left(-\frac{1}{b^2}\left(x - \frac{b\,s}{\sqrt{2}}\right)^2\right).\end{aligned}$$

Some rearrangement leads to the final form for the eigenstate-generating function:

$$f(x,\,s) = \sum_{n=0}^{\infty} \frac{s^n}{\sqrt{n!}}\,\psi_n(x) = \psi_0(x)\exp\left(\frac{\sqrt{2}}{b}\,s\,x - \frac{1}{2}\,s^2\right). \tag{2.14}$$

Before leaving this example it is useful to explore some applications of the generating function equation (2.14). First, since the quantum mechanical eigenvalue problem is based on a measurable quantity, and hence a Hermitian operator, it should be that the eigenfunctions are mutually orthogonal with respect to the inner product. To check this I evaluate the integral

$$\begin{aligned}&\int_{-\infty}^{\infty} f(x,\,t)^{*} f(x,\,s)\,dx\\ &= \frac{1}{\sqrt{\pi}b}\int_{-\infty}^{\infty}\exp\left(-\frac{b^2(s^2+t^2) - 2\sqrt{2}\,bx(s+t) + 2x^2}{2b^2}\right)dx.\end{aligned}$$

On completing the square in the exponent and integrating, the result is

$$\int_{-\infty}^{\infty} f(x,\, t)^* f(x,\, s)\, dx = e^{s\,t}.$$

The orthogonality is reflected in the fact that the integral depends only on the product st rather than s and t separately:

$$\sum_{m,n} \frac{s^m\, t^n}{\sqrt{m!n!}} \int_{-\infty}^{\infty} \psi_n^*(x)\psi_m(x)\, dx = \sum_n \frac{1}{n!} s^n t^n.$$

It follows from matching the coefficients that

$$\int_{-\infty}^{\infty} \psi_n^*(x)\psi_m(x)\, dx = \delta_{mn}$$

as expected.

Generating overlap integrals and matrix elements: To evaluate the set of overlap integrals of the form

$$S_{mn}(c) = \int_{-\infty}^{\infty} \psi_n^*(x)\psi_m(x - c)\, dx,$$

form a generating function,

$$F(s,\, t,\, c) = \int_{-\infty}^{\infty} f(x,\, t)^* f(x - c,\, s)\, dx = e^{-\frac{c^2}{4b^2} + \frac{c(t-s)}{\sqrt{2}b} + st}.$$

Thus,

$$F(s,\, t,\, c) = \sum_{m,n} \frac{s^m t^n}{\sqrt{m!n!}} S_{nm}(c) = e^{-\frac{c^2}{4b^2} + \frac{c(t-s)}{\sqrt{2}b} + st}.$$

The overlap integrals come from expanding in a double Taylor series. So, for instance,

$$\begin{aligned} S_{3,\,4}(c) &= \frac{1}{\sqrt{3!4!}} \left(\frac{\partial^7 F}{\partial s^4 \partial t^3} \right)_{s=0,\; t=0} \\ &= \frac{(-192b^6 c + 144b^4 c^3 - 24b^2 c^5 + c^7)e^{-\frac{c^2}{4b^2}}}{96\sqrt{2}\, b^7}. \end{aligned}$$

As a final example, one can obtain a generating function for matrix elements of a Gaussian function between two oscillator eigenstates. The matrix elements are defined as

$$V_{mn} = \int_{-\infty}^{\infty} \psi_n^*(x) e^{-q\,x^2} \psi_m(x)\, dx.$$

Let

$$G(s, t) = \int_{-\infty}^{\infty} f(x, t)^{*} e^{-q x^2} f(x, s)\, dx = \sum_{mn} \frac{t^m s^n}{\sqrt{m!n!}} V_{mn}.$$

The result is

$$G(s, t) = \frac{e^{\frac{2st - b^2 q(s^2+t^2)}{2b^2 q + 2}}}{\sqrt{b^2 q + 1}}.$$

In a similar way, one can compute generating functions for matrix elements of other families of functions or operators by evaluating the corresponding integrals with $f(x, s)$ of $f(x, t)$ in place of the oscillator eigenstates. This assumes first that one can perform the integration and then that the result is sufficiently simple so as to be of use. Fortunately, these conditions often turn out to be satisfied.

A side note—perhaps of interest—is that the eigenstate-generating function is also a coherent state of the oscillator. A coherent state is an eigenstate of the lowering operator a. This comes up in certain kinds of quantum mechanical problems. This property of $f(x, s)$ is seen as follows:

$$a\, f(x, s) = \sum_{n=0}^{\infty} \frac{s^n}{\sqrt{n!}}\, a\, \psi_n(x) = \sum_{n=0}^{\infty} \frac{s^n}{\sqrt{(n-1)!}}\, s\psi_{n-1}(x) = s\, f(x, s).$$

Thus $a\, f(x, s) = s\, f(x, s)$ for any s.

2.7 Bessel functions by factoring

The Bessel generating function has arisen is several places above. In this subsection, the Bessel equation is factored in a way somewhat similar to the one seen for the quantum oscillator. The development follows more or less the original one given by Weisner [3]. For a detailed study of operator factorization methods, see the famous paper of Infeld and Hull [16] and further developments in Miller [4]. In two subsequent subsections, Weisner's group theoretic method [2, 3] is first outlined in general and applied to finding the standard Bessel generating function. It is a good illustration of the general method.

Separation of the wave equation

$$\left(\nabla^2 - \frac{1}{c^2}\frac{\partial^2}{\partial t^2}\right)\phi = 0$$

in two spatial dimensions in polar coordinates by the substitution $\phi(r, \theta, t) = f(r)h(\theta)T(t)$, under the condition that T must remain bounded for all time, and that $h(\theta + 2\pi) = h(\theta)$ so that the solution is continuous and single valued, is an example separation of variables using canonical coordinates of a symmetry group. Canonical coordinates are discussed in sections below. The separated ordinary differential equation for $f(r)$ is

$$f''(r) + \frac{1}{r}f'(r) - \frac{m^2}{r^2}f(r) + \frac{\omega^2}{c^2}f(r) = 0. \tag{2.15}$$

Let $k^2 = \omega^2/c^2$ and $x = kr$. Finally, let $y(x) = f(r)$. Then

$$y''(x) + \frac{1}{x}y'(x) - \frac{m^2}{x^2}y(x) + y(x) = 0.$$

This is Bessel's differential equation in the case where m is an integer. It is really a set of equations, one for each value of m. Applying some of what one learns in a first course in differential equations shows that $x = 0$ is a regular singular point. The Frobenius theorem says that there should be at least one solution proportional to $x^s(1 + \mathcal{O}(x))$. On making this substitution, it appears that $s = m$ is a double root. So the second solution for any given m may be logarithmically singular. It is interesting to rewrite the latter equation as

$$\left(-\frac{d^2}{dx^2} - \frac{1}{x}\frac{d}{dx} + \frac{m^2}{x^2}\right)y_m = y_m. \tag{2.16}$$

The operator on the left is a sort of symmetry of y_m in that it leaves y_m invariant. Is it possible to factor this operator, as was done in the previous example of the quantum operator? The answer is yes, but not quite in the same way. I proceed, as before, by guessing a factored form and introducing two undetermined coefficients. I have the operator acting on a test function to remind myself that it is an operator:

$$\left(-\frac{d}{dx} + \frac{a}{x}\right)\left(\frac{d}{dx} + \frac{b}{x}\right)F = -\frac{d^2F}{dx^2} - \frac{(b-a)}{x}\frac{dF}{dx} + \frac{(a\,b - b)}{x^2}F.$$

In order to match equation (2.16) it must be that $b - a = 1$ and $a\,b - b = m^2$. Either $a = m - 1$ and $b = m$, or else $a = -(m + 1)$ and $b = -m$. Thus, again as in the oscillator case, there are two factored forms of the equation, namely

$$\left(-\frac{d}{dx} + \frac{m-1}{x}\right)\left(\frac{d}{dx} + \frac{m}{x}\right)y_m = y_m, \tag{2.17}$$

and

$$\left(\frac{d}{dx} + \frac{m+1}{x}\right)\left(-\frac{d}{dx} + \frac{m}{x}\right)y_m = y_m, \tag{2.18}$$

where a minus sign has been cancelled in the second one. Define two operators, $\mathcal{A}_m$ and $\mathcal{B}_m$, which are really two sets of operators indexed by m, in the following way:

$$\mathcal{A}_m = -\frac{d}{dx} + \frac{m}{x}, \tag{2.19}$$

$$\mathcal{B}_m = +\frac{d}{dx} + \frac{m}{x}. \tag{2.20}$$

In terms of these operators, the two factored forms of the Bessel equation are

$$\mathcal{A}_{m-1}\mathcal{B}_m y_m = y_m, \tag{2.21}$$

$$\mathcal{B}_{m+1}\mathcal{A}_m y_m = y_m. \tag{2.22}$$

Experience with the oscillator problem suggests operating on equation (2.21) with $\mathcal{B}_m$,

$$\mathcal{B}_m(\mathcal{A}_{m-1}\mathcal{B}_m y_m) = \mathcal{B}_m y_m,$$
$$(\mathcal{B}_m\mathcal{A}_{m-1})(\mathcal{B}_m y_m) = \mathcal{B}_m y_m.$$

If $\mathcal{B}_m y_m \neq 0$ then by comparing this line to equation (2.22),

$$\mathcal{B}_m y_m = k_{m-1} y_{m-1},$$

where k_m is some real constant for each m. Similarly, starting from equation (2.22) and operating with $\mathcal{A}_m$ leads to

$$\mathcal{A}_m y_m = \ell_m y_{m+1},$$

where ℓ_m is some other constant for each m. Then from equation (2.22) one has that $k_m \ell_m = 1$.

The regular solutions sought are the Bessel functions of type 1, $\{J_m(x)\}$. They are normalized by choosing $J_0(0) = 1$, since the regular solutions are such that $y_m(x) \sim x^m$, and then setting $k_m = 1$ for each m,

$$J_{m+1}(x) = \mathcal{A}_m J_m(x), \qquad J_{m-1}(x) = \mathcal{B}_m J_m(x). \tag{2.23}$$

At this point, if one knew $J_0(x)$ all the Bessel functions of the first kind for integer m would be known. One way to proceed would be to solve the differential equation with $m = 0$ subject to $y(0) = 1$, $y'(0) = 0$ by the method of Frobenius to get $J_0(x)$ and then to use the shift operators $\mathcal{A}_m$ and $\mathcal{B}_m$ explicitly to get the other Bessel functions. Instead, the plan is to compute the generating function directly using Weisner's method, to which we now turn.

Let m index the set of ordinary differential equations satisfied by the functions of interest, and suppose there may be also other, fixed parameters, indicated by k. An example would be the unscaled version of Bessel's equation appearing as (2.15) rewritten as

$$\left(\frac{d^2}{dx^2} + \frac{1}{x}\frac{d}{dx} - \frac{m^2}{x^2} + k^2\right) y_m(x, k) = 0, \tag{2.24}$$

which we generalize now to

$$\mathcal{L}\left(x, \frac{d}{dx}, m\right) y_m(x) = 0. \tag{2.25}$$

The linear differential operator $\mathcal{L}$ is a polynomial function of d/dx and m. It is a Laurent polynomial in x—meaning a linear combination of finitely many integer powers of x, positive or negative—and it depends in some fixed way also on k. In the general discussion k will be suppressed in the argument list of either the operator or $y_m(x)$. The latter, for instance, could be written as $y_m(x, k)$. The functions $y_m(x)$ are particular solutions of the linear homogeneous differential equations (2.25).

The basic strategy of the Weisner method is to shift one's attention from the set of functions $\{y_m(x)\}$ satisfying differential equations (2.25) to a single partial differential equation satisfied by the product $s^n y_n(x)$, for fixed but arbitrary n. One could view the latter expression as a trivial sort of generating function that generates only one function for one fixed n. But since $s^n y_n(x)$ is an eigenfunction of the operator $s(\partial/\partial s)$, such that

$$\mathcal{L}\left(x, \frac{\partial}{\partial x}, s\frac{\partial}{\partial s}\right)\left(s^n\, y_n(x)\right) = s^n\, \mathcal{L}\left(x, \frac{d}{dx}, n\right) y_n(x) = 0,$$

it satisfies an n-independent, partial differential equation. This is true for any real number n, which need not be an integer. In short notation, when $\mathcal{L}$ represents this partial differential operator,

$$\mathcal{L}\left(s^n y_n(x)\right) = 0.$$

Now suppose one can find a group $\{g_\alpha\}$ with generator U acting on functions of x and s such that U commutes with $\mathcal{L}$, and thus all powers of U commute with $\mathcal{L}$. Then so does $g_\alpha = \exp(\alpha U)$ commute with the partial differential operator,

$$0 = \exp(\alpha\, U)\left(\mathcal{L}\, s^n\, y_n(x)\right) = \mathcal{L}\left(\exp(\alpha\, U)\left(s^n\, y_n(x)\right)\right).$$

If the group action takes $s^n\, y_n(x)$ into some other, non-zero function $f(x, s, \alpha)$ and if this function can be expanded in powers of s, then

$$\mathcal{L}\left(x, \frac{\partial}{\partial x}, s\frac{\partial}{\partial s}\right) f(x, s, \alpha) = \sum_m A_m\, s^m\, \mathcal{L}\left(x, \frac{d}{dx}, m\right) u_m(x, \alpha) = 0.$$

2.8 Bessel function generator

In general, any method that produces solutions to the partial differential equation

$$\mathcal{L}\left(x, \frac{\partial}{\partial x}, s\frac{\partial}{\partial s}\right) f(x, s, \alpha) = 0$$

will result in some form of generating function. For the Bessel functions, one can take advantage of the operators $\mathcal{A}_m$ and $\mathcal{B}_m$, forming the partial differential operators:

$$\mathcal{A} = -\frac{\partial}{\partial x} + \frac{s}{x}\frac{\partial}{\partial s}, \qquad \mathcal{B} = +\frac{\partial}{\partial x} + \frac{s}{x}\frac{\partial}{\partial s}.$$

Notice that

$$\mathcal{A}\big(s^n\, y_n(x)\big) = s^n y_{n+1}(x), \qquad \mathcal{B}\big(s^n\, y_n(x)\big) = s^n y_{n-1}(x).$$

Then consider as a candidate the generating sum

$$f(x,\, s) = \sum_{m=-\infty}^{\infty} s^m\, y_m(x).$$

Then $f(x, s)$ is simultaneously an eigenfunction of $\mathcal{A}$ and $\mathcal{B}$, such that

$$s\,\mathcal{A}f(x,\, s) = f(x,\, s), \qquad \frac{1}{s}\,\mathcal{B}f(x,\, s) = f(x,\, s),$$

or

$$-\,s\frac{\partial f}{\partial x} + \frac{s^2}{x}\frac{\partial f}{\partial s} = f,$$
$$+\,\frac{1}{s}\frac{\partial f}{\partial x} + \frac{1}{x}\frac{\partial f}{\partial s} = f.$$

This implies that the operators $\mathcal{A}$ and $\mathcal{B}$ generate symmetries. The two differential equations may be solved for the partial derivatives,

$$\frac{\partial f}{\partial x} = \frac{1}{2}\left(s - \frac{1}{s}\right)f,$$
$$\frac{\partial f}{\partial s} = \frac{x}{2}\left(1 + \frac{1}{s^2}\right)f.$$

Integrating the first one gives

$$f(x,\, s) = A(s)\,\exp\left(\frac{x}{2}\left(s - \frac{1}{s}\right)\right).$$

Then the second differential equation shows that $A(s)$ is a constant A, independent of s. The value of the constant A is found from setting x to zero, and using that all the Bessel functions $\{J_m(x)\}$ are zero at $x = 0$ except for the case $J_0(0) = 1$. Thus $A = 1$. This gives the expected result:

$$f(x,\, s) = \exp\left(\frac{x}{2}\left(s - \frac{1}{s}\right)\right).$$

The Weisner method of creating a partial differential equation in the independent variables x and s using sums of terms proportional to $s^m\, y_m(x)$ has quite general applicability [2, 4]. Up to this point the simple Bessel example has shown only a hint of this. To go further leads into a brief discussion of multi-parameter groups and Lie algebras.

2.9 Multi-parameter Lie groups

For the most part, the groups of interest will appear naturally as groups of transformations acting on some space, as was seen above for the case of one-parameter groups. When it is more natural to think of the space as a vector space, then the transformations may be expressed as matrices, and when the space is a function space—which one can also think of generally as a vector space—then the group may correspond more naturally to differential operators. As before, it is often possible to write group elements as exponential expressions involving a set of infinitesimal generators. Here are two examples to illustrate some general points.

Euclidean group in the plane [4, 9]: In plane geometry, two figures are congruent if it is possible to make one coincide with the other by sliding it around in the plane and/or rotating it. For the moment, let me disallow reflections. So the lines, points or regions making up the figures are assumed to be equivalent, or essentially the same, under these two kinds of operations if one can be transformed exactly into the other. These operations, rigid translations and rotations about some point, have in common that they preserve the distances between points in the figures. Because they preserve distances, they preserve triangles and hence also angles. One can think of plane, Euclidean geometry as the study of properties of transformations that preserve distances in this way. The general length-preserving operations, not including reflections, are combinations of translations (in each of two independent directions) and rotations about any point. Some thought will show that in fact it is only necessary to think of rotations about a single center, since rotations about another point can be made up by combining translations with rotations about the fixed reference point. The latter can be taken to be the coordinate origin. So the group transformations are, for translation,

$$g_{(x),\,\alpha}\, x = \tilde{x} = x + \alpha, \tag{2.26}$$

$$g_{(x),\,\alpha}\, y = \tilde{y} = y. \tag{2.27}$$

$$g_{(y),\,\alpha}\, x = \tilde{x} = x, \tag{2.28}$$

$$g_{(y),\,\alpha}\, y = \tilde{y} = y + \alpha, \tag{2.29}$$

and for rotation about the origin

$$g_{(rot),\,\alpha}\, x = \tilde{x} = x\cos\alpha + y\sin\alpha, \tag{2.30}$$

$$g_{(rot),\,\alpha}\, y = \tilde{y} = y\cos\alpha - x\sin\alpha. \tag{2.31}$$

The Euclidean group must contain the two kinds of translation and rotation as subgroups. The infinitesimal generators of these three one-parameter subgroups are

$$U_x = \frac{\partial}{\partial x}, \qquad U_y = \frac{\partial}{\partial y}, \qquad U_{rot} = y\,\frac{\partial}{\partial x} - x\,\frac{\partial}{\partial y}.$$

Thinking of the Euclidean transformations as moving points in the plane, it seems clear that any point can be moved into any other point in an infinite number of ways. But it also appears that the order of application of the different subgroup actions makes a difference. Start with a point with coordinates $(a, 0)$. Rotating first counter-clockwise about the origin by $\pi/2$ and then translating by a in the x direction takes the initial point into (a, a). However first translating in the x direction by a and then rotating counter-clockwise by $\pi/2$ takes the same original point $(a, 0)$ into $(0, 2a)$. Reversing the order of operation therefore gives different results. This should not be a surprise. Groups need not be Abelean and group operations need not commute [5, 9].

Generally, the fact that the group transformations do not commute is reflected in the fact that the infinitesimal generators or Lie derivatives do not commute. So, for example, the fact that

$$e^{\alpha U_{rot}} e^{\beta U_x} \neq e^{\beta U_x} e^{\alpha U_{rot}}$$

can be traced to the fact that the two generators do not commute. Expanding the left side,

$$\begin{aligned} e^{\alpha U_{rot}} e^{\beta U_x} &= \left(1 + \alpha U_{rot} + \frac{1}{2}\alpha^2 U_{rot}^2 + \cdots\right)\left(1 + \beta U_x + \frac{1}{2}\beta^2 U_x^2 + \cdots\right) \\ &= 1 + \alpha U_{rot} + \beta U_x + \frac{1}{2}\alpha^2 U_{rot}^2 + \alpha\,\beta\,U_{rot}\,U_x + \frac{1}{2}\beta^2 U_x^2 + \cdots. \end{aligned}$$

Expanding the right side in the same way,

$$\begin{aligned} e^{\beta U_x} e^{\alpha U_{rot}} &= \left(1 + \beta U_x + \frac{1}{2}\beta^2 U_x^2 + \cdots\right)\left(1 + \alpha U_{rot} + \frac{1}{2}\alpha^2 U_{rot}^2 + \cdots\right) \\ &= 1 + \beta U_x + \alpha U_{rot} + \frac{1}{2}\beta^2 U_x^2 + \alpha\,\beta\,U_x\,U_{rot} + \frac{1}{2}\alpha^2 U_{rot}^2 + \cdots. \end{aligned}$$

The series are partially ordered according to the combined degree of α and β. There are two things to notice, both in this example and in general. First, up through terms of first order, the two series *do* agree, and the other is that they do not agree from second order on, precisely because the infinitesimal generators do not commute. In fact, in this example

$$[U_x,\ U_{rot}] = -U_y.$$

The commutator notation is the standard one used already with the quantum mechanical operators of the harmonic oscillator, with which the reader is probably familiar. Namely, if A and B are operators then $[A, B]$ is the operator such that

$$[A, B]f = A(Bf) - B(Af)$$

for any f in the domain of A and B. Apparently $[A, B] = -[B, A]$ always. Also one finds in general for commutators

$$[A, [B, C] + [B, [C, A]] + [C, [A, B]] = 0.$$

Continuing on in the Euclidean example, one can compute a commutator for each pair of generators,

$$[U_x, U_{rot}] = -U_y, \qquad [U_y, U_{rot}] = U_x, \qquad [U_x, U_y] = 0.$$

The other commutators follow from $[B, A] = -[A, B]$ and hence also $[A, A] = 0$. Notice that only the translation generators appear as commutators and that these in turn commute.

The rotation group in three dimensions: The rotations in three dimensions are generated by

$$U_{xy} = -y\frac{\partial}{\partial x} + x\frac{\partial}{\partial y}, \qquad U_{yz} = -z\frac{\partial}{\partial y} + y\frac{\partial}{\partial z}, \qquad U_{zx} = -x\frac{\partial}{\partial z} + z\frac{\partial}{\partial x}.$$

These generate right-handed rotations in the xy-plane (or around the z-axis), the yz-plane and the zx-plane. The distinct commutators are

$$[U_{xy}, U_{yz}] = -U_{zx}, \qquad [U_{yz}, U_{zx}] = -U_{xy}, \qquad [U_{zx}, U_{xy}] = -U_{yz}.$$

The others follow from anti-symmetry. So, in this case *each* of the generators appears as a commutator of two other generators.

The point of the examples was to illustrate that the collection of infinitesimal generators of a multi-parameter Lie group has some formal structure. The generators in each case form a basis for a Lie algebra. It will be best to begin to describe Lie algebras in a more general, more abstract way [5].

References

[1] Schwalm W 2017 *Lectures on Selected Topics in Mathematical Physics: Introduction to Lie Theory with Applications IOP Concise Physics* (San Rafael, CA: Morgan & Claypool)

[2] McBride E B 1971 *Obtaining Generating Functions* (Berlin: Springer)

[3] Weisner L 1955 Group-theoretic origin of certain generating functions *Pac J. Math.* **5** 1033–9

[4] Miller W 1968 *Lie Theory and Special Functions* (New York: Academic)

[5] Hamermesh M 1962 *Group Theory and Its Application to Physical Problems* (Reading, MA: Addison-Wesley)

[6] Tinkham M 1964 *Group Theory and Quantum Mechanics* (New York: McGraw-Hill)

[7] Lie S 1888 Klassifikation und integration von gewhnlichen differentialgleichungen zwischen x, y die eine gruppe von transformationen gestatten *Math. Ann.* **32** 213–81 (in German)

[8] Lie S 1891 *Vorlesungen über Differentialgleichunen mit Bekannten Infinitesimaln Transformationen* (Leipzig: Taubner) (in German)

[9] Cohen A 1911 *An Introduction to the Lie Theory of One-parameter Groups with Applications to the Solution of Differential Equations* (New York: Heath & Co)

[10] Cornell University Library 2016 A digital collection of historical math monographs http://ebooks.library.cornell.edu/m/math/
[11] Olver P J 1986 *Application of Lie Groups to Differential equations* (New York: Springer)
[12] Roman P 1975 *Some Modern Mathematics for Physicists and Other Outsiders* vol. 1 (Oxford: Pergamon)
[13] Mathews J and Walker R L 1970 *Mathematical Methods of Physics* 2nd edn (Reading, MA: Addison Wesley)
[14] Herman R 1966 *Lie Groups for Physicists* (New York: Benjamin)
[15] Schrödinger E 1940 *Proc. R. Irish Acad.* A **46** 9–16
[16] Infeld L and Hull T E 1951 The factorization method *Rev. Mod. Phys.* **23** 21–68
[17] Lipkin H J 2002 *Lie Groups for Pedestrians* (New York: Dover)
[18] Wybourne B G 1974 *Classical Groups for Physicists* (New York: Wiley)

Chapter 3

Lie algebras

An algebra is a vector space in which a vector product is defined between any two vectors. A Lie algebra is an algebra with a bracket product $[a, b]$ that generalizes the commutator that has arisen naturally in the context of group generators. In the sub sections below, after a brief and informal introduction to algebras focusing on the role played by a basis, a distinction is made between associative and non-associative algebras. There is a short discussion of associative algebras as matrix algebras, and then a physicist's introduction to just a few ideas about classification and structure of Lie algebras, concentrating on things that are apt to arise in physical calculations. The objective is to provide at least a bare, informal sketch of an abstract framework within which certain concepts or procedures can be discussed in general.

The introductory material on algebras in general is found in most abstract algebra books; see reference [1] and material therein. In concentrated form, although quite readable if one starts at the beginning, most of the following material on the abstract theory of associative, non-associative and Lie algebras is treated in Jacobson's book on Lie algebras [2]. The latter author has a preference for applying transformations from the right, which is different from the way they appear in other places, including most physics books, but this should cause no confusion. Material on semi-simple Lie algebras is found in an extremely clear form, especially from a physics perspective, in the small book by Cahn [3]. Both of the aforementioned are mathematically rigorous as well as generally accessible.

3.1 Algebras

An algebra is an abstract algebraic system, like a group or a ring. It is a vector space in which there is a vector product between vectors. Suppose V is an n-dimensional vector space over some field F. Indeed, n could be infinite, and when it is some complications can arise, but let us assume it is finite. In this book, the field will be either the complex or the real numbers. To construct an algebra based on V there needs to be a product, which I will denote for the time being by $\circ$. This takes two

doi:10.1088/2053-2571/aae789ch3

vectors into another vector. The basic question to answer is how this product will interact with the scalar multiplication and addition operations of the vector space. Let u, v and w be vectors in V, and c be any number from the field F. The minimum two requirements are

$$(cu) \circ v = c(u \circ v) = u \circ (cv) \tag{3.1}$$

$$(u + v) \circ w = (u \circ w) + (v \circ w). \tag{3.2}$$

After making these two requirements two things follow. The first is that all properties of the algebra are determined once the product is determined for each pair of basis vectors for some fixed basis of the space V. By a basis I mean here only a fixed set of n linearly independent vectors, which need not be orthogonal with respect to any inner product. Here n should be the maximum number of such vectors in any independent set, so that n is the dimension of the space. Any vector in the space can be written as a sum of basis vectors in a unique way. So, if $\{b_i\}$ is a basis in V, let

$$u = \sum_{i=1}^{n} u^i\, b_i, \qquad v = \sum_{j=0}^{n} v^j\, b_j,$$

then if $u \circ v = w$,

$$u \circ v = \sum_{i,j} u^i v^j (b_i \circ b_j).$$

One need only know what $b_i \circ b_j$ is for each pair of basis vectors. Since the result must again be a vector, these products of basis vectors can in general be specified by no more than n^3 numbers:

$$(b_i \circ b_j) = \sum_{k=1}^{n} c_{ij}^{\,k}\, b_k.$$

The numbers $\{c_{ij}^{\,k}\}$ are the structure constants, and they define the products completely. Finally,

$$w = \sum_{i,j,k} u^i v^j\; c_{ij}^{\,k} b_k.$$

Einstein sum convention: From here on, I will use the usual summation convention in which when a term in an expression contains the product of two indexed quantities indexed by the same letter, one superscripted and the other subscripted, a summation is implied. Thus in the previous discussion one has

$$u = u^i b_i, \qquad v = v^j b_j,$$

$$(b_i \circ b_j) = c_{ij}^{\,k} b_k,$$

and so on. This sum convention is standard and simplifies the notation.

In addition to shifting attention to the products of basis functions, the other thing that happens, given the properties (3.1) and (3.2), is that quite often the distinctions between products of numbers that appear in the field, or scalar multiplication of a vector by a scalar, or the vector product of two vectors, may be fairly clear from context. Sometimes a different symbol is used for the vector multiplication, as will be seen presently below. But in fact it is often possible without ambiguity to avoid an ornate symbol such as $\circ$ altogether and indicate the product by juxtaposition, while of course keeping the factors in proper order, since in general $b_j \circ b_i$ need not equal $b_i \circ b_j$.

Two properties, each of which an algebra may or may not have, are associativity, where the product satisfies

$$(u \circ v) \circ w = u \circ (v \circ w) \qquad \Leftrightarrow \qquad c_{ij}^{\ m} c_{mk}^{\ n} = c_{jk}^{\ m} c_{im}^{\ n}, \tag{3.3}$$

and commutativity, where the product is such that

$$u \circ v = v \circ u \qquad \Leftrightarrow \qquad c_{ij}^{\ m} = c_{ji}^{\ m}. \tag{3.4}$$

A commutative algebra is also called Abelian. A simple example of an Abelian algebra is generated over the field of real numbers by the diagonal matrices. The product is matrix multiplication, and matrix multiplication is associative. For instance, let

$$b_1 = \begin{pmatrix} 1 & 0 & 0 \\ 0 & 1 & 0 \\ 0 & 0 & 1 \end{pmatrix}, \qquad b_2 = \begin{pmatrix} 1 & 0 & 0 \\ 0 & -1 & 0 \\ 0 & 0 & 1 \end{pmatrix}, \qquad b_3 = \begin{pmatrix} 1 & 0 & 0 \\ 0 & 1 & 0 \\ 0 & 0 & -1 \end{pmatrix}$$

and

$$b_4 = \begin{pmatrix} 1 & 0 & 0 \\ 0 & -1 & 0 \\ 0 & 0 & -1 \end{pmatrix}.$$

In this case the non-zero structure constants are all +1,

$$c_{11}^{\ 1} = c_{22}^{\ 1} = c_{33}^{\ 1} = c_{44}^{\ 1} = 1, \quad c_{12}^{\ 2} = c_{21}^{\ 2} = c_{13}^{\ 3} = c_{31}^{\ 3} = c_{14}^{\ 4} = c_{41}^{\ 4} = 1,$$

$$c_{23}^{\ 4} = c_{32}^{\ 4} = 1, \quad c_{24}^{\ 3} = c_{42}^{\ 3} = 1 \qquad \text{and} \qquad c_{34}^{\ 2} = c_{43}^{\ 2} = 1.$$

So this algebra is both associative and commutative, or Abelean.

An example of an associative, non-Abelean algebra is the algebra of quaternions. Because it is usual to do so, I index the basis starting from 0 rather than 1. Also, in this paragraph I represent the components as they often appear with subscripted rather than superscripted variables to simplify the notation. The quaternions are of the form

$$q = q_0 b_0 + q_1 b_1 + q_2 b_2 + q_3 b_3,$$

where the basis objects satisfy $b_0^2 = 1$, $b_1^2 = b_2^2 = b_3^2 = -1$ and

$$b_1b_2 = -b_2b_1 = b_3, \qquad b_1b_3 = -b_3b_1 = -b_2, \qquad b_2b_3 = -b_3b_2 = b_1.$$

The non-zero structure constants are $c_{00}^0 = 1$, $c_{11}^0 = c_{22}^0 = c_{33}^0 = -1$,

$$c_{12}^3 = c_{23}^1 = c_{31}^2 = 1, \qquad \text{and} \qquad c_{32}^1 = c_{21}^3 = c_{13}^2 = -1.$$

Indeed the quaternions form a non-Abelean algebra. It is interesting that if the field F for the quaternions is taken to be the real numbers, then any non-zero quaternion q has a unique inverse (both left and right), namely

$$q^{-1} = \frac{q_0b_0 - q_1b_1 - q_2b_2 - q_3b_3}{q_0^2 + q_1^2 + q_2^2 + q_3^2}.$$

The 'real' quaternions form a division algebra. This does not work if F is taken to be the complex numbers, since then it possible for the denominator to vanish, even when not all the coefficients are zero.

3.2 Associative algebras are essentially matrix algebras

Any associative algebra is isomorphic to a matrix algebra. What does this mean, and why is it so?

To say that two algebraic structures are isomorphic is to say they have the same structure. One may recall that two groups $(G, \cdot)$ and $(H, *)$ are isomorphic if there is an invertible mapping from G onto H which preserves the operation. That is to say, if there is a bijection ϕ from G to H such that

$$g_1 \cdot g_2 = g_3 \qquad \Leftrightarrow \qquad \phi(g_1) * \phi(g_2) = \phi(g_3).$$

Then it is sometimes said that the groups are really the same group and that the mapping is merely a renaming of the elements. If one relaxes the condition that the mapping ϕ be one-to-one and requires only that it be onto the group H, then the operation conserving map ϕ is a homomorphism rather than an isomorphism.

Similarly, there may be an isomorphism or a homomorphism between algebras. Two algebras are isomorphic if there is a bijection from one onto the other that preserves the vector addition, the scalar multiply, the vector product and the distributive laws. In short, one could say that the isomorphism is a mapping that preserves both the vector space properties and the vector product. Then, because of properties (3.1) and (3.2), it follows that the bijection ϕ can be determined by what it does to the basis set $\{b_j\}$.

On the other hand, a homomorphism from algebra A to algebra B is like an isomorphism in that it preserves the vector addition, the scalar multiply, the vector product and the distributive laws, but it is more general in that it need not be one-to-one. So it need not be an invertible map.

Let $\{b_j\}$ be a basis for A. If ϕ is a homomorphism from A to B,

$$\phi(u) = \phi(u^j b_j) = u^j \phi(b_j).$$

This specifies completely how any vector u transforms. The product in A of two such vectors $u \cdot v = w$ becomes in B

$$\phi(u) * \phi(v) = u^i v^j \phi(b_i) * \phi(b_j).$$

On the other hand,

$$\phi(u) * \phi(v) = \phi(w) = w^k \phi(b_k) = c_{ij}^{\ k} u^i v^j \phi(b_k).$$

The structure constants would be the same if ϕ were an isomorphism defined in such a way that the new basis is $\{t_i\}$ where the image $t_i = \phi(b_i)$ of the old basis. Of course one can in addition perform a nonsingular change of basis vectors before or after the transformation and still have an isomorphism. Generally, a homomorphism implies a nontrivial, linear transformation of the basis vectors. But in any case knowing how the basis transforms under a homomorphism or isomorphism will always be sufficient for transforming any vector.

So suppose ϕ is an isomorphism from A to B and suppose

$$\phi(b_i) = \gamma_i^{\ m} t_m.$$

Then the isomorphism includes a change of basis. The structure constants must change to some different but equivalent set of structure constants. How exactly do they change, if ϕ is a general isomorphism? In B,

$$t_\ell * t_m = \tilde{c}_{\ell m}^{\ n} t_n.$$

Then,

$$\phi(b_i \cdot b_j) = \phi\left(c_{ij}^{\ k}\, b_k\right),$$

$$\phi(b_i) * \phi(b_j) = c_{ij}^{\ k} \phi(b_k),$$

$$\gamma_i^{\ \ell} \gamma_j^{\ m} t_\ell * t_m = c_{ij}^{\ k} \gamma_k^{\ n} t_n,$$

$$\gamma_i^{\ \ell} \gamma_j^{\ m}\ \tilde{c}_{\ell m}^{\ n} t_n = \gamma_k^{\ n} c_{ij}^{\ k} t_n.$$

So,

$$\tilde{c}_{\ell m}^{\ n} = (\gamma^{-1})_\ell^{\ i}\, (\gamma^{-1})_m^{\ j}\, \gamma_k^{\ n} c_{ij}^{\ k}.$$

When the basis and the structure constants are fixed, there is no particular need to express the vector product as * or $\cdot$ or $\circ$. When there is no ambiguity, it can be indicated by juxtaposition.

As seen in previous sections, a group applied to a physical system, or in the context of a physical formalism or a physical problem, is often represented naturally by the action of a set of linear differential operators on functions. In this way, the action of the infinitesimal generators forms a representation of an algebra. The natural kind of product becomes a composition of mappings where first one operator acts and then another. If one considers the product of differential operators as the vector product of the operator algebra, then it is associative. On the other hand, if the product is taken to be the commutator of operators, then it is anti-commutative and Jacobi non-associative. However, by a representation of an algebra one often means a matrix representation.

Representations: A (matrix) representation of an algebra A is a homomorphism from A onto an algebra of matrices. That is to say it is a mapping from a general algebra A onto an algebra of square matrices, which need not be one-to-one but which preserves the vector space properties, the vector product and the distributive laws. Matrix multiplication is associative, so if the product in A is to be represented as matrix multiplication, then A must be an associative algebra. As will become clear subsequently, other products may be defined on the matrices besides matrix multiplication. To wit, in a Lie algebra of matrices the appropriate product is the matrix commutator.

An *isomorphism* from a given algebra A onto an algebra in which the basis vectors are matrices is referred to as a *faithful* representation of the algebra. If A is associative, then the product in the matrix algebra can be matrix multiplication.

To show how at least one faithful matrix representation of an associative algebra can always be constructed consider a general associative algebra with a basis $\{b_j\}$. The components $\{u^j\}$ of u in the given basis can be viewed as forming a column matrix, i.e. column vector. In terms of the basis, the product $w = u \circ v$ of u with any other vector v is

$$w^m = c_{ij}^{\ m} u^i v^j.$$

Multiplying by $c_{mk}^{\ n}$ and summing on m,

$$c_{mk}^{\ n} w^m = c_{ij}^{\ m} c_{mk}^{\ n} u^i v^j.$$

While this is true in general, it does not fit the form of a matrix multiplication. However, the algebra is associative, so the structure constants need to obey the associativity condition (3.3). This gives

$$c_{mk}^{\ n} w^m = c_{jk}^{\ m} c_{im}^{\ n} u^i v^j.$$

Then rearranging,

$$w^m c_{mk}^{\ n} = u^i c_{im}^{\ n} v^j c_{jk}^{\ m}.$$

Let the matrix $M(w)$ corresponding to w be defined by its entries as

$$M(w)^n_k = w^m c_{mk}^{\ n},$$

where n is the row and k is the column index. Then,

$$M(w)^n_k = M(u)^n_m \, M(v)^m_k, \qquad \text{or} \qquad M(w) = M(u)\, M(v).$$

This kind of representation will be possible only for associative algebras, since matrix multiplication is associative. The matrix dimension in this particular representation (the regular representation) is the same as the dimensionality of the algebra, or in other words the same as the number of basis vectors. For a given algebra there will be faithful (i.e. isomorphic) representations. There will be non-faithful or homomorphic representations as well. Some of these may be in terms of smaller matrices, and it is always possible to form representations in terms of larger square matrices, for instance by using matrices that consist of smaller, non-overlapping square matrices located on the diagonal where the diagonal blocks are themselves representations.

As an example of a faithful representation with dimension smaller than the regular one, return to the quaternions over the reals. The algebra is four dimensional, so the regular representation is in terms of 4×4 matrices. However, a smaller faithful representation is found in terms of 2×2 complex matrices. So the basis matrices are complex, although the algebra is over the reals, so the vector components are real. In this case,

$$\phi(b_0) = I, \quad \phi(b_1) = -i\sigma_x, \quad \phi(b_2) = -i\sigma_y, \quad \phi(b_3) = -i\sigma_z,$$

with I the 2×2 identity matrix and

$$\sigma_x = \begin{pmatrix} 0 & 1 \\ 1 & 0 \end{pmatrix} \quad \sigma_y = \begin{pmatrix} 0 & -i \\ i & 0 \end{pmatrix} \quad \sigma_z = \begin{pmatrix} 1 & 0 \\ 0 & -1 \end{pmatrix}.$$

The vector components in this algebra are chosen from the real numbers. Choosing components from the complex numbers would also result in an algebra, but a different one. There are four basis matrices, so the dimension of the algebra over the reals is 4, but the basis matrices are 2×2 complex.

3.3 Lie algebras are commutator subalgebras

Lie algebras are non-commutative, non-associative algebras. An n-dimensional Lie algebra over a field F is an algebra with a product, usually indicated by square brackets, $[u, v]$ which, besides satisfying the two linearity properties (3.1), (3.2) is anti-commutative,

$$[v, u] = -[u, v], \tag{3.5}$$

and is non-associative according to the relation

$$[u, [v, w]] + [v, [w, u]] + [w, [u, v]] = 0. \tag{3.6}$$

In this book, F will be the real or the complex numbers. The dimension n need not be finite, but in the applications of interest here there will be a finite number n of basis vectors, so the dimension n will be finite. The corresponding Lie groups will be finitely generated.

The commutators of the infinitesimal group generators described in section 2.9 above were seen to have all the properties of the Lie bracket. Given an associative algebra, one can construct a Lie algebra in which the bracket of the Lie algebra is the commutator derived from the associative product. Conversely, it is a famous result [1, 3] that any Lie algebra is isomorphic to a commutator subalgebra of some associative algebra. That is to say, the set of all commutators of a given associative algebra defines a bracket product with respect to which it forms a Lie algebra, and every Lie algebra is isomorphic to a *subalgebra of* the commutator algebra of some associative algebra. So any Lie algebra is isomorphic to a space of matrices with a product that is the matrix commutator, rather than the ordinary, associative matrix product.

To show that the commutator algebra of an associative algebra is a Lie algebra is not difficult. To go the other way is a little more difficult, however one can once again construct a regular representation which will often be isomorphic. This time it will be a commutator algebra generated by a set of matrices. In fact, a regular representation of a Lie algebra A is again found from its structure constants. The first step is to construct a certain algebra of linear transformations which is associative, because the composition of mappings is defined to be associative. Then this gives a matrix algebra, the commutator subalgebra of which under certain conditions contains a faithful representation of A.

The adjoint map: Let A be a Lie algebra. The *adjoint* is a mapping from A, considered as a vector space, into the space of endomorphisms of A, i.e. the set of linear transformations or operators taking A into itself. Each vector u in the Lie algebra A corresponds to a linear mapping of A into A, namely the adjoint operator $\mathrm{ad}(u)$ defined as

$$\mathrm{ad}(u) : A \to A$$

$$\mathrm{ad}(u) : v \mapsto [u, v].$$

One can write

$$\mathrm{ad}(u)\, v = [u, v]. \tag{3.7}$$

It is a linear operator because the Lie bracket is bilinear. The set of operators forms an associative algebra where the product is a composition. The operator commutators form a Lie algebra homomorphic to A. To see this, suppose $[u, v] = w$ in A. Then let $x \in A$ be an arbitrary vector:

$$\mathrm{ad}(u)\mathrm{ad}(v)\, x - \mathrm{ad}(v)\mathrm{ad}(u)\, x = [u, [v, x]] - [v, [u, x]].$$

From anti-symmetry,

$$\mathrm{ad}(u)\mathrm{ad}(v)\, x - \mathrm{ad}(v)\mathrm{ad}(u)\, x = [u, [v, x]] + [v, [x, u]],$$

and the Jacobi property,

$$\mathrm{ad}(u)\mathrm{ad}(v)\, x - \mathrm{ad}(v)\mathrm{ad}(u)\, x = -[x, [u, v]],$$

and again anti-symmetry,

$$\mathrm{ad}(u)\mathrm{ad}(v)\, x - \mathrm{ad}(v)\mathrm{ad}(u)\, x = [[u, v], x] = \mathrm{ad}(w)\, x.$$

Altogether, since x is arbitrary,

$$[\mathrm{ad}(u), \mathrm{ad}(v)] = \mathrm{ad}(w).$$

So the adjoint map preserves the product. The adjoint map is a Lie algebra homomorphism. Is it an isomorphism? Suppose the adjoint takes two vectors in A into the same operator? Then for any $x \in A$,

$$[u, x] = [v, x], \qquad \text{so} \qquad [u - v, x] = 0.$$

That means u–v must belong to the subspace of A that commutes with every vector in A. If it exists, this subspace is an Abelean subalgebra of A. If A has no Abelean subalgebra, then the adjoint map gives an isomorphism from A to a commutator subalgebra.

The structure constants for the Lie bracket product $[u, v] = w$ in A with respect to a given basis $\{b_i\}$ are determined by the basis set. Expand both u and v as usual, $u = u^i b_i$, $v = v^j b_j$, then from

$$[b_i,\ b_j] = c_{ij}^k b_k$$

one finds

$$\mathrm{ad}(u)\, v = [u, v] = u^i v^j c_{ij}^k b_k.$$

The commutators satisfy by definition the same anti-symmetry and Jacobi non-associativity properties as any Lie bracket. In terms of the structure constants,

$$c_{ji}^\ell = -c_{ij}^\ell, \qquad \text{and} \qquad c_{jk}^\ell c_{i\ell}^m + c_{ki}^\ell c_{j\ell}^m + c_{ki}^\ell c_{j\ell}^m = 0.$$

Then one has

$$\mathrm{ad}(u)v = [u, v] = c_{ij}^\ell u^i v^j.$$

This suggests the definition of a matrix representation of ad(u) acting on vector components, namely it prompts one to define an $n \times n$ matrix $M(u)$ by

$$M(u)_j^\ell = c_{ij}^\ell u^i.$$

It is worth considering what 'ad' actually is. Given a vector u in the Lie algebra A, ad(u) is a linear transformation acting on all the vectors in A, and that transformation, when expressed in components, is the matrix $M(u)$. The set of matrices $M = \{M(u)\}$ for all $u \in A$ forms an associative algebra with respect to matrix multiplication. It is a sub algebra of $gl(n)$, the nonsingular linear transformations acting on A as a vector space. Since the composition of mappings ad(u) ad(v) or matrix multiplication $M(u)M(v)$ is associative, it cannot represent the bracket operation in A. What represents the general bracket in A is the commutator:

$$\mathrm{ad}([u, v]) = [\mathrm{ad}(u), \mathrm{ad}(v)].$$

The brackets on the right and on the left in the latter expression are different. On the left, $[u, v]$ is the abstract Lie bracket in the algebra A, while on the right the bracket represents a commutator,

$$[\mathrm{ad}(u), \mathrm{ad}(v)] = \mathrm{ad}(u)\mathrm{ad}(v) - \mathrm{ad}(v)\mathrm{ad}(u).$$

A representation is usually realized in terms of either the differential generators or Lie derivatives, or in terms of matrices. There are other concrete representations in terms of the differential operators or matrices, where the Lie operation is a commutator. For the matrices it is a matrix commutator,

$$[M(u),\, M(v)] = M(u)M(v) - M(v)M(u).$$

So we have seen at least one way in which an abstract Lie algebra can be expressed as a commutator algebra of matrices. As long as A contains no Abelean subalgebra, the regular representation in terms of the structure constants is a faithful $n \times n$ representation, where n is the dimension of A. In such a case it is isomorphic to A, where more generally one will be interested in both faithful and non-faithful $m \times m$ matrix representations that are homomorphic images of A.

For a given Lie algebra, one can find matrix representations. But the representations that often arise naturally are the commutator algebras of infinitesimal generators. From the operators, one can construct a basis and then calculate structure constants. From the structure constants, one may form a regular matrix representation.

There are other ways to form matrix representations, but instead of exploring this topic, let us construct some further examples of algebras that come directly from groups of function substitutions that give rise to infinitesimal generators. The infinitesimal generators of a multi-parameter group form the basis of a Lie algebra under commutation. Given the basis, one can work backwards and reconstruct the group. Strictly speaking, this works for a local Lie group, or in other words for the part of the Lie group continuously connected to the identity. There can be other parts of the group that are not connected to the identity.

Group from an invertible polynomial transformation: Begin with the invertible polynomial transformation

$$u = x - y - \frac{x^2}{a}, \qquad v = \frac{1}{2}\left(x + y + \frac{x^2}{a}\right),$$

and its inverse,

$$x = \frac{1}{2}(u + 2v) \quad y = \frac{1}{2}(2v - u) - \frac{(u + 2v)^2}{4\,a}.$$

In this pair of transformations a is not a group parameter. In physical terms, if x and y are lengths, then a is a length inserted to maintain dimensional consistency. The transformation is just one map, or coordinate transformation, and its inverse. To create a group action I begin with (x, y), express these in terms of (u, v), then replace u by $u + \alpha a$ and transform back. The result is

$$g_\alpha x = \tilde{x} = x + \alpha\frac{a}{2}, \quad g_\alpha y = \tilde{y} = y - \frac{\alpha((2 + \alpha)a + 4x)}{4}.$$

This is a group of transformations on functions of x and y, each one of which is nonlinear. The identity transformation is where $\alpha = 0$, as usual. Imagine starting with an initial point (x, y) and letting α vary from $-\infty$ to $+\infty$, the point $(\tilde{x}, \tilde{y})$ moves along a smooth curve, namely the *orbit* of the point (x, y), which in this case happens to be a parabola,

$$\tilde{x} + \tilde{y} + \frac{\tilde{x}^2}{a} = x + y + \frac{x^2}{a} = \text{a constant}.$$

As one can see, every point in the plane belongs to one and only one such orbit. To lie on the same orbit is an equivalence relation, so the set of orbits divides the points of the plane into equivalence classes. A tangent vector at (x, y) is a vector tangent to the orbit and has components proportional to

$$\xi = \frac{1}{2}, \qquad \eta = -\frac{1}{2} - \frac{x}{a}.$$

The infinitesimal generator is

$$U = \frac{1}{2}\frac{\partial}{\partial x} - \left(\frac{1}{2} + \frac{x}{a}\right)\frac{\partial}{\partial y}.$$

This is viewed as a tangent vector with basis vectors $b_1 = \partial/\partial x$ and $b_2 = \partial/\partial y$, having the corresponding components ξ and η.

The reader has probably noticed that, since the group action in terms of the u and v coordinates is a simple translation along u holding v fixed, the group must be equivalent to a simple translation, modulo a change of coordinates. The algebra, of course, reflects this since both the simple u translation of (u, v) and the resulting group action on (x, y), which is translation along the parabolic orbits, are Abelean groups with parameter range $\alpha \in (-\infty, \infty)$. Indeed this is true algebraically, but as one can see the action on the functions of x and y is very different from simple translation.

Since the action of a one-parameter group such as the one under consideration is to transform a general point into other points along the same orbit, a question arises naturally: when are two points on the same orbit? In other words, given points with coordinates (x_1, y_1) and (x_2, y_2), under what condition is there an element of the group that takes the first into the second? In terms of the parameter α, one has in general

$$\tilde{x}(x_1, y_1, \alpha) = x_2 \qquad \text{and} \qquad \tilde{y}(x_1, y_1, \alpha) = y_2.$$

If this relation is satisfied by some α the two points are on the same orbit. In this case eliminating α leads back to the requirement that the two points lie on the same parabola. The orbits are the parabolic curves.

Now suppose the group is extended by adding another generator,

$$V = \frac{\partial}{\partial x},$$

so that V generates a translation in the x direction. The algebra has to contain the subspace generated by U and V and also any other generators that arise from taking brackets:

$$[U, V] = \frac{1}{a}\frac{\partial}{\partial y},$$

which means the subspace generated by y translations with generator

$$W = \frac{\partial}{\partial y}$$

is also included. Then one finds closure,

$$[U, V] = \frac{1}{a}W, \qquad [V, W] = 0, \qquad [W, U] = 0.$$

A general vector in algebra generated by including both U and V is

$$X = \alpha U + \beta V + \gamma W.$$

Both $\beta V + \gamma W$ and $\alpha U + \gamma W$ generate two-dimensional subgroups. Starting from any point (x_1, y_1), any other point (x_2, y_2) can be reached by the action of either of these two subgroups. The set of points reached through the group action starting from a certain point (x, y) is still called the orbit of that point, even when the group has more than one parameter, so in this case, the orbit of any one point (x, y) under any of these two subgroups, or of course under the whole group, is the entire xy-plane, or the entire space where the group acts. So one says that these groups *act transitively* on the space. The one-parameter group generated only by U acts transitively on each parabolic orbit. One also says that each orbit is a *homogeneous space* of the group action. The entire xy-plane is a homogeneous space of the full group or of the two-parameter subgroup generated by $\beta V + \gamma W$ or by $\alpha U + \gamma W$.

Conformal subgroup: Next, consider a group of transformations of the plane that preserves angles between curves. Such transformations are conformal, and so this will be a subgroup of the conformal group in two dimensions. The full conformal group in two dimensions contains all the transformations defined by real and imaginary parts of any analytic function, so it is very large. Of interest at the moment will be just a subgroup. Certainly, all the Euclidean group preserves angles, so translation and rotation generators will be included. Uniform dilation preserves angles too. Finally, there is a special conformal transformation composed of an inversion followed by a translation, followed by another inversion. Clearly the inversions by themselves would not be part of a continuous group, but the combination of three operations makes a continuous one-parameter subgroup. Let T_1 be inversion with respect to the circle of radius a, and let $T_2(\alpha)$ be a translation in the x direction. Notice that T_1 is its own inverse. First invert

$$T_1\,x = \frac{a^2 x}{x^2 + y^2}, \qquad T_1\,y = \frac{a^2 y}{x^2 + y^2},$$

then translate by α along x,

$$T_2(\alpha)\, T_1 x = \frac{x + \alpha a}{(x + \alpha a)^2 + y^2 a^2}, \qquad T_2(\alpha)\, T_1 y = \frac{a^2 y}{(x + \alpha a)^2 + y^2 a^2}.$$

Finally, inverting once more,

$$g_{(xc)\alpha}\, x = \tilde{x} = T_1\, T_2(\alpha)\, T_1 x = \frac{a(ax + \alpha(x^2 + y^2))}{(a + \alpha x)^2 + \alpha^2 y^2},$$
$$g_{(xc)\alpha}\, y = \tilde{y} = T_1\, T_2(\alpha)\, T_1 y = \frac{a^2 y}{(a + \alpha x)^2 + \alpha^2 y^2}.$$

The infinitesimal generators are constructed in the usual way, first for the special conformal subgroup

$$\xi = \left(\frac{\partial \tilde{x}}{\partial \alpha}\right)_{\alpha \to 0} = \frac{y^2 - x^2}{a}, \qquad \eta = \left(\frac{\partial \tilde{y}}{\partial \alpha}\right)_{\alpha \to 0} = -2\frac{xy}{a}.$$

So the generator is

$$U_{cx} = (y^2 - x^2)\frac{1}{a}\frac{\partial}{\partial x} - 2xy\frac{1}{a}\frac{\partial}{\partial y}.$$

In order to simplify notation, rescale the variables x and y to make them dimensionless. Thus in the expressions above x and y are replaced by ax and ay, respectively. This is the same as rescaling the circle of inversion form radius a to a unit circle of radius $a = 1$. In other words, simply set a to 1. The infinitesimal generator is then just

$$U_{cx} = (y^2 - x^2)\frac{\partial}{\partial x} - 2xy\frac{\partial}{\partial y}.$$

Similarly, the infinitesimal generators for the x and y translation are

$$U_x = \frac{\partial}{\partial x}, \qquad U_y = \frac{\partial}{\partial y}.$$

For rotation and dilation, one has

$$U_{rot} = -y\frac{\partial}{\partial x} + x\frac{\partial}{\partial y}, \qquad U_{dil} = x\frac{\partial}{\partial x} + y\frac{\partial}{\partial y}.$$

The commutator of U_{cx} with rotation generator U_{rot} gives a new vector,

$$[U_{cx}, U_{rot}] = 2xy\frac{\partial}{\partial x} + (y^2 - x^2)\frac{\partial}{\partial y}.$$

This will be U_{cy}. The set of six vectors U_x, U_y, U_{rot}, U_{dil}, U_{cx} and U_{cy} is closed under commutation and so generate a Lie algebra. The commutators between basis vectors are shown in table 3.1.

Table 3.1. Commutator table.

$[A, B]$	U_x	U_y	U_{rot}	U_{dil}	U_{cx}	U_{cy}
U_x	0	0	U_y	U_x	$-2U_{dil}$	$2U_{rot}$
U_y	0	0	$-U_x$	U_y	$-2U_{rot}$	$-2U_{dil}$
U_{rot}	$-U_y$	U_x	0	0	$-U_{cy}$	U_{cx}
U_{dil}	$-U_x$	$-U_y$	0	0	U_{cx}	U_{cy}
U_{cx}	$2U_{dil}$	$2U_{rot}$	U_{cy}	$-U_{cx}$	0	0
U_{cy}	$-2U_{rot}$	$2U_{dil}$	$-U_{cx}$	$-U_{cy}$	0	0

There are 24 non-zero structure constants, defined by

$$[U_i, U_j] = c_{ij}^k U_k.$$

The reader should be able to see what these are.

In order to demonstrate the process of reconstructing the group action, consider the subgroup generated by U_{cy}:

$$U_{cy} = \xi\frac{\partial}{\partial x} + \eta\frac{\partial}{\partial y} = 2xy\frac{\partial}{\partial x} + (y^2 - x^2)\frac{\partial}{\partial y}.$$

Then the action of the subgroup generated by U_{cy} is recovered by integrating the autonomous system

$$\frac{d\tilde{x}}{d\alpha} = 2\tilde{x}\tilde{y}, \qquad \frac{d\tilde{y}}{d\alpha} = \tilde{y}^2 - \tilde{x}^2.$$

These differential equations solve easily through the complex substitution $\tilde{z} = \tilde{x} + i\tilde{y}$. Then,

$$\frac{d\tilde{z}}{d\alpha} = -i\tilde{z}^2.$$

Integrating this and applying the boundary conditions, one finds

$$\tilde{x} = \frac{x}{\alpha^2(x^2 + y^2) \pm 2\alpha y + 1}, \qquad \tilde{y} = \frac{y \pm \alpha(x^2 + y^2)}{\alpha^2(x^2 + y^2) \pm 2\alpha y + 1}.$$

The choice of the $\pm$ sign is unimportant, since it just reflects two different, equivalent parameterizations related by $\alpha \to -\alpha$. The circle of inversion can be brought back to radius a, if desired, by substituting

$$x \to x/a \qquad \text{and} \qquad y \to y/a.$$

Then the standard action of the subgroup generated by U_{cy} is

$$g_{(cy)\alpha}\, x = \tilde{x} = \frac{a^2 x}{\alpha^2 x^2 + (\alpha y + a)^2}, \qquad g_{(cy)\alpha}\, y = \tilde{y} = \frac{a(ay + \alpha(x^2 + y^2))}{\alpha^2 x^2 + (\alpha y + a)^2}.$$

The action of the most general element g_α of the full six-parameter group, where $\alpha = \{\alpha^1, \alpha^2, \dots, \alpha^6\}$, is

$$g_\alpha f(x, y) = e^{\alpha^k U_k} f(x, y),$$

where

$$\alpha^k U_k = \alpha^1 U_x + \alpha^2 U_y + \alpha^3 U_{rot} + \alpha^4 U_{dil} + \alpha^5 U_{cx} + \alpha^6 U_{cy},$$

with all independent choices of the six parameters. It is possible also to represent the same general group transformation as

$$g_\beta f(x, y) = e^{\beta^1 U_x} e^{\beta^2 U_2} e^{\beta^3 U_{rot}} \cdots e^{\beta^6 U_{cy}} f(x, y),$$

where the β parameter values would need to be chosen differently from the corresponding α values in order to index the same group element, owing to the fact that the factors from the different one-parameter subgroups do not commute.

3.4 Ideals and classification of complex Lie algebras

If A is an algebra, and B is a subset of A that is also an algebra with respect to the same vector product, then B is a subalgebra of A. Thus, if B is a subspace of A and has a basis that is closed under the vector product of A, then B is a subalgebra of A, since B inherits other properties from A.

An ideal is a special kind of subalgebra. Suppose A is a Lie algebra and J is a subalgebra of A with the property that, for any two vectors $u \in A$ and $v \in J$ where u may or may not be in J, then

$$[u, v] \in J.$$

In a sense, J is a subalgebra that captures vectors from A. Another name for an ideal is an invariant subalgebra.

Consider the notation

$$u + J,$$

where u is a vector in A that may or may not be in J, and J is an ideal of A. This expression will denote the set of vectors of the form $u + v$ where v is any vector in J. Then, of course, in the special case that u also happens to be in J, this set would just be the same as J. But in general it is not the same. One may call $u + J$ the coset of u in A with respect to the ideal J, or sometimes u modulo J. Clearly, since the zero vector is in J, J being a subspace, it follows that u is one of the vectors in the set $u + J$. Notice that if u does not belong to J, then the zero vector does not belong to $u + J$ so $u + J$ is not a subalgebra of A because it is not a subspace.

Two vectors u_1 and u_2 may be in the same coset $u + J$. For this to happen there must be two vectors v_1 and v_2 in J such that $u_1 = u + v_1$ and $u_2 = u + v_2$. Then $u_2 - u_1 = v_2 - v_1$ belongs to J. To be 'in the same coset' modulo J is to differ only by some vector in the ideal J. The reader can show that two cosets $u_1 + J$ and $u_2 + J$ are either the same coset or else they have no vectors in common. This means that each vector in the algebra A is in one and only one coset modulo J, and so the cosets partition the algebra A. On thinking of cosets as sets of vectors, it appears that what is important in the coset $u + J$ is the component of u that is in the orthogonal complement of the subspace J in the algebra A.

It is possible to define a new, somewhat more abstract Lie algebra in which the vectors are the cosets of A with respect to any fixed ideal J. The operations in this 'quotient algebra' of A modulo J are defined as follows:

$$(u_1 + J) + (u_2 + J) = u_1 + u_2 + J,$$

and

$$[(u_1 + J), (u_2 + J)] = [u_1, u_2] + J.$$

The latter expressions define operations on sets in terms of representatives, because the expression $u + J$ stands for a set of vectors that is the same as the set $u + v + J$ when v belongs to J. One needs to be sure the result of the operation will be the same in each case, no matter which representative of the coset is used to compute the sum or the vector product. Consider first the sum expressed differently but using the same rule,

$$(u_1 + v_1 + J) + (u_2 + v_2 + J) = u_1 + u_2 + (v_1 + v_2) + J = u_2 + u_2 + J,$$

since $v_1 + v_2$ belongs to J, J being a subspace. This shows that defining the sum in terms of representatives is unambiguous. Then consider the vector product:

$$[(u_1 + v_1 + J), (u_2 + v_2 + J)] = +[u_1, u_2] + [u_1, v_2] + [v_1, u_2] + [v_1, v_2] + J.$$

However, each of the products $[u_1, v_2]$, $[v_1, u_2]$ and $[v_1, v_2]$ belongs to J, J being an ideal. So indeed,

$$[(u_1 + v_1 + J), (u_2 + v_2 + J)] = [u_1, u_2] + J,$$

and the result is the same, independent of which way one elects to represent the cosets. The algebra of cosets will be referred to below as the quotient algebra of A modulo J. The conclusion is that whenever a Lie algebra A contains a subalgebra J that is an ideal, i.e. which 'captures products', then it is possible in some respects to deal with the quotient algebra, which will often be simpler. Of course, every Lie algebra contains two trivial ideals, namely the one comprising only the zero vector $\{0\}$ and the other being the algebra A itself. These trivial ideals are of no help in simplifying anything, so usually one disregards them.

The possibility that A might contain a nontrivial ideal and thus of forming a quotient leads to some ways of classifying a Lie algebra A.

Abelean Lie algebras: The most primitive Lie algebra is one in which every product is zero, or $[u, v] = 0$ for each u and v. This satisfies the requirements of anti-symmetry and the Jacobi property by default. Referring to the bracket operation loosely as a commutator, one may say that all the vectors 'commute'. Then the algebra is said to be Abelean. It was noted earlier, for example, that all one-dimensional algebras are Abelean.

Simple Lie algebras: If a non-Abelean Lie algebra has no nontrivial ideals, which is to say it has no ideals other than itself and the zero ideal, then it is called *simple*.

Notice that the requirement that it be non-Abelean automatically excludes a one-dimensional algebra from being simple, since these algebras are all Abelean. An example of a simple, three-dimensional Lie algebra would be the algebra *so*(3), which is the algebra corresponding to the group *SO*(3) of 3×3 rotation matrices. It is isomorphic to the algebra generated by the infinitesimal rotation generators U_{xy}, U_{yz} and U_{zx} derived above in the introduction to multi-parameter groups in section 2.9. The group *SO*(3) can be defined over the real or over the complex numbers, the parameter values thus taking on either real or complex values. In detail these cases are denoted *SO*(3, *R*) or *SO*(3, *C*) and the corresponding algebras are *so* (3, *R*), and *so*(3, *C*) where the vectors have real or complex coefficients. It is a standard notation that when a Lie group is abbreviated using an upper case acronym, then the corresponding Lie algebra is denoted by the same abbreviation in lower case.

Semi-simple Lie algebras: A Lie algebra is semi-simple if it has no non-zero Abelean ideals. A semi-simple Lie algebra is a direct sum of simple Lie algebras. What is a direct sum?

Let A and B be two Lie algebras. Suppose they are two distinct, invariant subalgebras (*two ideals*) of some Lie algebra L. They have only the zero vector in common. The direct sum $A \oplus B$ of the Lie algebras is the vector space comprising vectors of the form $a + b$ where a belongs to A and b belongs to B, with the obvious product

$$[a_1 + b_1, a_2 + b_2] = [a_1, a_2] + [a_1, b_2] + [a_2, b_1] + [b_1, b_2],$$

but where, because A and B are distinct ideals, both $[a_1, b_2] = 0$ and $[a_2, b_1] = 0$, so the product in the direct sum $A \oplus B$ is

$$[a_1 + b_1, a_2 + b_2] = [a_1, a_2] + [b_1, b_2].$$

The direct sum of two ideals is a vector space direct sum equipped with a bracket product that acts independently in each of the two ideals.

Notice that, since an n-dimensional semi-simple Lie algebra contains no Abelean ideal, then as was seen above, the commutators of the matrices formed from the adjoint map, which is to say from the structure constants, is a faithful $n \times n$ representation. In other words, the algebra is isomorphic to the commutator algebra of the structure constant matrices.

Moreover, since a Lie algebra is always isomorphic a commutator algebra of matrices, or in other words it has matrix representations, it will turn out that a semi-simple Lie algebra, when written as $n \times n$ matrices, can by a similarity transformation always be written in block-diagonal form in which the blocks on the diagonal represent the simple factor algebras. Some short comments may help show why this is the case.

In the following, let $\{L_k\}$ for $k = 1, 2, \ldots, m$ represent subspaces of a Lie algebra L. Let the expression $[L_i, L_j]$ represent the subspace generated by all bracket products between pairs of vectors, the first from L_i and the second from L_j. Because of the anti-symmetry of the product, the spaces $[L_i, L_j]$ and $[L_j, L_i]$ are the same. Then, for example, an Abelean Lie algebra A is one for which $[A, A] = 0$.

Everything commutes. To be a subalgebra of Lie algebra L, a subspace L_j need only be closed under the bracket product. The assertion that a subalgebra J be an ideal of L can be written now as

$$[J, L] \subset J.$$

Suppose L_i and L_j are simple ideals of L having only the zero vector in common. Let the dimensions of L_i and L_j be n_i and n_j. A matrix representation of $L_i \oplus L_j$ can be constructed using a catenated basis of the form

$$\{a_1, a_2, a_3, \ldots a_{n_i}, b_1, b_2, \ldots, b_{n_j}\},$$

where the a vectors span L_i and the b vectors span L_j. Each matrix in this representation will be block diagonal, the upper diagonal block of each will be $n_i \times n_i$ and the lower diagonal block $n_j \times n_j$. The upper blocks taken together comprise a matrix representation of L_i and the lower blocks make up a representation of L_j. If there is a finite collection of disjoint ideals of L such that

$$L = \sum_j L_j, \qquad \text{meaning} \qquad L = L_1 \oplus L_2 \oplus L_3 \oplus L_4 \oplus \cdots \oplus L_m,$$

where the ideals have only the zero vector in common, then L is semi-simple. The expression on the right-hand side is a direct sum, meaning that as a vector space it is a direct sum such that each L_k is a subspace orthogonal to each other, and also that the bracket product operates independently in each subspace:

$$[L_i, L_j] \subseteq L_i \quad \text{if} \quad i = j \quad \text{otherwise} \quad [L_i, L_j] = \{0\}.$$

Of course, the decomposition into diagonal blocks can be hidden by applying a fixed similarity transformation $Q^{-1}M(u)Q$ to each matrix in the representation, the same nonsingular, square matrix Q for each u. Thus a matrix representation of a semi-simple Lie algebra can always be put into block-diagonal form by a similarity transformation, though it need not appear in such a form.

To summarize, a semi-simple Lie algebra is one that has no non-zero Abelean ideals. As a result, it turns out that a semi-simple Lie algebra is the direct sum of simple ideals, although we have not shown it. The matrix representations can decompose by an appropriate similarity transformation into block-diagonal form, where the blocks can be studied separately as simple Lie algebras.

Nilpotent Lie algebras: Because it seems natural to speak of the space spanned by set of products $[A, B]$ of two Lie algebras as the product of A and B, one may think of the space L^p defined by the recursion

$$L^p = [L, L^{p-1}],$$

with $L^1 = L$ as the 'pth power' of L. By definition, L^p is a subspace of L^{p-1} and is closed under the bracket operation so L^p is a subalgebra of L^{p-1}. The fact that $[L^p, L^{p-1}] \subseteq L^{p-1}$ shows also that L^p is an ideal in L^{p-1}, so that in the nested sequence

$$L \supseteq L^2 \supseteq L^3 \supseteq \cdots$$

each L^p is an ideal in L^{p-1}. The sequence is referred to as the *descending central series of ideals*. If it happens to terminate with $L^p = \{0\}$ for some power p, then L is a nilpotent Lie algebra.

Solvable Lie algebras: The definition of a solvable algebra is straightforward. However, before stating it, a comment to show some motivation may be useful. Recall that in some ways the bracket product is like a derivative:

$$\text{ad}(u)\, v = [u, v].$$

As an operator, $ad(u)$ is linear with respect to addition and multiplication by constants,

$$\text{ad}(u)\,(av + bw) = [u,\ av + bw] = a\,[u,\ v] + b\,[u,\ w] = a\,\text{ad}(u)v + b\,\text{ad}(u)w.$$

The Jacobi property of the bracket is also a sort of Leibniz property for $\text{ad}(u)$,

$$\begin{aligned}\text{ad}(u)\,[v, w] &= [u, [v, w]] = -[v, [w, u] - [w, [u, v]] = [[u, v], w] + [v, [u, w]] \\ &= [(\text{ad}(u)v), w] + [v, (\text{ad}(u)w)].\end{aligned}$$

One may say that the linear operator $\text{ad}(u)$ acts symmetrically with respect to the (anti-symmetric) bracket product. In that sense it is a derivative, or a 'derivation'. The adjoint maps, when considered as derivations, make up the *inner* derivations. For any Lie algebra L, the ideal $L^2 = [L, L]$, which is the span of Lie products in L, is also the span of inner derivations, $L^2 = L^{(1)}$. In the sequence

$$L \supseteq L^{(1)} \supseteq L^{(2)} \supseteq \cdots,$$

where $L^{(d+1)} = [L^{(d)}, L^{(d)}]$ and $L^{(0)} = L$, each $L^{(d)}$ is again an ideal in $L^{(d-1)}$, and it is the span of derivations in $L^{(d-1)}$. $L^{(d)}$ is the dth *derived* algebra, and the latter sequence is the *series of derived algebras* in L. If there is an n such that $L^{(n)} = \{0\}$, then the Lie algebra L is solvable. It is fairly easy to show by induction that a nilpotent Lie algebra is also solvable.

3.5 Levi's decomposition

In this subsection, the discussion is specialized to the case of complex Lie algebras. The algebras are vector spaces over the field of complex numbers. One may think of them as deriving from Lie groups in which the parameters α are complex. This makes the classification easier. The case for the real algebras is more complicated, but of course these situations are related.

Levi's theorem asserts that a finite dimensional, complex Lie algebra A decomposes as

$$A = R\ \oplus_s\ (L_1 \oplus L_2 \oplus L_3 \oplus \cdots \oplus L_m), \tag{3.8}$$

where the terms in the direct sum in parentheses are simple, so that the parenthetical sum altogether represents a semi-simple Lie algebra. To be a direct sum means to be expressible in the block form described above in which blocks on the diagonal correspond to terms in the sum. The other term R is a solvable Lie algebra, and the symbol $\oplus_s$ indicates a semi-direct sum. Thus equation (3.8) can be summarized by writing that a finite dimensional, complex Lie algebra A is the semi-direct sum of two subalgebras,

$$A = R \oplus_s L,$$

where R and L are solvable and semi-simple Lie algebras, respectively.

What is a semi-direct sum?

A semi-direct sum $J \oplus_s K$ between two subalgebras of a Lie algebra A, in which J is an ideal, but K is a non-invariant subalgebra is, first of all, a direct sum $J + K$ with respect to vector space properties, as in the case of the direct sum. But the bracket in the semi-direct case no longer acts independently on each subspace. In the semi-direct case one has for the product

$$[J, J] \subseteq J, \qquad [J, K] \subseteq J, \qquad [K, K] \subseteq K.$$

Each vector in A is the unique sum of two components, one in J and one in K, such that with respect to the bracket in A:

$$[a_1, a_2] = [j_1 + k_1, j_2 + k_2] = \left([j_1, j_2] + [j_1, k_2] + [k_1, j_2]\right) + [k_1, k_2].$$

Notice that when R is absent, then the algebra A is semi-simple and each subalgebra L_j is an ideal in A. But when R is non-zero, the simple subalgebras are no longer ideals in A.

3.6 The Killing form

Recall that the adjoint map associates each element u of a Lie algebra with a linear transformation, $\mathrm{ad}(u)$ such that

$$\mathrm{ad}(u)v = [u, v].$$

It associates, therefore, a matrix $M(u)$ with u and hence a matrix $M(b_j)$ with each element of a basis set b_j, defined with respect to a particular basis by the structure constants

$$\mathrm{ad}(b_j)b_k = [b_j, b_k] = c_{ij}^{\ k} b_k,$$

so that

$$M(b_j)_i^k = c_{ij}^{\ k}.$$

Both $\mathrm{ad}(u)$ and $M(u)$ represent u, but the subtle difference is that the matrix $M(u)$, or in particular the entries $\{M_{ij}(u)\}$, refer to a particular basis, while the adjoint is coordinate free.

In order to fix ideas, recall that an elementary norm can be applied to a vector space of $n \times n$ square matrices. Suppose for a moment, just for simplicity, that the vector space is over the real numbers and the matrix entries are real. This norm is not the operator norm, but the sum of the squares of the matrix entries,

$$|M| = \sum_\alpha |M_\alpha|^2 = \sum_{ij} M_i^{\ j} M_i^{\ j} = \mathrm{trace}(M^T M).$$

It is just the Euclidean norm when you think of the matrix entries as components of an $n \times n$-dimensional vector such that the ordinary, real inner product is

$$\vec{A} \cdot \vec{B} = c = \sum_{ij} A_i^{\,j} B_i^{\,j} = c.$$

But, thinking of the matrices as matrices, this is

$$\mathrm{trace}(A^T B) = c.$$

These observations motivate the definition of a bilinear form on the Lie algebra A. It is not an inner product but it is a bilinear map from $A \times A$ to the coefficient field. The *Killing form* is defined as

$$K(u, v) = \mathrm{trace}(\mathrm{ad}(u)\mathrm{ad}(v)).$$

If I introduce a basis $\{b_j\}$ it becomes a matrix,

$$K_{ij} = K(b_i, b_j).$$

In terms of the structure constants

$$K_{ij} = \mathrm{trace}\big(M(b_i)M(b_j)\big) = c_{ip}^{q} c_{jq}^{p},$$

where by the Einstein convention the indices p and q are summed.

The Killing form is symmetric:

$$K(v, u) = \mathrm{trace}(\mathrm{ad}(v)\mathrm{ad}(u)) = \mathrm{trace}(\mathrm{ad}(u)\mathrm{ad}(v)) = K(u, v),$$

and so $K_{ji} = K_{ij}$. When the structure constants are real, the matrix K is a real symmetric matrix. But if not, then it is complex symmetric, rather than Hermitian. Let u and v be any two vectors in the Lie algebra A.

$$K(u, v) = K(u^i b_i, v^j b_j) = K_{ij} u^i v^j$$

is a number in the coefficient field.

The matrix K appears in the role of a metric tensor. It is symmetric, but the trouble is that it may be singular. A square matrix will be singular if and only if it takes some non-zero vector in the space into the zero vector. In that case it has no inverse. It is useful to apply this criterion to the general case of $K(u, v)$. The form K is *degenerate* on the Lie algebra A if there is a non-zero vector w in A such that $K(u, w) = 0$ for every vector u in A, including w. In order to be useful for constructing a metric tensor, K has to be non-degenerate. Notice that K will be non-degenerate if the determinant of the matrix $\{K_{ij}\}$ is non-zero.

Here is a significant application of the Killing form, a set of results due to Èlie Cartan and others, which again in keeping with the informal nature of the current presentation will not be proven here.

Cartan's criterion: A Lie algebra is semi-simple if and only if its Killing form is non-degenerate.

It follows from between this result and the Levi decomposition that if K is non-degenerate, then A has no solvable part, and thus that A is semi-simple. It is the direct sum of simple ideals. One can think of studying the ideals separately. The discussion from here on in this section will be directed toward the simple Lie subalgebras, as these form a set of building blocks and can be studied somewhat separately.

Since the Killing form is non-degenerate in a semi-simple Lie algebra, it can be viewed as a metric tensor. The bilinear sum

$$K(u, v) = K_{ij}u^i v^j$$

defines a pseudo-inner product. It is not an inner product, because it need not be positive definite, thus in particular for a non-zero vector v, $K(v, v)$ can be zero. This is a familiar state of affairs in the theory of relativity, where the flat-space metric tensor is nonsingular but is not positive definite, so there can be non-zero, 'light-like vectors', for which the pseudo-inner product of such a vector with itself is zero. In relativity, the value of $g_{ij}u^i u^j$ can be positive, zero or negative.

As an operator on the Lie algebra, ad(u) acts skew-symmetrically with respect to the (symmetric) Killing form, as can be seen as follows. Consider

$$\gamma(u, v, w) = K(u, \mathrm{ad}(v)w) = K(u, [v, w]),$$

as a mapping from $L \times L \times L$ to the base field. Since $\mathrm{ad}[v, w] = [\mathrm{ad}(v), \mathrm{ad}(w)]$, where the second bracket is a commutator,

$$\gamma(u, v, w) = \mathrm{trace}(\mathrm{ad}(u)(\mathrm{ad}(v)\mathrm{ad}(w) - \mathrm{ad}(w)\mathrm{ad}(v))).$$

Then using cyclic invariance of the trace,

$$\gamma(u, v, w) = \mathrm{trace}(\mathrm{ad}(u)\mathrm{ad}(v)\mathrm{ad}(w) - \mathrm{ad}(v)\mathrm{ad}(u)\mathrm{ad}(w))),$$

whence

$$\gamma(u, v, w) = K([u, v], w) = -K(\mathrm{ad}(v)\, u, w).$$

Indeed, the sign is reversed when ad(v) acts on the other argument:

$$K(u, \mathrm{ad}(v)\, w) = -K(\mathrm{ad}(v)\, u, w) \qquad \text{or} \qquad \gamma(u, v, w) = -\gamma(v, u, w).$$

Since $[w, v] = -[v, w]$ one also has $\gamma(u, w, v) = -\gamma(u, v, w)$. Taken together these two reflection properties imply the third $\gamma(u, w, v) = -\gamma(w, v, u)$. Altogether, γ is anti-symmetric with respect to interchange of any pair of arguments:

$$K(u, \mathrm{ad}(v)w) = u^i v^j w^k K_{im} c^{\,m}_{jk} = \gamma_{ijk} u^i v^j w^k,$$

where $\gamma_{ijk} = \gamma_o \varepsilon_{ijk}$ is fully anti-symmetric.

3.7 Cartan subalgebra

Every Lie algebra has a Cartan subalgebra. In a semi-simple Lie algebra it is a set of mutually commuting vectors that is maximal in some sense. In the context of quantum theory, for example, where the Lie brackets are commutators, one can think of a collection of mutually commuting operators that thus represent simultaneously measurable physical quantities. In the quantum system one may be able to find a set of ladder operators, or raising and lowering operators, that transform from one mutual eigenfunction of the commuting operators to another, as do the raising and lowering operator in the simple oscillator problem or in the case of angular momentum. Or in classical mechanics, the Cartan subalgebra may represent a set of mutually conserved quantities such that the Poisson brackets vanish in pairs. The following material generalizes these ideas. The root vectors to be introduced below are abstractions of the ladder operators in the context of a Lie algebra.

Definition: A Cartan subalgebra H of an n-dimensional Lie algebra A is a nilpotent subalgebra of dimension d for which there exists a basis

$$\{h_1, h_2, h_3, \dots, h_d\}$$

that can be extended to a basis on A,

$$\{h_1, h_2, \dots, h_d, e_1, e_2, \dots, e_{n-d}\},$$

in such a way that each e_α is an eigenvector of each ad (h_k). Thus

$$\mathrm{ad}(h_k)\, e_\alpha = [h_k, e_\alpha] = \lambda_\alpha(h_k)\, e_\alpha. \tag{3.9}$$

H is also maximal in the sense that it is not properly contained in a larger subalgebra with the same property.

Cartan's theorem guarantees that every finite dimensional Lie algebra over the complex numbers, whether semi-simple or not, has such a subalgebra H. The Lie algebra A may have more than one Cartan subalgebra, but it is guaranteed to have at least one. Moreover, if A is semi-simple, then H is Abelean, meaning that all the vectors in H commute with one another.

A brief comparison to a quantum mechanical system: A physics student can step back for a moment and look at this construction in the following way. From a physical standpoint, it is a familiar situation. Let us interpret the Lie brackets as commutators between operators in an associative algebra, which we may always do. When there is a quantum mechanical system with a set of simultaneously observable quantities represented, in the quantum mechanical sense, by Hermitian operators, acting in turn on the space of possible state functions of the physical system, these operators must commute with one another in the same way as the vectors $\{h_k\}$ of the Cartan subalgebra do for a semi-simple Lie algebra. The quantum operators belong to an associative algebra. Suppose each h_k in the Cartan subalgebra corresponds to a quantum observable $\mathcal{T}_k$ in the simultaneously measurable set. They represent the Lie algebra in that the brackets correspond to commutators. Similarly, let each e_α correspond to an operator $\mathcal{A}_\alpha$. The brackets in this quantum mechanical example may be either matrix commutators or commutators between differential operators. Suppose the physical system is in a quantum state represented by the function ψ_w. If one measures the quantity represented by $\mathcal{T}_k$, the possible results are the eigenvalues $\tau(w, k)$ such that $\mathcal{T}_k\, \psi_w = \tau(w, k)\, \psi_w$. Then from equation (3.9), interpreting the bracket as a commutator of operators in the usual associative algebra of quantum mechanics, one finds

$$\begin{aligned}\mathcal{T}_k(\mathcal{A}_\alpha\psi_w) &= (\mathcal{A}_\alpha\mathcal{T}_k + [\mathcal{T}_k, \mathcal{A}_\alpha])\psi_w \\ &= (\tau(w,k) + \lambda_\alpha(h_k))(\mathcal{A}_\alpha\psi_w).\end{aligned}$$

This implies that, as a quantum mechanical operator, either $\mathcal{A}_\alpha\psi_w = 0$ and the result is just zero, or else

$$\mathcal{A}_\alpha\psi_w, = (\text{constant})\psi_{w'},$$

where $\psi_{w'}$ is another state function of the system, with the new eigenvalue corresponding to another value for the measurement of the quantity represented by $\mathcal{T}_k$,

$$\tau(w', k) = \tau(w, k) + \lambda_\alpha(h_k),$$

the constant being included to preserve the probability interpretation of quantum mechanics. Thus $\mathcal{A}_\alpha$ acts as a raising or lowering operator with respect to measurement values of $\mathcal{T}_k$. At the same time, this new state has in general new eigenvalues for each other observable quantity $\mathcal{T}_j$, this being shifted by $\lambda_\alpha(h_j)$ and so on. Note however that the amount of the shift does not depend on w, i.e. on other parameters of the eigenfunctions themselves, rather only on $\mathcal{A}_\alpha$ and $\mathcal{T}_k$.

Return to a general discussion of the Cartan subalgebra: When presented with the basic relation (3.9) in this form,

$$\operatorname{ad}(h_k)\, e_\alpha = \lambda_\alpha(h_k)\, e_\alpha,$$

one is apt to think of $\operatorname{ad}(h_k)$ as acting on the eigenvector e_α to produce the eigenvalue times eigenvector on the right-hand side. Certainly this is a valid point of view, but it will be useful to shift the perspective somewhat. Notice that the eigenvalue λ itself is a linear function of both h_k and e_α, since the bracket $[a, b]$ is linear in both its arguments. This is nothing new, since one is used to combining linear maps linearly, as for instance with linear operators. So

$$\begin{aligned}&(c_1\operatorname{ad}(h_1) + c_2\operatorname{ad}(h_2) + \cdots + c_d\operatorname{ad}(h_d))e_\alpha \\ &= (c_1\lambda_\alpha(h_1) + c_2\lambda_\alpha(h_2) + \cdots + c_d\lambda_\alpha(h_d))e_\alpha,\end{aligned}$$

which I can write as

$$\lambda_\alpha(c_1h_1 + c_2h_2 + \cdots + c_dh_d)e_\alpha.$$

So it is fairly clear that the eigenvalue λ_α is a linear functional, or in other words a linear, scalar valued map from H to the complex numbers. Often in applications or for formal development it is useful to consider H as a real vector space and to restrict attention to a mapping into the reals. It should be clear also that, as always, the norm of each eigenvector e_α is undetermined, but that the definition of the dual vectors does not depend on this fact. Let $\Phi = \{\lambda_\alpha\}$ be the set of $n - d$ linear functionals thus defined.

It turns out to be the case that if a particular λ_α belongs to Φ, then $-\lambda_\alpha$ also belongs to Φ. Looking back at the situation in the quantum mechanical system, it means that whenever there is a raising operator for a certain measurable quantity there should be a lowering operator, and vice versa.

The eigenvalue mappings $\{\lambda_\alpha\}$ are called *roots*, and for each α the vector e_α corresponding to λ_α is a *root vector*. The linear space of vectors over the complex numbers spanned by the roots is the dual H^* of the Cartan subalgebra H, where H^* is the space of all linear functionals defined on H.

Notice two things, however: the first is that not every vector in the dual H^* is a root. The roots are only the particular linear functional in the set $\Phi = \{\lambda_\alpha\}$ that correspond to the root vectors $\{e_\alpha\}$ as eigenvalues in equation (3.9). Thus there is the dual space H^*, and within it there is the finite set of roots. The other point is that there are $n - d$ independent root vectors e_α in the basis, each of which defines a root. The dual space H^* of H should have the dimension of H, which is d. So generally, the root set Φ, although its linear combinations span H^*, cannot serve as a basis. The root set Φ is a subset of H^*, but it is not a basis set, because in general the roots are not linearly independent. The proof [3] that if $\lambda \in \Phi$ then $-\lambda \in \Phi$ relates to the anti-symmetry of the adjoint with respect to the Killing form K, as demonstrated above.

A notation often in use, when the root vector e_α gives the root λ_α and another root vector $e_{\alpha'}$ corresponds to $\lambda_{\alpha'} = -\lambda_\alpha$, then $e_{\alpha'}$ can be re-designated as $e_{-\alpha}$ and $\lambda_{\alpha'}$ as $\lambda_{-\alpha}$. It is easy to see that λ_α and $\lambda_{-\alpha} = -\lambda_\alpha$ are not independent.

It is possible to select from the root set Φ, a subset Π of *fundamental roots* $\{\pi_1, \pi_2, \dots, \pi_d\}$ that does serve as a basis for H^* and which has the following properties.

- $\Pi \subset \Phi$ are linearly independent.
- For each root $\lambda \in \Phi$ there is a set of non-negative integers $\{n_1, n_2 \cdots, n_d\}$ and an overall sign $\epsilon \in \{+1, -1\}$, such that $\lambda = \epsilon(n_1\pi_1 + n_2\pi_2 + \cdots + n_d\pi_d)$.
- $\mathrm{span}_{\mathbb{C}}(\Pi) = H^*$.

The vectors in Π are linearly independent, so now they can serve as a basis for all of H^*. If the coefficients are taken from the non-negative integers, then to within an overall sign, each of the roots λ_α can be written as expansion with non-negative integer coefficients as shown. If, on the other hand, one forms all complex linear combinations of the basis vectors in Π, the result is the whole dual H^* of the Cartan subalgebra H.

Inner product of roots: The Killing form is not generally an inner product on H because it need not be positive semi-definite. However in a semi-simple Lie algebra it induces a positive definite form and an inner product on H^*. For any $x \in H^*$ there is a unique $h_x \in H$ such that $x(y) = K(h_x, y)$ for any other $y \in H$. To find h_x one may use components with respect to the basis in H,

$$y = y^j h_j, \quad h_x = h_x^i h_i.$$

Since x is linear,

$$x(y) = x(y^j h_j) = y^j\, x(h_j) = K(h_x^i\, h_i,\, y^j h_j),$$

$$y^j x(h_j) = K_{ij} h_x^i y^j.$$

This must hold for any y, so cancelling

$$x(h_j) = K_{ij} h_x^i.$$

The inverse matrix $K^{-1} = \{K^{ij}\}$ exists because the Killing form is not degenerate, so the components of h_x are determined,

$$h_x^i = K^{ij} x(h_j), \qquad \text{so} \qquad h_x = K^{ij} x(h_j)\, h_k.$$

Then if u and u belong to H^*, let

$$\langle u,\, v\rangle = K(h_u,\, h_v) = K^{im} K^{jn} u(h_i),\, v(h_j) K_{mn},$$

or

$$\langle u,\, v\rangle = K^{ij} u(h_i) v(h_j).$$

The inverse Killing matrix thus serves as a metric tensor in H^*.

3.8 Root geometry, Weyl group and brief comment on classification

We have seen that a semi-simple Lie algebra is the direct sum of simple Lie algebras. The simple lie algebras over the complex numbers are classified completely. As with other important topics, this classification will not be discussed in depth. However, here is a brief look at some of it.

Up to this point there is the set Π of fundamental roots, which is a subset of the root set Φ. These are both subsets of the dual space H^* of linear functions defined on the complex Cartan subalgebra H. It is also instructive to consider a version of H and of H^* where the coefficient field is taken to be the real numbers.

When the algebra is semi-simple, and the coefficient field is taken as real, then $\langle u, v\rangle$ is real and $\langle u, u\rangle \geqslant 0$, being 0 only when $u = 0$, so the pseudo-inner product becomes a true inner product. The real vector space generated by Π over the reals can be denoted H_R^*. Clearly $H_R^* \subset H^*$. In H_R^* one can define lengths and angles. The basic classification of simple Lie algebras is in terms of geometry of the root set Φ which in turn is determined by the set Π of fundamental roots. Notice that, unlike the root vectors, $\{e_\alpha\}$, which are eigenvectors in the basic equations (3.9) and hence have an unspecified length, the root vectors $\{\lambda_\alpha\}$ and hence also $\{\pi_\alpha\}$ have rigidly specified components, and hence both the length and direction is specified for each one. The squared length of π_α is $\langle \pi_\alpha,\, \pi_\alpha\rangle$ and the cosine of the angle between vectors π_α and π_β is

$$\cos \vartheta_{\alpha\beta} = \frac{\langle \pi_\alpha,\, \pi_\beta\rangle}{\sqrt{\langle \pi_\alpha,\, \pi_\alpha\rangle \langle \pi_\beta,\, \pi_\beta\rangle}}.$$

Of prime interest is the symmetry group of Φ and the reconstruction of the root set Φ from the subset Π. The symmetry group of Φ is the Weyl group, the finite group of permutations of the roots that leaves Φ unchanged. It is generated by a set of reflections.

To review the way a vector can define a reflection operation, consider first the simple example of vectors in a three-dimensional Euclidean space. If $\vec{a}$ is some fixed vector and $\vec{r}$ is any point on the plane passing through the tip of $\vec{a}$ and perpendicular to $\vec{a}$, then

$$\vec{a} \cdot \vec{r} = \vec{a} \cdot \vec{a}.$$

This means that the component of $\vec{r}$ along $\vec{a}$ is constrained to be the same as the component of $\vec{a}$ along itself. Now consider instead a plane through the origin that is perpendicular to the vector $\vec{a}$. When the vector $\vec{r}$ lies in this plane, it is perpendicular to $\vec{a}$, so $\vec{a} \cdot \vec{r} = 0$. A general vector $\vec{r}$ resolves into a component $\vec{r}_{\|}$ parallel to the plane (hence perpendicular to $\vec{a}$) plus a component $\vec{r}_{\perp}$ perpendicular to the plane (hence along $\vec{a}$):

$$\vec{r} = \vec{r}_{\|} + \vec{r}_{\perp},$$

where the part perpendicular to the plane is

$$\vec{r}_{\perp} = \frac{\vec{a} \cdot \vec{r}}{\vec{a} \cdot \vec{a}} \vec{a},$$

so the other part must be

$$\vec{r}_{\|} = \vec{r} - \vec{r}_{\perp} = \vec{r} - \frac{\vec{a} \cdot \vec{r}}{\vec{a} \cdot \vec{a}} \vec{a}.$$

The process of reflecting $\vec{r}$ through the plane defined by $\vec{a}$ is a matter of changing the sign of the component $\vec{r}_{\perp}$ perpendicular to the mirror. So $\vec{r} = \vec{r}_{\|} + \vec{r}_{\perp}$ is taken into $\vec{r}_{\|} - \vec{r}_{\perp}$. That means, associated with any vector $\vec{a}$ there is a reflection operator $\sigma_{\vec{a}}$ that acts on an arbitrary vector $\vec{r}$ by reflecting through the mirror plane defined by $\vec{a}$. Namely,

$$\sigma_{\vec{a}}(\vec{r}) = \vec{r} - 2\frac{\vec{a} \cdot \vec{r}}{\vec{a} \cdot \vec{a}} \vec{a}.$$

Now returning to Φ, the Weyl group W is the group of all permutations of Φ that each leave the constellation of roots, or the *root system* geometrically the same, preserving the lengths of the roots and angles between them. It is generated by the set of reflections $\{\sigma_k\}$ where

$$\sigma_k(h) = k - 2\frac{\langle \pi_k, h \rangle}{\langle \pi_k, \pi_k \rangle} \pi_k \qquad \text{(no summation implied)}$$

for all $\pi_k \in \Pi$. The Weyl group takes each fundamental root in Φ into a star-like set of symmetry-related roots in Φ all of the same length. Starting from points in the set Π, the action is such that by applying a succession of group elements, any other point in the set Φ can be reached. All elements of the Weil group are symmetries of Φ meaning that they permute the roots. So, starting from Π the generator elements of W are obtained, which define all of W. At the same time, the action of W on H_R^* takes roots from Π onto all the other roots in Φ. Then W is also the permutation group of symmetries of Φ.

The possible structures of W have famous diagrammatic representations, the Dynkin diagrams, in terms of which all possible simple Lie algebras have been enumerated. Rather than continuing on with a discussion of this topic, it seems best to return to the realm of physical applications. In so doing, one is reminded that in addition to the abstract structure of the Lie algebra, the characteristics of the group action on a given space or of a particular representation of the group as commutator structure in an associative algebra of operators are also important.

3.9 Representations and Casimir operators

Recall that a representation is a homomorphism from a Lie algebra to the commutator subalgebra of some associative algebra, usually of matrices of differential operators, such that the bracket product of the Lie algebra becomes the commutator. The Lie algebras of greatest interest come from the symmetry groups of physical systems, or of sets of equations relating to physical systems. There are other possibilities too. For instance, in classical mechanical systems there are Lie algebras represented by Poisson brackets operating on functions of the dynamical variables. In classical thermodynamics the Poisson brackets arise again in projecting response functions onto a manifold of intensive variables, and so on. For definiteness, the focus will often be on quantum mechanical systems or on problems in electromagnetism where linear operators arise. That is, it will often be on Lie group representations pertaining either to geometrical transformations, or on calculations involving the Schrödinger equation or the Helmholtz equation. The important point is that because these systems are described by operator algebra that is associative, there will be an associative product AB in the representations, as well as the commutator $[A, B]$. There is only the bracket $[a, b]$ in the Lie algebra itself.

To reiterate some of what has gone before, a quantum mechanical operator $\mathcal{Q}$ operates on a function space of complex valued state functions, such as ψ. If $\mathcal{Q}$ represents (in the physical sense) a measurable quantity Q, then making a measurement of Q with an idealized apparatus that introduces no extra uncertainty into the measurement must result in a measured value that is an eigenvalue of $\mathcal{Q}$. That is how quantum theory works. So the eigenvalues should be real numbers. This is one reason that physically measurable $\mathcal{Q}$ needs to be a Hermitian operator, meaning

$$\langle \psi_a, \mathcal{Q}\, \psi_b \rangle = \langle \mathcal{Q}\, \psi_a, \psi_b \rangle,$$

where $\langle f, g \rangle$ is an appropriate inner product between functions. Another reason $\mathcal{Q}$ should be Hermitian is that it needs to have a complete, orthogonal set of eigenfunctions corresponding to the different possible values (the eigenvalues) of physical measurements of quantity Q.

To support the probability interpretation of quantum theory, the norm of a state function ψ is fixed in some way. A typical condition is $\langle\psi, \psi\rangle = 1$. A symmetry transformation of a quantum system that preserves this normalization and also preserves squared matrix elements such as $|\langle\psi_1, \mathcal{Q}\,\psi_2\rangle|^2$ is a unitary operator, $\mathcal{K}$,

$$\mathcal{K}^\dagger\mathcal{K} = 1.$$

In other words, the Hermitian conjugate of $\mathcal{K}$ is the same as its inverse. If $\mathcal{K}$ belongs to a local, one-parameter Lie group of unitary transformations, then

$$\mathcal{K} = \exp(\alpha U) = 1 + \alpha U + \cdots$$

and

$$\mathcal{K}^\dagger = \exp(\alpha^* U^\dagger) = 1 + \alpha^* U^\dagger + \cdots = \exp(-\alpha U) = 1 - \alpha\ U + \cdots.$$

Assuming the group parameter α is real, one finds that

$$U^\dagger = -U$$

so U is skew Hermitian. It is usual to write the generator U in terms of a Hermitian operator,

$$\mathcal{K} = \exp(\pm i\mathcal{M})$$

for instance, so that $U = \pm i\ \mathcal{M}$ where $\mathcal{M}^\dagger = \mathcal{M}$. This leaves open the possibility that $\mathcal{M}$ may represent some physical observable.

Symmetries: Roughly speaking, a symmetry is a transformation that may correspond to rotating or shifting the coordinates or to some more abstract operation, which may change the form of the state functions or of the physically observable operators or both, but does so in such a way that the squared matrix elements do not change. If $\mathcal{K}$ is a symmetry with respect to $\mathcal{Q}$ then

$$\mathcal{K}^\dagger\mathcal{Q}\mathcal{K} = \mathcal{Q}.$$

Expanding in s and equating term by term it follows that

$$\mathcal{K} = e^{is\mathcal{M}}$$

is a symmetry with respect to $\mathcal{Q}$, or that $\mathcal{M}$ generates a symmetry, if and only if

$$[\mathcal{M}, \mathcal{Q}] = 0.$$

Also, it was seen above that if $\mathcal{A}_\alpha$ acts as a ladder operator with respect to eigenfunctions of an observable T corresponding to $\mathcal{T}$ then

$$[\mathcal{T}, \mathcal{A}_\alpha] = \tau_\alpha\,\mathcal{A}_\alpha.$$

Then $\mathcal{A}_\alpha$ acts to transform eigenfunctions of $\mathcal{T}$ into other eigenfunctions with eigenvalue shifted by τ_α. So without going into detail, it should seem plausible at least that while the quantum mechanical operators naturally belong to an associative algebra, the commutator subalgebras to which they also belong will serve as representations for certain Lie algebras. So one studies these two things together, the Lie algebras and the representations in terms of operators.

Casimir operators: Generally, a Casimir is a quadratic expression formed from operators in some associative algebra in which there is a commutator representation of the Lie algebra, and the quadratic form is such that it commutes with the representations of all vectors in the Lie algebra. For example it was seen previously that rotations in three dimensions are generated by

$$U_{xy} = -y\frac{\partial}{\partial x} + x\frac{\partial}{\partial y}, \qquad U_{yz} = -z\frac{\partial}{\partial y} + y\frac{\partial}{\partial z}, \qquad U_{zx} = -x\frac{\partial}{\partial z} + z\frac{\partial}{\partial x},$$

which generate right-handed rotations in the xy-plane, the yz-plane and the zx-plane. The distinct commutators are

$$[U_{xy}, U_{yz}] = -U_{zx}, \qquad [U_{yz}, U_{zx}] = -U_{xy}, \qquad [U_{zx}, U_{xy}] = -U_{yz}.$$

Form using the associative product the operator

$$U^2 = U_{xy}^2 + U_{yz}^2 + U_{zx}^2.$$

Then, for instance,

$$[U^2, U_{xy}] = [U_{yz}^2 + U_{zx}^2, U_{xy}].$$

If one uses linearity and the derivation property, namely

$$[AB, C] = [A, B]C + A[B, C],$$

and then calculates in the associative algebra, the result is

$$\begin{aligned}[U^2, U_{xy}] &= U_{yz}[U_{yz}, U_{xy}] + [U_{yz}, U_{xy}]U_{yz} + U_{zx}[U_{zx}, U_{xy}] + [U_{zx}, U_{xy}]U_{zx} \\ &= +U_{yz}U_{zx} + U_{zx}U_{xz} - U_{zx}U_{yz} - U_{yz}U_{zx} = 0.\end{aligned}$$

Thus U^2 is an example of a quadratic Casimir operator. In the context of quantum mechanics, the associated Hermitian operators,

$$L_x = -i\hbar U_{yz}, \; L_y = -i\hbar U_{zx}, \; L_z = -i\,\hbar U_{xy},$$

are components of angular momentum and the Casimir operator is L^2, the angular momentum squared.

How can one construct such an operator in a representation of a general, semi-simple Lie algebra? Consider a representation where the Lie algebra basis vectors

$$\{b_1, b_2, b_3, \dots, b_n\}$$

are represented by operators

$$\{\mathcal{B}_1, \mathcal{B}_2, \mathcal{B}_3, \dots, \mathcal{B}_n\}.$$

One Casimir operator that can be formed from the complete basis in a simple way is the following one:

$$\mathcal{C} = K^{ij}\mathcal{B}_i\mathcal{B}_j,$$

where $\{K^{ij}\} = K^{-1}$ is the inverse of the Killing matrix $K_{ij} = K(b_i, b_j)$. To show this operator commutes with each operator $\mathcal{B}_k$ it may help to review some things.

Starting from a representation in terms of a commutator subalgebra, we saw by exploiting cyclical invariance of the trace that for any vectors u, v and w in the Lie algebra,

$$K(u, [v, w]) = K([u, v], w).$$

Then using the structure constants

$$[b_i, b_j] = c_{ij}^{\,k} b_k$$

and the definition

$$K_{ij} = K(b_i, b_j)$$

one finds

$$c_{jk}^{\,\ell}\, K_{i\ell} = c_{ij}^{\,\ell} K_{\ell k}.$$

The Killing form matrix is symmetric and the structure constants are skew-symmetric with respect to interchanging the lower indices. So, from this expression it follows, by looking first at the right side and then at the left, that the quantity

$$\gamma_{ijk} \equiv c_{ij}^{\,\ell} K_{\ell k},$$

changes sign under exchange of any pair of its three free indices. Then we had concluded above that

$$c_{ij}^{\,\ell} K_{\ell k} = \gamma_{ijk} = \gamma_o\, \varepsilon_{ijk},$$

for some constant γ_o, where ε_{ijk} is the Levi Civita anti-symmetric symbol. Thus one can solve for the structure constant in terms of the Levi Civita symbol:

$$K^{pk} c_{ij}^{\,\ell} K_{\ell k} = c_{ij}^{\,p} = \gamma_o\, K^{pk}\epsilon_{ijk}. \tag{3.10}$$

Now consider the commutator of $\mathcal{C}$ with any one of the basis operators. Using the Leibniz rule, i.e. the derivation property

$$[\mathcal{C}, \mathcal{B}_k] = K^{ij}\big[\mathcal{B}_i\mathcal{B}_j, \mathcal{B}_k\big] = K^{ij}\big(\mathcal{B}_i[\mathcal{B}_j, \mathcal{B}_k] + [\mathcal{B}_i, \mathcal{B}_k]\, \mathcal{B}_j\big).$$

Then, using again the structure constants:

$$[\mathcal{C}, \mathcal{B}_k] = K^{ij}\big(c_{jk}^{\,n}\, \mathcal{B}_i\, \mathcal{B}_n + c_{ik}^{\,n}\, \mathcal{B}_n\, \mathcal{B}_j\big).$$

In the second term I can interchange the dummy indices i and j, making use of symmetry of K^{ij}. The result is

$$[\mathcal{C}, \mathcal{B}_k] = K^{ij}\, c_{jk}^{\,n}(\mathcal{B}_i\, \mathcal{B}_n + \mathcal{B}_n\, \mathcal{B}_i).$$

Observing that the operator part of the expression is symmetric with respect to exchanging i and n, my plan for proving that it is zero is to show the tensor it is contracted with is skew-symmetric with respect to exchanging i and n. That way the sum will be zero. From equation (3.10),

$$c_{jk}^{\,n} = \gamma_o\, K^{mn}\epsilon_{jkm},$$

so

$$[\mathcal{C}, \mathcal{B}_k] = \gamma_o\, K^{ij}\, K^{mn}\epsilon_{jkm}(\mathcal{B}_i\, \mathcal{B}_n + \mathcal{B}_n\, \mathcal{B}_i) = \gamma_o\, \epsilon_{jkm}(\mathcal{B}^j\, \mathcal{B}^m + \mathcal{B}^m\, \mathcal{B}^j).$$

If one exchanges the summation index labels m and j, the operator expression in parentheses remains the same while the Levi Civita symbol changes sign. The commutator equals minus itself, and so it is zero: $[\mathcal{C}, \mathcal{B}_k] = 0$. Thus as predicted, $\mathcal{C}$ commutes with every operator in the representation.

References

[1] Roman P 1975 *Some Modern Mathematics for Physicists and Other Outsiders* vol 1 (Oxford: Pergamon)

[2] Jacobson N 1962 *Lie Algebras* (New York: Dover)

[3] Cahn R 1984 *Semi-simple Lie Algebras and Their Representations* (New York: Benjamin)

Chapter 4

Examples and applications

In order to round out the development, it is useful to have some more detailed examples. In this section the algebra *so*(5,R) of the five-dimensional (5D) rotation group and the quantum mechanical problem of a charged two-dimensional (2D) oscillator in a magnetic field are taken up. Finally, as another example, two generating functions for the spherical harmonics are obtained using Weisner's method.

4.1 The algebra *so*(5)

Here we select a Cartan subalgebra, construct root vectors and work out the root system for *so*(5), which is the simple Lie algebra associated with the group *SO*(5) of 5D rotations.

Consider a Euclidean 5D space with Cartesian coordinates. Because the coordinates are Cartesian in this example, I find it convenient to index them with subscripts rather than superscripts. The Lie group *SO*(5,*R*) is the group of rotations in a real 5D space. It is characterized by its action on this space. It transforms the coordinates in such a way that it preserves the sum

$$r^2 = x_1^2 + x_2^2 + x_3^2 + x_4^2 + x_5^2.$$

It might seem natural to begin by considering a matrix representation, but I elect to start from a set of infinitesimal generators. One can proceed directly by analogy with the three-dimensional (3D) case to find the infinitesimal generators for the local Lie group, i.e. the part connected continuously to the origin. The natural one-dimensional (1D) subgroups each act in a particular plane. In dimensions higher than three they are not characterized as rotations about a given axis, since there are several—in this case three—different axes perpendicular to a given plane. In the $x_i x_j$-plane, the rotation generator is

$$U_{ij} = x_i \frac{\partial}{\partial x_j} - x_j \frac{\partial}{\partial x_i} = x_i \partial_j - x_j \partial_i.$$

There are 10 of these. The reader should check to see that this annihilates r^2. One can make use of bilinearity and the derivation property to evaluate commutators

doi:10.1088/2053-2571/aae789ch4

involving associative products. To form commutators between the generators, first consider

$$[x_p\partial_q,\ x_r\partial_s] = x_p\delta_{qr}\partial_s - x_r\delta_{ps}\partial_q.$$

So,

$$[U_{pq},\ U_{rs}] = [x_p\partial_q - x_q\partial_p,\ x_r\partial_s - x_s\partial_r] = x_p\delta_{qr}\partial_s - x_r\delta_{ps}\partial_q - x_q\delta_{pr}\partial_s + x_r\delta_{qs}\partial_p \\ - x_p\delta_{qs}\partial_r + x_s\delta_{pr}\partial_q + x_q\delta_{ps}\partial_r - x_s\delta_{qr}\partial_p,$$

or

$$[U_{pq},\ U_{rs}] = \delta_{qr}U_{ps} + \delta_{ps}U_{qr} - \delta_{pr}U_{qs} - \delta_{qs}U_{pr}.$$

This holds modulo $U_{ps} = -U_{sp}$ and so on.

Physicists represent the generators by Hermitian operators, for the reasons mentioned above. In the case of the more familiar $so(3)$ one has $L_z = L_{21} = -i\,\hbar\,U_{12}$ so generalizing, $L_{jk} = -i\hbar U_{jk}$. It is more convenient to drop the $\hbar$. It can be reinserted if needs be. Let

$$U_{jk} = i\,L_{jk}.$$

Notice, though, that this substitution makes the Lie infinitesimal generator $i\,L_{jk}$ rather than L_{jk}, so the sign of the Killing form will be different and the structure constants will differ by factors of complex i if L_{jk} is used to calculate them instead of U_{jk}. Then the commutators become

$$[L_{pq},\ L_{rs}] = i\delta_{qr}L_{ps} + i\delta_{ps}L_{qr} - i\delta_{pr}L_{qs} - i\delta_{qs}L_{pr}.$$

The bracket is non-zero if and only if the two indices share one and only one digit. That is to say, if one and only one member of the set $\{p,q\}$ matches an element of the set $\{r,\ s\}$. Altogether, there are 30 different non-zero structure constants, counting the pair $c_{ij}^{k} = -c_{ji}^{k}$ as one, with values of ± 1, (or $\pm i$). From this, when the Killing form is calculated it is found to be diagonal, and in fact it is -6 times the identity matrix. The minus sign is important, and needs to be added if the L operators are used to calculate it. As in the 3D case, the Casimir operator is just L^2, the sum of squares. There are 14 different Cartan subalgebras, these being generated by pairs of generators with disjoint indices:

$$\{L_{12},\ L_{34}\},\quad \{L_{12},\ L_{45}\},\quad \{L_{12},\ L_{35}\},\quad \{L_{13},\ L_{24}\},\quad \cdots$$

Select $H = \text{span}\{L_{12},\ L_{34}\}$. To find the root vectors, a standard method is to make an arbitrary linear combination with undetermined coefficients, which are adjusted to satisfy the equations (3.9). Let the operator representing a general root vector e_α be written as A_α, or simply as A:

$$A = c_{12}\,L_{12} + c_{23}\,L_{23} + c_{34}\,L_{34} + c_{45}\,L_{45} + c_{51}\,L_{51} + c_{13}\,L_{13} + c_{35}\,L_{35} \\ + c_{52}\,L_{52} + c_{24}\,L_{24} + c_{41}\,L_{41}. \tag{4.1}$$

Due to the choice of the Cartan subalgebra H, the coefficients c_{12} and c_{34} are zero. The remaining space is eight-dimensional. The other coefficients are chosen in such a way that vector A satisfies the two eigenvalues

$$[L_{12}, A] = \mu A, \quad [L_{34}, A] = v A.$$

In other words, A should be an eigenvector simultaneously of $\mathrm{ad}(L_{12})$ and $\mathrm{ad}(L_{34})$. One expects eight different solutions.

Making the substitution (4.1) in the eigenvalue equation for $\mathrm{ad}(L_{12})$ yields

$$\begin{aligned}
&(x_3(c_{23} - ic_{13}\mu) + i\mu(c_{41}x_4 + c_{51}x_5) + c_{24}x_4 - c_{52}x_5)\partial_1 \\
&+(x_3(-(c_{13} + ic_{23}\mu)) + x_4(c_{41} - ic_{24}\mu) + x_5(c_{51} + ic_{52}\mu))\partial_2 \\
&-(ic_{13}\mu x_1 + ic_{23}\mu x_2 - ic_{35}\mu x_5 - c_{23}x_1 + c_{13}x_2)\partial_3 \\
&+(i(c_{24}\mu x_2 - c_{45}\mu x_5 + ic_{41}x_2) - x_1(c_{24} + ic_{41}\mu))\partial_4 \\
&+(-c_{51}(x_2 + i\mu x_1) + c_{52}(x_1 - i\mu x_2) + i\mu(c_{35}x_3 + c_{45}x_4))\partial_5 = 0.
\end{aligned}$$

Since the operator is zero the coefficient of each partial derivative is zero. Moreover, it has to vanish identically, i.e. at each point in space, so the coefficient of each monomial in the coordinates has to be zero. So for instance in the first term, or in the coefficient of ∂_1, the collected coefficient of x_3 is $c_{23} - i\mu\, c_{13}$. Thus $c_{23} - i\mu\, c_{13} = 0$. Setting the coefficient of each of the coordinate monomials to zero leads to a set of equations,

$$c_{23} - ic_{13}\mu = 0, \quad c_{24} + ic_{41}\mu = 0, \quad c_{51}\mu + ic_{52} = 0.$$

Similarly, for the coefficients of $\partial_2, \partial_3 \ldots$,

$$-c_{13} - ic_{23}\mu = 0, \quad c_{41} - ic_{24}\mu = 0, \quad c_{51} + ic_{52}\mu = 0,$$

$$c_{13}\mu + ic_{23} = 0, \quad c_{13} + ic_{23}\mu = 0, \quad c_{35}\mu = 0,$$

$$-c_{24} - ic_{41}\mu = 0, \quad c_{24}\mu + ic_{41} = 0, \quad c_{45}\mu = 0,$$

$$c_{52} - ic_{51}\mu = 0, \quad -c_{51} - ic_{52}\mu = 0, \quad c_{35}\mu = 0, \quad c_{45}\mu = 0.$$

There are three eigenvalues, and thus three families of possible root vectors. First, A may belong to the family

$$A_{-} = c_{13}\,(L_{13} - i\,L_{23}) + c_{24}\,(L_{24} - i\,L_{41}) + c_{51}(L_{51} - iL_{52}),$$

with eigenvalue $\mu = -1$, or A can belong to the family

$$A_0 = c_{35}\,L_{35} + c_{45}\,L_{45},$$

with eigenvalue $\mu = 0$, or else A can be of the form

$$A_+ = c_{13}(L_{13} + i\,L_{23}) + c_{24}(L_{24} + i\,L_{41}) + c_{51}(L_{51} + iL_{52}),$$

with $\mu = +1$.

For each of the three families, A_-, A_0, and A_+, the constants have thus been adjust to make each an eigenvector of $\mathrm{ad}(L_{12})$ with a definite μ eigenvalue. Now for each case, the remaining constants need to be adjusted via the same procedure to form an eigenvector also of $\mathrm{ad}(L_{23})$. The operators $\{A_\alpha\}$ corresponding to the root vectors $\{e_\alpha\}$ are found thus. The results are as follows:

$$A_1 = A_{(--)} = (L_{13} - i\,L_{23}) - (L_{24} - i\,L_{41}),$$
$$\lambda_1(L_{12}) = -1,\ \ \lambda_1(L_{34}) = -1.$$

$$A_2 = A_{(0,\,-)} = L_{35} - i\,L_{45},$$
$$\lambda_2(L_{12}) = 0,\ \ \lambda_2(L_{34}) = -1.$$

$$A_3 = A_{(+,\,-)} = (L_{13} + i\,L_{23}) + (L_{24} + i\,L_{41}),$$
$$\lambda_3(L_{12}) = +1,\ \ \lambda_3(L_{34}) = -1.$$

$$A_4 = A_{(-,\,0)} = L_{51} - i\,L_{52},$$
$$\lambda_4(L_{12}) = -1,\ \ \lambda_4(L_{34}) = 0.$$

$$A_5 = A_{(+,\,0)} = L_{51} + i\,L_{52},$$
$$\lambda_5(L_{12}) = +1,\ \ \lambda_5(L_{34}) = 0.$$

$$A_6 = L_{(-,\,+)} = (L_{13} - i\,L_{23}) + (L_{24} - i\,L_{41}),$$
$$\lambda_6(L_{12}) = -1,\ \ \lambda_6(L_{34}) = +1.$$

$$A_7 = A_{(0,\,+)} = L_{35} + i\,L_{45},$$
$$\lambda_7(L_{12}) = 0,\ \ \lambda_7(L_{34}) = +1.$$

$$A_8 = A_{(+,\,+)} = (L_{13} + i\,L_{23}) - (L_{24} + i\,L_{41}),$$
$$\lambda_8(L_{12}) = +1,\ \ \lambda_8(L_{34}) = +1.$$

The roots are shown in component form in figure 4.1.

What is the fundamental root set in this case? There must be two vectors in Π, since H and hence also H^* is 2D. One may take $\pi_1 = \lambda_5$ and $\pi_2 = \lambda_6$: (Why not take $\pi_2 = \lambda_7$?)

$$\lambda_5 = \pi_1, \quad \lambda_6 = \pi_2, \quad \lambda_7 = \pi_1 + \pi_2, \quad \lambda_8 = 2\pi_1 + \pi_2.$$

Then the other roots are just minus these:

$$\lambda_4 = -\pi_1, \quad \lambda_3 = -\pi_2, \quad \lambda_2 = -(\pi_1 + \pi_2), \quad \lambda_1 = -(2\pi_1 + \pi_2).$$

This illustrates the general fact described above that for each $\lambda \in \Phi$ there is a set of non-negative integers $\{n_1, n_2, \ldots\}$ and an over-all sign $\epsilon \in \{-1, +1\}$, which depends on λ, such that

$$\lambda = \epsilon(n_1\,\pi_1 + n_2\,\pi_2 + \cdots + n_8\,\pi_8).$$

The Weyl group is generated by σ_1 and σ_2 corresponding to reflections through planes perpendicular to $\pi_1 = \lambda_5$ and $\pi_2 = \lambda_6$. It is instructive to see how these two operations σ_1 and σ_2 can map π_1 and π_2 around to all the other roots in Φ. Notice that the roots form two stars, each comprising roots of the same length. Starting from $\pi_1 = \lambda_5$ one finds

$$\lambda_7 = \sigma_2(\lambda_5), \quad \lambda_4 = \sigma_1(\lambda_5), \quad \lambda_2 = \sigma_2(\sigma_1(\lambda_5)).$$

Then, for the other star one can write similarly,

$$\lambda_2 = \sigma_1(\lambda_6), \quad \lambda_3 = \sigma_2(\lambda_6), \quad \lambda_1 = \sigma_1(\sigma_2(\lambda_6)).$$

Recasting *so*(5) as represented by quantum operators: It is often useful to think of the vectors in the Lie algebra that have come from the infinitesimal generators, which we have already expressed in Hermitian form, as quantum operators acting on a space of possible quantum states of some physical system. To this end, let eight

root	L_{12}	L_{34}
λ_1	−1	−1
λ_2	0	−1
λ_3	+1	−1
λ_4	−1	0
λ_5	+1	0
λ_6	−1	+1
λ_7	0	+1
λ_8	+1	+1

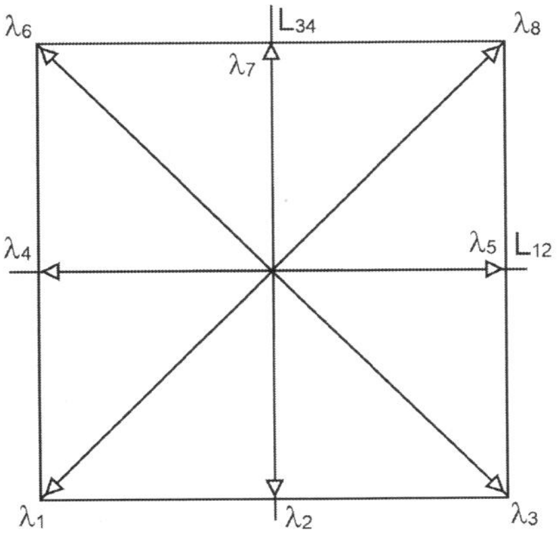

Figure 4.1. Roots for so(5) Lie algebra.

operators outside of the Cartan subalgebra, the root vectors, be represented as A, $A^\dagger$, B, $B^\dagger$, D, $D^\dagger$, F and $F^\dagger$, where

$$A = A_{(-,\,-)},\quad A^\dagger = A_{(+,\,+)}\quad B = A_{(-,\,+)},\quad B^\dagger = A_{(+,\,-)},$$
$$D = A_{(-,\,0)},\quad D^\dagger = A_{(+,\,0)}\quad F = A_{(0,\,-)},\quad F^\dagger = A_{(0,\,+)}.$$

A, $A^\dagger$, B, $B^\dagger$, D, $D^\dagger$, F and $F^\dagger$ are written in a way that emphasizes the Hermitian conjugate relation between raising and lowering operators. The conjugate, as in $A^\dagger$, is with respect to the inner product induced on the 4-sphere. For complex functions of five real variables serving as Cartesian coordinates,

$$\langle\langle f, g\rangle\rangle = \int_{all\ space} f^*(x)g(x)d^5x.$$

When there is $SO(5)$ symmetry, the orbit of a point a distance r from the origin is a 4-sphere of radius r. One can introduce polar coordinates, $\{r, \theta_1, \theta_2, \theta_3, \theta_4\}$. In the special case when f and g separate into products of radial times angular parts,

$$\langle\langle f, g\rangle\rangle = (f_{rad}, g_{rad})\langle f_{ang}, g_{ang}\rangle.$$

Then as usual,

$$\langle u, A\, v\rangle = \langle A^\dagger\, u, v\rangle.$$

A Casimir operator can be formed from these operators in a natural way from the associative products. One way to proceed, since the transformation involved is linear, is to express the basis operators in terms of the set $\{A, B, D, F\}$ and Hermitian conjugates. Doing this and then forming sums of squares,

$$\mathcal{C} = \mathcal{L}^2 = L_{12}^2 + L_{23}^2 + L_{34}^2 + L_{45}^2 + L_{51}^2 + L_{13}^2 + L_{35}^2 + L_{52}^2 + L_{24}^2 + L_{41}^2,$$

I find

$$L_{13}^2 + L_{23}^2 + L_{24}^2 + L_{41}^2 = \frac{1}{4}(A\,A^\dagger + A^\dagger A + B\,B^\dagger + B^\dagger B),$$
$$L_{35}^2 + L_{45}^2 = \frac{1}{2}(D\,D^\dagger + D^\dagger D),$$
$$L_{51}^2 + L_{52}^2 = \frac{1}{2}(F\,F^\dagger + F^\dagger F).$$

The Casimir operator found in this way is

$$\mathcal{C} = H + \frac{1}{4}(A\,A^\dagger + A^\dagger A + B\,B^\dagger + B^\dagger B) + \frac{1}{2}(D\,D^\dagger + D^\dagger D + F\,F^\dagger + F^\dagger F),$$

where

$$H = L_{12}^2 + L_{34}^2.$$

4.2 Two-dimensional oscillator in a magnetic field

The harmonic oscillator is a famously solvable problem in either classical or quantum mechanics because its dynamical equations are linear. Any physical system that can be approximated by an oscillator—for instance by restricting attention to the vicinity of an equilibrium point—can be analyzed in a very satisfactory way. This linearity correlates with the fact that the Hamiltonian function for a harmonic oscillator is a quadratic form in the dynamical variables. If an electromagnetic field is applied, linear perturbations of the Hamiltonian result, but the Hamiltonian can still be made into a homogeneous quadratic form by a linear transformation. The problem remains exactly solvable [2].

To make this section self-contained for a general reader, here is a short review.

The 1D oscillator was introduced in one of the earlier sections above. For this system, the coordinate x is associated with a momentum p. In quantum theory, these are operators:

$$x\colon\ f(x) \mapsto x\, f(x), \quad p\colon\ f(x) \mapsto -i\, \hbar f'(x).$$

This means that x (the operator) takes $f(x)$ into x (the coordinate variable) *times* $f(x)$, while the operator p takes $f(x)$ into an expression proportional to its derivative. The complex unit i is included to make p Hermitian. These operate on a space of state functions such as $\psi(x)$ that represent probability amplitudes. The operator product is defined by composition. First one operates and then the other, and so this is an associative algebra. The commutator algebra will have the Lie structure of interest. Stationary states of a quantum system are solutions to an eigenvalue problem associated with a Hamiltonian operator H, namely

$$H\, \psi(x) = E\, \psi(x),$$

the Schrödinger equation. For the 1D oscillator it was seen above that

$$H = \frac{1}{2m}p^2 + \frac{1}{2}\kappa x^2.$$

The spring constant κ is also $\kappa = m\, \omega_0^2$ where the constant ω_0 is an angular frequency.

In this section the oscillator will be altered in two ways: it is extended to two spatial dimensions with coordinates x and y and a magnetic field is applied.

In the following, x and y are either Hermitian operators or the coordinate variables considered as members of the function space. I hope that the distinction will be clear from context. A significant feature of quantum theory is that, as operators, the position–momentum pairs x and p_x and y and p_y do not commute: $[x_i, p_j] = i\, \hbar\, \delta_{ij}$. This concludes the review.

The Hamiltonian for the isotropic 2D oscillator without an applied field is

$$H_0 = \frac{1}{2m}\left(p_x^2 + p_y^2\right) + \frac{1}{2}m\, \omega_0^2 (x^2 + y^2).$$

Applying a magnetic field of strength B_o (constant in both time and space) in the z direction perpendicular to the plane changes H in such a way as to include the effect of the magnetic, Lorentz force, where one is assuming the oscillator is electrically charged. The operator H is formed from the classical Hamiltonian function by replacing the position and momentum variables by the operators according to a definite recipe. Generally, to apply a constant magnetic field, momentum p_i is replaced by $p_i - q\,A_i$. Here, q is the charge on the oscillator, and A_i is the ith component of the magnetic vector potential. For the case at hand one can apply the field by making the substitution [1],

$$p_x \rightarrow p_x + \frac{1}{2} q B_o y,$$

$$p_y \rightarrow p_y - \frac{1}{2} q B_o x.$$

The starting point for the physics problem is the Hamiltonian operator

$$H = \frac{1}{2m}\left(\left(p_x + \frac{1}{2} q B_o y\right)^2 + \left(p_y - \frac{1}{2} q B_o x\right)^2\right) + \frac{1}{2}\, m\, \omega_0^2 (x^2 + y^2),$$

acting on functions of x and y that are assumed to go to zero sufficiently rapidly at infinity. The first objective in this section is to find the stationary states of the system, meaning simultaneous eigenfunctions of H and any operator that commutes with H, and then to construct a generating functions.

The associative algebra is generated by the operators $\{x, p_x, y, p_y\}$. The Lie algebra of interest will be some subset of its commutator algebra. To simplify the problem, notice by comparing coefficients of x^2 or y^2 in the different terms in H that ω_0 and the ratio qB_o/m have the same physical dimensions. Therefore, let the symbol ω_1 replace $qB_o/2m$. This is a second, independent parameter of the physical problem. It is half of the cyclotron frequency. The frequency parameter ω_0 depends on the strength of the spring, and the new frequency parameter ω_1 depends on the strength of the magnetic field. Expanding and simplifying, one finds

$$H = \frac{1}{2m}\left(p_x^2 + p_y^2\right) - \omega_1 L + \frac{1}{2} m\left(\omega_0^2 + \omega_1^2\right)(x^2 + y^2),$$

where

$$L = x\, p_y - y\, p_x .$$

This is the z component of the angular momentum operator. From the fact that the momenta are represented as partial derivatives, it is seen that L is proportional to the infinitesimal rotation generator,

$$L = -i\,\hbar\left(x\frac{\partial}{\partial y} - y\frac{\partial}{\partial x}\right) = i\,\hbar\, U_{xy}.$$

One can write

$$H = H_0 - \omega_1 L,$$

where

$$H_0 = \frac{1}{2m}\left(p_x^2 + p_y^2\right) + \frac{1}{2} m\, \Omega^2(x^2 + y^2),$$

where $\Omega = \sqrt{(\omega_0^2 + \omega_1^2)}$. The Lie algebra of interest contains

$$x, \quad y, \quad p_x, \quad p_y, \quad L, \quad H_0, \quad \text{and} \quad 1.$$

It must contain 1 because $[x, p_x] = i\,\hbar$, which is a constant. Experience with physical systems should suggest that L and H_0 commute. The reason is that U_{xy} generates a rotation of the xy-plane. So for any function $f(x, y)$,

$$e^{\alpha U_{xy}}((x^2 + y^2)f(x, y)) = (\tilde{x}^2 + \tilde{y}^2)f(\tilde{x}, \tilde{y}).$$

But a circle is invariant under rotation, so $(\tilde{x}^2 + \tilde{y}^2) = (x^2 + y^2)$. If I think of $(x^2 + y^2)$ as an operator for 'multiplying by' $(x^2 + y^2)$ then

$$e^{\alpha U_{xy}}((x^2 + y^2)f(x, y)) = (\tilde{x}^2 + \tilde{y}^2)f(\tilde{x}, \tilde{y}) = (x^2 + y^2)e^{\alpha U_{xy}}f(x, y)$$

shows that

$$[e^{\alpha U_{xy}}, (x^2 + y^2)] = 0.$$

Expanding in a Taylor series in α and comparing terms shows that $[U_{xy}, (x^2 + y^2)] = 0$. The operator U_{xy} does not act on functions of p_x or p_y so it would seem to commute with all of H_0. But, there is one important reservation. The problem is that a transformation acting on x and y automatically effects a transformation acting on p_x and p_y, because these operators are derivatives with respect to x and y. It is not very difficult to work out what this induced transformation is, and then to see that in this case the full rotation transformation still commutes with both $(x^2 + y^2)$ and $(p_x^2 + p_y^2)$. Of course $(p_x^2 + p_y^2)$ looks to be rotation invariant, and indeed it is. However, there is a much easier way to show that L and H_0 commute. That is simply to work out the commutators directly:

$$\begin{aligned}[L, x^2 + y^2] &= [x\, p_y - y\, p_x, x^2 + y^2] = [x\, p_y, y^2] - [y\, p_x, x^2] \\ &= x\, y\, [p_y, y] + x\, [p_y, y]\, y - y\, [p_x, x]\, x - y\, [p_x, x]\, y\, x = 2\, i\, \hbar\, (x\, y - y\, x) = 0.\end{aligned}$$

Similarly,

$$\left[L, p_x^2 + p_y^2\right] = \left[x\, p_y - y\, p_x, p_x^2 + p_y^2\right] = 2\, i\, \hbar\, p_y\, p_x - 2\, i\, \hbar\, p_x\, p_y = 0.$$

Thus indeed $[L, H_0] = 0$. There are three mutually commuting operators: $\{H_0, L, 1\}$. Now, to be sure the algebra is closed, several more commutators need to be found:

$$[x, H_0] = \frac{1}{2m}\left[x, p_x^2\right] = \frac{i\,\hbar}{m}\,p_x, \quad [y, H_0] = \frac{i\,\hbar}{m}\,p_y,$$
$$[p_x, H_0] = -i\,\hbar m\,\Omega^2\,x, \quad \text{and} \quad [p_y, H_0] = -i\,\hbar\,m\,\Omega^2\,y.$$

Then

$$[x, L] = -i\,\hbar\,y, \quad [y, L] = +i\,\hbar\,x, \quad [p_x, L] = -i\,\hbar\,p_y, \quad [p_y, L] = +i\,\hbar\,p_x.$$

Finding the root vectors: From a physical point of view, the root vectors, which are eigenvectors of ad(L) and of ad(H_0), are ladder operators connecting a set of mutual eigenfunctions of the energy and angular momentum. These are of central importance in a system of this type. Considering $\{H_0, L, 1\}$ as a basis for a Cartan subalgebra, one should set about finding root vectors. To this end, let

$$A = c_1\,x + c_2\,y + c_3\,p_x + c_4\,p_y.$$

The constants are to be adjusted so that $[L, A] = \mu\,A$ and $[H_0, A] = v\,A$.

$$[L, A] = c_1(i\,\hbar\,y) + c_2(-i\,\hbar\,x) + c_3(i\,\hbar\,p_y) + c_4(-i\,\hbar\,p_x),$$

so

$$i\hbar c_1 = \mu c_2, \qquad -i\hbar c_2 = \mu c_1, \qquad i\hbar c_3 = \mu c_4, \qquad i\hbar c_4 = \mu c_3.$$

From the first two equations, $\mu = \pm\hbar$. So

$$\mu = \pm\hbar, \qquad c_2 = \pm i\,c_1, \qquad c_4 = \pm i\,c_3.$$

Then

$$[H_0, A] = c_1\!\left(-\frac{i\,\hbar}{m}\,p_x,\right) + c_2\!\left(-\frac{i\,\hbar}{m}\,p_y,\right) + c_3\,(i\,\hbar m\,\Omega^2\,x) + c_4\,(i\,\hbar m\,\Omega^2\,y),$$

so

$$-\frac{i\,\hbar}{m}\,c_1 = v\,c_3, \quad -\frac{i\,\hbar}{m}\,c_2 = v\,c_4,$$
$$i\,\hbar m\,\Omega^2\,c_3 = v\,c_1, \quad i\,\hbar m\,\Omega^2\,c_4 = v\,c_2.$$

The first and third equation give $v = \pm\hbar\Omega$. So,

$$v = \pm\,\hbar\,\Omega \quad c_1 = \pm i\,m\,\Omega\,c_3, \qquad c_2 = \pm i\,m\,\Omega\,c_4.$$

The ladder operators corresponding to root vectors, scaled to dimensionless form and normalized such that $[a, a^\dagger] = [b, b^\dagger] = 1$, are listed in table 4.1. The dagger notation indicates that $a^\dagger$ and $b^\dagger$ are Hermitian conjugates of a and b respectively. The physical difference between a and b is that the y coordinate is revered. Hence p_y is also reversed. This reflection has the effect of interchanging left-hand and right-hand circulation. Classically it would reverse the angular momentum.

Table 4.1. Lader operators for charged two-dimensional oscillator in magnetic field.

v	v	$v - \omega_1\,\mu$	Ladder operator
$-\hbar$	$-\hbar\,\Omega$	$-\hbar(\Omega - \omega_1)$	$a = \frac{1}{2}\sqrt{\frac{m\,\Omega}{\hbar}}(x - i\,y) + \frac{1}{2\,\sqrt{m\,\hbar\,\Omega}}(+i\,p_x + p_y)$
$+\hbar$	$-\hbar\,\Omega$	$+\hbar(\Omega + \omega_1)$	$b = \frac{1}{2}\sqrt{\frac{m\,\Omega}{\hbar}}(x + i\,y) + \frac{1}{2\,\sqrt{m\,\hbar\,\Omega}}(+i\,p_x - p_y)$
$-\hbar$	$+\hbar\,\Omega$	$-\hbar(\Omega + \omega_1)$	$b^\dagger = \frac{1}{2}\sqrt{\frac{m\,\Omega}{\hbar}}(x - i\,y) + \frac{1}{2\,\sqrt{m\,\hbar\,\Omega}}(-i\,p_x - p_y)$
$+\hbar$	$+\hbar\,\Omega$	$+\hbar(\Omega - \omega_1)$	$a^\dagger = \frac{1}{2}\sqrt{\frac{m\,\Omega}{\hbar}},\,(x + i\,y) + \frac{1}{2\,\sqrt{m\,\hbar\,\Omega}}(-i\,p_x + p_y)$

With the associative multiplication it is possible to re-express the physical operators H_0 and L in terms of a, $a^\dagger$, b and $b^\dagger$:

$$a^\dagger\,a = \frac{m\,\Omega}{4\,\hbar}\,(x^2 + y^2) + \frac{1}{4\,m\,\hbar\,\Omega}\left(p_x^2 + p_y^2\right) + \frac{1}{2\,\hbar}\,(L - \hbar),$$
$$b^\dagger\,b = \frac{m\,\Omega}{4\,\hbar}\,(x^2 + y^2) + \frac{1}{4\,m\,\hbar\,\Omega}\left(p_x^2 + p_y^2\right) - \frac{1}{2\,\hbar}\,(L + \hbar).$$

Then,

$$H_o = \hbar\,\Omega(a^\dagger a + b^\dagger b + 1) \quad \text{and} \quad L = \hbar(a^\dagger a - b^\dagger b).$$

Putting these together gives for the full Hamiltonian operator

$$H = H_o - \omega_1\,L = \hbar(\Omega - \omega_1)a^\dagger\,a + \hbar\,(\Omega + \omega_1)\,b^\dagger\,b + \hbar\,\Omega.$$

Comparing the operators in table 4.1 with the ladder operators a and $a^\dagger$ in the introductory material on the 1D oscillator (quantum oscillator example), one can see that

$$a = \frac{1}{\sqrt{2}}\left(a_x - i\,a_y\right) \quad \text{and} \quad b = \frac{1}{\sqrt{2}}\left(a_x + i\,a_y\right).$$

These two commute. Then also

$$a^\dagger = \frac{1}{\sqrt{2}}\left(a_x^\dagger + i\,a_y^\dagger\right) \quad \text{and} \quad b^\dagger = \frac{1}{\sqrt{2}}\left(a_x^\dagger - i\,a_y^\dagger\right)$$

commute, and the transformation is unitary with determinant $+i$. The inverse is

$$a_x = \frac{1}{\sqrt{2}}(a + b), \qquad a_y = \frac{i}{\sqrt{2}}(a - b).$$

It can be seen in the table that the roots form a stretched $\times$. Because of the stretch, which depends on the model parameters, there cannot be a Weyl symmetry. Reflecting one of the roots through a line perpendicular to another would result in a new dual vector not in the root set. This should not be surprising, as the algebra is not semi-simple. It contains x and p_x and hence 1, and 1 by itself certainly generates an Abelean ideal.

The length scale in the physical oscillator system considered currently depends on the field strength as well as the force constant of the spring. It can be represented by

$$c = \sqrt{\frac{\hbar}{m\,\Omega}} = \sqrt{\frac{\hbar}{m\sqrt{\omega_0^2 + \omega_1^2}}} = \sqrt{\frac{\hbar}{m\sqrt{\omega_0^2 + \left(\frac{q\,B_o}{2\,m}\right)^2}}}.$$

In terms of this length scale,

$$a_x = \frac{1}{\sqrt{2}}\left(\frac{x}{c} + i\,\frac{p_x}{m\,\Omega\,c}\right) = \frac{1}{\sqrt{2}}\left(\frac{x}{c} + c\,\frac{\partial}{\partial x}\right), \quad \text{etc.}$$

From the root table it appears that the allowed, normalizable eigenvalues of energy and of angular momentum depend on integer quantum numbers. Consider dimensionless products $a^\dagger\,a$ and $b^\dagger\,b$ in the associative algebra. These are in the Cartan subalgebra of the commutator algebra. Suppose $\psi(x, y)$ is an eigenfunction of $a^\dagger\,a$,

$$a^\dagger\,a\,\psi(x, y) = \lambda\,\psi(x, y).$$

Then

$$\begin{aligned}(a^\dagger\,a)(a^\dagger\psi_{n,k}) &= a^\dagger\,(a\,a^\dagger)\psi_{n,k}(x, y) = a^\dagger([a, a^\dagger] + a^\dagger\,a)\psi_{n,k}(x, y),\\ &= (1 + \lambda)\psi_{n,k}.\end{aligned}$$

Thus, as in the case of the simple oscillator, starting from $n = 0$, continued application of $a^\dagger$ transforms $\psi(x, y)$ into an infinite succession of other eigenfunctions with eigenvalues increasing by 1 at each step, starting from the unique ground state annihilated by a and by b where $\lambda = 0$. The eigenvalues are integers $n = 0, 1, 2 \ldots$. A similar argument holds for b, implying another integer quantum number k. The a and b operators are constructed so that either $a\psi_{n,k}$ is zero, or it is proportional to another eigenfunction $\psi_{n-1,k}$. Similarly, either $b\psi_{n,k}$ is zero, or proportional to $\psi_{n,k-1}$. There are no normalizable solutions to $a^\dagger\psi_{n,k} = 0$, or to $b^\dagger\psi_{n,k} = 0$, so $a^\dagger\psi_{n,k}$ is proportional to $\psi_{n+1,k}$ and $b^\dagger\psi_{n,k}$ is proportional to $\psi_{n,k+1}$. Following the same procedure as in the case of the 1D oscillator, the ground state eigenfunction $\psi_{0,\,0}(x, y)$ should be annihilated by both a and b, and hence by both a_x and a_y. This fact leads to a separable differential equation, and the normalized solution is

$$\psi_{0,0}(x, y) = \frac{1}{c\,\sqrt{\pi}}\exp\left(-\frac{1}{2}\left(\frac{x^2 + y^2}{c^2}\right)\right).$$

So the mutual eigenfunctions of $a^\dagger a$ and $b^\dagger b$ are such that

$$a^\dagger a\,\psi_{n,k}(x, y) = n\psi_{n,k}(x, y) \quad b^\dagger b\,\psi_{n,k}(x, y) = k\psi_{n,k}(x, y).$$

From the root vectors and their roles as ladder operators, the spectra of allowed stationary-state energy and angular momentum eigenvalues are now determined. These energy and angular momentum eigenvalues are

$$\tilde{E}_{n,k} = \hbar(\Omega - \omega_1)n + \hbar\,(\Omega + \omega_1)\,k + \hbar\,\Omega, \quad \text{and} \quad \hbar\,\ell = \hbar\,(n - k). \tag{4.2}$$

It is instructive to reorganize these in terms of a principal energy quantum number $N = n + k$ by comparing to the case with the magnetic field shut off:

$$E = \hbar\,\Omega\,(N + 1) - \hbar\,\omega_1\,\ell.$$

One finds

$$n = \frac{N + \ell}{2}, \qquad k = \frac{N - \ell}{2}.$$

Thus N and ℓ must both be even or both be odd. For a fixed value of N, ℓ takes on values from 0 to N,

$$\ell = -N + 2\,k \quad \text{for} \quad k \in \{0,\, 1,\, 2,\, \dots\,,\, N\},$$

and the accessible values of N for fixed ℓ are

$$N = \ell + 2\,k \quad \text{for} \quad k\ \in\ \{0,\, 1,\, 2,\, 3,\, \dots\}.$$

The raising operator $a^\dagger$ takes $\psi_{n,k}$ into a function proportional to $\psi_{n+1,k}$:

$$a^\dagger\psi_{n,k}(x,\, y) = C_n\psi_{n+1,k}.$$

Normalization of all the eigenfunctions such that $\langle\psi_{n,k},\, \psi_{n,k}\rangle = 1$ lets one evaluate the proportionality constant:

$$\langle a^\dagger\psi_{n,k},\, a^\dagger\psi_{n,k}\rangle = |C_n|^2\,\langle\psi_{n+1,k},\, \psi_{n+1,k}\rangle = \langle\psi_{n,k},\, a\, a^\dagger\psi_{n,k}\rangle,$$

and using $[a,\, a^\dagger] = 1$,

$$|C_n|^2\,\langle\psi_{n+1,k},\, \psi_{n+1,k}\rangle = (n + 1)\,\langle\psi_{n,k},\, \psi_{n,k}\rangle\ .$$

Thus $|C_n| = \sqrt{n + 1}$ as in the case of the 1D oscillator. A similar result obtains for $b^\dagger$. Finally, bearing in mind that a and b commute,

$$\psi_{n,k}(x,\, y) = \frac{1}{\sqrt{n!\; k!}}\,(a^\dagger)^n(b^\dagger)^k\psi_{0,0}(x,\, y).$$

Generating function for the eigenfunctions: To derive a generating function I can use roughly the same method as was used for the 1D oscillator. That was to construct a group transformation taking a trivial generating function into a non-trivial one. Consider

$$f(x,\, y,\, s,\, t) = \sum_{n,k} \frac{s^n\, t^k}{\sqrt{n!\; k!}}\psi_{n,k}(x,\, y).$$

Making use of the commutativity,

$$f(x, y, s, t) = \exp(s\, a^{\dagger}) \exp(t\, b^{\dagger}) \psi_{0,0}(x, y).$$

The difficulty in evaluating this is that the root vectors $a^{\dagger}$ and $b^{\dagger}$ are not in the form of infinitesimal generators. This sort of problem was seen earlier in the case of the 1D oscillator. The successful approach was to extend the group action. Now the same strategy can be applied here.

Starting from

$$a^{\dagger} = \frac{1}{\sqrt{2}}\left(a_x^{\dagger} + i\, a_y^{\dagger}\right) = \frac{1}{2}\left(\frac{x + i\, y}{c} - c\left(\frac{\partial}{\partial x} + i\, \frac{\partial}{\partial y}\right)\right)$$

and

$$b^{\dagger} = \frac{1}{\sqrt{2}}\left(a_x^{\dagger} - i\, a_y^{\dagger}\right) = \frac{1}{2}\left(\frac{x + i\, y}{c} - c\left(\frac{\partial}{\partial x} - i\, \frac{\partial}{\partial y}\right)\right)$$

one finds that both $a^{\dagger}$ and $b^{\dagger}$ annihilate

$$u(x, y) = \exp\left(\frac{x^2 + y^2}{2\, c^2}\right).$$

Suppose the function $u(x, y)$ is annihilated by $a^{\dagger}$. Then consider how the operator $a^{\dagger}$ acts on the product $u(x, y)\, f(x, y)$ for a general $f(x, y)$:

$$a^{\dagger}(u(x, y)\, f(x, y)) = (a^{\dagger}\, u(x, y)) f(x, y) + u(x, y) \frac{c}{2}\left(\frac{\partial}{\partial x} + i\, \frac{\partial}{\partial y}\right) f(x, y).$$

The first term on the right-hand side vanishes, and then in brief notation

$$a^{\dagger}\, u\, f = u\, \frac{c}{2}\left(\partial_x + i\, \partial_y\right) f.$$

As an operator,

$$a^{\dagger} = \frac{c}{2}\, u\left(\partial_x + i\, \partial_y\right) \frac{1}{u}.$$

Similarly, changing y to $-y$,

$$b^{\dagger} = \frac{c}{2}\, u\left(\partial_x - i\, \partial_y\right) \frac{1}{u}.$$

The operators will be exponentiated. When exponentiated, the resulting operators define a group action. The action can be made into a simple translation when the physical coordinates x and y are expressed in terms of canonical variables of the group, which simplify the form of the Taylor series. First notice that

$$\frac{1}{2}\left(\partial_x + i\, \partial_y\right)(x + i\, y) = 0, \quad \frac{1}{2}\left(\partial_x + i\, \partial_y\right)(x - i\, y) = 1,$$

and also that

$$\frac{1}{2}(\partial_x - i\,\partial_y)(x + i\,y) = 1, \quad \frac{1}{2}(\partial_x - i\,\partial_y)(x - i\,y) = 0.$$

If I introduce the canonical variables

$$\xi = \frac{x - i\,y}{c}, \quad \eta = \xi^* = \frac{x + i\,y}{c},$$

then

$$a^\dagger = -u\,\frac{\partial}{\partial\xi}\,\frac{1}{u}, \quad b^\dagger = -u\,\frac{\partial}{\partial\eta}\,\frac{1}{u}.$$

The fact that $a^\dagger$ and $b^\dagger$ commute is now expressed simply in terms of Clairaut's theorem on the equality of mixed partials for sufficiently well behaved functions. Application of these expressions simplifies calculation of the generating function, working first in the variables ξ and η and then transforming back to x and y:

$$f(x, y, s, t) = \exp(s\,a^\dagger)\exp(t\,b^\dagger)\psi_{0,0}(x, y).$$

Let

$$\hat{\psi}(\xi, \eta) = \psi_{0,0}(x(\xi, \eta), y(\xi, \eta)), \quad \hat{u}(\xi, \eta) = u(x(\xi, \eta), y(\xi, \eta)).$$

Then

$$\begin{aligned}
\hat{f}(\xi, \eta, s, t) &= \exp\left(-s\,\hat{u}\,\frac{\partial}{\partial\xi}\,\frac{1}{\hat{u}}\right)\exp\left(-t\,\hat{u}\,\frac{\partial}{\partial\eta}\,\frac{1}{\hat{u}}\right)\hat{\psi}(\xi, \eta)\\
&= \hat{u}(\xi, \eta)\exp\left(-s\,\frac{\partial}{\partial\xi}\right)\exp\left(-t\,\frac{\partial}{\partial\eta}\right)\frac{1}{\hat{u}(\xi, \eta)}\hat{\psi}(\xi, \eta).\\
\hat{f}(\xi, \eta, s, t) &= \hat{u}(\xi, \eta)\frac{\hat{\psi}(\xi - s, \eta - t)}{\hat{u}(\xi - s, \eta - t)}.
\end{aligned}$$

This is easy to evaluate:

$$\hat{\psi} = \frac{1}{c\,\sqrt{\pi}}\exp\left(-\frac{1}{2}\,\xi\,\eta\right), \quad \hat{u} = \exp\left(+\frac{1}{2}\xi\,\eta\right).$$

On substituting and simplifying,

$$\hat{f}(\xi, \eta, s, t) = \hat{\psi}(\xi, \eta)e^{-s\,t + (s\,\eta + t\,\xi)}.$$

Finally, substituting back to the coordinates (x, y), one finds that the generating function for eigenstates of a charged quantum oscillator in a magnetic field has a simple form, to wit

$$\begin{aligned}
f(x, y, s, t) &= \sum_{n,k}\frac{s^n\,t^k}{\sqrt{n!\,k!}}\psi_{n,k}(x, y) \qquad (4.3)\\
&= \psi_{0,\,0}(x, y)\exp\left(s\frac{x + i\,y}{c} + t\frac{x - i\,y}{c} - s\,t\right).
\end{aligned}$$

This contains quite a bit of information. An important thing to check is the set of all normalization integrals. It should be the case that

$$\langle \psi^*_{n_1,\ell_1}, \psi_{n_2,\ell_2} \rangle = \delta_{n_1,n_2}\delta_{\ell_1,\ell_2}. \tag{4.4}$$

To check this, one can start from the eigenfunction generating function:

$$\int f(x, -y, s, t) f(x, y, u, v)\, dx\, dy =$$

$$\frac{1}{\pi\, c^2} \int \exp\left(-(st + uv) + \frac{x}{c}(s + t + u + v) - i\frac{y}{c}(s - t - u + v) - \frac{x^2 + y^2}{c^2} \right) dx\, dy.$$

This kind of Gaussian integral appears often. To evaluate it one transforms linearly from x and y to variables that make the cross terms in the quadratic part of the exponent vanish, and then completes the square to incorporate the linear terms The result is

$$\int f(x, -y, s, t) f(x, y, u, v)\, dx\, dy = e^{s\,u+t\,v}.$$

Complex conjugation of the eigenfunction on the left accounts for the sign reversal of y in the generator. Since in the Maclaurin series of this expression, non-vanishing terms contain equal powers of u and s and equal powers of v and t, it implies orthogonality of the eigenfunctions. Expanding

$$e^{s\,u+t\,v} = e^{s\,u}\, e^{t\,v} = \sum_{n,k} \frac{1}{n!\,k!} (s\,u)^n (t\,v)^k.$$

Comparing with

$$\int f(x, -y, s, t) f(x, y, u, v)\, dx\, dy = \sum_{n,k} \frac{(s\,u)^n\,(t\,v)^k}{n!\,k!} \langle \psi_{n,k}, \psi_{n,k} \rangle$$

equation (4.4) follows.

Sombrero-shaped potential in a magnetic field: To demonstrate the use of the formal results obtained up to this point, it is interesting to consider a perturbation to the Hamiltonian. Suppose one were to add a repulsive Gaussian potential that would push the states of the physical system away from the coordinate origin. That is to say, suppose in addition to the magnetic field and the oscillator potential one were to include a perturbation

$$V = V_0 e^{-r^2/b^2},$$

where b is a length controlling the width of the Gaussian hump and V_0 is its height at the origin. This combined potential takes the shape of a sombrero or of the bottom of a wine bottle as illustrated in figure 4.2. Suppose one is interested in finding the

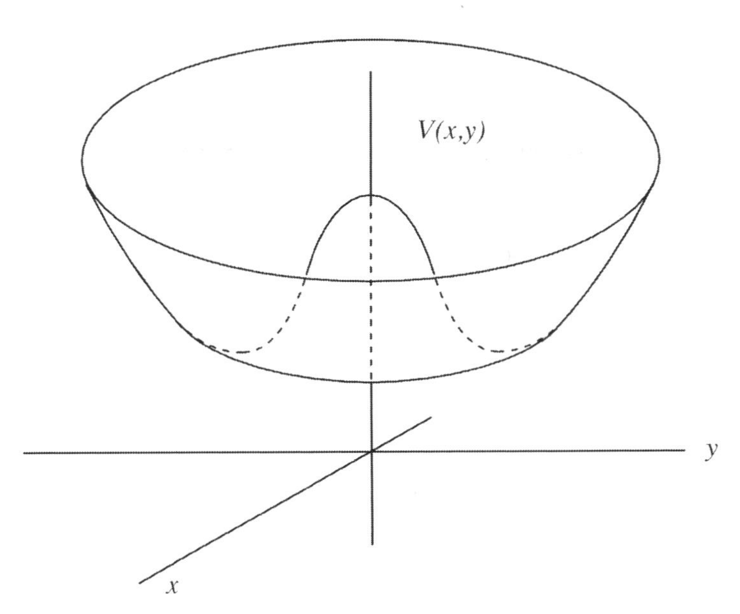

Figure 4.2. Sombrero potential. With appropriate parameter values, the combined parabolic oscillator potential and repulsive Gaussian hump form a potential energy function of the shape shown. The height of the central peak is V_o and its width is controlled by the length parameter b.

ground state and first few excited states of a single charged particle moving under the influence of this potential and an applied magnetic field.

The purpose of this brief interlude is not so much to explore the physical properties by analysing the approximate solutions as it is to show interested readers a sketch of a fairly general way to proceed. Upon adding the perturbation, I no longer know how to find the stationary states of the problem in closed form, if that is possible, but a rather good variational approximation is available, especially for the low-lying states.

The plan for performing this variational calculation is the usual Rayleigh–Ritz method. The eigenfunctions of the perturbed problem are assumed to be expandable in the basis of unperturbed eigenvectors, which can be chosen to be orthonormal. The approximation is to construct a model problem in which the Hamiltonian operator K of the full problem, with

$$K = H + V = (H_0 - \omega_1 L) + V$$

being projected onto a finite set of eigenfunctions of the unperturbed Hamiltonian

$$H = H_0 - \omega_1 L.$$

Thus an arbitrary eigenfunction of K is expanded approximately as a finite sum of eigenfunctions of H,

$$\psi = \sum_{n,k} c_{nk}\, \psi_{nk}.$$

The sum is over a finite subset of eigenfunctions of the oscillator in a magnetic field as represented by H, which may be selected as having in common a specific eigenvalue $\hbar\ell$ with respect to L, since both H and K commute with L. Whether or not this simplification of reducing the problem based on the rotation symmetry is made use of, a variational problem that approximates the original one is to find stationary values of the functional

$$F = \langle \psi, K\, \psi \rangle$$

with respect to the expansion coefficients $\{c_{nk}\}$ subject to the normalization constraint

$$\Phi = \langle \psi, \psi \rangle = 1.$$

So, one finds eigenvalues of the matrix K and the corresponding normalized eigenvectors. All of this may be familiar from basic quantum mechanics.

The matrix entries in K are (with no summation implied on the repeated indices)

$$K_{(n_1k_1)(n_2k_2)} = \langle \psi_{n_1k_1}, K\psi_{n_2k_2} \rangle = E^{(0)}_{n_1,\, k_1} \delta_{n_1n_2} \delta_{k_1k_2} + \langle \psi_{n_1k_1}, V\, \psi_{n_2k_2} \rangle,$$

where from equation (4.2)

$$E^{(0)}_{n,k} = \hbar(\Omega - \omega_1)n + \hbar\,(\Omega + \omega_1)\,k + \hbar\,\Omega.$$

This leads to the main task, which is to devise an efficient method for calculating the matrix entries of the V matrix, or in essence for calculating

$$M_{(n_1k_1)(n_2k_2)} = \int \psi^*_{n_1k_1}(x, y) \exp\left(-\frac{x^2 + y^2}{b^2}\right) \psi_{n_2k_2}(x, y)\, dx\, dy.$$

Having found already the generating function f for the eigenfunctions $\{\psi_{nk}\}$ one is in position to construct a generating function G for these matrix entries. Multiplying the previous expression by appropriate powers of the auxiliary counting variables s, t, u and v, as before, and summing,

$$\begin{aligned} G(s, t, u, v) &= \sum_{n_1,k_1,n_2,k_2} \frac{s^{n_1} t^{k_1} u^{n_2} v^{k_2}}{\sqrt{n_1! k_1! n_2! k_2!}} M_{(n_1k_1)(n_2k_2)} \\ &= \int f(x, -y, s, t) \exp\left(-\frac{x^2 + y^2}{b^2}\right) f(x, y, u, v)\, dx\, dy. \end{aligned}$$

Once again the integral is a standard Gaussian one. The result is

$$G(s, t, u, v) = \frac{b^2}{b^2 + c^2} \exp\left(\frac{b^2(su + tv) - c^2(st + uv)}{b^2 + c^2}\right). \tag{4.5}$$

The physical problem has rotation symmetry. Rotating the oscillator around the field axis leaves its Hamiltonian operator K and its Schrödinger equation invariant. By Nöther's theorem, the angular momentum is conserved. It should also be the case that, since L commutes with all of K, matrix entries of K should vanish between basis functions with different angular momentum quantum numbers ℓ, where $\ell = n - k$. That means there should be a selection rule $\ell_1 = \ell_2$, or

$$k_2 = k_1 - n_1 + n_2.$$

Then

$$s^{n_1}\, t^{k_1}\, u^{n_2}\, v^{k_2} = s^{n_1}\, t^{k_1}\, u^{n_2} v^{k_1 - n_1 + n_2} = \left(\frac{s}{v}\right)^{n_1} (t\, v)^{k_1}\, (u\, v)^{n_2}.$$

Respect for the selection rule requires that G be a function, really, of only three variables, the ratio s/v, and the products $t\, v$ and $u\, v$. This is true, because referring back to equation (4.5), one sees that the important conditions are met already in the two expressions,

$$(s\, u + t\, v) = \left(\frac{s}{v}\, u\, v + t\, v\right), \quad \text{and} \quad (s\, t + u\, v) = \left(\frac{s}{v}\, t\, v + u\, v\right).$$

To form a symmetry-reduced K matrix for a given value of $\ell \geqslant 0$, note that $n = k + \ell$ and that both n and k are non-negative integers. So to fill a symmetry-reduced K matrix for any given value of ℓ and of any size $k_{max} \times k_{max}$,

$$K_\ell(k_1, k_2) = E^{(0)}_{\ell+k_1-1,\, k_1-1}\delta_{k_1,\, k_2} - V_o M_{(\ell+k_1-1,\, k_1-1),\, (\ell+k_2-1,\, k_2-1)},$$

with

$$k_1 \text{ and } k_2 \in \{1, 2, 3 \cdots k_{max}\}.$$

This same matrix can be used for both $+\ell$ and $-\ell$, since changing the sign of ℓ corresponds to reversing the field direction, or in other words, it corresponds to changing the sign of ω_1.

Figure 4.3 shows an example calculation performed by constructing a 7×7 version of $K_{\ell=1}$. Only the lower five levels are shown as these are assumed to be represented more accurately in the approximate basis. The solid curves show variation of the levels as a function of applied field when $\ell = +1$ and angular momentum is parallel to the applied field. The dashed curves represent either one of two different things. They could represent the same levels when the field is reversed, so that in a sense the abscissa is folded to show both $+B_o$ and $-B_o$ superimposed. Alternatively, for the corresponding states with $\ell = -1$, the meanings of the dashed and solid curves would be exchanged.

4.3 Generating functions for spherical harmonics

The spherical harmonics arise from spherical symmetry. These are mutual eigenfunctions of the angular momentum operators L_z and L^2. In classical physics, the factors of $\hbar$ can be scaled out if they are not relevant.

For example, the Helmholtz partial differential equation with spherically symmetric boundary conditions decomposes in terms of angular momentum. So

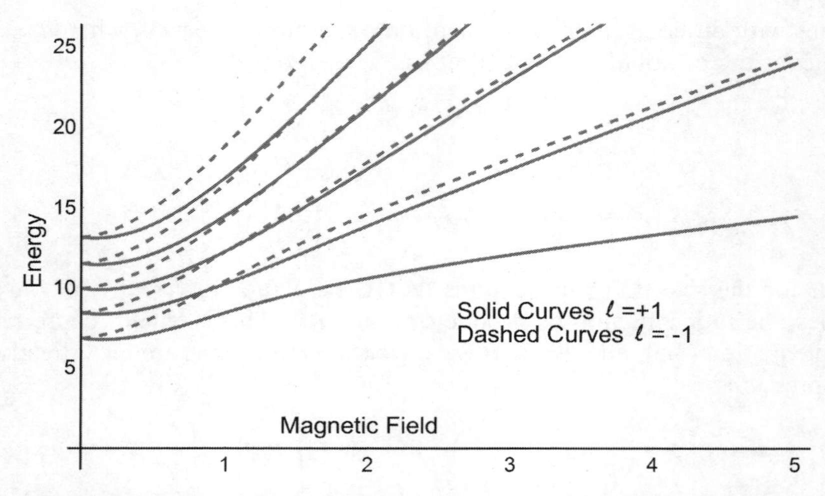

Figure 4.3. Magnetic field effect on an oscillator perturbed by a Gaussian hump. The lowest-lying energy levels of the perturbed oscillator in a magnetic with $\ell = \pm 1$. The energy is in units of $\hbar\,\omega_0$ and magnetic field in units of $2m\omega_0/q$. The dashed curves represent either the case $\ell = -1$ for field in the same, upward direction, or $\ell = +1$ with the field direction reversed. Gaussian hump: $V_0 = 10\hbar\,\omega_0$, $b = \sqrt{\hbar/(m\omega_0)}$.

$\vec{p} = -i\nabla$. A physics student might think of it this way. Imagine the operator $\vec{p}$ as an ordinary vector, the momentum of a particle. It has a radial component and a component perpendicular to the position vector $\vec{r}$:

$$\vec{p} = \vec{p}_r + \vec{p}_\perp.$$

Thus,

$$p^2 = p_r^2 + p_\perp^2,$$

and

$$L^2 = r^2\, p_\perp^2.$$

So,

$$-\nabla^2 = p_r^2 + \frac{1}{r^2} L^2.$$

On comparing with the expression for ∇^2 in spherical coordinates, the Laplacian operator can also be written

$$\nabla^2 = \frac{1}{r}\,\partial_r^2\, r - \frac{1}{r^2}\, L^2. \tag{4.6}$$

It is fairly clear that if $Y(\theta,\phi)$ is an eigenfunction of L^2 with eigenvalue λ, then

$$\nabla^2(Y(\theta,\phi) f(r)) = Y(\theta,\phi)\left(\frac{1}{r}\partial_r^2(r\, f(r)) - \frac{\lambda}{r^2} f(r)\right).$$

So there should exist separation solutions. Because ∇^2 with appropriate boundary conditions is Hermitian, there should exist a set of mutually orthogonal eigenfunctions for L^2, and so certain problems will become tractable. This suggests a motivation for studying these eigenfunctions. One can do so in the context of the Lie algebra $so(3)$ of proper rotations.

The group $SO(3R)$ is the group of proper rotations in R^3, the special orthogonal transformations. Each element of this group preserves the quadratic form

$$r^2 = x^2 + y^2 + z^2,$$

and considered as a matrix it has determinant +1. The algebra $so(3R)$ of corresponding special orthogonal group of rotations in R^5 was treated briefly above. It is easy to see that each of the infinitesimal generators

$$U_{12} = x\partial_y - y\partial_x, \qquad U_{23} = y\partial_z - z\partial_y, \qquad U_{31} = z\partial_x - x\partial_z$$

annihilates r^2 when r^2 is considered as an operand in the function space where they act, or that they commute with r^2 when r^2 is considered as an operator on the same space that multiplies a given function by $x^2 + y^2 + z^2$. From either point of view one sees that the group generated by these operators leaves r^2 unchanged. The standard, Hermitian version of this basis set for the Lie algebra $so(3)$ is encountered in quantum mechanics as the components of orbital angular momentum, which are taken normally as

$$L_x = -i\,\hbar\, U_{23}, \qquad L_y = -i\,\hbar\, U_{31}, \qquad L_z = -i\,\hbar\, U_{12},$$

which satisfy the commutator relations

$$[L_x, L_y] = i\,\hbar\, L_z, \qquad [L_y, L_z] = i\,\hbar\, L_x, \qquad [L_z, L_x] = i\,\hbar\, L_y.$$

The constant $\hbar$ is Planck's constant, which enters through the standard recipe in quantum theory for writing operators to represent measurable, physical quantities that involve momentum. As mentioned above, it is possible to re-scale the coordinates in such a way as to eliminate factors of $\hbar$. It is preferable to do this in classical physics calculations where $\hbar$ does not enter naturally. From here on, I set $\hbar$ to 1, meaning that it has been scaled out of the formalism. It is not difficult to restore factors of $\hbar$ if they are needed. The quadratic Casimir operator constructed in the usual way in the associative algebra is just the square of total angular momentum,

$$L^2 = L_x^2 + L_y^2 + L_z^2.$$

Thus,

$$[L^2, L_x] = [L^2, L_y] = [L^2, L_z] = 0.$$

As it happens, these basic commutation relations and the structure constants they imply are isomorphic to those of the Lie algebra for the special unitary group $SO(2)$ of 2×2 unitary matrices with determinant +1, although the groups are different. This fact was noted earlier in the context of the quaternions. From the brackets of the generators it is clear that the Lie algebra is simple. It contains no ideal at all. Any of the three generators in the basis can be chosen as generating the Cartan subalgebra. One often chooses L_z. In the context of quantum mechanics, it is useful to include L^2 in the Cartan subalgebra, as well, as one is interested in simultaneous measurements of the two operators L_z and L^2. Although the reader may be quite familiar with the basic construction of the spherical harmonics as simultaneous eigenfunctions of L_z and L^2, it is outlined here to make the presentation self-contained. A reader who is completely familiar with construction of the eigenstates for the angular momentum may wish to skip down to the problem of finding the generating function, and then refer back to this section if desired.

Using undetermined coefficients, as in the case of $so(5)$, one constructs a pair of root vectors

$$L_\pm = L_x \pm i\, L_y,$$

such that

$$[L_x, L_\pm] = \pm\, L_\pm.$$

Of course, both of these commute with L^2. The next objective is to derive a set $\{y_{\lambda,\mu}\}$ of normalized, simultaneous eigenfunctions of L_z and L^2. That this is possible depends on the fact that the group manifold of $SO(3,\ R)$ is a compact set. Every infinite sequence of points in the parameter space contains a subsequence that converges to a point in the space. In such a case, as in the case of rotations in two dimensions where the single parameter range is finite and homeomorphic to a circle, a discrete set of eigenfunctions can be found for the Cartan subalgebra that is countable and such that the eigenfunctions are in a sense orthogonal. Furthermore, it is a famous result by H Weyl that a Lie semi-simple connected Lie group is compact if the Killing form in its Lie algebra is negative definite. That is to say, that the Killing form (calculated using structure constants derived from the infinitesimal generators U_{12}, U_{23} and U_{31}) can be diagonalized and that its diagonal form can be scaled to minus the identity. The special orthogonal groups $SO(n)$ have this property.

Review of basic construction of L^2, L_z eigenstates: A standard construction of the spherical harmonics proceeds as follows. The reader who is familiar with this may wish to skip to the material following equation (4.9), where the topic of constructing generating functions is taken up, and to refer back if necessary.

Because $[L^2, L_z] = 0$, I know from general principles that there is a set of functions (normalizable with respect to integrating over polar angles in the standard inner product). Let us call them $y_{\ell m}(x,\ y,\ z)$. A more usual set of spherical harmonics will relate in a simple way to these functions when the Cartesian coordinates are changed to polar form. The functions $y_{\ell m}(x,\ y,\ z)$ are to be eigenfunctions simultaneously of the commuting Hermitian operators L^2 and L_z,

$$L^2 y_{\ell,m} = \lambda_\ell y_{\ell,m},$$
$$L_z y_{\ell,m} = \mu_m y_{\ell,m}.$$

Since the operators are measurable quantities, hence Hermitian, the eigenvalues are real and the functions are orthonormal in pairs,

$$\left\langle y_{\ell,m},\ y_{\ell',m'}\right\rangle = \delta_{\ell,\ell'}\delta_{m,m'}.$$

Using undetermined coefficients I constructed two ladder operators L_+, L_- (root vectors) such that

$$[L_z, L_+] = +L_+,$$
$$[L_z, L_-] = -L_-.$$

These two operators both commute with L^2. Now one is in a familiar place. For example, since

$$[L_z, L_+] = +L_+,$$
$$L_z L_+ y_{\ell,m} - L_+ L_z y_{\ell,m} = L_+ y_{\ell,m},$$

one sees that

$$L_z\big(L_+ y_{\ell,m}\big) = \mu_m L_+ y_{\ell,m} + L_+ y_{\ell,m} \quad \Rightarrow \quad L_z\big(L_+ y_{\ell,m}\big) = (\mu_m + 1)\big(L_+ y_{\ell,m}\big).$$

Apparently either $L_+ y_{\ell,m} = 0$ or else $L_+ y_{\ell,m}$ is proportional to $y_{\ell,m'}$ such that $\mu_{m'} = \mu_m + 1$. So, if I have one eigenfunction $y_{\ell,m}$ and eigenvalue μ_m, either I can form another one with higher eigenvalue $\mu_{m+1} = \mu_m + 1$, or else $L_+ y_{\ell,m} = 0$.

Now it turns out that, starting from a given $y_{\ell,m}$, and creating a series of eigenfunctions with higher and higher μ values, one must eventually find an eigenfunction with a maximum eigenvalue $\bar{\mu}$, and thus the eigenfunction must be annihilated by L_+. The reason is as follows.

Consider the fact that

$$\left\langle L_+ y_{\ell,m}, L_+ y_{\ell,m} \right\rangle \geqslant 0.$$

Using $(L_+)^\dagger = L_-$,

$$\left\langle L_+ y_{\ell,m}, L_+ y_{\ell,m} \right\rangle = \left\langle y_{\ell,m}, L_- L_+ y_{\ell,m} \right\rangle \geqslant 0,$$
$$L_- L_+ = L_x^2 + L_y^2 - L_z = L^2 - L_z^2 - L_z.$$

Consequently,

$$\left\langle L_+ y_{\ell,m}, L_+ y_{\ell,m} \right\rangle = \lambda_\ell - \mu_m^2 - \mu_m \geqslant 0.$$

So, as one constructs more and more eigenfunctions with higher and higher μ values (with the same λ value, since L_+ commutes with L^2), the expression that must to be non-negative would eventually become negative. This would be a contradiction. There must exist a certain eigenfunction $y_{\ell,m_{max}}$ with L_z eigenvalue $\bar{\mu}$ such that $L_+ y_{\ell,m_{max}} = 0$. This implies

$$\lambda_\ell - \bar{\mu}^2 - \bar{\mu} = 0.$$

Similarly starting with the lowering operator L_- one must arrive at a state $y_{\ell,m_{min}}$ with minimum eigenvalue $\underline{\mu}$ such that $L_- y_{\ell,m_{min}} = 0$. Working in the same way as with the raising operator, one finds

$$\lambda_\ell - \underline{\mu}^2 + \underline{\mu} = 0.$$

If I subtract the latter two results, ℓ cancels,

$$\bar{\mu}^2 + \bar{\mu} - \underline{\mu}^2 + \underline{\mu} = 0 = (\bar{\mu} + \underline{\mu})(\bar{\mu} - \underline{\mu} + 1).$$

It must be that $\underline{\mu} = -\bar{\mu}$.

Both $y_{\ell,m_{max}}$ and $y_{\ell,m_{min}}$ are unique for a given ℓ. Starting from $y_{\ell,m_{min}}$ and applying L_+ some integer N number of times must produce finally $y_{\ell,m_{max}}$:

$$N = \bar{\mu} - \underline{\mu} = 2\,\bar{\mu}.$$

The maximum $\bar{\mu}$ is some integer or odd half-integer value,

$$\bar{\mu} = \frac{N}{2},$$

where N depends only on ℓ. This value is normally called j:

$$\bar{\mu} = j \in \left\{0, \frac{1}{2}, 1, \frac{3}{2}, 2, \ldots\right\}.$$

So μ values for a given ℓ are at integer steps between $-j$ and j. Also, from above,

$$\lambda = j(j + 1).$$

The number j depends uniquely on the index value ℓ. When j is any general half-integer, j is used commonly as an index rather than ℓ.

Now there are two more points.

First, how do I know that there is not some other state with a quantum number $\tilde{\mu}$ in between the values we have just found? The answer is that, as in fact the argument above shows, if there were a state with such an eigenvalue, then in some integer number of steps one would arrive at a case where the norm that cannot be negative would become negative, unless somewhere along the way acting with L_+ would annihilate a state. But the system of PDEs

$$L^2 y = \lambda y, \quad L_z y = \mu y, \quad L_+ y = 0$$

has a unique solution. So for a given value of λ and hence of ℓ or j there is only one such unique state. So this shows that the original $\tilde{\mu}$ must be a member of the sequence already found, after all.

The second thing is this. One finds, by separating the Laplace PDE in polar coordinates, that the eigenvalues of L^2 have to be of the form $\ell(\ell + 1)$ where ℓ is strictly an integer. The integer comes about by solving the separated radial equation. The other, half-integer values of j correspond to spinor solutions. These have meaning as the intrinsic spin of certain particles, namely the Fermions, e.g. electrons. But the half-integer j solutions do not correspond to the smooth spherical harmonic solutions one will usually encounter. For these one has $\lambda_\ell = \ell(\ell + 1)$ with ℓ being a non-negative integer.

Going back to

$$\left\langle L_+ y_{\ell,m}, L_+ y_{\ell,m} \right\rangle = \lambda_\ell - \mu_m^2 - \mu_m = \ell(\ell + 1) - \mu(\mu + 1),$$

one may let $\mu = m$ with m ranging from $-\ell$ to ℓ. Then for normalization it must be that

$$L_+ y_{\ell,m} = \sqrt{\ell(\ell+1) - m(m+1)}\, y_{\ell,m+1}.$$

Consistent with this is that $L_+ y_{\ell,\,\ell} = 0$. Similarly for the lowering operator one finds

$$L_- y_{\ell,m} = \sqrt{\ell(\ell+1) - m(m-1)}\, y_{\ell,m-1}.$$

Finally, each of these has a factored form that will be very useful in the calculations to follow:

$$L^2 y_{\ell,m} = \ell(\ell+1) y_{\ell,m} \tag{4.7}$$

$$L_+ y_{\ell,m} = \sqrt{(\ell - m)(\ell + m + 1)}\, y_{\ell,m+1}, \tag{4.8}$$

$$L_- y_{\ell,m} = \sqrt{(\ell + m)(\ell - m + 1)}\, y_{\ell,m-1}. \tag{4.9}$$

Finding $y_{\ell,m}$ generating functions: There is some disagreement in definition of the spherical harmonics. This is a difference in the choice of phase in the normalization. There are several ways to construct a generating function for the set of $2\ell + 1$ functions $\{y_{\ell,m}\}$ for fixed ℓ:

1. One can show, based on the paragraphs directly above and the earlier section on Legendre polynomials, that the latter polynomials $\{P_\ell(\cos\theta)\}$ are each proportional to the corresponding spherical harmonic $y_{\ell,\,0}$ for the same ℓ. After demonstrating this it is a matter of finding the constant in each case by doing the normalization integral over the sphere, and then using the L_+ operator to find $y_{\ell,\,m}$ for positive m (and L_- to get $y_{\ell,m}$ for negative m). Then a generating relation can be found by exponentiating a form of L_+ or L_-.
2. The facts that $L_+ y_{\ell,\ell} = 0$ and $L_z y_{\ell,\ell} = \ell y_{\ell,\ell}$ allow us to solve for $y_{\ell,\,\ell}$ and give an alternative set of differential equations, which together with normalization define $y_{\ell,\,\ell}$ up to a phase, and then using L_- one can find the general $y_{\ell,m}$. A generating function is thus found by exponentiating a form of L_-.
3. Correspondingly, one could start from $L_- y_{\ell,-\ell} = 0$ and $L_z y_{\ell,-\ell} = -\ell y_{\ell,-\ell}$ to find $y_{\ell,-\ell}$. Then by exponentiating some form of L_+ another generating function for $y_{\ell,m}$ in the set for fixed ℓ can be found.

In addition to the three cases enumerated, one can find generating functions for sets of m and different ℓ. The different methods suggest different natural choices for the phases of $\{y_{\ell,m}\}$, since the natural choice in one instance is less natural in another.

Consider the first strategy in the list. The claim that $P_\ell(\mu) = ky_{\ell,0}(x, y, z)$ needs to be proven. It was seen previously that $r^\ell P_\ell(cos\theta)$ satisfies the Laplace equation. Using equation (4.6),

$$\nabla^2 r^\ell P_\ell = \left(\frac{1}{r}\partial_r^2 r - \frac{1}{r^2}L^2\right) r^\ell P_\ell = 0.$$

$$= \left(\frac{1}{r}\partial_r^2 r^{(\ell+1)} - r^{(\ell-2)}L^2\right) P_\ell,$$

so

$$r^{(\ell-2)}(\ell(\ell+1) - L^2)P_\ell = 0.$$

Indeed, $P_\ell(\cos\theta)$ is an eigenfunction of L^2:

$$L^2 P_\ell(\cos\theta) = \ell(\ell+1) P_\ell(\cos\theta).$$

Also, since L_z commutes with any function of r, and it contains derivatives with respect to x and y, but not z, and since $\cos\theta = z/r$,

$$L_z P_\ell(\cos\theta) = -i(x\,\partial_y - y\,\partial_x)P_\ell(z/r) = 0.$$

This $P_\ell(\cos\theta)$ is also an eigenfunction of L_z with $m = 0$, which shows that $P_\ell(z/r)$ is proportional to $y_{\ell,m=0}(x, y, z)$:

$$y_{\ell,0}(x, y, z) = k\, P_\ell(z/r).$$

The proportionality constant k is not very difficult to determine. From

$$\int_{\text{unit sphere}} |y_{\ell,0}|^2\, dA = 1,$$

and the proportionality

$$P_\ell(\mu) = ky_{\ell,0}(x, y, z)$$

with integral of the squared Legendre polynomial, obtained from the Legendre generating function

$$\int_{-1}^{1} P_\ell(x)^2\, dx = \frac{2}{2\ell+1},$$

one has that

$$k^2\int_{-\pi}^{\pi}\int_0^{\pi} \left| y_{\ell,0} \right|^2 \sin\theta d\theta\, d\phi = k^2 = 2\pi\int_0^{\pi} |P_\ell(\cos)|^2 \sin\theta d\theta$$

$$k^2 = 2\pi \int_{-1}^{1} |P_\ell(\mu)|^2 \, d\mu = \frac{4\pi}{2\ell + 1}.$$

Thus the relationship is

$$y_{\ell,0}(x, y, z) = \sqrt{\frac{2\ell + 1}{4\pi}} P_\ell\left(\frac{z}{r}\right).$$

Notice that I have selected the positive square root, which is a choice of phase. This choice will not necessarily agree with sign obtained naturally by other methods of determining the $\{y_{\ell,0}\}$.

Next, apply the raising operator repeatedly to $y_{\ell,0}$ in order to arrive at $y_{\ell,m}$. From

$$L_+ y_{\ell,m} = \sqrt{\ell(\ell + 1) - m(m + 1)}\, y_{\ell,m+1} = \sqrt{(\ell - m)(\ell + m + 1)}\, y_{\ell,m+1},$$

I find the first several results:

$$L_+ y_{\ell,0} = \sqrt{\ell(\ell + 1)}\, y_{\ell,1},$$

$$L_+^2 y_{\ell,0} = \sqrt{\ell(\ell + 1)}\, L_+ y_{\ell,1} = \sqrt{\ell(\ell + 1)}\, \sqrt{(\ell - 1)(\ell + 2)}\, y_{\ell,2},$$

$$L_+^3 y_{\ell,0} = \sqrt{\ell(\ell + 1)}\, \sqrt{(\ell - 1)(\ell + 2)}\, \sqrt{(\ell - 2)(\ell + 3)}\, y_{\ell,3}, \qquad \text{etc.}$$

The pattern is clear:

$$L_+^m y_{\ell,0} = \sqrt{\ell(\ell + 1)}\, \sqrt{(\ell - 1)(\ell + 2)} \cdots \sqrt{(\ell - m + 1)(\ell + m)}\, y_{\ell,m}$$

$$= \sqrt{\ell(\ell - 1)(\ell - 2)\cdots(\ell - m + 1)}\, \sqrt{(\ell + 1)(\ell + 2)\cdots(\ell + m)}\, y_{\ell,m}.$$

Rearranging a bit,

$$L_+^m y_{\ell,0} = \sqrt{\frac{\ell!}{(\ell - m)!}} \sqrt{\frac{(\ell + m)!}{\ell!}}\, y_{\ell,m}.$$

It seems that for non-negative m one can write

$$y_{\ell,m} = \sqrt{\frac{(\ell - m)!}{(\ell + m)!}}\, L_+^m y_{\ell,0} = \sqrt{\frac{2\ell + 1}{4\pi}} \sqrt{\frac{(\ell - m)!}{(\ell + m)!}}\, L_+^m P_\ell\left(\frac{z}{r}\right).$$

Similarly, for non-positive $\bar{m} = -m$,

$$y_{\ell,\bar{m}} = \sqrt{\frac{2\ell + 1}{4\pi}} \sqrt{\frac{(\ell - m)!}{(\ell + m)!}}\, L_-^m P_\ell\left(\frac{z}{r}\right).$$

In either case, $m \geqslant 0$ or $m < 0$, the coefficient involves $|m|$.

The plan now would be to use the group action generated by $L+$ or L_- to deform the generating function for the Legendre polynomials into one for the spherical harmonics. For definiteness, consider $m \geqslant 0$. This suggests a change of variables from the Cartesian x, y and z to canonical coordinates ξ, η and ζ for the L_+ operator, such that

$$L_+ = \frac{\partial}{\partial \xi}.$$

Then $L_+\xi = 1$, $L_+\eta = 0$ and $L_+\zeta = 0$. Thus η and ζ are independent solutions of the differential equation

$$L_+\zeta = 0.$$

$$z\left(\frac{\partial \zeta}{\partial x} + i\frac{\partial \zeta}{\partial y}\right) - (x + i\,y)\frac{\partial \zeta}{\partial z} = 0.$$

One solution is immediate from the fact that all of the rotation generators preserve r. Let $\zeta = r$. Then it is easy to see that L_+ also annihilates $x + i\,y$. So let $\eta = x + i\,y$. Finally, I find

$$L_+z = -(x + i\,y).$$

The expression $\eta = x + i\,y$ is annihilated by L_+ so it acts as a constant. Thus,

$$\xi = -\frac{z}{x + i\,y}, \qquad \eta = x + i\,y, \qquad \zeta = \sqrt{x^2 + y^2 + z^2} = r. \tag{4.10}$$

The latter transformation has a nice inverse, namely

$$x = \frac{\zeta^2 - \eta^2\xi^2 + \eta^2}{2\eta}, \qquad y = \frac{\mathrm{i}(\zeta^2 - \eta^2\xi^2 - \eta^2)}{2\eta}, \qquad z = -\xi\,\eta. \tag{4.11}$$

Very well. Now begin with the Legendre generating relation (2.5),

$$f(\cos\theta, s) = \frac{1}{\sqrt{1 - 2\,s\cos\theta + s^2}} = \sum_{\ell=0}^{\infty} s^\ell\, P_\ell(\cos\theta),$$

or

$$\frac{1}{\sqrt{1 - 2\,s\,\frac{z}{r} + s^2}} = \sum_{\ell=0}^{\infty} s^\ell \sqrt{\frac{4\pi}{2\ell + 1}}\, y_{\ell,0}(x, y, z).$$

The transformation of the Legendre generating function into the canonical coordinates results in

$$\tilde{f}(\xi, \eta, \zeta) = \frac{1}{\sqrt{1 - \frac{2\,s\,\xi\,\eta}{\zeta} + s^2}}.$$

Then, to generate $y_{\ell,\,m}$ for non-negative m,

$$\exp\left(t\,\frac{\partial}{\partial\xi}\right)\tilde{f}(\xi,\,\eta,\,\zeta) = \frac{1}{\sqrt{1-\frac{2\,s\,(\xi+t)\,\eta}{\zeta}+s^2}}.$$

Then transforming back to Cartesian form,

$$\exp\,(tL_+)f\left(\frac{z}{r},\,s\right) = \frac{1}{\sqrt{1-\frac{2\,s}{r}(z-t(x+i\,y))+s^2}}.$$

On the other hand,

$$\exp\,(tL_+)f\left(\frac{z}{r},\,s\right) = \sum_{\ell,m\geqslant 0}\frac{s^\ell t^m}{m!}\sqrt{\frac{(\ell+m)!}{(\ell-m)!}}\sqrt{\frac{4\,\pi}{2\ell+1}}\,y_{\ell,m}.$$

Putting these together gives a generating relation for spherical harmonics with $m \geqslant 0$,

$$\frac{1}{\sqrt{1-\frac{s}{r}(z-t(x+i\,y))+s^2}} = \sum_{\ell,m\geqslant 0}\frac{s^\ell t^m}{m!}\sqrt{\frac{(\ell+m)!}{(\ell-m)!}}\sqrt{\frac{4\,\pi}{2\ell+1}}\,y_{\ell,\,m}. \tag{4.12}$$

This contains a great deal of information about the spherical harmonics. To complete the expression, one should use L_- in the same way to find the corresponding sum for the cases $m \leqslant 0$, add the two results and then subtract the Legendre generator in order not to count the case $m = 0$ twice. The other generating functions that follow either from starting at the bottom, where $m = -\ell$ for fixed ℓ, and working up using L_+, or starting at the top, where $m = +\ell$, and working down using L_- are handled in a similar way. A result obtained by the latter computation, for instance, is

$$\begin{aligned}\frac{1}{\sqrt{4\pi}}\frac{\sqrt{(2\ell+1)!}}{2^\ell \ell!}\frac{1}{r^\ell}\left(t(x+iy)-2z-\frac{1}{t}(x-iy)\right)^\ell \\ = \sum_{m=-\ell}^{\ell}\sqrt{\frac{(2\ell)!}{(\ell-m)!(\ell+m)!}}\,t^m y_{\ell,m}.\end{aligned} \tag{4.13}$$

This result is more commonly seen. However, as noted in the preamble above, the phases (factors of powers of $\pm i$) of the spherical harmonics $y_{\ell,m}$ will *not* agree between the latter two expressions, equations (4.12, 4.13), since in the first case the signs are chosen such that $y_{\ell,0}$ is always positive, while in the latter expression, $y_{\ell,\ell}$ is required to be positive. Of course this can be fixed, but one must keep it in mind.

References

[1] Landau L D and Lifshitz E M 1958 *Quantum Mechanics, Non-relativistic Theory* (Oxford: Pergamon)

[2] Schwalm W, Nitschke C and Reis P 2005 Generating two-dimensional oscillator matrix elements sorted by angular momentum *J. Phys. A: Math. Gen.* **38** 9565–73

Chapter 5

Concluding remarks

In the sections above, there have been primary two goals. The first was to transmit in an informal way some general ideas about Lie theory and indeed a sketch of some of the content. The point of this was to suggest a more abstract context and a vocabulary into which certain physical ideas generalize, and with which one who happened to have been at the right place could now go on to read more. The other goal was to show how to use some of the general ideas to solve certain kinds of physical problems. The examples were chosen mostly to do this. Quite a bit is available about the connections between symmetry groups, algebras and special functions. In [1] it is shown that much of what is known about many special functions, including the factorization methods classified by Infeld and Hull [2], evolve from symmetries of the equations they satisfy with respect to the Euclidean group, or complex generalizations. In this brief monograph only a glance at each of a few topics could be accomplished. One thing not covered at all is the systematic construction of representations, reducibility and the reduction methods that are of great practical use in many kinds of applications. These are treated in other places, such as the books by Tinkham [3] or Hamermesh [4], and even though they are of central importance, especially in application, the space available above has been devoted to other topics that are less well covered elsewhere. One of these was the direct use of generating functions. Another thing that would be covered in a longer work is the construction of Lie algebra representations as combinations of second quantized operators for fermions or bosons. This topic is presented informally in a stimulating, short book by Lipkin [5]. Another important area is the application of the Lie formalism to uncoupling systems of nonlinear difference equations that appear, for instance, in real-space renormalization schemes, and other places [6–8].

References

[1] Miller W 1968 *Lie Theory and Special Functions* (New York: Academic)

[2] Infeld L and Hull T E 1951 The factorization method *Rev. Mod. Phys.* **23** 21–68

[3] Tinkham M 1964 *Group Theory and Quantum Mechanics* (New York: McGraw-Hill)

doi:10.1088/2053-2571/aae789ch5

[4] Hamermesh M 1962 *Group Theory and Its Application to Physical Problems* (Reading, MA: Addison-Wesley)
[5] Lipkin H J 2002 *Lie Groups for Pedestrians* (New York: Dover)
[6] Maeda S 1980 Canonical structure and symmetries for discrete systems *Math. Japonica* **25** 405–20
[7] Quispel G R W, Roberts J A G and Thompson C J 1989 Integrable mappings and soliton equations II *Physica* D **34** 183–92
[8] Moritz B, Schwalm W and Uherka D 1998 Finding Lie groups that reduce the order of discrete dynamical systems *J. Phys. A: Math. Gen.* **31** 7379–402

◎

编辑手记

世界著名物理学家恩斯特·马赫(Ernst Mach)曾感叹:

> 也许听起来奇怪,数学的力量在于它规避了一切不必要的思考和它惊人地节省了脑力活动.

本书是一部介绍数学在物理学中的应用的英文专著,中文书名或可译为《数学物理精选专题讲座:李理论的进一步应用》.

本书作者为威廉·A. 施瓦姆(William A. Schwalm),美国人,北达科他大学讲师,他的研究方向是凝聚态理论和物理问题的数学方法应用.他还著有《数学物理选修课:椭圆函数和椭圆积分》.施瓦姆博士获得了两项杰出的教学奖,即犹他大学物理学优秀本科生导师奖(1979 年)和北达科他大学麦克德莫特卓越教学奖(1995 年).

据作者在前言中介绍:

> 本著作致力于对李群、李代数和生成函数进行介绍,既适用于标准的特殊函数,也适用于解决某些类型的物理问题.本书绝不是完整的,也不具有数学严谨性.本书是对这些主题的非正式处理,适用于物理专业研究生或其他具有物理背景的人阅读,这些人希望本书以相对熟悉的风格和词汇对主题进行简短和非正式的介绍.本书是本系列[1]中专门讨论李群和微分方程的著作的续集.

在下面的介绍中,本书详细地回顾了李连续群的主题(在之前的卷中也被提到过),特别是局部群的李的无穷小生成元的构造和使用.例如,Euclid 群、二维中的特殊保形子群、Cremona 变换群、四元数、高维旋转群,以及特定问题中连续对称性的几个特定示例;构建问题的生成函数,比如用于标准弹簧和砝码振动的机械 Green 函数.对称群适用人们在与对称相关的问题和从一个形式结果到另一个形式结果之间进行导航,或者在相关问题与从一个特殊案例到一个一般解决方案之间进行导航.本书还简要讨论了与寻找群生成元的问题互补的重要问题,即当给定一组生成元时,重建局部李群的问题.

群生成元的一个显著特征是作为算子它们不能互相交换.换位子的生成元的向量空间成为一个李代数.量子算子或力学量或热力学量的 Poisson 括号形成了具有李结构的系统.从某种程度上说,本书处理的结合代数和非结合代数、换位子和李括号积、矩阵表示和作为微分算子的表示、李代数分类的主题比其他一般的物理书要更详细.尽管此种处理始终是一种非正式的对话式的处理方式,但我希望这些讨论足够广泛,可以揭示一些基本结构,这些结构可以成为对更抽象的处理方法的有用的介绍.

本书特别对 Cartan 子代数、根向量和根的概念,以及在某些物理问题中如何构造根向量及其与阶梯算子的关系进行了一些详细的讨论并举例说明.每一个李代数都有一个 Cartan 子代数.在一个半单李代数中,这是一组在某种意义上来说最大的相互交换的向量.例如,在量子理论的背景下,李括号是换位子,人们可以想到一组相互交换的算子,它们同时还表示可测量的物理量.在量子系统中,人们可能能够找到一组阶梯算子,或升降算子,它们从交换算子的一个相互特征函数转换为另一个相互特征函数,就像简单振子问题中的升降算子情况或角动量的情况一样.或者在经典力学中,Cartan 子代数可以表示一组相互守恒的量,使得 Poisson 括号成对消失.接下来的材料归纳了这些观点.下面要介绍的根向量是在李代数背景中阶梯算子的抽象.本书开发并详细说明了用于查找根向量或粗略地用于查找系统初等阶梯算子的基本策略,其处理包括五维旋度的代数 $SO(5)$ 和带外加磁场的二维谐振子.本书没有涉及,但在其他书籍中经常提到的非常重要且有用的话题就是矩阵表示的可约性.作者决定在本书中集中讨论一些不太容易在其他地方进行初级处理的主题和思想.

生成函数的概念,无论是用于操作特殊函数还是作为解决物理问题的一种方法,都是一个反复出现的主题.处理无穷集应用的一般情况,或找到无限类问题中的所有解决方案,或找到某个算子的所有矩阵元素等,通常比处理一些特殊情况更容易.在下面的处理中,李群始终用于将给定问题的平凡解变形为无穷集解的生成函数.例如,孤立点电荷的库仑势通过连续群变换而变形,首先转化为 Legendre 多项式的生成函数,然后通过李群 $SO(3)$ 转化为整个球谐函数集的生成函数.再举一个例子,一维振子的基态波函数以一种说明 Weisner 群论方法的方式转换为所有特征函数的生成函数.在这种情况下,生成函数变成了相干态.扩展李群行为的 Weisner 方法还为磁场中带电谐振子的角动量解析本征态制造了生成函数,进而给出了矩阵元素生成函数,其他问题也被解决了.作为一个说明性示例,在磁场的影响下,这些矩阵元素被用在一个 Sombrero 势上带电粒子移动的变分研究草图中.

本书目录为:

数学物理从某种意义上说就是数学建模. 北京十一中的青年数学教师朱浩楠多年来一直从事中学生数学建模的教学工作,是其中的佼佼者. 他将数学建模分为广义和狭义两种:狭义的理解,就是利用数学寻求现实问题的多、快、好、省的解决方案;广义的理解,整个数学都是数学建模的产物,是对数学结构的挖掘和发明. 不严格地说:数学建模就是将数学与生活的世界关联起来的思想方法.

本书就是此领域的一部杰作.

本书是我们数学工作室"极其庞大"的引进计划中的一部,原因是我们现在书虽多,但好书少. 据《中国出版年鉴》显示,从出版机构数量看,1980 年全国有出版社 192 家,2010 年增至 581 家;从出书品种看,1980 年出书 2.162 1 万种,2016 年增至 49.988 4 万种. 在相当长的一段时间里,全国年出书总量都在 50 万种左右,导致重复出版、跟风出版、同质化出版的现象比较严重,有数量、缺质量,有高原、缺高峰,没有生命力的"翻过即扔"的图书比比皆是,书多好少,图书总量严重过剩.

所以校枉必须过正!

刘培杰
2024 年 4 月 27 日
于哈工大

国外优秀数学著作
原版丛书（第三十辑）

工程师和科学家应用数学概论

Essential of Applied Mathematics for Engineers and Scientists(Second Edition)

（第二版）
（英文）

[美] 罗伯特·G.瓦特（Robert G. Watts） 著

HITP
哈爾濱工業大學出版社
HARBIN INSTITUTE OF TECHNOLOGY PRESS

黑版贸登字 08-2021-032 号

图书在版编目(CIP)数据

工程师和科学家应用数学概论:第二版=Essential of Applied Mathematics for Engineers and Scientists: Second Edition:英文/(美)罗伯特·G. 瓦特(Robert G. Watts)著. —哈尔滨:哈尔滨工业大学出版社,2024. 10
(国外优秀数学著作原版丛书. 第三十辑)
ISBN 978-7-5767-1340-4

Ⅰ. ①工… Ⅱ. ①罗… Ⅲ. ①应用数学-英文 Ⅳ. ①O29

中国国家版本馆 CIP 数据核字(2024)第 073691 号

GONGCHENGSHI HE SHUXUEJIA YINGYONG SHUXUE GAILUN: DI-ER BAN

策划编辑　刘培杰　杜莹雪
责任编辑　刘家琳　张嘉芮
封面设计　孙茵艾
出版发行　哈尔滨工业大学出版社
社　　址　哈尔滨市南岗区复华四道街 10 号　邮编 150006
传　　真　0451-86414749
网　　址　http://hitpress. hit. edu. cn
印　　刷　哈尔滨圣铂印刷有限公司
开　　本　889 mm×1 194 mm　1/16　印张 45.5　字数 641 千字
版　　次　2024 年 10 月第 1 版　2024 年 10 月第 1 次印刷
书　　号　ISBN 978-7-5767-1340-4
定　　价　252.00 元(全 4 册)

Contents

CHAPTER 1

Partial Differential Equations in Engineering

1.1 INTRODUCTORY COMMENTS

This book covers the material presented in a course in applied mathematics that is required for first-year graduate students in the departments of Chemical and Mechanical Engineering at Tulane University. A great deal of material is presented, covering boundary value problems, complex variables, and Fourier transforms. Therefore the depth of coverage is not as extensive as in many books. Our intent in the course is to introduce students to methods of solving linear partial differential equations. Subsequent courses such as conduction, solid mechanics, and fracture mechanics then provide necessary depth.

The reader will note some similarity to the three books, *Fourier Series and Boundary Value Problems, Complex Variables and Applications*, and *Operational Mathematics*, originally by R. V. Churchill. The first of these has been recently updated by James Ward Brown. The current author greatly admires these works, and studied them during his own tenure as a graduate student. The present book is more concise and leaves out some of the proofs in an attempt to present more material in a way that is still useful and is acceptable for engineering students.

First we review a few concepts about differential equations in general.

1.2 FUNDAMENTAL CONCEPTS

An *ordinary differential equation* expresses a dependent variable, say u, as a function of one independent variable, say x, and its derivatives. The *order* of the differential equation is given by the order of the highest derivative of the dependent variable. A boundary value problem consists of a differential equation that is defined for a given range of the independent variable (*domain*) along with conditions on the boundary of the domain. In order for the boundary value problem to have a unique solution the number of boundary conditions must equal the order of the differential equation. If the differential equation and the boundary conditions contain only terms of first degree in u and its derivatives the problem is *linear*. Otherwise it is *nonlinear*.

A *partial differential equation* expresses a dependent variable, say u, as a function of more than one independent variable, say x, y, and z. Partial derivatives are normally written as $\partial u/\partial x$. This is the first-order derivative of the dependent variable u with respect to the independent variable x. Sometimes we will use the notation u_x or when the derivative is an ordinary derivative we use u'. Higher order derivatives are written as $\partial^2 u/\partial x^2$ or u_{xx}. The order of the differential equation now depends on the orders of the derivatives of the dependent variables in terms of each of the independent variables. For example, it may be of order m for the x variable and of order n for the y variable. A boundary value problem consists of a partial differential equation defined on a domain in the space of the independent variables, for example the x, y, z space, along with conditions on the boundary. Once again, if the partial differential equation and the boundary conditions contain only terms of first degree in u and its derivatives the problem is linear. Otherwise it is nonlinear.

A differential equation or a boundary condition is *homogeneous* if it contains only terms involving the dependent variable.

Examples

Consider the ordinary differential equation

$$a(x)u'' + b(x)u = c(x), \quad 0 < x < A. \tag{1.1}$$

Two boundary conditions are required because the order of the equation is 2. Suppose

$$u(0) = 0 \quad \text{and} \quad u(A) = 1. \tag{1.2}$$

The problem is linear. If $c(x)$ is not zero the differential equation is nonhomogeneous. The first boundary condition is homogeneous, but the second boundary condition is nonhomogeneous.

Next consider the ordinary differential equation

$$a(u)u'' + b(x)u = c \qquad 0 < x < A \tag{1.3}$$

Again two boundary conditions are required. Regardless of the forms of the boundary conditions, the problem is nonlinear because the first term in the differential equations is not of first degree in u and u'' since the leading coefficient is a function of u. It is homogeneous only if $c = 0$.

Now consider the following three partial differential equations:

$$u_x + u_{xx} + u_{xy} = 1 \tag{1.4}$$

$$u_{xx} + u_{yy} + u_{zz} = 0 \tag{1.5}$$

$$uu_x + u_{yy} = 1 \tag{1.6}$$

The first equation is linear and nonhomogeneous. The third term is a *mixed partial derivative.* Since it is of second order in x two boundary conditions are necessary on x. It is first order in y, so that only one boundary condition is required on y. The second equation is linear and homogeneous and is of second order in all three variables. The third equation is nonlinear because the first term is not of first degree in u and u_x. It is of order 1 in x and order 2 in y.

In this book we consider only linear equations. We will now derive the partial differential equations that describe some of the physical phenomena that are common in engineering science.

Problems

Tell whether the following are linear or nonlinear and tell the order in each of the independent variables:

$$u'' + xu' + u^2 = 0$$

$$\tan(y)u_y + u_{yy} = 0$$

$$\tan(u)u_y + 3u = 0$$

$$u_{yyy} + u_{yx} + u = 0$$

1.3 THE HEAT CONDUCTION (OR DIFFUSION) EQUATION

1.3.1 Rectangular Cartesian Coordinates

The conduction of heat is only one example of the diffusion equation. There are many other important problems involving the diffusion of one substance in another. One example is the diffusion of one gas into another if both gases are motionless on the macroscopic level (no convection). The diffusion of heat in a motionless material is governed by Fourier's law which states that heat is conducted per unit area in the negative direction of the temperature gradient in the (vector) direction **n** in the amount $\partial u/\partial n$, that is

$$q^n = -k\partial u/\partial n \tag{1.7}$$

where q^n denotes the heat flux in the n direction (not the nth power). In this equation u is the local temperature and k is the thermal conductivity of the material. Alternatively u could be the partial fraction of a diffusing material in a host material and k the diffusivity of the diffusing material relative to the host material.

Consider the diffusion of heat in two dimensions in rectangular Cartesian coordinates. Fig. 1.1 shows an element of the material of dimension Δx by Δy by Δz. The material has a specific heat c and a density ρ. Heat is generated in the material at a rate q per unit volume. Performing a heat balance on the element, the time (t) rate of change of thermal energy within the element, $\rho c\, \Delta x \Delta y \Delta z \partial u/\partial t$ is equal to the rate of heat generated within the element

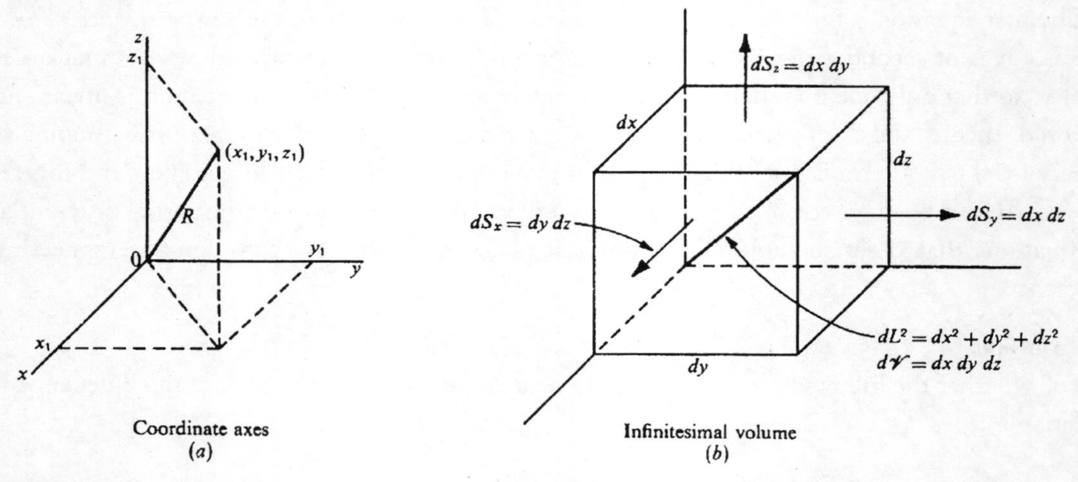

FIGURE 1.1: An element in three dimensional rectangular Cartesian coordinates

$q'''\Delta x\Delta y\Delta z$ minus the rate at which heat is conducted out of the material. The flux of heat conducted into the element at the x face is denoted by q^x while at the y face it is denoted by q^y. At $x + \Delta x$ the heat flux (i.e., per unit area) leaving the element in the x direction is $q^x + \Delta q^x$ while at $y + \Delta y$ the heat flux leaving in the y direction is $q^y + \Delta q^y$. Similarly for q^z. Expanding the latter three terms in Taylor series, we find that $q^x + \Delta q^x = q^x + q^x_x \Delta x + (1/2) q^x_{xx} (\Delta x)^2$ + terms of order $(\Delta x)^3$ or higher order. Similar expressions are obtained for $q^y + \Delta q^y$ and $q^z + \Delta q^z$ Completing the heat balance

$$\begin{aligned}\rho c\,\Delta x\Delta y\Delta z\partial u/\partial t = {} & q'''\Delta x\Delta y\Delta z + q^x\Delta y\Delta z + q^y\Delta x\Delta z \\ & - (q^x + q^x_x\Delta x + (1/2)q^x_{xx}(\Delta x)^2 + \cdots)\Delta y\Delta z \\ & - (q^y + q^y_y\Delta y + (1/2)q^y_{yy}(\Delta y)^2 + \cdots)\Delta x\Delta z \\ & - (q^z + q^z_z\Delta z + (1/2)q^z_{zz}(\Delta z)^2 + \cdots)\Delta x\Delta y\end{aligned} \tag{1.8}$$

The terms $q^x\Delta y\Delta z$, $q^y\Delta x\Delta z$, and $q^z\Delta x\Delta y$ cancel. Taking the limit as Δx, Δy, and Δz approach zero, noting that the terms multiplied by $(\Delta x)^2$, $(\Delta y)^2$, and $(\Delta z)^2$ may be neglected, dividing through by $\Delta x\Delta y\Delta z$ and noting that according to Fourier's law $q^x = -k\partial u/\partial x$, $q^y = -k\partial u/\partial y$, and $q^z = -k(\partial u/\partial z)$ we obtain the time-dependent heat conduction equation in three-dimensional rectangular Cartesian coordinates:

$$\rho c\,\partial u/\partial t = k(\partial^2 u/\partial x^2 + \partial^2 u/\partial y^2) + q \tag{1.9}$$

The equation is first order in t, and second order in both x and y. If the property values ρ, c and k and the heat generation rate per unit volume q are independent of the dependent

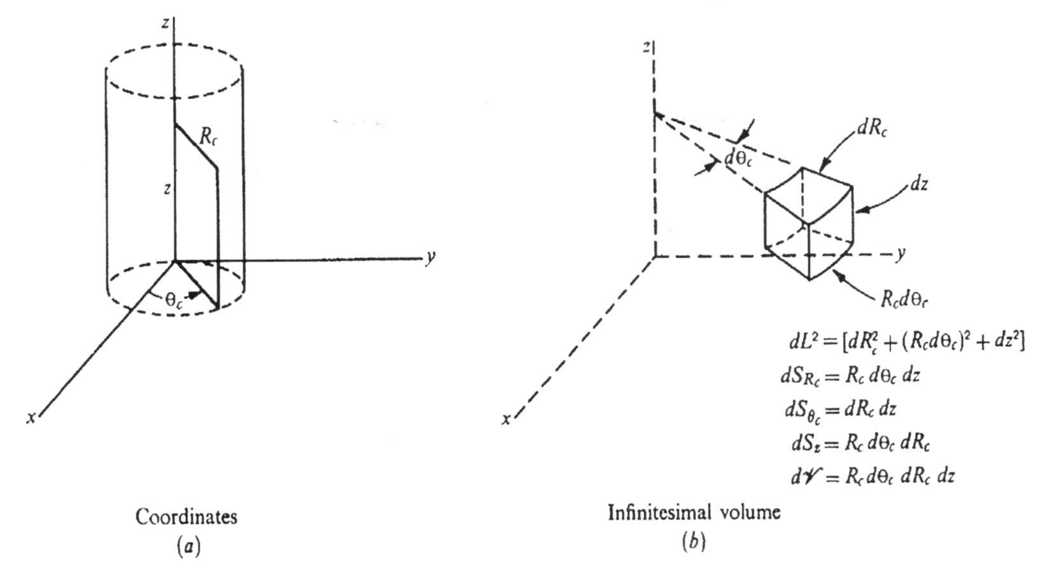

FIGURE 1.2: An element in cylindrical coordinates

variable, temperature the partial differential equation is linear. If q is zero, the equation is homogeneous. It is easy to see that if a third dimension, z, were included, the term $k\partial^2 u/\partial z^2$ must be added to the right-hand side of the above equation.

1.3.2 Cylindrical Coordinates

A small element of volume $r\Delta\Theta\Delta r\Delta z$ is shown in Fig. 1.2.

The method of developing the diffusion equation in cylindrical coordinates is much the same as for rectangular coordinates except that the heat conducted into and out of the element depends on the area as well as the heat flux as given by Fourier's law, and this area varies in the r-direction. Hence the heat conducted into the element at r is $q^r r\Delta\Theta\Delta z$, while the heat conducted out of the element at $r + \Delta r$ is $q^r r\Delta\Theta\Delta z + \partial(q^r r\Delta\Theta\Delta z)/\partial r(\Delta r)$ when terms of order $(\Delta r)^2$ are neglected as Δr approaches zero. In the z- and θ-directions the area does not change. Following the same procedure as in the discussion of rectangular coordinates, expanding the heat values on the three faces in Tayor series', and neglecting terms of order $(\Delta\Theta)^2$ and $(\Delta z)^2$ and higher,

$$\rho c r\Delta\theta\Delta r\Delta z\partial u/\partial t = -\partial(q^r r\Delta\theta\Delta z)/\partial r\Delta r - \partial(q^\theta\Delta r\Delta z)/\partial\theta\Delta\theta - \partial(q^z r\Delta\theta\Delta r)/\partial z\Delta z + qr\Delta\theta\Delta r\Delta z \quad (1.10)$$

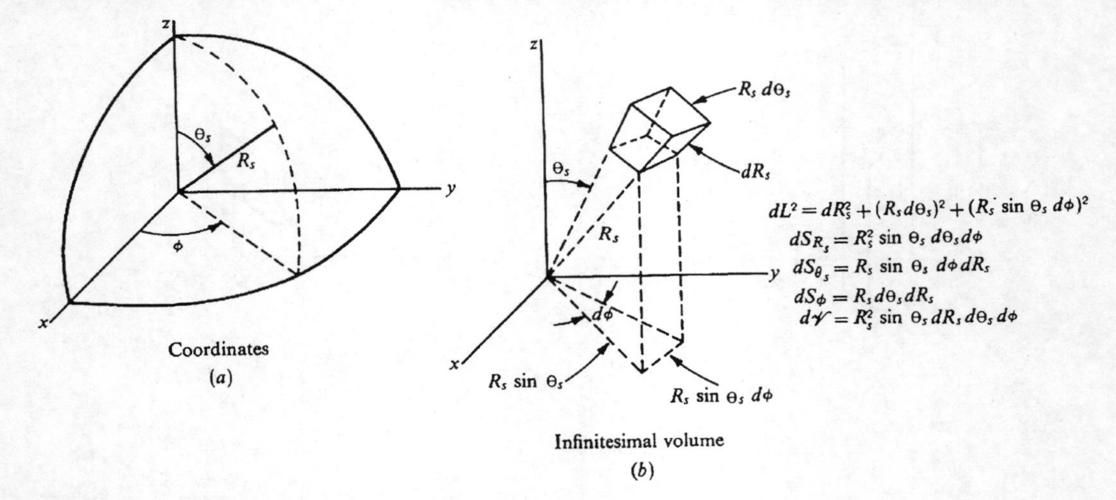

FIGURE 1.3: An element in spherical coordinates

Dividing through by the volume, we find after using Fourier's law for the heat fluxes

$$\rho c\, \partial u/\partial t = (1/r)\partial(r\,\partial u/\partial r)/\partial r + (1/r^2)\partial^2 u/\partial\theta^2 + \partial^2 u/\partial z^2 + q \tag{1.11}$$

1.3.3 Spherical Coordinates

An element in a spherical coordinate system is shown in Fig. 1.3. The volume of the element is $r\sin\theta\,\Delta\Phi\,\Delta r\, r\,\Delta\theta = r^2\sin\theta\,\Delta r\,\Delta\theta\,\Delta\Phi$. The net heat flows out of the element in the r, θ, and Φ directions are respectfully

$$q^r r^2 \sin\theta\,\Delta\theta\,\Delta\Phi \tag{1.12}$$

$$q^\theta r \sin\theta\,\Delta r\,\Delta\Phi \tag{1.13}$$

$$q^\Phi r\,\Delta\theta\,\Delta r \tag{1.14}$$

It is left as an exercise for the student to show that

$$\begin{aligned}\rho c\,\partial u/\partial t = k[&(1/r^2)\partial/\partial r(r^2\partial u/\partial r) + (1/r^2\sin^2\theta)\partial^2 u/\partial\Phi^2 \\ &+ (1/r^2\sin\theta)\partial(\sin\theta\,\partial u/\partial\theta)/\partial\theta + q\end{aligned} \tag{1.15}$$

The Laplacian Operator

The linear operator on the right-hand side of the heat equation is often referred to as the Laplacian operator and is written as ∇^2.

1.3.4 Boundary Conditions

Four types of boundary conditions are common in conduction problems.

a) Heat flux prescribed, in which case $k\partial u/\partial n$ is given.
b) Heat flux is zero (perhaps just a special case of (a)), in which case $\partial u/\partial n$ is zero.
c) Temperature u is prescribed.
d) Convection occurs at the boundary, in which case $k\partial u/\partial n = h(U - u)$.

Here n is a length in the direction normal to the surface, U is the temperature of the fluid next to the surface that is heating or cooling the surface, and h is the coefficient of convective heat transfer. Condition (d) is sometimes called Newton's law of cooling.

1.4 THE VIBRATING STRING

Next we consider a tightly stretched string on some interval of the x-axis. The string is vibrating about its equilibrium position so that its departure from equilibrium is $y(t, x)$. The string is assumed to be perfectly flexible with mass per unit length ρ.

Fig. 1.4 shows a portion of such a string that has been displaced upward. We assume that the tension in the string is constant. However the direction of the tension vector along the string varies. The tangent of the angle $\alpha(t, x)$ that the string makes with the horizontal is given by the slope of the wire, $\partial y/\partial x$,

$$V(x)/H = \tan\alpha(t, x) = \partial y/\partial x \tag{1.16}$$

If we assume that the angle α is small then the horizontal tension force is nearly equal to the magnitude of the tension vector itself. In this case the tangent of the slope of the wire

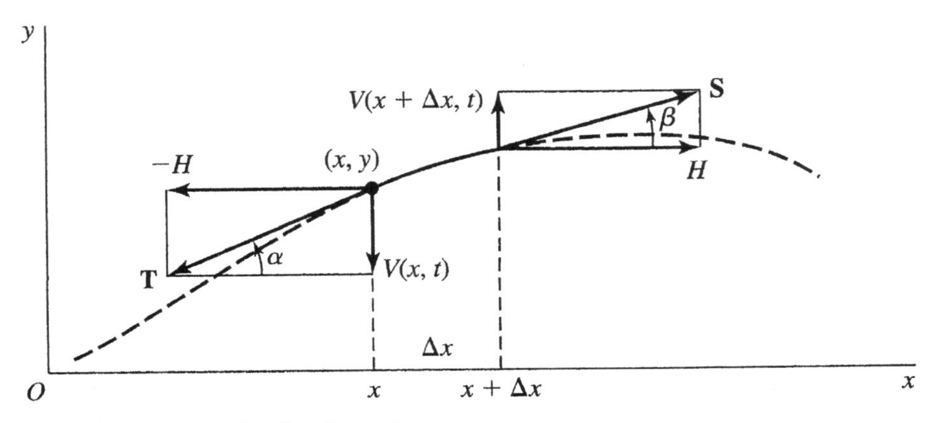

FIGURE 1.4: An element of a vibrating string

at $x + \Delta x$ is

$$V(x + \Delta x)/H = \tan\alpha(x + \Delta x) = \partial y/\partial x(x + \Delta x). \tag{1.17}$$

The vertical force V is then given by $H\partial y/\partial x$. The *net* vertical force is the difference between the vertical forces at x and $x + \Delta x$, and must be equal to the mass times the acceleration of that portion of the string. The mass is $\rho\Delta x$ and the acceleration is $\partial^2 y/\partial t^2$. Thus

$$\rho\Delta x\partial^2/\partial t^2 = H[\partial y/\partial x(x + \Delta x) - \partial y/\partial x(x)] \tag{1.18}$$

Expanding $\partial y/\partial x(x + \Delta x)$ in a Taylor series about $\Delta x = 0$ and neglecting terms of order $(\Delta x)^2$ and smaller, we find that

$$\rho y_{tt} = H y_{xx} \tag{1.19}$$

which is the wave equation. Usually it is presented as

$$y_{tt} = a^2 y_{xx} \tag{1.20}$$

where $a^2 = H/\rho$ is a wave speed term.

Had we included the weight of the string there would have been an extra term on the right-hand side of this equation, the acceleration of gravity (downward). Had we included a damping force proportional to the velocity of the string, another negative term would result:

$$\rho y_{tt} = H y_{xx} - b y_t - g \tag{1.21}$$

1.4.1 Boundary Conditions

The partial differential equation is linear and if the gravity term is included it is nonhomogeneous. It is second order in both t and x, and requires two boundary conditions (initial conditions) on t and two boundary conditions on x. The two conditions on t are normally specifying the initial velocity and acceleration. The conditions on x are normally specifying the conditions at the ends of the string, i.e., at $x = 0$ and $x = L$.

1.5 VIBRATING MEMBRANE

The partial differential equation describing the motion of a vibrating membrane is simply an extension of the right-hand side of the equation of the vibrating string to two dimensions. Thus,

$$\rho y_{tt} + b y_t = -g + \nabla^2 y \tag{1.22}$$

In this equation, ρ is the density per unit area and $\nabla^2 y$ is the *Laplacian operator* in either rectangular or cylindrical coordinates.

1.6 LONGITUDINAL DISPLACEMENTS OF AN ELASTIC BAR

The longitudinal displacements of an elastic bar are described by Eq. (1.20) except the in this case $a^2 = E/\rho$, where ρ is the density and E is Young's modulus.

FURTHER READING

V. Arpaci, *Conduction Heat Transfer*. Reading, MA: Addison-Wesley, 1966.

J. W. Brown and R. V. Churchill, *Fourier Series and Boundary Value Problems*. 6th edition. New York: McGraw-Hill, 2001.

P. V. O'Neil, *Advanced Engineering Mathematics*. 5th edition. Pacific Grove, CA: Brooks/Cole-Thomas Learning, 2003.

CHAPTER 2

The Fourier Method: Separation of Variables

In this chapter we will work through a few example problems in order to introduce the general idea of *separation of variables* and the concept of orthogonal functions before moving on to a more complete discussion of orthogonal function theory. We will also introduce the concepts of *nondimensionalization* and *normalization*.

The goal here is to use the three theorems stated below to walk the student through the solution of several types of problems using the concept of separation of variables and learn some early lessons on how to apply the method without getting too much into the details that will be covered later, especially in Chapter 3.

We state here without proof three fundamental theorems that will be useful in finding series solutions to partial differential equations.

Theorem 2.1. *Linear Superposition: If a group of functions* u_n, $n = m$ *through* $n = M$ *are all solutions to some linear differential equation then*

$$\sum_{n=m}^{M} c_n u_n$$

is also a solution.

Theorem 2.2. *Orthogonal Functions: Certain sets of functions* Ψ_n *defined on the interval* (a, b) *possess the property that*

$$\int_a^b \Psi_n \Psi_m dx = constant, \ n = m$$

$$\int_a^b \Psi_n \Psi_m dx = 0, \ n \neq m$$

These are called orthogonal functions. Examples are the sine and cosine functions. This idea is discussed fully in Chapter 3, particularly in connection with Sturm–Liouville equations.

Theorem 2.3. *Fourier Series: A piecewise continuous function $f(x)$ defined on (a, b) can be represented by a series of orthogonal functions $\Psi_n(x)$ on that interval as*

$$f(x) = \sum_{n=0}^{\infty} A_n \Psi_n(x)$$

where

$$A_n = \frac{\int_{x=a}^{b} f(x)\Psi_n(x)dx}{\int_{x=a}^{b} \Psi_n(x)\Psi_n(x)dx}$$

These properties will be used in the following examples to introduce the idea of solution of partial differential equations using the concept of separation of variables.

2.1 HEAT CONDUCTION

We will first examine how Theorems 1, 2, and 3 are systematically used to obtain solutions to problems in heat conduction in the forms of infinite series. We set out the methodology in detail, step-by-step, with comments on lessons learned in each case. We will see that the mathematics often serves as a guide, telling us when we make a bad assumption about solution forms.

Example 2.1. A Transient Heat Conduction Problem

Consider a flat plate occupying the space between $x = 0$ and $x = L$. The plate stretches out in the y and z directions far enough that variations in temperature in those directions may be neglected. Initially the plate is at a uniform temperature u_0. At time $t = 0^+$ the wall at $x = 0$ is raised to u_1 while the wall at $x = L$ is insulated. The boundary value problem is then

$$\rho c u_t = k u_{xx} \quad 0 < x < L \quad t > 0 \tag{2.1}$$

$$\begin{aligned} u(t, 0) &= u_1 \\ u_x(t, L) &= 0 \\ u(0, x) &= u_0 \end{aligned} \tag{2.2}$$

2.1.1 Scales and Dimensionless Variables

When it is possible it is always a good idea to write both the independent and dependent variables in such a way that they range from zero to unity. In the next few problems we shall show how this can often be done.

We first note that the problem has a fundamental length scale, so that if we define another space variable $\xi = x/L$, the partial differential equation can be written as

$$\rho c u_t = L^{-2} k u_{\xi\xi} \quad 0 < \xi < 1 \quad t < 0 \tag{2.3}$$

Next we note that if we define a dimensionless time-like variable as $\tau = \alpha t/L^2$, where $\alpha = k/\rho c$ is called the *thermal diffusivity*, we find

$$u_\tau = u_{\xi\xi} \tag{2.4}$$

We now proceed to *nondimensionalize* and *normalize* the dependent variable and the boundary conditions. We define a new variable

$$U = (u - u_1)/(u_0 - u_1) \tag{2.5}$$

Note that this variable is always between 0 and 1 and is dimensionless. Our boundary value problem is now devoid of constants.

$$U_\tau = U_{\xi\xi} \tag{2.6}$$

$$\begin{aligned} &U(\tau, 0) = 0 \\ &U_\xi(\tau, 1) = 0 \\ &U(0, \xi) = 1 \end{aligned} \tag{2.7}$$

All but one of the boundary conditions are homogeneous. *This will prove necessary in our analysis.*

2.1.2 Separation of Variables

Begin by assuming $U = \Gamma(\tau)\Phi(\xi)$. Insert this into the differential equation and obtain

$$\Phi(\xi)\Gamma_\tau(\tau) = \Gamma(\tau)\Phi_{\xi\xi}(\xi). \tag{2.8}$$

Next divide both sides by $U = \Phi\Gamma$,

$$\frac{\Gamma_\tau}{\Gamma} = \frac{\Phi_{\xi\xi}}{\Phi} = \pm\lambda^2 \tag{2.9}$$

The left-hand side of the above equation is a function of τ only while the right-hand side is a function only of ξ. This can only be true if both are constants since they are equal to each other. λ^2 is always positive, but we must decide whether to use the plus sign or the minus sign. We have two ordinary differential equations instead of one partial differential equation. Solution for Γ gives a constant times either $\exp(-\lambda^2\tau)$ or $\exp(+\lambda^2\tau)$. Since we know that U is always between 0 and 1, we see immediately that we must choose the minus sign. The second ordinary differential equation is

$$\Phi_{\xi\xi} = -\lambda^2\Phi \tag{2.10}$$

and we deduce that the two homogeneous boundary conditions are

$$\begin{aligned} \Phi(0) &= 0 \\ \Phi_\xi(1) &= 0 \end{aligned} \tag{2.11}$$

Solving the differential equation we find

$$\Phi = A\cos(\lambda\xi) + B\sin(\lambda\xi) \tag{2.12}$$

where A and B are constants to be determined. The first boundary condition requires that $A = 0$.

The second boundary condition requires that either $B = 0$ or $\cos(\lambda) = 0$. Since the former cannot be true (U is not zero!) the latter must be true. ξ can take on any of an infinite number of values $\lambda_n = (2n-1)\pi/2$, where n is an integer between negative and positive infinity. Equation (2.10) together with boundary conditions (2.11) is called a Sturm–Liouville problem. The solutions are called *eigenfunctions* and the λ_n are called *eigenvalues*. A full discussion of Sturm–Liouville theory will be presented in Chapter 3.

Hence the apparent solution to our partial differential equation is any one of the following:

$$U_n = B_n \exp[-(2n-1)^2\pi^2\tau/4)]\sin[\pi(2n-1)\xi/2]. \tag{2.13}$$

2.1.3 Superposition

Linear differential equations possess the important property that if each solution U_n satisfies the differential equation and the boundary conditions then the linear combination

$$\sum_{n=1}^{\infty} B_n \exp[-(2n-1)^2\pi^2\tau/4]\sin[\pi(2n-1)\xi/2] = \sum_{n=1}^{\infty} U_n \tag{2.14}$$

also satisfies them, as stated in Theorem 2. Can we build this into a solution that satisfies the one remaining boundary condition? The final condition (the nonhomogeneous initial condition) states that

$$1 = \sum_{n=1}^{\infty} B_n \sin(\pi(2n-1)\xi/2) \tag{2.15}$$

This is called a Fourier sine series representation of 1. The topic of Fourier series is further discussed in Chapter 3.

2.1.4 Orthogonality

It may seem hopeless at this point when we see that we need to find an infinite number of constants B_n. What saves us is a concept called *orthogonality* (to be discussed in a more general way in Chapter 3). The functions $\sin(\pi(2n-1)\xi/2)$ form an orthogonal set on the interval $0 < \xi < 1$, which means that

$$\begin{aligned}\int_0^1 \sin(\pi(2n-1)\xi/2)\sin(\pi(2m-1)\xi/2)d\xi &= 0 \text{ when } m \neq n \\ &= 1/2 \text{ when } m = n\end{aligned} \tag{2.16}$$

Hence if we multiply both sides of the final equation by $\sin(\pi(2m-1)\xi/2)d\xi$ and integrate over the interval, we find that all of the terms in which $m \neq n$ are zero, and we are left with one term, the general term for the nth B, B_n

$$B_n = 2\int_0^1 \sin(\pi(2n-1)\xi/2)d\xi = \frac{4}{\pi(2n-1)} \tag{2.17}$$

Thus

$$U = \sum_{n=1}^{\infty} \frac{4}{\pi(2n-1)} \exp[-\pi^2(2n-1)^2\tau/4]\sin[\pi(2n-1)\xi/2] \tag{2.18}$$

satisfies both the partial differential equation and the boundary and initial conditions, and therefore is a solution to the boundary value problem.

2.1.5 Lessons

We began by assuming a solution that was the product of two variables, each a function of only one of the independent variables. Each of the resulting ordinary differential equations was then solved. The two homogeneous boundary conditions were used to evaluate one of the constant coefficients and the separation constant λ. It was found to have an infinite number of values. These are called *eigenvalues* and the resulting functions $\sin\lambda_n\xi$ are called *eigenfunctions*. Linear superposition was then used to build a solution in the form of an infinite series. The infinite series was then required to satisfy the initial condition, the only nonhomogeneous condition. The coefficients of the series were determined using the concept of *orthogonality* stated in Theorem 3, resulting in a Fourier series. Each of these concepts will be discussed further in Chapter 3. *For now we state that many important functions are members of orthogonal sets.*

The method would not have worked had the differential equation not been homogeneous. (Try it.) It also would not have worked if more than one boundary condition had been nonhomogeneous. We will see how to get around these problems shortly.

Problems

1. Equation (2.9) could just as easily have been written as

$$\frac{\Gamma_\tau}{\Gamma} = \frac{\Phi_{\xi\xi}}{\Phi} = +\lambda^2$$

Show two reasons why this would reduce to the trivial solution or a solution for which Γ approaches infinity as τ approaches infinity, and that therefore the minus sign must be chosen.

2. Solve the above problem with boundary conditions

$$U_\xi(\tau, 0) = 0 \quad \text{and} \quad U(\tau, 1) = 0$$

using the steps given above.

Hint: $\cos(n\pi x)$ is an orthogonal set on (0, 1). *The result will be a Fourier cosine series representation of 1.*

3. Plot U versus ξ for $\tau = 0.001,\ 0.01,$ and 0.1 in Eq. (2.18). Comment.

Example 2.2. A Steady Heat Transfer Problem in Two Dimensions

Heat is conducted in a region of height a and width b. Temperature is a function of two space dimensions and independent of time. Three sides are at temperature u_0 and the fourth side is at temperature u_1. The formulation is as follows:

$$\frac{\partial^2 u}{\partial x^2} + \frac{\partial^2 u}{\partial y^2} = 0 \tag{2.19}$$

with boundary conditions

$$\begin{aligned} &u(0, x) = u(b, x) = u(y, a) = u_0 \\ &u(y, 0) = u_1 \end{aligned} \tag{2.20}$$

2.1.6 Scales and Dimensionless Variables

First note that there are two obvious length scales, a and b. We can choose either one of them to nondimensionalize x and y. We define

$$\xi = x/a \quad \text{and} \quad \eta = y/b \tag{2.21}$$

so that both dimensionless lengths are normalized.

To normalize temperature we choose

$$U = \frac{u - u_0}{u_1 - u_0} \tag{2.22}$$

The problem statement reduces to

$$U_{\xi\xi} + \left(\frac{a}{b}\right)^2 U_{\eta\eta} = 0 \tag{2.23}$$

$$U(0, \xi) = U(1, \xi) = U(\eta, 1) = 0$$

$$U(\eta, 0) = 1 \tag{2.24}$$

2.1.7 Separation of Variables

As before, we assume a solution of the form $U(\xi, n) = X(\xi)Y(\eta)$. We substitute this into the differential equation and obtain

$$Y(\eta)X_{\xi\xi}(\xi) = -X(\xi)\left(\frac{a}{b}\right)^2 Y_{\eta\eta}(\eta) \tag{2.25}$$

Next we divide both sides by $U(\xi, n)$ and obtain

$$\frac{X_{\xi\xi}}{X} = -\left(\frac{a}{b}\right)^2 \frac{Y_{nn}}{Y} = \pm\lambda^2 \tag{2.26}$$

In order for the function only of ξ on the left-hand side of this equation to be equal to the function only of η on the right-hand side, both must be constant.

2.1.8 Choosing the Sign of the Separation Constant

However in this case it is not as clear as the case of Example 1 what the sign of this constant must be. Hence we have designated the constant as $\pm\lambda^2$ so that for real values of λ the $\pm$ sign determines the sign of the constant. Let us proceed by choosing the negative sign and see where this leads.

Thus

$$X_{\xi\xi} = -\lambda^2 X$$

$$Y(\eta)X(0) = 1$$

$$Y(\eta)X(1) = 0 \tag{2.27}$$

or

$$X(0) = 1$$

$$X(1) = 0 \tag{2.28}$$

and

$$Y_{\eta\eta} = \mp\left(\frac{b}{a}\right)^2 \lambda^2 Y \tag{2.29}$$

$$X(\xi)Y(0) = X(\xi)Y(1) = 0$$

$$Y(0) = Y(1) = 0 \tag{2.30}$$

The solution of the differential equation in the η direction is

$$Y(\eta) = A\cosh(b\lambda\eta/a) + B\sinh(b\lambda\eta/a) \tag{2.31}$$

Applying the first boundary condition (at $\eta = 0$) we find that $A = 0$. When we apply the boundary condition at $\eta = 1$ however, we find that it requires that

$$0 = B\sinh(b\lambda/a) \tag{2.32}$$

so that either $B = 0$ or $\lambda = 0$. Neither of these is acceptable since either would require that $Y(\eta) = 0$ for all values of η.

We next try the positive sign. In this case

$$X_{\xi\xi} = \lambda^2 X \tag{2.33}$$

$$Y_{\eta\eta} = -\left(\frac{b}{a}\right)^2 \lambda^2 Y \tag{2.34}$$

with the same boundary conditions given above. The solution for $Y(\eta)$ is now

$$Y(\eta) = A\cos(b\lambda\eta/a) + B\sin(b\lambda\eta/a) \tag{2.35}$$

The boundary condition at $\eta = 0$ requires that

$$0 = A\cos(0) + B\sin(0) \tag{2.36}$$

so that again $A = 0$. The boundary condition at $\eta = 1$ requires that

$$0 = B\sin(b\lambda/a) \tag{2.37}$$

Since we don't want B to be zero, we can satisfy this condition if

$$\lambda_n = an\pi/b, \quad n = 0, 1, 2, 3, \ldots \tag{2.38}$$

Thus

$$Y(\eta) = B\sin(n\pi\eta) \tag{2.39}$$

Solution for $X(\xi)$ yields hyperbolic functions.

$$X(\xi) = C\cosh(\lambda_n \xi) + D\sinh(\lambda_n \xi) \tag{2.40}$$

The boundary condition at $\xi = 1$ requires that

$$0 = C\cosh(\lambda_n) + D\sinh(\lambda_n) \tag{2.41}$$

or, solving for C in terms of D,

$$C = -D\tanh(\lambda_n) \tag{2.42}$$

One solution of our problem is therefore

$$U_n(\xi, \eta) = K_n \sin(n\pi\eta)[\sinh(an\pi\xi/b) - \cosh(an\pi\xi/b)\tanh(an\pi/b)] \tag{2.43}$$

2.1.9 Superposition

According to the superposition theorem (Theorem 2) we can now form a solution as

$$U(\xi, \eta) = \sum_{n=0}^{\infty} K_n \sin(n\pi\eta)[\sinh(an\pi\xi/b) - \cosh(an\pi\xi/b)\tanh(an\pi/b)] \tag{2.44}$$

The final boundary condition (the nonhomogeneous one) can now be applied,

$$1 = -\sum_{n=1}^{\infty} K_n \sin(n\pi\eta)\tanh(an\pi/b) \tag{2.45}$$

2.1.10 Orthogonality

We have already noted that the sine function is an orthogonal function as defined on (0, 1). Thus, we multiply both sides of this equation by $\sin(m\pi\eta)d\eta$ and integrate over (0, 1), noting that according to the orthogonality theorem (Theorem 3) the integral is zero unless $n = m$. The result is

$$\int_{\eta=0}^{1} \sin(n\pi\eta)d\eta = -K_n \int_{\eta=0}^{1} \sin^2(n\pi\eta)d\eta \tanh(an\pi/b) \tag{2.46}$$

$$\frac{1}{n\pi}[1-(-1)^n] = -K_n \tanh(an\pi/b)\frac{1}{2} \tag{2.47}$$

$$K_n = -\frac{2[1-(-1)^n]}{n\pi\tanh(an\pi/b)} \tag{2.48}$$

The solution is represented by the infinite series

$$U(\xi, \eta) = \sum_{n=1}^{\infty} \frac{2[1-(-1)^n]}{n\pi \tanh(an\pi/b)} \sin(n\pi\eta) \times [\cosh(an\pi\xi/b)\tanh(an\pi/b) - \sinh(an\pi\xi/b)] \tag{2.49}$$

2.1.11 Lessons

The methodology for this problem is the same as in Example 1.

Example 2.3. A Steady Conduction Problem in Two Dimensions: Addition of Solutions

We now illustrate a problem in which two of the boundary conditions are nonhomogeneous. Since the problem and the boundary conditions are both linear we can simply break the problem into two problems and add them. Consider steady conduction in a square region L by L in size. Two sides are at temperature u_0 while the other two sides are at temperature u_1.

$$u_{xx} + u_{yy} = 0 \tag{2.50}$$

We need four boundary conditions since the differential equation is of order 2 in both independent variables.

$$u(0, y) = u(L, y) = u_0 \tag{2.51}$$

$$u(x, 0) = u(x, L) = u_1 \tag{2.52}$$

2.1.12 Scales and Dimensionless Variables

The length scale is L, so we let $\xi = x/L$ and $\eta = y/L$. We can make the first two boundary conditions homogeneous while normalizing the second two by defining a dimensionless temperature as

$$U = \frac{u - u_0}{u_1 - u_0} \tag{2.53}$$

Then

$$U_{\xi\xi} + U_{\eta\eta} = 0 \tag{2.54}$$

$$U(0, \eta) = U(1, \eta) = 0 \tag{2.55}$$

$$U(\xi, 0) = U(\xi, 1) = 1 \tag{2.56}$$

2.1.13 Getting to One Nonhomogeneous Condition

There are two nonhomogeneous boundary conditions, so we must find a way to only have one. Let $U = V + W$ so that we have two problems, each with one nonhomogeneous boundary condition.

$$W_{\xi\xi} + W_{\eta\eta} = 0 \tag{2.57}$$

$$W(0, \eta) = W(1, \eta) = W(\xi, 0) = 0 \tag{2.58}$$

$$W(\xi, 1) = 1$$

$$V_{\xi\xi} + V_{\eta\eta} = 0 \tag{2.59}$$

$$V(0, \eta) = V(1, \eta) = V(\xi, 1) = 0 \tag{2.60}$$

$$V(\xi, 0) = 1$$

(It should be clear that these two problems are identical if we put $V = W(1 - \eta)$. We will therefore only need to solve for W.)

2.1.14 Separation of Variables

Separate variables by letting $W(\xi, \eta) = P(\xi)Q(\eta)$.

$$\frac{P_{\xi\xi}}{P} = -\frac{Q_{\eta\eta}}{Q} = \pm\lambda^2 \tag{2.61}$$

2.1.15 Choosing the Sign of the Separation Constant

Once again it is not immediately clear whether to choose the plus sign or the minus sign. Let's see what happens if we choose the plus sign.

$$P_{\xi\xi} = \lambda^2 P \tag{2.62}$$

The solution is exponentials or hyperbolic functions.

$$P = A\sinh(\lambda\xi) + B\cosh(\lambda\xi) \tag{2.63}$$

Applying the boundary condition on $\xi = 0$, we find that $B = 0$. The boundary condition on $\xi = 1$ requires that $\underline{A}\sinh(\lambda) = 0$, which can only be satisfied if $A = 0$ or $\lambda = 0$, which yields a trivial solution, $W = 0$, and is unacceptable. The only hope for a solution is thus choosing the minus sign.

If we choose the minus sign in Eq. (2.61) then

$$P_{\xi\xi} = -\lambda^2 P \tag{2.64}$$

$$Q_{\eta\eta} = \lambda^2 Q \tag{2.65}$$

with solutions

$$P = A\sin(\lambda\xi) + B\cos(\lambda\xi) \tag{2.66}$$

and

$$Q = C\sinh(\lambda\eta) + D\cosh(\lambda\eta) \tag{2.67}$$

respectively. Remembering to *apply the homogeneous boundary conditions first*, we find that for $W(0, \eta) = 0$, $B = 0$ and for $W(1, \eta) = 0$, $\sin(\lambda) = 0$. Thus, $\lambda = n\pi$, our eigenvalues corresponding to the eigenfunctions $\sin(n\pi\xi)$. The last homogeneous boundary condition is $W(\xi, 0) = 0$, which requires that $D = 0$. There are an infinite number of solutions of the form

$$PQ_n = K_n \sinh(n\pi\eta)\sin(n\pi\xi) \tag{2.68}$$

2.1.16 Superposition

Since our problem is linear we apply superposition.

$$W = \sum_{n=1}^{\infty} K_n \sinh(n\pi\eta)\sin(n\pi\xi) \tag{2.69}$$

Applying the final boundary condition, $W(\xi, 1) = 1$

$$1 = \sum_{n=1}^{\infty} K_n \sinh(n\pi)\sin(n\pi\xi). \tag{2.70}$$

2.1.17 Orthogonality

Multiplying both sides of Eq. (2.70) by $\sin(m\pi\xi)$ and integrating over the interval (0, 1)

$$\int_0^1 \sin(m\pi\xi)d\xi = \sum_{n=0}^{\infty} K_n \sinh(n\pi) \int_0^1 \sin(n\pi\xi)\sin(m\pi\xi)d\xi \tag{2.71}$$

The orthogonality property of the sine eigenfunction states that

$$\int_0^1 \sin(n\pi\xi)\sin(m\pi\xi)d\xi = \begin{matrix} 0, & m \neq n \\ 1/2, & m = n \end{matrix} \tag{2.72}$$

Thus,

$$K_n = 2/\sinh(n\pi) \tag{2.73}$$

and

$$W = \sum_{n=0}^{\infty} \frac{2}{\sinh(n\pi)} \sinh(n\pi\eta)\sin(n\pi\xi) \tag{2.74}$$

Recall that

$$V = W(\xi, 1 - \eta) \quad \text{and} \quad U = V + W$$

2.1.18 Lessons

If there are two nonhomogeneous boundary conditions break the problem into two problems that can be added (since the equations are linear) to give the complete solution. If you are unsure of the sign of the separation constant just assume a sign and move on. *Listen to what the mathematics is telling you.* It will always tell you if you choose wrong.

Example 2.4. A Non-homogeneous Heat Conduction Problem

Consider now the arrangement above, but with a heat source, and with both boundaries held at the initial temperature u_0. The heat source is initially zero and is turned on at $t = 0^+$. *The exercise illustrates the method of solving the problem when the single nonhomogeneous condition is in the partial differential equation rather than one of the boundary conditions.*

$$\rho c u_t = k u_{xx} + q \tag{2.75}$$

$$\begin{aligned} u(0, x) &= u_0 \\ u(t, 0) &= u_0 \\ u(t, L) &= u_0 \end{aligned} \tag{2.76}$$

2.1.19 Scales and Dimensionless Variables

Observe that the length scale is still L, so we define $\xi = x/L$. Recall that $k/\rho c = \alpha$ is the diffusivity. How shall we nondimensionalize temperature? We want as many ones and zeros in coefficients in the partial differential equation and the boundary conditions as possible. Define $U = (u - u_0)/S$, where S stands for "something with dimensions of temperature" that we must find. Dividing both sides of the partial differential equation by q and substituting for x

$$\frac{L^2 S \rho c U_t}{q} = \frac{k S U_{\xi\xi}}{q} + 1 \tag{2.77}$$

Letting $S = q/k$ leads to one as the coefficient of the first term on the right-hand side. Choosing the same dimensionless time as before, $\tau = \alpha t/L^2$ results in one as the coefficient of the time derivative term. We now have

$$U_\tau = U_{\xi\xi} + 1 \tag{2.78}$$

$$U(0,\xi) = 0$$
$$U(\tau,0) = 0 \tag{2.79}$$
$$U(\tau,1) = 0$$

2.1.20 Relocating the Nonhomogeneity

We have only one nonhomogeneous condition, but it's in the wrong place. The differential equation won't separate. For example if we let $U(\xi,\tau) = P(\xi)G(\tau)$ and insert this into the partial differential equation and divide by PG, we find

$$\frac{G'(\tau)}{G} = \frac{P''(\xi)}{P} + \frac{1}{PG} \tag{2.80}$$

The technique to deal with this is to relocate the nonhomogenous condition to the initial condition. Assume a solution in the form $U = W(\xi) + V(\tau,\xi)$. We now have

$$V_\tau = V_{\xi\xi} + W_{\xi\xi} + 1 \tag{2.81}$$

If we set $W_{\xi\xi} = -1$, the differential equation for V becomes homogeneous. We then set both W and V equal to zero at $\xi = 0$ and 1 and $V(0,\xi) = -W(\xi)$

$$W_{\xi\xi} = -1 \tag{2.82}$$

$$W(0) = W(1) = 0 \tag{2.83}$$

and

$$V_\tau = V_{\xi\xi} \tag{2.84}$$

$$V(0,\xi) = -W(\xi)$$
$$V(\tau,0) = 0 \tag{2.85}$$
$$V(\tau,1) = 0$$

The solution for W is parabolic

$$W = \frac{1}{2}\xi(1-\xi) \tag{2.86}$$

2.1.21 Separating Variables

We now solve for V using separation of variables.

$$V = P(\tau)Q(\xi) \tag{2.87}$$

$$\frac{P_\tau}{P} = \frac{Q_{\xi\xi}}{Q} = \pm\lambda^2 \tag{2.88}$$

We must choose the minus sign once again (see Problem 1 above) to have a negative exponential for $P(\tau)$. (We will see later that it's not always so obvious.) $P = \exp(-\lambda^2\tau)$.

The solution for Q is once again sines and cosines.

$$Q = A\cos(\lambda\xi) + B\sin(\lambda\xi) \tag{2.89}$$

The boundary condition $V(\tau, 0) = 0$ requires that $Q(0) = 0$. Hence, $A = 0$. The boundary condition $V(\tau, 1) = 0$ requires that $Q(1) = 0$. Since B cannot be zero, $\sin(\lambda) = 0$ so that our eigenvalues are $\lambda = n\pi$ and our eigenfunctions are $\sin(n\pi\xi)$.

2.1.22 Superposition

Once again using linear superposition,

$$V = \sum_{n=0}^{\infty} B_n \exp(-n^2\pi^2\tau)\sin(n\pi\xi) \tag{2.90}$$

Applying the initial condition

$$\frac{1}{2}\xi(\xi - 1) = \sum_{n=1}^{\infty} B_n \sin(n\pi\xi) \tag{2.91}$$

This is a Fourier sine series representation of $\frac{1}{2}\xi(\xi - 1)$. We now use the orthogonality of the sine function to obtain the coefficients B_n.

2.1.23 Orthogonality

Using the concept of orthogonality again, we multiply both sides by $\sin(m\pi\xi)d\xi$ and integrate over the space noting that the integral is zero if m is not equal to n. Thus, since

$$\int_0^1 \sin^2(n\pi\xi)d\xi = \frac{1}{2} \tag{2.92}$$

$$B_n = \int_0^1 \xi(\xi - 1)\sin(n\pi\xi)d\xi \tag{2.93}$$

2.1.24 Lessons

When the differential equation is nonhomogeneous use the linearity of the differential equation to transfer the nonhomogeneous condition to one of the boundary conditions. Usually this will result in a homogeneous partial differential equation and an ordinary differential equation.

We pause here to note that while the method of separation of variables is straightforward in principle, a certain amount of intuition or, if you wish, cleverness is often required in order to put the equation and boundary conditions in an appropriate form. The student working diligently will soon develop these skills.

Problems

1. Using these ideas obtain a series solution to the boundary value problem

$$\begin{aligned} u_t &= u_{xx} \\ u(t, 1) &= 0 \\ u(t, 0) &= 0 \\ u(0, x) &= 1 \end{aligned}$$

2. Find a series solution to the boundary value problem

$$\begin{aligned} u_t &= u_{xx} + x \\ u_x(t, 0) &= 0 \\ u(t, 1) &= 0 \\ u(0, x) &= 0 \end{aligned}$$

2.2 VIBRATIONS

In vibrations problems the dependent variable occurs in the differential equation as a second-order derivative of the independent variable t. The methodology is, however, essentially the same as it is in the diffusion equation. We first apply separation of variables, then use the boundary conditions to obtain eigenfunctions and eigenvalues, and use the linearity and orthogonality principles and the single nonhomogeneous condition to obtain a series solution. Once again, if there are more than one nonhomogeneous condition we use the linear superposition principle to obtain solutions for each nonhomogeneous condition and add the resulting solutions. We illustrate these ideas with several examples.

Example 2.5. A Vibrating String

Consider a string of length L fixed at the ends. The string is initially held in a fixed position $y(0, x) = f(x)$, where it is clear that $f(x)$ must be zero at both $x = 0$ and $x = L$. The boundary value problem is as follows:

$$y_{tt} = a^2 y_{xx} \tag{2.94}$$

$$y(t, 0) = 0$$

$$y(t, L) = 0 \tag{2.95}$$

$$y(0, x) = f(x)$$

$$y_t(0, x) = 0$$

2.2.1 Scales and Dimensionless Variables

The problem has the obvious length scale L. Hence let $\xi = x/L$. Now let $\tau = ta/L$ and the equation becomes

$$y_{\tau\tau} = y_{\xi\xi} \tag{2.96}$$

One could now nondimensionalize y, for example, by defining a new variable as $f(x)/f_{\max}$, but it wouldn't simplify things. The boundary conditions remain the same except t and x are replaced by τ and ξ.

2.2.2 Separation of Variables

You know the dance. Let $y = P(\tau)Q(\xi)$. Differentiating and substituting into Eq. (2.96),

$$P_{\tau\tau} Q = P Q_{\xi\xi} \tag{2.97}$$

Dividing by PQ and noting that $P_{\tau\tau}/P$ and $Q_{\xi\xi}/Q$ cannot be equal to one another unless they are both constants, we find

$$P_{\tau\tau}/P = Q_{\xi\xi}/Q = \pm\lambda^2 \tag{2.98}$$

It should be physically clear that we want the minus sign. Otherwise both solutions will be hyperbolic functions. However if you choose the plus sign you will immediately find that the boundary conditions on ξ cannot be satisfied. Refer back to (2.63) and the sentences following.

The two ordinary differential equations and homogeneous boundary conditions are

$$P_{\tau\tau} + \lambda^2 P = 0 \tag{2.99}$$

$$P_\tau(0) = 0$$

and

$$Q_{\xi\xi} + \lambda^2 Q = 0 \tag{2.100}$$
$$Q(0) = 0$$
$$Q(1) = 0$$

The solutions are

$$P = A\sin(\lambda\tau) + B\cos(\lambda\tau) \tag{2.101}$$
$$Q = C\sin(\lambda\xi) + D\cos(\lambda\xi) \tag{2.102}$$

The first boundary condition of Eq. (2.100) requires that $D = 0$. The second requires that $C\sin(\lambda)$ be zero. Our eigenvalues are again $\lambda_n = n\pi$. The boundary condition at $\tau = 0$, that $P_\tau = 0$ requires that $A = 0$. Thus

$$PQ_n = K_n \sin(n\pi\xi)\cos(n\pi\tau) \tag{2.103}$$

The final form of the solution is then

$$y(\tau,\xi) = \sum_{n=0}^{\infty} K_n \sin(n\pi\xi)\cos(n\pi\tau) \tag{2.104}$$

2.2.3 Orthogonality

Applying the final (nonhomogeneous) boundary condition (the initial position).

$$f(\xi) = \sum_{n=0}^{\infty} K_n \sin(n\pi\xi) \tag{2.105}$$

In particular, if $f(x) = hx, \quad 0 < x < 1/2$

$$= h(1-x), \qquad 1/2 < x < 1 \tag{2.106}$$

$$\int_0^1 f(x)\sin(n\pi x)dx = \int_0^{1/2} hx\sin(n\pi x)dx + \int_{1/2}^1 h(1-x)\sin(n\pi x)dx$$
$$= \frac{2h}{n^2\pi^2}\sin\left(\frac{n\pi}{2}\right) = \frac{2h}{n^2\pi^2}(-1)^{n+1} \tag{2.107}$$

and

$$\int_0^1 K_n \sin^2(n\pi x)dx = K_n/2 \tag{2.108}$$

so that

$$y = \frac{4h}{\pi^2} \sum_{n=1}^{\infty} \frac{(-1)^{n+1}}{n^2} \sin(n\pi\xi)\cos(n\pi\tau) \tag{2.109}$$

2.2.4 Lessons

The solutions are in the form of infinite series. The coefficients of the terms of the series are determined by using the fact that the solutions of at least one of the ordinary differential equations are orthogonal functions. The orthogonality condition allows us to calculate these coefficients.

Problem

1. Solve the boundary value problem

$$\begin{aligned} u_{tt} &= u_{xx} \\ u(t, 0) &= u(t, 1) = 0 \\ u(0, x) &= 0 \\ u_t(0, x) &= f(x) \end{aligned}$$

   Find the special case when $f(x) = \sin(\pi x)$.

FURTHER READING

V. Arpaci, *Conduction Heat Transfer*. Reading, MA: Addison-Wesley, 1966.

J. W. Brown and R. V. Churchill, *Fourier Series and Boundary Value Problems*. 6th edition. New York: McGraw-Hill, 2001.

CHAPTER 3

Orthogonal Sets of Functions

In this chapter we elaborate on the concepts of orthogonality and Fourier series. We begin with the familiar concept of orthogonality of vectors. We then extend the idea to orthogonality of functions and the use of this idea to represent general functions as Fourier series—series of orthogonal functions.

Next we show that solutions of a fairly general linear ordinary differential equation—the Sturm–Liouville equation—are orthogonal functions. Several examples are given.

3.1 VECTORS

We begin our study of orthogonality with the familiar topic of orthogonal vectors. Suppose $\mathbf{u}(1)$, $\mathbf{u}(2)$, and $\mathbf{u}(3)$ are the three rectangular component vectors in an ordinary three-dimensional space. The norm of the vector (its length) $||\mathbf{u}||$ is

$$||\mathbf{u}|| = [u(1)^2 + u(2)^2 + u(3)^2]^{1/2} \tag{3.1}$$

If $||\mathbf{u}|| = 1$, $\mathbf{u}$ is said to be normalized. If $||\mathbf{u}|| = 0$, $\mathbf{u}(r) = 0$ for each r and $\mathbf{u}$ is the zero vector.

A linear combination of two vectors $\mathbf{u}_1$ and $\mathbf{u}_2$ is

$$\mathbf{u} = c_1\mathbf{u}_1 + c_2\mathbf{u}_2, \tag{3.2}$$

The scalar or inner product of the two vectors $\mathbf{u}_1$ and $\mathbf{u}_2$ is defined as

$$(\mathbf{u}_1, \mathbf{u}_2) = \sum_{r=1}^{3} u_1(r)u_2(r) = \|u_1\|\|u_2\|\cos\theta \tag{3.3}$$

3.1.1 Orthogonality of Vectors

If neither $\mathbf{u}_1$ nor $\mathbf{u}_2$ is the zero vector and if

$$(\mathbf{u}_1, \mathbf{u}_2) = 0 \tag{3.4}$$

then $\theta = \pi/2$ and the vectors are *orthogonal*. The norm of a vector $\mathbf{u}$ is

$$||\mathbf{u}|| = (\mathbf{u}, \mathbf{u})^{1/2} \tag{3.5}$$

3.1.2 Orthonormal Sets of Vectors

The vector $\mathbf{\Phi}_n = \mathbf{u}_n/||\mathbf{u}_n||$ has magnitude unity, and if $\mathbf{u}_1$ and $\mathbf{u}_2$ are orthogonal then $\mathbf{\Phi}_1$ and $\mathbf{\Phi}_2$ are orthonormal and their inner product is

$$(\mathbf{\Phi}_n, \mathbf{\Phi}_m) = \delta_{nm} = 0, m \neq n \qquad (3.6)$$
$$= 1, m = n$$

where δ_{nm} is called the Kronecker delta.

If $\mathbf{\Phi}_1$, $\mathbf{\Phi}_2$, and $\mathbf{\Phi}_3$ are three vectors that are mutually orthogonal to each other then every vector in three-dimensional space can be written as a linear combination of $\mathbf{\Phi}_1$, $\mathbf{\Phi}_2$, and $\mathbf{\Phi}_3$; that is,

$$\mathbf{f}(r) = c_1\mathbf{\Phi}_1 + c_2\mathbf{\Phi}_2 + c_3\mathbf{\Phi}_3 \qquad (3.7)$$

Note that due to the fact that the vectors $\mathbf{\Phi}_n$ form an orthonormal set,

$$(\mathbf{f}, \mathbf{\Phi}_1) = c_1, (\mathbf{f}, \mathbf{\Phi}_2) = c_2, (\mathbf{f}, \mathbf{\Phi}_3) = c_3 \qquad (3.8)$$

Simply put, suppose the vector **f** is

$$\mathbf{f} = 2\mathbf{\Phi}_1 + 4\mathbf{\Phi}_2 + \mathbf{\Phi}_3. \qquad (3.9)$$

Taking the inner product of **f** with $\mathbf{\Phi}_1$ we find that

$$(\mathbf{f}, \mathbf{\Phi}_1) = 2(\mathbf{\Phi}_1, \mathbf{\Phi}_1) + 4(\mathbf{\Phi}_1, \mathbf{\Phi}_2) + (\mathbf{\Phi}_1, \mathbf{\Phi}_3) \qquad (3.10)$$

and according to Eq. (3.8) $c_1 = 2$. Similarly, $c_2 = 4$ and $c_3 = 1$.

3.2 FUNCTIONS

3.2.1 Orthonormal Sets of Functions and Fourier Series

Suppose there is a set *of orthonormal functions* $\Phi_n(x)$ defined on an interval $a < x < b$ ($\sqrt{2}\sin(n\pi x)$ on the interval $0 < x < 1$ is an example). A set of orthonormal *functions* is defined as one whose inner product, defined as $\int_{x=a}^{b} \Phi_n(x)\Phi_m(x)dx$, is

$$(\Phi_n, \Phi_m) = \int_{x=a}^{b} \Phi_n \Phi_m \, dx = \delta_{nm} \qquad (3.11)$$

Suppose we can express a function as an infinite series of these orthonormal functions,

$$f(x) = \sum_{n=0}^{\infty} c_n\Phi_n \quad \text{on} \quad a < x < b \qquad (3.12)$$

Equation (3.12) is called a *Fourier series* of $f(x)$ in terms of the orthonormal function set $\Phi_n(x)$.

If we now form the inner product of Φ_m with both sides of Eq. (3.12) and use the definition of an orthonormal function set as stated in Eq. (3.11) we see that the inner product of $f(x)$ and $\Phi_n(x)$ is c_n.

$$c_n \int_{x=a}^{b} \Phi_n^2(\xi) d\xi = c_n = \int_{x=a}^{b} f(\xi)\, \Phi_n(\xi) d\xi \tag{3.13}$$

In particular, consider a set of functions Ψ_n that are orthogonal on the interval (a, b) so that

$$\begin{aligned} \int_{x=a}^{b} \Psi_n(\xi)\Psi_m(\xi) d\xi &= 0, \quad m \neq n \\ &= \|\Psi_n\|^2, \quad m = n \end{aligned} \tag{3.14}$$

where $\|\Psi_n\|^2 = \int_{x=a}^{b} \Psi_n^2(\xi) d\xi$ is called the square of the norm of Ψ_n. The functions

$$\frac{\Psi_n}{\|\Psi_n\|} = \Phi_n \tag{3.15}$$

then form an orthonormal set. We now show how to form the series representation of the function $f(x)$ as a series expansion in terms of the orthogonal (but not orthonormal) set of functions $\Psi_n(x)$.

$$f(x) = \sum_{n=0}^{\infty} \frac{\Psi_n}{\|\Psi_n\|} \int_{\xi=a}^{b} f(\xi) \frac{\Psi_n(\xi)}{\|\Psi_n\|} d\xi = \sum_{n=0}^{\infty} \Psi_n \int_{\xi=a}^{b} f(\xi) \frac{\Psi_n(\xi)}{\|\Psi_n\|^2} d\xi \tag{3.16}$$

This is called a Fourier series representation of the function $f(x)$.

As a concrete example, the square of the norm of the sine function on the interval $(0, \pi)$ is

$$\|\sin(nx)\|^2 = \int_{\xi=0}^{\pi} \sin^2(n\xi) d\xi = \frac{\pi}{2} \tag{3.17}$$

so that the corresponding orthonormal function is

$$\Phi = \sqrt{\frac{2}{\pi}} \sin(nx) \tag{3.18}$$

A function can be represented by a series of sine functions on the interval $(0, \pi)$ as

$$f(x) = \sum_{n=0}^{\infty} \sin(nx) \int_{\varsigma=0}^{\pi} \frac{\sin(n\varsigma)}{\pi/2} f(\varsigma) d\varsigma \tag{3.19}$$

This is a *Fourier sine series.*

3.2.2 Best Approximation

We next ask whether, since we can never sum to infinity, the values of the constants c_n in Eq. (3.13) give the most accurate approximation of the function. To illustrate the idea we return to the idea of *orthogonal vectors* in three-dimensional space. Suppose we want to approximate a three-dimensional vector with a two-dimensional vector. What will be the components of the two-dimensional vector that best approximate the three-dimensional vector?

Let the three-dimensional vector be $\mathbf{f} = c_1\mathbf{\Phi}_1 + c_2\mathbf{\Phi}_2 + c_3\mathbf{\Phi}_3$. Let the two-dimensional vector be $\mathbf{k} = a_1\mathbf{\Phi}_1 + a_2\mathbf{\Phi}_2$. We wish to minimize $||\mathbf{k} - \mathbf{f}||$.

$$||\mathbf{k} - \mathbf{f}|| = \left\{(a_1 - c_1)^2 + (a_2 - c_2)^2 + c_3^2\right\}^{1/2} \tag{3.20}$$

It is clear from the above equation (and also from Fig. 3.1) that this will be minimized when $a_1 = c_1$ and $a_2 = c_2$.

Turning now to the *orthogonal function* series, we attempt to minimize the difference between the function with an infinite number of terms and the summation only to some finite value m. The square of the error is

$$E^2 = \int_{x=a}^{b} (f(x) - K_m(x))^2 dx = \int_{x=a}^{b} \left[f^2(x) + K^2(x) - 2f(x)K(x)\right] dx \tag{3.21}$$

where

$$f(x) = \sum_{n=1}^{\infty} c_n \Phi_n(x) \tag{3.22}$$

and

$$K_m = \sum_{n=1}^{m} a_n \Phi_n(x) \tag{3.23}$$

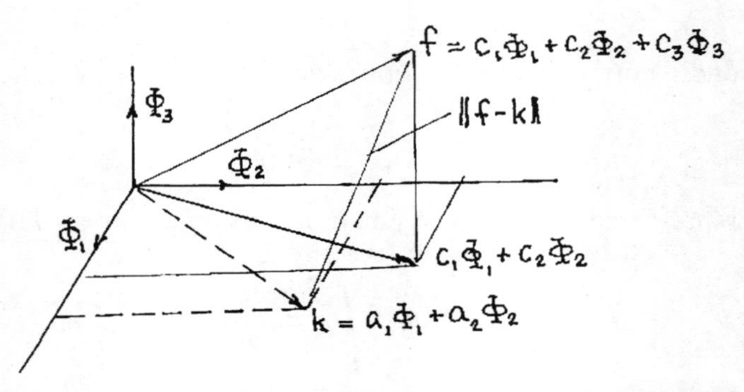

FIGURE 3.1: Best approximation of a three-dimensional vector in two dimensions

Noting that

$$\int_{x=a}^{b} K_m^2(x)dx = \sum_{n=1}^{m}\sum_{j=1}^{m} a_n a_j \int_{x=a}^{b} \Phi_n(x)\Phi_j(x)dx = \sum_{n=1}^{m} a_n^2 = a_1^2 + a_2^2 + a_3^2 + \cdots + a_m^2 \tag{3.24}$$

and

$$\int_{x=a}^{b} f(x)K(x)dx = \sum_{n=1}^{\infty}\sum_{j=1}^{m} c_n a_j \int_{x=a}^{b} \Phi_n(x)\Phi_j(x)dx$$

$$= \sum_{n=1}^{m} c_n a_n = c_1 a_1 + c_2 a_2 + \cdots + c_m a_m \tag{3.25}$$

$$E^2 = \int_{x=a}^{b} f^2(x)dx + a_1^2 + \cdots + a_m^2 - 2a_1c_1 - \cdots - 2a_m c_m \tag{3.26}$$

Now add and subtract $c_1^2, c_2^2, \ldots, c_m^2$. Thus Eq. (3.26) becomes

$$E^2 = \int_{x=a}^{b} f^2(x)dx - c_1^2 - c_2^2 - \cdots - c_m^2 + (a_1 - c_1)^2 + (a_2 - c_2)^2 + \cdots + (a_m - c_m)^2 \tag{3.27}$$

which is clearly minimized when $a_n = c_n$.

3.2.3 Convergence of Fourier Series

We briefly consider the question of whether the Fourier series actually converges to the function $f(x)$ for all values, say, on the interval $a \le x \le b$. The series will converge to the function if the value of E defined in (3.19) approaches zero as m approaches infinity. Suffice to say that this is true for functions that are continuous with piecewise continuous first derivatives, that is, most physically realistic temperature distributions, displacements of vibrating strings and bars. In each particular situation, however, one should use the various convergence theorems that are presented in most elementary calculus books. Uniform convergence of Fourier series is discussed extensively in the book *Fourier Series and Boundary Value Problems* by James Ward Brown and R. V. Churchill. In this chapter we give only a few physically clear examples.

3.2.4 Examples of Fourier Series

Example 3.1. Determine a Fourier sine series representation of $f(x) = x$ on the interval $(0, 1)$. The series will take the form

$$x = \sum_{j=0}^{\infty} c_j \sin(j\pi x) \tag{3.28}$$

since the $\sin(j\pi x)$ forms an orthogonal set on $(0, 1)$, multiply both sides by $\sin(k\pi x)dx$ and integrate over the interval on which the function is orthogonal.

$$\int_{x=0}^{1} x \sin(k\pi x)dx = \sum_{k=0}^{\infty} \int_{x=0}^{1} c_j \sin(j\pi x)\sin(k\pi x)dx \tag{3.29}$$

Noting that all of the terms on the right-hand side of (2.20) are zero except the one for which $k = j$,

$$\int_{x=0}^{1} x \sin(j\pi x)dx = c_j \int_{x=0}^{1} \sin^2(j\pi x)dx \tag{3.30}$$

After integrating we find

$$\frac{(-1)^{j+1}}{j\pi} = \frac{c_j}{2} \tag{3.31}$$

Thus,

$$x = \sum_{j=0}^{\infty} \frac{(-1)^{j+1}}{j\pi} 2\sin(j\pi x) \tag{3.32}$$

This is an alternating sign series in which the coefficients always decrease as j increases, and it therefore converges. The sine function is periodic and so the series must also be a periodic function beyond the interval $(0, 1)$. The series outside this interval forms the *periodic continuation* of the series. Note that the sine is an odd function so that $\sin(j\pi x) = -\sin(-j\pi x)$. Thus the periodic continuation looks like Fig. 3.2. The series converges everywhere, but at $x = 1$ it is identically zero instead of one. It converges to $1 - \varepsilon$ arbitrarily close to $x = 1$.

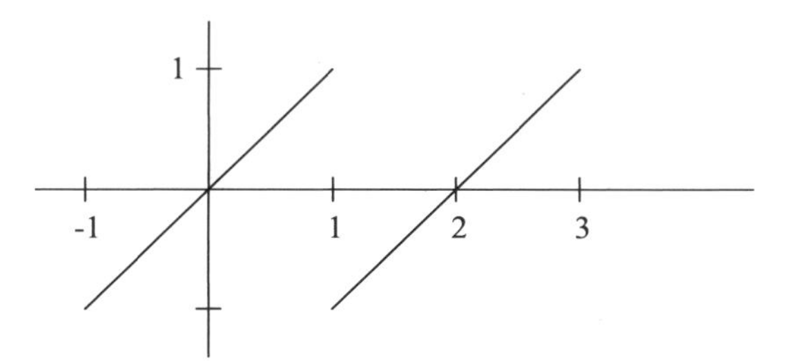

FIGURE 3.2: The periodic continuation of the function x represented by the sine series

Example 3.2. Find a Fourier cosine for $f(x) = x$ on the interval (0, 1). In this case

$$x = \sum_{n=0}^{\infty} c_n \cos(n\pi x) \tag{3.33}$$

Multiply both sides by $\cos(m\pi x)dx$ and integrate over (0, 1).

$$\int_{x=0}^{1} x \cos(m\pi x)dx = \sum_{n=0}^{\infty} c_n \int_{x=0}^{1} \cos(m\pi x)\cos(n\pi x)dx \tag{3.34}$$

and noting that $\cos(n\pi x)$ is an orthogonal set on (0, 1) all terms in (2.23) are zero except when $n = m$. Evaluating the integrals,

$$\frac{c_n}{2} = \frac{[(-1)^2 - 1]}{(n\pi)^2} \tag{3.35}$$

There is a problem when $n = 0$. Both the numerator and the denominator are zero there. However we can evaluate c_0 by noting that according to Eq. (3.26)

$$\int_{x=0}^{1} x dx = c_0 = \frac{1}{2} \tag{3.36}$$

and the cosine series is therefore

$$x = \frac{1}{2} + \sum_{n=1}^{\infty} 2\frac{[(-1)^n - 1]}{(n\pi)^2}\cos(n\pi x) \tag{3.37}$$

The series converges to x everywhere. Since $\cos(n\pi x) = \cos(-n\pi x)$ it is an even function and its *periodic continuation* is shown in Fig. 3.3. Note that the sine series is discontinuous at $x = 1$, while the cosine series is continuous everywhere. (Which is the better representation?)

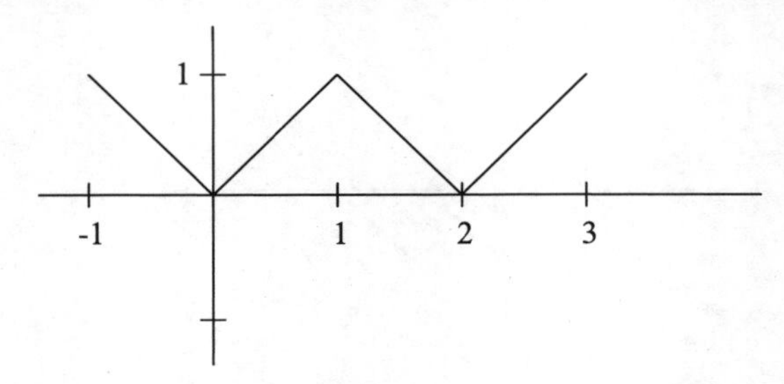

FIGURE 3.3: The periodic continuation of the series in Example 3.2

It should be clear from the above examples that in general a Fourier sine/cosine series of a function $f(x)$ defined on $0 \leq x \leq 1$ can be written as

$$f(x) = \frac{c_0}{2} + \sum_{n=1}^{\infty} c_n \cos(n\pi x) + \sum_{n=1}^{\infty} b_n \sin(n\pi x) \tag{3.38}$$

where

$$c_n = \frac{\int_{x=0}^{1} f(x)\cos(n\pi x)dx}{\int_{x=0}^{1} \cos^2(n\pi x)dx} \qquad n = 0, 1, 2, 3, \ldots$$

$$b_n = \frac{\int_{x=0}^{1} f(x)\sin(n\pi x)dx}{\int_{x=0}^{1} \sin^2(n\pi x)dx} \qquad n = 1, 2, 3, \ldots \tag{3.39}$$

Problems

1. Show that

$$\int_{x=0}^{\pi} \sin(nx)\sin(mx)dx = 0$$

when $n \neq m$.

2. Find the Fourier sine series for $f(x) = 1 - x$ on the interval (0, 1). Sketch the periodic continuation. Sum the series for the first five terms and sketch over two periods. Discuss convergence of the series, paying special attention to convergence at $x = 0$ and $x = 1$.
3. Find the Fourier cosine series for $1 - x$ on (0, 1). Sketch the periodic continuation. Sum the first two terms and sketch. Sum the first five terms and sketch over two periods. Discuss convergence, paying special attention to convergence at $x = 0$ and $x = 1$.

3.3 STURM–LIOUVILLE PROBLEMS: ORTHOGONAL FUNCTIONS

We now proceed to show that solutions of a certain ordinary differential equation with certain boundary conditions (called a Sturm–Liouville problem) are *orthogonal functions with respect to a weighting function*, and that therefore a well-behaved function can be represented by an infinite series of these orthogonal functions (called eigenfunctions), as in Eqs. (3.12) and (3.16).

Recall that the problem

$$X_{xx} + \lambda^2 X = 0, X(0) = 0, X(1) = 0 \qquad 0 \le x \le 1 \tag{3.40}$$

has solutions only for $\lambda = n\pi$ and that the solutions, $\sin(n\pi x)$ are orthogonal on the interval (0, 1). The sine functions are called eigenfunctions and $\lambda = n\pi$ are called eigenvalues.

As another example, consider the problem

$$X_{xx} + \lambda^2 X = 0 \tag{3.41}$$

with boundary conditions

$$\begin{aligned} X(0) &= 0 \\ X(1) + HX_x(1) &= 0 \end{aligned} \tag{3.42}$$

The solution of the differential equation is

$$X = A\sin(\lambda x) + B\cos(\lambda x)) \tag{3.43}$$

The first boundary condition guarantees that $B = 0$. The second boundary condition is satisfied by the equation

$$A[\sin(\lambda) + H\lambda\cos(\lambda)] = 0 \tag{3.44}$$

Since A cannot be zero, this implies that

$$-\tan(\lambda) = H\lambda. \tag{3.45}$$

The *eigenfunctions* are $\sin(\lambda x)$ and the *eigenvalues* are solutions of Eq. (3.45). This is illustrated graphically in Fig. 3.4.

We will generally be interested in the fairly general linear second-order differential equation and boundary conditions given in Eqs. (3.46) and (3.47).

$$\frac{d}{dx}\left[r(x)\frac{dX}{dx}\right] + [q(x) + \lambda p(x)]X = 0 \quad a \le x \le b \tag{3.46}$$

$$\begin{aligned} &a_1 X(a) + a_2 dX(a)/dx = 0 \\ &b_1 X(b) + b_2 dX(b)/dx = 0 \end{aligned} \tag{3.47}$$

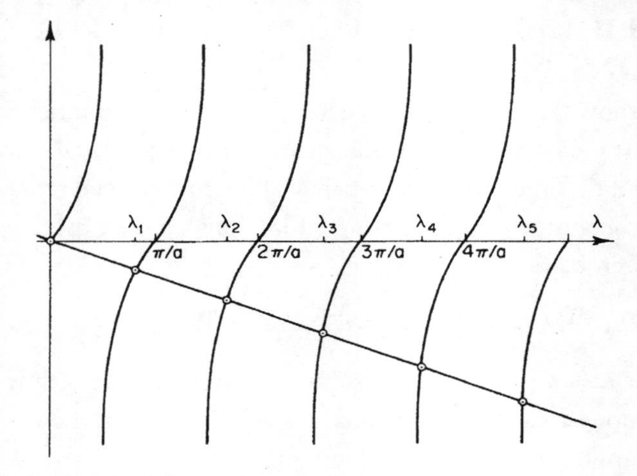

FIGURE 3.4: Eigenvalues of $-\tan(\lambda) = H\lambda$

Solutions exist only for discrete values λ_n the *eigenvalues.* The corresponding solutions $X_n(x)$ are the *eigenfunctions.*

3.3.1 Orthogonality of Eigenfunctions

Consider two solutions of (3.46) and (3.47), X_n and X_m corresponding to eigenvalues λ_n and λ_m. The primes denote differentiation with respect to x.

$$(rX'_m)' + qX_m = -\lambda_m pX_m \tag{3.48}$$

$$(rX'_n)' + qX_n = -\lambda_n pX_n \tag{3.49}$$

Multiply the first by X_n and the second by X_m and subtract, obtaining the following:

$$(rX_nX'_m - rX_mX'_n)' = (\lambda_n - \lambda_m)pX_mX_n \tag{3.50}$$

Integrating both sides

$$r(X'_mX_n - X'_nX_m)_a^b = (\lambda_n - \lambda_m)\int_a^b p(x)X_nX_m dx \tag{3.51}$$

Inserting the boundary conditions into the left-hand side of (3.51)

$$\begin{aligned} &X'_m(b)X_n(b) - X'_m(a)X_n(a) - X'_n(b)X_m(b) + X'_n(a)X_m(a) \\ &= -\frac{b_1}{b_2}X_m(b)X_n(b) + \frac{a_1}{a_2}X_m(a)X_n(a) - \frac{a_1}{a_2}X_n(a)X_m(a) + \frac{b_1}{b_2}X_m(b)X_n(b) = 0 \end{aligned} \tag{3.52}$$

Thus

$$(\lambda_n - \lambda_m) \int_a^b p(x) X_n X_m dx = 0, \quad m \neq n \tag{3.53}$$

Notice that X_m and X_n are *orthogonal with respect to the weighting function $p(x)$ on the interval (a, b)*. Obvious examples are the sine and cosine functions.

Example 3.3. Example 2.1 in Chapter 2 is an example in which the *eigenfunctions* are $\sin(\lambda_n \xi)$ and the *eigenvalues* are $(2n-1)\pi/2$.

Example 3.4. If the boundary conditions in Example 2.1 in Chapter 2 are changed to

$$\Phi'(0) = 0 \qquad \Phi(1) = 0 \tag{3.54}$$

we note that the general solution of the differential equation is

$$\Phi(\xi) = A\cos(\lambda\xi) + B\sin(\lambda\xi) \tag{3.55}$$

The boundary conditions require that $B = 0$ and $\cos(\lambda) = 0$. The values of λ can take on any of the values $\pi/2, 3\pi/2, 5\pi/2, \ldots, (2n-1)\pi/2$. The *eigenfunctions* are $\cos(\lambda_n \xi)$ and the *eigenvalue* are $\lambda_n = (2n-1)\pi/2$.

Example 3.5. Suppose the boundary conditions in the original problem (Example 1, Chapter 2) take on the more complicated form

$$\Phi(0) = 0 \qquad \Phi(1) + h\Phi'(1) = 0 \tag{3.56}$$

The first boundary condition requires that $B = 0$. The second boundary conditions require that

$$\sin(\lambda_n) + h\lambda_n \cos(\lambda_n) = 0, \text{ or} \tag{3.57}$$

$$\lambda_n = -\frac{1}{h}\tan(\lambda_n) \tag{3.58}$$

which is a transcendental equation that must be solved for the *eigenvalues*. The *eigenfunctions* are, of course, $\sin(\lambda_n x)$.

Example 3.6. A Physical Example: Heat Conduction in Cylindrical Coordinates

The heat conduction equation in cylindrical coordinates is

$$\frac{\partial u}{\partial t} = \frac{\partial^2 u}{\partial r^2} + \frac{1}{r}\frac{\partial u}{\partial r} \qquad 0 < r < 1 \tag{3.59}$$

with boundary conditions at $R = 0$ and $r = 1$ and initial condition $u(0, r) = f(r)$.

Separating variables as $u = R(r)T(t)$,

$$\frac{1}{T}\frac{dT}{dt} = \frac{1}{R}\frac{d^2R}{dr^2} + \frac{1}{rR}\frac{dR}{dr} = -\lambda^2 \qquad 0 \le r \le 1, \quad 0 \le t \tag{3.60}$$

(Why the minus sign?)

The equation for $R(r)$ is

$$(rR')' + \lambda^2 r R = 0, \tag{3.61}$$

which is a Sturm–Liouville equation *with weighting function* r. It is an eigenvalue problem with an infinite number of eigenfunctions corresponding to the eigenvalues λ_n. There will be two solutions $R_1(\lambda_n r)$ and $R_2(\lambda_n r)$ for each λ_n. The solutions are called Bessel functions, and they will be discussed in Chapter 4.

$$R_n(\lambda_n r) = A_n R_1(\lambda_n r) + B_n R_2(\lambda_n r) \tag{3.62}$$

The boundary conditions on r are used to determine a relation between the constants A and B. For solutions $R(\lambda_n r)$ and $R(\lambda_m r)$

$$\int_0^1 r R(\lambda_n r) R(\lambda_m r) dr = 0, \quad n \neq m \tag{3.63}$$

is the orthogonality condition.

The solution for $T(t)$ is the exponential $e^{-\lambda_n^2 t}$ for all n. Thus, the solution of (3.60), because of superposition, can be written as an infinite series in a form something like

$$u = \sum_{n=0}^{\infty} K_n e^{-\lambda_n^2} R(\lambda_n r) \tag{3.64}$$

and the orthogonality condition is used to find K_n as

$$K_n = \int_{r=0}^{1} f(r) R(\lambda_n r) r\,dr \Big/ \int_{r=0}^{1} f(r) R^2(\lambda_n r) r\,dr \tag{3.65}$$

Problems

1. For Example 2.1 in Chapter 2 with the new boundary conditions described in Example 3.2 above, find K_n and write the infinite series solution to the revised problem.

FURTHER READING

J. W. Brown and R. V. Churchill, *Fourier Series and Boundary Value Problems*, 6th edition. New York: McGraw-Hill, 2001.

P. V. O'Neil, *Advanced Engineering Mathematics*. 5th edition. Brooks/Cole Thompson, Pacific Grove, CA, 2003.

CHAPTER 4
Series Solutions of Ordinary Differential Equations

4.1 GENERAL SERIES SOLUTIONS

The purpose of this chapter is to present a method of obtaining solutions of linear second-order ordinary differential equations in the form of Taylor series'. The methodology is then used to obtain solutions of two special differential equations, Bessel's equation and Legendre's equation. Properties of the solutions—Bessel functions and Legendre functions—which are extensively used in solving problems in mathematical physics, are discussed briefly. Bessel functions are used in solving both diffusion and vibrations problems in cylindrical coordinates. The functions $R(\lambda_n r)$ in Example 3.4 at the end of Chapter 3 are called Bessel functions. Legendre functions are useful in solving problems in spherical coordinates. Associated Legendre functions, also useful in solving problems in spherical coordinates, are briefly discussed.

4.1.1 Definitions

In this chapter we will be concerned with linear second-order equations. A general case is

$$a(x)u'' + b(x)u' + c(x)u = f(x) \tag{4.1}$$

Division by $a(x)$ gives

$$u'' + p(x)u' + q(x)u = r(x) \tag{4.2}$$

Recall that if $r(x)$ is zero the equation is *homogeneous*. The solution can be written as the sum of a *homogeneous solution* $u_h(x)$ and a *particular solution* $u_p(x)$. If $r(x)$ is zero, $u_p = 0$. The nature of the solution and the solution method depend on the nature of the coefficients $p(x)$ and $q(x)$. If each of these functions can be expanded in a Taylor series about a point x_0 the point is said to be an *ordinary point* and the function is *analytic* at that point. If either of the coefficients is not analytic at x_0, the point is a *singular point*. If x_0 is a singular point and if $(x - x_0)p(x)$ and $(x - x_0)^2 q(x)$ are analytic, then the singularities are said to be *removable* and the singular point is a *regular singular point*. If this is not the case the singular point is *irregular*.

4.1.2 Ordinary Points and Series Solutions

If the point x_0 is an ordinary point the dependent variable has a solution in the neighborhood of x_0 of the form

$$u(x) = \sum_{n=0}^{\infty} c_n (x - x_0)^n \tag{4.3}$$

We now illustrate the solution method with two examples.

Example 4.1. Find a series solution in the form of Eq. (4.3) about the point $x = 0$ of the differential equation

$$u'' + x^2 u = 0 \tag{4.4}$$

The point $x = 0$ is an ordinary point so at least near $x = 0$ there is a solution in the form of the above series. Differentiating (4.3) twice and inserting it into (4.4)

$$u' = \sum_{n=0}^{\infty} n c_n x^{n-1}$$

$$u'' = \sum_{n=0}^{\infty} n(n-1) c_n x^{n-2}$$

$$\sum_{n=0}^{\infty} n(n-1) c_n x^{n-2} + \sum_{n=0}^{\infty} x^{n+2} c_n = 0 \tag{4.5}$$

Note that the first term in the u' series is zero while the first two terms in the u'' series are zero. We can shift the indices in both summations so that the power of x is the same in both series by setting $n - 2 = m$ in the first series.

$$\sum_{n=0}^{\infty} n(n-1) c_n x^{n-2} = \sum_{m=-2}^{\infty} (m+2)(m+1) c_{m+2} x^m = \sum_{m=0}^{\infty} (m+2)(m+1) c_{m+2} x^m \tag{4.6}$$

Noting that m is a "dummy variable" and that the first two terms in the series are zero the series can be written as

$$\sum_{n=0}^{\infty} (n+2)(n+1) c_{n+2} x^n \tag{4.7}$$

In a similar way we can write the second term as

$$\sum_{n=0}^{\infty} c_n x^{n+2} = \sum_{n=2}^{\infty} c_{n-2} x^n \tag{4.8}$$

We now have

$$\sum_{n=0}^{\infty}(n+2)(n+1)c_{n+2}x^n + \sum_{n=2}^{\infty}c_{n-2}x^n = 0 \tag{4.9}$$

which can be written as

$$2c_2 + 6c_3x + \sum_{n=2}^{\infty}[(n+2)(n+1)c_{n+2} + c_{n-2}]x^n = 0 \tag{4.10}$$

Each coefficient of x^n must be zero in order to satisfy Eq. (4.10). Thus c_2 and c_3 must be zero and

$$c_{n+2} = -c_{n-2}/(n+2)(n+1) \tag{4.11}$$

while c_0 and c_1 remain arbitrary.

Setting $n = 2$, we find that $c_4 = -c_0/12$ and setting $n = 3$, $c_5 = -c_1/20$. Since c_2 and c_3 are zero, so are c_6, c_7, c_{10}, c_{11}, etc. Also, $c_8 = -c_4/(8)(7) = c_0/(4)(3)(8)(7)$ and

$$c_9 = -c_5/(9)(8) = c_1/(5)(4)(9)(8).$$

The first few terms of the series are

$$u(x) = c_0(1 - x^4/12 + x^6/672 + \cdots) + c_1(1 - x^5/20 + x^9/1440 + \cdots) \tag{4.12}$$

The values of c_0 and c_1 may be found from appropriate boundary conditions. These are both alternating sign series with each term smaller than the previous term at least for $x \leq 1$ and it is therefore convergent at least under these conditions.

The constants c_0 and c_1 can be determined from boundary conditions. For example if $u(0) = 0$, $c_0 + c_1 = 0$, so $c_1 = -c_0$. If $u(1) = 1$,

$$c_0[-1/12 + 1/20 + 1/672 - 1/1440 + \cdots] = 1$$

Example 4.2. Find a series solution in the form of Eq. (4.3) of the differential equation

$$u'' + xu' + u = x^2 \tag{4.13}$$

valid near $x = 0$.

Assuming a solution in the form of (4.3), differentiating and inserting into (4.13),

$$\sum_{n=0}^{\infty}(n-1)nc_nx^{n-2} + \sum_{n=0}^{\infty}nc_nx^n + \sum_{n=0}^{\infty}c_nx^n - x^2 = 0 \tag{4.14}$$

Shifting the indices as before

$$\sum_{n=0}^{\infty}(n+2)(n+1)c_{n+2}x^n + \sum_{n=0}^{\infty} nc_n x^n + \sum_{n=0}^{\infty} c_n x^n - x^2 = 0 \tag{4.15}$$

Once again, each of the coefficients of x^n must be zero.

Setting $n = 0$, we see that

$$\begin{aligned} n &= 0 : 2c_2 + c_0 = 0, \qquad c_2 = -c_0/2 \\ n &= 1 : 6c_3 + 2c_1 = 0, \qquad c_3 = -c_1/3 \\ n &= 2 : 12c_4 + 3c_2 - 1 = 0, \quad c_4 = (1 + 3c_0/2)/12 \\ n &> 2 : c_{n+2} = \frac{c_n}{n+2} \end{aligned} \tag{4.16}$$

The last of these is called a *recurrence formula.*

Thus,

$$\begin{aligned} u &= c_0(1 - x^2/2 + x^4/8 - x^6/(8)(6) + \cdots) \\ &\quad + c_1(x - x^3/3 + x^5/(3)(5) - x^7/(3)(5)(7) + \cdots) \\ &\quad + x^4(1/12 - x^2/(12)(6) + \cdots) \end{aligned} \tag{4.17}$$

Note that the series on the third line of (4.17) is the *particular solution* of (4.13). The constants c_0 and c_1 are to be evaluated using the boundary conditions.

4.1.3 Lessons: Finding Series Solutions for Differential Equations with Ordinary Points

If x_0 is an ordinary point assume a solution in the form of Eq. (4.3) and substitute into the differential equation. Then equate the coefficients of equal powers of x. This will give a recurrence formula from which two series may be obtained in terms of two arbitrary constants. These may be evaluated by using the two boundary conditions.

Problems

1. The differential equation

$$u'' + xu' + xu = x$$

has ordinary points everywhere. Find a series solution near $x = 0$.

2. Find a series solution of the differential equation

$$u'' + (1 + x^2)u = x$$

near $x = 0$ and identify the particular solution.

3. The differential equation

$$(1 - x^2)u'' + u = 0$$

has singular points at $x = \pm 1$, but is analytic near $x = 0$. Find a series solution that is valid near $x = 0$ and discuss the radius of convergence.

4.1.4 Regular Singular Points and the Method of Frobenius

If x_0 is a singular point in (4.2) there may not be a power series solution of the form of Eq. (4.3). In such a case we proceed by assuming a solution of the form

$$u(x) = \sum_{n=0}^{\infty} c_n (x - x_0)^{n+r} \tag{4.18}$$

in which $c_0 \neq 0$ and r is any constant, not necessarily an integer. This is called the method of Frobenius and the series is called a Frobenius series. The Frobenius series need not be a power series because r may be a fraction or even negative. Differentiating once

$$u' = \sum_{n=0}^{\infty} (n + r)c_n (x - x_0)^{n+r-1} \tag{4.19}$$

and differentiating again

$$u'' = \sum_{n=0}^{\infty} (n + r - 1)(n + r)c_n (x - x_0)^{n+r-2} \tag{4.20}$$

These are then substituted into the differential equation, shifting is done where required so that each term contains x raised to the power n, and the coefficients of x^n are each set equal to zero. The coefficient associated with the lowest power of x will be a quadratic equation that can be solved for the index r. It is called an *indicial equation*. There will therefore be two roots of this equation corresponding to two series solutions. The values of c_n are determined as above by a *recurrence equation* for each of the roots. Three possible cases are important: (a) the roots are distinct and do not differ by an integer, (b) the roots differ by an integer, and (c) the roots are coincident, i.e., repeated. We illustrate the method by a series of examples.

Example 4.3 (distinct roots). Solve the equation

$$x^2 u'' + x(1/2 + 2x)u' + (x - 1/2)u = 0 \tag{4.21}$$

The coefficient of the u' term is

$$p(x) = \frac{(1/2 + 2x)}{x} \tag{4.22}$$

and the coefficient of the u'' term is

$$q(x) = \frac{(x - 1/2)}{x^2} \tag{4.23}$$

Both have singularities at $x = 0$. However multiplying $p(x)$ by x and $q(x)$ by x^2 the singularities are removed. Thus $x = 0$ is a regular singular point. Assume a solution in the form of the Frobenius series: $u = \sum_{n=0}^{\infty} c_n x^{n+r}$, differentiate twice and substitute into (4.21) obtaining

$$\sum_{n=0}^{\infty}(n+r)(n+r-1)x^{n+1} + \sum_{n=0}^{\infty}\frac{1}{2}(n+r)c_n x^{n+r} + \sum_{n=0}^{\infty} 2(n+r)c_n x^{n+r+1}$$
$$+ \sum_{n=0}^{\infty} c_n x^{n+r+1} - \sum_{n=0}^{\infty}\frac{1}{2}c_n x^{n+r} = 0 \tag{4.24}$$

The indices of the third and fourth summations are now shifted as in Example 4.1 and we find

$$\left[r(r-1) + \frac{1}{2}r - \frac{1}{2}\right]c_0 x^r + \sum_{n=1}^{\infty}\left[(n+r)(n+r-1) + \frac{1}{2}(n+r) - \frac{1}{2}\right]c_n x^{n+r}$$
$$+ \sum_{n=1}^{\infty}[2(n+r-1)+1]c_{n-1}x^{n+r} = 0 \tag{4.25}$$

Each coefficient must be zero for the equation to be true. Thus the coefficient of the c_0 term must be zero since c_0 itself cannot be zero. This gives a quadratic equation to be solved for r, and this is called an *indicial equation* (since we are solving for the index, r).

$$r(r-1) + \frac{1}{2}r - \frac{1}{2} = 0 \tag{4.26}$$

with $r = 1$ and $r = -1/2$. The coefficients of x^{n+r} must also be zero. Thus

$$[(n+r)(n+r-1) + 1/2(n+r) - 1/2]c_n + [2(n+r-1)+1]c_{n-1} = 0\,. \tag{4.27}$$

The *recurrence equation* is therefore

$$c_n = -\frac{2(n+r-1)+1}{(n+r)(n+r-1) + \frac{1}{2}(n+r) - \frac{1}{2}}c_{n-1} \tag{4.28}$$

For the case of $r = 1$

$$c_n = -\frac{2n+1}{n\left(n+\frac{3}{2}\right)}c_{n-1} \tag{4.29}$$

Computing a few of the coefficients,

$$c_1 = -\frac{3}{\frac{5}{2}}c_0 = -\frac{6}{5}c_0$$

$$c_2 = -\frac{5}{7}c_1 = -\frac{6}{7}c_0$$

$$c_3 = -\frac{7}{\frac{27}{2}}c_2 = -\frac{4}{9}c_0$$

etc. and the first Frobenius series is

$$u_1 = c_0\left(x - \frac{6}{5}x^2 + \frac{6}{7}x^3 - \frac{4}{9}x^4 + \cdots\right) \tag{4.30}$$

Setting $r = -1/2$ in the recurrence equation (4.26) and using b_n instead of c_n to distinguish it from the first case,

$$b_n = -\frac{2n-2}{n\left(n-\frac{3}{2}\right)}b_{n-1} \tag{4.31}$$

Noting that in this case $b_1 = 0$, all the following $b_n s$ must be zero and the second Frobenius series has only one term: $b_0 x^{-1/2}$. The complete solution is

$$u = c_0\left(x - \frac{6}{5}x^2 + \frac{6}{7}x^3 - \frac{4}{9}x^4 + \cdots\right) + b_0 x^{-1/2} \tag{4.32}$$

Example 4.4 (repeated roots). Next consider the differential equation

$$x^2u'' - xu' + (x+1)u = 0 \tag{4.33}$$

There is a regular singular point at $x = 0$, so we attempt a Frobenius series around $x = 0$.

Differentiating (4.17) and substituting into (4.30),

$$\sum_{n=0}^{\infty}(n+r-1)(n+r)c_n x^{n+r} - \sum_{n=0}^{\infty}(n+r)c_n x^{n+r} + \sum_{n=0}^{\infty}c_n x^{n+r} + \sum_{n=0}^{\infty}c_n x^{n+r+1} = 0 \tag{4.34}$$

or

$$[r(r-1) - r + 1]c_0x^r + \sum_{n=1}^{\infty}[(n+r-1)(n+r) - (n+r) + 1]c_n x^{n+r} + \sum_{n=1}^{\infty}c_{n-1}x^{n+r} = 0 \tag{4.35}$$

where we have shifted the index in the last sum.

The indicial equation is

$$r(r-1) - r + 1 = 0 \tag{4.36}$$

and the roots of this equation are both $r = 1$. Setting the last two sums to zero we find the recurrence equation

$$c_n = -\frac{1}{(n+r-1)(n+r)-(n+r)+1}c_{n-1} \tag{4.37}$$

and since $r = 1$,

$$\begin{aligned} c_n &= -\frac{1}{n(n+1)-(n+1)+1}c_{n-1} \\ c_1 &= -c_0 \\ c_2 &= \frac{-1}{6-3+1}c_1 = \frac{1}{4}c_0 \\ c_3 &= \frac{-1}{12-4+1}c_2 = \frac{-1}{9}c_1 = \frac{-1}{36}c_0 \end{aligned} \tag{4.38}$$

etc.

The Frobenius series is

$$u_1 = c_0\left(x - x^2 + \frac{1}{4}x^3 - \frac{1}{36}x^4 + \dots\right) \tag{4.39}$$

In this case there is no second solution in the form of a Frobenius series because of the repeated root. We shall soon see what form the second solution takes.

Example 4.5 (roots differing by an integer 1). Next consider the equation

$$x^2u'' - 2xu' + (x+2)u = 0 \tag{4.40}$$

There is a regular singular point at $x = 0$. We therefore expect a solution in the form of the Frobenius series (4.18). Substituting (4.18), (4.19), (4.20) into our differential equation, we obtain

$$\sum_{n=0}^{\infty}(n+r)(n+r-1)c_nx^{n+r} - \sum_{n=0}^{\infty}2(n+r)c_nx^{n+r} + \sum_{n=0}^{\infty}2c_nx^{n+r} + \sum_{n=0}^{\infty}c_nx^{n+r+1} = 0 \tag{4.41}$$

Taking out the $n = 0$ term and shifting the last summation,

$$[r(r-1)-2r+2]c_0x^r + \sum_{n=1}^{\infty}[(n+r)(n+r-1)-2(n+r)+2]c_nx^{n+r} + \sum_{n=1}^{\infty}c_{n-1}x^{n+r} = 0 \tag{4.42}$$

The first term is the indicial equation.

$$r(r-1) - 2r + 2 = 0 \tag{4.43}$$

There are two distinct roots, $r_1 = 2$ and $r_2 = 1$. However they differ by an integer.

$$r_1 - r_2 = 1.$$

Substituting $r_1 = 2$ into (4.39) and noting that each coefficient of x^{n+r} must be zero,

$$[(n+2)(n+1) - 2(n+2) + 2]c_n + c_{n-1} = 0 \tag{4.44}$$

The recurrence equation is

$$\begin{aligned} c_n &= \frac{-c_{n-1}}{(n+2)(n-1)+2} \\ c_1 &= \frac{-c_0}{2} \\ c_2 &= \frac{-c_1}{6} = c_0\frac{c_0}{12} \\ c_3 &= \frac{-c_2}{12} = \frac{-c_0}{144} \end{aligned} \tag{4.45}$$

The first Frobenius series is therefore

$$u_1 = c_0\left[x^2 - \frac{1}{2}x^3 + \frac{1}{12}x^4 - \frac{1}{144}x^5 + \dots\right] \tag{4.46}$$

We now attempt to find the Frobenius series corresponding to $r_2 = 1$. Substituting into (4.44) we find that

$$[n(n+1) - 2(n+1) + 2]c_n = -c_{n-1} \tag{4.47}$$

When $n = 1$, c_0 must be zero. Hence c_n must be zero for all n and the attempt to find a second Frobenius series has failed. This will not always be the case when roots differ by an integer as illustrated in the following example.

Example 4.6 (roots differing by an integer 2). Consider the differential equation

$$x^2u'' + x^2u' - 2u = 0 \tag{4.48}$$

You may show that the indicial equation is $r^2 - r - 2 = 0$ with roots $r_1 = 2$, $r_2 = -1$ and the roots differ by an integer. When $r = 2$ the recurrence equation is

$$c_n = -\frac{n+1}{n(n+3)}c_{n-1} \tag{4.49}$$

The first Frobenius series is

$$u_1 = c_0 x^2 \left[1 - \frac{1}{2}x + \frac{3}{20}x^2 - \frac{1}{30}x^3 + \dots\right] \tag{4.50}$$

When $r = -1$ the recurrence equation is

$$[(n-1)(n-2) - 2]b_n + (n-2)b_{n-1} = 0 \tag{4.51}$$

When $n = 3$ this results in $b_2 = 0$. Thus $b_n = 0$ for all $n \geq 2$ and the second series terminates.

$$u_2 = b_0 \left(\frac{1}{x} - \frac{1}{2}\right) \tag{4.52}$$

4.1.5 Lessons: Finding Series Solution for Differential Equations with Regular Singular Points

1. Assume a solution of the form

$$u = \sum_{n=0}^{\infty} c_n x^{n+r}, \; c_0 \neq 0 \tag{4.53}$$

Differentiate term by term and insert into the differential equation. Set the coefficient of the lowest power of x to zero to obtain a quadratic equation on r.

If the indicial equation yields two roots that do not differ by an integer there will always be two Frobenius series, one for each root of the indicial equation.

2. If the roots are the same (repeated roots) the form of the second solution will be

$$u_2 = u_1 \ln(x) + \sum_{n=1}^{\infty} b_n x^{n+r_1} \tag{4.54}$$

This equation is substituted into the differential equation to determine b_n.

3. If the roots differ by an integer, choose the largest root to obtain a Frobenius series for u_1. The second solution may be another Frobenius series. If the method fails assume a solution of the form

$$u_2 = u_1 \; \ln(x) + \sum_{n=1}^{\infty} b_n x^{n+r_2} \tag{4.55}$$

This equation is substituted into the differential equation to find b_n.

This is considered in the next section.

4.1.6 Logarithms and Second Solutions

Example 4.7. Reconsider Example 4.4 and assume a solution in the form of (4.54). Recall that in Example 4.4 the differential equation was

$$x^2u'' - xu' + (1+x)u = 0 \tag{4.56}$$

and the indicial equation yielded a double root at $r = 1$.

A single Frobenius series was

$$u_1 = x - x^2 + \frac{x^3}{4} - \frac{x^4}{36} + \cdots$$

Now differentiate Eq. (4.54).

$$u_2' = u_1' \ln x + \frac{1}{x}u_1 + \sum_{n=1}^{\infty}(n+r)b_n x^{n+r-1}$$

$$u_2'' = u_1'' \ln x + \frac{2}{x}u_1' - \frac{1}{x^2}u_1 + \sum_{n=1}^{\infty}(n+r-1)(n+r)b_n x^{n+r-2} \tag{4.57}$$

Inserting this into the differential equation gives

$$\begin{aligned}
&\ln(x)[x^2u_1'' - xu_1' + (1+x)u_1] + 2(xu_1' - u_1)\\
&+ \sum_{n=1}^{\infty}[b_n(n+r-1)(n+r)x^{n+r} - b_n(n+r)x^{n+r} + b_n x^{n+r}]\\
&+ \sum_{n=1}^{\infty} b_n x^{n+r+1} = 0
\end{aligned} \tag{4.58}$$

The first term on the left-hand side of (4.52) is clearly zero because the term in brackets is the original equation. Noting that $r = 1$ in this case and substituting from the Frobenius series for u_1, we find (c_0 can be set equal to unity without losing generality)

$$2\left[-x^2 + \frac{x^3}{3} - \frac{x^4}{12} + \cdots\right] + \sum_{n=1}^{\infty}[n(n+1) - (n+1) + 1]b_n x^{n+1} + \sum_{n=2}^{\infty} b_{n-1}x^{n+1} = 0 \tag{4.59}$$

or

$$-2x^2 + x^3 - \frac{x^4}{6} + \cdots + b_1x^2 + \sum_{n=2}^{\infty}\left[n^2b_n + b_{n-1}\right]x^{n+1} = 0 \tag{4.60}$$

Equating coefficients of x raised to powers we find that $b_1 = 2$

For $n \geq 2$

$$1 + 4b_2 + b_1 = 0 \qquad b_2 = -3/4$$
$$-\frac{1}{6} + 9b_3 + b_2 = 0 \qquad b_3 = \frac{11}{108}$$

etc.

$$u_2 = u_1 \ln x + \left(2x^2 - \frac{3}{4}x^3 + \frac{11}{108}x^4 - \cdots\right) \tag{4.61}$$

The complete solution is

$$u = [C_1 + C_2 \ln x]\, u_1 + C_2 \left[2x^2 - \frac{3}{4}x^3 + \frac{11}{108}x^4 - \cdots\right] \tag{4.62}$$

Example 4.8. Reconsider Example 4.5 in which a second Frobenius series could not be found because the roots of the indicial equation differed by an integer. We attempt a second solution in the form of (4.55).

The differential equation in Example 4.5 was

$$x^2u'' - 2xu' + (x+2)u = 0$$

and the roots of the indicial equation were $r = 2$ and $r = 1$, and are therefore separated by an integer. We found one Frobenius series

$$u_1 = x^2 - \frac{1}{2}x^3 + \frac{1}{12}x^4 - \frac{1}{144}x^5 + \cdots$$

for the root $r = 2$, but were unable to find another Frobenius series for the case of $r = 1$.

Assume a second solution of the form in Eq. (4.55). Differentiating and substituting into (4.40)

$$\begin{aligned}
&[x^2u_1'' - 2xu' + (x+2)u]\ln(x) + 2xu' - 3u_1 \\
&\quad + \sum_{n=1}^{\infty} b_n[(n+r)(n+r-1) - 2(n+r) + 2]x^{n+r} \\
&\quad + \sum_{n=1}^{\infty} b_n x^{n+r+1} = 0
\end{aligned} \tag{4.63}$$

Noting that the first term in the brackets is zero, inserting u_1 and u_1' from (4.50) and noting that $r_2 = 1$

$$x^2 - \frac{3}{2}x^3 + \frac{5}{12}x^4 - \frac{7}{144}x^5 + \ldots + b_0x^2 + \sum_{n=2}^{\infty}\{[n(n-1)]b_n + b_{n-1}\}x^{n+1} = 0 \tag{4.64}$$

Equating x^2 terms, we find that $b_0 = -1$. For higher order terms

$$\frac{3}{2} = 2b_2 + b_1 = 2b_2 + b_1$$

Taking $b_1 = 0$,

$$b_2 = \frac{3}{4}$$

$$-\frac{5}{12} = 6b_3 + b_2 = 6b_3 + \frac{3}{4}$$

$$b_3 = -\frac{7}{36}$$

The second solution is

$$u_2 = u_1 \ \ln(x) - \left(x - \frac{3}{4}x^3 + \frac{7}{36}x^4 - \ldots\right) \tag{4.65}$$

The complete solution is therefore

$$u = [C_1 + C_2 \ln x]\, u_1 - C_2 \left[x - \frac{3}{4}x^3 + \frac{7}{36}x^4 - \cdots\right] \tag{4.66}$$

Problems

1. Find two Frobenius series solutions

$$x^2u'' + 2xu' + (x^2 - 2)u = 0$$

2. Find two Frobenious series solutions

$$x^2u'' + xu' + \left(x^2 - \frac{1}{4}\right) u = 0$$

3. Show that the indicial equation for the differential equation

$$xu'' + u' + xu = 0$$

has roots $s = -1$ and that the differential equation has only one Frobenius series solution. Find that solution. Then find another solution in the form

$$u = \ln \sum_{n=0}^{\infty} c_n x^{n+s} + \sum_{m=0}^{\infty} a_n x^{s+m}$$

where the first summation above is the first Frobenius solution.

4.2 BESSEL FUNCTIONS

A few differential equations are so widely useful in applied mathematics that they have been named after the mathematician who first explored their theory. Such is the case with Bessel's equation. It occurs in problems involving the Laplacian $\nabla^2 u$ in cylindrical coordinates when variables are separated. Bessel's equation is a Sturm–Liouville equation of the form

$$\rho^2 \frac{d^2u}{d\rho^2} + \rho \frac{du}{d\rho} + (\lambda^2\rho^2 - \nu^2)u = 0 \tag{4.67}$$

Changing the independent variable $x = \lambda\rho$, the equation becomes

$$x^2u'' + xu' + (x^2 - \nu^2)u = 0 \tag{4.68}$$

4.2.1 Solutions of Bessel's Equation

Recalling the standard forms (4.1) and (4.2) we see that it is a linear homogeneous equation with variable coefficients and with a regular singular point at $x = 0$. We therefore assume a solution of the form of a Frobenius series (4.17).

$$u = \sum_{j=0}^{\infty} c_j x^{j+r} \tag{4.69}$$

Upon differentiating twice and substituting into (4.68) we find

$$\sum_{j=0}^{\infty}[(j+r-1)(j+r) + (j+r) - \nu^2]c_j x^{j+r} + \sum_{j=0} c_j x^{j+r+2} = 0 \tag{4.70}$$

In general ν can be any real number. We will first explore some of the properties of the solution when ν is a nonnegative integer, 0, 1, 2, 3, ... First note that

$$(j+r-1)(j+r) + (j+r) = (j+r)^2 \tag{4.71}$$

Shifting the exponent in the second summation and writing out the first two terms in the first

$$(r-n)(r+n)c_0 + (r+1-n)(r+1+n)c_1 x$$
$$+ \sum_{j=2}^{\infty}[(r+j-n)(r+j+n)c_j + c_{j-2}]x^j = 0 \tag{4.72}$$

In order for the coefficient of the x^0 term to vanish $r = n$ or $r = -n$. (This is the indicial equation.) In order for the coefficient of the x term to vanish $c_1 = 0$. For each term in the summation to vanish

$$c_j = \frac{-1}{(r+j-n)(r+j+n)}c_{j-2} = \frac{-1}{j(2n+j)}c_{j-2}, \qquad r=n \qquad j=2,3,4,\cdots \quad (4.73)$$

This is the recurrence relation. Since $c_1 = 0$, all $c_j = 0$ when j is an odd number. It is therefore convenient to write $j = 2k$ and note that

$$c_{2k} = \frac{-1}{2^2 k(r+k)}c_{2k-2} \tag{4.74}$$

so that

$$c_{2k} = \frac{(-1)^k}{k!(n+1)(n+2)\dots(n+k)2^{2k}}c_0 \tag{4.75}$$

The Frobenius series is

$$u = c_0 x^n \left[1 + \sum_{k=1}^{\infty} \frac{(-1)^k}{k!(n+1)(n+2)\dots.(n+k)}\left(\frac{x}{2}\right)^{2k}\right] \tag{4.76}$$

Now c_0 is an arbitrary constant so we can choose it to be $c_0 = 1/n!2^n$ in which case the above equation reduces to

$$J_n = u = \sum_{k=0}^{\infty} \frac{(-1)^k}{k!(n+k)!}\left(\frac{x}{2}\right)^{n+2k} \tag{4.77}$$

The usual notation is J_n and the function is called a *Bessel function of the first kind of order n*. Note that we can immediately conclude from (4.77) that

$$J_n(-x) = (-1)^n J_n(x) \tag{4.78}$$

Note that the roots of the indicial equation differ by an integer. When $r = -n$ (4.72) does not yield a useful second solution since the denominator is zero for $j = 0$ or $2n$. In any case it is easy to show that $J_n(x) = (-1)^n J_{-n}$, so when r is an integer the two solutions are not independent.

A second solution is determined by the methods detailed above and involves natural logarithms. The details are very messy and will not be given here. The result is

$$Y_n(x) = \frac{2}{\pi}\left\{J_n(x)\left[\ln\left(\frac{x}{2}\right) + \gamma\right] + \sum_{k=1}^{\infty}\frac{(-1)^{k+1}[\phi(k)+\phi(k+1)]}{2^{2k+n+1}k!(k+n)!}x^{2k+n}\right\}$$

$$-\frac{2}{\pi}\sum_{k=0}^{n-1}\frac{(n-k-1)!}{2^{2k-n+1}k!}x^{2k-n} \tag{4.79}$$

In this equation $\Phi(k) = 1 + 1/2 + 1/3 + \cdots + 1/k$ and γ is Euler's constant 0.5772156649 ……

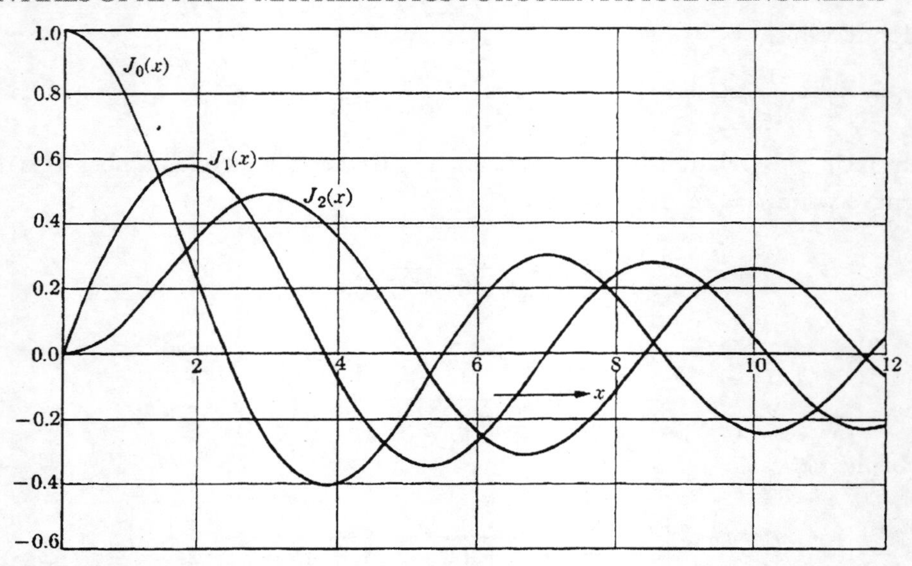

FIGURE 4.1: Bessel functions of the first kind

Bessel functions of the first and second kinds of order zero are particularly useful in solving practical problems (Fig. 4.1). For these cases

$$J_0(x) = \sum_{k=0}^{\infty} \frac{(-1)^k}{(k!)^2} \left(\frac{x}{2}\right)^{2k} \tag{4.80}$$

and

$$Y_0 = J_0(x)\ln(x) + \sum_{k=1}^{\infty} \frac{(-1)^{k+1}}{2^{2k}(k!)^2} \phi(k) x^{2k} \tag{4.81}$$

The case of $\nu \neq n$. Recall that in (4.70) if ν is not an integer, a part of the denominator is

$$(1+\nu)(2+\nu)(3+\nu)\ldots(n+\nu) \tag{4.82}$$

We were then able to use the familiar properties of *factorials* to simplify the expression for $J_n(x)$. If $\nu \neq n$ we can use the properties of the *gamma function* to the same end. The gamma function is defined as

$$\Gamma(\nu) = \int_0^{\infty} t^{\nu-1} e^{-t} dt \tag{4.83}$$

Note that

$$\Gamma(\nu+1)=\int_0^\infty t^\nu e^{-t}dt \tag{4.84}$$

and integrating by parts

$$\Gamma(\nu+1)=[-t\,\nu e^{-t}]_0^\infty+\nu\int_0^\infty t^{\nu-1}e^{-t}dt=\nu\Gamma(\nu) \tag{4.85}$$

and (4.82) can be written as

$$(1+\nu)(2+\nu)(3+\nu)\ldots.(n+\nu)=\frac{\Gamma(n+\nu+1)}{\Gamma(\nu+1)} \tag{4.86}$$

so that when ν is not an integer

$$J_\nu(x)=\sum_{n=0}^{\infty}\frac{(-1)^n}{2^{2n+\nu}n!\Gamma(n+\nu+1)}x^{2n+\nu} \tag{4.87}$$

Fig. 4.3 is a graphical representation of the gamma function.

Here are the rules

1. If 2ν is not an integer, J_ν and $J_{-\nu}$ are linearly independent and the general solution of Bessel's equation of order ν is

$$u(x)=AJ_\nu(x)+BJ_{-\nu}(x) \tag{4.88}$$

   where A and B are constants to be determined by boundary conditions.

2. If 2ν is an odd positive integer J_ν and $J_{-\nu}$ are still linearly independent and the solution form (4.88) is still valid.
3. If 2ν is an even integer, $J_\nu(x)$ and $J_{-\nu}(x)$ are not linearly independent and the solution takes the form

$$u(x)=AJ_\nu(x)+BY_\nu(x) \tag{4.89}$$

Bessel functions are tabulated functions, just as are exponentials and trigonometric functions. Some examples of their shapes are shown in Figs. 4.1 and 4.2.

Note that both $J_\nu(x)$ and $Y_\nu(x)$ have an infinite number of zeros and we denote them as λ_j, $j=0,1,2,3,\ldots$

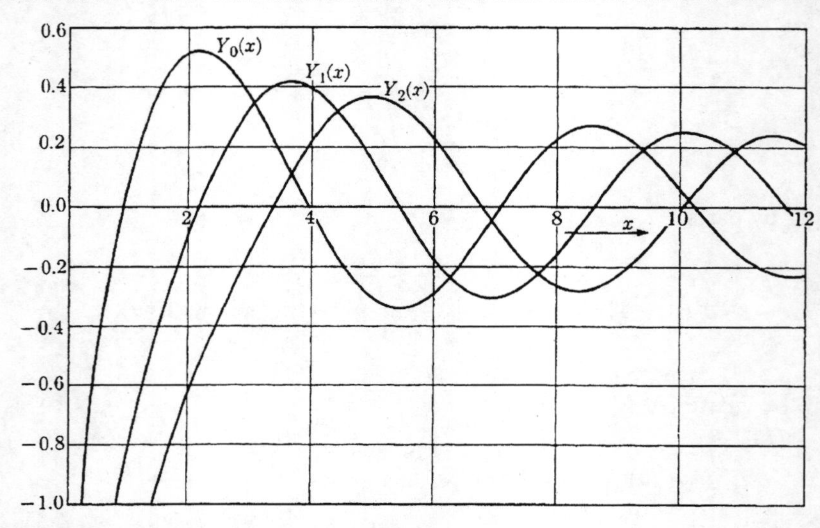

FIGURE 4.2: Bessel functions of the second kind

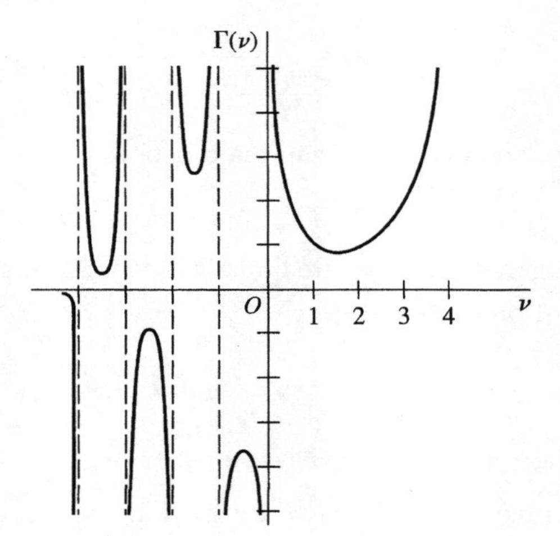

FIGURE 4.3: The gamma function

Some important relations involving Bessel functions are shown in Table 4.1. We will derive only the first, namely

$$\frac{d}{dx}(x^{\nu}J_{\nu}(x)) = x^{\nu}J_{\nu-1}(x) \tag{4.90}$$

$$\frac{d}{dx}(x^{\nu}J_{\nu}(x)) = \frac{d}{dx}\left[\sum_{n=0}^{\infty}\frac{(-1)^{n}}{2^{2n+\nu}n!\Gamma(n+\nu+1)}x^{2n+2\nu}\right] \tag{4.91}$$

TABLE 4.1: Some Properties of Bessel Functions
1. $[x^{\nu}J_{\nu}(x)]' = x^{\nu}J_{\nu-1}(x)$
2. $[\vec{x}^{-\nu}J_{\nu}(x)]' = -x^{-\nu}J_{\nu+1}(x)$
3. $J_{\nu-1}(x) + J_{\nu+1}(x) = 2\nu/x[J_{\nu}(x)]$
4. $\mathrm{J}_{\nu-1}(x) - J_{\nu+1}(x) = 2J_{\nu}(x)\prime$
5. $\int x^{\nu}J_{\nu-1}(x)dx = x^{\nu}J_{\nu} + \mathit{constant}$
6. $\int x^{-\nu}J_{\nu+1}(x)dx = x^{-\nu}J_{\nu}(x) + \mathit{constant}$

$$= \sum_{n=0}^{\infty} \frac{(-1)^n 2(n+\nu)}{2^{2n+\nu} n!(n+\nu)\Gamma(n+\nu)} x^{2n+2\nu-1} \tag{4.92}$$

$$= x^{\nu} \sum_{n=0}^{\infty} \frac{(-1)^n}{2^{2n+\nu-1} n!\Gamma(n+\nu)} x^{2n+2\nu-1} = x^{\nu}J_{\nu-1}(x) \tag{4.93}$$

These will prove important when we begin solving partial differential equations in cylindrical coordinates using separation of variables.

Bessel's equation is of the form (4.138) of a Sturm–Liouville equation and the functions $J_n(x)$ are orthogonal with respect to a weight function ρ (see Eqs. (3.46) and (3.53), Chapter 3).

Note that Bessel's equation (4.67) with $\nu = n$ is

$$\rho^2 J_n'' + \rho J_n' + (\lambda^2\rho^2 - n^2)J_n = 0 \tag{4.94}$$

which can be written as

$$\frac{d}{d\rho}(\rho J_n')^2 + (\lambda^2\rho^2 - n^2)\frac{d}{d\rho}J_n^2 = 0 \tag{4.95}$$

Integrating, we find that

$$[(\rho J')^2 + (\lambda^2\rho^2 - n^2)J^2]_0^1 - 2\lambda^2 \int_{\rho=0}^{1} \rho J^2 d\rho = 0 \tag{4.96}$$

Thus,

$$2\lambda^2 \int_{\rho=0}^{1} \rho J_n^2 d\rho = \lambda^2[J_n'(\lambda)]^2 + (\lambda^2 - n^2)[J_n(\lambda)]^2 \tag{4.97}$$

Thus, we note from that if the eigenvalues are λ_j, the roots of $J_\nu(\lambda_j \rho) = 0$ the orthogonality condition is, according to Eq. (3.53) in Chapter 3

$$\int_0^1 \rho J_n(\lambda_j \rho) J_n(\lambda_k \rho) d\rho = 0, \qquad j \neq k$$

$$= \frac{1}{2}[J_{n+1}(\lambda_j)]^2, \quad j = k \tag{4.98}$$

On the other hand, if the eigenvalues are the roots of the equation

$$HJ_n(\lambda_j) + \lambda_j J_n'(\lambda_j) = 0$$

$$\int_0^1 \rho J_n(\lambda_j \rho) J_n(\lambda_k \rho) d\rho = 0, \quad j \neq k$$

$$= \frac{(\lambda_j^2 - n^2 + H^2)[J_n(\lambda_j)]^2}{2\lambda_j^2}, \quad j = k \tag{4.99}$$

Using the equations in the table above and integrating by parts it is not difficult to show that

$$\int_{s=0}^{x} s^n J_0(s) ds = x^n J_1(x) + (n-1)x^{n-1} J_0(x) - (n-1)^2 \int_{s=0}^{x} s^{n-2} J_0(s) ds \tag{4.100}$$

4.2.2 Fourier–Bessel Series

Owing to the fact that Bessel's equation with appropriate boundary conditions is a Sturm–Liouville system it is possible to use the orthogonality property to expand any piecewise continuous function on the interval $0 < x < 1$ as a series of Bessel functions. For example, let

$$f(x) = \sum_{n=1}^{\infty} A_n J_0(\lambda_n x) \tag{4.101}$$

Multiplying both sides by $x J_0(\lambda_k x) dx$ and integrating from $x = 0$ to $x = 1$ (recall that the weighting function x must be used to insure orthogonality) and noting the orthogonality property we find that

$$f(x) = \sum_{j=1}^{\infty} \frac{\int_{x=0}^{1} x f(x) J_0(\lambda_j x) dx}{\int_{x=0}^{1} x [J_0(\lambda_j x)]^2 dx} J_0(\lambda_j x) \tag{4.102}$$

Example 4.9. Derive a Fourier–Bessel series representation of 1 on the interval $0 < x < 1$. We note that with $J_0(\lambda_j) = 0$

$$\int_{x=0}^{1} x[J_0(\lambda_j x)]^2 dx = \frac{1}{2}[J_1(\lambda_j)]^2 \tag{4.103}$$

and

$$\int_{x=0}^{1} x J_0(\lambda_j x) dx = J_1(\lambda_j) \tag{4.104}$$

Thus

$$1 = 2\sum_{j=1}^{\infty} \frac{J_0(\lambda_j x)}{\lambda_j J_1(\lambda_j)} \tag{4.105}$$

Example 4.10 (A problem in cylindrical coordinates). A cylinder of radius r_1 is initially at a temperature u_0 when its surface temperature is increased to u_1. It is sufficiently long that variation in the z direction may be neglected and there is no variation in the θ direction. There is no heat generation. From Chapter 1, Eq. (1.11)

$$u_t = \frac{\alpha}{r}(r u_r)_r \tag{4.106}$$

$$u(0, r) = u_0$$

$$u(t, r_1) = u_1$$

$$u \text{ is bounded} \tag{4.107}$$

The length scale is r_1 and the time scale is r_1^2/α. A dimensionless dependent variable that normalizes the problem is $(u - u_1)/(u_0 - u_1) = U$. Setting $\eta = r/r_1$ and $\tau = t\alpha/r_1^2$,

$$U_\tau = \frac{1}{\eta}(\eta U_\eta)_\eta \tag{4.108}$$

$$U(0, \eta) = 1$$

$$U(\tau, 1) = 0 \tag{4.109}$$

$$U \text{ is bounded}$$

Separate variables as $T(\tau)R(\eta)$. Substitute into the differential equation and divide by TR.

$$\frac{T_\tau}{T} = \frac{1}{R\eta}(\eta R_\eta)_\eta = \pm\lambda^2 \tag{4.110}$$

where the minus sign is chosen so that the function is bounded. The solution for T is exponential and we recognize the equation for R as Bessel's equation with $\nu = 0$.

$$\frac{1}{\eta}(\eta R_\eta)_\eta + \lambda^2 R = 0 \tag{4.111}$$

The solution is a linear combination of the two Bessel functions of order 0.

$$C_1 J_0(\lambda\eta) + C_2 Y_0(\lambda\eta) \tag{4.112}$$

Since we have seen that Y_0 is unbounded as η approaches zero, C_2 must be zero. Furthermore, the boundary condition at $\eta = 1$ requires that $J_0(\lambda) = 0$, so that our eigenfunctions are $J_0(\lambda\eta)$ and the corresponding eigenvalues are the roots of $J_0(\lambda_n) = 0$.

$$U_n = K_n e^{-\lambda_n^2 \tau} J_0(\lambda_n \eta), \qquad n = 1, 2, 3, 4, \ldots \tag{4.113}$$

Summing (linear superposition)

$$U = \sum_{n=1}^{\infty} K_n e^{-\lambda_n^2 \tau} J_0(\lambda_n \eta) \tag{4.114}$$

Using the initial condition,

$$1 = \sum_{n=1}^{\infty} K_n J_0(\lambda_n \eta) \tag{4.115}$$

Bessel functions are orthogonal *with respect to weighting factor* η since theyare solutions to a Sturm–Liouville system. Therefore when we multiply both sides of this equation by $\eta J_0(\lambda_m \eta) d\eta$ and integrate over (0, 1) all of the terms in the summation are zero except when $m = n$. Thus,

$$\int_{\eta=0}^{1} J_0(\lambda_n \eta)\eta d\eta = K_n \int_{\eta=0}^{1} J_0^2(\lambda_n \eta)\eta d\eta \tag{4.116}$$

but

$$\int_{\eta=0}^{1} \eta J_0^2(\lambda_n \eta) d\eta = \frac{J_1^2(\lambda_n)}{2}$$

$$\int_{\eta=0}^{1} \eta J_0(\lambda_n \eta) d\eta = \frac{1}{\lambda_n} J_1(\lambda_n) \tag{4.117}$$

Thus

$$U(\tau,\eta)=\sum_{n=0}^{\infty}\frac{2}{\lambda_n J_1(\lambda_n)}\,e^{-\lambda_n^2\tau}J_0(\lambda_n\eta) \tag{4.118}$$

Example 4.11 (Heat generation in a cylinder). Reconsider the problem of heat transfer in a long cylinder but with heat generation and at a normalized initial temperature of zero.

$$u_\tau=\frac{1}{r}(ru_r)_r+q_0 \tag{4.119}$$

$$u(\tau,1)=u(0,r)=0,\ \ u\text{ bounded} \tag{4.120}$$

Our experience with the above example hints that the solution maybe of the form

$$u=\sum_{j=1}^{\infty}A_j(\tau)J_0(\lambda_j r) \tag{4.121}$$

This equation satisfies the boundary condition at $r=1$ and $A_j(\tau)$ is to be determined. Substituting into the partial differential equation gives

$$\sum_{j=1}^{\infty}A_j'(\tau)J_0(\lambda_j)=\sum_{j=1}^{\infty}A_j(\tau)\frac{1}{r}\frac{d}{dr}\left[r\frac{dJ_0}{dr}\right]+q_0 \tag{4.122}$$

In view of Bessel's differential equation, the first term on the right can be written as

$$\sum_{j=1}^{\infty}-\lambda_j^2J_0(\lambda_j r)A_j(\tau) \tag{4.123}$$

The second term can be represented as a Fourier–Bessel series as follows:

$$q_0=q_0\sum_{j=1}^{\infty}\frac{2J_0(\lambda_j r)}{\lambda_j J_1(\lambda_j)} \tag{4.124}$$

as shown in Example 4.9 above.

Equating coefficients of $J_0(\lambda_j r)$ we find that $A_j(\tau)$ must satisfy the ordinary differential equation

$$A'(\tau)+\lambda_j^2A(\tau)=q_0\frac{2}{\lambda_j J_1(\lambda_j)} \tag{4.125}$$

with the initial condition $A(0)=0$.

Solution of this simple first-order linear differential equations yields

$$A_j(\tau)=\frac{2q_0}{\lambda_j^3J_1(\lambda_j)}+C\exp(-\lambda_j^2\tau) \tag{4.126}$$

After applying the initial condition

$$A_j(\tau) = \frac{2q_0}{\lambda_j^3 J_1(\lambda_j)} \left[1 - \exp(-\lambda_j^2 \tau)\right] \tag{4.127}$$

The solution is therefore

$$u(\tau, r) = \sum_{j=1}^{\infty} \frac{2q_0}{\lambda_j^3 J_1(\lambda_j)} \left[1 - \exp(-\lambda_j^2 \tau)\right] J_0(\lambda_j r) \tag{4.128}$$

Example 4.12 (Time dependent heat generation). Suppose that instead of constant heat generation, the generation is time dependent, $q(\tau)$. The differential equation for $A(\tau)$ then becomes

$$A'(\tau) + \lambda_j^2 A(\tau) = \frac{2q(\tau)}{\lambda_j J_1(\lambda_j)} \tag{4.129}$$

An integrating factor for this equation is $\exp(\lambda_j^2 \tau)$ so that the equation can be written as

$$\frac{d}{d\tau}\left[A_j \exp(\lambda_j^2 \tau)\right] = \frac{2q(\tau)}{\lambda_j J_1(\lambda_j)} \exp(\lambda_j^2 \tau) \tag{4.130}$$

Integrating and introducing as a dummy variable t

$$A_j(\tau) = \frac{2}{\lambda_j J_1(\lambda_j)} \int_{t=0}^{\tau} q(t) \exp(-\lambda_j^2(\tau - t)) dt \tag{4.131}$$

Problems

1. By differentiating the series form of $J_0(x)$ term by term show that

$$J_0'(x) = -J_1(x)$$

2. Show that

$$\int x J_0(x) dx = x J_1(x) + constant$$

3. Using the expression for $\int_{s=0}^{x} s^n J_0(s) ds$ show that

$$\int_{s=0}^{x} s^5 J_0(s) ds = x(x^2 - 8)[4x J_0(x) + (x^2 - 8) J_1(x)]$$

4. Express $1 - x$ as a Fourier–Bessel series.

4.3 LEGENDRE FUNCTIONS

We now consider another second-order linear differential that is common for problems involving the Laplacian in spherical coordinates. It is called Legendre's equation,

$$(1 - x^2)u'' - 2xu' + ku = 0 \tag{4.132}$$

This is clearly a Sturm–Liouville equation and we will seek a series solution near the origin, which is a regular point. We therefore assume a solution in the form of (4.3).

$$u = \sum_{j=0}^{\infty} c_j x^j \tag{4.133}$$

Differentiating (4.133) and substituting into (4.132) we find

$$\sum_{j=0}^{\infty} [j(j-1)c_j x^{j-2}(1-x^2) - 2jc_j x^j + n(n+1)c_j x^j] \tag{4.134}$$

or

$$\sum_{j=0}^{\infty} \{[k - j(j+1)]c_j x^j + j(j-1)c_j x^{j-2}\} = 0 \tag{4.135}$$

On shifting the last term,

$$\sum_{j=0}^{\infty} \{(j+2)(j+1)c_{j+2} + [k - j(j+1)]c_j\} x^j = 0 \tag{4.136}$$

The recurrence relation is

$$c_{j+2} = -\frac{j(j+1) - k}{(j+1)(j+2)} c_j \tag{4.137}$$

There are thus two independent Frobenius series. It can be shown that they both diverge at $x = 1$ unless they terminate at some point. It is easy to see from (4.137) that they do in fact terminate if $k = n(n+1)$.

Since n and j are integers it follows that $c_{n+2} = 0$ and consequently c_{n+4}, c_{n+6}, etc. are all zero. Therefore the solutions, which depend on n (i.e., the eigenfunctions) are polynomials, series that terminate at $j = n$. For example, if $n = 0$, $c_2 = 0$ and the solution is a constant. If

$n = 1$ $c_n = 0$ when $n \geq 1$ and the polynomial is x. In general

$$u = P_n(x) = c_n \left[x^n - \frac{n(n-1)}{2(2n-1)} x^{n-2} + \frac{n(n-1)(n-2)(n-3)}{2(4)(2n-1)(2n-3)} x^{n-4} - \dots \right]$$
$$= \frac{1}{2^k} \sum_{k=0}^{m} \frac{(-1)^k}{k!} \frac{(2n-2k)!}{(n-2k)!(n-k)!} x^{n-2k} \tag{4.138}$$

where $m = n/2$ if n is even and $(n-1)/2$ if n is odd.

The coefficient c_n is of course arbitrary. It turns out to be convenient to choose it to be

$$c_0 = 1$$
$$c_n = \frac{(2n-1)(2n-3)\cdots 1}{n!} \tag{4.139}$$

the first few polynomials are

$$P_0 = 1,\ P_1 = x,\ P_2 = (3x^2 - 1)/2,\ P_3 = (5x^3 - 3x)/2,\ P_4 = (35x^4 - 30x^2 + 3)/8,$$

Successive Legendre polynomials can be generated by the use of Rodrigues' formula

$$P_n(x) = \frac{1}{2^n n!} \frac{d^n}{dx^n}(x^2 - 1)^n \tag{4.140}$$

For example

$$P_5 = (63x^5 - 70x^3 + 15x)/8$$

Fig. 4.4 shows graphs of several Legendre polynomials.

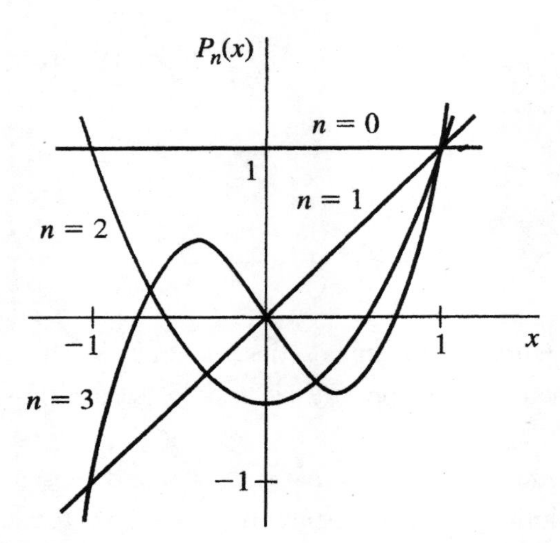

FIGURE 4.4: Legendre polynomials

The second solution of Legendre's equation can be found by the method of variation of parameters. The result is

$$Q_n(x) = P_n(x) \int \frac{d\zeta}{P_n^2(\zeta)(1-\zeta^2)} \tag{4.141}$$

It can be shown that this generally takes on a logarithmic form involving $\ln[(x+1)/(x-1)]$ which goes to infinity at $x = 1$. In fact it can be shown that the first two of these functions are

$$Q_0 = \frac{1}{2}\ln\frac{1+x}{1-x} \quad \text{and} \quad Q_1 = \frac{x}{2}\ln\frac{1+x}{1-x} - 1 \tag{4.142}$$

Thus the complete solution of the Legendre equation is

$$u = AP_n(x) + BQ_n(x) \tag{4.143}$$

where $P_n(x)$ and $Q_n(x)$ are Legendre polynomials of the first and second kind. If we require the solution to be finite at $x = 1$, B must be zero.

Referring back to Eqs. (3.46) through (3.53) in Chapter 3, we note that the eigenvalues $\lambda = n(n+1)$ and the eigenfunctions are $P_n(x)$ and $Q_n(x)$. We further note from (3.46) and (3.47) that the weight function is one and that the orthogonality condition is

$$\int_{-1}^{1} P_n(x)P_m(x)dx = \frac{2}{2n+1}\delta_{mn} \tag{4.144}$$

where δ_{mn} is Kronecker's delta, 1 when $n = m$ and 0 otherwise.

Example 4.13. Steady heat conduction in a sphere

Consider heat transfer in a solid sphere whose surface temperature is a function of θ, the angle measured downward from the z-axis (see Fig. 1.3 Chapter 1). The problem is steady and there is no heat source.

$$\begin{aligned} &r\frac{\partial^2}{\partial r^2}(ru) + \frac{1}{\sin\theta}\frac{\partial}{\partial\theta}\left(\sin\theta\frac{\partial u}{\partial\theta}\right) = 0 \\ &u(r=1) = f(\theta) \\ &u \text{ is bounded} \end{aligned} \tag{4.145}$$

Substituting $x = \cos\theta$,

$$r\frac{\partial^2}{\partial r^2}(ru) + \frac{\partial}{\partial x}\left[(1-x^2)\frac{\partial u}{\partial x}\right] = 0 \tag{4.146}$$

We separate variables by assuming $u = R(r)X(x)$. Substitute into the equation and divide by RX and find

$$\frac{r}{R}(rR)'' = -\frac{[(1-x^2)X']'}{X} = \pm\lambda^2 \tag{4.147}$$

or

$$\begin{aligned} r(rR)'' \mp \lambda^2 R = 0 \\ [(1-x^2)X']' \pm \lambda^2 X = 0 \end{aligned} \tag{4.148}$$

The second of these is Legendre's equation, and we have seen that it has bounded solutions at $r = 1$ when $\lambda^2 = n(n+1)$. The first equation is of the Cauchy–Euler type with solution

$$R = C_1 r^n + C_2 r^{-n-1} \tag{4.149}$$

Noting that the constant C_2 must be zero to obtain a bounded solution at $r = 0$, and using superposition,

$$u = \sum_{n=0}^{\infty} K_n r^n P_n(x) \tag{4.150}$$

and using the condition at $fr = 1$ and the orthogonality of the Legendre polynomial

$$\int_{\theta=0}^{\pi} f(\theta)P_n(\cos\theta)d\theta = \int_{\theta=0}^{\pi} K_n P_n^2(\cos\theta)d\theta = \frac{2K_n}{2n+1} \tag{4.151}$$

4.4 ASSOCIATED LEGENDRE FUNCTIONS

Equation (1.15) in Chapter 1 can be put in the form

$$\frac{1}{\alpha}\frac{\partial u}{\partial t} = \left\{\frac{\partial^2 u}{\partial r^2} + \frac{2}{r}\frac{\partial u}{\partial r}\right\} + \frac{1}{r^2}\frac{\partial}{\partial\mu}\left\{(1-\mu^2)\frac{\partial u}{\partial\mu}\right\} + \frac{1}{r^2(1-\mu^2)}\frac{\partial^2 u}{\partial\Phi^2} \tag{4.152}$$

by substituting $\mu = \cos\theta$.

We shall see later that on separating variables in the case where u is a function of r, θ, Φ, and t, we find the following differential equation in the μ variable:

$$\frac{d}{d\mu}\left\{(1-\mu^2)\frac{df}{d\mu}\right\}+\left\{n(n+1)-\frac{m^2}{1-\mu^2}\right\}f=0 \tag{4.153}$$

We state without proof that the solution is the associated Legendre function $P_n^m(\mu)$. The associated Legendre polynomial is given by

$$P_n^m=(1-\mu^2)^{1/2m}\frac{d^m}{d\mu^m}P_n(\mu) \tag{4.154}$$

The orthogonality condition is

$$\int_{-1}^{1}[P_n^m(\mu)]^2 d\mu=\frac{2(n+m)!}{(2n+1)(n-m)!} \tag{4.155}$$

and

$$\int_{-1}^{1}P_n^m P_{n'}^m d\mu=0 \qquad n\neq n' \tag{4.156}$$

The associated Legendre function of the second kind is singular at $x=\pm 1$ and may be computed by the formula

$$Q_n^m(x)=(1-x^2)^{m/2}\frac{d^m Q_n(x)}{dx^m} \tag{4.157}$$

Problems

1. Find and carefully plot P_6 and P_7.
2. Perform the integral above and show that

$$Q_0(x)=CP_0(x)\int_{\xi=0}^{x}\frac{d\xi}{(1-\xi^2)P_0(\xi)}=\frac{C}{2}\ln\left(\frac{1+x}{1-x}\right)$$

and that

$$Q_1(x)=Cx\int_{\xi=0}^{x}\frac{d\xi}{\xi^2(1-\xi^2)}=\frac{Cx}{2}\ln\left(\frac{1+x}{1-x}\right)-1$$

3. Using the equation above find $Q_0^0(x)$ and $Q_1^1(x)$

FURTHER READING

J. W. Brown and R. V. Churchill, *Fourier Series and Boundary Value Problems*. New York: McGraw-Hill, 2001.

C. F. Chan Man Fong, D. DeKee, and P. N. Kaloni, *Advanced Mathematics for Engineering and Science*. 2nd edition. Singapore: World Scientific, 2004.

P. V. O'Neil, *Advanced Engineering Mathematics*. 5th edition. Brooks/Cole Thompson, Pacific Grove, CA, 2003.

CHAPTER 5

Solutions Using Fourier Series and Integrals

We have already demonstrated solution of partial differential equations for some simple cases in rectangular Cartesian coordinates in Chapter 2. We now consider some slightly more complicated problems as well as solutions in spherical and cylindrical coordinate systems to further demonstrate the Fourier method of separation of variables.

5.1 CONDUCTION (OR DIFFUSION) PROBLEMS

Example 5.1 (Double Fourier series in conduction). We now consider transient heat conduction in two dimensions. The problem is stated as follows:

$$\begin{aligned} u_t &= \alpha(u_{xx} + u_{yy}) \\ u(t,0,y) &= u(t,a,y) = u(t,x,0) = u(t,x,b) = u_0 \\ u(0,x,y) &= f(x,y) \end{aligned} \tag{5.1}$$

That is, the sides of a rectangular area with initial temperature $f(x, y)$ are kept at a constant temperature u_0. We first attempt to scale and nondimensionalize the equation and boundary conditions. Note that there are two length scales, a and b. We can choose either, but there will remain an extra parameter, either a/b or b/a in the equation. If we take $\xi = x/a$ and $\eta = y/b$ then (5.1) can be written as

$$\frac{a^2}{\alpha} u_t = \left(u_{\xi\xi} + \frac{a^2}{b^2} u_{\eta\eta} \right) \tag{5.2}$$

The time scale is now chosen as a^2/α and the dimensionless time is $\tau = \alpha t/a^2$. We also choose a new dependent variable $U(\tau, \xi, \eta) = (u - u_0)/(f_{\max} - u_0)$. The now nondimensionalized system is

$$\begin{aligned} U_\tau &= U_{\xi\xi} + r^2 U_{\eta\eta} \\ U(\tau,0,\eta) &= U(\tau,1,\eta) = U(\tau,\xi,0) = U(\tau,\xi,1) = 0 \\ U(0,\xi,\eta) &= (f - u_0)/(f_{\max} - u_0) = g(\xi,\eta) \end{aligned} \tag{5.3}$$

We now proceed by separating variables. Let

$$U(\tau, \xi, \eta) = T(\tau)X(\xi)Y(\eta) \tag{5.4}$$

Differentiating and inserting into (5.3) and dividing by (5.4) we find

$$\frac{T'}{T} = \frac{X''Y + r^2Y''X}{XY} \tag{5.5}$$

where the primes indicate differentiation with respect to the variable in question and $r = a/b$. Since the left-hand side of (5.5) is a function only of τ and the right-hand side is only a function of ξ and η both sides must be constant. If the solution is to be finite in time we must choose the constant to be negative, $-\lambda^2$. Replacing T'/T by $-\lambda^2$ and rearranging,

$$-\lambda^2 - \frac{X''}{X} = r\frac{Y''}{Y} \tag{5.6}$$

Once again we see that both sides must be constants. How do we choose the signs? It should be clear by now that if either of the constants is positive solutions for X or Y will take the form of hyperbolic functions or exponentials and the boundary conditions on ξ or η cannot be satisfied. Thus,

$$\frac{T'}{T} = -\lambda^2 \tag{5.7}$$

$$\frac{X''}{X} = -\beta^2 \tag{5.8}$$

$$r^2\frac{Y''}{Y} = -\gamma^2 \tag{5.9}$$

Note that X and Y are eigenfunctions of (5.8) and (5.9), which are Sturm–Liouville equations and β and γ are the corresponding eigenvalues.

Solutions of (5.7), (5.8), and (5.9) are

$$T = A\exp(-\lambda^2\tau) \tag{5.10}$$

$$X = B_1\cos(\beta\xi) + B_2\sin(\beta\xi) \tag{5.11}$$

$$Y = C_1\cos(\gamma\eta/r) + C_2\sin(\gamma\eta/r) \tag{5.12}$$

Applying the first homogeneous boundary condition, we see that $X(0) = 0$, so that $B_1 = 0$. Applying the third homogeneous boundary condition we see that $Y(0) = 0$, so that $C_1 = 0$. The second homogeneous boundary condition requires that $\sin(\beta) = 0$, or $\beta = n\pi$. The last homogeneous boundary condition requires $\sin(\gamma/r) = 0$, or $\gamma = m\pi r$. According to (5.6), $\lambda^2 = \beta^2 + \gamma^2$. Combining these solutions, inserting into (5.4) we have one solution in the form

$$U_{mn}(\tau, \xi, \eta) = K_{nm} e^{-(n^2\pi^2 + m^2\pi^2 r^2)\tau} \sin(n\pi\xi)\sin(m\pi\eta) \tag{5.13}$$

for all $m, n = 1, 2, 3, 4, 5, \ldots$

Superposition now tells us that

$$\sum_{n=1}^{\infty}\sum_{m=1}^{\infty} K_{nm} e^{-(n^2\pi^2 + m^2\pi^2 r^2)\tau} \sin(n\pi\xi)\sin(m\pi) \tag{5.14}$$

Using the initial condition

$$g(\xi, \eta) = \sum_{n=1}^{\infty}\sum_{m=1}^{\infty} K_{nm} \sin(n\pi\xi)\sin(m\pi\eta) \tag{5.15}$$

We have a double Fourier series, and since both $\sin(n\pi\xi)$ and $\sin(m\pi\eta)$ are members of orthogonal sequences we can multiply both sides by $\sin(n\pi\xi)\sin(m\pi\eta)\mathrm{d}\xi\,\mathrm{d}\eta$ and integrate over the domains.

$$\begin{aligned}
&\int_{\xi=0}^{1}\int_{\eta=0}^{1} g(\xi, \eta)\sin(n\pi\xi)\sin(m\pi\eta)d\xi\, d\eta \\
&\quad = K_{nm}\int_{\xi=0}^{1}\int_{\eta=0}^{1} \sin^2(n\pi\xi)d\xi\, \sin^2(m\pi\eta)d\eta \\
&\quad = \frac{K_{nm}}{4}
\end{aligned} \tag{5.16}$$

Our solution is

$$\sum_{n=1}^{\infty}\sum_{m=1}^{\infty} 4\int_{\xi=0}^{1}\int_{\eta=0}^{1} g(\xi, \eta)\sin(n\pi\xi)\sin(m\pi\eta)d\xi\, d\eta\; e^{-(n^2\pi^2 + m^2\pi^2 r^2)\tau} \sin(n\pi\xi)\sin(m\pi\eta) \tag{5.17}$$

Example 5.2 (A convection boundary condition). Reconsider the problem defined by (2.1) in Chapter 2, but with different boundary and initial conditions,

$$u(t, 0) = u_0 = u(0, x) \tag{5.18}$$

$$k u_x(t, L) - h[u_1 - u(t, L)] = 0 \tag{5.19}$$

The physical problem is a slab with conductivity k initially at a temperature u_0 suddenly exposed at $x = L$ to a fluid at temperature u_1 through a heat transfer coefficient h while the $x = 0$ face is maintained at u_0.

The length and time scales are clearly the same as the problem in Chapter 2. Hence, $\tau = t\alpha/L^2$ and $\xi = x/L$. If we choose $U = (u - u_0)/(u_1 - u_0)$ we make the boundary condition at $x = 0$ homogeneous but the condition at $x = L$ is not. We have the same situation that we had in Section 2.3 of Chapter 2. The differential equation, one boundary condition, and the initial condition are homogeneous. Proceeding, we find

$$\begin{aligned} &U_\tau = U_{\xi\xi} \\ &U(\tau, 0) = U(0, \xi) = 0 \\ &U_\xi(\tau, 1) + B[U(\tau, 1) - 1] = 0 \end{aligned} \tag{5.20}$$

where $B = hL/k$. It is useful to relocate the nonhomogeneous condition as the initial condition. As in the previous problem we assume $U(\tau, \xi) = V(\tau, \xi) + W(\xi)$.

$$\begin{aligned} &V_\tau = V_{\xi\xi} + W_{\xi\xi} \\ &W(0) = 0 \\ &W_\xi(1) + B[W(1) - 1] = 0 \\ &V(\tau, 0) = 0 \\ &V_\xi(\tau, 1) + BV(\tau, 1) = 0 \\ &V(0, \xi) = -W(\xi) \end{aligned} \tag{5.21}$$

Set $W_{\xi\xi} = 0$. Integrating twice and using the two boundary conditions on W,

$$W(\xi) = \frac{B\xi}{B+1} \tag{5.22}$$

The initial condition on V becomes

$$V(0, \xi) = -B\xi/(B+1)\,. \tag{5.23}$$

Assume $V(\tau, \xi) = P(\tau)Q(\xi)$, substitute into the partial differential equation for V, and divide by PQ as usual.

$$\frac{P'}{P} = \frac{Q''}{Q} = \pm\lambda^2 \tag{5.24}$$

We must choose the minus sign for the solution to be bounded. Hence,

$$\begin{aligned} &P = Ae^{-\lambda^2\tau} \\ &Q = C_1\sin(\lambda\xi) + C_2\cos(\lambda\xi) \end{aligned} \tag{5.25}$$

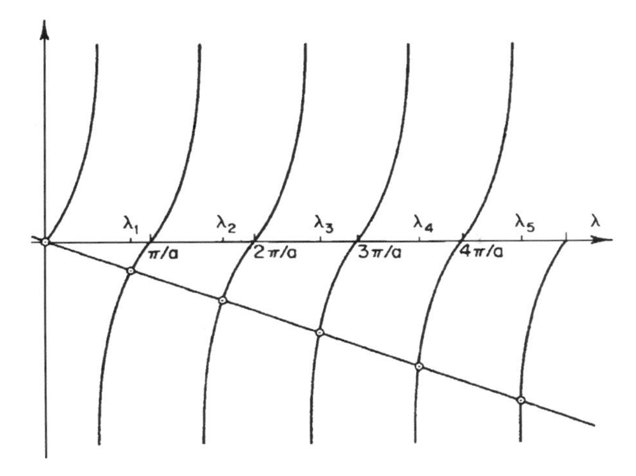

FIGURE 5.1: The eigenvalues of $\lambda_n = -B\tan(\lambda_n)$

Applying the boundary condition at $\xi = 0$, we find that $C_2 = 0$. Now applying the boundary condition on V at $\xi = 1$,

$$C_1\lambda\cos(\lambda) + C_1 B\sin(\lambda) = 0 \tag{5.26}$$

or

$$\lambda = -B\tan(\lambda) \tag{5.27}$$

This is the equation for determining the eigenvalues, λ_n. It is shown graphically in Fig. 5.1.

Example 5.3 (Superposition of several problems). We've seen now that in order to apply separation of variables the partial differential equation itself must be homogeneous and we have also seen a technique for transferring the inhomogeneity to one of the boundary conditions or to the initial condition. But what if several of the boundary conditions are nonhomogeneous? We demonstrate the technique with the following problem. We have a transient two-dimensional problem with given conditions on all four faces.

$$\begin{aligned}
u_t &= u_{xx} + u_{yy} \\
u(t,0,y) &= f_1(y) \\
u(t,a,y) &= f_2(y) \\
u(t,x,0) &= f_3(x) \\
u(t,x,b) &= f_4(x) \\
u(0,x,y) &= g(x,y)
\end{aligned} \tag{5.28}$$

The problem can be broken down into five problems. $u = u_1 + u_2 + u_3 + u_4 + u_5$.

$$\begin{aligned} &u_{1t} = u_{1xx} + u_{1yy} \\ &u_1(0, x, y) = g(x, y) \\ &u_1 = 0, \quad \text{all boundaries} \end{aligned} \tag{5.29}$$

$$\begin{aligned} &u_{2xx} + u_{2yy} = 0 \\ &u_2(0, y) = f_1(y) \\ &u_2 = 0 \quad \text{on all other boundaries} \end{aligned} \tag{5.30}$$

$$\begin{aligned} &u_{3xx} + u_{3yy} = 0 \\ &u_3(a, y) = f_2(y) \\ &u_3 = 0 \quad \text{on all other boundaries} \end{aligned} \tag{5.31}$$

$$\begin{aligned} &u_{4xx} + u_{4yy} = 0 \\ &u_4(x, 0) = f_3(x) \\ &u_4 = 0 \quad \text{on all other boundaries} \end{aligned} \tag{5.32}$$

$$\begin{aligned} &u_{5xx} + u_{5yy} = 0 \\ &u_5(x, b) = f_4(x) \\ &u_5 = 0 \quad \text{on all other boundaries} \end{aligned} \tag{5.33}$$

5.1.1 Time-Dependent Boundary Conditions

We will explore this topic when we discuss Laplace transforms.

Example 5.4 (A finite cylinder). Next we consider a cylinder of finite length $2L$ and radius r_1. As in the first problem in this chapter, there are two possible length scales and we choose r_1. The cylinder has temperature u_0 initially. The ends at $L = \pm L$ are suddenly insulated while the sides are exposed to a fluid at temperature u_1. The differential equation with no variation in the θ direction and the boundary conditions are

$$\begin{aligned} &u_t = \frac{\alpha}{r}(r u_r)_r + u_{zz} \\ &u_z(t, r, -L) = u_z(t, r, +L) = 0 \\ &k u_r(r_1) + h[u(r_1) - u_1(r_1)] = 0 \\ &u(0, r, z) = u_0 \\ &u \text{ is bounded} \end{aligned} \tag{5.34}$$

If we choose the length scale as r_1 then we define $\eta = r/r_1$, $\zeta = z/L$, and $\tau = \alpha t/r_1^2$. The normalized temperature can be chosen as $U = (u - u_1)(u_0 - u_1)$. With these we find that

$$\begin{aligned} U_\tau &= \frac{1}{\eta}(\eta U_\eta)_\eta + \left(\frac{r_1}{L}\right)^2 U_{\varsigma\varsigma} \\ U_\varsigma(\varsigma = \pm 1) &= 0 \\ U_\eta(\eta = 1) + BU(\eta = 1) &= 0 \\ U(\tau = 0) &= 1 \end{aligned} \tag{5.35}$$

where $B = hr_1/k$.

Let $U = T(\tau)R(\eta)Z(\zeta)$. Insert into the differential equation and divide by U.

$$\frac{T'}{T} = \frac{1}{\eta R}(\eta R')' + \left(\frac{r_1}{L}\right)^2 \frac{Z''}{Z} \tag{5.36}$$

$$\begin{aligned} Z_\varsigma(\varsigma = \pm 1) &= 0 \\ R_\eta(\eta = 1) + BR(\eta = 1) &= 0 \\ U(\tau = 0) &= 1 \end{aligned}$$

Again, the dance is the same. The left-hand side of Eq. (5.36) cannot be a function of η or ζ so each side must be a constant. The constant must be negative for the time term to be bounded.

Experience tells us that Z''/Z must be a negative constant because otherwise Z would be exponential functions and we could not simultaneously satisfy the boundary conditions at $\zeta = \pm 1$. Thus, we have

$$\begin{aligned} T' &= -\lambda^2 T \\ \eta^2 R'' + \eta R' + \beta^2 \eta^2 R &= 0 \\ Z'' &= -\gamma^2 \left(\frac{L}{r_1}\right)^2 Z \end{aligned} \tag{5.37}$$

with solutions

$$\begin{aligned} T &= Ae^{-\lambda^2 t} \\ Z &= C_1 \cos(\gamma L\varsigma/r_1) + C_2 \sin(\gamma L\varsigma/r_1) \\ R &= C_3 J_0(\beta\eta) + C_4 Y_o(\beta\eta) \end{aligned} \tag{5.38}$$

It is clear that C_4 must be zero always when the cylinder is not hollow because Y_0 is unbounded when $\eta = 0$. The boundary conditions at $\varsigma = \pm 1$ imply that Z is an even function, so that C_2

must be zero. The boundary condition at $\zeta = 1$ is

$$Z_\zeta = -C_1(\gamma L/r_1)\sin(\gamma L/r_1) = 0, \quad \text{or} \quad \gamma L/r_1 = n\pi \tag{5.39}$$

The boundary condition at $\eta = 1$ requires

$$\begin{aligned} &C_3[J_0'(\beta) + BJ_0(\beta)] = 0 \text{ or} \\ &BJ_0(\beta) = \beta J_1(\beta) \end{aligned} \tag{5.40}$$

which is the transcendental equation for finding β_m. Also note that

$$\lambda^2 = \gamma_n^2 + \beta_m^2 \tag{5.41}$$

By superposition we write the final form of the solution as

$$U(\tau, \eta, \varsigma) = \sum_{n=0}^{\infty}\sum_{m=0}^{\infty} K_{nm} e^{-(\gamma_n^2+\beta_m^2)\tau} J_0(\beta_m \eta)\cos(n\pi\varsigma) \tag{5.42}$$

K_{nm} is found using the orthogonality properties of $J_0(\beta_m \eta)$ and $\cos(n\pi\zeta)$ after using the initial condition.

$$\int_{r=0}^{1} rJ_0(\beta_m\eta)d\eta \int_{\varsigma=-1}^{1} \cos(n\pi\varsigma)d\varsigma = K_{nm}\int_{r=0}^{1} rJ_0^2(\beta_m\eta)d\eta \int_{\varsigma=-1}^{1} \cos^2(n\pi\varsigma)d\varsigma \tag{5.43}$$

Example 5.5 (Heat transfer in a sphere). Consider heat transfer in a solid sphere whose surface temperature is a function of θ, the angle measured downward from the z-axis (see Fig. 1.3, Chapter 1). The problem is steady and there is no heat source.

$$\begin{aligned} &r\frac{\partial^2}{\partial r^2}(ru) + \frac{1}{\sin\theta}\frac{\partial}{\partial\theta}\left(\sin\theta\frac{\partial u}{\partial\theta}\right) = 0 \\ &u(r=1) = f(\theta) \\ &u \text{ is bounded} \end{aligned} \tag{5.44}$$

Substituting $x = \cos\theta$,

$$r\frac{\partial^2}{\partial r^2}(ru) + \frac{\partial}{\partial x}\left[(1-x^2)\frac{\partial u}{\partial x}\right] = 0 \tag{5.45}$$

We separate variables by assuming $u = R(r)X(x)$. Substitute into the equation, divide by RX and find

$$\frac{r}{R}(r)'' = -\frac{[(1-x^2)X']'}{X} = \pm\lambda^2 \tag{5.46}$$

or

$$r(rR)'' \mp \lambda^2 R = 0$$
$$[(1 - x^2)X']' \pm \lambda^2 X = 0 \tag{5.47}$$

The second of these is Legendre's equation, and we have seen that it has bounded solutions at $r = 1$ when $\pm\lambda^2 = n(n+1)$. The first equation is of the Cauchy–Euler type with solution

$$R = C_1 r^n + C_2 r^{-n-1} \tag{5.48}$$

Noting that the constant C_2 must be zero to obtain a bounded solution at $r = 0$, and using superposition,

$$u = \sum_{n=0}^{\infty} K_n r^n P_n(x) \tag{5.49}$$

and using the condition at $f(r = 1)$ and the orthogonality of the Legendre polynomial

$$\int_{\theta=0}^{\pi} f(\theta) P_n(\cos\theta) d\theta = \int_{\theta=0}^{\pi} K_n P_n^2(\cos\theta) d\theta = \frac{2K_n}{2n+1} \tag{5.50}$$

$$K_n = \frac{2n+1}{2} \int_{\theta=0}^{\pi} f(\theta) P_n(\cos\theta) d\theta \tag{5.51}$$

5.2 VIBRATIONS PROBLEMS

We now consider some vibrations problems. In Chapter 2 we found a solution for a vibrating string initially displaced. We now consider the problem of a string forced by a sine function.

Example 5.6 (Resonance in a vibration problem). Equation (1.21) in Chapter 1 is

$$y_{tt} = a^2 y_{xx} + A\sin(\eta t) \tag{5.52}$$

Select a length scale as L, the length of the string, and a time scale L/a and defining $\xi = x/L$ and $\tau = ta/L$,

$$y_{\tau\tau} = y_{\xi\xi} + C\sin(\omega\tau) \tag{5.53}$$

where ω is a dimensionless frequency, $\eta L/a$ and $C = AL^2 a^2$.

The boundary conditions and initial velocity and displacement are all zero, so the boundary conditions are all homogeneous, while the differential equation is not. Back in Chapter 2 we saw one way of dealing with this. Note that it wouldn't have worked had q''' been a function of time. We approach this problem somewhat differently. From experience, we expect a solution of the form

$$y(\xi,\tau) = \sum_{n=1}^{\infty} B_n(\tau)\sin(n\pi\xi) \tag{5.54}$$

where the coefficients $B_n(\tau)$ are to be determined. Note that the equation above satisfies the end conditions. Inserting this series into the differential equation and using the Fourier sine series of C

$$C = \sum_{n=1}^{\infty} \frac{2C[1-(-1)^n]}{n\pi}\sin(n\pi\xi) \tag{5.55}$$

$$\sum_{n=1}^{\infty} B_n''(\tau)\sin(n\pi\xi) = \sum_{n=1}^{\infty}[-(n\pi)^2 B_n(\tau)]\sin(n\pi\xi) + C\sum_{n=1}^{\infty}\frac{2[1-(-1)^n]}{n\pi}\sin(n\pi\xi)\sin(\varpi\tau) \tag{5.56}$$

Thus

$$B_n'' = -(n\pi)^2 B_n + C\frac{2[1-(-1)^n]}{n\pi}\sin(\varpi\tau) \tag{5.57}$$

subject to initial conditions $y = 0$ and $y_\tau = 0$ at $\tau = 0$. When n is even the solution is zero. That is, since the right-hand side is zero when n is even,

$$B_n = C_1\cos(n\pi\tau) + C_2\sin(n\pi\tau) \tag{5.58}$$

But since both $B_n(0)$ and $B_n'(0)$ are zero, $C_1 = C_2 = 0$. When n is odd we can write

$$B_{2n-1}'' + [(2n-1)\pi]^2 B_{2n-1} = \frac{4C}{(2n-1)\pi}\sin(\omega\tau) \tag{5.59}$$

$(2n-1)\pi$ is the natural frequency of the system, ω_n. The homogeneous solution of the above equation is

$$B_{2n-1} = D_1\cos(\omega_n\tau) + D_2\sin(\omega_n\tau)\,. \tag{5.60}$$

To obtain the particular solution we assume a solution in the form of sines and cosines.

$$B_P = E_1\cos(\omega\tau) + E_2\sin(\omega\tau) \tag{5.61}$$

Differentiating and inserting into the differential equation we find

$$-E_1\omega^2\cos(\omega\tau) - E_2\omega^2\sin(\omega\tau) + \omega_n^2[E_1\cos(\omega\tau) + E_2\sin(\omega\tau)] = \frac{4C}{\omega_n}\sin(\omega\tau) \tag{5.62}$$

Equating coefficients of sine and cosine terms

$$\begin{aligned} E_1(\omega_n^2 - \omega^2)\cos(\omega\tau) &= 0 \qquad \omega \neq \omega_n \\ E_2(\omega_n^2 - \omega^2)\sin(\omega\tau) &= \frac{4C}{\omega_n}\sin(\omega\tau) \end{aligned} \tag{5.63}$$

Thus

$$E_1 = 0 \qquad E_2 = \frac{4C}{\omega_n(\omega_n^2 - \omega^2)} \qquad \omega \neq \omega_n \tag{5.64}$$

Combining the homogeneous and particular solutions

$$B_{2n-1} = D_1\cos(\omega_n\tau) + D_2\sin(\omega_n\tau) + \frac{4C}{\omega_n(\omega_n^2 - \omega^2)}\sin(\omega\tau) \tag{5.65}$$

The initial conditions at $\tau = 0$ require that

$$\begin{aligned} D_1 &= 0 \\ D_2 &= -\frac{4C(\omega/\omega_n)}{\omega_n(\omega_n^2 - \omega^2)} \end{aligned} \tag{5.66}$$

The solution for B_{2n-1} is

$$B_{2n-1} = \frac{4C}{\omega_n(\omega^2 - \omega_n^2)}\left(\frac{\omega}{\omega_n}\sin(\omega_n\tau) - \sin(\omega\tau)\right), \quad \omega \neq \omega_n \tag{5.67}$$

The solution is therefore

$$y(\xi, \tau) = 4C\sum_{n=1}^{\infty}\frac{\sin(\omega_n\xi)}{\omega_n(\omega^2 - \omega_n^2)}\left(\frac{\omega}{\omega_n}\sin(\omega_n\tau) - \sin(\omega\tau)\right) \tag{5.68}$$

When $\omega = \omega_n$ the above is not valid. The form of the particular solution should be chosen as

$$B_P = E_1\tau\cos(\omega\tau) + E_2\tau\sin(\omega\tau) \tag{5.69}$$

Differentiating and inserting into the differential equation for B_{2n-1}

$$[E_1\tau\omega_n^2 + 2E_2\omega_n - E_1\tau\omega_n^2]\cos(\omega_n\tau) + [E_2\tau\omega_n^2 - E_2\tau\omega_n^2 - 2E_1\omega_n]\sin(\omega_n\tau) = \frac{4C}{\omega_n}\sin(\omega_n\tau) \tag{5.70}$$

Thus

$$E_2 = 0 \qquad E_1 = -\frac{4C}{2\omega_n^2} \tag{5.71}$$

and the solution when $\omega = \omega_n$ is

$$B_{2n-1} = C_1 \cos(\omega_n \tau) + C_2 \sin(\omega_n \tau) - \frac{2C}{\omega_n^2} \tau \cos(\omega_n \tau) \tag{5.72}$$

The initial condition on position implies that $C_1 = 0$. The initial condition that the initial velocity is zero gives

$$\omega_n C_2 - \frac{2C}{\omega_n^2} = 0 \tag{5.73}$$

The solution for B_{2n-1} is

$$B_{2n-1} = \frac{2C}{\omega_n^3}[\sin(\omega_n \tau) - \omega_n \tau \cos(\omega_n \tau)] \tag{5.74}$$

Superposition now gives

$$y(\xi, \tau) = \sum_{n=1}^{\infty} \frac{2C}{\omega_n^3} \sin(\omega_n \xi)[\sin(\omega_n \tau) - \omega_n \tau \cos(\omega_n \tau)] \tag{5.75}$$

An interesting feature of the solution is that there are an infinite number of natural frequencies,

$$\eta = \frac{a}{L}[\pi, 3\pi, 5\pi, \ldots, (2n-1)\pi, \ldots] \tag{5.76}$$

If the system is excited at *any* of the frequencies, the magnitude of the oscillation will grow (theoretically) without bound. The smaller natural frequencies will cause the growth to be fastest.

Example 5.7 (Vibration of a circular membrane). Consider now a circular membrane (like a drum). The partial differential equation describing the displacement $y(t, r, \theta)$ was derived in Chapter 1.

$$a^{-2} \frac{\partial^2 y}{\partial t^2} = \frac{1}{r} \frac{\partial}{\partial r} \left(r \frac{\partial y}{\partial r} \right) + \frac{1}{r^2} \frac{\partial^2 y}{\partial \theta^2} \tag{5.77}$$

Suppose it has an initial displacement of $y(0, r, \theta) = f(r, \theta)$ and the velocity $y_t = 0$. The displacement at $r = r_1$ is also zero and the displacement must be finite for all r, θ, and t. The length scale is r_1 and the time scale is $r_1/a. r/r_1 = \eta$ and $ta/r_1 = \tau$.

We have

$$\frac{\partial^2 y}{\partial \tau^2} = \frac{1}{\eta} \frac{\partial}{\partial \eta} \left(\eta \frac{\partial y}{\partial \eta} \right) + \frac{1}{\eta^2} \frac{\partial^2 y}{\partial \theta^2} \tag{5.78}$$

Separation of variables as $y = T(\tau)R(\eta)S(\theta)$, substituting into the equation and dividing by TRS,

$$\frac{T''}{T} = \frac{1}{\eta R}(\eta R')' + \frac{1}{\eta^2}\frac{S''}{S} = -\lambda^2 \tag{5.79}$$

The negative sign is because we anticipate sine and cosine solutions for T.

We also note that

$$\lambda^2\eta^2 + \frac{\eta}{R}(\eta R')' = -\frac{S''}{S} = \pm\beta^2 \tag{5.80}$$

To avoid exponential solutions in the θ direction we must choose the positive sign. Thus we have

$$\begin{aligned} T'' &= -\lambda^2 T \\ S'' &= -\beta^2 S \\ \eta(\eta R')' + (\eta^2\lambda^2 - \beta^2)R &= 0 \end{aligned} \tag{5.81}$$

The solutions of the first two of these are

$$\begin{aligned} T &= A_1\cos(\lambda\tau) + A_2\sin(\lambda\tau) \\ S &= B_1\cos(\beta\theta) + B_2\sin(\beta\theta) \end{aligned} \tag{5.82}$$

The boundary condition on the initial velocity guarantees that $A_2 = 0$. β must be an integer so that the solution comes around to the same place after θ goes from 0 to 2π. Either B_1 and B_2 can be chosen zero because it doesn't matter where θ begins (we can adjust $f(r,\theta)$).

$$T(\tau)S(\theta) = AB\cos(\lambda\tau)\sin(n\theta) \tag{5.83}$$

The differential equation for R should be recognized from our discussion of Bessel functions. The solution with $\beta = n$ is the Bessel function of the first kind order n. The Bessel function of the second kind may be omitted because it is unbounded at $r = 0$. The condition that $R(1) = 0$ means that λ is the mth root of $J_n(\lambda_{mn}) = 0$. The solution can now be completed using superposition and the orthogonality properties.

$$y(\tau,\eta,\theta) = \sum_{n=0}^{\infty}\sum_{m=1}^{\infty} K_{nm}J_n(\lambda_{mn}\eta)\cos(\lambda_{mn}\tau)\sin(n\theta) \tag{5.84}$$

Using the initial condition

$$f(\eta,\theta) = \sum_{n=0}^{\infty}\sum_{m=1}^{\infty} K_{nm}J_n(\lambda_{mn}\eta)\sin(n\theta) \tag{5.85}$$

and the orthogonality of $\sin(n\theta)$ and $J_n(\lambda_{mn}\eta)$

$$\int_{\theta=0}^{2\pi}\int_{\eta=0}^{1} f(\eta,\theta)\eta J_n(\lambda_{mn}\eta)\sin(n\theta)d\theta d\eta = K_{nm}\int_{\theta=0}^{2\pi}\sin^2(n\theta)d\theta\int_{r=0}^{1}\eta J_n^2(\lambda_{mn}\eta)d\eta \tag{5.86}$$
$$= \frac{K_{nm}}{4}J_{n+1}^2(\lambda_{mn})$$

$$K_{nm} = \frac{4}{J_{n+1}^2(\lambda_{nm})}\int_{\theta=0}^{2\pi}\int_{\eta=0}^{1} f(\eta,\theta)\eta J_n(\lambda_{nm}\eta)\sin(n\theta)d\theta d\eta \tag{5.87}$$

Problems

1. The conduction equation in one dimension is to be solved subject to an insulated surface at $x = 0$ and a convective boundary condition at $x = L$. Initially the temperature is $u(0, x) = f(x)$, a function of position. Thus

$$u_t = \alpha u_{xx}$$
$$u_x(t, 0) = 0$$
$$ku_x(t, L) = -h[u(t, L) - u_1]$$
$$u(0, x) = f(x)$$

First nondimensionalize and normalize the equations. Then solve by separation of variables. Find a specific solution when $f(x) = 1 - x^2$.

2. Consider the diffusion problem

$$u_t = \alpha u_{xx} + q(x)$$
$$u_x(t, 0) = 0$$
$$u_x(t, L) = -h[u(t, L) - u_1]$$
$$u(0, x) = u_1$$

Define time and length scales and define a u scale such that the initial value of the dependent variable is zero. Solve by separation of variables and find a specific solution for $q(x) = Q$, a constant. Refer to Problem 2.1 in Chapter 2.

3. Solve the steady-state conduction

$$u_{xx} + u_{yy} = 0$$
$$u_x(0, y) = 0$$
$$u(a, y) = u_0$$
$$u(x, 0) = u_1$$
$$u_y(x, b) = -h[u(x, b) - u_1]$$

Note that one could choose a length scale either a or b. Choose a. Note that if you choose

$$U = \frac{u - u_1}{u_0 - u_1}$$

there is only one nonhomogeneous boundary condition and it is normalized. Solve by separation of variables.

5.3 FOURIER INTEGRALS

We consider now problems in which one dimension of the domain is infinite in extent. Recall that a function defined on an interval $(-c, c)$ can be represented as a Fourier series

$$f(x) = \frac{1}{2c}\int_{\varsigma=-c}^{c} f(\varsigma)d\varsigma + \frac{1}{c}\sum_{n=1}^{\infty}\int_{\varsigma=-c}^{c} f(\varsigma)\cos\left(\frac{n\pi\varsigma}{c}\right)d\varsigma\cos\left(\frac{n\pi x}{c}\right)$$
$$+\frac{1}{c}\sum_{n=1}^{\infty}\int_{\varsigma=-c}^{c} f(\varsigma)\sin\left(\frac{n\pi\varsigma}{c}\right)d\varsigma\sin\left(\frac{n\pi x}{c}\right) \tag{5.88}$$

which can be expressed using trigonometric identities as

$$f(x) = \frac{1}{2c}\int_{\varsigma=-c}^{c} f(\varsigma)d\varsigma + \frac{1}{c}\sum_{n=1}^{\infty}\int_{\varsigma=-c}^{c} f(\varsigma)\cos\left[\frac{n\pi}{c}(\varsigma - x)\right]d\varsigma \tag{5.89}$$

We now formally let c approach infinity. If $\int_{\varsigma=-c}^{\infty} f(\varsigma)d\varsigma$ exists, the first term vanishes. Let $\Delta\alpha = \pi/c$. Then

$$f(x) = \frac{2}{\pi}\sum_{n=1}^{\infty}\int_{\varsigma=0}^{c} f(\varsigma)\cos[n\Delta\alpha(\varsigma - x)d\varsigma\,\Delta\alpha \tag{5.90}$$

or, with

$$g_c(n\Delta\alpha, x) = \int\limits_{\varsigma=0}^{c} f(\varsigma)\cos[n\Delta\,\alpha(\varsigma - x)]d\varsigma \tag{5.91}$$

we have

$$f(x) = \sum_{n=1}^{\infty} g_c(n\Delta\alpha, x)\Delta\alpha \tag{5.92}$$

As c approaches infinity we can imagine that $\Delta\alpha$ approaches dα and $n\Delta\alpha$ approaches α, whereupon the equation for $f(x)$ becomes an integral expression

$$f(x) = \frac{2}{\pi}\int\limits_{\varsigma=0}^{\infty}\int\limits_{\alpha=0}^{\infty} f(\varsigma)\cos[\alpha(\varsigma - x)]d\varsigma\, d\alpha \tag{5.93}$$

which can alternatively be written as

$$f(x) = \int\limits_{\alpha=0}^{\infty} [A(\alpha)\cos\alpha\, x + B(\alpha)\sin\alpha\, x]d\alpha \tag{5.94}$$

where

$$A(\alpha) = \frac{2}{\pi}\int\limits_{\varsigma=0}^{\infty} f(\varsigma)\cos\alpha\varsigma\, d\varsigma \tag{5.95}$$

and

$$B(\alpha) = \frac{2}{\pi}\int\limits_{\varsigma=0}^{\infty} f(\varsigma)\sin\alpha\varsigma\, d\varsigma \tag{5.96}$$

Example 5.8 (Transient conduction in a semi-infinite region). Consider the boundary value problem

$$\begin{aligned} u_t &= u_{xx} \qquad (x \geq 0,\ t \geq 0) \\ u(0, t) &= 0 \\ u(x, 0) &= f(x) \end{aligned} \tag{5.97}$$

This represents transient heat conduction with an initial temperature $f(x)$ and the boundary at $x = 0$ suddenly reduced to zero. Separation of variables as $T(t)X(x)$ would normally yield a solution of the form

$$B_n \exp(-\lambda^2 t) \sin\left(\frac{\lambda x}{c}\right) \tag{5.98}$$

for a region of x on the interval $(0, c)$. Thus, for x on the interval $0 \leq x \leq \infty$ we have

$$B(\alpha) = \frac{2}{\pi} \int_{\varsigma=0}^{\infty} f(\varsigma) \sin \alpha\varsigma \, d\varsigma \tag{5.99}$$

and the solution is

$$u(x, t) = \frac{2}{\pi} \int_{\lambda=0}^{\infty} \exp(-\lambda^2 t) \sin(\lambda x) \int_{s=0}^{\infty} f(s) \sin(\lambda s) ds \, d\alpha \tag{5.100}$$

Noting that

$$2 \sin \alpha\, s \;\; \sin \alpha\, x = \cos \alpha\,(s - x) - \cos \alpha(s + x) \tag{5.101}$$

and that

$$\int_{0}^{\infty} \exp(-\gamma^2 \alpha) \cos(\gamma\, b) d\gamma = \frac{1}{2}\sqrt{\frac{\pi}{\alpha}} \exp\left(-\frac{b^2}{4\alpha}\right) \tag{5.102}$$

we have

$$u(x, t) = \frac{1}{2\sqrt{\pi t}} \int_{0}^{\infty} f(s) \left\{ \exp\left[-\frac{(s - x)^2}{4t}\right] - \exp\left[-\frac{(s + x)^2}{4t}\right] \right\} ds \tag{5.103}$$

Substituting into the first of these integrals $\sigma^2 = \frac{(s-x)^2}{4t}$ and into the second integral

$$\sigma^2 = \frac{(s + x)^2}{4t} \tag{5.104}$$

$$\begin{aligned} u(x, t) = {} & \frac{1}{\sqrt{\pi}} \int_{-x/2\sqrt{t}}^{\infty} f(x + 2\sigma\sqrt{t}) e^{-\sigma^2} d\sigma \\ & - \frac{1}{\sqrt{\pi}} \int_{x/2\sqrt{t}}^{\infty} f(-x + 2\sigma\sqrt{t}) e^{-\sigma^2} d\sigma \end{aligned} \tag{5.105}$$

In the special case where $f(x) = u_0$

$$u(x, t) = \frac{2u_0}{\sqrt{\pi}} \int_0^{x/2\sqrt{t}} \exp(-\sigma^2) d\sigma = u_0 \operatorname{erf}\left(\frac{x}{2\sqrt{t}}\right) \tag{5.106}$$

where erf(p) is the Gauss error function defined as

$$\operatorname{erf}(p) = \frac{2}{\sqrt{\pi}} \int_0^p \exp(-\sigma^2) d\sigma \tag{5.107}$$

Example 5.9 (Steady conduction in a quadrant). Next we consider steady conduction in the region $x \geq 0,\ y \geq 0$ in which the face at $x = 0$ is kept at zero temperature and the face at $y = 0$ is a function of x: $u = f(x)$. The solution is also assumed to be bounded.

$$u_{xx} + u_{yy} = 0 \tag{5.108}$$

$$u(x, 0) = f(x) \tag{5.109}$$

$$u(0, y) = 0 \tag{5.110}$$

Since $u(0, y) = 0$ the solution should take the form $e^{-\alpha y} \sin \alpha\, x$, which is, according to our experience with separation of variables, a solution of the equation $\nabla^2 u = 0$. We therefore assume a solution of the form

$$u(x, y) = \int_0^\infty B(\alpha) e^{-\alpha\, y} \sin \alpha\, x d\alpha \tag{5.111}$$

with

$$B(\alpha) = \frac{2}{\pi} \int_0^\infty f(\varsigma) \sin\alpha\varsigma\, d\varsigma \tag{5.112}$$

The solution can then be written as

$$u(x, y) = \frac{2}{\pi} \int_{\varsigma=0}^\infty f(\varsigma) \int_{\alpha=0}^\infty e^{-\alpha\, y} \sin \alpha\, x \sin \alpha\, \varsigma\, d\, \alpha\, d\, \varsigma \tag{5.113}$$

Using the trigonometric identity for $2 \sin ax \sin a\varsigma = \cos a(\varsigma - x) - \cos a(\varsigma + x)$ and noting that

$$\int_0^\infty e^{-\alpha\, y} \cos a\beta\, d\, \alpha = \frac{y}{\beta^2 + y^2} \tag{5.114}$$

we find

$$u(x, y) = \frac{y}{\pi} \int_0^\infty f(\varsigma) \left[\frac{1}{(\varsigma - x)^2 + y^2} - \frac{1}{(\varsigma + x)^2 + y^2} \right] d\varsigma \tag{5.115}$$

Problem

Consider the transient heat conduction problem

$$u_t = u_{xx} + u_{yy} \quad x \geq 0,\ 0 \leq y \leq 1, \quad t \geq 0$$

with boundary and initial conditions

$$u(t, 0, y) = 0$$
$$u(t, x, 0) = 0$$
$$u(t, x, 1) = 0$$
$$u(0, x, y) = u_0$$

and $u(t, x, y)$ is bounded.

Separate the problem into two problems $u(t, x, y) = v(t, x)w(t, y)$ and give appropriate boundary conditions. Show that the solution is given by

$$u(t, x, y) = \frac{4}{\pi} \operatorname{erf}\left[\frac{x}{2\sqrt{t}}\right] \sum_{n=1}^{\infty} \frac{\sin(2n-1)\pi y}{2n-1} \exp[-(2n-1)^2 \pi^2 t]$$

FURTHER READING

V. Arpaci, *Conduction Heat Transfer*. Reading, MA: Addison-Wesley, 1966.

J. W. Brown and R. V. Churchill, *Fourier Series and Boundary Value Problems*. 6th edition. New York: McGraw-Hill, 2001.

CHAPTER 6

Integral Transforms: The Laplace Transform

Integral transforms are a powerful method of obtaining solutions to both ordinary and partial differential equations. They are used to change ordinary differential equations into algebraic equations and partial differential into ordinary differential equations. The general idea is to multiply a function $f(t)$ of some independent variable t (not necessarily time) by a Kernel function $K(t, s)$ and integrate over some t space to obtain a function $F(s)$ of s which one hopes is easier to solve. Of course one must then inverse the process to find the desired function $f(t)$. In general,

$$F(s) = \int_{t=a}^{b} K(t, s) f(t) dt \tag{6.1}$$

6.1 THE LAPLACE TRANSFORM

A useful and widely used integral transform is the Laplace transform, defined as

$$L[f(t)] = F(s) = \int_{t=0}^{\infty} f(t) e^{-st} dt \tag{6.2}$$

Obviously, the integral must exist. The function $f(t)$ must be sectionally continuous and of exponential order, which is to say $|f(t)| \leq Me^{kt}$ when $t > 0$ for some constants M and k. For example neither the Laplace transform of t^{-1} nor $\exp(t^2)$ exists.

The inversion formula is

$$L^{-1}[F(s)] = f(t) = \frac{1}{2\pi i} \frac{\lim}{L \to \infty} \int_{\gamma - iL}^{\gamma + iL} F(s) e^{ts} ds \tag{6.3}$$

in which $\gamma - iL$ and $\gamma + iL$ are complex numbers. We will put off using the inversion integral until we cover complex variables. Meanwhile, there are many tables giving Laplace transforms and inverses. We will now spend considerable time developing the theory.

6.2 SOME IMPORTANT TRANSFORMS

6.2.1 Exponentials

First consider the exponential function:

$$L[e^{-at}] = \int_{t=0}^{\infty} e^{-at} e^{-st} dt = \int_{t=0}^{\infty} e^{-(s=a)t} dt = \frac{1}{s+a} \tag{6.4}$$

If $a = 0$, this reduces to

$$L[1] = 1/s \tag{6.5}$$

6.2.2 Shifting in the s-domain

$$L[e^{at} f(t)] = \int_{t=0}^{\infty} e^{-(s-a)t} f(t) dt = F(s-a) \tag{6.6}$$

6.2.3 Shifting in the time domain

Consider a function defined as

$$f(t) = 0 \quad t < a \qquad f(t) = f(t-a) \quad t > a \tag{6.7}$$

Then

$$\int_{\tau=0}^{\infty} e^{-s\tau} f(\tau - a) d\tau = \int_{\tau=0}^{a} 0 d\tau + \int_{\tau=a}^{\infty} e^{-s\tau} f(\tau - a) d\tau \tag{6.8}$$

Let $\tau - a = t$. Then

$$\int_{t=0}^{\infty} e^{-s(t+a)} f(t) dt = F(s) e^{-as} = L[f(t-a)] \tag{6.9}$$

the shifted function described above.

6.2.4 Sine and cosine

Now consider the sine and cosine functions. We shall see in the next chapter (and you should already know) that

$$e^{ikt} = \cos(kt) + i\sin(kt) \tag{6.10}$$

Thus the Laplace transform is

$$L[e^{ikt}] = L[\cos(kt)] + iL[\sin(kt)] = \frac{1}{s - ik} = \frac{s + ik}{(s + ik)(s - ik)} = \frac{s}{s^2 + k^2} + i\frac{k}{s^2 + k^2} \tag{6.11}$$

so

$$L[\sin(kt)] = \frac{k}{s^2 + k^2} \tag{6.12}$$

$$L[\cos(kt)] = \frac{s}{s^2 + k^2} \tag{6.13}$$

6.2.5 Hyperbolic functions

Similarly for hyperbolic functions

$$L[\sinh(kt)] = L\left[\frac{1}{2}(e^{kt} - e^{-kt})\right] = \frac{1}{2}\left[\frac{1}{s-k} - \frac{1}{s+k}\right] = \frac{k}{s^2 - k^2} \tag{6.14}$$

Similarly,

$$L[\cosh(kt)] = \frac{s}{s^2 - k^2} \tag{6.15}$$

6.2.6 Powers of t: t^m

We shall soon see that the Laplace transform of t^m is

$$L[t^m] = \frac{\Gamma(m+1)}{s^{m+1}} \qquad m > -1 \tag{6.16}$$

Using this together with the s domain shifting results,

$$L[t^m e^{-at}] = \frac{\Gamma(m+1)}{(s+a)^{m+1}} \tag{6.17}$$

Example 6.1. Find the inverse transform of the function

$$F(s) = \frac{1}{(s-1)^3}$$

This is a function that is shifted in the s-domain and hence Eq. (6.6) is applicable. Noting that $L^{-1}(1/s^3) = t^2/\Gamma(3) = t^2/2$ from Eq. (6.16)

$$f(t) = \frac{t^2}{2}e^t$$

Or we could use Eq. (6.17) directly.

Example 6.2. Find the inverse transform of the function

$$F(s) = \frac{3}{s^2+4}e^{-s}$$

The inverse transform of

$$F(s) = \frac{2}{s^2+4}$$

is, according to Eq. (6.11)

$$f(t) = \frac{3}{2}\sin(2t)$$

The exponential term implies shifting in the time domain by 1. Thus

$$\begin{aligned} f(t) &= 0, \quad t < 1 \\ &= \frac{3}{2}\sin[2(t-1)], \quad t > 1 \end{aligned}$$

Example 6.3. Find the inverse transform of

$$F(s) = \frac{s}{(s-2)^2+1}$$

The denominator is shifted in the s-domain. Thus we shift the numerator term and write $F(s)$ as two terms

$$F(s) = \frac{s-2}{(s-2)^2+1} + \frac{2}{(s-2)^2+1}$$

Equations (6.6), (6.12), and (6.13) are applicable. The inverse transform of the first of these is a shifted cosine and the second is a shifted sine. Therefore each must be multiplied by $\exp(2t)$. The inverse transform is

$$f(t) = e^{2t}\cos(t) + 2e^{2t}\sin(t)$$

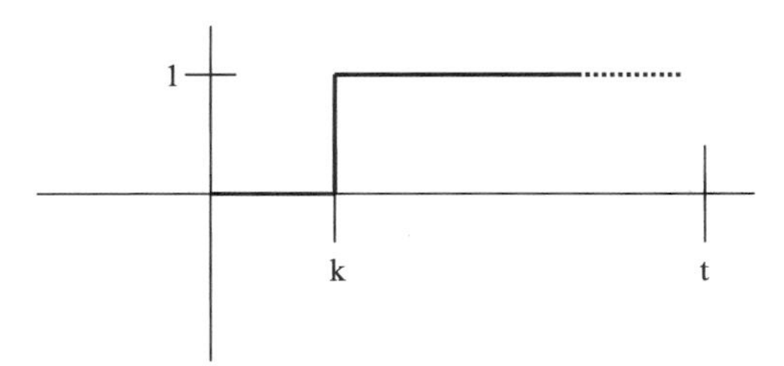

FIGURE 6.1: The Heaviside step

6.2.7 Heaviside step

A frequently useful function is the Heaviside step function, defined as

$$\begin{aligned} U_k(t) &= 0 \quad 0 < t < k \\ &= 1 \quad k < t \end{aligned} \tag{6.18}$$

It is shown in Fig. 6.1.

The Laplace transform is

$$L[U_k(t)] = \int_{t=k}^{\infty} e^{-st} dt = \frac{1}{s} e^{-ks} \tag{6.19}$$

The Heaviside step (sometimes called the unit step) is useful for finding the Laplace transforms of periodic functions.

Example 6.4 (Periodic functions). For example, consider the periodic function shown in Fig. 6.2.

It can be represented by an infinite series of shifted Heaviside functions as follows:

$$f(t) = U_0 - 2U_k + 2U_{2k} - 2U_{3k} + \cdots = U_0 + \sum_{n=1}^{\infty} (-1)^n 2U_{nk} \tag{6.20}$$

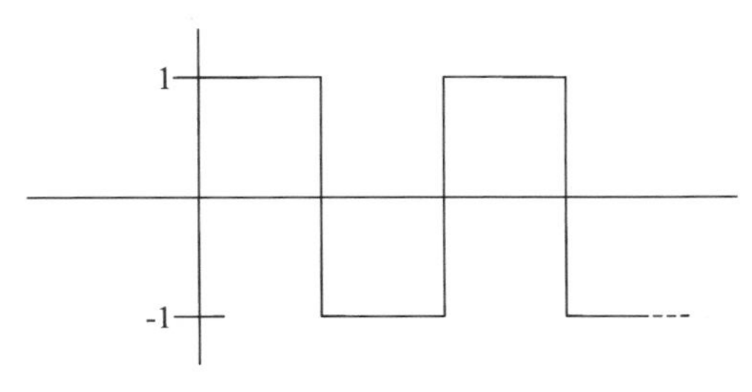

FIGURE 6.2: A periodic square wave

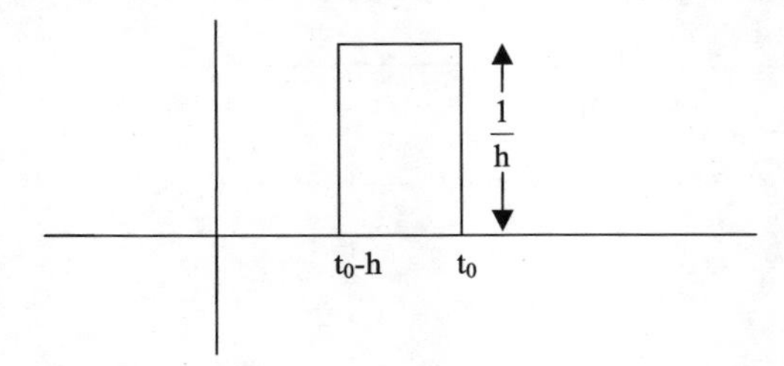

FIGURE 6.3: The Dirac delta function

The Laplace transform is found term by term,

$$L[f(t)] = \frac{1}{s}\{1 - 2e^{-sk}[1 - e^{-sk} + e^{-2sk} - e^{-3sk} \cdots]\}$$
$$= \frac{1}{s}\left\{1 - \frac{2e^{-sk}}{1 + e^{-sk}}\right\} = \frac{1}{s}\left(\frac{1 - e^{-sk}}{1 + e^{-sk}}\right) \tag{6.21}$$

6.2.8 The Dirac delta function

Consider a function defined by

$$\lim \frac{U_{t_0} - U_{t_0-h}}{h} = \delta(t_0) \qquad h \to 0 \tag{6.22}$$
$$L[\delta(t_0)] = e^{-st_0} \tag{6.23}$$

The function, without taking limits, is shown in Fig. 6.3.

6.2.9 Transforms of derivatives

$$L\left[\frac{df}{dt}\right] = \int_{t=0}^{\infty} \frac{df}{dt} e^{-st} dt = \int_{t=0}^{\infty} e^{-st} df \tag{6.24}$$

and integrating by parts

$$L\left[\frac{df}{dt}\right] = f(t)e^{-st}\Big|_0^{\infty} + s\int_{t=0}^{\infty} f(t)e^{-st} dt = sF(s) - f(0)$$

To find the Laplace transform of the second derivative we let $g(t) - f'(t)$. Taking the Laplace transform

$$L[g'(t)] = sG(s) - g(0)$$

and with

$$G(s) = L[f'(t)] = s\,F(s) - f(0)$$

we find that

$$L\left[\frac{d^2 f}{dt^2}\right] = s^2 F(s) - s f(0) - f'(0) \tag{6.25}$$

In general

$$L\left(\frac{d^n f}{dt^n}\right) = s^n F(s) - s^{n-1} f(0) - s^{n-2} f'(0) - \cdots - \frac{d^{n-1} f}{dt^{n-1}}(0) \tag{6.26}$$

The Laplace transform of t^m may be found by using the gamma function,

$$L[t^m] = \int_0^\infty t^m e^{-st} dt \quad \text{and} \quad \text{let } x = st \tag{6.27}$$

$$L[t^m] = \int_{x=0}^\infty \left(\frac{x}{s}\right)^m e^{-x} \frac{dx}{s} = \frac{1}{s^{m+1}} \int_{x=0}^\infty x^m e^{-x} dx = \frac{\Gamma(m+1)}{s^{m+1}} \tag{6.28}$$

which is true for all $m > -1$ even for nonintegers.

6.2.10 Laplace Transforms of Integrals

$$L\left[\int_{\tau=0}^{t} f(\tau) d\tau\right] = L[g(t)] \tag{6.29}$$

where $dg/dt = f(t)$. Thus $L[dg/dt] = s\,L[g(t)]$. Hence

$$L\left[\int_{\tau=0}^{t} f(\tau) d\tau\right] = \frac{1}{s} F(s) \tag{6.30}$$

6.2.11 Derivatives of Transforms

$$F(s) = \int_{t=0}^{\infty} f(t) e^{-st} dt \tag{6.31}$$

so

$$\frac{dF}{ds} = -\int_{t=0}^{\infty} tf(t)e^{-st}dt \tag{6.32}$$

and in general

$$\frac{d^n F}{ds^n} = L[(-t)^n f(t)] \tag{6.33}$$

For example

$$L[t\sin(kt)] = -\frac{d}{ds}\left(\frac{k}{s^2+k^2}\right) = \frac{2sk}{(s^2+k^2)^2} \tag{6.34}$$

6.3 LINEAR ORDINARY DIFFERENTIAL EQUATIONS WITH CONSTANT COEFFICIENTS

Example 6.5. A homogeneous linear ordinary differential equation

Consider the differential equation

$$\begin{aligned} &y'' + 4y' + 3y = 0 \\ &y(0) = 0 \\ &y'(0) = 2 \end{aligned} \tag{6.35}$$

$$L[y''] = s^2Y - sy(0) - y'(0) = s^2Y - 2 \tag{6.36}$$

$$L[y'] = sY - y(0) = sY \tag{6.37}$$

Therefore

$$(s^2 + 4s + 3)Y = 2 \tag{6.38}$$

$$Y = \frac{2}{(s+1)(s+3)} = \frac{A}{s+1} + \frac{B}{s+3} \tag{6.39}$$

To solve for A and B, note that clearing fractions,

$$\frac{A(s+3) + B(s+1)}{(s+1)(s+3)} = \frac{2}{(s+1)(s+3)} \tag{6.40}$$

Equating the numerators, or

$$A + B = 0 \quad 3A + B = 2: \quad A = 1 \quad B = -1 \tag{6.41}$$

and from Eq. (6.8)

$$Y = \frac{1}{s+1} - \frac{1}{s+3}$$
$$y = e^{-t} - e^{-3t} \tag{6.42}$$

6.4 SOME IMPORTANT THEOREMS

6.4.1 Initial Value Theorem

$$\lim_{s\to\infty} \int_{t=0}^{\infty} f'(t)e^{-st}dt = sF(s) - f(0) = 0 \tag{6.43}$$

Thus

$$\lim_{s\to\infty} sF(s) = \lim_{t\to 0} f(t) \tag{6.44}$$

6.4.2 Final Value Theorem

As s approaches zero the above integral approaches the limit as t approaches infinity minus $f(0)$. Thus

$$\lim_{s\to 0} sF(s) = \lim_{t\to\infty} f(t) \tag{6.45}$$

6.4.3 Convolution

A very important property of Laplace transforms is the convolution integral. As we shall see later, it allows us to write down solutions for very general forcing functions and also, in the case of partial differential equations, to treat both time dependent forcing and time dependent boundary conditions.

Consider the two functions $f(t)$ and $g(t)$. $F(s) = L[f(t)]$ and $G(s) = L[g(t)]$. Because of the time shifting feature,

$$e^{-s\tau}G(s) = L[g(t-\tau)] = \int_{t=0}^{\infty} e^{-st}g(t-\tau)dt \tag{6.46}$$

$$F(s)G(s) = \int_{\tau=0}^{\infty} f(\tau)e^{-s\tau}G(s)d\tau \tag{6.47}$$

But

$$e^{-s\tau}G(s) = \int_{t=0}^{\infty} e^{-st}g(t-\tau)dt \tag{6.48}$$

so that

$$F(s)G(s) = \int_{t=0}^{\infty} e^{-st} \int_{\tau=0}^{t} f(\tau)g(t-\tau)d\tau\, dt \tag{6.49}$$

where we have used the fact that $g(t-\tau) = 0$ when $\tau > t$. The inverse transform of $F(s)G(s)$ is

$$L^{-1}[F(s)G(s)] = \int_{\tau=0}^{t} f(\tau)g(t-\tau)d\tau \tag{6.50}$$

6.5 PARTIAL FRACTIONS

In the example differential equation above we determined two roots of the polynomial in the denominator, then separated the two roots so that the two expressions could be inverted in forms that we already knew. The method of separating out the expressions $1/(s+1)$ and $1/(s+3)$ is known as the method of partial fractions. We now develop the method into a more user friendly form.

6.5.1 Nonrepeating Roots

Suppose we wish to invert the transform $F(s) = p(s)/q(s)$, where $p(s)$ and $q(s)$ are polynomials. We first note that the inverse exists if the degree of $p(s)$ is lower than that of $q(s)$. Suppose $q(s)$ can be factored and a *nonrepeated root* is a.

$$F(s) = \frac{\phi(s)}{s-a} \tag{6.51}$$

According to the theory of partial fractions there exists a constant C such that

$$\frac{\phi(s)}{s-a} = \frac{C}{s-a} + H(s) \tag{6.52}$$

Multiply both sides by $(s-a)$ and take the limit as $s \to a$ and the result is

$$C = \phi(a) \tag{6.53}$$

Note also that the limit of

$$p(s)\frac{s-a}{q(s)} \tag{6.54}$$

as s approaches a is simply $p(s)/q'(s)$.

If $q(s)$ has no repeated roots and is of the form

$$q(s) = (s-a_1)(s-a_2)(s-a_3)\cdots(s-a_n) \tag{6.55}$$

then

$$L^{-1}\left[\frac{p(s)}{q(s)}\right] = \sum_{m=1}^{n} \frac{p(a_m)}{q'(a_m)} e^{a_m t} \tag{6.56}$$

Example 6.6. Find the inverse transform of

$$F(s) = \frac{4s+1}{(s^2+s)(4s^2-1)}$$

First separate out the roots of $q(s)$

$$q(s) = 4s(s+1)(s+1/2)(s-1/2)$$
$$q(s) = 4s^4 + 4s^3 - s^2 - s$$
$$q'(s) = 16s^3 + 12s^2 - 2s - 1$$

Thus

$$q'(0) = -1 \qquad p(0) = 1$$
$$q'(-1) = -3 \qquad p(-1) = -3$$
$$q'(-1/2) = 1 \qquad p(-1/2) = -1$$
$$q'(1/2) = 3 \qquad p(1/2) = 3$$
$$f(t) = e^{-t} - e^{-t/2} + e^{t/2} - 1$$

Example 6.7. Solve the differential equation

$$y'' - y = 1 - e^{3t}$$

subject to initial conditions

$$y'(0) = y(0) = 0$$

Taking the Laplace transform

$$(s^2-1)Y=\frac{1}{s}-\frac{1}{s-3}$$

$$Y(s)=\frac{1}{s(s^2-1)}-\frac{1}{(s-3)(s^2-1)}=\frac{1}{s(s+1)(s-1)}-\frac{1}{(s-3)(s+1)(s-1)}$$

First find the inverse transform of the first term.

$$q=s^3-s$$
$$q'=3s^2-1$$
$$q'(0)=-1 \qquad p(0)=1$$
$$q'(1)=2 \qquad p(1)=1$$
$$q'(-1)=2 \qquad p(-1)=1$$

The inverse transform is

$$-1+1/2e^t+1/2\,e^{-t}$$

Next consider the second term.

$$q=s^3-3s^2-s+3$$
$$q'=3s^2-6s-1$$
$$q'(-3)=44 \qquad p(-3)=1$$
$$q'(1)=-4 \qquad p(1)=1$$
$$q'(-1)=8 \qquad p(-1)=1$$

The inverse transform is

$$\frac{1}{44}e^{-3t}-\frac{1}{4}e^t+\frac{1}{8}e^{-t}$$

Thus

$$y(t)=\frac{1}{4}e^t+\frac{5}{8}e^{-t}+\frac{1}{44}e^{-3t}-1$$

6.5.2 Repeated Roots

We now consider the case when $q(s)$ has a repeated root $(s + a)^{n+1}$. Then

$$F(s) = \frac{p(s)}{q(s)} = \frac{\phi(s)}{(s-a)^{n+1}} \qquad n = 1, 2, 3, \ldots$$
$$= \frac{A_a}{(s-a)} + \frac{A_1}{(s-a)^2} + \cdots + \frac{A_n}{(s-a)^{n+1}} + H(s) \tag{6.57}$$

It follows that

$$\phi(s) = A_0(s-a)^n + \cdots + A_m(s-a)^{n-m} + \cdots + A_n + (s-a)^{n+1}H(s) \tag{6.58}$$

By letting $s \to a$ we see that $A_n = \phi(a)$. To find the remaining A's, differentiate ϕ $(n-r)$ times and take the limit as $s \to a$.

$$\phi^{(n-r)}(a) = (n-r)!A_r \tag{6.59}$$

Thus

$$F(s) = \sum_{r=0}^{n} \frac{\phi^{(n-r)}(a)}{(n-r)!}\frac{1}{(s-a)^{r+1}} + H(s) \tag{6.60}$$

If the inverse transform of $H(s)$ (the part containing no repeated roots) is $h(t)$ it follows from the shifting theorem and the inverse transform of $1/s^m$ that

$$f(t) = \sum_{r=0}^{n} \frac{\phi^{(n-r)}(a)}{(n-r)!r!} t^r e^{at} + h(t) \tag{6.61}$$

Example 6.8. Inverse transform with repeated roots

$$F(s) = \frac{s}{(s+2)^3(s+1)} = \frac{A_0}{(s+2)} + \frac{A_1}{(s+2)^2} + \frac{A_2}{(s+2)^3} + \frac{C}{(s+1)}$$

Multiply by $(s+2)^3$.

$$\frac{s}{(s+1)} = A_0(s+2)^2 + A_1(s+2) + A_2 + \frac{C(s+2)^3}{(s+1)} = \phi(s)$$

Take the limit as $s \to -2$,

$$A_2 = 2$$

Differentiate once

$$\phi' = \frac{1}{(s+1)^2} \qquad \phi'(-2) = 1 = A_1$$

$$\phi'' = \frac{-2}{(s+1)^3} \qquad \phi''(-2) = 2 = A_0$$

To find C, multiply by $(s+1)$ and take $s = -1$ (in the original equation).

$$C = -1.$$

Thus

$$F(s) = \frac{2}{(s+2)} + \frac{1}{(s+2)^2} + \frac{2}{(s+2)^3} - \frac{1}{(s+1)}$$

and noting the shifting theorem and the theorem on t^m,

$$f(t) = 2e^{-2t} + te^{-2t} + 2t^2e^{-2t} + e^{-t}$$

6.5.3 Quadratic Factors: Complex Roots

If $q(s)$ has complex roots and all the coefficients are real this part of $q(s)$ can always be written in the form

$$(s-a)^2 + b^2 \tag{6.62}$$

This is a shifted form of

$$s^2 + b^2 \tag{6.63}$$

This factor in the denominator leads to sines or cosines.

Example 6.9. Quadratic factors

Find the inverse transform of

$$F(s) = \frac{2(s-1)}{s^2+2s+5} = \frac{2s}{(s+1)^2+4} - \frac{1}{(s+1)^2+4}$$

Because of the shifted s in the denominator the numerator of the first term must also be shifted to be consistent. Thus we rewrite as

$$F(s) = \frac{2(s+1)}{(s+1)^2+4} - \frac{3}{(s+1)^2+4}$$

The inverse transform of

$$\frac{2s}{s^2+4}$$

is

$$2\cos(2t)$$

and the inverse of

$$\frac{-3}{s^2+4} = -\frac{3}{2}\frac{2}{(s^2+4)}$$

is

$$-\frac{3}{2}\sin(2t)$$

Thus

$$f(t) = 2e^{-t}\cos(2t) - \frac{3}{2}e^{-t}\sin(2t)$$

Tables of Laplace transforms and inverse transforms can be found in many books such as the book by Arpaci and in the Schaum's Outline referenced below. A brief table is given here in Appendix A.

Problems

1. Solve the problem

$$y''' - 2y'' + 5y' = 0$$
$$y(0) = y'(0) = 0 \qquad y''(0) = 1$$

   using Laplace transforms.
2. Find the general solution using Laplace transforms

$$y'' + k^2 y = a$$

3. Use convolution to find the solution to the following problem for general $g(t)$. Then find the solution for $g(t) = t^2$.

$$y'' + 2y' + y = g(t)$$
$$y'(0) = y(0) = 0$$

4. Find the inverse transforms.

   (a)
$$F(s) = \frac{s+c}{(s+a)(s+b)^2}$$

   (b)
$$F(s) = \frac{1}{(s^2+a^2)s^3}$$

(c)
$$F(s) = \frac{(s^2 - a^2)}{(s^2 + a^2)^2}$$

5. Find the periodic function whose Laplace transform is

$$F(s) = \frac{1}{s^2}\left[\frac{1 - e^{-s}}{1 + e^{-s}}\right]$$

and plot your results for $f(t)$ for several periods.

FURTHER READING

M. Abramowitz and I. A. Stegun, Eds., *Handbook of Mathematical Functions with Formulas, Graphs, and Mathematical Tables*. New York: Dover Publications, 1974.

V. S. Arpaci, *Conduction Heat Transfer*. Reading, MA: Addison-Wesley, 1966.

R. V. Churchill, *Operational Mathematics*, 3rd edition. New York: McGraw-Hill, 1972.

I. H. Sneddon, *The Use of Integral Transforms*. New York: McGraw-Hill, 1972.

CHAPTER 7

Complex Variables and the Laplace Inversion Integral

7.1 BASIC PROPERTIES

A *complex number* z can be defined as an ordered pair of real numbers, say x and y, where x is the real part of z and y is the real value of the imaginary part:

$$z = x + iy \tag{7.1}$$

where $i = \sqrt{-1}$

I am going to assume that the reader is familiar with the elementary properties of addition, subtraction, multiplication, etc. In general, complex numbers obey the same rules as real numbers. For example

$$(x_1 + iy_1)(x_2 + iy_2) = x_1x_2 - y_1y_2 + i(x_1y_2 + x_2y_1) \tag{7.2}$$

The *conjugate* of z is

$$\bar{z} = x - iy \tag{7.3}$$

It is often convenient to represent complex numbers on Cartesian coordinates with x and y as the axes. In such a case, we can represent the complex number (or variable) z as

$$z = x + iy = r(\cos\theta + i\sin\theta) \tag{7.4}$$

as shown in Fig. 7.1. We also define the exponential function of a complex number as $\cos\theta + i\sin\theta = e^{i\theta}$ which is suggested by replacing x in series $e^x = \sum_{n=0}^{\infty} \frac{x^n}{n!}$ by $i\theta$.

Accordingly,

$$e^{i\theta} = \cos\theta + i\sin\theta \tag{7.5}$$

and

$$e^{-i\theta} = \cos\theta - i\sin\theta \tag{7.6}$$

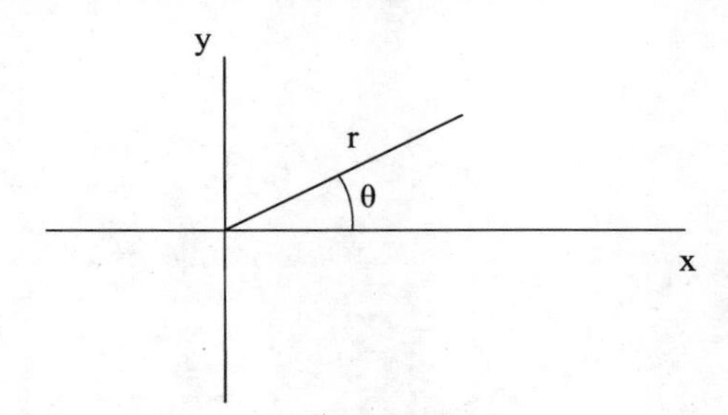

FIGURE 7.1: Polar representation of a complex variable z

Addition gives

$$\cos\theta = \frac{e^{i\theta} + e^{-i\theta}}{2} = \cosh(i\theta) \tag{7.7}$$

and subtraction gives

$$\sin\theta = \frac{e^{i\theta} - e^{-i\theta}}{2i} = -i\sinh(i\theta) \tag{7.8}$$

Note that

$$\begin{aligned}\cosh z &= \frac{1}{2}\left(e^{x+iy} + e^{-x-iy}\right) = \frac{1}{2}\left(e^{x}\left[\cos y + i\sin y\right] + e^{-x}\left[\cos y - i\sin y\right]\right)\\ &= \frac{e^{x} + e^{-x}}{2}\cos y + i\frac{e^{x} - e^{-x}}{2}\sin y\\ &= \cosh x\cos y + i\sinh x\sin y\end{aligned} \tag{7.9}$$

The reader may show that

$$\sinh z = \sinh x\cos y + i\cosh x\sin y. \tag{7.10}$$

Trigonometric functions are defined in the usual way:

$$\sin z = \frac{e^{iz} - e^{-iz}}{2i} \qquad \cos z = \frac{e^{iz} + e^{-iz}}{2} \qquad \tan z = \frac{\sin z}{\cos z} \tag{7.11}$$

Two complex numbers are equal if and only if their real parts are equal and their imaginary parts are equal.

Noting that

$$\begin{aligned} z^2 &= r^2(\cos^2\theta - \sin^2\theta + i2\sin\theta\cos\theta) \\ &= r^2\left[\frac{1}{2}(1+\cos 2\theta) - \frac{1}{2}(1-\cos 2\theta) + i\sin 2\theta\right] \\ &= r^2[\cos 2\theta + i\sin 2\theta] \end{aligned}$$

We deduce that

$$z^{1/2} = r^{1/2}(\cos\theta/2 + i\sin\theta/2) \tag{7.12}$$

In fact in general

$$z^{m/n} = r^{m/n}[\cos(m\theta/n) + i\sin(m\theta/n)] \tag{7.13}$$

Example 7.1. Find $i^{1/2}$.

Noting that when $z = I$, $r = 1$ and $\theta = \pi/2$, with $m = 1$ and $n = 2$. Thus

$$i^{1/2} = 1^{1/2}[\cos(\pi/4) + i\sin(\pi/4)] = \frac{1}{\sqrt{2}}(1+i)$$

Note, however, that if

$$w = \cos\left(\frac{\pi}{4} + \pi\right) + i\sin\left(\frac{\pi}{4} + \pi\right)$$

then $w^2 = i$. Hence $\frac{1}{\sqrt{2}}(-1 - i)$ is also a solution. The roots are shown in Fig. 7.2.

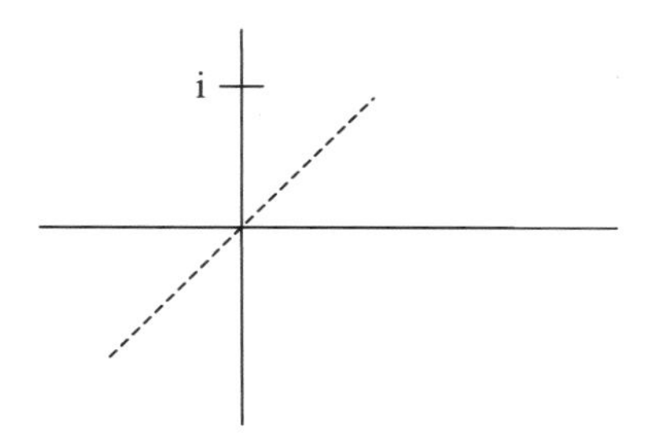

FIGURE 7.2: Roots of $i^{1/2}$

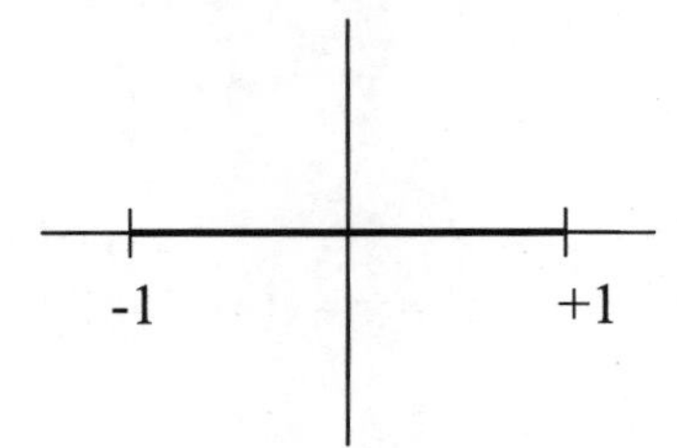

FIGURE 7.3: The roots of $1^{1/2}$

In fact in this example θ is also $\pi/2 + 2k\pi$. Using the fact that

$$z = re^{-i(\theta+2k\pi)} \qquad k = 1, 2, 3, \ldots$$

it is easy to show that

$$z^{1/n} = \sqrt[n]{r}\left[\cos\left(\frac{\theta + 2\pi k}{n}\right) + i\sin\left(\frac{\theta + 2\pi k}{n}\right)\right] \tag{7.14}$$

This is *De Moivre's theorem*. For example when $n = 2$ there are two solutions and when $n = 3$ there are three solutions. These solutions are called *branches* of $z^{1/n}$. A region in which the function is single valued is indicated by forming a *branch cut*, which is a line stretching from the origin outward such that the region between the positive real axis and the line contains only one solution. In the above example, a branch cut might be a line from the origin out the negative real axis.

Example 7.2. Find $1^{1/2}$ and represent it on the polar diagram.

$$1^{1/2} = 1\left[\cos\left(\frac{\theta}{2} + k\pi\right) + i\sin\left(\frac{\theta}{2} + k\pi\right)\right]$$

and since $\theta = 0$ in this case

$$1^{1/2} = \cos k\pi + i\sin k\pi$$

There are two distinct roots at $z = +1$ for $k = 0$ and -1 for $k = 1$. The two values are shown in Fig. 7.3. The two solutions are called branches of $\sqrt{1}$, and an appropriate branch cut might be from the origin out the positive imaginary axis, leaving as the single solution 1.

Example 7.3. Find the roots of $(1 + i)^{1/4}$.

Making use of Eq. (7.13) with $m = 1$ and $n = 4$, $r = \sqrt{2}$, $\theta = {}^{\pi}/_{4}$, we find that

$$(1 + i)^{1/4} = (\sqrt{2})^{1/4}\left[\cos\left(\frac{\pi}{16} + \frac{2k\pi}{4}\right) + i\sin\left(\frac{\pi}{16} + \frac{2k\pi}{4}\right)\right] \qquad k = 0, 1, 2, 3$$

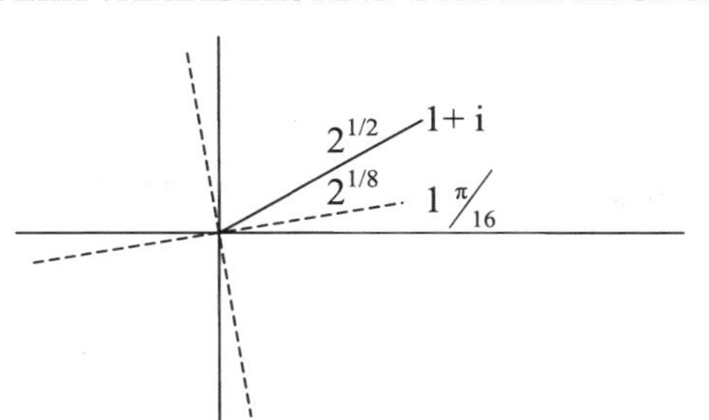

FIGURE 7.4: The roots of $(1+i)^{1/4}$

Hence, the four roots are as follows:

$$
\begin{aligned}
(1+i)^{1/4} &= 2^{1/8}\left[\cos\left(\frac{\pi}{16}\right) + i\sin\left(\frac{\pi}{16}\right)\right] \\
&= 2^{1/8}\left[\cos\left(\frac{\pi}{16}+\frac{\pi}{2}\right) + i\sin\left(\frac{\pi}{16}+\frac{\pi}{2}\right)\right] \\
&= 2^{1/8}\left[\cos\left(\frac{\pi}{16}+\pi\right) + i\sin\left(\frac{\pi}{16}+\pi\right)\right] \\
&= 2^{1/8}\left[\cos\left(\frac{\pi}{16}+\frac{3\pi}{2}\right) + i\sin\left(\frac{\pi}{16}+\frac{3\pi}{2}\right)\right]
\end{aligned}
$$

The locations of the roots are shown in Fig. 7.4.

The natural logarithm can be defined by writing $z = re^{i\theta}$ for $-\pi \leq \theta < \pi$ and noting that

$$\ln z = \ln r + i\theta \tag{7.15}$$

and since z is not affected by adding $2n\pi$ to θ this expression can also be written as

$$\ln z = \ln r + i\,(\theta + 2n\pi) \quad \text{with} \quad n = 0, 1, 2, \ldots \tag{7.16}$$

When $n = 0$ we obtain the *principal branch*. All of the single valued branches are analytic for $r > 0$ and $\theta_0 < \theta < \theta_0 + 2\pi$.

7.1.1 Limits and Differentiation of Complex Variables: Analytic Functions

Consider a *function of a complex variable* $f(z)$. We generally write

$$f(z) = u(x, y) + iv(x, y)$$

where u and v are real functions of x and y. The derivative of a complex variable is defined as follows:

$$f' = \lim_{\Delta z \to 0} \frac{f(z+\Delta z) - f(z)}{\Delta z} \tag{7.17}$$

or

$$f'(z) = \lim_{\Delta x, \Delta y \to 0} \frac{u(x+\Delta x, y+\Delta y) + iv(x+\Delta x, y+\Delta y) - u(x, y) - iv(x, y)}{\Delta x + i\Delta y} \tag{7.18}$$

Taking the limit on Δx first, we find that

$$f'(z) = \lim_{\Delta y \to 0} \frac{u(x, y+\Delta y) + iv(x, y+\Delta y) - u(x, y) - iv(x, y)}{i\Delta y} \tag{7.19}$$

and now taking the limit on Δy,

$$f'(z) = \frac{1}{i}\frac{\partial u}{\partial y} + \frac{\partial v}{\partial y} = \frac{\partial v}{\partial y} - i\frac{\partial u}{\partial y} \tag{7.20}$$

Conversely, taking the limit on Δy first,

$$\begin{aligned} j'(z) &= \lim_{\Delta x \to 0} \frac{u(x+\Delta x, y) + iv(x+\Delta x, y) - u(x, y) - iv(x, y)}{\Delta x} \\ &= \frac{\partial u}{\partial x} + i\frac{\partial v}{\partial x} \end{aligned} \tag{7.21}$$

The derivative exists only if

$$\frac{\partial u}{\partial x} = \frac{\partial v}{\partial y} \quad \text{and} \quad \frac{\partial u}{\partial y} = -\frac{\partial v}{\partial x} \tag{7.22}$$

These are called the *Cauchy—Riemann conditions*, and in this case the function is said to be *analytic*. If a function is analytic for all x and y it is *entire*.

Polynomials are entire as are trigonometric and hyperbolic functions and exponential functions. We note in passing that analytic functions share the property that both real and imaginary parts satisfy the equation $\nabla^2 u = \nabla^2 v = 0$ in two-dimensional space. It should be obvious at this point that this is important in the solution of the steady-state diffusion equation

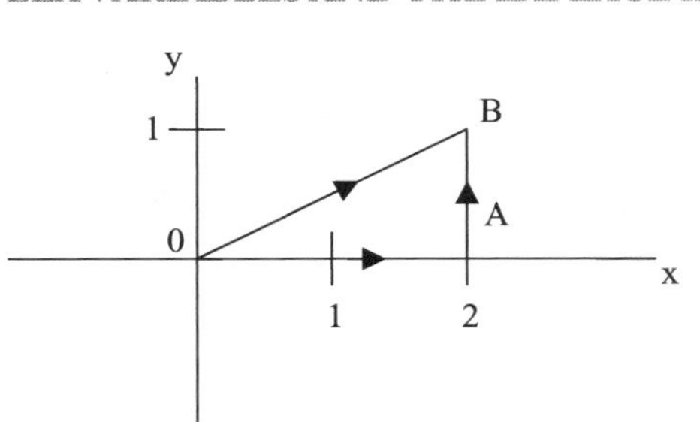

FIGURE 7.5: Integration of an analytic function along two paths

in two dimensions. We mention here that it is also important in the study of incompressible, inviscid fluid mechanics and in other areas of science and engineering. You will undoubtedly meet with it in some of you clurses.

Example 7.4.

$$f = z^2 \quad f' = 2z$$
$$f = \sin z \quad f' = \cos z$$
$$f = e^{az} \quad f' = ae^{az}$$

Integrals

Consider the line integral along a curve C defined as $x = 2y$ from the origin to the point $x = 2,\ y = 1$, path OB in Fig. 7.5.

$$\int_C z^2 dz$$

We can write

$$z^2 = x^2 - y^2 + 2ixy = 3y^2 + 4y^2 i$$

and $dz = (2 + i)dy$

Thus

$$\int_{y=0}^{1} (3y^2 + 4y^2 i)(2 + i)dy = (3 + 4i)(2 + i) \int_{y=0}^{1} y^2 dy = \frac{2}{3} + \frac{11}{3} i$$

On the other hand, if we perform the same integral along the x axis to $x = 2$ and then along the vertical line $x = 2$ to the same point, path OAB in Fig. 7.5, we find that

$$\int_{x=0}^{2} x^2 dx + \int_{y=0}^{1} (2+iy)^2 i dy = \frac{8}{3} + i \int_{y=0}^{1} (4 - y^2 + 4iy) dy = \frac{2}{3} + \frac{11}{3} i$$

This happened because the function z^2 is *analytic within the region between the two curves*. In general, if a function is analytic in the region contained between the curves, the integral

$$\int_C f(z) dz \tag{7.23}$$

is *independent of the path of* C. Since any two integrals are the same, and since if we integrate the first integral along BO only the sign changes, we see that the integral around the closed contour is zero.

$$\oint_C f(z) dz = 0 \tag{7.24}$$

This is called the *Cauchy–Goursat theorem* and is true as long as the region R within the closed curve C is *simply connected* and the function is analytic everywhere within the region. A *simply connected region* R is one in which every closed curve within it encloses only points in R.

The theorem can be extended to allow for multiply connected regions. Fig. 7.6 shows a doubly connected region. The method is to make a cut through part of the region and to integrate counterclockwise around C_1, along the path C_2 through the region, clockwise around the interior curve C_3, and back out along C_4. Clearly, the integral along C_2 and C_4 cancels, so that

$$\oint_{C_1} f(z) dz + \oint_{C_3} f(z) dz = 0 \tag{7.25}$$

where the first integral is counterclockwise and second clockwise.

7.1.2 The Cauchy Integral Formula

Now consider the following integral:

$$\oint_C \frac{f(z) dz}{(z - z_0)} \tag{7.26}$$

If the function $f(z)$ is analytic then the integrand is also analytic at all points except $z = z_0$. We now form a circle C_2 of radius r_0 around the point $z = z_0$ that is small enough to fit inside

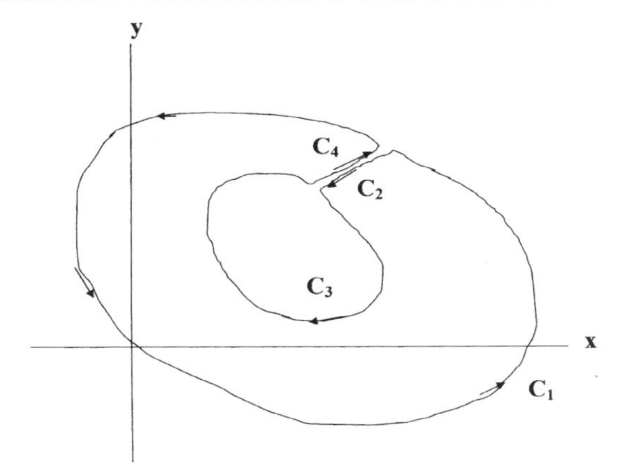

FIGURE 7.6: A doubly connected region

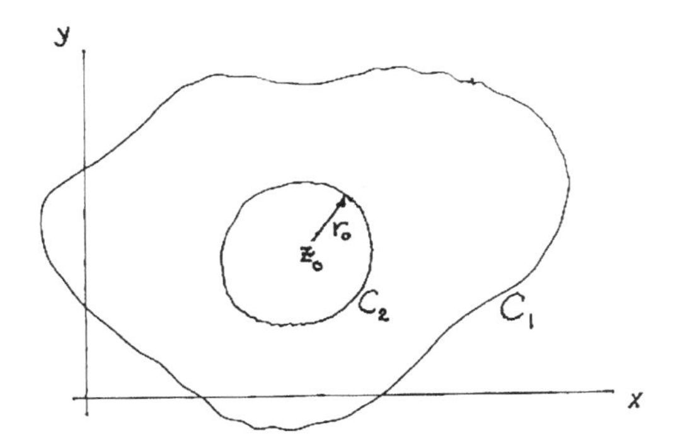

FIGURE 7.7: Derivation of Cauchy's integral formula

the curve C_1 as shown in Fig. 7.7. Thus we can write

$$\oint_{C_1} \frac{f(z)}{z-z_0} dz - \oint_{C_2} \frac{f(z)}{z-z_0} dz = 0 \tag{7.27}$$

where both integrations are counterclockwise. Let r_0 now approach zero so that in the second integral z approaches z_0, $z - z_0 = r_0 e^{i\theta}$ and $dz = r_0 i e^{i\theta} d\theta$. The second integral is as follows:

$$\oint_{C_2} \frac{f(z_0)}{r_0 e^{i\theta}} r_0 i e^{i\theta} d\theta = -f(z_0) i \int_{\theta=0}^{2\pi} d\theta = -2\pi i f(z_0)$$

Thus, *Cauchy's integral formula* is

$$f(z_0) = \frac{1}{2\pi i} \oint_C \frac{f(z)}{z-z_0} dz \tag{7.28}$$

where the integral is taken counterclockwise and $f(z)$ is analytic inside C.

We can formally differentiate the above equation n times with respect to z_0 and find an extension as

$$f^{(n)}(z_0) = \frac{n!}{2\pi i} \oint_C \frac{f(z)}{(z - z_0)^{n+1}} dz \tag{7.29}$$

Problems

1. Show that

   (a) $\sinh z = \sinh x \cos y + i \cosh x \sin y$

   (b) $\cos z = \cos x \cosh y - i \sin x \sinh y$

   and show that each is *entire*.

2. Find all of the values of

   (a) $(-1 + i\sqrt{3})^{\frac{3}{2}}$

   (b) $8^{\frac{1}{6}}$

3. Find all the roots of the equation

   $\sin z = \cosh 4$

4. Find all the zeros of

   (a) $\sinh z$

   (b) $\cosh z$

CHAPTER 8

Solutions with Laplace Transforms

In this chapter, we present detailed solutions of some boundary value problems using the Laplace transform method. Problems in both mechanical vibrations and diffusion are presented along with the details of the inversion method.

8.1 MECHANICAL VIBRATIONS

Example 8.1. Consider an elastic bar with one end of the bar fixed and a constant force F per unit area at the other end acting parallel to the bar. The appropriate partial differential equation and boundary and initial conditions for the displacement $y(x, t)$ are as follows:

$$
\begin{aligned}
&y_{\tau\tau} = y_{\zeta\zeta}, 0 < \zeta < 1, \quad t > 0 \\
&y(\zeta, 0) = y_t(\zeta, 0) = 0 \\
&y(0, \tau) = 0 \\
&y_\varsigma(1, \tau) = F/E = g
\end{aligned}
$$

We obtain the Laplace transform of the equation and boundary conditions as

$$
\begin{aligned}
&s^2 Y = Y_{\varsigma\varsigma} \\
&Y(s, 0) = 0 \\
&Y_\varsigma(s, 1) = g/s
\end{aligned}
$$

Solving the differential equation for $Y(s, \zeta)$,

$$Y(s) = (A \sinh \varsigma s + B \cosh \varsigma\, s)$$

Applying the boundary conditions we find that $B = 0$ and

$$
\begin{aligned}
&\frac{g}{s} = As \cosh s \\
&A = \frac{g}{s^2 \cosh s} \\
&Y(s) = \frac{g \sinh \varsigma\, s}{s^2 \cosh s}
\end{aligned}
$$

Since the function $\frac{1}{s}\sinh\varsigma s = \varsigma + \frac{s^2\varsigma^3}{3!} + \frac{s^4\varsigma^5}{5!} + \ldots$ the function

$$\frac{1}{s}\sinh\varsigma s$$

is analytic and $Y(s)$ can be written as the ratio of two analytic functions

$$Y(s) = \frac{\frac{1}{s}\sinh\varsigma s}{s\cosh s}$$

$Y(s)$ therefore has a simple pole at $s = 0$ and the residue there is

$$R(s=0) = \lim_{s\to 0} sY(s) = \lim_{s\to 0}\frac{\varsigma + \frac{s^2\varsigma^3}{3!} + \ldots}{\cosh s} = g\varsigma$$

The remaining poles are the singularities of $\cosh s$. But $\cosh s = \cosh x\cos y + i\sinh x\sin y$, so the zeros of this function are at $x = 0$ and $\cos y = 0$.

Hence, $s_n = i(2n-1)\pi/2$. The residues at these points are

$$R(s=s_n) = \lim_{s\to s_n}\left[\frac{g\sinh\varsigma s}{s\frac{d}{ds}(s\cosh s)}\right]e^{s\tau} = \frac{g}{s_n^2}\frac{\sinh\varsigma s_n}{\sinh s_n}e^{s_n\tau}\,(n = \pm 1, \pm 2, \pm 3\ldots)$$

Since

$$\sinh\left[i\frac{2n-1}{2}(\pi\varsigma)\right] = i\sin\left[\frac{2n-1}{2}(\pi\varsigma)\right]$$

we have

$$R(s=s_n) = \frac{gi\sin\left[\frac{2n-1}{2}(\pi\varsigma)\right]}{-\left[\frac{2n-1}{2}\pi\right]^2 i\sin\left[\frac{2n-1}{2}\pi\right]}\exp\left[i\frac{2n-1}{2}\pi\tau\right]$$

and

$$\sin\left[\frac{2n-1}{s}\pi\right] = (-1)^{n+1}$$

The exponential function can be written as

$$\exp\left[i\frac{2n-1}{2}\pi\tau\right] = \cos\left[\frac{2n-1}{2}\pi\tau\right] + i\sin\left[\frac{2n-1}{2}\pi\tau\right]$$

Note that for the poles on the negative imaginary axis ($n < 0$) this expression can be written as

$$\exp\left[i\frac{2m-1}{2}\pi\tau\right] = \cos\left[\frac{2m-1}{2}\pi\tau\right] - i\sin\left[\frac{2m-1}{2}\pi\tau\right]$$

where $m = -n > 0$. This corresponds to the conjugate poles.

Thus for *each* of the sets of poles we have

$$R(s = s_n) = \frac{4g(-1)^n}{\pi^2(2n-1)^2} \sin \frac{(2n-1)\pi\varsigma}{2} \exp\left[\frac{(2n-1)\pi\tau i}{2}\right]$$

Now adding the residues corresponding to each pole and its conjugate we find that the final solution is as follows:

$$y(\varsigma, \tau) = g\left[\varsigma + \frac{8}{\pi^2}\sum_{n=1}^{\infty} \frac{(-1)^n}{(2n-1)^2} \sin\frac{(2n-1)\pi\varsigma}{2} \cos\frac{(2n-1)\pi\tau}{2}\right]$$

Suppose that instead of a constant force at $\zeta = 1$, we allow g to be a function of τ. In this case, the Laplace transform of $y(\zeta, \tau)$ takes the form

$$Y(\varsigma, s) = \frac{G(s)\sinh(\varsigma s)}{s \cosh s}$$

The simple pole with residue $g\zeta$ is not present. However, the other poles are still at the same s_n values. The residues at each of the conjugate poles of the function

$$F(s) = \frac{\sinh(\varsigma s)}{s \cosh s}$$

are

$$\frac{2(-1)^n}{\pi(2n-1)} \sin\frac{(2n-1)\pi\varsigma}{2} \sin\frac{(2n-1)\pi\tau}{2} = f(\varsigma, \tau)$$

According to the convolution theorem

$$y(\varsigma, \tau) = \int_{\tau'=0}^{\tau} y(\tau - \tau')g(\tau')d\tau'$$

$$y(\varsigma, \tau) = \frac{4}{\pi}\sum_{n=0}^{\infty} \frac{(-1)^n}{(2n-1)} \sin\frac{(2n-1)\pi\varsigma}{2} \int_{\tau'}^{\tau} g(\tau - \tau')\sin\frac{(2n-1)\pi\tau'}{2} d\tau'.$$

In the case that $g =$ constant, integration recovers the previous equation.

Example 8.2. An infinitely long string is initially at rest when the end at $x = 0$ undergoes a transverse displacement $y(0, t) = f(t)$. The displacement is described by the differential equation and boundary conditions as follows:

$$\frac{\partial^2 y}{\partial t^2} = \frac{\partial^2 y}{\partial x^2}$$

$$y(x, 0) = y_t(x, 0) = 0$$

$$y(0, t) = f(t)$$

$$y \text{ is bounded}$$

Taking the Laplace transform with respect to time and applying the initial conditions yields

$$s^2 Y(x, s) = \frac{d^2 Y(x, s)}{dx^2}$$

The solution may be written in terms of exponential functions

$$Y(x, s) = Ae^{-sx} + Be^{sx}$$

In order for the solution to be bounded $B = 0$. Applying the condition at $x = 0$ we find

$$A = F(s)$$

where $F(s)$ is the Laplace transform of $f(t)$.

Writing the solution in the form

$$Y(x, s) = s\,F(s)\frac{e^{-sx}}{s}$$

and noting that the inverse transform of e^{-sx}/s is the Heaviside step $U_x(t)$ where

$$U_x(t) = 0 \quad t < x$$

$$U_x(t) = 1 \quad t > x$$

and that the inverse transform of $s\,F(s)$ is $f'(t)$, we find using convolution that

$$y(x, t) = \int_{\mu=0}^{t} f'(t-\mu) U_x(\mu) d\mu = f(t-x) \quad x < t$$

$$= 0 \quad x > t$$

For example, if $f(t) = \sin \omega t$

$$y(x, t) = \sin \omega(t - x) \quad x < t$$

$$= 0 \quad x > t$$

Problems

1. Solve the above vibration problem when

$$y(0, \tau) = 0$$
$$y(1, \tau) = g(\tau)$$

Hint: To make use of convolution see Example 8.3.

2. Solve the problem

$$\frac{\partial^2 y}{\partial t^2} = \frac{\partial^2 y}{\partial x^2}$$
$$y_x(0, t) = y(x, 0) = y_t(x, 0) = 0$$
$$y(1, t) = h$$

using the Laplace transform method.

8.2 DIFFUSION OR CONDUCTION PROBLEMS

We now consider the conduction problem

Example 8.3.

$$u_\tau = u_{\varsigma\varsigma}$$
$$u(1, \tau) = f(\tau)$$
$$u(0, \tau) = 0$$
$$u(\varsigma, 0) = 0$$

Taking the Laplace transform of the equation and boundary conditions and noting that $u(\varsigma, 0) = 0$,

$$sU(s) = U_{\varsigma\varsigma}$$

solution yields

$$U = A \sinh \sqrt{s}\,\varsigma + B \cosh \sqrt{s}\,\varsigma$$
$$U(0, s) = 0$$
$$U(1, s) = F(s)$$

The first condition implies that $B = 0$ and the second gives

$$F(s) = A \sinh \sqrt{s}$$

and so $U = F(s)\frac{\sinh \sqrt{s}\,\varsigma}{\sinh \sqrt{s}}$.

If $f(\tau) = 1$, $F(s) = 1/s$, a particular solution, V, is

$$V = \frac{\sinh \sqrt{s}\,\varsigma}{s \sinh \sqrt{s}}$$

where

$$v = L^{-1} V(s)$$

Now,

$$\frac{\sinh \sqrt{s}\,\varsigma}{\sinh \sqrt{s}} = \frac{\varsigma \sqrt{s} + \frac{(\varsigma\sqrt{s})^3}{3!} + \frac{(\varsigma\sqrt{s})^5}{5!} + \dots}{\sqrt{s} + \frac{(\sqrt{s})^3}{3!} + \frac{(\sqrt{s})^5}{5!} + \dots}$$

and so there is a simple pole of $Ve^{s\tau}$ at $s = 0$. Also, since when $\sinh\sqrt{s} = 0$, $\sinh\varsigma\sqrt{s}$ not necessarily zero, there are simple poles at $\sinh \sqrt{s} = 0$ or $s = -n^2\pi^2$. The residue at the pole $s = 0$ is

$$\lim_{s \to 0} s\, V(s) e^{s\tau} = \varsigma$$

and since $V(s)\, e^{s\tau}$ has the form $P(s)/Q(s)$ the residue of the pole at $-n^2\pi^2$ is

$$\frac{P(\varsigma, -n^2\pi^2)}{Q'(-n^2\pi^2)} e^{-n^2\pi^2\tau} = \left. \frac{\sinh \varsigma \sqrt{s}\, e^{-n^2\pi^2\tau}}{\frac{\sqrt{s}}{2} \cosh \sqrt{s} + \sinh \sqrt{s}} \right]_{s=-n^2\pi^2} = 2 \frac{\sin(n\pi\,\varsigma)}{n\pi \cos(n\pi)} e^{-n^2\pi^2\tau}$$

The solution for $v(\zeta, \tau)$ is then

$$v(\varsigma, \tau) = \varsigma + \sum_{n=1}^{\infty} \frac{2(-1)^n}{n\pi} e^{-n^2\pi^2\tau} \sin(n\pi\,\varsigma)$$

The solution for the general case as originally stated with $u(1, \tau) = f(\tau)$ is obtained by first differentiating the equation for $v(\zeta, \tau)$ and then noting the following:

$$U(\varsigma, s) = s\, F(s) \frac{\sinh \varsigma \sqrt{s}}{s \sinh \sqrt{s}}$$

and

$$L\left[f'(\tau)\right] = s\, F(s) - f(\tau = 0)$$

so that

$$U(\varsigma, s) = f(\tau = 0) V(\varsigma, s) + L\left[f'(s)\right] V(\varsigma, s)$$

Consequently

$$u(\varsigma,\tau) = f(\tau=0)v(\varsigma,\tau) + \int\limits_{\tau'=0}^{\tau} f'(\tau-\tau')v(\varsigma,\tau')d\tau'$$

$$= \varsigma\, f(\tau) + \frac{2f(0)}{\pi}\sum_{n=1}^{\infty}\frac{(-1)^n}{n}e^{-n^2\pi^2\tau}\sin(n\pi\,\varsigma)$$

$$+\frac{2}{\pi}\sum_{n=1}^{\infty}\frac{(-1)^n}{n}\sin(n\pi\,\varsigma)\int\limits_{\tau'=0}^{\tau} f'(\tau-\tau')e^{-n^2\pi^2\tau'}d\tau'$$

This series converges rapidly for large values of τ. However for small values of τ, it converges slowly. *There is another form of solution that converges rapidly for small* τ.

The Laplace transform of $v(\zeta,\tau)$ can be written as

$$\frac{\sinh\varsigma\sqrt{s}}{s\,\sinh\sqrt{s}} = \frac{e^{\varsigma\sqrt{s}}-e^{-\varsigma\sqrt{s}}}{s(e^{\varsigma\sqrt{s}}-e^{-\sqrt{s}})} = \frac{1}{s\,e^{\sqrt{s}}}\frac{e^{\varsigma\sqrt{s}}-e^{-\varsigma\sqrt{s}}}{1-e^{-2\sqrt{s}}}$$

$$= \frac{1}{s\,e^{\sqrt{s}}}\left[e^{\varsigma\sqrt{s}}-e^{-\varsigma\sqrt{s}}\right]\left[1+e^{-2\sqrt{s}}+e^{-4\sqrt{s}}+e^{-6\sqrt{s}}+\ldots\right]$$

$$= \frac{1}{s}\sum_{n=0}^{\infty}\left[e^{-(1+2n-\varsigma)\sqrt{s}}-e^{-(1+2n+\varsigma)\sqrt{s}}\right]$$

The inverse Laplace transform of $\frac{e^{=k\sqrt{s}}}{s}$ is the complimentary error function, defined by

$$\operatorname{erfc}(k/2\sqrt{\tau}) = 1-\frac{2}{\sqrt{\pi}}\int\limits_{x=0}^{k/2\sqrt{\tau}} e^{-x^2}\,dx$$

Thus we have

$$v(\varsigma,\tau) = \sum_{n=0}^{\infty}\left[\operatorname{erfc}\left(\frac{1+2n-\varsigma}{2\sqrt{\tau}}\right)-\operatorname{erfc}\left(\frac{1+2n+\varsigma}{2\sqrt{\tau}}\right)\right]$$

and this series converges rapidly for small values of τ.

Example 8.4. Next we consider a conduction problem with a convective boundary condition:

$$\begin{aligned}
&u_\tau = u_{\varsigma\varsigma}\\
&u(\tau,0) = 0\\
&u_\varsigma(\tau,1) + Hu(\tau,1) = 0\\
&u(0,\varsigma) = \varsigma
\end{aligned}$$

Taking the Laplace transform

$$sU - \varsigma = U_{\varsigma\varsigma}$$
$$U(s, 0) = 0$$
$$U_{\varsigma}(s, 1) + HU(s, 1) = 0$$

The differential equation has a homogeneous solution

$$U_h = A\cosh(\sqrt{s}\,\varsigma) + B\sinh(\sqrt{s}\,\varsigma)$$

and a particular solution

$$U_p = \frac{\varsigma}{s}$$

so that

$$U = \frac{\varsigma}{s} + A\cosh(\sqrt{s}\,\varsigma) + B\sinh(\sqrt{s}\,\varsigma)$$

Applying the boundary conditions, we find $A = 0$

$$B = -\frac{1 + H}{s\left[\sqrt{s}\cosh(\sqrt{s}) + H\sinh(\sqrt{s})\right]}$$

The Laplace transform of the solution is as follows:

$$U = \frac{\varsigma}{s} - \frac{(1 + H)\sinh(\sqrt{s}\,\varsigma)}{s\left[\sqrt{s}\cosh(\sqrt{s}) + H\sinh(\sqrt{s})\right]}$$

The inverse transform of the first term is simply ζ. For the second term, we must first find the poles. There is an isolated pole at $s = 0$. To obtain the residue of this pole note that

$$\frac{\lim}{s \to 0} - \frac{(1 + H)\sinh\varsigma\sqrt{s}}{\sqrt{s}\cosh\sqrt{s} + H\sinh\sqrt{s}}e^{s\tau} = \frac{\lim}{s \to 0} - \frac{(1 + H)(\varsigma\sqrt{s} + \cdots)}{\sqrt{s} + H(\sqrt{s} + \cdots)} = -\varsigma$$

canceling the first residue. To find the remaining residues let $\sqrt{s} = x + iy$. Then

$$(x + iy)\left[\cosh x\cos y + i\sinh x\sin y\right] + H\left[\sinh x\cos y + i\cosh x\sin y\right] = 0$$

Setting real and imaginary parts equal to 0 yields

$$x\cosh x\cos y - y\sinh x\sin y + H\sinh x\cos y = 0$$

and

$$y\cosh x\cos y + x\sinh x\sin y + H\cosh x\sin y = 0$$

which yields

$$x = 0$$

$$y \cos y + H \sin y = 0$$

The solution for the second term of U is

$$\frac{\lim}{s \to iy} \frac{(s - iy)(1 + H) \sinh(\sqrt{s}\varsigma)e^{s\tau}}{s\left[\sqrt{s}\cosh(\sqrt{s}) + H \sinh\sqrt{s})\right]}$$

or

$$\left[\frac{P(\varsigma, s)e^{s\tau}}{Q'(\varsigma, s)}\right]_{s=-y^2}$$

where

$$Q = s\left[\sqrt{s}\cosh\sqrt{s} + H \sinh\sqrt{s}\right]$$

$$Q' = \sqrt{s}\cosh\sqrt{s} + H\sinh\sqrt{s} + s\left[\frac{1}{2\sqrt{s}}\cosh\sqrt{s} + \frac{1}{2}\sinh\sqrt{s} + \frac{H}{2\sqrt{s}}\cosh\sqrt{s}\right]$$

$$Q' = \frac{\sqrt{s}(1 + H)}{2}\cosh\sqrt{s} + \frac{s}{2}\sinh\sqrt{s}$$

$$Q' = \left[\frac{\sqrt{s}(1 + H)}{2} - \frac{s\sqrt{s}}{2H}\right]\cosh\sqrt{s}$$

$$Q'(s = -y^2) = \left[\frac{H(H+1) + y^2}{2H}\right] iy\cos(y)$$

while

$$P(s = -y^2) = (1 + H)i\sin(y\varsigma)e^{-y^2\tau}$$

$$u_n(\varsigma, \tau) = \frac{-(1 + H)\sin(y_n\varsigma)e^{-y^2\tau}}{\left[\frac{H(H+1)+y^2}{2H}\right] y_n\cos(y_n)} = \frac{-2H(H+1)\sin(y_n\varsigma)e^{-y^2\tau}}{[H(H+1) + y^2]y_n\cos(y_n)}$$

$$= \left[\frac{2(H+1)}{H(H+1) + y^2}\right]\frac{\sin\varsigma y_n}{\sin y_n}e^{-y_n^2\tau}$$

The solution is therefore

$$u(\varsigma, \tau) = \sum_{n=1}^{\infty} \frac{2(H+1)}{H(H+1) + y^2}\frac{\sin\varsigma y_n}{\sin y_n}e^{-y_n^2\tau}$$

Note that as a partial check on this solution, we can evaluate the result when $H \to \infty$ as

$$u(\varsigma, \tau) = \sum_{n=1}^{\infty} \frac{-2}{y_n \cos y_n} \sin \varsigma\, y_n e^{-y_n^2 \tau} = \sum_{n=1}^{\infty} \frac{2(-1)^{n+1}}{n\pi} \sin(n\pi\,\varsigma) e^{-n^2\pi^2\tau}$$

in agreement with the separation of variables solution. Also, letting $H \to 0$ we find

$$u(\varsigma, \tau) = \sum_{n=1}^{\infty} \frac{2}{y_n^2} \frac{\sin(y_n \varsigma)}{\sin(y_n)} e^{y_n^2 \tau}$$

with $y_n = \frac{2n-1}{2}\pi$ again in agreement with the separation of variables solution.

Example 8.5. Next we consider a conduction (diffusion) problem with a transient source $q(\tau)$. (Nondimensionalization and normalization are left as an exercise.)

$$\begin{aligned} u_\tau &= u_{\varsigma\varsigma} + q(\tau) \\ u(\varsigma, 0) &= 0 = u_\varsigma(0, \tau) \\ u(1, \tau) &= 1 \end{aligned}$$

Obtaining the Laplace transform of the equation and boundary conditions we find

$$\begin{aligned} sU &= U_{\varsigma\varsigma} + Q(s) \\ U_\varsigma(0, s) &= 0 \\ U(1, s) &= \frac{1}{s} \end{aligned}$$

A particular solution is

$$U_P = \frac{Q(s)}{s}$$

and the homogeneous solution is

$$U_H = A \sinh(\varsigma\sqrt{s}) + B \cosh(\varsigma\sqrt{s})$$

Hence the general solution is

$$U = \frac{Q}{s} + A \sinh(\varsigma\sqrt{s}) + B \cosh(\varsigma, \sqrt{s})$$

Using the boundary conditions

$$U_\varsigma(0, s) = 0, \quad A = 0$$

$$U(1, s) = \frac{1}{s} = \frac{Q}{s} + B\cosh(\sqrt{s}) \quad B = \frac{1 - Q}{s\cosh(\sqrt{s})}$$

$$U = \frac{Q}{s} + \frac{1 - Q}{s}\frac{\cosh(\varsigma\sqrt{s})}{\cosh(\sqrt{s})}$$

The poles are (with $\sqrt{s} = x + iy$)

$$\cosh\sqrt{s} = 0 \quad \text{or} \quad \cos y = 0 \quad \sqrt{s} = \pm\frac{2n-1}{2}\pi\, i$$

$$s = -\left(\frac{2n-1}{2}\right)^2 \pi^2 = -\lambda_n^2 \quad n = 1, 2, 3, \ldots$$

or when $s = 0$.

When $s = 0$ the residue is

$$\text{Res} = \frac{\lim}{s \to 0} s\, U(s) e^{s\tau} = 1$$

The denominator of the second term is $s\cosh\sqrt{s}$ and its derivative with respect to s is

$$\cosh\sqrt{s} + \frac{\sqrt{s}}{2}\sinh\sqrt{s}$$

When $s = -\lambda_n^2$, we have for the residue of the second term

$$\frac{\lim}{s \to -\lambda_n^2}\left[\frac{(1 - Q)\cosh(\varsigma\sqrt{s})}{\cosh\sqrt{s} + \frac{\sqrt{s}}{2}\sinh\sqrt{s}}\right] e^{s\tau}$$

and since

$$\sinh\sqrt{s} = i\sin\left(\frac{2n-1}{2}\right)\pi = i(-1)^{n+1}$$

and

$$\cosh(\varsigma\sqrt{s}) = \cos\left(\frac{2n-1}{2}\right)\varsigma\pi$$

we have

$$L^{-1}\frac{\cosh(\varsigma\sqrt{s})}{s\cosh\sqrt{s}} = \frac{\cos\left(\frac{2n-1}{2}\varsigma\pi\right)}{\left(\frac{2n-1}{2}\right)\pi\, i^2(-1)^{n+1}} e^{-\left(\frac{2n-1}{2}\right)^2\pi^2\tau} = \frac{2(-1)^n\cos\left(\frac{2n-1}{2}\varsigma\pi\right)}{(2n-1)\pi} e^{-\left(\frac{2n-1}{2}\right)^2\pi^2\tau}$$

We now use the convolution principle to evaluate the solution for the general case of $q(\tau)$. We are searching for the inverse transform of

$$\frac{1}{s}\frac{\cosh(\varsigma\sqrt{s})}{\cosh\sqrt{s}}+\frac{Q(s)}{s}\left(1-\frac{\cosh(\varsigma\sqrt{s})}{\cosh\sqrt{s}}\right)$$

The inverse transform of the first term is given above. As for the second term, the inverse transform of $Q(s)$ is simply $q(\tau)$ and the inverse transform of the second term, absent $Q(s)$ is

$$1-\frac{2(-1)^{n+1}\cos\left(\frac{2n-1}{2}\varsigma\pi\right)}{(2n-1)\pi}e^{-\left(\frac{2n-1}{2}\right)^2\pi^2\tau}$$

According to the convolution principle, and summing over all poles

$$u(\varsigma,\tau)=\sum_{n=1}^{\infty}\frac{2(-1)^{n+1}\cos\left(\frac{2n-1}{2}\varsigma\pi\right)}{(2n-1)\pi}e^{-\left(\frac{2n-1}{2}\right)^2\pi^2\tau}$$
$$+\sum_{n=1}^{\infty}\int_{\tau'=0}^{\tau}\left[1-\frac{2(-1)^{n+1}\cos\left(\frac{2n-1}{2}\varsigma\pi\right)}{(2n-1)\pi}\right]e^{-\left(\frac{2n-1}{2}\right)^2\pi^2\tau}q(\tau-\tau')d\tau'$$

Example 8.6. Next consider heat conduction in a semiinfinite region $x>0$, $t>0$. The initial temperature is zero and the wall is subjected to a temperature $u(0,t)=f(t)$ at the $x=0$ surface.

$$u_t=u_{xx}$$
$$u(x,0)=0$$
$$u(0,t)=f(t)$$

and u is bounded.

Taking the Laplace transform and applying the initial condition

$$sU=U_{xx}$$

Thus

$$U(x,s)=A\sinh x\sqrt{s}+B\cosh x\sqrt{s}$$

Both functions are unbounded for $x\to\infty$. Thus it is more convenient to use the equivalent solution

$$U(x,s)=Ae^{-x\sqrt{s}}+Be^{x\sqrt{s}}=Ae^{-x\sqrt{s}}$$

in order for the function to be bounded. Applying the boundary condition at $x=0$

$$F(s)=A$$

Thus we have

$$U(x, s) = F(s)e^{-x\sqrt{s}}$$

Multiplying and dividing by s gives

$$U(x, s) = s\,F(s)\frac{e^{-x\sqrt{s}}}{s}$$

The inverse transform of $e^{-x\sqrt{s}}/s$ is

$$L^{-1}\left[\frac{e^{-x\sqrt{s}}}{s}\right] = \text{erfc}\left(\frac{x}{2\sqrt{t}}\right)$$

and we have seen that

$$L\{f'\} = s\,F(s) - f(0)$$

Thus, making use of convolution, we find

$$u(x, t) = f(0)\text{erfc}\left(\frac{x}{2\sqrt{t}}\right) + \int_{\mu=0}^{t} f'(t-\mu)\,\text{erfc}\frac{x}{2\sqrt{\mu}}d\mu$$

Example 8.7. Now consider a problem in cylindrical coordinates. An infinite cylinder is initially at dimensionless temperature $u(r, 0) = 1$ and dimensionless temperature at the surface $u(1, t) = 0$. We have

$$\frac{\partial u}{\partial t} = \frac{1}{r}\frac{\partial}{\partial r}\left(r\frac{\partial u}{\partial r}\right)$$
$$u(1, t) = 0$$
$$u(r, 0) = 1$$
$$u \text{ bounded}$$

The Laplace transform with respect to time yields

$$s\,U(r, s) - 1 = \frac{1}{r}\frac{d}{dr}\left(r\frac{dU}{dr}\right)$$

with

$$U(1, s) = \frac{1}{s}$$

Obtaining the homogeneous and particular solutions yields

$$U(r, s) = \frac{1}{s} + AJ_0(i\sqrt{s}r) + BY_0(i\sqrt{s}r)$$

The boundedness condition requires that $B = 0$, while the condition at $r = 1$

$$A = -\frac{1}{sJ_0(i\sqrt{s})}$$

Thus

$$U(r, s) = \frac{1}{s} - \frac{J_0(i\sqrt{s}r)}{sJ_0(i\sqrt{s})}$$

The inverse transform is as follows:

$$u(r, t) = 1 - \sum \text{Residues of } \left[e^{st}\frac{J_0(i\sqrt{s}r}{sJ_0(i\sqrt{s})}\right]$$

Poles of the function occur at $s = 0$ and $J_0(i\sqrt{s}) = 0$ or $i\sqrt{s} = \lambda_n$, the roots of the Bessel function of the first kind order are zero. Thus, they occur at $s = -\lambda_n^2$. The residues are

$$\frac{\lim}{s \to 0}\left[e^{st}\frac{J_0(i\sqrt{s}r)}{J_0(i\sqrt{s})}\right] = 1$$

and

$$\frac{\lim}{s \to -\lambda_n^2}\left[e^{st}\frac{J_0(i\sqrt{s}r)}{sJ_0'(i\sqrt{s})}\right] = \frac{\lim}{s \to -\lambda_n^2}\left[e^{st}\frac{J_0(i\sqrt{s}r)}{-J_1(i\sqrt{s})\,i/2\sqrt{s}}\right] = e^{-\lambda_n^2 t}\left[\frac{J_0(\lambda_n r)}{-\frac{1}{2}\lambda_n J_1(\lambda_n)}\right]$$

The two unity residues cancel and the final solution is as follows:

$$u(r, t) = \sum_{n=1}^{\infty} e^{-\lambda_n^{2t}}\frac{J_0(\lambda_n r)}{\lambda_n J_1(\lambda_n)}$$

Problems

1. Consider a finite wall with initial temperature zero and the wall at $x = 0$ insulated. The wall at $x = 1$ is subjected to a temperature $u(1, t) = f(t)$ for $t > 0$. Find $u(x, t)$.
2. Consider a finite wall with initial temperature zero and with the temperature at $x = 0$ $u(0, t) = 0$. The temperature gradient at $x = 1$ suddenly becomes $u_x(1, t) = f(t)$ for $t > 0$. Find the temperature when $f(t) = 1$ and for general $f(t)$.
3. A cylinder is initially at temperature $u = 1$ and the surface is subject to a convective boundary condition $u_r(t, 1) + Hu(t, 1) = 0$. Find $u(t, r)$.

8.3 DUHAMEL'S THEOREM

We are now prepared to solve the more general problem

$$\nabla^2 u + g(r, t) = \frac{\partial u}{\partial t} \tag{8.1}$$

where r may be considered a vector, that is, the problem is in three dimensions. The general boundary conditions are

$$\frac{\partial u}{\partial n_i} + h_i u = f_i(r, t) \text{ on the boundary } S_i \tag{8.2}$$

and

$$u(r, 0) = F(r) \tag{8.3}$$

initially. Here $\frac{\partial u}{\partial n_i}$ represents the normal derivative of u at the surface. We present Duhamel's theorem without proof.

Consider the auxiliary problem

$$\nabla^2 P + g(r, \lambda) = \frac{\partial P}{\partial t} \tag{8.4}$$

where λ is a timelike constant with boundary conditions

$$\frac{\partial P}{\partial n_i} + h_i P = f_i(r, \lambda) \text{ on the boundary } S_i \tag{8.5}$$

and initial condition

$$P(r, 0) = F(r) \tag{8.6}$$

The solution of Eqs. (8.1), (8.2), and (8.3) is as follows:

$$u(x, y, z, t) = \frac{\partial}{\partial t} \int_{\lambda=0}^{t} P(x, y, z, \lambda, t-\lambda) d\lambda = F(x, y, z) + \int_{\lambda=0}^{t} \frac{\partial}{\partial t} P(x, y, z, \lambda, t-\lambda) d\lambda \tag{8.7}$$

This is Duhamel's theorem. For a proof, refer to the book by Arpaci.

Example 8.8. Consider now the following problem with a time-dependent heat source:

$$u_t = u_{xx} + xe^{-t}$$
$$u(0, t) = u(1, t) = 0$$
$$u(x, 0) = 0$$

We first solve the problem

$$P_t = P_{xx} + xe^{-\lambda}$$
$$P(0, t) = P(1, t) = 0$$
$$P(x, 0) = 0$$

while holding λ constant.

Recall from Chapter 2 that one technique in this case is to assume a solution of the form

$$P(x, \lambda, t) = X(x) + W(x, \lambda, t)$$

so that

$$W_t = W_{xx}$$
$$W(0, \lambda, t) = W(1, \lambda, t) = 0$$
$$W(x, \lambda, 0) = -X(x, \lambda)$$

and

$$X_{xx} + xe^{-\lambda} = 0$$
$$X(0) = X(1) = 0$$

Separating variables in the equation for $W(x, t)$, we find that for $W(x, \lambda, t) = S(x)Q(t)$

$$\frac{Q_t}{Q} = \frac{S_{xx}}{S} = -\beta^2$$

The minus sign has been chosen so that Q remains bounded. The boundary conditions on $S(x)$ are as follows:

$$S(0) = S(1) = 0$$

The solution gives

$$S = A\sin(\beta x) + B\cos(\beta x)$$
$$Q = Ce^{-\beta t}$$

Applying the boundary condition at $x = 0$ requires that $B = 0$ and applying the boundary condition at $x = 1$ requires that $\sin(\beta) = 0$ or $\beta = n\pi$.

Solving for $X(x)$ and applying the boundary conditions gives

$$X = \frac{x}{6}(1 - x^2)e^{-\lambda} = -W(x, \lambda, 0)$$

The solution for $W(x, t)$ is then obtained by superposition:

$$W(x, t) = \sum_{n=0}^{\infty} K_n e^{-n^2\pi^2 t} \sin(n\pi x)$$

and using the orthogonality principle

$$e^{-\lambda} \int_{x=0}^{1} \frac{x}{6}(x^2 - 1) \sin(n\pi x) dx = K_n \int_{n=0}^{1} \sin^2(n\pi x) dx = \frac{1}{2} K_n$$

so

$$W(x, t) = \sum_{n=1}^{\infty} e^{-\lambda} \int_{x=0}^{1} \frac{x}{3}(x^2 - 1) \sin(n\pi x) dx \, e^{-n^2\pi^2 t} \sin(n\pi x)$$

$$P(x, \lambda, t) = \left\{ \frac{x}{6}(1 - x^2) + \sum_{n=1}^{\infty} \int_{x=0}^{1} \frac{x}{3}(x^2 - 1) \sin(n\pi x) dx \, \sin(n\pi x) \, e^{-n^2\pi^2 t} \right\} e^{-\lambda}$$

and

$$P(x, \lambda, t - \lambda) = \left\{ \frac{x}{6}(1 - x^2) e^{-\lambda} \right\} + \sum_{n=1}^{\infty} \int_{x=0}^{1} \frac{x}{3}(x^2 - 1) \sin(n\pi x) dx \, \sin(n\pi x) \, e^{-n^2\pi^2 t} e^{n^2\pi^2 \lambda - \lambda}$$

$$\frac{\partial}{\partial t} P(x, \lambda, t - \lambda) = \sum_{n=1}^{\infty} n^2\pi^2 \int_{x=0}^{1} \frac{x}{3}(1 - x^2) \sin(n\pi x) dx e^{-n^2\pi^2 t} e^{(n^2\pi^2 - 1)\lambda}$$

According to Duhamel's theorem, the solution for $u(x, t)$ is then

$$u(x, t) = \sum_{n=1}^{\infty} \int_{x=0}^{1} \frac{x}{3}(1 - x^2) n^2\pi^2 \sin(n\pi x) dx \, \sin(n\pi x) \int_{\lambda=0}^{t} e^{-n^2\pi^2(t-\lambda) - \lambda} d\lambda$$

$$= \sum_{n=1}^{\infty} \frac{n^2\pi^2}{n^2\pi^2 - 1} \int_{x=0}^{1} \frac{x}{3}(1 - x^2) \sin(n\pi x) dx \, [e^{-t} - e^{-n^2\pi^2 t}] \sin(n\pi x)$$

Example 8.9. Reconsider Example 8.6 in which $u_t = u_{xx}$ on the half space, with

$$u(x, 0) = 0$$

$$u(0, t) = f(t)$$

To solve this using Duhamel's theorem, we first set $f(t) = f(\lambda)$ with λ a timelike constant.

Following the procedure outlined at the beginning of Example 8.6, we find

$$U(x, s) = f(\lambda)\frac{e^{-x\sqrt{s}}}{s}$$

The inverse transform is as follows:

$$u(x, t, \lambda) = f(\lambda)\,\mathrm{erfc}\left(\frac{x}{2\sqrt{t}}\right)$$

Using Duhamel's theorem,

$$u(x, t) = \int_{\lambda=0}^{t} \frac{\partial}{\partial t}\left[f(\lambda)\mathrm{erfc}\left(\frac{x}{2\sqrt{t-\lambda}}\right)\right] d\lambda$$

which is a different form of the solution given in Example 8.6.

Problems

1. Show that the solutions given in Examples 8.6 and 8.9 are equivalent.
2. Use Duhamel's theorem along with Laplace transforms to solve the following conduction problem on the half space:

$$u_t = u_{xx}$$
$$u(x, 0) = 0$$
$$u_x(0, t) = f(t)$$

3. Solve the following problem first using separation of variables:

$$\frac{\partial u}{\partial t} = \frac{\partial^2 u}{\partial x^2} + \sin(\pi x)$$
$$u(t, 0) = 0$$
$$u(t, 1) = 0$$
$$u(0, x) = 0$$

4. Consider now the problem

$$\frac{\partial u}{\partial t} = \frac{\partial^2 u}{\partial x^2} + \sin(\pi x)te^{-t}$$

with the same boundary conditions as Problem 7. Solve using Duhamel's theorem.

FURTHER READING

V. S. Arpaci, *Conduction Heat Transfer*, Reading, MA: Addison-Wesley, 1966.

R. V. Churchill, *Operational Mathematics*, 3rd ed. New York: McGraw-Hill, 1972.

I. H. Sneddon, *The Use of Integral Transforms*, New York: McGraw-Hill, 1972.

CHAPTER 9

Sturm–Liouville Transforms

Sturm–Liouville transforms include a variety of examples of choices of the kernel function $K(s, t)$ that was presented in the general transform equation at the beginning of Chapter 6. We first illustrate the idea with a simple example of the Fourier sine transform, which is a special case of a Sturm–Liouville transform. We then move on to the general case and work out some examples.

9.1 A PRELIMINARY EXAMPLE: FOURIER SINE TRANSFORM

Example 9.1. Consider the boundary value problem

$$u_t = u_{xx} \qquad x \leq 0 \leq 1$$

with boundary conditions

$$u(0, t) = 0$$
$$u_x(1, t) + Hu(1, t) = 0$$

and initial condition

$$u(x, 0) = 1$$

Multiply both sides of the differential equation by $\sin(\lambda x)dx$ and integrate over the interval $x \leq 0 \leq 1$.

$$\int_{x=0}^{1} \sin(\lambda x)\frac{d^2u}{dx^2}\,dx = \frac{d}{dt}\int_{x=0}^{1} u(x, t)\sin(\lambda x)dx$$

Integration of the left hand side by parts yields

$$\int_{x=0}^{1} \frac{d^2}{dx^2}[\sin(\lambda x)]u(x, t)dx + \left[\sin(\lambda x)\frac{du}{dx} - u\frac{d}{dx}[\sin(\lambda x)]\right]_0^1$$

and applying the boundary conditions and noting that

$$\frac{d^2}{dx^2}[\sin(\lambda x)] = -\lambda^2 \sin(\lambda x)$$

we have

$$-\lambda^2 \int_{x=0}^{1} \sin(\lambda x) u(x,t) dx + [u_x \sin(\lambda x) - \lambda u \cos(\lambda x)]_0^1$$
$$= -\lambda^2 U(\lambda, t) - u(1)[\lambda \cos\lambda + H \sin\lambda]$$

Defining

$$S_\lambda\{u(x,t)\} = \int_{x=0}^{1} u(x,t) \sin(\lambda x) dx = U(\lambda, t)$$

as the Fourier sine transform of $u(x, t)$ and setting

$$\lambda \cos\lambda + H \sin\lambda = 0$$

we find

$$U_t(\lambda, t) = -\lambda^2 U(\lambda, t)$$

whose solution is

$$U(\lambda, t) = A e^{-\lambda^2 t}$$

The initial condition of the transformed function is

$$U(\lambda, 0) = \int_{x=0}^{1} \sin(\lambda x) dx = \frac{1}{\lambda}[1 - \cos(\lambda)]$$

Applying the initial condition we find

$$U(\lambda, t) = \frac{1}{\lambda}[1 - \cos(\lambda)] e^{-\lambda^2 t}$$

It now remains to find from this the value of $u(x, t)$.

Recall from the general theory of Fourier series that any odd function of x defined on $0 \le x \le 1$ can be expanded in a Fourier sine series in the form

$$u(x,t) = \sum_{n=1}^{\infty} \frac{\sin(\lambda_n x)}{\|\sin(\lambda_n)\|^2} \int_{\xi=0}^{1} u(\xi, t) \sin(\lambda_n \xi) d\xi$$

and this is simply

$$u(x,t)=\sum_{n=1}^{\infty}\frac{\sin(\lambda_n x)}{\left\|\sin(\lambda_n)\right\|^2}U(\lambda_n,t)$$

with λ_n given by the transcendental equation above. The final solution is therefore

$$u(x,t)=\sum_{n=1}^{\infty}\frac{2(1-\cos\lambda_n)}{\lambda_n-\frac{1}{2}\sin(2\lambda_n)}\sin(\lambda_n x)e^{-\lambda_n^2 t}$$

9.2 GENERALIZATION: THE STURM–LIOUVILLE TRANSFORM: THEORY

Consider the differential operator D

$$D[f(x)]=A(x)f''+B(x)f'+C(x)f \qquad a\le x\le b \tag{9.1}$$

with boundary conditions of the form

$$\begin{aligned}N_\alpha[f(x)]_{x=a}&=f(a)\cos\alpha+f'(a)\sin\alpha\\ N_\beta[f(x)]_{x=b}&=f(b)\cos\beta+f'(b)\sin\beta\end{aligned} \tag{9.2}$$

where the symbols N_α and N_β are differential operators that define the boundary conditions. For example the differential operator might be

$$D[f(x)]=f_{xx}$$

and the boundary conditions might be defined by the operators

$$N_\alpha[f(x)]_{x=a}=f(a)=0$$

and

$$N_\beta[f(x)]_{x=b}=f(b)+Hf'(b)=0$$

We define an integral transformation

$$T[f(x)]=\int_a^b f(x)K(x,\lambda)dx=F(\lambda) \tag{9.3}$$

We wish to transform these differential forms into algebraic forms. First we write the differential operator in standard form. Let

$$
\begin{aligned}
r(x) &= \exp \int_a^x \frac{B(\xi)}{A(\xi)} d\xi \\
p(x) &= \frac{r(x)}{A(x)} \\
q(x) &= -p(x)C(x)
\end{aligned}
\tag{9.4}
$$

Then

$$
D[f(x)] = \frac{1}{p(x)} \left[(rf')' - qf\right] = \frac{1}{p(x)} \Re[f(x)] \tag{9.5}
$$

where $\Re$ is the Sturm–Liouville operator.

Let the kernel function $K(x, \lambda)$in Eq. (9.3) be

$$
K(x, \lambda) = p(x)\Phi(x, \lambda) \tag{9.6}
$$

Then

$$
\begin{aligned}
T[D[f(x)]] &= \int_a^b \Phi(x, \lambda)\Re[f(x)]dx \\
&= \int_a^b f(x)\Re[\Phi(x, \lambda)]dx + [(\Phi f_x - \Phi_x f)r(x)]_a^b
\end{aligned}
\tag{9.7}
$$

while

$$
\begin{aligned}
N_\alpha[f(a)] &= f(a) \cos \alpha + f(a) \sin \alpha \\
N'_\alpha[f(a)] &= \frac{d}{d\alpha} \left(f(a) \cos \alpha + f'(a) \sin \alpha\right) \\
&= -f(a) \sin \alpha + f'(a) \cos \alpha
\end{aligned}
\tag{9.8}
$$

so that

$$
\begin{aligned}
f(a) &= N_\alpha[f(a)] \cos \alpha - N'_\alpha[f(a)] \sin \alpha \\
f'(a) &= N'_\alpha[f(a)] \cos \alpha + N_\alpha[f(a)] \sin \alpha
\end{aligned}
\tag{9.9}
$$

where the prime indicates differentiation with respect to α.

The lower boundary condition at $x = a$ is then

$$[\Phi(a,\lambda)f'(a) - \Phi'(a,\lambda)f(a)]\,r(a) = \left[\begin{array}{l}\Phi(a,\lambda)N'_\alpha[f(a)]\cos\alpha + \Phi(a,\lambda)N_\alpha[f(a)]\sin\alpha - \Phi'(a,\lambda)N_\alpha[f(a)]\cos\alpha \\ +\Phi'(a,\lambda)N'_\alpha[f(a)]\sin\alpha\end{array}\right] r(a) \tag{9.10}$$

But if $\Phi(x, \lambda)$ is chosen to satisfy the Sturm–Liouville equation and the boundary conditions then

$$\begin{aligned} N_\alpha[\Phi(x,\lambda)]_{x=a} &= \Phi(a,\lambda)\cos\alpha + \Phi'(a,\lambda)\sin\alpha \\ N_\beta[\Phi(x,\lambda)]_{x=b} &= \Phi(b,\lambda)\cos\beta + \Phi'(b,\lambda)\sin\beta \end{aligned} \tag{9.11}$$

and

$$\begin{aligned} \Phi(a,\lambda) &= N_\alpha[\Phi(a,\lambda)]\cos\alpha - N'_\alpha[\Phi(a,\lambda)]\sin\alpha \\ \Phi'(a,\lambda) &= N'_\alpha[\Phi(a,\lambda)]\cos\alpha + N_\alpha[\Phi(a,\lambda)]\sin\alpha \end{aligned} \tag{9.12}$$

and we have

$$\begin{aligned} &[(N'_\alpha[\Phi(a,\lambda)]\cos\alpha + N_\alpha[f(a)]\sin\alpha)(N_\alpha[\Phi(a,\lambda)]\cos\alpha + N'_\alpha[\Phi(a,\lambda)]\sin\alpha) \\ &\quad - (N'_\alpha[\Phi(a,\lambda)]\cos\alpha + N_\alpha[\Phi(a,\lambda)]\sin\alpha)(N'_\alpha[f(a)]\cos\alpha \\ &\quad - N_\alpha[f(a)]\sin\alpha)]r(a) \\ &= \{N'_\alpha[f(a)]N_\alpha[\Phi(a,\lambda)] - N_\alpha[f(a)]N'_\alpha[\Phi(a,\lambda)]\}r(a) \end{aligned} \tag{9.13}$$

If the kernel function is chosen so that $N_\alpha[\Phi(a, \lambda)] = 0$, for example, the lower boundary condition is

$$-N_\alpha[f(a)]N'_\alpha[\Phi(a,\lambda)]r(a) \tag{9.14}$$

Similarly, at $x = b$

$$\left[\Phi(b,\lambda)f'(b) - \Phi'(b,\lambda)f(b)\right] r(b) = -N_\beta[f(b)]N'_\beta[\Phi(b,\lambda)]r(b) \tag{9.15}$$

Since $\Phi(x, \lambda)$ satisfies the Sturm–Liouville equation, there are n solutions forming a set of orthogonal functions with weight function $p(x)$ and

$$\Re\Phi_n(x,\lambda_n) = -\lambda_n^2 p(x)\Phi_n(x,\lambda_n) \tag{9.16}$$

so that

$$T\{D[f(x)]\} = -\lambda^2 \int_{x=a}^{b} p(x)f(x)\Phi_n(x,\lambda)dx + N_\alpha[f(a)]N'_\alpha[\Phi_n(a,\lambda)]r(a) - N_\beta[f(b)]N'_\beta[\Phi_n(b,\lambda)]r(b) \tag{9.17}$$

where

$$\lambda_n^2 \int_a^b p(x)f_n(x)\Phi_n(x,\lambda_n)dx = \lambda_n^2 F_n(\lambda_n) \tag{9.18}$$

9.3 THE INVERSE TRANSFORM

The great thing about Sturm–Liouville transforms is that the inversion is so easy. Recall that the generalized Fourier series of a function $f(x)$ is

$$f(x) = \sum_{n=1}^{\infty} \frac{\Phi_n(x,\lambda_n)}{\|\Phi_n\|} \int_a^b f_n(\xi)p(\xi)\frac{\Phi_n(\xi,\lambda_n)}{\|\Phi_n\|}d\xi = \sum_{n=1}^{\infty} \frac{\Phi_n(x)}{\|\Phi_n\|^2}F(\lambda_n) \tag{9.19}$$

where the functions $\Phi_n(x,\lambda_n)$form an orthogonal set with respect to the weight function $p(x)$.

Example 9.2 (The cosine transform). Consider the diffusion equation

$$y_t = y_{xx} \qquad 0 \le x \le 1 \qquad t > 0$$
$$y_x(0,t) = y(1,t) = 0$$
$$y(x,0) = f(x)$$

To find the proper kernel function $K(x,\lambda)$ we note that according to Eq. (9.16) $\Phi_n(x,\lambda_n)$ must satisfy the Sturm–Liouville equation

$$\Re[\Phi_n(x,\lambda)] = -p(x)\Phi_n(x,\lambda)$$

where for the current problem

$$\Re[\Phi_n(x,\lambda)] = \frac{d^2}{dx^2}[\Phi_n(x,\lambda)] \quad \text{and} \quad p(x) = 1$$

along with the boundary conditions (9.11)

$$N_\alpha[\Phi(x,\lambda)]_{x=a} = \Phi_x(0,\lambda) = 0$$
$$N_\beta[\Phi(x,\lambda)]_{x=b} = \Phi(1,\lambda) = 0$$

Solution of this differential equation and applying the boundary conditions yields an infinite number of functions (as any Sturm–Liouville problem)

$$\Phi(x, \lambda_n) = A\cos(\lambda_n x)$$

with

$$\cos(\lambda_n) = 0 \qquad \lambda_n = \frac{(2n-1)}{2}\pi$$

Thus, the appropriate kernel function is $K(x, \lambda_n) = \cos(\lambda_n x)$ with $\lambda_n = (2n-1)\frac{\pi}{2}$. Using this kernel function in the original partial differential equation, we find

$$\frac{dY}{dt} = -\lambda_n^2 Y$$

where $C_\lambda\left\{y(x, t)\right\} = Y(t, \lambda_n)$ is the cosine transform of $y(t, x)$. The solution gives

$$Y(t, \lambda_n) = Be^{-\lambda^2 t}$$

and applying the cosine transform of the initial condition

$$B = \int_{x=0}^{1} f(x)\cos(\lambda_n x)dx$$

According to Eq. (9.19) the solution is as follows:

$$y(x, t) = \sum_{n=0}^{\infty} \frac{\cos(\lambda_n x)}{\left\|\cos(\lambda_n x)\right\|^2} \int_{x-0}^{1} f(x)\cos(\lambda_n x)dxe^{-\lambda_n^2 t}$$

Example 9.3 (The Hankel transform). Next consider the diffusion equation in cylindrical coordinates.

$$u_t = \frac{1}{r}\frac{d}{dr}\left(r\frac{du}{dr}\right)$$

Boundary and initial conditions are prescribed as

$$u_r(t, 0) = 0$$

$$u(t, 1) = 0$$

$$u(0, r) = f(r)$$

First we find the proper kernel function

$$\Re[\Phi(r, \lambda_n)] = \frac{d}{dr}\left(r\frac{d\Phi_n}{dr}\right) = -\lambda_n^2 r\,\Phi$$

with boundary conditions

$$\Phi_r(\lambda_n, 0) = 0$$
$$\Phi(\lambda_n, 1) = 0$$

The solution is the Bessel function $J_0(\lambda_n r)$ with λ_n given by $J_0(\lambda_n) = 0$. Thus the transform of $u(t, r)$ is as follows:

$$H_\lambda \{u(t, r)\} = U(t, \lambda_n) = \int_{r=0}^{1} r J_0(\lambda_n r) u(t, r) dr$$

This is called a *Hankel transform*. The appropriate differential equation for $U(t, \lambda_n)$ is

$$\frac{dU_n}{dt} = -\lambda_n^2 U_n$$

so that

$$U_n(t, \lambda_n) = B e^{-\lambda_n^2 t}$$

Applying the initial condition, we find

$$B = \int_{r=0}^{1} r f(r) J_0(\lambda_n r) dr$$

and from Eq. (9.19)

$$u(t, r) = \sum_{n=0}^{\infty} \frac{\int_{r=0}^{1} r f(r) J_0(\lambda_n r) dr}{\|J_0(\lambda_n r)\|^2} J_0(\lambda_n r) e^{-\lambda_n^2 t}$$

Example 9.4 (The sine transform with a source). Next consider a one-dimensional transient diffusion with a source term $q(x)$:

$$u_t = u_{xx} + q(x)$$
$$y(0, x) = y(t, 0) = t(t, \pi) = 0$$

First we determine that the sine transform is appropriate. The operator $\Re$ is such that

$$\Re\Phi = \Phi_{xx} = \lambda\Phi$$

and according to the boundary conditions we must choose $\Phi = \sin(nx)$ and $\lambda = -n^2$. The sine transform of $q(x)$ is $Q(\lambda)$.

$$U_t = -n^2 U + Q(\lambda)$$
$$U = U(\lambda, t)$$

The homogeneous and particular solutions give

$$U_n = Ce^{-n^2 t} + \frac{Q_n}{n^2}$$

when $t = 0$, $U = 0$ so that

$$C = -\frac{Q_n}{n^2}$$

where Q_n is given by

$$Q_n = \int_{x=0}^{\pi} q(x)\sin(nx)dx$$

Since $U_n = \frac{Q_n}{n^2}[1 - e^{-n^2 t}]$ the solution is

$$u(x, t) = \sum_{n=1}^{\infty} \frac{Q_n}{n^2}[1 - e^{-n^2 t}]\frac{\sin(nx)}{\left\|\sin(nx)\right\|^2}$$

Note that Q_n is just the nth term of the Fourier sine series of $q(x)$. For example, if $q(x) = x$,

$$Q_n = \frac{\pi}{n}(-1)^{n+1}$$

Example 9.5 (A mixed transform). Consider steady temperatures in a half cylinder of infinite length with internal heat generation, $q(r)$ that is a function of the radial position. The appropriate differential equation is

$$u_{rr} + \frac{1}{r}u_r + \frac{1}{r^2}u_{\theta\theta} + u_{zz} + q(r) = 0 \quad 0 \le r \le 1 \quad 0 \le z \le \infty \quad 0 \le \theta \le \pi$$

with boundary conditions

$$u(1, \theta, z) = 1$$
$$u(r, 0, z) = u(r, \pi, z) = u(r, \theta, 0) = 0$$

Let the sine transform of u be denoted by $S_n\{u(r,\theta,z)\} = U_n(r,n,z)$ with respect to θ on the interval $(0,\pi)$. Then

$$\frac{\partial^2 U_n}{\partial r^2} + \frac{1}{r}\frac{\partial U_n}{\partial r} - \frac{n^2}{r^2}U_n + \frac{\partial^2 U_n}{\partial z^2} + q(r)S_n(1) = 0$$

where $S_n(1)$ is the sine transform of 1, and the boundary conditions for $u(r,\theta,z)$ on θ have been used.

Note that the operator on Φ in the r coordinate direction is

$$\Re\left[\Phi(r,\mu_j)\right] = \frac{1}{r}\frac{d}{dr}\left(r\frac{d\Phi}{dr}\right) - \frac{n^2}{r^2}\Phi = -\mu_j^2\Phi$$

With the boundary condition at $r = 1$ chosen as $\Phi(1,\mu_j) = 0$ this gives the kernel function as $\Phi = rJ_n(r,\mu_j)$ with eigenvalues determined by $J_n(1,\mu_j) = 0$

We now apply the finite Hankel transform to the above partial differential equation and denote the Hankel transform of U_n by U_{jn}.

After applying the boundary condition on r we find, after noting that

$$N_\beta[U_n(z,1)] = S_n(1)$$
$$N'_\beta[\Phi(1,z)] = -\mu_j J_{n+1}(\mu_j)$$

$-\mu_j^2 U_{jn} + \mu_j J_{n+1}(\mu_j)S_n(1) + \frac{d^2U_{jn}}{dz^2} + Q_j(\mu_j)S_n(1) = 0$. Here $Q_j(\mu_j)$ is the Hankel transform of $q(r)$.

Solving the resulting ordinary differential equation and applying the boundary condition at $z = 0$,

$$U_{jn}(\mu_j,n,z) = S_n(1)\frac{Q_j(\mu_j) + \mu_j J_{n+1}(\mu_j)}{\mu_j^2}[1 - \exp(-\mu_j z)]$$

We now invert the transform for the sine and Hankel transforms according to Eq. (9.19) and find that

$$u(r,\theta,z) = \frac{4}{\pi}\sum_{n=1}^{\infty}\sum_{j=1}^{\infty}\frac{U_{jn}(\mu_j,n,z)}{[J_{n+1}(\mu_j)]^2}J_n(\mu_j r)\sin(n\theta)$$

Note that

$$S_n(1) = [1 - (-1)^n]/n$$

Problems

Use an appropriate Sturm–Liouville transform to solve each of the following problems:

1. Chapter 3, Problem 1.
2. Chapter 2, Problem 2.
3. Chapter 3, Problem 3.

$$\frac{\partial u}{\partial t} = \frac{1}{r}\frac{\partial}{\partial r}\left(r\frac{\partial u}{\partial r}\right) + G(\text{constant } t)$$

4. $u(r, 0) = 0$

   $u(1, t) = 0$

   u bounded

5. Solve the following using an appropriate Sturm–Liouville transform:

$$\frac{\partial^2 u}{\partial x^2} = \frac{\partial u}{\partial t}$$

$$u(t, 0) = 0$$

$$u(t, 1) = 0$$

$$u(0, x) = \sin(\pi x)$$

6. Find the solution for general $\rho(t)$:

$$\frac{\partial u}{\partial t} = \frac{\partial^2 u}{\partial x^2}$$

$$u(t, 0) = 0$$

$$u(t, 1) = \rho(t)$$

$$u(0.x) = 0$$

FURTHER READING

V. S. Arpaci, *Conduction Heat Transfer*, Reading, MA: Addison-Wesley, 1966.

R. V. Churchill, *Operational Mathematics*, 3rd ed. New York: McGraw-Hill, 1972.

I. H. Sneddon, *The Use of Integral Transforms*, New York: McGraw-Hill, 1972.

CHAPTER 10

Introduction to Perturbation Methods

Perturbation theory is an approximate method of solving equations which contain a parameter that is small in some sense. The method should result in an approximate solution that may be termed "precise" in the sense that the error (the difference between the approximate and exact solutions) is understood and controllable and can be made smaller by some rational technique. Perturbation methods are particularly useful in obtaining solutions to equations that are nonlinear or have variable coefficients. In addition it is important to note that if the method yields a simple, accurate approximate solution of any problem it may be more useful than an exact solution that is more complicated.

10.1 EXAMPLES FROM ALGEBRA

We begin with examples from algebra in order to introduce the ideas of regular perturbations and singular perturbations. We start with a problem of extracting the roots of a quadratic equation that contains a small parameter $\varepsilon << 1$.

10.1.1 REGULAR PERTURBATION

Example 10.1 Consider, for example the equation

$$x^2 + \varepsilon x - 1 = 0 \tag{10.1}$$

The exact solution for the roots is, of course, simply obtained from the quadratic formula.

$$x = -\frac{\varepsilon}{2} \pm \sqrt{1 + \frac{\varepsilon^2}{4}} \tag{10.2}$$

which for $\varepsilon = 0.1$ yields exact solutions

$$x = 0.962422837$$
and
$$x = -1.062422837$$

Eq. (10.2) can be expanded for small values of ε in the rapidly convergent series

$$x = 1 - \frac{\varepsilon}{2} + \frac{\varepsilon^2}{8} - \frac{\varepsilon^4}{128} + \ldots\ldots. \tag{10.3}$$

or

$$x = -1 - \frac{\varepsilon}{2} - \frac{\varepsilon^2}{8} + \frac{\varepsilon^4}{128} - \ldots\ldots \tag{10.4}$$

To apply perturbation theory we first note that if $\varepsilon = 0$ the two roots of the equation, which we will call the zeroth order solutions, are $x_0 = \pm 1$. We assume a solution of the form

$$x = x_0 + a_1\varepsilon + a_2\varepsilon^2 + a_3\varepsilon^3 + a_4\varepsilon^4 + \ldots.. \tag{10.5}$$

Substituting (10.5) into (10.1)

$$1 + (2a_1 + 1)\varepsilon + (a_1^2 + 2a_2 + a_1)\varepsilon^2 + (2a_1a_2 + 2a_3 + a_2)\varepsilon^3 + \ldots\ldots - 1 = 0 \tag{10.6}$$

where we have substituted $x_0 = 1$. Each of the coefficients of ε^n must be zero. Solving for a_n we find

$$\begin{aligned} a_1 &= -\tfrac{1}{2} \\ a_2 &= \tfrac{1}{8} \\ a_3 &= 0 \end{aligned} \tag{10.7}$$

So that the approximate solution for the root near $x = 1$ is

$$x = 1 - \frac{\varepsilon}{2} + \frac{\varepsilon^2}{8} + O\left(\varepsilon^4\right) \tag{10.8}$$

The symbol $O(\varepsilon^4)$ means that the next term in the series is of order ε^4.

Performing the same operation with $x_0 = -1$

$$1 - (1 + 2a_1)\varepsilon + (a_1^2 - 2a_2 + a_1)\varepsilon^2 + (2a_1a_2 - 2a_3 + a_2)\varepsilon^3 + \ldots. - 1 = 0 \tag{10.9}$$

Again setting the coefficients of ε^n equal to zero

$$\begin{aligned} a_1 &= -\tfrac{1}{2} \\ a_2 &= -\tfrac{1}{8} \\ a_3 &= 0 \end{aligned} \tag{10.10}$$

so that the root near $x_0 = -1$ is

$$x = -1 - \frac{\varepsilon}{2} - \frac{\varepsilon^2}{8} + O\left(\varepsilon^4\right) \tag{10.11}$$

The first three terms in (10.8) give $x = 0.951249219$, accurate to within 1.16% of the exact value while (10.11) gives the second root as $x = -1.051249219$, which is accurate to within 1.05%.

Example 10.2 Next suppose the small parameter occurs multiplied by the squared term.

$$\varepsilon x^2 + x - 1 = 0 \tag{10.12}$$

Using the quadratic formula gives the exact solution.

$$x = -\frac{1}{2\varepsilon} \pm \sqrt{\frac{1}{4\varepsilon^2} + \frac{1}{\varepsilon}} \tag{10.13}$$

If $\varepsilon = 0.1$ (10.13) gives two solutions:

$$x = 0.916079783$$

and

$$x = -10.91607983$$

We attempt to follow the same procedure that we used in Example 10.1, which we call a *regular perturbation*, to obtain an approximate solution. If $\varepsilon = 0$ identically, $x_0 = 1$. Using (10.5) with $x_0 = 1$ and substituting into (10.12) we find

$$(1 + a_1)\varepsilon + (2a_1 + a_2)\varepsilon^2 + (2a_2 + a_1^2 + a_3)\varepsilon^3 + ... = 0 \tag{10.14}$$

Setting the coefficients of $\varepsilon^n = 0$, solving for a_n and substituting into (10.5)

$$x = 1 - \varepsilon + 2\varepsilon^2 - 5\varepsilon^3 + ... \tag{10.15}$$

gives $x = 0.915$, close to the exact value. However, Eq. (10.12) clearly has two roots, and the method cannot give an approximation for the second root.

The essential problem is that the second root is not small. In fact (10.13) shows that as $\varepsilon \to 0$, $|x| \to \frac{1}{2\varepsilon}$ so that the term εx^2 is never negligible.

10.1.2 SINGULAR PERTURBATION

Arranging (10.12) in normal form

$$x^2 + \frac{x-1}{\varepsilon} = 0 \tag{10.12a}$$

and the equation is said to be singular as $\varepsilon \to 0$. If we set $x\varepsilon = u$ we find an equation for u as

$$u^2 + u - \varepsilon = 0 \tag{10.16}$$

With ε identically zero, $u = 0$ or -1. Assuming that u may be approximated by a series like (10.5) we find that

$$(-a_1 - 1)\varepsilon + (a_1^2 - a_2)\varepsilon^2 + (2a_1a_2 - a_3)\varepsilon^3 + ... = 0 \tag{10.17}$$

$$\begin{aligned} a_1 &= -1 \\ a_2 &= 1 \\ a_3 &= -2 \end{aligned} \tag{10.18}$$

so that

$$x = -\frac{1}{\varepsilon} - 1 + \varepsilon - 2\varepsilon^2 + ... \tag{10.19}$$

The three term approximation of the negative root is therefore $x = -10.92$, within .03% of the exact solution.

Example 10.3 As a third algebraic example consider

$$x^2 - 2\varepsilon x - \varepsilon = 0 \tag{10.20}$$

This at first seems like a harmless problem that appears at first glance to be amenable to a regular perturbation expansion since the x^2 term is not lost when $\varepsilon \to 0$. We proceed optimistically by taking

$$x = x_0 + a_1\varepsilon + a_2\varepsilon^2 + a_3\varepsilon^3 + ... \tag{10.21}$$

Substituting into (10.20) we find

$$x_0^2 + (2x_0a_1 - 2x_0 - 1)\varepsilon + (a_1^2 + 2x_0a_2 - 2a_1)\varepsilon^2 + ... = 0 \tag{10.22}$$

from which we find

$$\begin{aligned} &x_0 = 0 \\ &2x_0a_1 - 2x_0 - 1 = 0 \\ &a_1^2 + 2x_0a_2 - 2a_1 = 0 \end{aligned} \tag{10.23}$$

From the second of these we conclude that either $x_0 = -1$ or that there is something wrong. That is, (10.21) is not an appropriate expansion in this case.

Note that (10.20) tells us that as $\varepsilon \to 0$, $x \to 0$. Moreover, in writing (10.21) we have essentially assumed that $\varepsilon \to 0$ in such a manner that $\frac{x}{\varepsilon} \to$ constant. Let us suppose instead that as $\varepsilon \to 0$

$$\frac{x(\varepsilon)}{\varepsilon^p} \to \text{constant} \tag{10.24}$$

We then define a new variable

$$x = \varepsilon^p v(\varepsilon) \tag{10.25}$$

such that $v(0) \neq 0$. Substitution into (10.20) yields

$$\varepsilon^{2p}v^2 - 2\varepsilon^{p+1}v - \varepsilon = Q \tag{10.26}$$

where Q must be *identically* zero. Note that $\frac{Q}{\varepsilon}$ must also be zero no matter how small ε becomes, as long as it is not identically zero.

Now, if $p > 1/2$, $2p - 1 > 0$ and in the limit as $\varepsilon - 0$, $\varepsilon^{2p-1}v^2(\varepsilon) - 2\varepsilon^p v(\varepsilon) - 1 \to -1$, which cannot be true given that $Q = 0$ identically.

Next suppose p <1/2. Again, $\frac{Q}{\varepsilon^{2p}}$ is identically zero for all ε including in the limit as $\varepsilon \to 0$. In the limit as $\varepsilon \to 0$, $v(\varepsilon)^2 - \varepsilon^{1-p}v(\varepsilon) - \varepsilon^{1-2p} \to v(0)^2 \neq 0$. Thus, $p = 1/2$ is the only possibility left, so we attempt a solution with this value. Hence,

$$x = \varepsilon^{1/2}v(\varepsilon) \tag{10.27}$$

Substitution into (10.20) gives

$$v^2 - 2\sqrt{\varepsilon}v - 1 = 0 \tag{10.28}$$

and this can now be solved by a regular perturbation assuming $\beta = \sqrt{\varepsilon} << 1$. Hence,

$$v = v_0 + a_1\beta + a_2\beta^2 + a_3\beta^3 + \ldots.. \tag{10.29}$$

Inserting this into (10.28) with $\beta = \sqrt{\varepsilon}$

$$v_0 - 1 + (2v_0a_1 - 2v_0)\beta + (a_1^2 + 2v_0a_2 - 2a_1)\beta^2 + \ldots\ldots = 0 \tag{10.30}$$

Thus,

$$\begin{aligned} v_0 &= \pm 1 \\ a_1 &= 1 \\ a_2 &= +\tfrac{1}{2} \quad \text{or} \quad -\tfrac{1}{2} \end{aligned} \tag{10.31}$$

Thus, the two solutions are

$$\nu = 1 + \sqrt{\varepsilon} + \frac{\varepsilon}{2} + \ldots. \tag{10.32}$$

and

$$\nu = -1 + \sqrt{\varepsilon} - \frac{\varepsilon}{2} + \ldots. \tag{10.33}$$

The approximate solutions are

$$x = \sqrt{\varepsilon} + \varepsilon + \frac{1}{2}\varepsilon\sqrt{\varepsilon} + \ldots \tag{10.34}$$

and

$$x = -\sqrt{\varepsilon} + \varepsilon - \frac{1}{2}\varepsilon\sqrt{\varepsilon} + \ldots\ldots. \tag{10.35}$$

If $\beta = 0.1$, approximate roots are 0.1105 and -0.0805 whereas the exact solution is, .12499 and $-.0805$. Note that when $\beta = 0.1$, $\varepsilon = 0.36623$, so that ε is not particularly small.

Example 10.4 Finally, consider the third order algebraic equation,

$$\varepsilon x^3 + x - 2 = 0 \tag{10.36}$$

While the exact solution of the quadratic equation is easy to find, the solution to the cubic is not so easy. Let us see what singular perturbation theory can provide as an approximation.

First we try a regular perturbation expansion of the form

$$x = x_0 + x_1 \varepsilon + \varepsilon^2 x_2 + \ldots. \tag{10.37}$$

Substituting into Eq. (10.36) gives the following set of equations:

$$\begin{aligned} \varepsilon^0 &: x_0 = 2 \\ \varepsilon^1 &: x_1 = -x_0^3 \\ \varepsilon^2 &: x_2 = -3x_0^2 x_1 \end{aligned} \tag{10.38}$$

Thus, the regular perturbation yields only one root, which is approximated by

$$x = 2 - 8\varepsilon + 96\varepsilon^2 + O\left(\varepsilon^3\right) \tag{10.39}$$

The problem is that, once again, as $\varepsilon \to 0$ the first term is lost unless x becomes very large. But if x is very large, what must we do about the second term? Let us again explore what happens if we let

$$x = \varepsilon^p \nu(\varepsilon) \tag{10.40}$$

Equation (10.36) becomes

$$\varepsilon^{1+3p} \nu^3(\varepsilon) + \varepsilon^p \nu(\varepsilon) - 2 = 0 \tag{10.41}$$

As $\varepsilon \to 0$ the first two terms retain the same orders of magnitude only if $p = -1/2$. In this case,

$$\nu^3 + \nu - 2\varepsilon^{1/2} = 0 \tag{10.42}$$

We try an expansion of the form

$$\nu = \left(a_0 + \varepsilon^{1/2} a_1 + \varepsilon a_2 + \ldots..\right) \tag{10.43}$$

Substituting into ....,

$$\left(a_0^3 + a_0\right) + \varepsilon^{1/2}\left(3a_1 a_0^2 + a_1 - 2\right) + \varepsilon\left(3a_0^2 a_2 + 3a_1^2 a_0 + a_2\right) + O\left(\varepsilon^{3/2}\right) = 0 \tag{10.44}$$

Equating terms of the same order,

$$\begin{aligned} a_0 &= 0 \text{ or } \pm i \\ a_1 &= -1 \\ a_2 &= \pm \tfrac{3}{2} i \end{aligned} \tag{10.45}$$

The yields two complex roots:

$$\begin{aligned} & i\left(\varepsilon^{-1/2} + \tfrac{3}{2}\varepsilon^{1/2}\right) - 1 \\ & -i\left(\varepsilon^{-1/2} + \tfrac{3}{2}\varepsilon^{1/2}\right) - 1 \end{aligned} \tag{10.46}$$

Problems

Using appropriate perturbation analysis and for small ε

1. Find the approximate roots of

   $x^2 - \varepsilon x - 1 = 0$

2. Find the approximate roots of

   $\varepsilon x^2 + 2x + 1 = 0$

3. Find the approximate roots of

   $\varepsilon x^3 - x^2 + 1 = 0$

   and evaluate for $\varepsilon = 0.1$ and $\varepsilon = .01$

   Answer:

$$\varepsilon = .1, -0.995, 1.057, 9.898$$
$$\varepsilon = .01, -0.995, 1.005, 99.990$$

10.2 EXAMPLES FROM ORDINARY DIFFERENTIAL EQUATIONS

We now introduce some ideas about approximate solutions of ordinary differential equations using perturbation methods. We first consider two examples of problems that are solvable by regular perturbation methods. Then we illustrate the failure of the regular perturbation method to yield an approximate solution in two simple equations before moving on in Chapter 11 to more fully discuss singular methods.

Example 10.5 Without going into the details of the physical derivation, the following equation and initial condition describe the cooling by convection of a small object that has a heat capacity that varies slightly with temperature.

$$\begin{aligned} (1 + \varepsilon x)\tfrac{dx}{dt} + x &= 0 \\ x(0) &= 1 \end{aligned} \tag{10.47}$$

(The physical heat transfer derivation can be found in the book by Aziz and Na at the end of this chapter.)

One might think of x as a dimensionless temperature departure from that of the environment. The initial dimensionless temperature is 1 and ε is a small dimensionless parameter. As is usual in the regular perturbation approach we begin by taking

$$x = x_0 + \varepsilon x_1 + \varepsilon^2 x_2 + O\left(\varepsilon^3\right) \tag{10.48}$$

Inserting this into (10.47) results in

$$\frac{dx_0}{dt}+x_0+\varepsilon\left(\frac{dx_1}{dt}+x_1+x_0\frac{dx_0}{dt}\right)+\varepsilon^2\left(\frac{dx_2}{dt}+x_2+x_0\frac{dx_1}{dt}+x_1\frac{dx_0}{dt}\right)+O\left(\varepsilon^3\right)=0 \tag{10.49}$$

Thus,

$$\begin{aligned}&\tfrac{dx_0}{dt}+x_0=0,\, x_0(0)=1\\&\tfrac{dx_1}{dt}+x_1+x_0\tfrac{dx_0}{dt}=0,\, x_1(0)=0\\&\tfrac{dx_2}{dt}+x_2+x_0\tfrac{dx_1}{dt}+x_1\tfrac{dx_0}{dt}=0.\;\; x_2(0)=0\end{aligned} \tag{10.50}$$

The solution of the first of these is

$$x_0 = e^{-t} \tag{10.51}$$

The second equation (associated with the ε term) is

$$x_1 = e^{-t} - e^{-2t} \tag{10.52}$$

In the case of the third equation we note that

$$x_2 = e^{-t} - 2e^{-2t} + e^{-3t} \tag{10.53}$$

The approximate solution is then

$$x = e^{-t} + \varepsilon\left(e^{-t} - e^{-2t}\right) + \varepsilon^2\left(e^{-t} - 2e^{-2t} + e^{-3t}\right) + O\left(\varepsilon^3\right) \tag{10.54}$$

Note that the equation can be easily solved by separation of variables and the exact solution is

$$\ln x + \varepsilon(x - 1) = -t \tag{10.55}$$

(We comment here that in (10.55), x is a rather complicated implicit function of t. Equations like (10.54) are often easier to work with and, as long as they are sufficiently accurate, preferred. The author strongly feels that a good approximate solution that is simple is often more useful than a complicated exact solution.)

Example 10.6 Next consider the case of a spring mass damper system with a small spring constant. The mass is initially displaced by a non-dimensional amount $x(0) = 1$ and with a non-dimensional velocity $\frac{dx}{dt}(0) = 0$. The differential equation and initial conditions are as follows:

$$\begin{aligned}&\tfrac{d^2x}{dt_2} + \tfrac{dx}{dt} + \varepsilon x = 0\\&x(0) = 1\\&\tfrac{dx}{dt}(0) = 0\end{aligned} \tag{10.56}$$

We attempt a regular perturbation solution in the form

$$x = x_0 + \varepsilon x_1 + \varepsilon^2 x_2 + O\left(\varepsilon^3\right) \tag{10.57}$$

Substituting in the usual way we find

$$\begin{aligned} &\frac{d^2x_0}{dt^2} + \frac{dx_0}{dt} + \varepsilon\left(\frac{d^2x_1}{dt^2} + \frac{dx_1}{dt} + x_0\right) + O\left(\varepsilon^2\right) = 0 \\ &x_0(0) = 1, \quad \frac{dx_0}{dt}(0) = 0 \\ &x_1(0) = \frac{dx_1}{dt}(0) = 0 \end{aligned} \tag{10.58}$$

(It is clear to the reader at this point that it is easy to solve the original differential equation, since it is linear and rather easy to solve by Laplace transforms or other means. We continue merely to illustrate the perturbation method.)

Solving,

$$\begin{aligned} x_0 &= 1 \\ x_1 &= 2 - t - e^{-t} \end{aligned} \tag{10.59}$$

We can see a problem arising here that often arises when using approximate methods. As long as t is sufficiently small, the solution,

$$x = 1 - \varepsilon\left(t + e^{-t} - 2\right) + O\left(\varepsilon^2\right) \tag{10.60}$$

is accurate. However, when t is larger than, perhaps, ε, the result is not useful. In fact, when t approaches infinity, x approaches negative infinity, which clearly cannot be true. Keeping the ε^2 term helps, but not much.

Example 10.7 Next consider the case of the mass, spring damper system in which the mass is small. Initially the mass is at rest, $x(0) = \frac{dx}{dt} = 0$, and a sudden force, a dimensionless impulsive force of F(0) = $\delta(0)$, is applied. We represent this force as $1/\varepsilon$. The appropriate differential equation and initial conditions are

$$\begin{aligned} &\varepsilon\frac{d^2x}{dt^2} + \frac{dx}{dt} + x = 0 \\ &x(0) = 0 \\ &\frac{dx}{dt}(0) = {}^1\!/_\varepsilon \end{aligned} \tag{10.61}$$

Physically, when ε is very small, this corresponds to an impulsive force acting on the mass, giving it a large initial speed. From experience with the algebraic equations, we might expect that, with the small term multiplying the second derivative term, a regular perturbation approach might not work. Such will herewith be demonstrated.

Begin by assuming a regular perturbation approach with

$$x = x_0 + \varepsilon x_1 + O\left(\varepsilon^2\right) \tag{10.62}$$

$$\varepsilon \frac{d^2 x_0}{dt^2} + \varepsilon^2 \frac{d^2 x_1}{dt^2} + \frac{dx_0}{dt} + \varepsilon \frac{dx_1}{dt} + x_0 + \varepsilon x_1 = 1 \tag{10.63}$$

The first order equation, the equation when $\varepsilon = 0$, is

$$\frac{dx_0}{dt} + x_0 = 0 \tag{10.64}$$

The second derivative term has been eliminated, so that the solution cannot satisfy the initial conditions, which require both the terms on the left hand side of the equation to be zero at $t = 0$. Clearly, the regular perturbation approach will not work.

We now turn to an exposition of the classical use of singular perturbation methods in the useful approximate solutions of boundary value problems.

Problems

Use regular perturbation theory to find an approximate solution for small ε

10.4. $y'' + \varepsilon y' + 1 = 0, \; y(0) = 0, \; y'(0) = 1$

10.5. $y' + y + \varepsilon y^2 = x, \; y(1) = 1$

REFERENCES

[1] Aziz, A. and T. Y. Na, *Perturbation Methods in Heat Transfer*, Hemisphere Pub. Co., New York. 1984. Cited on page(s)

[2] Simmonds, J. G. and J. E. Mann, *A First Look at Perturbation Theory*, Robert E. Krieger Pub. Co., Melebar, Florida, 1986. Cited on page(s)

CHAPTER 11

Singular Perturbation Theory of Differential Equations

A singular perturbation approach is usually required when a small coefficient is multiplied by the highest derivative term, so that in some region of the domain, the dependent variable changes so rapidly that this term, although multiplied by a small parameter, cannot be considered to be negligible. This often (but not always) occurs at a boundary of the domain. The approach is usually the following:

The independent variable, say t, is replaced by a variable that depends on the small parameter, say ε, by

$$\eta = \vartheta(\varepsilon)\, x \tag{11.1}$$

and $\varphi(\varepsilon)$ is chosen to remove the small parameter from the highest derivative term. This should allow a regular perturbation expansion to be performed within the region in which the independent variable changes rapidly, which is called the *inner region*. Next, a regular perturbation is performed with the result generally being appropriate for the region outside that where the dependent variable changes rapidly. This is the *outer region*. An intermediate region is now identified in which the inner and outer solutions form a *common solution* in some sense (best demonstrated by the examples below). This is referred to as the *matching problem*. A *uniformly valid solution* can now be obtained as the sum of the inner and outer solutions minus the common solution.

We begin our illustrative examples by reconsidering the problem introduced in Example 10.7 in the last chapter, which is often used as an introduction to some of the fundamental problems in the concept of singular perturbation theory.

Example 11.1 We begin with Eq. (10.61). The regular perturbation method has failed because the first approximation yielded a first order differential equation whose boundary conditions could not both be satisfied. If we change the independent variable to

$$t = \varepsilon\, \tau \tag{11.2}$$

Equation (10.61) becomes

$$\begin{aligned} &\frac{d^2x}{d\tau^2} + \frac{dx}{d\tau} + \varepsilon\, x = 0 \\ &x(0) = 0 \\ &\frac{dx}{d\tau} = 1 \end{aligned} \tag{11.3}$$

Now let

$$x = x_0 + \varepsilon x_1 + O\left(\varepsilon^2\right) \tag{11.4}$$

Equating terms multiplied by powers of ε yields a first approximation

$$\begin{aligned} &\frac{d^2x_0}{d\tau^2} + \frac{dx_0}{d\tau} = 0 \\ &x_0(0) = 0 \\ &\frac{dx_0}{d\tau}(0) = 1 \end{aligned} \tag{11.5}$$

The solution is

$$x_0 = 1 - e^{-\tau} \tag{11.6}$$

The differential equation for x_1 is

$$\begin{aligned} &\frac{d^2x_1}{d\tau^2} + \frac{dx_1}{d\tau} + x_0 = 0 \\ &\frac{d^2x_1}{d\tau^2} + \frac{dx_1}{d\tau} = e^{-\tau} - 1 \\ &x_1(0) = \frac{dx_1}{d\tau} = 0 \end{aligned} \tag{11.7}$$

Solution gives

$$x_1 = 2\left(1 - e^{-\tau}\right) - \tau\left(1 + e^{-\tau}\right) \tag{11.8}$$

The solution is good for $t = \varepsilon\tau$ small. It is called the *inner solution.*

When t is large, such that τ is of order $1/\varepsilon$ or larger, the second derivative becomes increasingly negligible, so that a regular perturbation expansion is appropriate. Hence, referring back to (10.63) and (10.64), and solving for the first two approximations,

$$\begin{aligned} x_0 &= A_0 e^{-t} \\ x_1 &= A_1 e^{-t} - A_0 e^{-t} \end{aligned} \tag{11.9}$$

This is referred to as the *outer solution*, the solution that is appropriate when t is large.

It is possible to obtain a *uniformly valid solution* that satisfies both the inner solution and the outer solution. This is called the *matching problem.*

We proceed in the following manner.

It seems reasonable that the inner and outer solutions should agree in some since in some overlap region that is intermediate between the inner and outer regions. If t is of order $t = O(\varepsilon)$ then t is within the inner solution region, while if t is of order $t = O(1)$ the t is within the outer solution region. We might expect, then, that an overlap region will be characterized by values of t for which $t = O(\sqrt{\varepsilon})$. Since $\sqrt{\varepsilon}$ approaches zero less rapidly than does ε, we introduce an intermediate time scale

$$\eta = \frac{t}{\sqrt{\varepsilon}} \tag{11.10}$$

To obtain a solution that is common between the inner and outer solutions we require that

$$\lim_{\varepsilon \to 0} x_{\text{inner}}\left(\sqrt{\varepsilon}\eta\right) = \lim_{\varepsilon \to 0} x_{\text{outer}}\left(\sqrt{\varepsilon}\eta\right) \tag{11.11}$$

This results in

$$\lim_{\varepsilon \to 0}\left(1 - e^{\eta/\sqrt{\varepsilon}}\right) = \lim_{\varepsilon \to 0} A_0 e^{\eta\sqrt{\varepsilon}}, \quad \text{so that } A_0 = 1 \tag{11.12}$$

The uniform solution is obtained by summing the inner and outer solutions and subtracting the overlap.

$$x = 1 - e^{-t/\varepsilon} + e^{-t} - 1 = e^{-t} - e^{-t/\varepsilon} \tag{11.13}$$

Obtaining the second approximation is left as an exercise.

Example 11.2 Next consider the differential equation

$$\begin{aligned} &\varepsilon u'' - (2 - x^2)u = -1, \ 0 \le x \le 1 \\ &\varepsilon \langle\langle 1 \\ &u(0) = 0, \ u(1) = 1 \end{aligned} \tag{11.14}$$

It should be clear to the reader that a regular perturbation will not produce a useful solution because the small term ε is multiplied by the second derivative term, which would be eliminated if the regular perturbation were used, so the boundary conditions could not be satisfied. We search for a useful singular perturbation approach.

Let

$$\begin{aligned} &x = \varphi(\varepsilon)\,\varsigma \\ &\frac{\varepsilon}{\vartheta^2}\frac{d^2u}{d\varsigma^2} - \left(2 - \vartheta^2\varsigma^2\right)u = -1 \end{aligned} \tag{11.15}$$

If we choose $\vartheta = \varepsilon^{1/2}$ the differential equation becomes

$$\frac{d^2u}{d\varsigma^2} - (2 - \varepsilon\varsigma^2)u = -1 \tag{11.16}$$

with boundary conditions

$$u(0) = 0, \quad u\left(\varsigma = \varepsilon^{-1/2}\right) = 1 \tag{11.17}$$

For the inner solution

$$u_{\text{inner}} = f_0 + \varepsilon f_1 + O\left(\varepsilon^2\right) \tag{11.18}$$

$$f_0'' + \varepsilon f_1'' + \ldots.. - 2f_0 + 2\varepsilon\varsigma^2 f_0 - 2\varepsilon f_1 + \ldots. = -1 \tag{11.19}$$

Thus,

$$f_0'' - 2f_0 = -1 \tag{11.20}$$

This is a two-point boundary value problem, so we must be careful. The solution for f_0 needs only to satisfy the boundary condition at $\varsigma = 0$.

The general solution for f_0 is

$$f_0 = \frac{1}{2} + Ae^{\sqrt{2}\varsigma} + Be^{-\sqrt{2}\varsigma} \tag{11.21}$$

To satisfy the boundary condition at $x = 0$ we must have $A + B = -1/2$.

For the solution when the second derivative term in the original equation is small (i.e., when x is large enough that $\sqrt{\varepsilon}x$ is no longer small), the outer solution is

$$u_{\text{outer}} = g = \frac{1}{2 - x^2} \tag{11.22}$$

We now need to match the solutions at some intermediate value of x, and then obtain a uniformly valid approximate solution. In the limit as x approaches zero, g, the outer solution, approaches $1/2$, while when $x = 1$, the boundary condition at $x = 1$, $u(1) = 1$, is satisfied. If we choose $A = 0$ and $B = -1/2$, the limit as $\varepsilon \to 0$ for the inner solution and $\varepsilon \to 0$ for the outer solution yields

$$\frac{\lim}{\varepsilon \to 0}\left(\frac{1}{2} - \frac{1}{2}e^{-\sqrt{2/\varepsilon^x}}\right) = \frac{\lim}{x \to 0}\frac{1}{2 - x^2} = \frac{1}{2} \tag{11.23}$$

The uniformly valid solution is given by the sum of the inner and outer solutions minus the common part.

$$u_0 = f_0 + g_0 - \frac{1}{2} = \frac{1}{2 - x^2} - \frac{1}{2}e^{-x\sqrt{2/\varepsilon}} \tag{11.24}$$

Example 11.3 (a problem from Carrier): Consider now the differential equation

$$\begin{aligned} &\varepsilon\frac{d^2u}{dx^2} - (2 - x^2)u = -1 \\ &u(-1) = u(1) = 0 \end{aligned} \tag{11.25}$$

As in the above example, if ε were identically zero, the solution would be

$$u = \frac{1}{2 - x^2} \tag{11.26}$$

However, this solution cannot satisfy either of the two boundary conditions and, therefore, cannot be an accurate solution near either of the two boundaries. We conclude that there are boundary layers near the boundaries where the solution changes rapidly in order to satisfy the boundary conditions. Taking the above function to be an outer solution that is valid away from the boundaries, we assume that an approximate solution takes the form

$$u_0 + w(x, \varepsilon) + v(x, \varepsilon) \tag{11.27}$$

where $w(x, \varepsilon)$ vanishes rapidly away from $x = -1$ and $v(x, \varepsilon)$ vanishes rapidly away from $x = +1$. Near $x = -1$ it makes sense to define a new variable

$$\xi = (1 + x)\vartheta(\varepsilon) \tag{11.28}$$

and

$$w = W(\xi) \tag{11.29}$$

(The choice shifts the origin of the variable such that $x = -1$ corresponds to $\xi = 0$.)

Similarly, we define

$$\eta = (1 - x)\psi(\varepsilon) \tag{11.30}$$

and

$$v = V(\eta) \tag{11.31}$$

so that

$$u = u_0(x) + W(\xi) + V(\eta) \tag{11.32}$$

$$\varepsilon u_{0_{xx}} + \varepsilon\vartheta^2(\varepsilon)W_{\xi\xi} + \varepsilon\psi^2(\varepsilon)V_{\eta\eta} - (2 - x^2)\left[u_0 + W(\xi) + V(\eta)\right] = -1 \tag{11.33}$$

Near $\xi = 0$, and $\eta = 0$, $\varepsilon u_{0xx} = O(\varepsilon)$ and if we choose

$$\vartheta(\varepsilon) = \psi(\varepsilon) = \sqrt{\varepsilon} \tag{11.34}$$

We find that near $\xi = 0$ or $\eta = 0$

$$W_{\xi\xi} - (2 - x^2)W(\xi) + V_{\eta\eta} - (2 - x^2)V(\eta) = 0 \tag{11.35}$$

to order ε.

Near $x = -1$ or $x = +1(2 - x^2)$ is of order 1.

Near $\xi = 0$, we consider the equation

$$W_{\xi\xi} - W = 0 \tag{11.36}$$

with solution satisfying the boundary condition $W(0) = -1$, with a bounded solution. The result is

$$W = -e^{-(1+x)/\sqrt{\varepsilon}} \tag{11.37}$$

By symmetry,

$$V = -e^{-(1-x)/\sqrt{\varepsilon}} \tag{11.38}$$

The complete solution is then

$$u = \frac{1}{2 - x^2} - \exp\left(-\frac{x + 1}{\sqrt{\varepsilon}}\right) - \exp\left(-\frac{1 - x}{\sqrt{\varepsilon}}\right) \tag{11.39}$$

Example 11.4 Consider now a slightly more general linear second order differential equation

$$\varepsilon y'' + a(x)y' + b(x)y = 0, \quad 0 < x < 1, \quad y(0) = 0, \quad y(1) = 1 \tag{11.40}$$

where $\varepsilon << 1$ and both $a(x)$ and $b(x)$ are continuously differentiable.

It should be clear to the reader that a regular perturbation approach would lead to elimination of the second derivative term, and would not lead to a useful result. Example 11.1 hints that we should choose a new variable as

$$\xi = \varepsilon^{-1} g(x), \quad g(0) = 0 \tag{11.41}$$

Thus,

$$\begin{aligned} y &= y(x, \xi, \varepsilon) \\ y' &= y_x + y_\xi \varepsilon^{-1} g'^2 \\ y'' &= y_{xx} + 2y_{x\xi}\varepsilon^{-1} g' + y_\xi \varepsilon^{-1} g'' + y_{\xi\xi}\varepsilon^{-2} g'^2 \end{aligned} \tag{11.42}$$

Here the subscripts represent differentiation with respect to the subscripted variable.

Substituting into (11.40),

$$\begin{aligned} &g'^2 y_{\xi\xi} + ag' y_\xi + \varepsilon(2y_{x\xi} g' + y_\xi g'' + ay_x + +by) + \varepsilon^2 y_{xx} = 0 \\ &y(0, 0, \varepsilon) = 0, \, y\left(1, \varepsilon^{-1} g(1), \varepsilon\right) = 1 \end{aligned} \tag{11.43}$$

Now we assume that

$$y(x, \xi, \varepsilon) = Y_0(x, \xi) + \varepsilon Y_1(x, \xi) + \ldots \tag{11.44}$$

Equating to zero terms of successive powers of ε we find

$$g'^2 Y_{0,\xi\xi} + ag' Y_{0,\xi} = 0, \quad Y_0(0, 0) = 0, \quad Y_0(1, \varepsilon^{-1} g(1)) = 1 \tag{11.45}$$

$$\begin{aligned} &g'^2 Y_{1,\xi\xi} + ag' Y_{1,\xi} + 2g' Y_{0,x\xi} + g'' Y_{0,\xi} + aY_{0,x} + bY_0 = 0, \\ &Y_1(0, 0) = 0, \quad Y_1(1, \varepsilon^{-1} g(1)) = 0 \end{aligned} \tag{11.46}$$

The zeroth order solutions simplifies if we let

$$a(x) = g', \quad g(x) = \int_0^x a(x)\, dx \tag{11.47}$$

Solving (11.45),

$$Y_0(x, \xi) = A_0(x) + B_0(x) \exp(-\xi) \tag{11.48}$$

The first boundary condition in (11.45) gives

$$A_0(0) + B_0(0) = 0 \tag{11.49}$$

At $x = 1$, $\exp(-\varepsilon^{-1}\int_0^1 a(x)dx)$, so that $\exp(-\xi)$ is transcendentally small. Thus,

$$A_0(1) = 1 \tag{11.49a}$$

The differential equation for Y_1 can now be written

$$Y_{1,\xi\xi} + Y_{1,\xi} = \exp^{-\xi}\left[\frac{B_0'}{a} + \frac{(a'-b)\,B_0}{a^2}\right] - \left(\frac{A_0'}{a} + \frac{bA_0}{a^2}\right) \tag{11.50}$$

If the right-hand side of this equation is nonzero, the solution will involve terms like $F(x)\xi e - \xi + G(x)\xi$. These are called resonant terms, and since $\xi = g(x)/\varepsilon$ and we are interested in the case where $\varepsilon \to 0$ the right-hand side of (11.50) must be identically zero.

Thus,

$$\begin{aligned} &aA_0' + bA_0 = 0 \\ &\text{and} \\ &aB_0' + (a' - b)B_0 = 0 \end{aligned} \tag{11.51}$$

The first of these has the solution

$$A_0(x) = \exp\int_x^1 [b(t)a(t)]\,dt \tag{11.52}$$

while the second yields

$$B_0(x) = [C_0/a(x)]\exp\left[\int_0^x (b(t)/a(t))\,dt\right] \tag{11.53}$$

C_0 can be determined by applying (11.49) with A_0 given by (11.52) and we find that

$$C_0 = -a(0)\exp\int_0^1 [b(t)/a(t)]\,dt \tag{11.54}$$

The zeroth order solution is

$$\begin{aligned} Y_0(x,\xi) = &\exp\int_x^1 [b(t)/a(t)]\,dt \\ &- \frac{a(0)}{a(x)}e^{-\xi}\exp\int_0^1 [b(t)/a(t)]\,dt\,\exp\left[\int_0^x [b(t)/a(t)]\,dt\right] \end{aligned} \tag{11.55}$$

Example 11.5 In some cases boundary layers occur in the interior of the solution domain. Such is the case with the following equation. Consider

$$\varepsilon^2 y'' + 2xy' - 2y = 0, \quad -1 < x < 1, \quad y(-1) = A, \quad y(1) = B \tag{11.56}$$

With $\varepsilon = 0$ identically, the reduced equation yields $y = Cx$, and the boundary condition at $x = -1$ requires that $C = -A$. On the other hand, applying the boundary condition at $x = 1$ gives $C = B$. The implication is that

$$\begin{aligned} y &= -Ax, \quad x < 0 \\ y &= Bx, \quad x > 0 \end{aligned} \tag{11.57}$$

From this we deduce that at $x = 0$ there is a cusp, a point where the first derivative is discontinuous. We look for a solution near $x = 0$ in which the leading term is not negligible. Let

$$\xi = x/\varepsilon \tag{11.58}$$

Then

$$\frac{d^2y}{d\xi^2} + 2\xi\frac{dy}{d\xi} - 2y = 0, \quad -\varepsilon^{-1} < \xi < \varepsilon^{-1} \tag{11.59}$$

One solution of this equation is clearly $y = \xi$. The other is then of the form (using variation of parameters)

$$y = \xi u(\xi) \tag{11.60}$$

Substituting into (11.59)

$$\xi u'' + 2(1 + \xi^2)u' = 0 \tag{11.61}$$

and integrating once,

$$u' = c_1\xi^{-2}\exp\left(-\xi^2\right) \tag{11.62}$$

A second integration gives

$$u = c_2 + c_1\int_\xi^\infty \frac{\exp(-t^2)}{t^2}\,dt \tag{11.63}$$

Integration by parts gives

$$\begin{aligned} u &= C + D\left[\frac{1}{\sqrt{\pi}}\exp\left(-\xi^2\right) + \frac{2}{\sqrt{\pi}}\int_\xi^\infty \exp\left(-t^2\right)\,dt\right] \\ &= C + D\left[\frac{1}{\sqrt{\pi}}\exp\left(-\xi^2\right) + \operatorname{erf}(\xi)\right] \end{aligned} \tag{11.64}$$

where $\operatorname{erf}(\xi)$ is the error function.

Now

$$y = Ex + F\left[\frac{\varepsilon}{\sqrt{\pi}}\exp\left(-x^2/\varepsilon^2\right) + x\operatorname{erf}(x/\varepsilon)\right] \tag{11.65}$$

Imposing the boundary conditions given in (11.56) and neglecting terms multiplied by ε,

$$y = x\left[\frac{B-A}{2} + \frac{B+A}{2}\operatorname{erf}(x/\varepsilon)\right] - \frac{B+A}{2\sqrt{\pi}}\varepsilon\exp\left(-x^2/\varepsilon^2\right) \tag{11.66}$$

Example 11.6 Finally, we consider a partial differential equation that describes the motion of the fluid in semicircular can of the human ear. The symbol u represents the velocity of the fluid inside the semicircular can relative to the wall of the canal. For the case of an impulse response (a sudden movement of the head) the equation describing the velocity of the fluid can be written

$$\frac{\partial u}{\partial t} + \left(1 + \frac{\gamma}{\beta}\right)\delta(t) = \frac{1}{r}\frac{\partial}{\partial r}\left(r\left(\frac{\partial u}{\partial r}\right)\right) - \varepsilon \int_0^t \int_0^1 u\, r\, dr\, dt \tag{11.67}$$

The symbols γ and β represent properties of the canal whereas ε also contains properties of the canal fluid. If you are interested in the derivation of (11.67) you can find it in the article by van Buskirk, Watts and Liu found at the end of this chapter. For the human ear γ/β is of order unity and ε is very small, of order 0.02.

The boundary and initial conditions state that the velocity is zero at the wall ($r = 1$), the slope of the velocity is zero at the midpoint of the channel ($r = 0$) and the velocity is initially zero ($t = 0$).

$$u(1, t) = u(r, 0) = \frac{\partial u}{\partial r}(0, t) = 0 \tag{11.68}$$

For *very small values of time* the regular expansion

$$u = u_0 + \varepsilon u_1 + \varepsilon^2 u_2 + \ldots. \tag{11.69}$$

leads to the following partial differential equation for u_0.

$$\frac{\partial u_o}{\partial t} + \left(1 + \frac{\gamma}{\beta}\right)\delta\,(t) = \frac{1}{r}\frac{\partial}{\partial r}\left(r\,\frac{\partial u_0}{\partial r}\right) \tag{11.70}$$

The solution is easily obtained by Laplace transforms.

$$u_0 = -2\left(1 + \frac{\gamma}{\beta}\right)\sum_{n=1}^{\infty}\frac{\exp\left(-\lambda_n^2\, t\right) J_0\,(\lambda_n\, r)}{\lambda_n\, J_1\,(\lambda_n)} \tag{11.71}$$

It is useful to determine the volume flow rate integral as

$$V = \int_0^t \int_0^1 u\, r\, dr\, dt \tag{11.72}$$

This quantity represents the total displacement of the fluid. In the present case the integration is easily performed with the result:

$$V_0 = -2\left(1 + \frac{\gamma}{\beta}\right)\sum_{n=1}^{\infty}\frac{1 - \exp\left(-\lambda_n^2\, t\right)}{\lambda_n^4} \tag{11.73}$$

Note now why the present system is singular. It is simply because for small values of time (and ε) the integral term in (11.67) can be neglected as a zeroth approximation, while for larger times, regardless of the (non-zero) value of ε, the integral term can be dominant.

For *large times* the velocity will be changing very slowly. We then let $\tau = \varepsilon t$, the stretched independent variable. Substituting into (11.67) we find

$$\varepsilon \frac{\partial u}{\partial \tau} + \left(1 + \frac{\gamma}{\beta}\right) \delta\,(\tau/\varepsilon) = \frac{1}{r}\left(r \frac{\partial u}{\partial r}\right) - \int_0^{\tau} \int_0^1 ur\,dr\,d\tau \tag{11.74}$$

The integral can now be written

$$\int_0^{\nu} \int_0^1 u\,r\,dr\,d\tau + \int_{\nu}^{\tau} \int_0^1 u\,r\,dr\,d\tau \tag{11.75}$$

When $\varepsilon << \nu(\varepsilon) << 1$, and if ε is sufficiently small then for all practical purposes the first of these integrals can be written

$$\int_0^{\nu} \int_0^1 u\,r\,dr\,d\tau = -2\left(1 + \frac{\gamma}{\beta}\right) \varepsilon \sum_{n=1}^{\infty} \frac{1}{\lambda_n^4} \tag{11.76}$$

so that to a zeroth approximation (11.74) can be written

$$\frac{1}{r}\frac{\partial}{\partial r}\left(r \frac{\partial u_0}{\partial r}\right) = \int_{\nu(\varepsilon)}^{\tau} \int_0^1 u_0\,r\,dr\,d\tau - 2\left(1 + \frac{\gamma}{\beta}\right) \varepsilon \sum_{n=1}^{\infty} \left(\frac{1}{\lambda_n^4}\right) \tag{11.77}$$

Equation (11.77) is easily solved using Laplace transforms and the result is

$$u_0 = \frac{\varepsilon}{4} \sum_{n=1}^{\infty} \left(\frac{2\,(1 + \gamma/\beta)}{\lambda_n^4}\right) \left(1 - r^2\right) \exp\,(-\,\varepsilon\,t\,/\,16) \tag{11.78}$$

The corresponding volume displacement is

$$V_0 = \int_{\nu(\varepsilon)}^{t} \int_0^1 u_0\,r dr\,dt = \left[\exp(-\,\nu(\varepsilon)\,) - \exp(\,t\varepsilon/16)\right] \sum_{n=1}^{\infty} \frac{2\,(1 + \gamma/\beta)}{\lambda_n^4} \tag{11.79}$$

and since we have assumed that $\nu(\varepsilon) << 1$

$$V_0 = \left[1 - \exp(t\varepsilon/16)\right] \sum_{n=1}^{\infty} \frac{2(1 + \gamma/\beta)}{\lambda_n^4} \tag{11.80}$$

This is the solution for large values of time. Equations (11.73) and (11.80) can now be combined to give the uniformly valid zeroth approximation.

$$V_0 = \sum_{n=1}^{\infty} \frac{2\,(1 + \gamma/\beta)}{\lambda_n^2} \left[\exp\,(-\lambda_n^2\,t) - \exp(-\,t\varepsilon/16\,)\right] \tag{11.81}$$

Problems

1. Find an approximate solution for small ε.

   (a) $\varepsilon y' + y = 0, \quad y(0) = 1$

   (b) $\varepsilon y'' + (1+\varepsilon)y' + y = 0, \quad y(0) = 0, \quad y(1) = 1$

   (c) Show that the exact solution of (11.25) is

   (d) Consider

$$\varepsilon^2 y'' + y' + y + x = 1, \quad 0 < x < 1, \quad y(0) = 0, \quad y(1) = 0.$$

   Noting that when ε is identically zero the solution is $y = 2 - x$, neither boundary condition can be satisfied. Singularities (boundary layers) occur at both $x = 0$ and $x = 1$. Find the zeroth order approximate solution.

   (e) Graph Equations (11.39) and (11.66)

REFERENCES

[1] Carrier, G. F. and C. E, Pearson, *Ordinary Differential Equations*, Blaisdell Pub. Co., Waltham, Mass. Cited on page(s)

[2] Van Buskirk, W. C., R. G. Watts, and Y. K. Liu, "The Fluid Mechanics of the Semicircular Canals," *J. Fluid Mech.*, vol. 78, Part 1, pp. 87–98, 1976. Cited on page(s)

[3] Simmonds, J. G. and J. E. Mann, *A First Look at Perturbation Theory*, Robert E. Krieger Pub. Co., Melebar, Florida, 1986. Cited on page(s)

[4] Ali Hasan Nayfeh, *Perturbation Methods*, Wiley, Weinham, Germany, 2007, 420 pp. Cited on page(s)

[5] Alan W. Bush, *Perturbation Methods for Engineers and Scientists*, CRC Press, 1992, 303 pp. Cited on page(s)

Appendix A: The Roots of Certain Transcendental Equations

TABLE A.1: The first six roots, † α_n, of

$$\alpha \tan \alpha + C = 0.$$

C	α_1	α_2	α_3	α_4	α_5	α_6
0	0	3.1416	6.2832	9.4248	12.5664	15.7080
0.001	0.0316	3.1419	6.2833	9.4249	12.5665	15.7080
0.002	0.0447	3.1422	6.2835	9.4250	12.5665	15.7081
0.004	0.0632	3.1429	6.2838	9.4252	12.5667	15.7082
0.006	0.0774	3.1435	6.2841	9.4254	12.5668	15.7083
0.008	0.0893	3.1441	6.2845	9.4256	12.5670	15.7085
0.01	0.0998	3.1448	6.2848	9.4258	12.5672	15.7086
0.02	0.1410	3.1479	6.2864	9.4269	12.5680	15.7092
0.04	0.1987	3.1543	6.2895	9.4290	12.5696	15.7105
0.06	0.2425	3.1606	6.2927	9.4311	12.5711	15.7118
0.08	0.2791	3.1668	6.2959	9.4333	12.5727	15.7131
0.1	0.3111	3.1731	6.2991	9.4354	12.5743	15.7143
0.2	0.4328	3.2039	6.3148	9.4459	12.5823	15.7207
0.3	0.5218	3.2341	6.3305	9.4565	12.5902	15.7270
0.4	0.5932	3.2636	6.3461	9.4670	12.5981	15.7334
0.5	0.6533	3.2923	6.3616	9.4775	12.6060	15.7397
0.6	0.7051	3.3204	6.3770	9.4879	12.6139	15.7460
0.7	0.7506	3.3477	6.3923	9.4983	12.6218	15.7524
0.8	0.7910	3.3744	6.4074	9’5087	12.6296	15.7587

TABLE A.1: (*continue*)

$\alpha \tan \alpha + C = 0.$						
C	α_1	α_2	α_3	α_4	α_5	α_6
0.9	0.8274	3.4003	6.4224	9.5190	12.6375	15.7650
1.0	0.8603	3.4256	6.4373	9.5293	12.6453	15.7713
1.5	0.9882	3.5422	6.5097	9.5801	12.6841	15.8026
2.0	1.0769	3.6436	6.5783	9.6296	12.7223	15.8336
3.0	1.1925	3.8088	6.7040	9.7240	12.7966	15.8945
4.0	1.2646	3.9352	6.8140	9.8119	12.8678	15.9536
5.0	1.3138	4.0336	6.9096	9.8928	12.9352	16.0107
6.0	1.3496	4.1116	6.9924	9.9667	12.9988	16.0654
7.0	1.3766	4.1746	7.0640	10.0339	13.0584	16.1177
8.0	1.3978	4.2264	7.1263	10.0949	13.1141	16.1675
9.0	1.4149	4.2694	7.1806	10.1502	13.1660	16.2147
10.0	1.4289	4.3058	7.2281	10.2003	13.2142	16.2594
15.0	1.4729	4.4255	7.3959	10.3898	13.4078	16.4474
20.0	1.4961	4.4915	7.4954	10.5117	13.5420	16.5864
30.0	1.5202	4.5615	7.6057	10.6543	13.7085	16.7691
40.0	1.5325	4.5979	7.6647	10.7334	13.8048	16.8794
50.0	1.5400	4.6202	7.7012	10.7832	13.8666	16.9519
60.0	1.5451	4.6353	7.7259	10.8172	13.9094	17.0026
80.0	1.5514	4.6543	7.7573	10.8606	13.9644	17.0686
100.0	1.5552	4.6658	7.7764	10.8871	13.9981	17.1093
∞	1.5708	4.7124	7.8540	10.9956	14.1372	17.2788

† The roots of this equation are all real if $C > 0$.

TABLE A.2: The first six roots, † α_n, of

$$\alpha \cot\alpha + C = 0.$$

C	α_1	α_2	α_3	C	α_1	α_2
−1.0	0	4.4934	7.7253	10.9041	14.0662	17.2208
−0.995	0.1224	4.4945	7.7259	10.9046	14.0666	17.2210
−0.99	0.1730	4.4956	7.7265	10.9050	14.0669	17.2213
−0.98	0.2445	4.4979	7.7278	10.9060	14.0676	17.2219
−0.97	0.2991	4.5001	7.7291	10.9069	14.0683	17.2225
−0.96	0.3450	4.5023	7.7304	10.9078	14.0690	17.2231
−0.95	0.3854	4.5045	7.7317	10.9087	14.0697	17.2237
−0.94	0.4217	4.5068	7.7330	10.9096	14.0705	17.2242
−0.93	0.4551	4.5090	7.7343	10.9105	14.0712	17.2248
−0.92	0.4860	4.5112	7.7356	10.9115	14.0719	17.2254
−0.91	0.5150	4.5134	7.7369	10.9124	14.0726	17.2260
−0.90	0.5423	4.5157	7.7382	10.9133	14.0733	17.2266
−0.85	0.6609	4.5268	7.7447	10.9179	14.0769	17.2295
−0.8	0.7593	4.5379	7.7511	10.9225	14.0804	17.2324
−0.7	0.9208	4.5601	7.7641	10.9316	14.0875	17.2382
−0.6	1.0528	4.5822	7.7770	10.9408	14.0946	17.2440
−0.5	J.1656	4.6042	7.7899	10.9499	14.1017	17.2498
−0.4	1.2644	4.6261	7.8028	10.9591	14.1088	17.2556
−0.3	1.3525	4.6479	7.8156	10.9682	14.1159	17.2614
−0.2	1.4320	4.6696	7.8284	10.9774	14.1230	17.2672
−0.1	1.5044	4.6911	7.8412	10.9865	14.1301	17.2730
0	1.5708	4.7124	7.8540	10.9956	14.1372	17.2788
0.1	1.6320	4.7335	7.8667	11.0047	14.1443	17.2845
0.2	1.6887	4.7544	7.8794	11.0137	14.1513	17.2903
0.3	1.7414	4.7751	7.8920	11.0228	14.1584	17.2961
0.4	1.7906	4.7956	7.9046	11.0318	14.1654	17.3019

TABLE A.2: (*continue*)

$\alpha \cot\alpha + C = 0.$						
C	α_1	α_2	α_3	C	α_1	α_2
0.5	1.8366	4.8158	7.9171	11.0409	14.1724	17.3076
0.6	1.8798	4.8358	7.9295	11.0498	14.1795	17.3134
0.7	1.9203	4.8556	7.9419	11.0588	14.1865	17.3192
0.8	1.9586	4.8751	7.9542	11.0677	14.1935	17.3249
0.9	1.9947	4.8943	7.9665	11.0767	14.2005	17.3306
1.0	2.0288	4.9132	7.9787	11.0856	14.2075	17.3364
1.5	2.1746	5.0037	8.0385	1 J.1296	14.2421	17.3649
2.0	2.2889	5.0870	8.0962	1 J.1727	14.2764	17.3932
3.0	2.4557	5.2329	8.2045	11.2560	14.3434	17.4490
4.0	2.5704	5.3540	8.3029	11.3349	14.4080	17.5034
5.0	2.6537	5.4544	8.3914	11.4086	14.4699	17.5562
6.0	2.7165	5.5378	8.4703	11.4773	14.5288	17.6072
7.0	2.7654	5,6078	8.5406	11.5408	14.5847	17.6562
8.0	2.8044	5.6669	8.6031	11.5994	14.6374	17.7032
9.0	2.8363	5.7172	8.6587	11.6532	14.6870	17.7481
10.0	2.8628	5.7606	8.7083	11.7027	14.7335	17.7908
15.0	2.9476	5.9080	8.8898	11.8959	14.9251	17.9742
20.0	2.9930	5.9921	9.0019	12.0250	15.0625	18.1136
30.0	3.0406	6.0831	9.1294	12.1807	15.2380	18.3018
40.0	3.0651	6.1311	9.1987	12.2688	15.3417	18.4180
50.0	3.0801	6.1606	9.2420	12.3247	15.4090	18.4953
60.0	3.0901	6.1805	9.2715	12.3632	15.4559	18.5497
80.0	3.1028	6.2058	9.3089	12.4124	15.5164	18.6209
100.0	3.1105	6.2211	9.3317	12.4426	15.5537	18.6650
∞	3.1416	6.2832	9.4248	12.5664	15.7080	18.8496

† The roots of this equation are all real if $C > -1$. These negative values of C arise in connection with the sphere, §9.4.

TABLE A.3: The first six roots α_n, of

$$\alpha J_1(\alpha) - C J_0(\alpha) = 0$$

C	α_1	α_2	α_3	α_4	α_5	α_6
0	0	3.8317	7.0156	10.1735	13.3237	16.4706
0.01	0.1412	3.8343	7.0170	10.1745	13.3244	16.4712
0.02	0.1995	3.8369	7.0184	10.1754	13.3252	16.4718
0.04	0.2814	3.8421	7.0213	10.1774	13.3267	16.4731
0.06	0.3438	3.8473	7.0241	10.1794	13.3282	16.4743
0.08	0.3960	3.8525	7.0270	10.1813	13.3297	16.4755
0.1	0.4417	3.8577	7.0298	10.1833	13.3312	16.4767
0.15	0.5376	3.8706	7.0369	10.1882	13.3349	16.4797
0.2	0.6170	3.8835	7.0440	10.1931	13.3387	16.4828
0.3	0 7465	3.9091	7.0582	10.2029	13.3462	16.4888
0.4	0.8516	3.9344	7.0723	10.2127	13.3537	16.4949
0.5	0.9408	3.9594	7.0864	10.2225	13.3611	16.5010
0.6	1.0184	3.9841	7.1004	10.2322	13.3686	16.5070
0.7	1.0873	4.0085	7.1143	10.2419	13.3761	16.5131
0.8	1.1490	4.0325	7.1282	10.2516	13.3835	16.5191
0.9	1.2048	4.0562	7.1421	10.2613	13.3910	16.5251
1.0	1.2558	4.0795	7.1558	10.2710	13.3984	16.5312
1.5	1.4569	4.1902	7.2233	10.3188	13.4353	16.5612
2.0	1.5994	4.2910	7.2884	10.3658	13.4719	16.5910
3.0	1.7887	4.4634	7.4103	10.4566	13.5434	16.6499
4.0	1.9081	4.6018	7.5201	10.5423	13.6125	16.7073
5.0	1.9898	4.7131	7.6177	10.6223	13.6786	16.7630
6.0	2.0490	4.8033	7.7039	10.6964	13.7414	16.8168
7.0	2.0937	4.8772	7.7797	10.7646	13.8008	16.8684
8.0	2.1286	4.9384	7.8464	10.8271	13.8566	16.9179
9.0	2.1566	4.9897	7.9051	10.8842	13.9090	16.9650
10.0	2.1795	5.0332	7.9569	10.9363	13.9580	17.0099
15.0	2.2509	5.1773	8.1422	11.1367	14.1576	17.2008
20.0	2.2880	5.2568	8.2534	11.2677	14.2983	17.3442
30.0	2.3261	5.3410	8.3771	11.4221	14.4748	17.5348
40.0	2.3455	5.3846	8.4432	11.5081	14.5774	17.6508
50.0	2.3572	5.4112	8.4840	11.5621	14.6433	17.7272
60.0	2.3651	5.4291	8.5116	11.5990	14.6889	17.7807
80.0	2.3750	5.4516	8.5466	11.6461	14.7475	17.8502
100.0	2.3809	5.4652	8.5678	11.6747	14.7834	17.8931
∞	2.4048	5.5201	8.6537	11.7915	14.9309	18.0711

Appendix B

In this table $q = (p/a)^{1/2}$; a and x are positive real; α, β, γ are unrestricted; k is a finite integer; n is a finite integer or zero; v is a fractional number; $1 \cdot 2 \cdot 3 \cdots n = n!$; $1 \cdot 3 \cdot 5 \cdots (2n-1) = (2n-1)!!$ $n\Gamma(n) = \Gamma(n+1) = n!$; $\Gamma(1) = 0! = 1$; $\Gamma(v)\Gamma(1-v) = \pi/\sin v\pi$; $\Gamma(\frac{1}{2}) = \pi^{1/2}$

NO.	TRANSFORM	FUNCTION
1	$\frac{1}{p}$	1
2	$\frac{1}{p^2}$	t
3	$\frac{1}{p^k}$	$\frac{t^{k-1}}{(k-1)!}$
4	$\frac{1}{p^{1/2}}$	$\frac{1}{(\pi t)^{1/2}}$
5	$\frac{1}{p^{3/2}}$	$2\left(\frac{t}{\pi}\right)^{\frac{1}{2}}$
6	$\frac{1}{p^{k+1/2}}$	$\frac{2^k}{\pi^{1/2}(2k-1)!!}t^{k-1/2}$
7	$\frac{1}{p^v}$	$\frac{t^{v-1}}{\Gamma(v)}$
8	$p^{1/2}$	$-\frac{1}{2\pi^{1/2}t^{5/2}}$
9	$p^{3/2}$	$\frac{3}{4\pi^{1/2}t^{5/2}}$
10	$p^{k-1/2}$	$\frac{(-1)^k(2k-1)!!}{2^k\pi^{1/2}t^{k+1/2}}$
11	p^{n-v}	$\frac{t^{v-n-1}}{\Gamma(v-n)}$
12	$\frac{1}{p+\alpha}$	$e^{-\alpha t}$
13	$\frac{1}{(p+\alpha)(p+\beta)}$	$\frac{e^{-\beta t}-e^{-\alpha t}}{\alpha-\beta}$
14	$\frac{1}{(p+\alpha)^2}$	$te^{-\alpha t}$

15	$\frac{1}{(p+\alpha)(p+\beta)(p+\gamma)}$	$\frac{(\gamma-\beta)e^{-\alpha t}+(\alpha-\gamma)e^{-\beta t}+(\beta-\alpha)e^{-\gamma t}}{(\alpha-\beta)(\beta-\gamma)(\gamma-\alpha)}$
16	$\frac{1}{(p+\alpha)^2(p+\beta)}$	$\frac{e^{-\beta t}-e^{-\alpha t}[1-(\beta-\alpha)t]}{(\beta-\alpha)^2}$
17	$\frac{1}{(p+\alpha)^3}$	$\frac{1}{2}t^2e^{-\alpha t}$
18	$\frac{1}{(p+\alpha)^k}$	$\frac{t^{k-1}e^{-\alpha t}}{(k-1)!}$
19	$\frac{p}{(p+\alpha)(p+\beta)}$	$\frac{\alpha e^{-\alpha t}-\beta e^{-\beta t}}{\alpha-\beta}$
20	$\frac{p}{(p+\alpha)^2}$	$(1-\alpha t)e^{-\alpha t}$
21	$\frac{p}{(p+\alpha)(p+\beta)(p+\gamma)}$	$\frac{\alpha(\beta-\gamma)e^{-\alpha t}+\beta(\gamma-\alpha)e^{-\beta t}+\gamma(\alpha-\beta)e^{-\gamma t}}{(\alpha-\beta)(\beta-\gamma)(\gamma-\alpha)}$
22	$\frac{p}{(p+\alpha)^2(p+\beta)}$	$\frac{[\beta-\alpha(\beta-\alpha)t]e^{-\alpha t}-\beta e^{-\beta t}}{(\beta-\alpha)^2}$
23	$\frac{p}{(p+\alpha)^3}$	$t\left(1-\frac{1}{2}\alpha t\right)e^{-\alpha t}$
24	$\frac{\alpha}{p^2+\alpha^2}$	$\sin\alpha t$
25	$\frac{p}{p^2+\alpha^2}$	$\cos\alpha t$
26	$\frac{\alpha}{p^2-\alpha^2}$	$\sinh\alpha t$
27	$\frac{p}{p^2-\alpha^2}$	$\cosh\alpha t$
28	e^{-qx}	$\frac{x}{2(\pi\alpha t^3)^{1/2}}e^{-x^2/4\alpha t}$
29	$\frac{e^{-qx}}{q}$	$\left(\frac{\alpha}{\pi t}\right)^{1/2}e^{-x^2/4\alpha t}$
30	$\frac{e^{-qx}}{p}$	$\operatorname{erfc}\left[\frac{x}{2(\alpha t)^{1/2}}\right]$
31	$\frac{e^{-qx}}{qp}$	$2\left(\frac{\alpha t}{\pi}\right)^{1/2}e^{-x^2/4\alpha t}-x\operatorname{erfc}\left[\frac{x}{2(\alpha t)^{1/2}}\right]$
32	$\frac{e^{-qx}}{p^2}$	$\left(t+\frac{x^2}{2\alpha}\right)\operatorname{erfc}\left[\frac{x}{2(\alpha t)^{1/2}}\right]-x\left(\frac{t}{\alpha\pi}\right)^{1/2}e^{-x^2/4\alpha t}$
33	$\frac{e^{-qx}}{p^{1+n/2}}$	$\frac{(\gamma-\beta)e^{-\alpha t}+(\alpha-\gamma)e^{-\beta t}+(\beta-\alpha)e^{-\gamma t}}{(\alpha-\beta)(\beta-\gamma)(\gamma-\alpha)}$

34	$\dfrac{e^{-qx}}{p^{3/4}}$	$\dfrac{e^{-\beta t} - e^{-\alpha t}[1-(\beta-\alpha)t]}{(\beta-\alpha)^2}$
35	$\dfrac{e^{-qx}}{q+\beta}$	$\dfrac{1}{2}t^2 e^{-\alpha t}$
36	$\dfrac{e^{-qx}}{q(q+\beta)}$	$\dfrac{t^{k-1}e^{-\alpha t}}{(k-1)!}$
37	$\dfrac{e^{-qx}}{p(q+\beta)}$	$\dfrac{\alpha e^{-\alpha t} - \beta e^{-\beta t}}{\alpha - \beta}$
38	$\dfrac{e^{-qx}}{qp(q+\beta)}$	$(1-\alpha t)e^{-\alpha t}$
39	$\dfrac{e^{-qx}}{q^{n+1}(q+\beta)}$	$\dfrac{\alpha(\beta-\gamma)e^{-\alpha t} + \beta(\gamma-\alpha)e^{-\beta t} + \gamma(\alpha-\beta)e^{-\gamma t}}{(\alpha-\beta)(\beta-\gamma)(\gamma-\alpha)}$
40	$\dfrac{e^{-qx}}{(q+\beta)^2}$	$\dfrac{[\beta-\alpha(\beta-\alpha)t]e^{-\alpha t} - \beta e^{-\beta t}}{(\beta-\alpha)^2}$
41	$\dfrac{e^{-qx}}{p(q+\beta)^2}$	$t\left(1-\dfrac{1}{2}\alpha t\right)e^{-\alpha t}$
42	$\dfrac{e^{-qx}}{p-\gamma}$	$\sin \alpha t$
43	$\dfrac{e^{-qx}}{q(p-\gamma)}$	$\frac{1}{2}e^{\gamma t}\left(\dfrac{\alpha}{\gamma}\right)^{1/2}\left\{\begin{array}{l} e^{-x(\gamma/\alpha)^{1/2}}\operatorname{erfc}\left[\dfrac{x}{2(\alpha t)^{1/2}} - (\gamma t)^{1/2}\right] \\ +e^{x(\gamma/\alpha)^{1/2}}\operatorname{erfc}\left[\dfrac{x}{2(\alpha t)^{1/2}} + (\gamma t)^{1/2}\right]\end{array}\right\}$
44	$\dfrac{e^{-qx}}{(p-\gamma)^2}$	$\frac{1}{2}e^{\gamma t}\left\{\begin{array}{l}\left[t - \dfrac{x}{2(\alpha t)^{1/2}}\right]e^{-x(\gamma/\alpha)^{1/2}}\operatorname{erfc}\left[\dfrac{x}{2(\alpha t)^{1/2}} - (\gamma t)^{1/2}\right] \\ +\left[t + \dfrac{x}{2(\alpha t)^{1/2}}\right]e^{x(\gamma/\alpha)^{1/2}}\operatorname{erfc}\left[\dfrac{x}{2(\alpha t)^{1/2}} + (\gamma t)^{1/2}\right]\end{array}\right\}$
45	$\dfrac{e^{-qx}}{(p-\gamma)(q+\beta)}$, $\gamma \neq \alpha\beta^2$	$\frac{1}{2}e^{\gamma t}\left\{\begin{array}{l}\dfrac{\alpha^{1/2}}{\alpha^{1/2}\beta+\gamma^{1/2}}e^{-x(\gamma/\alpha)^{1/2}}\operatorname{erfc}\left[\dfrac{x}{2(\alpha t)^{1/2}} - (\gamma t)^{1/2}\right] \\ +\dfrac{\alpha^{1/2}}{\alpha^{1/2}\beta-\gamma^{1/2}}e^{x(\gamma/\alpha)^{1/2}}\operatorname{erfc}\left[\dfrac{x}{2(\alpha t)^{1/2}} + (\gamma t)^{1/2}\right]\end{array}\right\}$ $-\dfrac{\alpha\beta}{\alpha\beta^2-\gamma}e^{\beta x+\alpha\beta^2 t}\operatorname{erfc}\left[\dfrac{x}{2(\alpha t)^{1/2}} + \beta(\alpha t)^{1/2}\right]$
46	$e^{x/p} - 1$	$\left(\dfrac{x}{t}\right)^{1/2} I_1\left[2(xt)^{1/2}\right]$
47	$\dfrac{1}{p}e^{x/p}$	$I_0\left[2(xt)^{1/2}\right]$
48	$\dfrac{1}{p^{y}}e^{x/p}$	$\left(\dfrac{t}{x}\right)^{(v-1)/2} I_{v-1}\left[2(xt)^{1/2}\right]$

49	$K_0(qx)$	$\frac{1}{2t}e^{-x^2/4\alpha t}$
50	$\frac{1}{p^{1/2}}K_{2v}(qx)$	$\frac{1}{2(\pi t)^{1/2}}e^{-x^2 8\alpha t}K_v\left(\frac{x^2}{8\alpha t}\right)$
51	$p^{v/2-1}K_v(qx)$	$x^{-v}\alpha^{v/2}2^{v-1}\int_{x^2/4\alpha t}^{\infty}e^{-u}u^{v-1}du$
52	$p^{v/2}K_v(qx)$	$\frac{x^v}{\alpha^{v/2}(2t)^{v+1}}e^{-x^2/4\alpha t}$
53	$\left[p-(p^2-x^2)^{1/2}\right]^v$	$v\frac{x^v}{t}I_v(xt)$
54	$e^{x\left[(p+\alpha)^{1/2}-(p+\beta)^{1/2}\right]^z}-1$	$\frac{x(\alpha-\beta)e^{-(\alpha+\beta)t/2}I_1\left[\frac{1}{2}(\alpha-\beta)t^{1/2}(t+4x)^{1/2}\right]}{t^{1/2}(t+4x)^{1/2}}$
55	$\frac{e^{x\left[p-(p+\alpha)^{1/2}(p+\beta)^{1/2}\right]}}{(p+\alpha)^{1/2}(p+\beta)^{1/2}}$	$e^{-(\alpha+\beta)(t+x)/2}I_0\left[\frac{1}{2}(\alpha-\beta)t^{1/2}(t+2x)^{1/2}\right]$
56	$\frac{e^{x\left[(p+\alpha)^{1/2}-(p+\beta)^{1/2}\right]^2}}{(p+\alpha)^{1/2}(p+\beta)^{1/2}\left[(p+\alpha)^{1/2}+(p+\beta)^{1/2}\right]^{2v}}$	$\frac{t^{v/2}e^{-(\alpha+\beta)t/2}I_v\left[\frac{1}{2}(\alpha-\beta)t^{1/2}(t+4x)^{1/2}\right]}{(\alpha-\beta)^v(t+4x)^{v/2}}$

Author Biography

Dr. Robert G. Watts is the Cornelia and Arthur L. Jung Professor of Mechanical Engineering at Tulane University. He holds a BS (1959) in mechanical engineering from Tulane, an MS(1960) in nuclear engineering from the Massachusetts Institute of Technology and a PhD (1965) from Purdue University in mechanical engineering. He spent a year as a Postdoctoral associate studying atmospheric and ocean science at Harvard University. He has taught advanced applied mathematics and thermal science at Tulane for most of his 43 years of service to that university.

Dr. Watts is the author of *Keep Your Eye on the Ball: The Science and Folklore of Baseball* (W. H. Freeman) and the editor of *Engineering Response to Global Climate Change* (CRC Press) and *Innovative Energy Strategies for CO2 Stabilization* (Cambridge University Press) as well as many papers on global warming, paleoclimatology energy and the physic of sport. He is a Fellow of the American Society of Mechanical Engineers.

◎ 编辑手记

世界著名数学家 P. R. Halmos 曾指出：

> 一个数学革新的来源在哪里？有时在数学之外，但绝不是总在数学之外. 正如数学家对工程学、物理学、心理学、经济学和其他学科做出了贡献一样，这些领域反过来也使数学保持了有生气的创造力，其方式就是向数学提出令人兴奋的问题，引导其沿着新的方向发展，有时还为表达数学概念建议某种语言. 曾经发生过这样一件事：当一个物理学家需要一个数学定理时，它已经放在那儿了，等待着人们取用. 但是更经常发生的事则是，当人们需要某些新东西时，这个消息要经过几十年或者更长的时间才能渗透到象牙塔内，而且差不多要再经过这么长的时间，其答案才能从塔内传达出来.

本书是一部讲应用数学的教程，中文书名可译为《工程师和科学家应用数学概论（第二版）》.

本书的作者为罗伯特 · G. 瓦特（Robert G. Watts），美国人，杜兰大学机械工程的教授. 他拥有杜兰大学机械工程学士学位（1959 年）、麻省理工学院核工程硕士学位（1960 年）和普渡大学机械工程博士学位（1965 年）. 他在哈佛大学担任了一年的博士后研究员，研究大气和海洋科学，在他任职的 43 年中，他大部分时间都在教授高级应用数学和热力学的相关课程.

这本面向工程师的实用数学畅销书的第二版包括了关于摄动方法和理论的新章节和扩展章节.本书是关于线性偏微分方程的书籍,该主题在工程学与物理科学中很常见.本书几乎对所有工程领域的研究生和高年级本科生都非常有用,还对希望学习高等数学是如何在其专业中应用的物理学、化学、地球物理学和其他物理科学的专业工程师都有用.读者将会了解有关传热、流体流动和机械振动的应用.本书着重讲解了物理问题的解决方法和应用,还详细展现了很多例子,并全面解释了它们与真实世界的关系.本书还给出了建议读者进行扩展阅读的参考文献.本书的主题还包括经典的变量分离和正交函数、拉普拉斯变换、复变量和斯图姆-刘维尔变换.第二版增加了关于摄动方法、微分方程的奇异摄动理论两个章节,还对部分章节进行了修订.

本书的目录为:

6. 积分变换:拉普拉斯变换

6.1　拉普拉斯变换

6.2　一些重要的变换

6.3　常系数的线性常微分方程

6.4　一些重要理论

6.5　部分分式

7. 复变量与拉普拉斯反演积分

7.1　基本性质

8. 带拉普拉斯变换的解

8.1　机械振动

8.2　扩散或传导问题

8.3　杜阿梅尔定理

9. 斯图姆–刘维尔变换

9.1　一个初等例子:傅里叶正弦变换

9.2　泛论:斯图姆–刘维尔变换——理论

9.3　逆变换

10. 摄动方法介绍

10.1　代数中的例子

11. 微分方程的奇异摄动理论

可能是由于作者本身就是热力学专家,所以本书的核心内容就是热传导方程.关于此方程的课后练习,笔者最推崇 J. 巴斯 1965 年在 Academic Press Inc 出版的 *Exercises in Mathematics*①中提供的习题,由浅入深,难度适中,梯度合理,且多用傅里叶级数方法,与本书甚合,兹摘录几题如下:

问题 1　考虑热传导方程

$$\frac{\partial y}{\partial t}=a^2\frac{\partial^2 y}{\partial x^2} \tag{1}$$

这里 a 是一个给定的实常数.

A. 求形如

$$y=f(x)g(t)$$

的解.

B. 设

① 摘自《数学习题》,J. 巴斯著,徐信之译,上海科学技术出版社,1986.

$$\exp(\omega x + a^2\omega^2 t) = \sum_{n=0}^{\infty} \frac{\omega^n}{n!} V_n(x,t)$$

证明函数 $V_n(x,t)$ 都是式(1) 的解,并且证明,当 $t < 0$ 时,它们可以用由

$$\exp(-u^2 + 2us) = \sum_{n=0}^{\infty} \frac{1}{n!} u^n P_n(s)$$

所定义的埃尔米特多项式 $P_n(s)$ 表示.

C. 求式(1) 的解 y,它是 x 的以 2π 为周期的周期函数,当 $t \to \infty$ 时,它是有界的;当 $t = 0$ 时,它化为一个给定的函数 $y_0(x)$.

D. 当 $y_0(x) = u_1(0 < x < \alpha)$ 和 $y_0(x) = u_2(\alpha < x < 2\pi)$ 时,通过计算求解,这里 u_1 与 u_2 是给定的常数. 当 $t \to \infty$ 时,$y(x,t)$ 的极限是什么?

解 A. 我们有

$$fg' = a^2 f'' g \quad 或 \quad \frac{1}{a^2}\frac{g'}{g} = \frac{f''}{f}$$

后一个方程的左端是 t 的函数,而右端是 x 的函数. 仅当它们有一个公共常数值时才可能相等. 如果这个常数值是负的,那么有

$$f'' + \omega^2 f = 0, \quad \frac{g'}{g} = -a^2\omega^2$$

由此可得初等解

$$f = e^{i\omega x}, \quad g = \exp(-a^2\omega^2 t)$$

和

$$y = \exp(i\omega x - a^2\omega^2 t)$$

化为实值,我们有两个解,即

$$y = \exp(-a^2\omega^2 t)\cos \omega x$$

与

$$y = \exp(-a^2\omega^2 t)\sin \omega x$$

如果常数是正的,那么有

$$f'' - \omega^2 f = 0, \quad \frac{g'}{g} = a^2\omega^2$$

由此可得

$$y = \exp(\omega x + a^2\omega^2 t)$$

B. 关于 x 二次可微及关于 t 一次可微的形如

$$\sum c_n \exp(i\omega_n x - a^2\omega_n^2 t)$$

的每一个级数都是热传导方程的一个解(θ 函数就是一个例子).

在下面的问题中,我们将更详细地讨论这一类型的解. 这里我们要考察一族与初等解 $\exp(\omega x + a^2\omega^2 t)$ 有关的解. 当把 $\exp(\omega x + a^2\omega^2 t)$ 考虑为复变量 ω 的函数时,它是全纯的,因此,可以在任何圆内把它展为幂级数.

我们写出

$$\exp(\omega x + a^2\omega^2 t) = \sum_{n=0}^{\infty} \frac{\omega^n}{n!} V_n(x,t)$$

然后限定 ω 取实值. 用 A 表示算子

$$\frac{\partial}{\partial t} - a^2 \frac{\partial^2}{\partial x^2}$$

根据假设,我们有

$$A\left(\sum_{n=0}^{\infty} \frac{\omega^n}{n!} V_n(x,t)\right) = 0$$

假定这个级数关于 x 或 t 可逐项微商,则有

$$\sum_{n=0}^{\infty} \frac{\omega^n}{n!} A V_n(x,t) = 0$$

因为这个幂级数的极限不管 ω 取什么值都是零,所以,它所有的系数一定都是零,因此,$AV_n(x,t) = 0$. 对于每一个 n,函数 $V_n(x,t)$ 是热传导方程的一个解.

现在我们用

$$\exp(-u^2 + 2us) = \sum_{n=0}^{\infty} \frac{1}{n!} u^n P_n(s)$$

定义了埃尔米特多项式 $P_n(s)$.

设

$$u^2 = -a^2\omega^2 t, \quad 2us = \omega x$$

或

$$u = a\omega\sqrt{-t}, \quad s = \frac{x}{2a\sqrt{-t}}$$

并假设 $t < 0$,我们看到

$$\exp(a^2\omega^2 t + \omega x) = \sum \frac{1}{n!}(-t)^{n/2} a^n \omega^n P_n\left(\frac{x}{2a\sqrt{-t}}\right)$$

因此

$$V_n(x,t) = a^n(-t)^{n/2} P_n\left(\frac{x}{2a\sqrt{-t}}\right)$$

这个关系式指出了函数 V_n 与埃尔米特多项式的联系.

例

$$P_2(s) = 4s^2 - 2, \quad V_2(x,t) = x^2 + 2a^2 t$$

$$P_3(s) = 8s^3 - 12s, \quad V_3(x,t) = x^3 + 6xa^2 t$$

多项式 $x^2 + 2a^2 t$ 与 $x^3 + 6xa^2 t$ 都是方程

$$\frac{\partial y}{\partial t} = a^2 \frac{\partial^2 y}{\partial x^2}$$

的解.

C. 我们希望$y(x,t)$是x的以2π为周期的周期函数,当$t\to+\infty$时,它是有界的.这样一来,我们需要选择某些初等解

$$\exp(-a^2\omega^2t)\cos\omega x\quad\text{与}\quad\exp(-a^2\omega^2t)\sin\omega x$$

然后把所选择的解组合,形如

$$y=\exp(-a^2\omega^2t)(A\cos\omega x+B\sin\omega x)$$

的一个解,这个解依赖于两个待定的新常数A与B.

要函数$f(x)$以2π为周期,需要ω取整数值.对n的每一个值,我们指定A与B的值为A_n与B_n.这样一来,我们构造了一个解,它在形式上由下面的级数表示

$$y=\sum_{n=0}^{\infty}\exp(-a^2n^2t)(A_n\cos nx+B_n\sin nx)\tag{2}$$

由初始条件知,当$t=0$时,y等于给定的函数$y_0(x)$

$$y_0(x)=\sum_{n=0}^{\infty}(A_n\cos nx+B_n\sin nx)$$

我们来计算A_n与B_n,现在它们是函数$y_0(x)$在区间$[0,2\pi]$上的傅里叶系数

$$A_n=\frac{1}{\pi}\int_0^{2\pi}y_0(s)\cos ns\mathrm{d}s$$

$$B_n=\frac{1}{\pi}\int_0^{2\pi}y_0(s)\sin ns\mathrm{d}s$$

$$A_0=\frac{1}{2\pi}\int_0^{2\pi}y_0(s)\mathrm{d}s$$

$$A_0\cos nx+B_0\sin nx=\frac{1}{\pi}\int_0^{2\pi}y_0(s)\cos n(x-s)\mathrm{d}s$$

因此,我们有

$$y=\frac{1}{2\pi}\int_0^{2\pi}y_0(s)\mathrm{d}s+\frac{1}{\pi}\sum_{n=1}^{\infty}\exp(-a^2n^2t)\int_0^{2\pi}y_0(s)\cos n(x-s)\mathrm{d}s$$

以

$$\frac{1}{\pi}\exp(-a^2n^2t)y_0(s)\cos n(x-s)$$

为通项的级数对所有的$t>0$,有以$(M/\pi)\exp(-a^2n^2t)$为通项的上级数,这里M是$|y_0(s)|$的上确界.因此,这个级数关于s是一致收敛的,可以对它逐项积分.若设

$$h(x,t)=\frac{1}{\pi}\left[\frac{1}{2}+\sum_{n=1}^{\infty}\exp(-a^2n^2t)\cos nx\right]$$

则有

$$y=\int_0^{2\pi}h(t,x-s)y_0(s)\mathrm{d}s$$

如果级数(2)关于t可逐项微商一次,关于x可逐项微商两次,也就是说,如果通项为

$$n^2\exp(-a^2n^2t)\int_0^{2\pi}y_0(s)\cos n(x-s)\,\mathrm{d}s$$

的级数关于t与x是一致收敛的,那么函数$y(x,t)$的确是热传导方程的解.现在,后一个级数以通项为

$$2\pi Mn^2\exp(-a^2n^2t)$$

的级数作为上级数;而对于一切$t\geqslant t_0>0$,这个级数又以通项为

$$2\pi Mn^2\exp(-a^2n^2t_0)$$

的级数为上级数,由这个级数的收敛性可推出前面级数的一致收敛性.

D.例(图1).

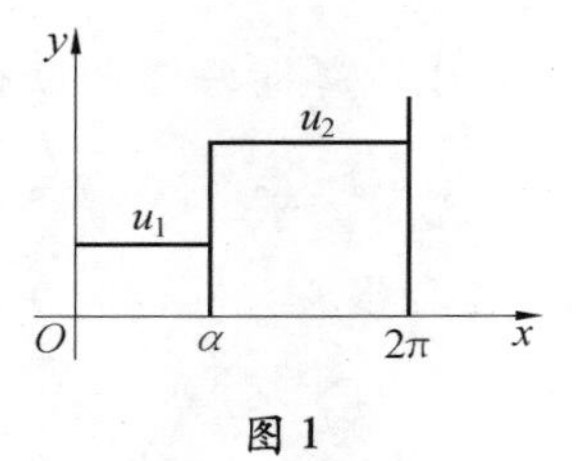

图1

$$y_0(x)=\begin{cases}u_1, & 0<x<\alpha\\ u_2, & \alpha<x<2\pi\end{cases}$$

这里我们有

$$A_n=\frac{u_1}{\pi}\int_0^{\alpha}\cos ns\mathrm{d}s+\frac{u_2}{\pi}\int_{\alpha}^{2\pi}\cos ns\mathrm{d}s$$

$$B_n=\frac{u_1}{\pi}\int_0^{\alpha}\sin ns\mathrm{d}s+\frac{u_2}{\pi}\int_{\alpha}^{2\pi}\sin ns\mathrm{d}s$$

或

$$A_n=\frac{u_1-u_2}{\pi}\cdot\frac{\sin n\alpha}{n}$$

$$A_0=\frac{\alpha u_1}{2\pi}+(2\pi-\alpha)\frac{u_2}{2\pi}$$

$$B_n=\frac{u_1-u_2}{n\pi}(1-\cos n\alpha)$$

$$A_n\cos nx+B_n\sin nx=2\frac{u_1-u_2}{n\pi}\sin\frac{n\alpha}{2}\cos n\left(x-\frac{\alpha}{2}\right)$$

问题2 考虑偏微分方程

$$\frac{\partial y}{\partial t}=a^2\frac{\partial^2 y}{\partial x^2}\tag{1}$$

求一个定义在 $0 \leqslant x \leqslant l, t > 0$ 上的解，它满足下述条件：

(1) $\partial y/\partial t + Ky = 0, x = l, t$ 是任意的（K 是一个给定的常数）.

(2) $\partial y/\partial x = 0, x = 0, t$ 是任意的.

(3) $\lim\limits_{t \to \infty} y(x,t) = 0, x$ 是任意的.

(4) $y(x,0) = f(x), f(x)$ 是定义在 $0 \leqslant x \leqslant l$ 上的已知函数.

A. 求形如

$$y = e^{-a^2\omega^2 t}\cos \omega x$$

的解，使之满足条件(1)(2)(3)，并证明 ω 是方程

$$\omega \tan \omega l = K$$

的一个解.

B. 设 $\omega_1, \omega_2, \cdots, \omega_n, \cdots$ 是上述方程的正根. 试证明，函数

$$g_n(x) = \cos \omega_n x$$

在区间 $[0,l]$ 上构成一个正交级数，并计算它们的范数.

C. 借助形如

$$\sum_n A_n \exp(-a^2\omega_n^2 t)\cos \omega_n x$$

的级数，表出式(1) 的满足条件(1)(2)(3)(4) 的解（不需证明计算的合理性）. 利用 $f(x)$ 计算系数 A_n. 处理 $K = 0$ 及 $K = \infty$ 的特殊情况.

解 A. 我们看到

$$y(x,t) = \exp(-a^2\omega^2 t)(A\cos \omega x + B\sin \omega x)$$

满足热传导方程，这里 A, B, ω 都是实常数.

我们写出边界条件

$$\frac{\partial y}{\partial x} = \exp(-a^2\omega^2 t)(-\omega A\sin \omega x + \omega B\cos \omega x)$$

条件

$$\frac{\partial y}{\partial x}(0,t) = 0$$

蕴含着 $B = 0$. 因此

$$\frac{\partial y}{\partial x} + Ky = \exp(-a^2\omega^2 t)(KA\cos \omega x - \omega A\sin \omega x)$$

条件

$$\left(\frac{\partial y}{\partial x} + Ky\right)(l,t) = \exp(-a^2\omega^2 t)(KA\cos \omega l - \omega A\sin \omega l) = 0$$

或者蕴含着 $A = 0$（这个解不能接受，因为它蕴含着 $y(x,t) = 0$），或者蕴含着 $\omega \tan \omega l = K$.

函数

$$y(x,t) = A\exp(-a^2\omega^2 t)\cos \omega x$$

（其中 $\omega \tan \omega l = K$）是传热导方程的一个满足边界条件的初等解. 当 $t \to$

∞ 时,它趋向于零. 这就证明了后面选择正的实常数 ω^2 的合理性.

B. 特征函数解的正交性. 方程 $\omega\tan\omega l = K$ 可以写为

$$\tan\omega l = \frac{Kl}{\omega l}$$

由图(图2)可以看到,在0与$\pi/2$之间,π与$3\pi/2$之间,…,共有无穷多个解 $\omega_1,\omega_2,\cdots,\omega_n,\cdots$.

这些解是我们所考虑的边值问题的特征值. 这个问题在于求微分方程

$$g''(x)+\omega^2 g(x)=0$$

的解 $\cos\omega x$,使得

$$g'(x)=0,\quad x=0$$
$$g'(x)+Kg(x)=0,\quad x=l$$

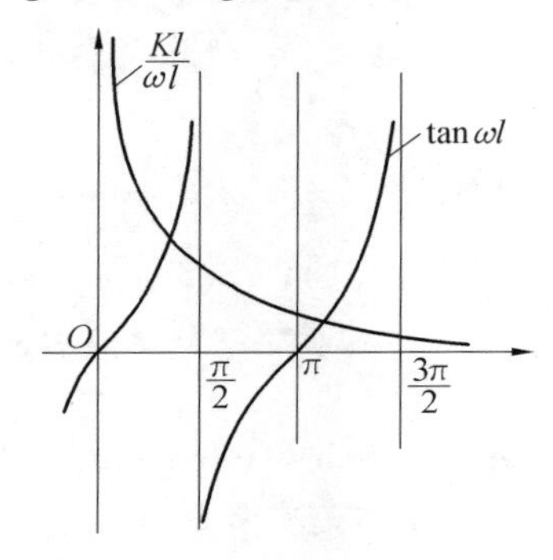

图2

这些边界条件确定了一个在区间$[0,l]$上的函数的向量空间 Ω. 我们已经知道,在这个空间中,特征函数是正交的. 为了证明这一点,考虑两个相应于特征函数 g_q 与 g_p 的特征值 ω_q 与 ω_p. 我们有

$$g''_q+\omega_q^2 g_q=0,\quad g''_p+\omega_p^2 g_p=0$$

由此可得

$$g_q g''_p - g''_q g_p + (\omega_p^2-\omega_q^2)g_q g_p = 0$$

如果从0到 l 积分,并注意到

$$g_q g''_p - g''_q g_p = \frac{\mathrm{d}}{\mathrm{d}x}(g_q g'_p - g'_q g_p)$$

那么有

$$[g_q g'_p - g'_q g_p]_0^l + (\omega_p^2-\omega_q^2)\int_0^l g_q g_p \mathrm{d}x = 0$$

由边界条件,方括号中的表达式在下限 $x=0$ 处取零值. 类似地,当 $x=l$ 时,我们有

$$g'(l)=-Kg(l)$$

由此可得

$$g_q(l)g'_p(l) - g'_q(l)g_p(l) = -Kg_q(l)g_p(l) + Kg_q(l)g_p(l) = 0$$

因此,整个方括号的表达式取零值,由此,因为 $\omega_p^2 - \omega_q^2 \neq 0$,所以有

$$\int_0^l g_q g_p \mathrm{d}x = 0$$

这样一来,函数 g_q 与 g_p 在区间$[0,l]$上的确是正交的.

这一结果可以利用 $g(x)$ 的表达式通过初等积分计算来证明. 我们定义

$$I = \int_0^l \cos \omega_p x \cos \omega_q x \mathrm{d}x, \quad p \neq q$$

于是

$$\begin{aligned} 2I &= \int_0^l [\cos(\omega_p - \omega_q)x + \cos(\omega_p - \omega_q)x] \mathrm{d}x \\ &= \frac{\sin(\omega_p - \omega_q)l}{\omega_p - \omega_q} + \frac{\sin(\omega_p + \omega_q)l}{\omega_p + \omega_q} \\ &= \frac{2\omega_p \tan \omega_p l - 2\omega_q \tan \omega_q l}{\omega_p^2 - \omega_q^2} \end{aligned}$$

因为 $\omega_p \tan \omega_p l$ 的值与 p 无关,所以 $I=0$. 这样一来,函数序列 $\cos \omega_n(x)$ 在$[0,l]$上是正交的.

最后,我们来计算 $\cos \omega_n x$ 的范数

$$\begin{aligned} 2\|\cos \omega_n x\|^2 &= 2\int_0^l \cos^2 \omega_n x \mathrm{d}x = l + \frac{\sin 2\omega_n l}{2\omega_n} \\ &= l + \frac{2\tan \omega_n l}{2\omega_n(1 + \tan^2 \omega_n l)} \\ &= l + \frac{K}{K^2 + \omega_n^2} \end{aligned}$$

$$\|\cos \omega_n x\| = \frac{1}{\sqrt{2}}\left(l + \frac{K}{K^2 + \omega_n^2}\right)^{1/2}$$

对应于正交序列 $\cos \omega_n x$ 的规格化的正交序列以 $\cos \omega_n x / \|\cos \omega_n x\|$ 为通项,所以

$$\frac{\sqrt{2}\cos \omega_n x}{\{l + [K/(K^2 + \omega_n^2)]\}^{1/2}}$$

为通项.

C. 假定函数 $y(x,t)$ 可以表示为形如

$$y = \sum_{n=0}^{\infty} A_n \exp(-a^2\omega_n^2 t) \cos \omega_n x \tag{2}$$

的级数.

现在写出 $t=0$ 时的初始条件 $y=f(x)$,我们有

$$\sum_{n=0}^{\infty} A_n \cos \omega_n x = f(x), \quad 0 < x < l$$

这样一来,这个函数 $f(x)$ 是三角级数的和. 但是 ω_n 不都是 ω_0 的倍数,不可能推出 $\omega_n = n\omega_0$;或者更一般地,存在一个依赖于 n 的整数 $\varphi(n)$,使得 $\omega_n = \omega_0 \varphi(n)$. 级数(2) 可能不是周期的傅里叶级数.

我们不需寻求这个假设的正确性的证明,而来证明可利用关系式(2) 计算系数 A_n. 为了使符号简单起见,设 $\cos \omega_n x = g_n(x)$,并写出

$$\sum_{n=0}^{\infty} A_n g_n = f$$

作这个方程的每一项与 g_p 的内积(在函数空间 Ω 的意义下),假定这个运算可以逐项进行,由于 g_n 的正交性,只剩下

$$A_p(g_p, g_p) = (f, g_p) \quad 或 \quad A_p = \frac{(f, g_p)}{\| g_p \|^2}$$

范数 $\| g_p \|$ 在问题 B 中计算过;分子的内积是

$$(f, g_p) = \int_0^l f(x) \cos \omega_p x \mathrm{d}x$$

这样一来,A_p 就确定了.

于是我们有

$$y = \sum_{n=0}^{\infty} A_n \exp(-a^2 \omega_n^2 t) \cos \omega_n x, \quad t > 0$$

我们证明 y 关于 t 一次可微,关于 x 二次可微,然后推出 y 的确是热传导方程的解.

整个问题在于证明,级数

$$\sum_{n=0}^{\infty} \omega_n^2 A_n \exp(-a^2 \omega_n^2 t) \cos \omega_n x$$

是一致收敛的. 当 $t \geqslant t_0 > 0$ 时,这个级数以级数

$$\sum_{n=0}^{\infty} \omega_n^2 | A_n | \exp(-a^2 \omega_n^2 t_0)$$

为上级数. 根据施瓦兹不等式,有

$$| A_n | = \frac{|(f, g_n)|}{\| g_n \|^2} \leqslant \frac{\| f \| \cdot \| g_n \|}{\| g_n \|^2} = \frac{\| f \|}{\| g_n \|}$$

现在

$$\| g_n \| = \frac{1}{\sqrt{2}} \left(l + \frac{K}{K^2 + \omega_n^2} \right)^{1/2} > \frac{1}{\sqrt{2}} \cdot \frac{\sqrt{K}}{(K^2 + \omega_n^2)^{1/2}}$$

我们可以证明

$$\omega_n^2 | A_n | \leqslant \| f \| \frac{\omega_n^2}{\| g_n \|} \leqslant \left(\frac{2}{K} \right)^{1/2} \omega_n^2 (K^2 + \omega_n^2)^{1/2} \| f \|$$

上级数的一般项与

$$\omega_n^2\exp(-a^2\omega_n^2t_0)$$

以同样的方式趋向于零. 因为 $\omega_n>(n-1)\pi/l$,所以指数部分保证了这个级数的收敛性. 由此得到关于 y 求导数所得到的级数是一致收敛的.

特殊情况. 若 $K=0$,则有 $\tan\omega l=0$. 在这种情况下,$\omega_n=n\pi/l$,于是级数(2)是普通的周期的傅里叶级数. 在 $K=\infty$ 时,我们有类似的结论.

注意:由施瓦兹不等式所得到的上估计 $\|A_n\|\leqslant\|f\|/\|g_n\|$ 对于证明 y 的可微性而言是足够的. 但是,这个方法较为笨拙. 只需证明 $\|A_n\|$ 是有界的就可以了,因为当 $n\to\infty$ 时,$\|g_n\|$ 有极限. 现在我们知道,当 $\omega_n\to\infty$ 时,积分

$$\int_0^l f(x)\cos\omega_nx\mathrm{d}x$$

趋向于零,也就是 $A_n\to0$. 这还不足以证明级数 $\sum A_n\cos\omega_nx$ 是收敛的. 恰如讨论普通的傅里叶级数的收敛性一样,我们还需要研究函数 $f(x)$ 的精确性质.

考虑热传导方程

$$\frac{\partial y}{\partial t}=a^2\frac{\partial^2y}{\partial x^2},\quad a>0$$

我们知道这个方程恰好有一个解. 当 $t=0$ 时,它化为函数 $f(x)$,这个函数对一切 x 有定义,并从 $-\infty$ 到 $+\infty$ 绝对可积.

这个解是

$$y=\frac{1}{2a(\pi t)^{1/2}}\int_{-\infty}^{+\infty}\exp\left[-\frac{(x-s)^2}{4a^2t}\right]f(s)\mathrm{d}s\tag{3}$$

在 t 的地方写成 $t-t'$,就可以用任意的值 t' 代替值 $t=0$.

非齐次热传导方程

$$\frac{\partial y}{\partial t}=a^2\frac{\partial^2y}{\partial x^2}+\varphi(x,t)$$

并证明,在 $t=0$ 时的解可以用类似于(3)的公式来表示. 我们可以从这个公式推出,当 $t=0$ 时化为给定函数 $f(x)$ 的解.

问题3　考虑非齐次热传导方程

$$\frac{\partial y}{\partial t}=a^2\frac{\partial^2y}{\partial x^2}+\varphi(x,t)\tag{1}$$

这里 $\varphi(x,t)$ 是一个给定的函数,对每一个固定的 t 值,它从 $x=-\infty$ 到 $x=+\infty$ 是绝对可积的.

试证明,当 $t=0$ 时,这个方程取零值的解 y_0 是

$$y_0=\int_0^t u_0(x,t,t')\mathrm{d}t'$$

这里 u_0 是齐次方程 $\partial u/\partial t = a^2\partial^2 u/\partial x^2$ 的解，当 $t=t'$ 时，它化为 $\varphi(x,t')$. 然后求(1) 在 $t=0$ 时化为给定函数 $f(x)$ 的解.

解 设 $u_0(x,t,t')$ 是齐次方程

$$\frac{\partial u}{\partial t} = a^2\frac{\partial^2 u}{\partial x^2}$$

的解，当 $t=t'$ 时，它化为 $\varphi(x,t')$. 假定已知这个解的存在性与唯一性. 考虑函数

$$y_0(x,t) = \int_0^t u_0(x,t,t')\mathrm{d}t'$$

因为 $u_0(x,t,t')$ 是连续的，所以

$$\frac{\partial y_0}{\partial t} = u_0(x,t,t) + \int_0^t \frac{\partial u_0}{\partial t}(x,t,t')\mathrm{d}t' = \varphi(x,t) + \int_0^t \frac{\partial u_0}{\partial t}(x,t,t')\mathrm{d}t'$$

我们也可以在积分号下对 x 求微商

$$a^2\frac{\partial^2 y_0}{\partial x^2} = \int_0^t a^2\frac{\partial^2 u_0}{\partial x^2}(x,t,t')\mathrm{d}t' = \int_0^t \frac{\partial u_0}{\partial t}(x,t,t')\mathrm{d}t'$$

于是，关于

$$\int_0^t \frac{\partial u_0}{\partial t}(x,t,t')\mathrm{d}t'$$

我们有两个表达式，即

$$\frac{\partial y_0}{\partial t'} - \varphi(x,t) \quad 与 \quad a^2\frac{\partial^2 y_0}{\partial x^2}$$

因此

$$\frac{\partial y_0}{\partial t} = a^2\frac{\partial^2 y_0}{\partial x^2} + \varphi(x,t)$$

函数 y_0 的确是非齐次热传导方程的解，当 $t=0$ 时，它取零值.

因为我们知道

$$u_0(x,t,t') = \frac{1}{2a[\pi(t-t')]^{1/2}}\int_{-\infty}^{+\infty}\exp\left[-\frac{(x-s)^2}{4a^2(t-t')}\right]\varphi(s,t')\mathrm{d}s$$

所以解 y_0 正是所要求的.

从而容易构造非齐次方程的解 y，使得当 $t=0$ 时，它取非零值 $f(x)$. 我们有

$$\frac{\partial y}{\partial t} = a^2\frac{\partial^2 y}{\partial x^2} + \varphi(x,t), \quad \frac{\partial y_0}{\partial t} = a^2\frac{\partial^2 y_0}{\partial x^2} + \varphi(x,t)$$

于是 $y-y_0$ 是非齐次方程

$$\frac{\partial}{\partial t}(y-y_0) = a^2\frac{\partial^2}{\partial x^2}(y-y_0)$$

的解. 现在，在 $t=0$ 处，函数 $y-y_0$ 与 y_0 取同样的值 $f(x)$. 因此有

$$y-y_0=y_1 \quad 或 \quad y=y_0+y_1$$

其中

$$y_1=\frac{1}{2a(\pi t)^{1/2}}\int_{-\infty}^{+\infty}\exp\left[-\frac{(x-s)^2}{4a^2t}\right]f(s)\,\mathrm{d}s$$

借助线性积分变换将 $f(s)$ 变到 y_1. 把这个变换写成下述公式的形式

$$y_1=\int_{-\infty}^{+\infty}K(x,s,t)f(s)\,\mathrm{d}s$$

其中

$$K(x,s,t)=\frac{1}{2a(\pi t)^{1/2}}\exp\left[-\frac{(x-s)^2}{4a^2t}\right]$$

这样一来,我们有

$$y=\int_0^t\mathrm{d}t'\left[\int_{-\infty}^{+\infty}K(x,s,t-t')\varphi(s,t')\,\mathrm{d}s\right]+\int_{-\infty}^{+\infty}K(x,s,t)f(s)\,\mathrm{d}s$$

下面我们要再介绍一个热传导方程关于变量 x 的柯西问题,即:求一个解,使得当 $x=0$ 时,在 t 轴上的一个给定的区间 S 上,它化为一个已知函数;而它的导数 $\partial y/\partial x$ 化为另一个已知函数.

根据柯西-柯瓦列夫斯基定理,这个问题恰有一个定义在 xt-平面上的线段 S 的邻域内的解.

当然,当 S 是有限线段时,这是合理的.

在这个问题中,我们把 S 取为 t 的整个正半轴. 我们要在整个半平面 $t>0$ 内确定这个解. 在这种情况下,不能给出一个存在及唯一性定理. 我们将只满足于形式上构造一个满足已知条件的解. 这个解用积分表示,而不用傅里叶级数表示,这是因为要把已知条件用到无界集上.

问题 4 A. 求热传导方程

$$\frac{\partial y}{\partial t}=a^2\frac{\partial^2 y}{\partial x^2}$$

满足下列条件的解:

当 $x=0$ 时

$$y=\begin{cases}0, & t<0\\ f(t), & t>0\end{cases},\quad \frac{\partial y}{\partial x}=\begin{cases}0, & t<0\\ g(t), & t>0\end{cases}$$

这里 f,g 是两个定义在 $t>0$ 上的已知函数. 假定计算是合理的,用形如

$$y=\int_{-\infty}^{+\infty}[A(\omega)\cos\omega x+B(\omega)\sin\omega x]\exp(-a^2\omega^2t)\,\mathrm{d}\omega$$

的积分表示 y,其中 $A(\omega)$ 是偶函数;$B(\omega)$ 是奇函数. 通过设 $a^2\omega^2=s$ 将这一问题化为一个拉普拉斯变换的问题.

B. 当 $f(t)=1/\sqrt{t}$,$g(t)=0$ 时,做出具体计算.

解 考虑热传导方程

$$\frac{\partial y}{\partial t}=a^2\frac{\partial^2 y}{\partial x^2}$$

我们求满足下述条件的解:

当 $x=0$ 时

$$y=\begin{cases}0, & t\leqslant 0\\ f(t), & t>0\end{cases},\quad \frac{\partial y}{\partial x}=\begin{cases}0, & t\leqslant 0\\ g(t), & t>0\end{cases}$$

已知的条件是边界条件,而不是初始条件.这一问题是定义在整个半轴 $t>0$ 上的柯西问题.

因为当 $t<0$ 时,$y=0$ 是一个解,所以只需考虑 t 的正值.

与前面的练习一样,我们要求一个傅里叶积分形式的解,而不是傅里叶级数形式的解

$$y(x,t)=\int_{-\infty}^{+\infty}[A(\omega)\cos\omega x+B(\omega)\sin\omega x]\exp(-a^2\omega^2 t)\,\mathrm{d}\omega$$

这里 $A(\omega)$ 是一个偶函数,而 $B(\omega)$ 是一个奇函数.当 $x=0$ 时,我们有

$$\begin{aligned}y(0,t)&=\int_{-\infty}^{+\infty}A(\omega)\exp(-a^2\omega^2 t)\,\mathrm{d}\omega\\&=2\int_0^{+\infty}A(\omega)\exp(-a^2\omega^2 t)\,\mathrm{d}\omega\end{aligned}$$

这是因为 $A(\omega)$ 是偶函数,所以我们还有

$$\begin{aligned}\frac{\partial y}{\partial x}(0,t)&=\int_{-\infty}^{+\infty}\omega B(\omega)\exp(-a^2\omega^2 t)\,\mathrm{d}\omega\\&=2\int_0^{+\infty}\omega B(\omega)\exp(-a^2\omega^2 t)\,\mathrm{d}\omega\end{aligned}$$

这是因为 $B(\omega)$ 是奇函数.

设 $a^2\omega^2=s$,于是

$$f(t)=\int_0^{+\infty}\frac{A(\sqrt{s}/a)}{a\sqrt{s}}\mathrm{e}^{-st}\,\mathrm{d}s$$

$$g(t)=\int_0^{+\infty}\frac{B(\sqrt{s}/a)}{a^2}\mathrm{e}^{-st}\,\mathrm{d}s$$

函数 $f(t)$ 与 $g(t)$ 是函数

$$\frac{A(\sqrt{s}/a)}{a\sqrt{s}}\quad 与\quad \frac{1}{a^2}B\left(\frac{\sqrt{s}}{a}\right)$$

的拉普拉斯变换.因此,如果 F 与 G 分别是 f 与 g 的原函数,那么有

$$A(\omega)=a^2\omega F(a^2\omega^2),\quad B(\omega)=a^2G(a^2\omega^2)$$

它们就确定了函数 y.

例 $f(t)=1/\sqrt{t}$,$g(t)=0$.首先,我们有

$$B(\omega)=0$$

其次,$1/\sqrt{t}$ 是 $1/\sqrt{\pi s}$ 的拉普拉斯变换. 因此,$A(\omega)=a/\sqrt{\pi}$. 由此可得 $y(x,t)$ 的表达式

$$y(x,t)=\int_{-\infty}^{+\infty}\frac{a}{\sqrt{\pi}}\cos\omega x\exp(-a^2\omega^2t)\mathrm{d}\omega$$

或

$$y(x,t)=\int_{-\infty}^{+\infty}\frac{a}{\sqrt{\pi}}\mathrm{e}^{\mathrm{i}\omega x}\exp(-a^2\omega^2t)\mathrm{d}\omega$$

当把后一个积分中的指数部分写成三角函数的形式时,包含奇函数 $\sin\omega x$ 那一项的积分为零.

设 $a^2\omega^2t=s^2/2$,于是

$$y(x,t)=\int_{-\infty}^{+\infty}\frac{1}{(2\pi t)^{1/2}}\exp\left(\mathrm{i}s\frac{x}{a(2t)^{1/2}}\right)\exp\left(-\frac{s^2}{2}\right)\mathrm{d}s$$

今设 $x/a(2t)^{1/2}=z$,则

$$y(x,t)=\int_{-\infty}^{+\infty}\frac{1}{(2\pi t)^{1/2}}\mathrm{e}^{\mathrm{i}sz}\exp\left(-\frac{s^2}{2}\right)\mathrm{d}s$$

除去差一个因子 $1/(2\pi t)^{1/2}$ 外,右端的积分是 $\exp(-s^2/2)$ 的傅里叶变换,也就是 $(2\pi)^{1/2}\exp(-z^2/2)$ 的傅里叶变换.

因此

$$g(x,t)=\frac{1}{\sqrt{t}}\exp\left(-\frac{x^2}{4a^2t}\right),\quad t>0$$

注意,当 t 从右边趋向于零时,y 趋向于零. 因此,我们找到的这个解,当 $t<0$ 时,可以延拓为零函数. 对 $\partial y/\partial x$ 来说,也是如此.

在本书即将出版之际,世界局势动荡,经济形势恶化,国内各行各业内卷严重,出版行业压力山大,故此我们工作室的图书定价会有小幅上调,其实这种情况在中国出版史上早就出现过.

20 世纪 20 年代末到 30 年代初期,由于受到大萧条的冲击,世界金价、银价剧烈波动,带累国内物价出现大幅度上涨. 为了减少损失,上海书业界对书价进行了调整,普遍按照定价上调 10%. 贴在一些图书版权页上标有“同业公议照码加一成”的小纸条便成了这段历史的见证.

刘培杰

2024 年 9 月 1 日

于哈工大

国外优秀数学著作
原版丛书（第三十辑）

Fast Start Advanced Calculus

高等微积分快速入门

（英文）

［美］丹尼尔·阿什洛克（Daniel Ashlock） 著

哈爾濱工業大學出版社
HARBIN INSTITUTE OF TECHNOLOGY PRESS

黑版贸登字 08-2021-031 号

图书在版编目(CIP)数据

高等微积分快速入门=Fast Start Advanced Calculus:英文/(美)丹尼尔·阿什洛克(Daniel Ashlock)著.—哈尔滨:哈尔滨工业大学出版社,2024.10
(国外优秀数学著作原版丛书.第三十辑)
ISBN 978-7-5767-1340-4

Ⅰ.①高… Ⅱ.①丹… Ⅲ.①微积分-英文 Ⅳ.①O172

中国国家版本馆 CIP 数据核字(2024)第 073689 号

GAODENG WEIJIFEN KUAISU RUMEN

策划编辑 刘培杰 杜莹雪
责任编辑 刘家琳 张嘉芮
封面设计 孙茵艾
出版发行 哈尔滨工业大学出版社
社　　址 哈尔滨市南岗区复华四道街 10 号 邮编 150006
传　　真 0451-86414749
网　　址 http://hitpress. hit. edu. cn
印　　刷 哈尔滨圣铂印刷有限公司
开　　本 889 mm×1 194 mm 1/16 印张 45.5 字数 641 千字
版　　次 2024 年 10 月第 1 版 2024 年 10 月第 1 次印刷
书　　号 ISBN 978-7-5767-1340-4
定　　价 252.00 元(全 4 册)

Contents

Preface

This text covers multi-variable integral and differential calculus, presuming familiarity with the single variable techniques from the precursor texts *Fast Start Differential Calculus* and *Fast Start Differential Calculus*. The texts were developed for a course that arose from a perennial complaint by the physics department at the University of Guelph that the introductory calculus courses covered topics roughly a year after they were needed. In an attempt to address this concern, a multi-disciplinary team created a two-semester integrated calculus and physics course. This book covers the integral calculus topics from that course as well a material on the behavior of polynomial function. The philosophy of the course was that the calculus will be delivered before it is needed, often just in time, and that the physics will serve as a substantial collection of motivating examples that will anchor the student's understanding of the mathematics.

The course has run three times before this text was started, and it was used in draft form for the fourth offering of the course, and then for two additional years. There is a good deal of classroom experience and testing behind this text. There is also enough information to confirm our hypothesis that the course would help students. The combined drop and flunk rate for this course is consistently under 3%, where 20% is more typical for first-year university calculus. Co-instruction of calculus and physics works. It is important to note that we did not achieve these results by watering down the math. The topics covered, in two semesters, are about half again as many as are covered by a standard first-year calculus course. That's the big surprise: covering more topics faster increased the average grade and reduced the failure rate. Using physics as a knowledge anchor worked even better than we had hoped.

This text, and its two companion volumes, *Fast Start Differential Calculus* and *Fast Start Integral Calculus*, make a number of innovations that have caused mathematical colleagues to raise objections. In mathematics it is traditional, even dogmatic, that math be taught in an order in which no thing is presented until the concepts on which it rests are already in hand. This is correct, useful dogma for mathematics students. It also leads to teaching difficult proofs to students who are still hungover from beginning-of-semester parties. This text neither emphasizes nor neglects theory, but it does move theory away from the beginning of the course in acknowledgment of the fact that this material is philosophically difficult and intellectually challenging. The course also presents a broad integrated picture as soon as possible. The text also emphasizes cleverness and computational efficiency. Remember that "mathematics is the art of avoiding calculation."

It is important to state what was sacrificed to make this course and this text work the way they do. This is not a good text for math majors, unless they get the theoretical parts of calculus later in a real analysis course. The text is relatively informal, almost entirely example driven, and application motivated. The author is a math professor with a CalTech Ph.D. and three decades of experience teaching math at all levels from 7th grade (as a volunteer) to graduate education including having supervised a dozen successful doctoral students. The author's calculus credits include calculus for math and engineering, calculus for biology, calculus for business, and multivariate and vector calculus.

Daniel Ashlock
August 2019

Acknowledgments

This text was written for a course developed by a team including my co-developers Joanne M. O'Meara of the Department of Physics and Lori Jones and Dan Thomas of the Department of Chemistry, University of Guelph. Andrew McEachern, Cameron McGuinness, Jeremy Gilbert, and Amanda Saunders have served as head TAs and instructors for the course over the last six years and had a substantial impact on the development of both the course and this text. Martin Williams, of the Department of Physics at Guelph, has been an able partner on the physics side delivering the course and helping get the integration of the calculus and physics correct. I also owe six years of students thanks for serving as the test bed for the material. Many thanks to all these people for making it possible to decide what went into the text and what didn't. I also owe a great debt to Wendy Ashlock and Cameron McGuinness at Ashlock and McGuinness Consulting for removing a large number of errors and making numerous suggestions to enhance the clarity of the text. This is the fourth edition that corrects several mistakes and adds a very modest number of topics.

Daniel Ashlock
August 2019

CHAPTER 1

Advanced Derivatives

This chapter deals with the issues of derivatives of functions that have multiple independent variables. In our earlier studies we learned to take derivatives of functions of the form $y = f(x)$. Now we will learn to work with functions of the form $z = f(x, y)$, functions that graph as surfaces in a space like the one shown in Figure 1.1.

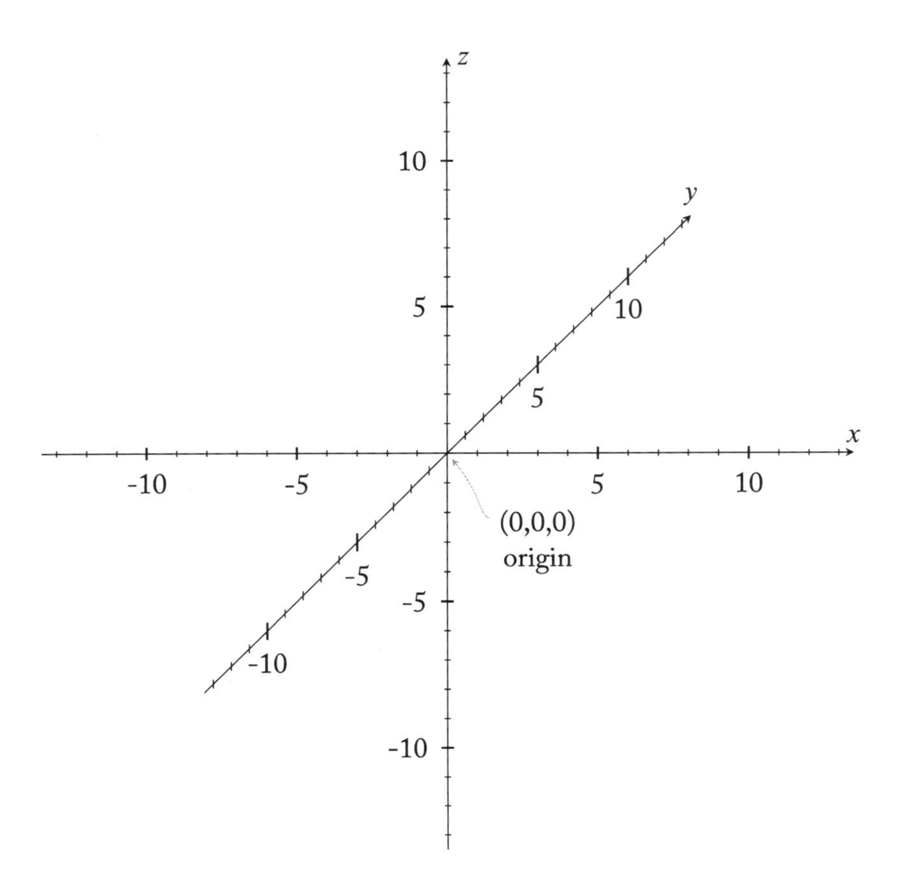

Figure 1.1: Three-dimensional coordinate system.

Where before we had points (x, y), we now have points (x, y, z). We've looked at the graph of functions like $y = x^2$ over and over. Now let's look at the three-dimensional analog:

$$z = x^2 + y^2$$

for $-4 \leq x, y \leq 4$. A graph of the function is shown in Figure 1.2.

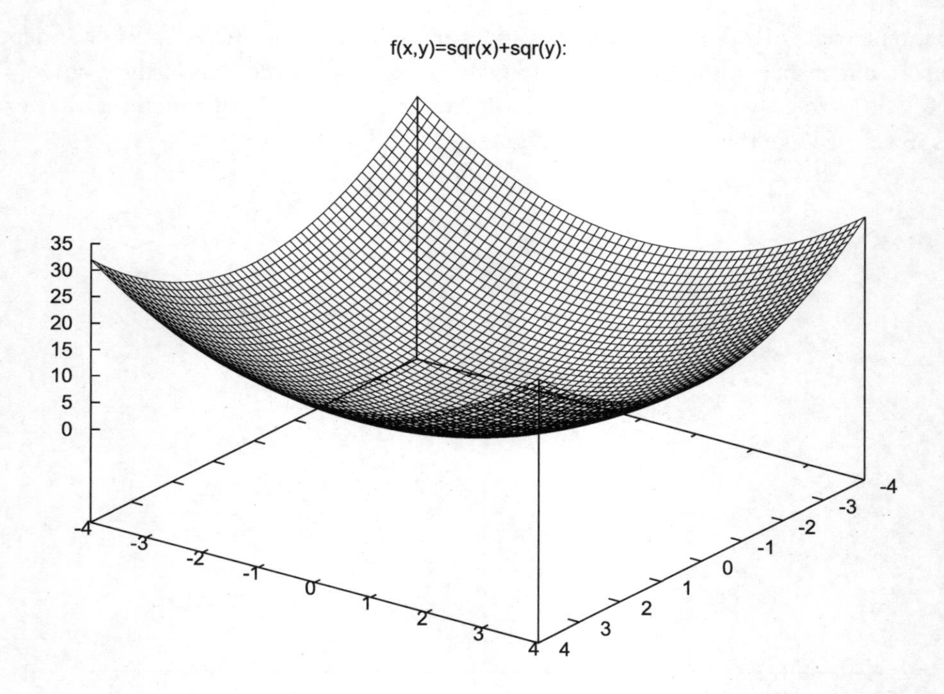

Figure 1.2: A graph of $f(x, y) = x^2 + y^2$ for $-4 \leq x, y \leq 4$.

1.1 PARTIAL DERIVATIVES

So far in this text we have had one independent and one dependent variable. Now we have two independent variables, which has a huge effect on derivatives. The number of directions on the number line is two – left and right or plus and minus. As we learned in *Fast Start Integral Calculus*, there are an infinite number of directions when there are two variables to choose directions among. Each unit vector starting at the origin points in a different direction. Every single vector can be written as:

$$\vec{v} = (a, b) = a \cdot (1, 0) + b \cdot (0, 1)$$

In other words, each vector is a combination of the two fundamental vectors $\vec{e}_1 = (1, 0)$ and $\vec{e}_2 = (0, 1)$.

On the surface of a function $z = f(x, y)$, the function has a rate of change in every direction at every point. Pick a direction – the slope in that direction is the rate of change in that direction. The basis for all of these rates of change are the *partial derivatives* – derivatives in the direction of x and y.

Knowledge Box 1.1

Partial Derivatives

If $z = f(x, y)$ is a function of two variables, then there are two fundamental derivatives

$$z_x = \frac{\partial f}{\partial x} \text{ and } z_y = \frac{\partial f}{\partial y}$$

These are called the **partial derivatives** *of z (or f) with respect to x and y, respectively.*

In order to take the partial derivative of a function with respect to one variable, all other variables are treated as constants. This is most easily understood through examples.

Example 1.1 Suppose that $z = x^2 + 3xy + y^2$. Find z_x and z_y.

Solution:

$$z_x = 2x + 3y \text{ and } z_y = 3x + 2y$$

To see this, notice that, when we are taking the derivative with respect to x, the derivative of $3xy$ is $3y$, *because y is treated as a constant.* Similarly, the derivative of $3xy$ with respect to y is $3x$. The derivatives for x^2 and y^2 are $2x$ and $2y$ when they are the active variable and zero when the other variable is active, because the derivative of a constant is zero.

◊

It is important to remember that every derivative rule we have learned so far applies when we are taking partial derivatives – the product, quotient, and chain rules and all the individual formulas for functions.

Example 1.2 Find $\frac{\partial f}{\partial x}$ if

$$f(x, y) = \frac{x}{x^2 + y^2}$$

Solution:

This problem requires the quotient rule. Since a "prime" hash-mark doesn't carry its identity (with respect to x or with respect to y) we used the symbol $\frac{\partial}{\partial x}$ to mean "derivative with respect to x."

This means that

$$\begin{aligned}\frac{\partial}{\partial x}\left(\frac{x}{x^2 + y^2}\right) &= \frac{\frac{\partial}{\partial x}x \cdot (x^2 + y^2) - \frac{\partial}{\partial x}(x^2 + y^2) \cdot x}{(x^2 + y^2)^2} \\ &= \frac{1 \cdot (x^2 + y^2) - 2x \cdot x}{(x^2 + y^2)^2} \\ &= \frac{y^2 - x^2}{(x^2 + y^2)^2}\end{aligned}$$

◊

The next example uses the chain rule. It also uses the additional notation "f_y" as another way of saying "the partial derivative of $f(x, y)$ with respect to y."

Example 1.3 Find $f_y(x, y)$ if

$$f(x, y) = \sin(2xy + 1)$$

Solution:

$$f_y(x, y) = \cos(2xy + 1) \cdot 2x$$

◊

As long as we remember that y is the active variable and x is treated as a constant, this is not difficult.

Example 1.4 Find the partial derivatives with respect to x and y of

$$f(x, y) = (3x + 4y)^5$$

Solution:

$$f_x(x, y) = 5(3x + 4y)^4 \cdot 3 \qquad f_y(x, y) = 5(3x + 4y)^4 \cdot 4$$

The only point where things are different is the way the chain rule acts depending on if x or y is the active variable.

◊

Example 1.5 Find the partial derivatives with respect to x and y for

$$f(x, y) = \frac{x}{y}$$

Solution:

$$f_x(x, y) = \frac{1}{y} \qquad f_y(x, y) = \frac{-x}{y^2}$$

For f_x, $\frac{1}{y}$ is effectively a constant, while, for f_y, we employ the reciprocal rule while x plays the part of a constant.

◊

1.1.1 IMPLICIT PARTIAL DERIVATIVES

Since natural laws are often stated in the form of equations that are not in functional form, implicit derivatives are very useful in physics. It turns out that implicit partial derivatives are a lot like standard partial derivatives – as long as you remember which variable is active.

Example 1.6 Find z_x and z_y if $x^2 + y^2 + z^2 = 16$.

Solution:

z is the dependent variable and so gets a z_x or a z_y each time we take a derivative of it, while x and y take turns being the active variable and a constant, respectively. So,

$$2x + 2z \cdot z_x = 0 \text{ and } 2y + 2z \cdot z_y = 0$$

Simplifying we get that

$$z_x = -\frac{x}{z} \text{ and } z_y = -\frac{y}{z}$$

◊

Example 1.7 Find z_y if

$$(xyz + 1)^3 = 6y$$

Solution:

In this example z is the dependent variable, y is the active variable, and x is acting like a constant. So:

$$\begin{aligned} 3(xyz+1)^2(xz + xy \cdot z_y) &= 6 \\ xz + xy \cdot z_y &= \frac{2}{(xyz+1)^2} \\ z_y &= \frac{2}{xy \cdot (xyz+1)^2} - \frac{z}{y} \end{aligned}$$

◊

1.1.2 HIGHER-ORDER PARTIAL DERIVATIVES

In earlier chapters, when we wanted the second derivative of a function we just took the derivative again. The fact that we have multiple choices of which derivative to take complicates this. If the first derivative is with respect to x or y and so is the second, then we get four possible nominal second derivatives:

$$f_{xx} \qquad f_{xy} \qquad f_{yx} \qquad f_{yy}$$

A useful fact saves us from having the number of higher-order partial derivatives explode.

Knowledge Box 1.2

When working with $f(x, y)$,

$$f_{xy}(x, y) = f_{yx}(x, y)$$

and in general the order in which partial derivatives are taken does not affect the result of taking them.

Example 1.8 If

$$f(x, y) = x^3 + 3xy + y^2$$

find f_{xx}, f_{xy}, f_{yx}, and f_{yy}, and verify that $f_{xy} = f_{yx}$.

Solution:

Start by finding f_x and f_y and then keep going.

$$f_x = 3x^2 + 3y$$

$$f_y = 3x + 2y$$

$$f_{xx} = 6x$$

$$f_{xy} = 3$$

$$f_{yx} = 3$$

$$f_{yy} = 2$$

And we see that $f_{xy} = 3 = f_{yx}$, achieving the desired verification. From this point on we will only compute one of the two mixed partials f_{xy} and f_{yx}.

◊

The "order doesn't matter" rule means, for example, that $f_{xxy} = f_{xyx} = f_{yxx}$. So the degree to which this rule reduces the number of higher-order derivatives increases with the order of the derivative.

Example 1.9 Find f_{xx}, f_{xy}, and f_{yy} for

$$f(x, y) = x^2 \sin(y).$$

Solution:
Compute the first partials first and keep going.

$$f_x = 2x\sin(y)$$

$$f_y = x^2\cos(y)$$

$$f_{xx} = 2\sin(y)$$

$$f_{xy} = 2x\cos(y)$$

$$f_{yy} = -x^2\sin(y)$$

◊

Example 1.10 Find f_{xy} for

$$f(x,y) = \frac{xy}{x^2+1}.$$

Solution:

The fact that we get the same result by computing the partials with respect to x and y in either order means that we may choose the order to minimize our work. The order y then x is easier, because y vanishes.

$$f(x,y) = y \cdot \frac{x}{x^2+1}$$

$$f_y = \frac{x}{x^2+1}$$

$$f_{xy} = \frac{(x^2+1)(1) - x(2x)}{(x^2+1)^2} \qquad \text{Quotient rule.}$$

$$= \frac{1-x^2}{(x^2+1)^2}$$

◊

PROBLEMS

Problem 1.11 For each of the following functions find f_x and f_y.

1. $f(x, y) = 2x^2 + 3xy + y^2 + 4x + 2y + 7$
2. $g(x, y) = \sin(xy)$
3. $h(x, y) = x^3y^3$
4. $r(x, y) = \ln(x^2 + y^2 + 1)$
5. $s(x, y) = \dfrac{1}{x^2 + y^2 + 1}$
6. $q(x, y) = \dfrac{x - y}{x + y}$
7. $a(x, y) = e^{x \sin(y)}$
8. $b(x, y) = (x^2 + xy + 1)^6$

Problem 1.12 For each of the following functions find f_{xx}, f_{xy}, and f_{yy}.

1. $f(x, y) = x^2 - 5xy + 2y^2 + 3x - 6y + 11$
2. $g(x, y) = \tan^{-1}(xy)$
3. $h(x, y) = (x^3 + 1)(y^3 + 1)$
4. $r(x, y) = e^{x^2+y^2-5}$
5. $s(x, y) = \dfrac{1}{x^2 + y^2 + 1}$
6. $q(x, y) = \dfrac{2x - 3y}{5x + y}$
7. $a(x, y) = \ln(x \sin(y))$
8. $b(x, y) = (2x^2 - xy)^4$

Problem 1.13 Find z_x and z_y if $(xyz + 2)^3 = 4$.

Problem 1.14 Find z_x and z_y if $\dfrac{x}{yz} = 1$.

Problem 1.15 Find z_x and z_y if $\dfrac{x + z}{y + z} = 4$.

Problem 1.16 Find z_x and z_y if $\dfrac{3x + z}{2z} = 4xy$.

Problem 1.17 Find z_x and z_y if $\cos(x + y + z) = \dfrac{\sqrt{2}}{2}$.

Problem 1.18 Find z_x and z_y if $\tan^{-1}(z - xy) = 1$.

Problem 1.19 Find z_x and z_y if $(xy + xz + yz)^3 = 16$.

Problem 1.20 For each of the following functions find f_{xy}.

1. $f(x, y) = x^2y + x^3y + y\sin(x)$
2. $g(x, y) = x\sin(y) + x\tan^{-1}(x)$
3. $h(x, y) = xy^2 + x^2y + xy$
4. $r(x, y) = \left(x(y + 1)^5 + 1\right)^2$
5. $s(x, y) = x \cdot \sin(xy)$
6. $q(x, y) = y \cdot \cos(xy)$
7. $a(x, y) = \dfrac{x}{y} + \dfrac{y}{x} + y^3$
8. $b(x, y) = (1 + x + y + xy)^4$

Problem 1.21 Find f_x, f_y, f_{xx}, f_{xy}, and f_{yy} if $f(x, y) = \dfrac{x^2}{x^2 + y^2}$.

Problem 1.22 Find g_x, g_y, g_{xx}, g_{xy}, and g_{yy} if $g(x, y) = \tan^{-1}(xy + 1)$.

Problem 1.23 Rind f_{xx}, f_{xxy}, and f_{xxyy} if $h(x, y) = \sin(xy)$.

Problem 1.24 Find z_x and z_y if $z = (x^2 + 1)^y$.

Problem 1.25 Find z_x and z_y if $z^{xy} = 2$.

Problem 1.26 Find z_x and z_y if $z = \dfrac{x^3(x + 1)^2(x - 1)^3}{y^2(y - 1)^3(y + 1)^5}$.

1.2 THE GRADIENT AND DIRECTIONAL DERIVATIVES

We are now ready to look at some of the opportunities that are available once we understand partial derivatives. At any point on the surface that forms the graph of $z = f(x, y)$ there are an infinite number of directions and so an infinite number of rates at which the function is changing. Pick a direction, and the function has a rate of change *in that direction.*

It turns out that there is some order to this richness of directions and rates of growth, in the form of a simple formula for the direction in which the function is growing fastest.

Knowledge Box 1.3

> *If $z = f(x, y)$ is a function of two variables, then*
>
> $$\nabla f(x, y) = (f_x, f_y)$$
>
> *is called the* **gradient** *of $f(x, y)$. The gradient of a function points in the direction it is growing most quickly; the rate of growth is the magnitude of the gradient.*

Example 1.27 Find the gradient of the function $f(x, y) = x^2 + y^2 + 3xy$.

Solution:

Using the formula given, $\nabla f(x, y) = (2x + 3y, 3x + 2y)$.

$$\diamondsuit$$

We can ask much more complex questions about the gradient than simply computing its value.

Example 1.28 At what points is the function

$$g(x, y) = \sin(x) + \cos(y)$$

changing the fastest in its direction of maximum increase?

Solution:

This question wants us to maximize the magnitude of the gradient.

First compute the gradient:

$$\nabla g(x, y) = (\cos(x), -\sin(y))$$

The magnitude of this is

$$\sqrt{\cos^2(x) + \sin^2(y)}$$

Since x and y vary independently, the answer is simply those points that make $\cos^2(x)$ and $\sin^2(y)$ both one.

So, the answer is those points (x, y) such that

$$x = n\pi \text{ and } y = \frac{2m + 1}{2}\pi$$

where n and m are whole numbers.

◊

The graph in Figure 1.3 might help you understand Example 1.28.

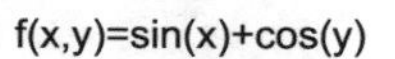

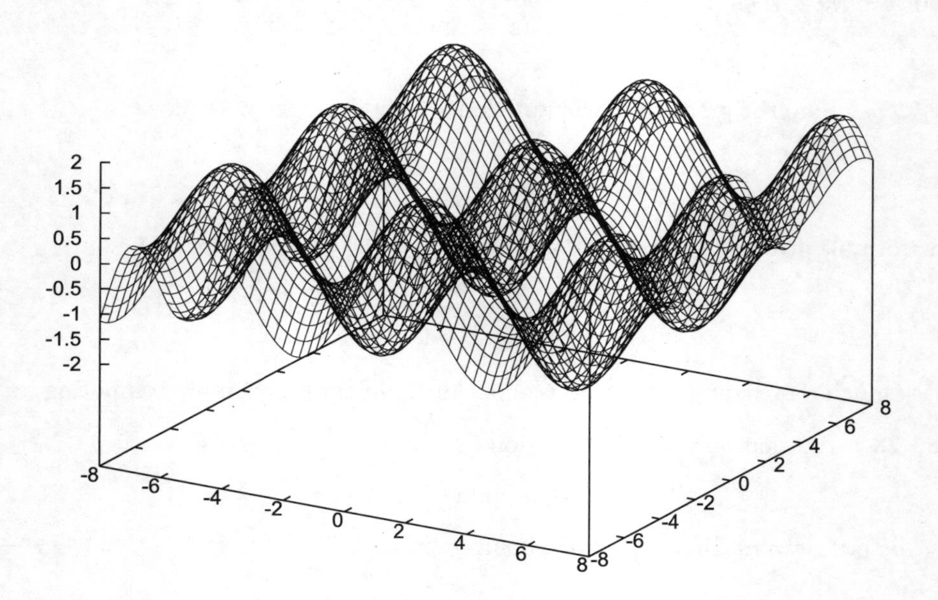

Figure 1.3: A graph of $z = \sin(x) + \cos(y)$ for $-8 \leq x, y \leq 8$.

Example 1.29 Find the gradient of

$$z = x^y$$

Solution:

The partial derivative with respect to x is not hard because y is treated as a constant – so $z_x = yx^{y-1}$. The partial derivative with respect to y is trickier. It uses the formula for the derivative of a constant to a variable power which gives $z_y = x^y \cdot \ln(x)$.

This makes the gradient

$$\nabla z = \left(yx^{y-1}, x^y \cdot \ln(x)\right)$$

◊

What is the physical meaning of the gradient? We have already noted that it points in the direction in which a surface grows fastest away from the point where the gradient is computed – the steepest uphill slope away from the point. The magnitude of the gradient also gives of the *rate* of fastest growth. Another fact is that the negative of the function

$$-\nabla f(x, y)$$

is the steepest downhill slope away from the point.

In other words, the negative of the gradient is the direction that a ball, starting at rest, will roll.

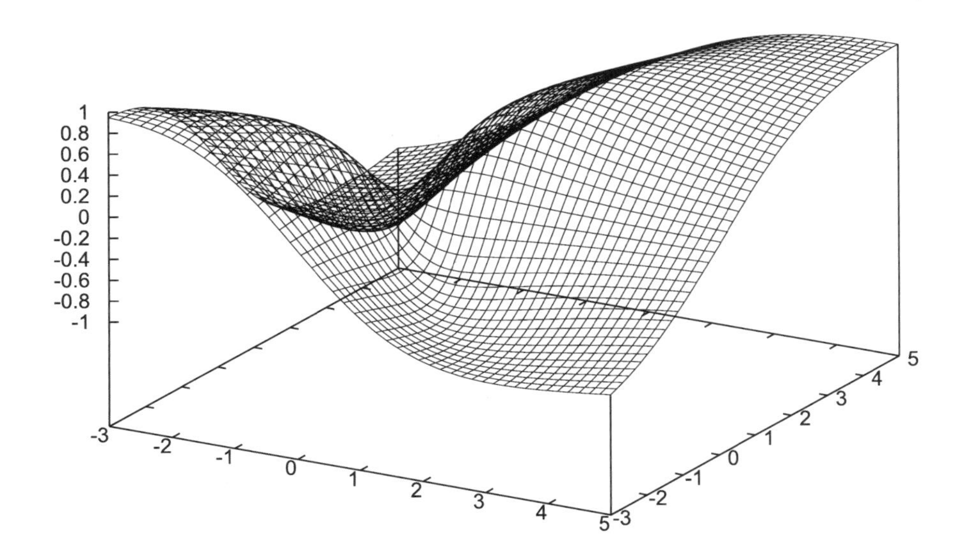

Figure 1.4: A graph of $H(x, y) = \dfrac{2x + y}{x^2 + y^2 + 1}$ for $-3 \le x, y \le 5$

Example 1.30 Suppose that the function $H(x, y) = \dfrac{2x + y}{x^2 + y^2 + 1}$ (Figure 1.4) describes the height of a surface. Which direction will a ball placed at the point $(1, 2)$ roll?

Solution:

We need to compute the negative of the gradient of $H(x, y)$.

$$H_x(x, y) = \frac{(x^2 + y^2 + 1)(2) - (2x + y)(2x)}{(x^2 + y^2 + 1)^2}$$

$$H_x(1, 2) = \frac{12 - 8}{36} = 1/9$$

$$H_y(x, y) = \frac{(x^2 + y^2 + 1)(1) - (2x + y)(2y)}{(x^2 + y^2 + 1)^2}$$

$$H_y(1, 2) = \frac{6 - 8}{36} = -1/18$$

So the ball rolls in the direction of the vector

$$-\nabla H(1, 2) = \left(\frac{-1}{9}, \frac{1}{18}\right)$$

◊

Notice that, if we were designing a game, then we could use a well-chosen equation to give us a height map for the surface, and the gradient could be used to tell which ways balls would roll and water would flow.

Visualization helps us understand functions. This leads to the question: what does a gradient look like? The gradient of a function $f(x, y)$ assigns a vector to each point in space. That means that we could get an idea of what a gradient looks like by plotting the vectors of the gradient on a grid of points.

Example 1.31 Let $f(x, y) = x^2 + y^2$. For all points (x, y) with coordinates in the set $\{\pm 2, \pm 1.5, \pm 1, \pm 0.5, 0\}$, plot the point and the vector starting at that point in the direction $\nabla f(x, y)$.

Solution:

Since $\nabla f(x, y) = (2x, 2y)$ the vectors are:

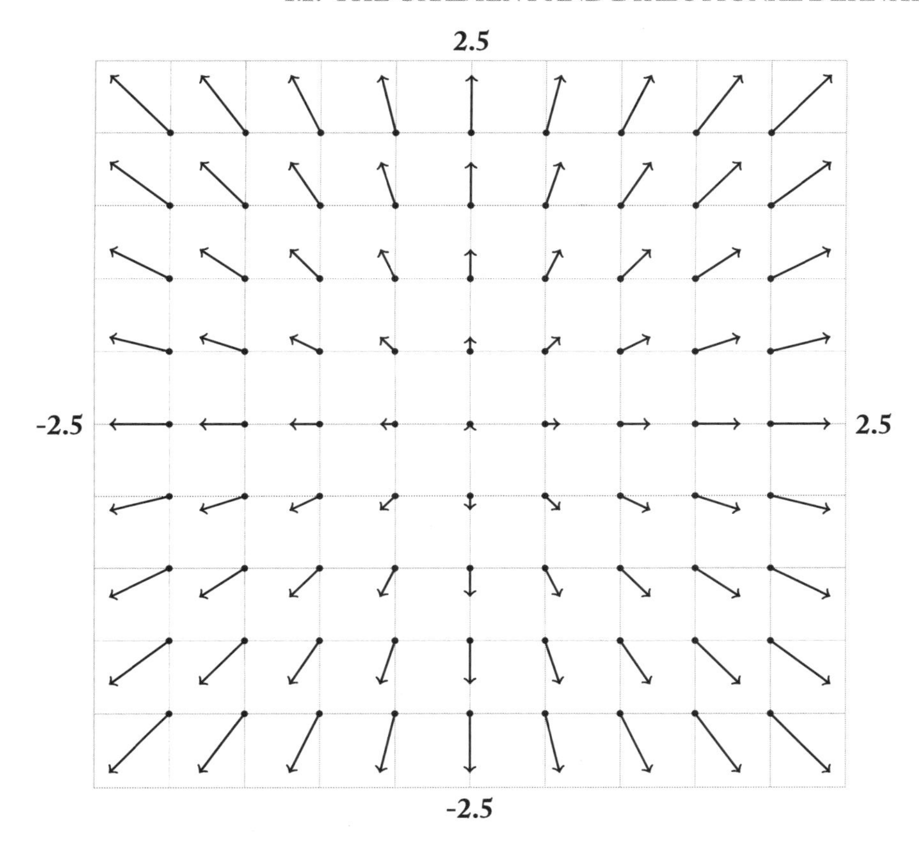

Notice that all the gradient vectors point directly away from the origin. This corresponds to our notion that the gradient is the direction of fastest growth.

◊

The function $\dfrac{2x+y}{x^2+y^2+1}$ from Example 1.30, the rolling ball question, will probably have a more interesting gradient than the simple paraboloid in Example 1.31.

Example 1.32 Let $H(x,y) = \dfrac{2x+y}{x^2+y^2+1}$. For all points (x,y) with coordinates in the set $\{\pm 3, \pm 2.5, \pm 2, \pm 1.5, \pm 1, \pm 0.5, 0\}$, plot the point and the vector starting at that point in the direction $\nabla H(x,y)$.

Solution:

We are plotting the vectors drawn from the gradient

$$\nabla H(x, y) = \left(\frac{2y^2 - 2x^2 - 2xy + 2}{(x^2 + y^2 + 1)^2}, \frac{x^2 - y^2 - 4xy + 1}{(x^2 + y^2 + 1)^2} \right)$$

which yields the picture:

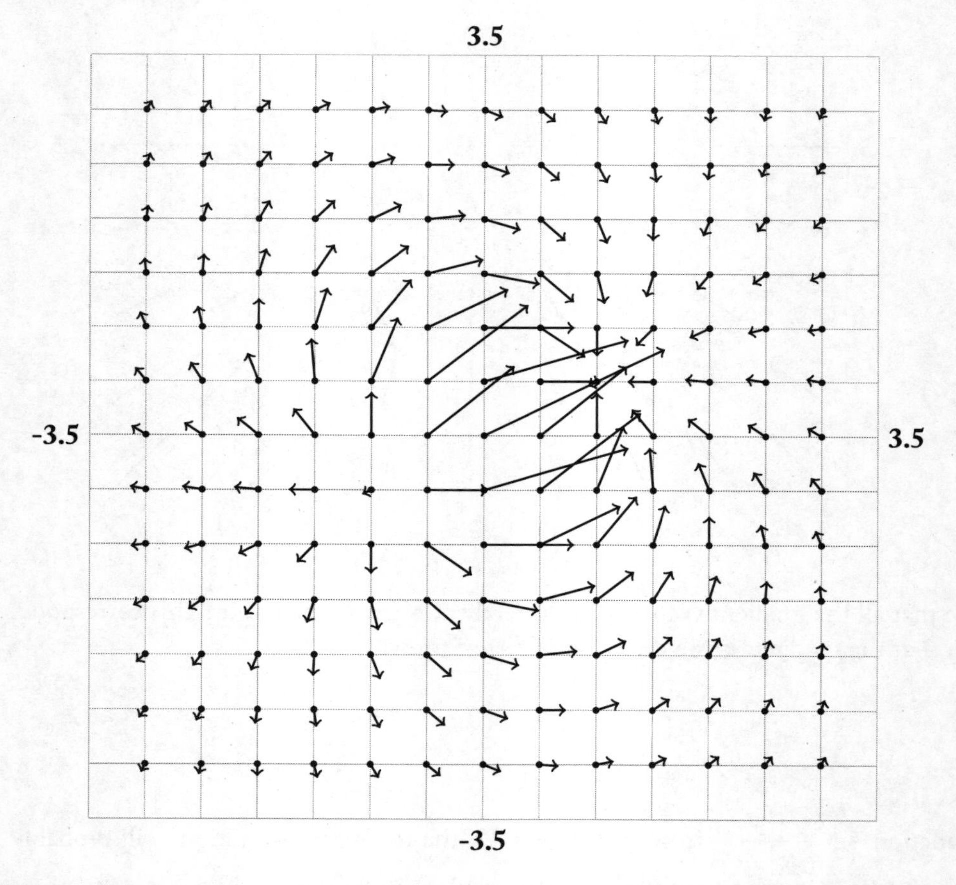

The gradient vectors point in many directions. A ball rolling on this surface could potentially have a very complex path.

◊

Example 1.33 Find a reasonable sketch of the vector field associated with the gradient of $g(x, y) = \sin(x) + \cos(y)$ from Example 1.28. Use grid points with coordinates $\pm n$ for $n = 0, 1, \ldots 8$.

Solution:
We are plotting the vectors drawn from the gradient

$$\nabla g(x, y) = (\cos(x), -\sin(y))$$

which yields the picture:

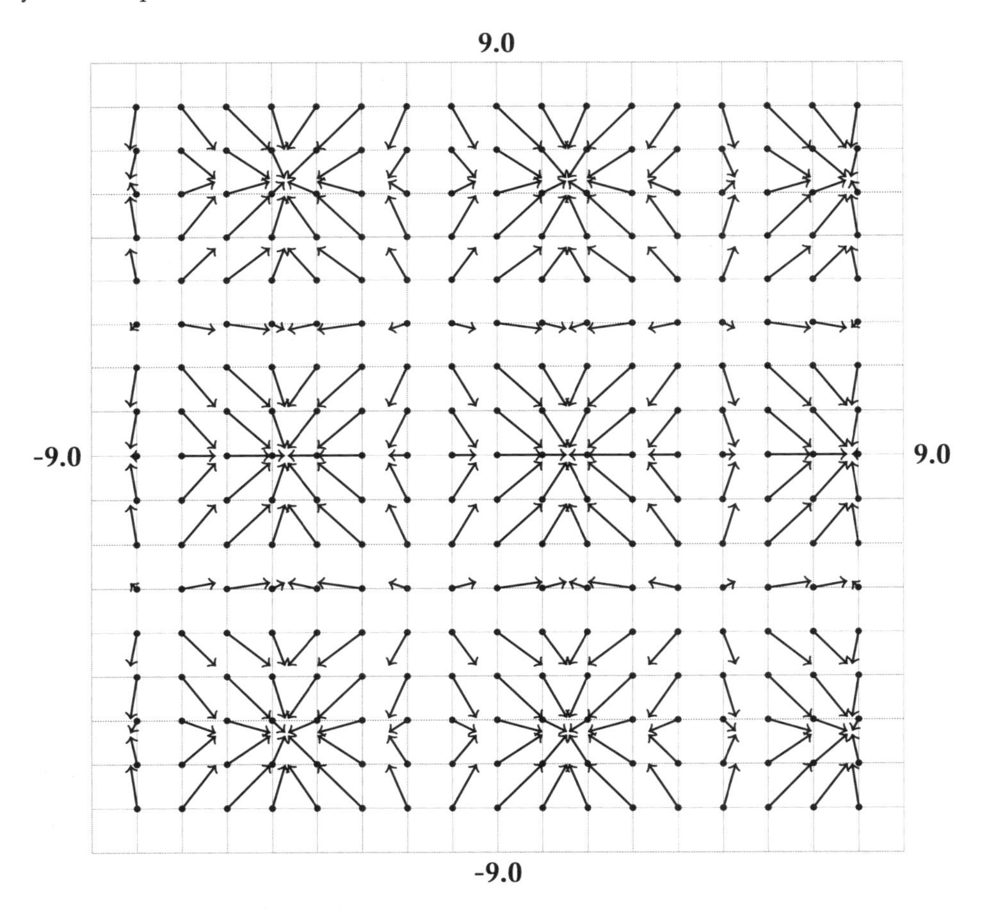

The periodicity of the vector field is easy to see. The different cells of the periodicity are slightly different because the periods are multiples of π, and the grid used to display the vectors is sized in multiples of one.

◊

Now that we have the gradient, it is possible to define the derivative in a particular direction.

Knowledge Box 1.4

If $z = f(x, y)$ is a differentiable function of two variables, and $\vec{u} = (a, b)$ is a unit vector, then the **derivative of $f(x, y)$ in the direction of $\vec{u}$ is:**

$$\nabla_{\vec{u}} f(x, y) = \vec{u} \cdot \nabla f(x, y) = af_x + bf_y$$

If $\vec{v} = (r, s)$ is any vector, then the **derivative of $f(x, y)$ in the direction of $\vec{v}$ is:**

$$\nabla_{\vec{v}} f(x, y) = \frac{\vec{v}}{|\vec{v}|} \cdot \nabla f(x, y)$$

Notice that we are continuing the practice of using unit vectors to designate directions – even when computing the derivative of a function in the direction of a general vector, we first coerce it to be a unit vector.

It is worth mentioning that, when computing directional derivatives, we start with a scalar quantity – the function $f(x, y)$. When we compute the gradient, we get the vector quantity $\nabla f(x, y) = (f_x, f_y)$ but then return to a scalar function of two variables $\nabla_{\vec{u}} f(x, y)$. It is important to keep track of the type of object – scalar or vector – that you are working with.

Example 1.34 Find the derivative of

$$f(x, y) = x^2 + y^2$$

in the direction of $\vec{u} = (1/2, \sqrt{3}/2)$.

Solution:

The vector $\vec{u}$ is a unit vector so, starting with $f_x = 2x$, $f_y = 2y$, we get:

$$\nabla_{\vec{u}} f(x, y) = \frac{1}{2} 2x + \frac{\sqrt{3}}{2} 2y = x + \sqrt{3} y$$

◊

The directional derivative occasionally comes up in the natural course of trying to solve a problem. There is one very natural application: finding level curves. First let's define level curves.

Knowledge Box 1.5

Level Curves

If $z = f(x, y)$ defines a surface, then the **level curve of height c** *of $f(x, y)$ is the set of points that solve the equation*

$$f(x, y) = c.$$

Example 1.35 Plot the level curves for $c \in \{1, 2, 3, 4, 5, 6\}$ for

$$f(x, y) = x^2 + 2y^2$$

Solution:

The equation $x^2 + 2y^2 = c$ is an ellipse that is $\sqrt{2}$ times as far across in the x direction as the y directions. Solving for the points where $x = 0$ or $y = 0$ gives us the extreme points of the ellipse for each value of c, and we get the following picture.

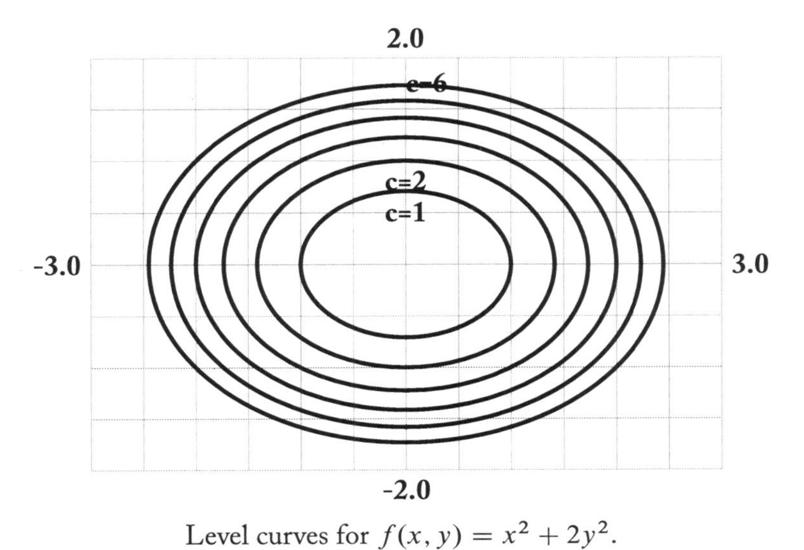

Level curves for $f(x, y) = x^2 + 2y^2$.

Since the values of c used to derive the level curves are equally spaced, the fact that the curves are getting closer together gives a sense of how steeply the graph of $f(x, y) = x^2 + 2y^2$ is sloped.

◊

In Example 1.35 we can plot the level curves because we recognize them as ellipses. In general, that's not going to work. If we supply a graphics system with information about the gradient of a curve, then it can sketch level curves by following the correct directional derivative.

Knowledge Box 1.6

If $z = f(x, y)$ defines a surface, then the level curve at the point (a, b) proceeds in the direction of the unit vector $\vec{u}$ such that

$$|\nabla_{\vec{u}} f(a, b)| = 0.$$

In other words, it proceeds in the direction such that the height of the graph of $f(x, y)$ is not changing.

Example 1.36 Find, in general, the direction of the level curves of

$$f(x, y) = x^2 + 3y^2.$$

Solution:

To find the general direction of level curves, we need to solve this equation for $\vec{u}$.

$$|\nabla_{\vec{u}} f(a, b)| = 0$$

Let $\vec{u} = (a, b)$.

$$\nabla\left(x^2 + 3y^2\right) \cdot (a, b) = 0$$

$$(2x, 6y) \cdot (a, b) = 0$$

$$2xa + 6yb = 0$$

$$2xa = -6yb$$

$$a = -3\left(\frac{y}{x}\right) b$$

Giving us the direction in which the level curves go.

If we set $b = 1$, then, at a given point (x, y) in space, the level curve at that point is in the direction $\left(-\frac{3y}{x}, 1\right)$. This encodes the direction.

To be thorough, let's turn this into a unit vector.

$$\left|\left(-\frac{3y}{x}, 1\right)\right| = \sqrt{\left(-\frac{3y}{x}\right)^2 + 1^2} = \sqrt{\frac{9y^2}{x^2} + 1} = \frac{1}{x}\sqrt{9y^2 + x^2}$$

So, the direction, as a unit vector is:

$$\left(-\frac{3y}{\sqrt{9y^2 + x^2}}, \frac{x}{\sqrt{9y^2 + x^2}}\right)$$

Notice that if we had chosen $b = -1$, we would have gotten

$$\left(\frac{3y}{\sqrt{9y^2 + x^2}}, -\frac{x}{\sqrt{9y^2 + x^2}}\right)$$

which is an equally valid solution.

The level curve points in two opposite directions.

A natural question at this point is: what do the vectors we found in Example 1.36 look like?

Example 1.37 Using the solution to Example 1.36 plot the directions of the level curves as a vector field.

Solution:

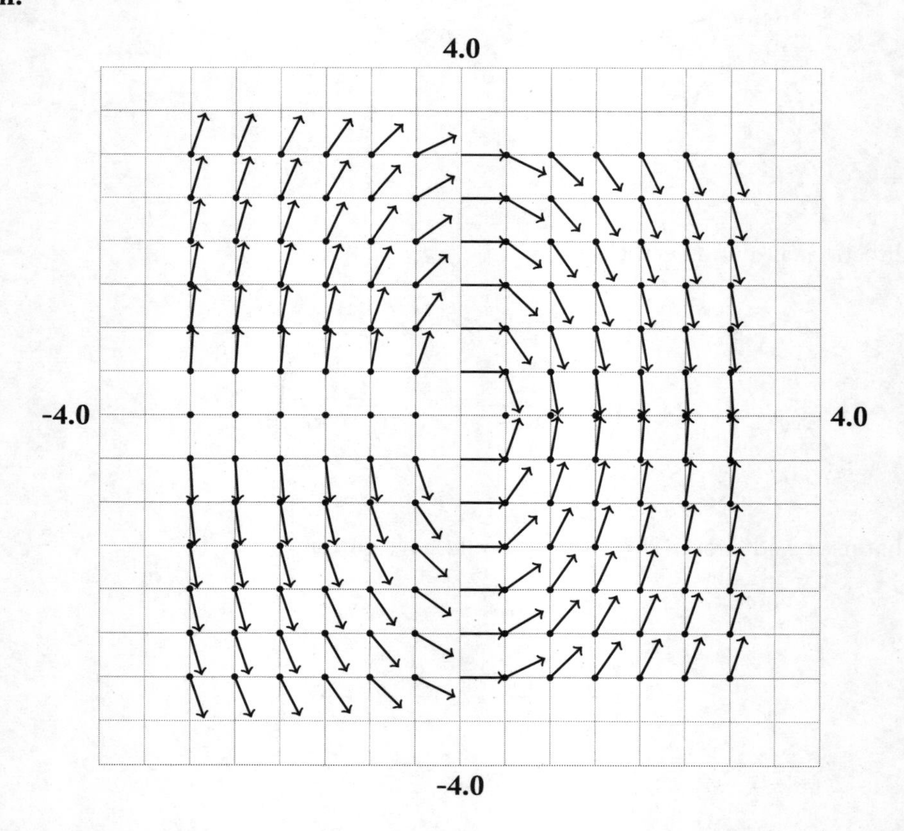

At each point, there are actually two vectors that point in the direction of the level curve. If $\vec{u}$ points in the direction of the level curve, then so does $-\vec{u}$. The figure above sometimes shows one and sometimes the other. The curves that these vectors are following track the vertically long ellipses that one would expect as level curves of:

$$f(x, y) = x^2 + 3y^2$$

◊

PROBLEMS

Problem 1.38 For each of the following functions, find their gradient.

1. $f(x, y) = x^2 + 4xy + 7y^2 + 2x - 5y + 1$
2. $g(x, y) = \sin(xy)$
3. $h(x, y) = \dfrac{x^2}{y^2 + 1}$
4. $r(x, y) = \left(x^2 + y^2\right)^{3/2}$
5. $s(x, y) = \sin(x) + \cos(y)$
6. $q(x, y) = x^5 y + xy^5 + x^3 y^3 + 1$
7. $a(x, y) = \dfrac{x}{x^2 + y^2}$
8. $b(x, y) = x^y$

Problem 1.39 For the function

$$g(x, y) = x^2 + 3xy + y^2,$$

remembering that directions should be reported as unit vectors, find:

1. The greatest rate of growth of the curve in any direction at the point $(1, 1)$
2. The direction of greatest growth at $(1, 1)$
3. The greatest rate of growth of the curve in any direction at the point $(-1, 2)$
4. The direction of greatest growth at $(-1, 2)$
5. The greatest rate of growth of the curve in any direction at the point $(0, 3)$
6. The direction of greatest growth at $(0, 3)$

Problem 1.40 Suppose we are trying to find level curves. Is it possible to find points where there are more than two directions in which the surface does not grow? Either explain why this cannot happen or give an example where it does.

Problem 1.41 For each of the following functions, sketch the gradient vector field of the function for the points (x, y) with

$$x, y \in \{0,\ \pm 1,\ \pm 2,\ \pm 3\}.$$

Exclude points, if any, where the gradient does not exist. This is a problem where a spreadsheet may be useful for performing routine computation.

1. $f(x, y) = x^2 + y^2 + 1$
2. $g(x, y) = \sin(xy)$
3. $h(x, y) = \dfrac{x}{x + y}$
4. $r(x, y) = \sqrt{x^2 + y^2}$
5. $s(x, y) = \sin(x) + \cos(y)$
6. $q(x, y) = x^5 y + x y^5 + x^3 y^3 + 1$

Problem 1.42 Find the directional derivative of

$$f(x, y) = x^2 + y^3$$

in the direction of (1,1) at (0,0).

Problem 1.43 Find the directional derivative of

$$g(x, y) = x^2 - y^2$$

in the direction of $(\sqrt{3}/2, 1/2)$ at (2,1).

Problem 1.44 Find the directional derivative of

$$h(x, y) = (y - 2x)^3$$

in the direction of (0,-1) at (-1,1).

Problem 1.45 As in Example 1.36 find the general direction for level curves at (x, y) for the function

$$H(x, y) = \frac{1}{x^2 + y^2}$$

and plot the vector field for all points with whole number coordinates in the range $-3 \leq x, y \leq 3$.

Problem 1.46 As in Example 1.36 find the general direction for level curves at (x, y) for the function

$$H(x, y) = \frac{2}{x^2 - 2x + y^2 - 2y + 2}$$

and plot the vector field for all points with whole number coordinates in the range $-5 \leq x, y \leq 5$.

Problem 1.47 Which way will a ball placed at (1,1) on the surface given by the function in Problem 1.42 roll?

1.3 TANGENT PLANES

One of the first things we built after developing skill with the derivative was tangent lines to a curve. With a function $z = f(x, y)$ the analogous object is a tangent *plane*. There are several ways to specify a plane.

Knowledge Box 1.7

Formulas for planes

If a, b, c, and d are constants, all of the following formulas specify planes:

$$z = ax + by + c$$

$$ax + by + cz = d.$$

$$(a, b, c) \cdot (x, y, z) = d$$

Note that the third formula, while phrased in terms of a dot product, is actually the same as the second.

Let's get some practice with converting between the different possible forms of a plane.

Example 1.48 If

$$(3, -2, 5) \cdot (x, y, z) = 2$$

find the other two forms of the plane.

Solution:

$$\begin{aligned} (3, -2, 5) \cdot (x, y, z) &= 2 \\ 3x - 2y + 5z &= 2 & \text{Second form} \\ 5z &= -3x + 2y + 2 \\ z &= -0.6x + 0.4y + 0.4 & \text{First form} \end{aligned}$$

◊

The most common way to find a tangent line for a function is to take the point of tangency – which must be on the line – together with the slope of the line found by computing the derivative and use the point slope form to find the formula for the line. It turns out that there is a way of specifying planes that is similar to the point slope form of a line. Remember that if $\vec{v}$ and $\vec{w}$ are vectors that are at right angles to one another, then $\vec{v} \cdot \vec{w} = 0$.

Knowledge Box 1.8

Formula for a plane at right angles to a vector

Suppose that $\vec{v}$ is a vector at right angles to a plane in three dimensions and that (a, b, c) is a point on that plane. Then a formula for the plane is

$$\vec{v} \cdot (x - a, y - b, z - c) = 0.$$

Notice that (a, b, c) is constructively on the plane and that our knowledge of the dot product tells us it is at right angles to $\vec{v}$.

Example 1.49 Find the plane at right angles to $\vec{v} = (1, 2, 1)$ through the point $(2, -1, 5)$.

Solution:

Simply substitute into the formula given in Knowledge Box 1.8.

$$\begin{aligned} (1,2,1) \cdot (x-2, y+1, z-5) &= 0 \\ 1(x-2) + 2(y+1) + 1(z-5) &= 0 \\ x + 2y + z - 2 + 2 - 5 &= 0 \\ x + 2y + z &= 5 \end{aligned}$$

◊

Knowledge Box 1.8 seems very special purpose, but it turns out to be very useful in light of another fact. Suppose that

$$f(x, y, z) = c$$

specifies a surface. We need to expand our definition of the gradient just a bit to

$$\nabla f(x, y, z) = \big(f_x(x, y, z), f_y(x, y, z), f_z(x, y, z)\big).$$

In this case the vector $\nabla f(a, b, c)$ points directly outward from the surface $f(x, y, z) = c$ at (a, b, c) and *it is at right angles to the tangent plane.*

Knowledge Box 1.9

Formula for the tangent plane to a surface

If $f(x, y, z) = c$ defines a surface, and (a, b, c) is a point on the surface, then a formula for the tangent plane to that surface at (a, b, c) is:

$$\left(f_x(a,b,c),\, f_y(a,b,c),\, f_z(a,b,c)\right) \cdot (x-a, y-b, z-c) = 0.$$

Defining a surface in the form $f(x, y, z) = c$ is a little bit new – but in fact this is another version of level curves, just one dimension higher. Let's practice.

Example 1.50 Find the tangent plane to the surface $x^2 + y^2 + z^2 = 3$ at the point $(1, -1, 1)$.

Solution:

Check that the point is on the surface: $(1)^2 + (-1)^2 + (1)^2 = 3$ – so it is. Next find the gradient

$$\nabla x^2 + y^2 + z^2 = (2x, 2y, 2z).$$

This means that the gradient at $(1, -1, 1)$ is $\vec{v} = (2, -2, 2)$. This makes the plane

$$\begin{aligned} (2,-2,2)\cdot(x-1, y+1, z-1) &= 0 \\ 2x - 2y + 2z - 2 - 2 - 2 &= 0 \\ 2x - 2y + 2z &= 6 \\ x - y + z &= 3 \end{aligned}$$

Notice that we simplified the form of the plane; this is not required but it does make for neater answers.

◊

The problem with finding the tangent plane to a surface $f(x, y, z) = c$ is that it does not solve the original problem – finding tangent planes to $z = f(x, y)$. A modest amount of algebra solves this problem. If $z = f(x, y)$ then $g(x, y, z) = z - f(x, y) = 0$ is in the correct form for our surface techniques. This gives us a new way of finding tangent planes to a function that defines a surface in 3-space.

Knowledge Box 1.10

Formula for the tangent plane to a functional surface

If $z = f(x, y)$ defines a surface, then the tangent plane to the surface at (a, b) may be obtained as

$$(-f_x(a,b), -f_y(a,b), 1) \cdot (x-a, y-b, z-f(a,b)) = 0.$$

This is the result of applying the gradient-of-a-surface formula to the surface $z - f(x, y) = 0$ at the point $(a, b, f(a, b))$.

Example 1.51 Find the tangent plane to $f(x, y) = x^2 - y^3$ at the point $(2, -1)$.

Solution:
Assemble the pieces and plug into Knowledge Box 1.10.

$$f_x(x, y) = 2x$$

$$f_x(2, -1) = 4$$

$$f_y(x, y) = -3y^2$$

$$f_y(2, -1) = -3$$

$$f(2, -1) = 4 - (-1) = 5$$

Put the plane together

$$(-4, 3, 1) \cdot (x - 2, y + 1, z - 5) = 0$$

$$-4x + 8 + 3y + 3 + z - 5 = 0$$

$$-4x + 3y + z = -6$$

Which is the tangent plane desired.

◇

Example 1.52 Find the tangent plane to

$$g(x, y) = xy^2$$

at (3,1).

Solution:

Assemble the pieces and plug into Knowledge Box 1.10.

$$f_x(x, y) = y^2$$

$$f_x(3, 1) = 1$$

$$f_y(x, y) = 2xy$$

$$f_y(3, 1) = 6$$

$$f(3, 1) = 3$$

Put the plane together

$$(-1, -6, 1) \cdot (x - 3, y - 1, z - 3) = 0$$

$$-x + 3 - 6y + 6 + z - 3 = 0$$

$$-x - 6y + z = -6$$

◊

PROBLEMS

Problem 1.53 Find the plane through (1,-1,1) at right angles to $\vec{v} = (2, 2, -1)$.

Problem 1.54 Find the plane through (2,0,5) at right angles to $\vec{v} = (1, 1, 1)$.

Problem 1.55 Find the plane through (3,2,1) at right angles to $\vec{v} = (1, -1, 2)$.

Problem 1.56 Find, in the form

$$ax + by + cz = d,$$

the tangent planes to the following curves at the indicated points.

1. $f(x, y) = x^2 + y^2 - 1$ at $(2, 2)$
2. $g(x, y) = 3x^2 + 2xy + 4^2 - 1$ at $(-1, 1)$
3. $h(x, y) = (x + y + 1)^2$ at $(0, 4)$
4. $r(x, y) = y \cdot \ln(x^2 + 1)$ at $(3, -1)$
5. $s(x, y) = e^{x^2+y^2}$ at $(0, 1)$
6. $q(x, y) = \sin(x)\cos(y)$ at $(\pi/3, \pi/6)$

Problem 1.57 Suppose that

$$x^2 + y^2 + z^2 = 12.$$

Find the tangent plane at each of the following points.

1. $p = (2, 2, 2)$
2. $q = (2, -2, 2)$
3. $r = (-2, -2, -2)$
4. $u = (1, 1, \sqrt{10})$
5. $v = (-\sqrt{7}, 1, 2)$
6. $w = (-1/2, 2.5, \sqrt{22}/2)$

Problem 1.58 Find all tangent planes to

$$x^2 + y^2 + z^2 = 75$$

that are at right angles to the vector $(2, 4, 6)$.

Problem 1.59 Find the tangent plane at $(-1, \pi/3)$ to

$$g(x, y) = x\cos(y)$$

Problem 1.60 Find the tangent plane to each of the following surfaces at the indicated point.

1. Surface $x + y + z^2 = 6$ at (1,1,2)
2. Surface $x^2 - y^3 + 5z = 19$ at (4,-2,-1)
3. Surface $3x + y^2 + z^2 = 8$ at (2,1,1)
4. Surface $(x - y)^3 + 2z = 14$ at (2,0,3)
5. Surface $xyz + x + y + z = 4$ at (1,1,1)
6. Surface $xy + yz + xz = 0$ at (-1,1,-2)
7. Surface $x^2 + y^2 + z^3 = 6$ at (2,1,1)
8. Surface $xy + xz + yz = 7$ at (1,1,3)

Problem 1.61 Find the tangent plane at $(0, 0)$ to

$$g(x, y) = e^{-(x^2+y^2)}$$

Problem 1.62 Find the tangent plane at $(\pi/4, \pi/4, \pi/4)$ to

$$\cos(xyz) = 0$$

Problem 1.63 If

$$x^2 + y^2 + z^2 = r^2$$

is a sphere of radius r centered at the origin $(0, 0, 0)$ (it is), show that a sphere has a tangent plane at right angles to any non-zero vector.

Problem 1.64 If you want to find the tangent plane to a point on a sphere, what is the simplest method? Explain.

Problem 1.65 If P and Q are planes that are both at right angles to a vector $\vec{v}$, and they are not equal, what can be said about the intersection of P and Q?

Problem 1.66 Suppose that $\vec{u}$ and $\vec{v}$ are vectors so that $\vec{v} \cdot \vec{u} = 0$. If P is a plane at right angles to $\vec{u}$, and Q is a plane at right angles to $\vec{v}$ in three-dimensional space, then what is the most that can be said about the intersection of the planes?

CHAPTER 2

Multivariate and Constrained Optimization

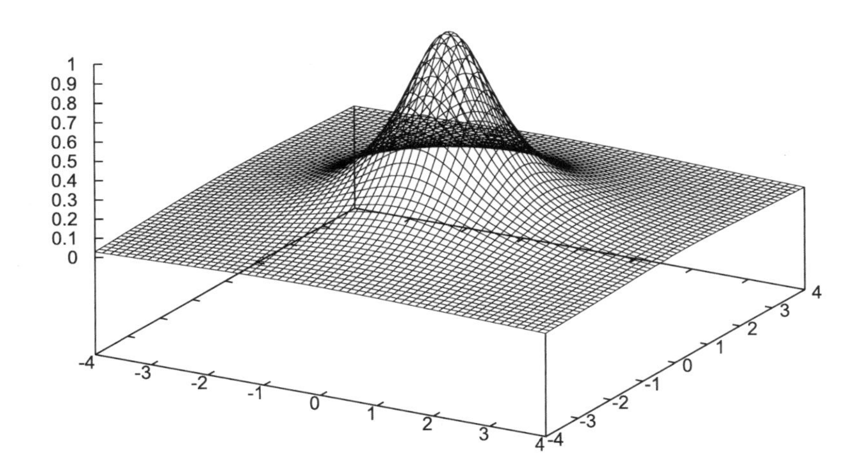

Figure 2.1: A graph of $f(x, y) = \dfrac{1}{x^2 + y^2 + 1}$ on $-4 \leq x, y \leq 4$.

A big application of derivatives in our earlier work was optimizing functions. In this chapter we learn to do this for multiple independent variables. In some ways we will be doing pretty much the same thing – but as always, more variables makes the process more complicated.

2.1 OPTIMIZATION WITH PARTIAL DERIVATIVES

Examine the hill-shaped graph in Figure 2.1. This function has a single optimum at $(x, y) = (0, 0)$. It is also positive everywhere. It is a good, simple function for demonstration purposes. The level curves are circles centered on the origin – except at the origin on top of the optimum, where we get a "level curve" consisting of a single point. This sort of point – where the directional derivatives are all zero – is the three-dimensional equivalent of a critical point.

Knowledge Box 2.1

Critical points for surfaces

If $z = f(x, y)$ defines a surface, then the **critical points** *of the function are at the points (x, y) that solve the equations:*

$$f_x(x, y) = 0 \text{ and } f_y(x, y) = 0.$$

As with our original optimization techniques, local and global optima occur at critical points or at the boundaries of the domain of optimization.

The information in Knowledge Box 2.1 gives us a good start on finding optima of multivariate functions – but it also contains a land mine. The phrase *the boundaries of the domain of optimization* is more fearsome when there are more dimensions. The next section of this chapter is about dealing with some of the kinds of boundaries that arise when optimizing surfaces.

Example 2.1 Demonstrate that the function in Figure 2.1, $f(x, y) = \frac{1}{x^2 + y^2 + 1}$, in fact has a critical point at (0,0) by solving for the partial derivatives equal to zero.

Solution:
The partial derivatives are:

$$f_x(x, y) = \frac{2x}{(x^2 + y^2 + 1)^2} \text{ and } f_y(x, y) = \frac{2y}{(x^2 + y^2 + 1)^2}$$

Remembering that a fraction is zero only when its numerator is zero – and noting the denominators of these partial derivative are never zero – we see we are solving the very difficult system of simultaneous equations:

$$2x = 0 \text{ and } 2y = 0$$

So we verify a single critical point at (0,0).

◇

Example 2.2 Find the critical point(s) of

$$g(x, y) = x^2 + 2y^2 + 4xy - 6x - 8y + 2$$

Solution:

Start by computing the partials.

$$\begin{aligned} g_x(x, y) &= 2x + 4y - 6 \\ g_y(x, y) &= 4y + 4x - 8 \end{aligned}$$

This gives us the simultaneous system:

$$\begin{aligned} 2x + 4y - 6 &= 0 \\ 4x + 4y - 8 &= 0 \end{aligned}$$

$$\begin{aligned} 2x + 4y &= 6 \\ 4x + 4y &= 8 \\ 2x &= 2 && \text{Second line minus first.} \\ x &= 1 \\ 2 + 4y &= 6 && \text{Plug x=1 into first line.} \\ 4y &= 4 \\ y &= 1 \end{aligned}$$

So we find a single critical point at $(x, y) = (1, 1)$.

◊

Now that we can locate critical points of surfaces, we have to deal with figuring out if the point is a local maximum, a local minimum, or something else. Let's begin by understanding the option of "something else." Examine the function in Figure 2.2. The partial derivatives are $f_x = 2x$ and $f_y = 2y$. So, it has a critical point at (0,0) in the center of the graph, but it *does not* have an optimum. This is a type of critical point called a **saddle point**.

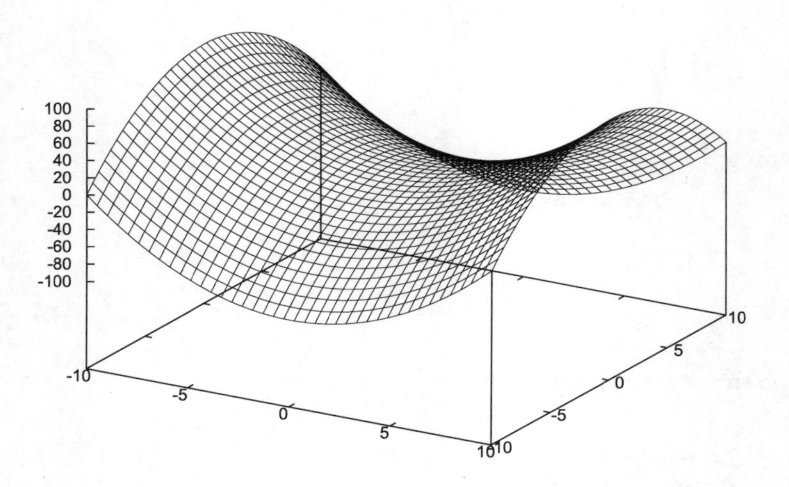

Figure 2.2: A graph of $f(x, y) = x^2 - y^2$ on $-10 \leq x, y \leq 10$. This graph exemplifies a saddle point.

Knowledge Box 2.2

Types of critical points for surfaces

When we find critical points by solving $f_x = 0$ and $f_y = 0$, there are three possible outcomes:

- *if all nearby points are lower, the point is a* **local maximum,**
- *if all nearby points are higher, the point is a* **local minimum,**
- *if there are nearby points that are both higher and lower, the point is a* **saddle point.**

If we have a good graph of a function, then we can often look at a critical point to determine its character. This is not always practical. The sign chart technique from *Fast Start Differential Calculus* is not possible – it only makes sense in a one-dimensional setting. There *is* an analog to the second derivative test. Look at the functions in Figures 2.1 and 2.2. If we slice them along the x and y axes, we see that the hill in Figure 2.1 is concave down along both axes, while the saddle in Figure 2.2 is concave down along the y axis and concave up along the x axis. This type of observation can be generalized into a type of second derivative test.

Knowledge Box 2.3

Classifying critical points for surfaces

For a function $z = f(x, y)$ that defines a surface, define

$$D(x, y) = f_{xx}(x, y) f_{yy}(x, y) - f_{xy}^2(x, y).$$

The quantity D is called the **discriminant** *of the system. Then for a critical point (a, b)*

- *If $D(a, b) > 0$ and $f_{xx}(a, b) > 0$ or $f_{yy}(a, b) > 0$, then the point is a local minimum.*
- *If $D(a, b) > 0$ and $f_{xx}(a, b) < 0$ or $f_{yy}(a, b) < 0$, then the point is a local maximum.*
- *If $D(a, b) < 0$, then the point is a saddle point.*
- *If $D(a, b) = 0$, then the test yields no information about the type of the critical point.*

The fourth outcome – no information – forces you to examine the function in other ways. This situation is usually the result of having functions with repeated roots or high powers of both variables. The function

$$q(x, y) = x^3 + y^3,$$

for example, has a critical point at $(0, 0)$, but all of f_{xx}, f_{yy}, and $f_{x,y}$ are zero. The critical point is a saddle point, but that conclusion follows from careful examination of the graph of the function, not from the second derivative test.

A part of the structure of this multivariate version of the second derivative test is that *either* f_{xx} or f_{yy} may be used to check if a critical point is a minimum or maximum. A frequent question by students is "which should I use?" The answer is "whichever is easier." The reason for this is that the outcome of the discriminant test that tells you a critical point is an optimum of some sort *forces* the signs of f_{xx} and f_{yy} to agree.

Examine the formula for $D(a, b)$ and see if you can tell why this is true.

Example 2.3 Find and classify the critical point of

$$h(x, y) = 4 + 3x + 5y - 3x^2 + xy - 2y^2$$

Solution:

The partials are $f_x = 3 - 6x + y$ and $f_y = 5 + x - 4y$. Setting these equal to zero we obtain the simultaneous system of equations:

$$\begin{aligned} 6x - y &= 3 \\ -x + 4y &= 5 \end{aligned}$$

$$\begin{aligned} &\text{Solve:} \\ -6x + 24y &= 30 \\ 23y &= 33 \\ y &= 33/23 \\ -x + 132/23 &= 115/23 \\ -x &= -17/23 \\ x &= 17/23 \end{aligned}$$

Now compute $D(x, y)$. We see $f_{xx} = -6$, $f_{yy} = -4$ and $f_{xy} = 1$. So,

$$D = (-6)(-4) - 1^2 = 23.$$

Since $D > 0$, the critical point is an optimum of some sort. The fact that $f_{xx} < 0$ or the fact that $f_{yy} < 0$ suffices to tell us this point is a maximum.

◊

Notice that in Example 2.3 the equation is a quadratic with large negative squared terms and a small mixed term (xy). The fact that its sole critical point is a maximum means that it is a paraboloid opening downward. In the next example we look at a quadratic with a large mixed term.

Example 2.4 Find and classify the critical points of

$$q(x, y) = x^2 + 6xy + y^2 - 14x - 10y + 3.$$

Solution:

Start by computing the needed partials and the discriminant.

$$f_x(x, y) = 2x + 6y - 14$$

$$f_y(x, y) = 6x + 2y - 10$$

$$f_{xx} = 2$$

$$f_{yy} = 2$$

$$f_{xy} = 6$$

$$D(x, y) = 2 \cdot 2 - 6^2 = -32$$

These calculations show that any critical point we find will be a saddle point since $D < 0$. To find the critical point we solve the first partials equal to zero obtaining the system

$$2x + 6y = 14$$

$$6x + 2y = 10$$

$$6x + 18y = 42 \qquad \text{3x line one}$$

$$16y = 32 \qquad \text{Line 3 minus line two}$$

$$y = 2$$

$$2x + 12 = 14 \qquad \text{Substitute}$$

$$2x = 2$$

$$x = 1$$

So we see that the function has, as its sole critical point, a saddle point at (1,2).

◊

The last two examples have been bivariate quadratic equations with a constant discriminant. This is a feature of all bivariate quadratics. All functions of this kind have a single critical point and a constant discriminant. For the next example we will tackle a more challenging example.

Example 2.5 Find and classify the critical points of $f(x, y) = x^4 + y^4 - 16xy + 6$.

Solution:

Start by computing the needed partials and the discriminant.

$$f_x(x, y) = 4x^3 - 16y$$

$$f_y(x, y) = 4y^3 - 16x$$

$$f_{xx}(x, y) = 12x^2$$

$$f_{yy}(x, y) = 12y^2$$

$$f_{xy}(x, y) = -16$$

$$D(x, y) = f_{xx} f_{yy} - f_{xy}^2$$

$$D(x, y) = 144x^2y^2 - 256$$

Next, we need to find the critical points by solving the system

$$4x^3 = 16y$$

$$4y^3 = 16x$$

or

$$x^3 = 4y$$

$$y^3 = 4x$$

Solve the second equation to get $y = \sqrt[3]{4x}$ and plug into the first, obtaining:

$$x^3 = 4\sqrt[3]{4x}$$

$$x^9 = 256x \qquad \text{Cube both sides}$$

$$x^9 - 256x = 0$$

$$x(x^8 - 256) = 0$$

$$x = 0, \pm\sqrt[8]{256}$$

$$x = 0, \pm 2$$

Since the equation system is symmetric in x and y we may deduce that $y = 0, \pm 2$ as well. Referring back to the original equations, it is not hard to see that, when $x = 0$, $y = 0$; when $x = 2$ so must y; when $x = -2$ so must y. This gives us three critical points: $(-2, -2)$, $(0, 0)$, and $(2, 2)$. Next, we check the discriminant at the critical points.

$$D(-2, -2) = 2304 - 256 = 2048 > 0$$

Since $f_{xx}(-2, -2) > 0$, this point is a minimum.

$$D(2, 2) = 2048 > 0$$

Another minimum, and

$$D(0, 0) = -256,$$

so this point is a saddle point.

One application of multivariate optimization is to minimize distances. For that we should state, or re-state, the definition of distance.

Knowledge Box 2.4

The definition of distance

If $p = (x_0, y_0)$ and $q = (x_1, y_1)$ are points in the plane, then the distance between p and q is

$$d(p, q) = \sqrt{(x_0 - x_1)^2 + (y_0 - y_1)^2}.$$

If $r = (x_0, y_0, z_0)$ and $s = (x_1, y_1, z_1)$ are points in space, then the distance between r and s is

$$d(r, s) = \sqrt{(x_0 - x_1)^2 + (y_0 - y_1)^2 + (z_0 - z_1)^2}.$$

The definition in Knowledge Box 2.4 can be extended to any number of dimensions – the distance between two points is the square root of the sum of the squared differences of the individual coordinates of the point. For this text only two- and three-dimensional distances are required. With the definition of distance in place, we can now pose a standard type of problem.

Example 2.6 What point on the plane $z = 3x + 2y + 4$ is closest to the origin?

Solution:

The origin is the point $(0, 0, 0)$, while a general point (x, y, z) on the plane has the form $(x, y, 3x + 2y + 4)$. That means that the distance we are trying to minimize is given by

$$d = \sqrt{(x-0)^2 + (y-0)^2 + (3x + 2y + 4 - 0)^2}$$

$$= \sqrt{10x^2 + 12xy + 5y^2 + 24x + 16y + 16}$$

The partial derivatives of this function are:

$$d_x = \frac{20x + 12y + 24}{2\sqrt{10x^2 + 12xy + 5y^2 + 24x + 16y + 16}}$$

$$d_y = \frac{12x + 10y + 16}{2\sqrt{10x^2 + 12xy + 5y^2 + 24x + 16y + 16}}$$

We need to solve the simultaneous equation in which each of these partials is zero. Remember a fraction is zero only where its numerator is zero. This maxim gives us the simultaneous equations:

$$20x + 12y + 24 = 0$$

$$12x + 10y + 16 = 0$$

Simplify

$$5x + 3y = -6$$

$$6x + 5y = -8$$

Solve:

$$30x + 18y = -36$$

$$30x + 25y = -40$$

$$7y = -4$$

$$y = -4/7$$

$$5x - 12/7 = -42/7$$

$$5x = -30/7$$

$$x = -6/7$$

So we find a single critical point $(-6/7, -4/7)$ Considering the geometry of a plane, we see that it has a unique closest approach to the origin. So, computing $z = 3x + 2y + 4 = -18/7 - 8/7 + 28/7 = 2/7$, gives us that the point on the plane closest to the origin is

$$(-6/7, -4/7, 2/7).$$

◊

Clever students will have noticed that when we were minimizing

$$\sqrt{10x^2 + 12xy + 5y^2 + 24x + 128 + 16}$$

the numerators of the partial derivatives were exactly the *partial derivatives* of $10x^2 + 12xy + 5y^2 + 24x + 128 + 16$. This observation is an instance of a more general shortcut.

Knowledge Box 2.5

Minimization of distance—a shortcut

Suppose that we are optimizing a function $g(x, y) = \sqrt{f(x, y)}$ *or* $g(x, y, z) = \sqrt{f(x, y, z)}$. *Then the critical points for optimization of* g *and* f *are the same, and the second derivative tests agree. The actual values of the function are not, which means care is required.*

Why does this shortcut work? It is because if $0 < a < b$, then $0 < \sqrt{a} < \sqrt{b}$. So optimizing the value finds where the optimum of the square root of the value is as well.

Example 2.7 Minimize:

$$g(x, y) = \sqrt{x^2 + y^2 + 3}$$

Solution:

In this case $g(x, y) = \sqrt{f(x, y)}$ where $f(x, y) = x^2 + y^2 + 3$, so the shortcut applies. Finding the relevant partials we see that

$$f_x(x, y) = 2x$$
$$f_y(x, y) = 2y$$

which is easy to see has a critical point at (0,0). The second derivative test shows that:

$$f_{xx} f_{yy} - f_{xy}^2 = 2 \cdot 2 - 0 = 4 > 0$$

So the critical point is an optimum. Since $f_{xx} = 2 > 0$, it is a minimum. This means the minimum value of $g(x, y)$ is at the point $(0, 0, \sqrt{3})$.

◊

Definition 2.1 *A function $m(x)$ is* **monotone increasing** *if, whenever $a < b$ and $m(x)$ exists on $[a, b]$, then $m(a) < m(b)$.*

Notice that $m(x) = \sqrt{x}$ is monotone increasing. In fact the shortcut in Knowledge Box 2.5 works for any monotone increasing function. These functions include: e^x, $\ln(x)$, $\tan^{-1}(x)$, x^n when n is odd, and $\sqrt[n]{x}$.

Knowledge Box 2.6

A test for a function being monotone increasing

We know that a function is increasing if its first derivative is positive. If a function exists and has a positive derivative on the interval [a,b] it is a monotone increasing function on [a,b].

Example 2.8 Show that $y = \ln(x)$ is monotone increasing where it exists.

Solution:
According to Knowledge Box 2.6, a function is increasing when its first derivative is positive. The function $y = \ln(x)$ only exists on the interval $(0, \infty)$. Its derivative is $y' = 1/x$ which is positive for any positive x. Thus, $\ln(x)$ is increasing everywhere that it exists and so is a monotone increasing function.

Using an extension of the shortcut from Knowledge Box 2.5 with these other monotone functions will make the homework problems much easier. Let's practice.

Example 2.9 Find the point of closest approach to the origin on the plane

$$3x - 4y - z = 4.$$

Solution:

First find the distance between the origin $(0, 0, 0)$ and a generic point $(x, y, 3x - 4y - 4)$ on the plane. We get that

$$d(x, y) = \sqrt{x^2 + y^2 + 9x^2 - 24xy + 16y^2 - 32y - 24x + 16}$$

$$d(x, y) = \sqrt{10x^2 - 24xy + 17y^2 - 32y - 24x + 16}$$

Using the distance minimization shortcut, find the critical points of

$$d^2(x, y) = 10x^2 - 24xy + 17y^2 - 32y - 24x + 16$$

which is the relatively simple quadratic case.

$$f_x(x, y) = 20x - 24y - 24$$

$$f_y(x, y) = -24x + 34y - 32$$

$$f_{xx}(x, y) = 20$$

$$f_{yy}(x, y) = 34$$

$$f_{x,y}(x, y) = -24$$

$$D(x, y) = 20 \cdot 34 - (-24)^2 = 104$$

So we see that $D > 0$ and $f_{xx} > 0$ meaning we will find a minimum distance. Now we must solve for the critical point.

$$20x - 24y = 24$$

$$-24x + 34y = 32$$

$$5x - 6y = 6$$

$$-12x + 17y = 16$$

$$60x - 72y = 72$$

$$-60x + 85y = 80$$

$$13y = 152$$

$$y = 152/13$$

$$5x - 912/13 = 78/13$$

$$5x = 990/13$$

$$x = 198/13$$

So the critical point is roughly at $x = 15.23$, $y = 11.69$, making the point of closest approach $(15.23, 11.69, 2.93)$. Not a lot easier – but we avoided all sorts of square roots.

PROBLEMS

Problem 2.10 Find the critical points for each of the following functions, and use the second derivative test to classify them as maxima, minima, or saddle points. Also, find the value of the function at the critical point.

1. $f(x, y) = x^2 + y^2 - 6x - 4y + 4$
2. $g(x, y) = 3x^2y + y^3 - 3x^2 - 3y^2 + 2$
3. $h(x, y) = 12 - x^2 + 3xy - y^2 + 2x - y + 1$
4. $q(x, y) = x^2 + xy + y^2 + 4x - 5y + 2$
5. $r(x, y) = x^4 + y^4 - 4xy + 1$
6. $s(x, y) = x^3 + 2y^3 - 4xy$

Problem 2.11 Show that

$$f(x, y) = x^4 + y^4$$

is an example of a function where the second derivative test yields no useful information.

Problem 2.12 Suppose that

$$f(x, y) = ax^2 + bxy + cy^2 + dx + ey + f.$$

Show that this function has at most one critical point. When it does have a critical point, derive rules based on the constants a-f for classifying that critical point.

Problem 2.13 Find the point on the plane

$$x + y + z = 4$$

closest to the origin.

Problem 2.14 Find the point on the plane

$$z = -x/3 + y/4 + 1$$

closest to the origin.

Problem 2.15 Find the point on the tangent plane of

$$f(x, y) = x^2 + y^2 + 1$$

at $(2, 1, 6)$ that is closest to the origin.

Problem 2.16 Find the critical points for each of the following functions, and use the second derivative test to classify them as maxima, minima, or saddle points. Also, find the value of the function at the critical point.

1. $f(x, y) = \ln(x^2 + y^2 + 4)$
2. $g(x, y) = e^{x^4+y^4-36xy+6}$
3. $h(x, y) = \tan^{-1}(x^3 + 2y^3 - 4xy)$
4. $r(x, y) = (x^2 + xy + 3y^2 - 4x + 3y + 1)^5$
5. $s(x, y) = \sqrt{x^2 + y^2 + 14}$
6. $q(x, y) = \tan^{-1}\left(e^{x^2+xy+y^2-2x-3y+1}\right)$

Problem 2.17 For each of the following functions either demonstrate it is not monotone increasing on the given interval or show that it is.

1. $y = e^x$ on $(-\infty, \infty)$
2. $y = x^3 - x$ on $(-\infty, \infty)$
3. $y = \dfrac{x}{x^2+1}$ on $(-1, 1)$
4. $y = \tan^{-1}(x)$ on $(-\infty, \infty)$
5. $y = \dfrac{x^2}{x^2+1}$ on $(0, \infty)$
6. $y = \dfrac{2x}{x^2+1}$ on $(-\infty, \infty)$
7. $y = \dfrac{e^x}{e^x+1}$ on $(-\infty, \infty)$

Problem 2.18 Suppose that we are going to cut a line of length L into three pieces. Find the division into pieces that maximizes the sum of the squares of the lengths. This can be phrased as a multivariate optimization problem.

Problem 2.19 Find the largest interval on which the function

$$\frac{x}{x^2+16}$$

is monotone increasing.

Problem 2.20 Find the largest interval on which the function

$$\frac{x^2-1}{x^2+4}$$

is monotone increasing.

2.2 THE EXTREME VALUE THEOREM REDUX

The extreme value theorem, last seen in *Fast Start Differential Calculus*, says that *The global maximum and minimum of a continuous, differentiable function must occur at critical points or at the boundaries of the domain where optimization is taking place.* This is just as true for optimizing a function $z = f(x, y)$. So what's changed? The largest change is that the boundary can now have a very complex shape.

In the initial section of this chapter we carefully avoided optimizing functions with boundaries, thus avoiding the issue with boundaries. In the next couple of examples we will demonstrate techniques for dealing with boundaries.

Example 2.21 Find the global maximum of

$$f(x, y) = (xy + 1)e^{-x-y}$$

for $x, y \geq 0$.

Solution:

This function is to be optimized only over the first quadrant. We begin by finding critical points in the usual fashion

$$f_x(x, y) = ye^{-x-y} + (xy + 1)e^{-x-y}(-1) = (-xy + y - 1)e^{-x-y}$$

$$f_y(x, y) = xe^{-x-y} + (xy + 1)e^{-x-y}(-1) = (-xy + x - 1)e^{-x-y}$$

Since the extreme value theorem tells us that the optima occur at boundaries or critical points, we won't need a second derivative test – we just compare values. This means out next step is to solve for any critical points. Remember that powers of e cannot be zero, giving us the system of equations:

$$-xy + y - 1 = 0$$

$$-xy + x - 1 = 0$$

$$xy = x - 1 = y - 1 \qquad \text{so } x = y$$

$$-x^2 + x - 1 = 0$$

$$x^2 - x + 1 = 0$$

$$x = \frac{1 \pm \sqrt{1-4}}{2} \qquad \text{There are no critical points!}$$

This means that the extreme value occurs on the boundaries – the positive x and y axes where either $x = 0$ or $y = 0$. So, we need the largest value of $f(0, y) = \mathrm{e}^{-y}$ for $y \geq 0$ and $f(x, 0) = \mathrm{e}^{-x}$ for $x \geq 0$. Since e^{-x} is largest (for non-negative x) at $x = 0$, we get that the global maximum is $f(0, 0) = 1$.

While the extreme value theorem lets us make decisions without the second derivative test, it forces us to examine the boundaries, which can be hard. It is sometimes possible to solve the problem by adopting a different point of view. The next example will use a transformation to a parametric curve to solve the problem.

Example 2.22 Maximize

$$g(x, y) = x^4 + y^4$$

for those points $\{(x, y) : x^2 + y^2 \leq 4\}$.

Solution:

This curve has a single critical point at $(x, y) = (0, 0)$ which is extremely easy to find. The boundary for the optimization domain is the circle $x^2 + y^2 = 4$, a circle of radius 2 centered at the origin. This boundary is also the parametric curve $(2\cos(t), 2\sin(t))$. This means that *on the boundary* the function is

$$g(2\cos(t), 2\sin(t)) = 16\cos^4(t) + 16\sin^4(t).$$

This means we can treat the location of optima on the boundary as a single-variable optimization task of the function $g(t) = 16\cos^4(t) + 16\sin^4(t)$. We see that

$$g'(t) = 64\cos^3(t) \cdot (-\sin(t)) + 64\sin^3(t)\cos(t) = 64\sin(t)\cos(t)(\sin^2(t) - \cos^2(t))$$

Solve:

$$\sin(t) = 0 \qquad t = (2n+1)\frac{\pi}{2}$$

$$\cos(t) = 0 \qquad t = n\pi$$

$$\sin^2(t) - \cos^2(t) = 0$$

$$\sin^2(t) = \cos^2(t) \qquad t = \pm(2n+1)\frac{\pi}{4}$$

If we plug these values of t back into the parametric curve, the points on the boundary that may be optima are: $(0, \pm 2)$, $(\pm 2, 0)$, and $(\pm\sqrt{2}, \pm\sqrt{2})$. Plugging the first four of these into the curve we get that $g(x, y) = 16$. Plugging the last four into the function we get $g(x, y) = 2 \cdot (\sqrt{2})^4 = 2 \cdot 4 = 8$. At the critical point (0,0) we see $g(x, y) = 0$. This means that the maximum value of $g(x, y)$ on the optimization domain is 16 at any of $(0, \pm 2)$, $(\pm 2, 0)$.

◊

Notice that the solutions to the parametric version of the boundary gave us an infinite number of solutions. But, when we returned to the (x, y, z) domain, this infinite collection of solutions were just the eight candidate points repeated over and over. This means that our failure to put a bound on the parameter t was not a problem; bounding t was not necessary.

A problem with the techniques developed in this section is that all of them are special purpose. We will develop general purpose techniques in Section 2.3 – the understanding of which is substantially aided by the practice we got in this section.

Example 2.23 If

$$h(x, y) = x^2 + y^2$$

find the minimum value of $h(x, y)$ among those points (x, y) on the line $y = 2x - 4$.

Solution:

The function $h(x, y)$ is a paraboloid that opens upward. If we look at the points (x, y, z) on $h(x, y)$ that happen to lie on a line, that line will slice a parabolic shape out of the surface defined by $h(x, y)$. First note that $h(x, y)$ has a single critical point at $(0, 0)$ which is *not in* the domain of optimization. This means we may consider only those points on $y = 2x - 4$, in other words on the function

$$h(x, 2x - 4) = x^2 + (2x - 4)^2 = 5x^2 - 16x + 16$$

This means that the problem consists of finding the minimum of $f(x) = 5x^2 - 16x + 16$.

$$f'(x) = 10x - 16 = 0$$

$$10x = 16$$

$$x = 8/5$$

$$y = 16/5 - 20/5 = -4/5$$

Since $f(x)$ opens upward, we see that this point is a minimum; there are no boundaries to the line that is constraining the values of (x, y) so the minimum value of $h(x, y)$ on the line is

$$h(8/5, -4/5) = \frac{64}{25} + \frac{16}{25} = \frac{80}{25} = \frac{16}{5}$$

◊

Example 2.23 did not really use the extreme value theorem. For that to happen we would need to optimize over a line segment instead of a full line.

Example 2.24 If

$$h(x, y) = x^2 + y^2$$

find the minimum and maximum value of $h(x, y)$ among those points (x, y) on the line $y = x + 1$ for $-4 \leq x \leq 4$.

Solution:

This problem is very similar to Example 2.23. The ends of the line segment are (-4,-3) and (4,5), found by substituting into the formula for the line. On the line $h(x, y)$ becomes

$$h(x, x+1) = x^2 + (x+1)^2 = 2x^2 + 2x + 1$$

So we get a critical point at $4x + 2 = 0$ or $x = -1/2$ which is the point $(-1/2, 1/2)$.

Plug in and we get

$$h(-4, -3) = 25$$

$$h(-1/2, 1/2) = 1/2$$

$$h(4, 5) = 41$$

This means the minimum value is 1/2 at the critical point and that the maximum value is 41 at one of the endpoints of the domain of optimization.

Making the domain a line segment is one of the simplest possible options. Let's look at another example with a more complex domain of optimization.

Example 2.25 If

$$s(x, y) = 2x + y - 4$$

find the minimum and maximum value of $s(x, y)$ among those points (x, y) on the curve $y = x^2 - 3$ for $-2 \leq x \leq 2$.

Solution:

In this problem we are cutting a parabolic segment out of the plane $s(x, y) = 2x + y - 4$. We get that the ends of the domain of optimization are $(2, 1)$ and $(-2, 1)$ by plugging in the ends of the interval in x to the formula for the parabolic segment. Substituting the parabolic segment into the plane yields

$$s(x, x^2 - 3) = 2x + x^2 - 3 - 4 = x^2 + 2x - 7$$

This means our critical point appears at $2x + 2 = 0$ or $x = -1$, making the candidate point $(-1, -2)$. Plugging the candidate points into $s(x, y)$ yields:

$$s(-2, 1) = -7$$

$$s(-1, -2) = -8$$

$$s(2, 1) = 1$$

This means that the maximum value is 1 at $(2, 1)$, and the minimum is -8 at $(-1, -2)$.

$\diamond$

Some sets of boundaries are simple enough that we can use geometric reasoning to avoid needing to use calculus on the boundaries.

Example 2.26 Find the maximum value of

$$h(x, y) = x^2 + y^2$$

for $-2 \leq x, y \leq 3$.

Solution:

First of all, we know that this surface has a single critical point at $(0, 0)$ – this surface is an old friend (see graph in Figure 1.2).

The domain of optimization is a square with corners $(-2, 2)$ and $(3, 3)$. Along each side of the square, we see that the boundary is a line – and so has extreme values at its ends. This means that we need only check the corners of the square; the interior of the edges – viewed as line segments – cannot attain maximum or minimum values *by the extreme value theorem*. This means we need only add the points $(-2, -2)$, $(-2, 3)$, $(3, -2)$, and $(3, 3)$ to our candidates.

Plugging in the candidate points we obtain:

$$h(-2-2) = 8$$

$$h(0,0) = 0$$

$$h(-2,3) = 13$$

$$h(3,-2) = 13$$

$$h(3,3) = 18$$

So the maximum is 18 at $(3,3)$, and the minimum is 0 at $(0,0)$.

The goal for this section was to set up LaGrange Multipliers – the topic of Section 2.3. The take-home message from this section is that the extreme value theorem implies that optimizing on a boundary is the difficult added portion of optimizing on a bounded domain.

The techniques that we will develop in the next section require that the boundary itself be a differentiable curve. Some of the boundaries in this section are made of several differentiable curves, meaning that the techniques in this section may be easier for those problems. If we have a boundary that is not differentiable, then the techniques in this section are the only option.

PROBLEMS

Problem 2.27 If we look at the points on

$$h(x,y) = x^2 + y^2 + 4x + 4$$

that fall on a line, then there is a minimum somewhere on the line. Find that minimum value for the following lines.

1. $y = 3x + 1$
2. $y = 4 - x$
3. $x + y = 6$
4. $2x - 7y = 24$
5. $x + y = \sqrt{3}/2$
6. $-3x + 5x = 7$

Problem 2.28 Find the maximum of

$$f(x, y) = (x^2 + y^2)\,\mathrm{e}^{-xy}$$

with (x, y) in the first quadrant where $0 < x, y$.

Hint: this function is symmetric. Use this fact.

Problem 2.29 Find the maximum and minimum of

$$q(x, y) = \frac{1}{x^2 + y^2 - 2x - 4y + 6}$$

in the first quadrant: $0 < x, y$.

Problem 2.30 Find the maximum of

$$g(x, y) = (x^4 + y^4)$$

on the set of points

$$\{(x, y) : x^2 + y^2 \leq 25\}.$$

Problem 2.31 Maximize

$$h(x) = x^4 + y^4$$

on the set of points $\{(x, y) : x^2 + y^2 \leq 25\}$.

Problem 2.32 If

$$g(x, y) = 2x^2 + 3y^2,$$

find the minimum of $g(x, y)$ for those points (x, y) on the line $y = x - 1$.

Problem 2.33 For the function

$$s(x, y) = x^2 + y^2 - 2x + 4y + 1$$

with (x, y) on the following line segments, find the minimum and maximum values.

1. $y = 2x - 2$ on $-2 \leq x \leq 2$
2. $y = x + 7$ on $-1 \leq x \leq 4$
3. $y = 6 - 5x$ on $0 \leq x \leq 6$
4. $x + y = 10$ on $4 \leq x \leq 10$
5. $2x - 7y = 8$ on $-10 \leq x \leq 10$
6. $2x - y = 13$ on $1 \leq x \leq 5$

Problem 2.34 Suppose that

$$p(x, y) = 3x - 5y + 2$$

Find the maximum and minimum values on the following parametric curves.

1. $(3t + 1, 5 - t)$; $-5 \leq t \leq 5$
2. $(\cos(t), \sin(t))$
3. $(\sin(t), 3\cos(t))$
4. $(\sin(2t), \cos(t))$
5. $(t\cos(t), t\sin(t))$; $0 \leq t \leq 2\pi$
6. $(3\sin(t), 2\sin(t))$

Problem 2.35 Find the maximum and minimum of

$$q(x, y) = \frac{1}{x^2 + y^2 + 4x - 12y + 45}$$

in the first quadrant: $0 < x, y$.

Problem 2.36 Suppose that $P = f(x, y)$ is a plane and that we are considering the points where $x^2 + y^2 = r^2$ for some constant r. If $f(x, y)$ is not equal to a constant, explain why there is a unique minimum and a unique maximum value.

Problem 2.37 If

$$f(x, y) = x^2 + y^2,$$

find the minimum of $f(x, y)$ for those points (x, y) on the line $y = mx + b$.

2.3 LAGRANGE MULTIPLIERS

This section introduces **constrained optimization** using a technique called **Lagrange multipliers**. The basic idea is that we want to optimize a function $f(x, y)$ at those points where $g(x, y) = c$. The function $g(x, y)$ is called the **constraint**.

The proof that Lagrange multipliers work is beyond the scope of this text, so we will begin by just stating the technique.

Knowledge Box 2.7

The method of LaGrange Multipliers with two variables

Suppose that $z = f(x, y)$ *defines a surface and that* $g(x, y) = c$ *specifies points of interest. Then the optima of* $f(x, y)$*, subject to the constraint that* $g(x, y) = c$*, occur at solutions to the system of equations*

$$f_x(x, y) = \lambda \cdot g_x(x, y)$$

$$f_y(x, y) = \lambda \cdot g_y(x, y)$$

$$g(x, y) = c$$

where λ *is an* **auxiliary variable.**

The variable λ is new and strange – it is the "multiplier" – and, as we will see, correct solutions to the system of equations that arise from Lagrange multipliers typically use λ in a fashion that causes it to drop out.

If the constraint $g(x, y) = c$ is thought of as the boundary of the domain of optimization, then using Lagrange multipliers gives us a tool for resolving the boundary as a source of optima as per the extreme value theorem.

With that context, let's practice our Lagrange multipliers.

Example 2.38 Find the maxima and minima of

$$f(x, y) = x + 4y - 2$$

on the ellipse $2x^2 + 3y^2 = 36$.

Solution:

The equations arising from the Lagrange multiplier technique are:

$$1 = \lambda \cdot 4x$$

$$4 = \lambda \cdot 6y$$

$$2x^2 + 3y^2 = 36$$

Solving:

$$x = \frac{1}{4\lambda}$$

$$y = \frac{2}{3\lambda}$$

Plug into the constraint and we get:

$$\frac{2}{16\lambda^2} + \frac{12}{9 \cdot \lambda^2} = 36$$

$$\frac{1}{8} + \frac{4}{3} = 36\lambda^2$$

$$35/24 = 36\lambda^2$$

$$35/864 = \lambda^2$$

or

$$\lambda = \pm\sqrt{35/864}$$

Which yields candidate points:

$$x = \pm\frac{1}{4}\sqrt{\frac{864}{35}} = \pm\sqrt{\frac{54}{35}}$$

$$y = \pm\frac{2}{3}\sqrt{\frac{864}{35}} = \pm\sqrt{\frac{384}{35}}$$

Since the equation of $f(x, y)$ is a plane that gets larger as x and y get larger, we see that the maximum is

$$f\left(\sqrt{54/35}, \sqrt{384/35}\right) \cong 12.5$$

and the minimum is

$$f\left(-\sqrt{54/35}, -\sqrt{384/35}\right) \cong -16.5$$

◇

Notice that in the calculations in Example 2.38, the auxiliary variable λ served as an informational conduit that let us discover the candidate points. The next example is much simpler.

Example 2.39 Find the point of closest approach of the line $2x + 5y = 3$ to the origin (0,0).

Solution:

This problem is like the closest approach of a plane to the origin problems, but in a lower dimension. The tricky part of this problem is phrasing it as a function to be minimized and a constraint.

Since we are minimizing distance we get that the function is the distance of a point (x, y) from the origin:

$$d(x, y) = \sqrt{(x-0)^2 - (y-0)^2} = \sqrt{x^2 + y^2}$$

As per the monotone function shortcut, we can instead optimize the function

$$d^2(x, y) = x^2 + y^2.$$

The constraint function is the line. We can phrase the line as a constraint by saying:

$$g(x, y) = 2x + 5y = 3$$

With the parts in place, we can extract the Lagrange multiplier equations.

$$2x = \lambda 2$$

$$2y = \lambda 5$$

$$2x + 5y = 3$$

Solve:

$$x = \lambda$$

$$y = \frac{5}{2} \cdot \lambda$$

$$2(\lambda) + 5\left(\frac{5}{2} \cdot \lambda\right) = 3$$

$$\frac{29}{2} \cdot \lambda = 3$$

$$\lambda = \frac{6}{29}$$

Yielding: $gx = \frac{6}{29}$ and $y = \frac{15}{29}$

This makes the point on $2x + 5y = 3$ closest to the origin $\left(\frac{6}{29}, \frac{15}{29}\right)$.

Example 2.40 Find the maximum value of

$$f(x, y) = x^2 + y^3$$

subject to the constraint $x^2 + y^2 = 9$.

Solution:

This problem is already in the correct form for Lagrange multipliers, so we may immediately derive the system of equations.

$$2x = 2x\lambda$$

$$2y = 3y^2\lambda$$

$$x^2 + y^2 = 9$$

The first equation yields the useful information that $\lambda = 1$ but that x may take on any value, unless $x = 0$ in which case λ may be anything.

Given that $\lambda = 1$, the second equation tells us $y = 0$ or $y = 2/3$. If λ is free and $x = 0$ then y may be anything.

Plugging the values for y into the constraint that says the points lie on a circle, we can retrieve values for x: when $y = 0$, $x = \pm 3$; when $y = 2/3$, $x = \pm\sqrt{9 - 4/9} = \pm\sqrt{77/9}$.

If $x = 0$ and y is free the constraint yields $y = \pm 3$.

This gives us the candidate points $(\pm 3, 0)$, $(0, \pm 3)$, $(2/3, \sqrt{77}/3)$, and $(2/3, -\sqrt{77}/3)$.

The sign of the x-coordinate is unimportant because $f(x, y)$ depends on x^2; since $f(x, y)$ depends on y^3, positive values yield larger values of f, and negative ones yield smaller values of f.

Since $f(3, 0) = 9$, $f(2/3, \sqrt{77}/3) \cong 25.47$, and $f(0, 3) = 27$, we get that the maximum value of the function on the circle is:

$$f(0, 3) = 27$$

◊

In Section 2.2 we found the minimum of a line on a quadratic surface that opens upward (Example 2.23). The next example lets us try a problem like this using the formalism of Lagrange multipliers.

Example 2.41 Find the minimum of $f(x, y) = x^2 - xy + y^2$ on the line $5x + 7y = 18$.

Solution:

The constraint is the line – so $g(x, y) = 5x + 7y = 18$. With this detail we can apply the Lagrange multiplier technique:

$$2x - y = 5\lambda$$

$$2y - x = 7\lambda$$

$$5x + 7y = 18$$

Solve the constraint for y and we get $y = \dfrac{18 - 5x}{7}$ so

$$2x - \frac{18 - 5x}{7} = 5\lambda$$

$$14x - 18 + 5x = 35\lambda$$

$$19x - 35\lambda = 18$$

and

$$2\left(\frac{18 - 5x}{7}\right) - x = 7\lambda$$

$$36 - 10x - 7x = 49\lambda$$

$$17x + 49\lambda = -36$$

$$\lambda = -(17x + 36)/49$$

$$19x + 35 \cdot (17x + 36)/49 = 18$$

$$1526x = -378$$

$$x = -189/763 \cong -0.25$$

$$y = 2097/763 \cong 2.75$$

Making the minimum

$$\left(\frac{-189}{763}\right)^2 - \frac{-189 \cdot 2097}{74^2} + \left(\frac{2097}{763}\right)^2 \cong 8.30$$

◊

This problem actually got harder when we used the Lagrange multiplier formalism to solve it. This is another example of how different tools are good for different problems.

At this point, we introduce the 3-space version of Lagrange multipliers. Or, we could say that this is the version of Lagrange multipliers that uses three independent variables. This version of the Lagrange multiplier technique widens the variety of problems we can work with.

Knowledge Box 2.8

The method of LaGrange Multipliers with three variables

Suppose that $w = f(x, y, z)$ defines a surface and that $g(x, y, z) = c$ specifies points of interest. Then the optima of $f(x, y, z)$, subject to the constraint that $g(x, y, z) = c$, occur at solutions to the system of equations

$$f_x(x, y, z) = \lambda \cdot g_x(x, y, z)$$

$$f_y(x, y, z) = \lambda \cdot g_y(x, y, z)$$

$$f_z(x, y, z) = \lambda \cdot g_z(x, y, z)$$

$$g(x, y, z) = c$$

where λ is an auxiliary variable.

Example 2.42 Find the point of intersection of the function $w = 2x + y - z$ and the sphere $x^2 + y^2 + z^2 = 8$ that has the largest value of w.

Solution:

The problem is already in the correct form to apply Lagrange multipliers.

$$2x = 2\lambda$$

$$2y = \lambda$$

$$2z = -\lambda$$

$$x^2 + y^2 + z^2 = 8$$

Solving the first three equations tell us that

$$x = \lambda \quad y = \frac{1}{2}\lambda \quad z = -\frac{1}{2}\lambda$$

Plugging these into the last equation tells us that

$$\lambda^2 + (1/4)\lambda^2 + (1/4)\lambda^2 = 8$$

$$(3/2)\lambda^2 = 8$$

$$\lambda^2 = 16/3$$

$$\lambda = \pm 4/\sqrt{3}$$

So we see that: $x = \pm\frac{4}{\sqrt{3}} \quad y = \pm\frac{2}{\sqrt{3}} \quad z = \pm\frac{2}{\sqrt{3}}$

Looking at the formula for w, we see that w grows as x, y, and $-z$.

So the point that maximizes w is: $(x, y, z) = \left(\frac{4}{\sqrt{3}}, \frac{2}{\sqrt{3}}, -\frac{2}{\sqrt{3}}\right)$.

◊

Example 2.43 Suppose we divide a rope of length 6 m into three pieces. What size of pieces maximizes the product of the lengths?

Solution:

Name the length of the pieces x, y, z. That means the function we are maximizing is $f(x, y, z) = xyz$ and the constraint is $x + y + z = 6$.

Having put the problem into the form for Lagrange multipliers, we can move to the system of equations.

$$yz = \lambda$$

$$xz = \lambda$$

$$xy = \lambda$$

$$x + y + z = 6$$

So $x = \lambda/z = y$ and $z = \lambda/x = y$ making $x = y = z$. Since they sum to 6, we see $x = y = z = 2$ is the sole candidate point. Testing constrained points near $(2, 2, 2)$, like $(1.9, 2.1, 2)$ and $(2, 1.95, 2.05)$, shows that this point is a maximum.

PROBLEMS

Problem 2.44 For each of the following sets of functions and constraints, write out but do not solve the Lagrange multiplier equations.

1. $f(x, y) = \cos(x)\sin(y)$ constrained by $x^2 + y^2 = 8$
2. $g(x, y) = xe^y$ constrained by $2x - 3y = 12$
3. $h(x, y) = \dfrac{x^2}{y^2 + 1}$ constrained by $3x^2 + y^2 = 48$
4. $r(x, y) = 2xy$ constrained by $x^2 + y^2 = 8$
5. $s(x, y) = \ln(x^2 + y^2 + 4)$ constrained by $xy = 9$
6. $q(x, y) = \tan^{-1}(xy + 1)$ constrained by $x^2 - y^2 = 4$

Problem 2.45 For each of the following lines, find the point on the line closest to the origin using the method of Lagrange multipliers.

1. $y = 2x + 1$
2. $y = 4 - x$
3. $2x + 3y = 17$
4. $x + y = 12$
5. $y = 5x + 25$
6. $2x + 4y = 8$

Problem 2.46 Using Lagrange multipliers, find the maximum and minimum values of

$$f(x, y) = x^4 + y^4$$

subject to the constraint that

$$x^2 + y^2 = 1.$$

Problem 2.47 Suppose that

$$z = 3x - 5y + 2$$

For each of the following constraints, find the maximum and minimum values of z subject to the constraint, if any.

1. $x^2 + y^2 = 4$
2. $x^2 + y^2 = 0.25$
3. $2x^2 + 6y^2 = 64$
4. $x^2 - xy + y^2 = 1$
5. $x^2 - y^2 = 4$
6. $xy = 16$

Problem 2.48 If

$$f(x, y) = x^3 - y^2$$

and $x^2 + y^2 = 25$, find the maximum and minimum values $f(x, y)$ can take on.

Problem 2.49 Use Lagrange multipliers to find the closest approach of the plane

$$f(x, y) = 3x - y + 2$$

to the origin.

Problem 2.50 If

$$f(x, y, z) = x^2 + 4y^2 + 9z^2$$

and $x + y + z = 60$, find the values of x, y, and z that maximize and minimize f.

Problem 2.51 If

$$h(x, y, z) = x^2 - y^2 + z^2$$

and $x + y + z = 6$, find the values of x, y, and z that maximize and minimize h.

Problem 2.52 If

$$q(x, y, z) = x^2 - y^2 - 2z^2$$

and $x + y + z = 300$, find the values of x, y, and z that maximize and minimize q.

Problem 2.53 Find the largest point in intersection of

$$x + 2y + 3z = 4$$

and a cylinder of radius 2 centered on the z-axis.

CHAPTER 3

Advanced Integration

This chapter covers various sorts of integration that compute volumes and areas. The first, volumes of revolution, is a small twist on the integration methods we have already studied. The other techniques involve multivariate integration, which is both newer and more difficult. This latter subject permits us to compute the volume under a surface as we computed the area under the curve earlier.

3.1 VOLUMES OF ROTATION

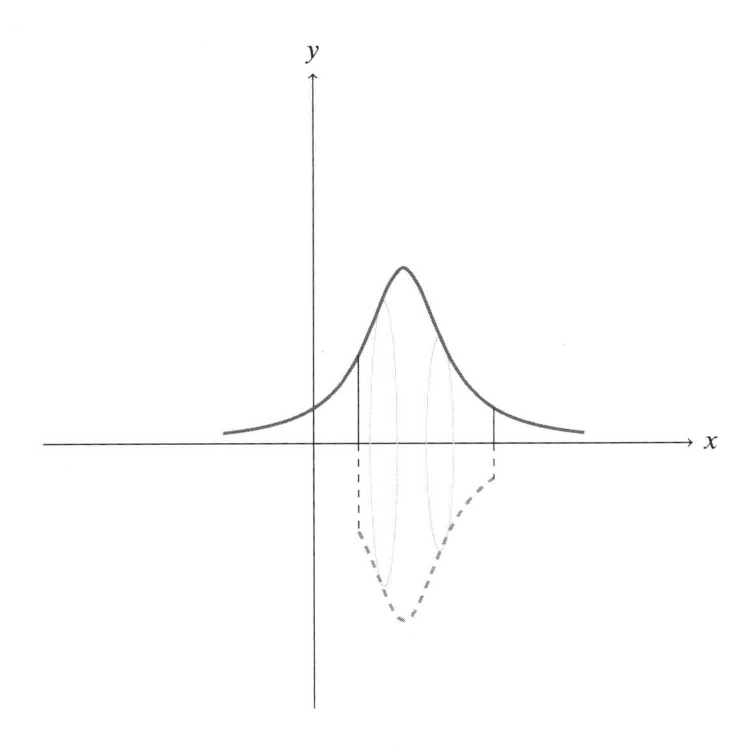

Figure 3.1: A portion of a curve rotated about the x-axis.

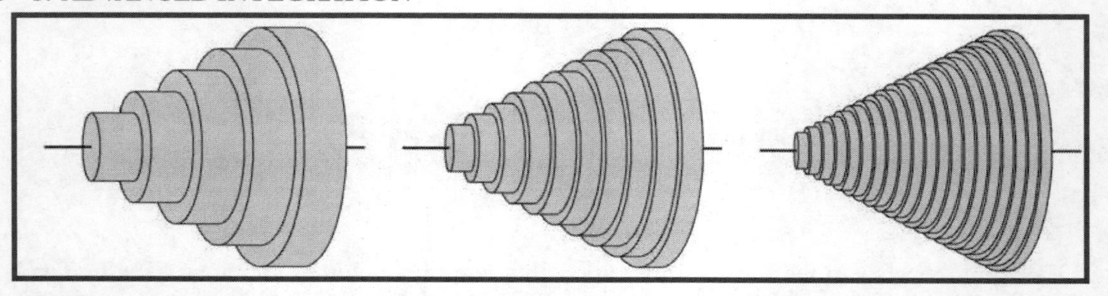

Figure 3.2: Approximations of a rotated volume using smaller and larger numbers of disks.

Figure 3.1 shows a portion of a curve and its shadow reflected about the x-axis. This is meant to evoke revolving the curve about the axis between the two vertical lines. When the curve is rotated, it encloses a volume – our goal is to compute that volume.

When integration was defined in *Fast Start Integral Calculus*, we first approximated the area represented by the integral with rectangles and then let the number of rectangles go to infinity, obtaining both the area and the integral as a limit. For volumes of rotation, we use a similar technique with *disks* instead of rectangles. Figure 3.2 shows how a conical volume is approximated using five, ten, and twenty disks. The more disks we use, the closer the sum of the volumes of the disks gets to the volume enclosed by rotating a curve about the x axis.

The final version of the area integral added up functional heights – rectangles of infinite thinness. Recall that the area under a curve from $x = a$ to $x = b$ was

$$\text{Area} = \int_a^b f(x) \cdot dx.$$

Given this, the analogous integral for volume of rotation is

$$\text{Volume} = \int_a^b \text{Area of Circles} \cdot dx = \int_a^b \pi r^2 \cdot dx$$

The radius of the circular slices of the volume we are trying to compute is given by the height of the function being rotated about the axis, giving us the final formula for *volume of rotation of a function about the x-axis.*

Knowledge Box 3.1

Formula for volume of rotation about the x axis

The volume V enclosed by rotating the function $f(x)$ about the x-axis from $x = a$ to $x = b$ is:

$$V = \pi \int_a^b f(x)^2 \cdot dx.$$

(Notice that we took the constant π out in front of the integral sign.)

Example 3.1 Find the volume enclosed by rotating $y = x^2$ about the x axis from $x = 0$ to $x = 2$.

Solution:

Start by sketching the situation:

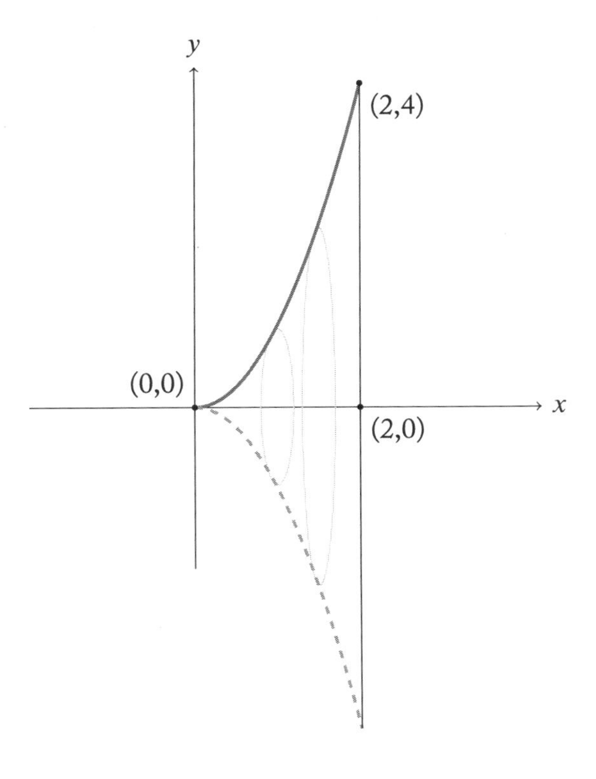

Using the equation in Knowledge Box 3.1 we get:

$$\begin{aligned}
V &= \pi \int_0^2 f(x)^2 \cdot dx \\
&= \pi \int_0^2 (x^2)^2 \cdot dx \\
&= \pi \int_0^2 x^4 \cdot dx \\
&= \pi \frac{1}{5} x^5 \Big|_0^2 \\
&= \pi \left(\frac{32}{5} - 0 \right) \\
&= \frac{32\pi}{5} \text{units}^3
\end{aligned}$$

And so we see the volume of the broad trumpet-shaped solid enclosed by rotating $y = x^2$ about the x-axis from $x = 0$ to $x = 2$ has a volume of $32\pi/5 \cong 503$ units^3.

◊

The next example is similar except that it uses a more difficult integral. If you're not comfortable with integration by parts, please review it in *Fast Start Integral Calculus*.

Example 3.2 Find the volume enclosed by rotating $y = \ln(x)$ about the x axis from $x = 1$ to $x = 4$.

Solution:

Start again with a sketch.

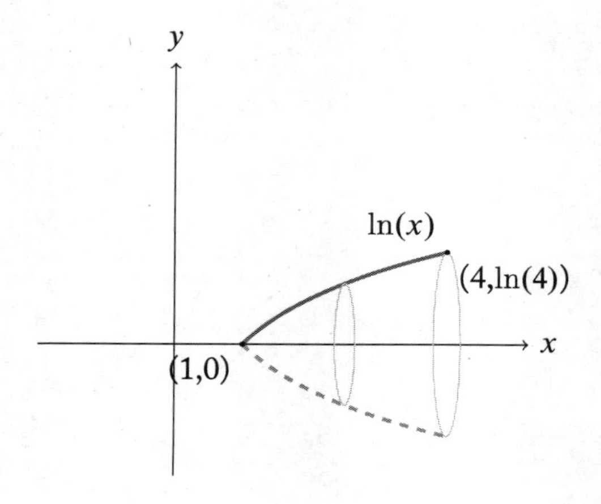

Applying the formula from Knowledge Box 3.1 we get:

$$\text{Volume} = \pi \int_1^4 \ln(x)^2 \cdot dx$$

$$U = \ln(x)^2 \qquad V = x$$

$$= \pi \left(x\ln(x)^2 - \int_1^4 2\ln(x)dx \right) \qquad dU = \frac{2\ln(x)}{x} \cdot dx \qquad dV = dx$$

$$U = \ln(x) \qquad V = x$$

$$= \pi \left(x\ln(x)^2 - 2x\ln(x) + 2\int_1^4 dx \right) \qquad dU = \frac{dx}{x} \qquad dV = dx$$

$$= \left(\pi x\ln(x)^2 - 2\pi x\ln(x) + 2\pi x\right)\Big|_1^4$$

$$= 4\pi\ln(4)^2 - 8\pi\ln(4) + 8\pi - 0 + 0 - 2\pi$$

$$\cong 8.16 \text{ units}^3$$

◊

Now that we have a formula for rotating objects about the x-axis, the next logical step is to rotate them about the y-axis. This turns out to be a little trickier.

There are two basic techniques.

1. We can figure out a new function (the inverse of the original one) that gives us disks along the y-axis, and integrate with respect to dy, or

2. We can use a different type of slice, the *cylindrical shell.* Instead of slicing the shape into disks and integrating them, we slice it into cylinders.

The radius of the disks for rotation about the x-axis was simply the value of the function, $f(x)$, but the distance from a graph to the y axis is just x. We need to get the information about the y-distance in somehow. We will start with Method 2, the method that uses a different type of slice (cylinders centered on the y-axis). Figure 3.3 shows the result of rotating a line about the y axis.

The surface area of a cylinder, not counting the top and bottom, is $2\pi rh$ where r is its radius and h is its height. The radius, as already noted, is x, and the height of the cylinder is $y = f(x)$.

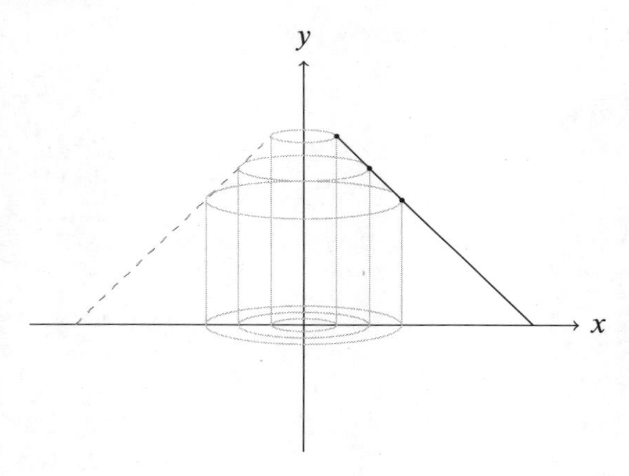

Figure 3.3: Cylindrical shells for computing a volume of revolution about the y-axis.

Thus, we have the area of our cylindrical slices

$$\text{Area} = 2\pi \cdot r \cdot h = 2\pi \cdot x \cdot f(x)$$

Knowledge Box 3.2

Volume of rotation with cylindrical shells about the y-axis

If we rotate a function $f(x)$ about the y-axis, then the volume enclosed by the curve between $x = a$ and $x = b$ is:

$$V = 2\pi \int_{x=a}^{x=b} x \cdot f(x) \cdot dx.$$

(Notice that we took the constant 2π out in front of the integral sign.)

One tricky thing about this is that the limits of integration are the range of x-values that the radii of the cylindrical shells span. The y-distances only come in via the participation of $f(x)$. The formula in Knowledge Box 3.2 assumes we are rotating the area *below the curve* around the y axis. Finding other areas may require a more complicated setup where we need to figure out the height of the cylinders.

Example 3.3 Find the volume of rotation of the area below the curve $f(x) = 1/x$ about the y-axis from $x = 0.5$ to $x = 2.0$.

Solution:

Sketch the situation.

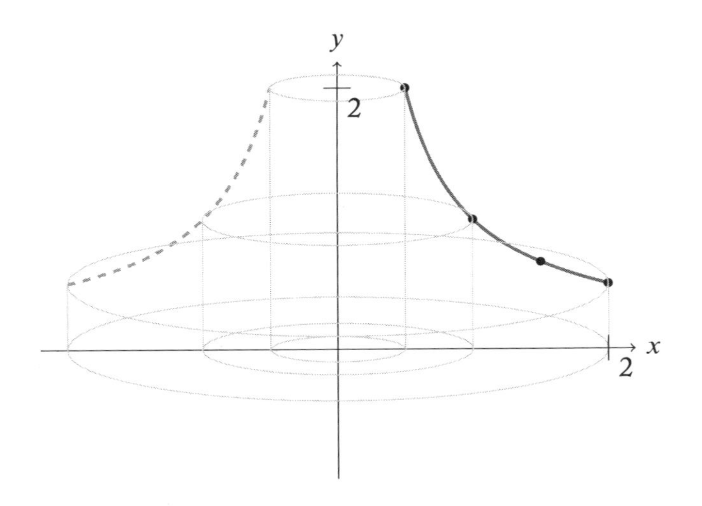

Applying the equation in Knowledge Box 3.2 we get:

$$\begin{aligned}
\text{Volume} &= 2\pi \int_{0.5}^{2} x \cdot f(x) \cdot dx \\
&= 2\pi \int_{0.5}^{2} x \cdot \frac{1}{x} \cdot dx \\
&= 2\pi \int_{0.5}^{2} dx \\
&= 2\pi x \Big|_{0.5}^{2} \\
&= 2\pi(2 - 0.5) \\
&= 3\pi \text{ units}^3
\end{aligned}$$

◊

In the next example we will try and find the volume enclosed by rotating a curve about the y axis. This means we will need to be much more careful about the heights of the cylinders.

Example 3.4 Find the volume enclosed by rotating the area bounded by $f(x) = x^{2/3}$, $x = 1$, and $y = 3^{2/3}$ about the y-axis from $x = 1$ to $x = 3$.

Solution:

Start with the traditional sketch.

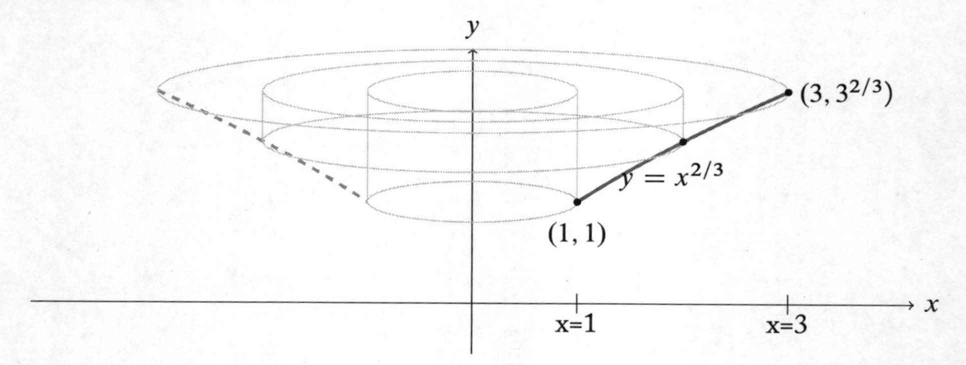

This example is one where the height of the cylinder is not just $f(x)$, so we need to find an expression for the height of the cylinder. This expression goes from $f(x)$ (at the bottom) to $y = 3^{2/3}$ (at the top). So, we get a formula of

$$h = 3^{2/3} - f(x) = 3^{2/3} - x^{2/3}$$

for the height of the cylinder. The area of a cylinder is thus $2\pi x \cdot (3^{2/3} - x^{2/3})$ which makes the volume:

$$\begin{aligned}
\text{Volume} &= 2\pi \int_1^3 x \cdot \left(3^{2/3} - x^{2/3}\right) dx \\
&= 2\pi \int_1^3 \left(x \cdot 3^{2/3} - x \cdot x^{2/3}\right) dx \\
&= 2\pi \int_1^3 \left(3^{2/3}x - x^{5/3}\right) dx \\
&= 2\pi \left(\frac{3^{2/3}}{2}x^2 - \frac{3}{8}x^{8/3}\right)\Bigg|_1^3 \\
&\cong 10.52 \text{ units}^3
\end{aligned}$$

◊

Example 3.4 demonstrates that the formula in Knowledge Box 3.2 only covers one type of rotation about the y axis. In general, it is necessary to remember that you are adding up cylindrical shells and to carefully figure out the height and radius, plugging into $A = 2\pi rh$ to get the formula to integrate. Care is also needed in figuring out the limits of integration. Next, we provide an example of a problem where we use disks to rotate about the y axis.

In order to do this we need to use inverse functions. These were defined in *Fast Start Differential Calculus*. Examples of inverse functions include the following.

- For $x \geq 0$, when $f(x) = x^2$ we have $f^{-1}(x) = \sqrt{x}$.
- If $g(x) = e^x$, we have $g^{-1}(x) = \ln(x)$.
- If $h(x) = \tan(x)$, we have $h^{-1}(x) = \tan^{-1}(x)$.

The notation for "inverse" and "negative first power" are identical and can only be told apart by examining context. Be careful!

Example 3.5 Rotation about the y-axis with disks: Find the volume of rotation when the area bounded by $y = x^2$, the y-axis, and $y = 4$ is rotated about the y axis.

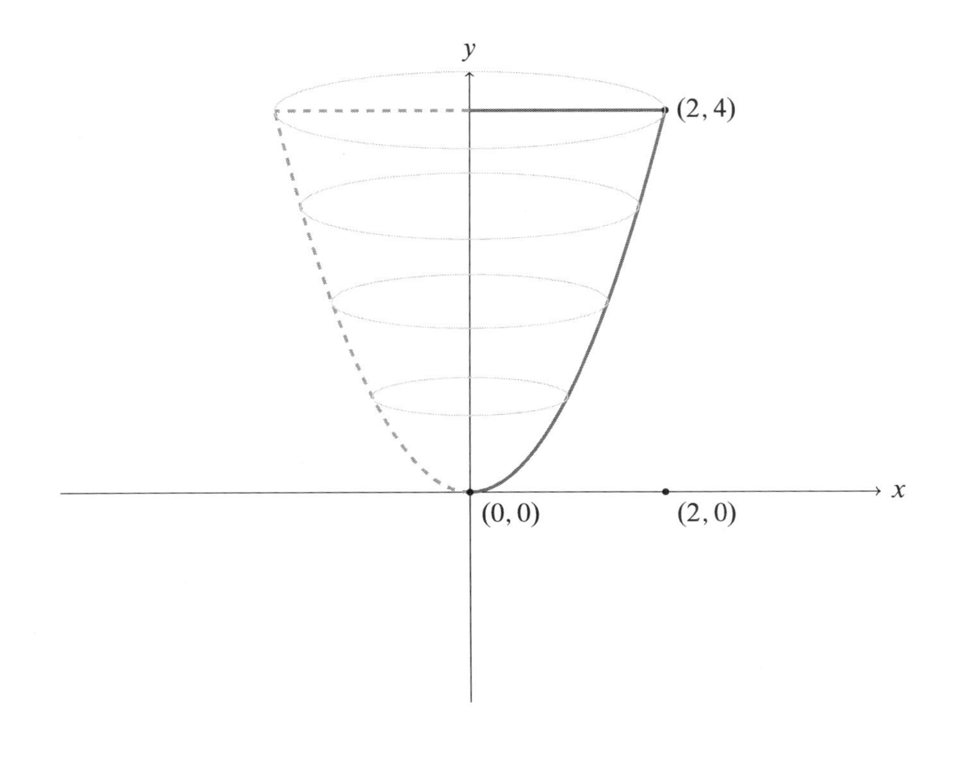

Solution:

For this problem, the radius of the disks is going from the y-axis to the curve $y = x^2$. We need to solve this equation for x which is not too difficult:

$$x = g(y) = \sqrt{y}$$

We ignore the negative square root because we obviously need a positive radius. This means that we have a disk radius of $x = g(y)$.

The volume formula becomes

$$V = \pi \int_{y_o}^{y_1} g(y)^2 \cdot dy$$

The function $x = g(y)$ is the *inverse function* of $y = f(x)$. We can now do the calculations for volume:

$$\begin{aligned}
\text{Volume} &= \pi \int_0^4 (\sqrt{y})^2 \cdot dy \\
&= \pi \int_0^4 y \cdot dy \\
&= \frac{\pi}{2} y^2 \Big|_0^4 \\
&= \frac{\pi}{2}(16 - 0) \\
&= 8\pi \text{ units}^3
\end{aligned}$$

To see that is it possible, let's set this up (but not calculate the integral) with cylinders.

The height of a cylinder is from $y = x^2$ to $y = 4$ meaning that $h = 4 - x^2$. So, the volume integral is

$$V = 2\pi \int_{x=0}^{x=2} x(4 - x^2) \cdot dx$$

Not a terribly hard integral, but one that is harder than the disk integral.

◊

Knowledge Box 3.3

Disks or Cylinders?

Both methods for finding volume of rotation involve adding up areas with integration to find a volume. The method of disks adds up circular disks and the method of cylinders adds up, well, cylinders. How do you tell which method to use?

You use whichever method you can set up and, if you can set up both, you use the one that yields the easier integral.

The next example is another one that lets us practice with the method of cylinders. It is an example where the calculations to find the radius values needed to use the method of disks is too hard.

Example 3.6 Compute the volume obtained by rotating the area bounded by the curve

$$y = 2x^2 - x^3$$

and the x-axis around the y-axis.

Solution:

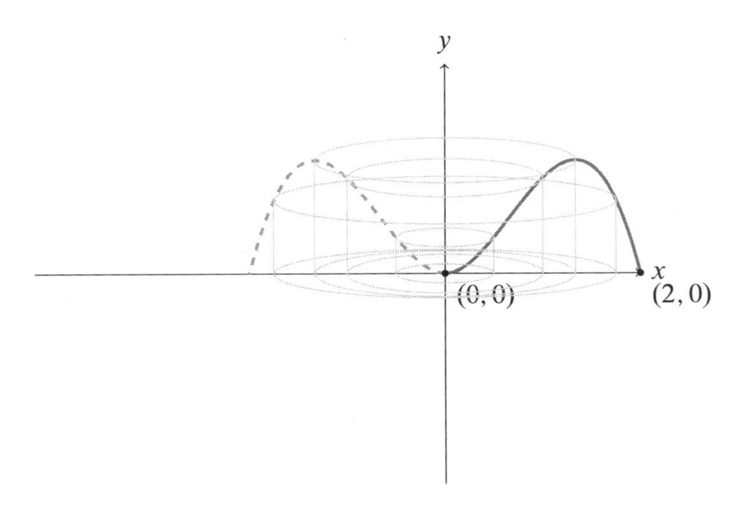

Let's do the calculations. The points where $y = 0$ are $x = 0$ and $x = 2$, giving us the limits of integration. In this case the radius of the cylinders is just x, and the height is just y. So, applying the formula from Knowledge Box 3.2 we get:

$$\begin{aligned}
\text{Volume} &= 2\pi \int_0^2 x(2x^2 - x^3) \cdot dx \\
&= 2\pi \int_0^2 (2x^3 - x^4) \cdot dx \\
&= 2\pi \left(\frac{1}{2}x^4 - \frac{1}{5}x^5 \Big|_0^2 \right) \\
&= 2\pi \left(8 - \frac{32}{5} - 0 + 0 \right) \\
&= 2\pi \cdot \frac{8}{5} \\
&= \frac{16\pi}{5} \text{ units}^3
\end{aligned}$$

If we wanted to use the method of disks we would need to solve $y = 2x^2 - x^3$ for x to get the inverse function – a challenging piece of algebra.

Suppose that we want to rotate the area between two curves about the x-axis. Then we get a large disk for the outer curve and a small disk for the inner curve – meaning that we get *washers*. A rendering of a washer is shown in Figure 3.4. The area of a washer with outer radius r_1 and inner radius r_2 is the difference of the area of the overall disk and the missing inner disk:

$$A = \pi r_1^2 - \pi r_2^2 = \pi(r_1^2 - r_2^2)$$

This area formula forms the basis of the integration performed with the **method of washers**.

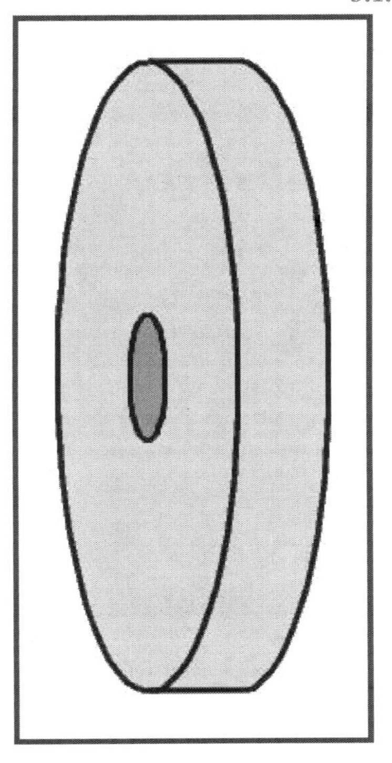

Figure 3.4: A washer or disk with a hole in the center. When we want to rotate the difference of two curves about the x-axis, this shape replaces the disks used when only one curve is rotated.

Knowledge Box 3.4

Formula for finding volume with the method of washers

If $y = f_1(x)$ and $y = f_2(x)$ are functions for which $f_1(x) \geq f_2(x)$ on the interval $[a, b]$, then the volume obtained by rotating the area between the curves about the x-axis is:

$$V = \pi \int_a^b \left(f_1(x)^2 - f_2(x)^2\right) \cdot dx.$$

(Notice that we took the constant π out in front of the integral sign.)

Example 3.7 Find the volume resulting from rotating the area between $f_1(x) = x^2$ and $f_2(x) = \sqrt{x}$ about the x-axis.

Solution:

To apply the method of washers, we need to know where the curves intersect and which one is on the outside edge of the washers. It is easy to see that they intersect at $(0, 0)$ and $(1, 1)$. So the limits of integration will be from $x = 0$ to $x = 1$. On the range $0 \leq x \leq 1$, it's easy to see that $x^2 \leq \sqrt{x}$. So, the outer curve is $f_2(x)$. This gives us enough information to set up the integral.

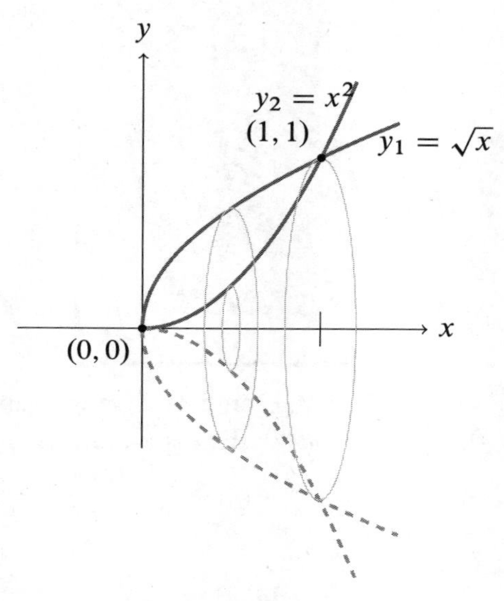

$$\text{Volume} = \pi \int_0^1 (f_1(x)^2 - f_2(x)^2) \cdot dx$$

$$= \pi \int_0^1 (x - x^4)\, dx$$

$$= \pi \left(\frac{1}{2}x^2 - \frac{1}{5}x^5 \right) \Big|_0^1$$

$$= \frac{3\pi}{10} \text{ units}^3$$

◇

There are a large number of different ways to rotate objects about the x- and y-axis. While formulas are presented in this section, it is a really good idea to identify the radius and height of the disks, cylindrical shells, or washers you are integrating to create a volume of rotation. Without that understanding, it is easy to pick the wrong formula. For Example 3.4 *none* of the formulas apply!

The formulas given in this section demonstrate how to add up infinitely thin disks, cylinders, and washers with an integral to compute a volume. These are relatively intuitive shapes with which we should already be familiar. Calculus is not constrained by familiarity: an integral can add up any sort of shapes. If you want a challenge, try to find the formula for the volume of a pyramid with a square base. The formula should be based on the side length of the square base and the height of the pyramid.

PROBLEMS

Problem 3.8 Find the volume of rotation about the x-axis for each of the functions below from $x = 0$ to $x = 1$.

1. $f(x) = 3x^5$
2. $g(x) = 0.5x^6$
3. $h(x) = 5x^7$
4. $r(x) = 12x^5$
5. $s(x) = 6x^{11}$
6. $q(x) = e^{-x}$
7. $a(x) = \cos(\pi x)$
8. $b(x) = (e^x + e^{-x})$

Problem 3.9 If we rotate $y = \sin(x)$ about the x-axis, the result is a string of beads. Find the volume of one bead.

Problem 3.10 Repeat Problem 3.9 for $f(x) = \sin(4x)$.

Problem 3.11 Recalculate Example 3.5 using the method of cylinders.

Problem 3.12 Verify the volume formula

$$V = \frac{4}{3}\pi r^3$$

for a sphere using the method of disks about the x-axis.

Problem 3.13 Find the value a so that rotating the curve $y = \ln(x)$ about the x-axis from $x = 0$ to $x = a$ yields a volume of 4 units^3.

Problem 3.14 Find the volume of rotation about the y axis of the area below each of the functions listed from $x = 0$ to $x = 1$.

1. $f(x) = 6x^9$
2. $g(x) = 16x^8$
3. $h(x) = 3x^{12}$
4. $r(x) = 5x^8$
5. $s(x) = 11x^8$
6. $q(x) = e^x$
7. $a(x) = \cos(\pi x)$
8. $b(x) = \frac{x}{x^2+1}$

Problem 3.15 We know the area under $y = 1/x$ on the interval $[1, \infty)$ is infinite. Find the volume of rotation of $y = 1/x$ on this interval.

Problem 3.16 For each of the following functions and starting and ending x values, find the volume of rotation of the function about the x-axis.

1. $f(x) = x^{-2/5}$ between $x = 3$ and $x = 7$
2. $f(x) = x^{-2/5}$ between $x = 3$ and $x = 5$
3. $f(x) = x^{2/7}$ between $x = 4$ and $x = 7$
4. $f(x) = x^{3/7}$ between $x = 5$ and $x = 7$
5. $f(x) = x^{-3}$ between $x = 5$ and $x = 9$
6. $f(x) = x^{-2}$ between $x = 8$ and $x = 9$

Problem 3.17 For each of the following pairs of functions, compute the volume obtained by rotating the area between the functions about the x-axis.

1. $f(x) = 3x^3$ and $g(x) = 9x^2$
2. $f(x) = x^3$ and $g(x) = 4x^2$
3. $f(x) = 3x^5$ and $g(x) = 48x$
4. $f(x) = 3x^4$ and $g(x) = 3x^3$
5. $f(x) = 2x^3$ and $g(x) = 6x^2$
6. $f(x) = x^3$ and $g(x) = x^2$

Problem 3.18 For each of the following pairs of functions, compute the volume obtained by rotating the area between the functions about the y-axis.

1. $f(x) = 7x^3$ and $g(x) = 28x^2$
2. $f(x) = 4x^6$ and $g(x) = 64x^2$
3. $f(x) = 2x^4$ and $g(x) = 18x^2$
4. $f(x) = 3x^6$ and $g(x) = 3x^2$
5. $f(x) = 7x^5$ and $g(x) = 21x^4$
6. $f(x) = 2x^3$ and $g(x) = 8x$

Problem 3.19 Using the method of disks, rotating about the x-axis, verify the formula for the volume of a cone of radius R and height H:

$$V = \frac{1}{3}\pi \cdot R^2 H$$

3.2 ARC LENGTH AND SURFACE AREA

In this section we will learn to compute the length of curves and, having done that, to find the surface area of figures of rotation.

A piece of a curve is called an **arc**. The key to finding the length of an arc is the **differential of arc length**.

In the past we have had quantities like dx and dy that measure infinitesimal changes in the directions of the variables x and y.

The differential of arc length is different – it does not point in a consistent direction, rather it points along a curve and so, by integrating it, we can find the length of a curve.

Examine Figure 3.5. The relationship between the change in x and y and the change in the length of the curve is Pythagorean, based on a right triangle.

If we take this relationship to the infinitesimal scale, we obtain a formula for the differential of arc length.

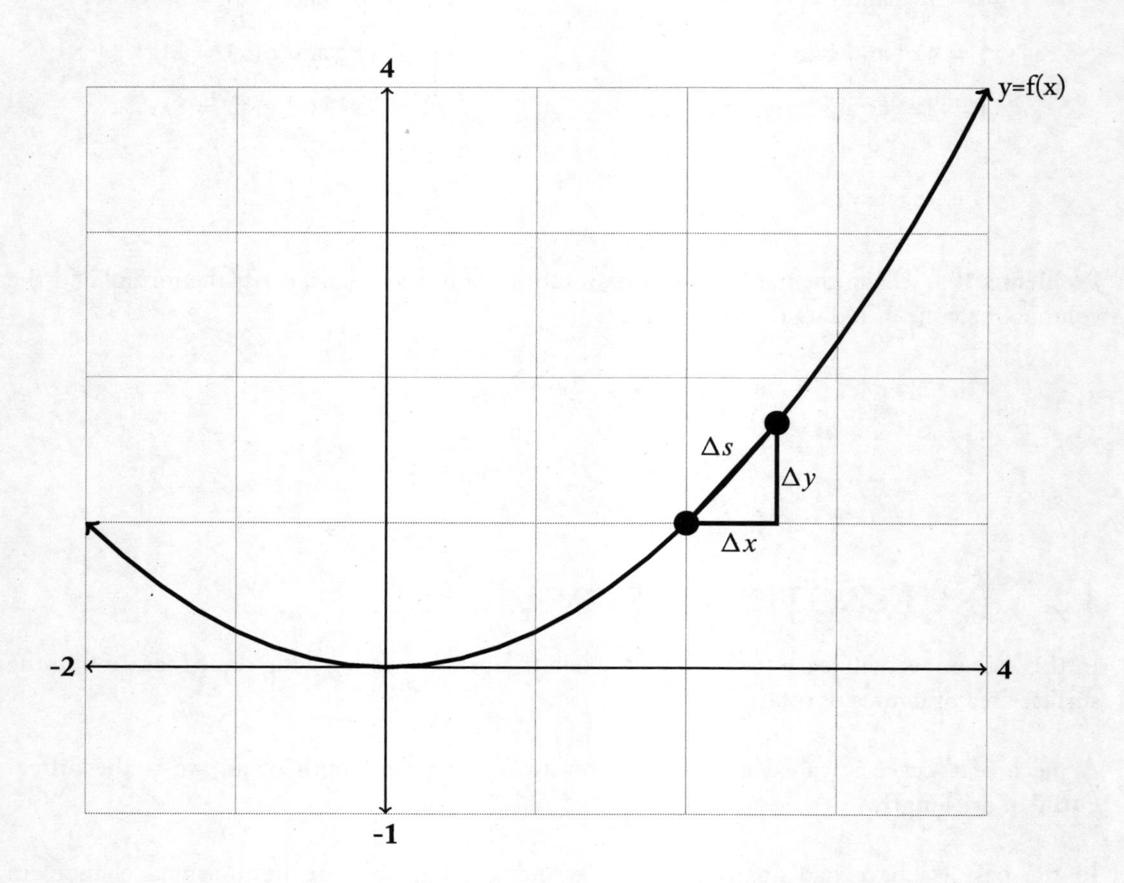

Figure 3.5: The triangle with sides Δx, Δy, and Δs shows how the change in the length of the curve is related to the changes in distance in the x and y directions.

Knowledge Box 3.5

The differential of arc length and arc length

If $y = f(x)$ is a continuous curve, then the rate at which the length of the graph of $f(x)$ changes is called the **differential of arc length,** *denoted by ds. The value of ds is:*

$$\begin{aligned} ds^2 &= dy^2 + dx^2 \\ ds &= \sqrt{dy^2 + dx^2} \\ &= \sqrt{\left(\frac{dy^2}{dx^2} + 1\right) \cdot dx^2} \\ &= \sqrt{(y')^2 + 1} \cdot dx. \end{aligned}$$

*The length S of a curve (***arclength** *of the curve) from $x = a$ to $x = b$ is:*

$$S = \int_a^b ds.$$

Example 3.20 Find the length of $y = 3x^{2/3}$ from $x = 1$ to $x = 8$.

Solution:

The first step in an arc length problem is to compute ds.

$$\begin{aligned} y &= 3x^{2/3} \\ y' &= 2x^{-1/3} \\ ds &= \sqrt{4x^{-2/3} + 1} \cdot dx \end{aligned}$$

This means that the desired length is

$$\begin{aligned} S &= \int_1^8 ds \\ &= \int_1^8 \sqrt{\frac{4}{x^{2/3}} + 1} \cdot dx \\ &= \int_1^8 \sqrt{\frac{4 + x^{2/3}}{x^{2/3}}} \cdot dx \\ &= \int_1^8 \sqrt{4 + x^{2/3}} \cdot \frac{dx}{x^{1/3}} \end{aligned}$$

Let $u = 4 + x^{2/3}$, then $du = \frac{2}{3}x^{-1/3} \cdot dx = \frac{2}{3}\frac{dx}{x^{1/3}}$

So, $\frac{3}{2}du = \frac{dx}{x^{1/3}}$.

Applying the substitution to the limits we see that the integral goes from $u = 5$ to $u = 8$. Transforming everything to u-space, the arc length is:

$$\begin{aligned} S &= \int_5^8 \sqrt{u} \cdot \frac{3}{2}du \\ &= \frac{3}{2}\int_5^8 u^{1/2}du \\ &= \frac{3}{2}\left(\frac{2}{3}u^{3/2}\right)\Bigg|_5^8 \\ &= 8^{3/2} - 5^{3/2} \\ &\cong 11.45 \text{ units}^2 \end{aligned}$$

◊

Alert students will have noticed that the function chosen to demonstrate arc length is not one of our usual go-to functions for demonstration. This is because the formula for ds yields some very difficult integrals. The next example is one such, but yields a formula we already know how to integrate.

Example 3.21 Find the arc length of $y = x^2$ from $x = 0$ to $x = 2$.

Solution:

Since $y' = 2x$, it is easy to find that $ds = \sqrt{4x^2 + 1} \cdot dx$, meaning our integral is:

$$S = \int_0^2 \sqrt{4x^2 + 1} \cdot dx$$

This is a trig-substitution integral. The triangle for this integral is

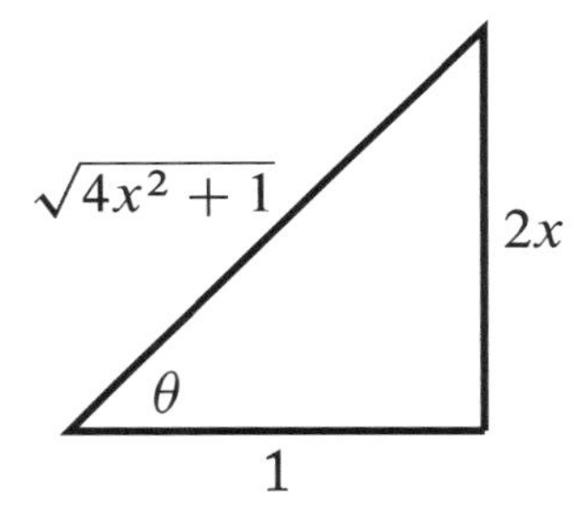

which means our substitutions are:

$$\sqrt{4x^2 + 1} = \sec(\theta)$$

$$2x = \tan(\theta)$$

$$x = (1/2)\tan(\theta)$$

$$dx = (1/2)\sec^2(\theta) \cdot d\theta$$

So we get:

$$\begin{aligned}
\int_0^2 \sqrt{4x^2+1}\cdot dx &= \int_?^? \sec(\theta)\cdot(1/2)\sec^2(\theta)\cdot d\theta \\
&= \frac{1}{2}\int_?^? \sec^3(\theta)\cdot d\theta \qquad \text{(known integral)} \\
&= \frac{1}{4}\left(\sec(\theta)\tan(\theta)+\ln|\sec(\theta)+\tan(\theta)|\right)\Big|_?^? \\
&= \frac{1}{4}\left(\sqrt{4x^2+1}\cdot 2x+\ln|\sqrt{4x^2+1}+2x|\right)\Big|_0^2
\end{aligned}$$

Now that we have performed the integral and transformed it back into x-space we can substitute in the limits and get the arc length.

$$\begin{aligned}
S &= \frac{1}{4}\left(\sqrt{17}\cdot 4+\ln|\sqrt{17}+4|-\sqrt{1}\cdot 0+\ln|\sqrt{1}+0|\right) \\
&= \frac{1}{4}\left(4\sqrt{17}+\ln(\sqrt{17}+4)\right) \cong 4.65 \text{ units}
\end{aligned}$$

Arc-lengths integrals are often challenging. Let's do one more example with a function chosen to keep the difficulty from getting out of hand. This example looks for a general formula for the arc length of a function.

Example 3.22 Find the length of $y = x^{3/2}$ from $x = 0$ to $x = a$.

Solution:

Since $y' = \frac{3}{2}x^{1/2}$ we see that

$$ds = \sqrt{\frac{9}{4}x+1}\cdot dx = \frac{1}{2}\sqrt{9x+4}\cdot dx$$

This means that the desired arc length is:

$$S = \int_0^a \frac{1}{2}\sqrt{9x+4} \cdot dx$$

$$= \frac{1}{2}\int_0^a \sqrt{9x+4} \cdot dx$$

Let $u = 9x + 4$ so that $\frac{1}{9}du = dx$

$$= \frac{1}{2}\int_?^? \sqrt{u} \cdot \frac{1}{9}du$$

$$= \frac{1}{18}\int_?^? u^{1/2} \cdot du$$

$$= \frac{1}{18}\left(\frac{2}{3}u^{3/2}\right)\Big|_?^?$$

Need to substitute back to x

$$= \frac{1}{27}(9x+4)^{3/2}\Big|_0^a$$

$$= \frac{(9a+4)^{3/2} - 8}{27}$$

which is the desired arc length formula.

We have already computed the volume of a solid that is enclosed by the graph of a function rotated about the x-axis. The solids defined in this fashion also have a surface area, the slices of which are circles.

Figure 3.6 shows examples of the circles that appear in such a rotation.

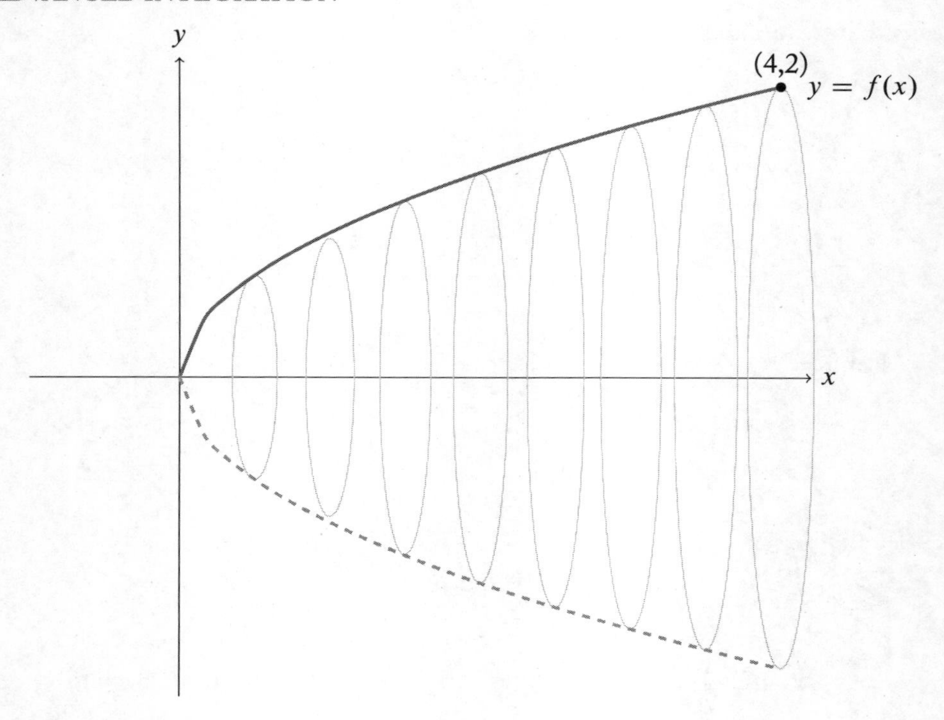

Figure 3.6: Shown are some circles that are the "slices" of the surface area obtained by rotating $f(x)$ about the x-axis.

Using Figure 3.6 as inspiration, we can compute surface area by integrating the circumference of circles with radius $f(x)$. Since these circle follow the arc of $f(x)$, the correct type of change is the differential of arc length, ds. This leads to the formula in Knowledge Box 3.6.

Knowledge Box 3.6

Surface area of rotation

If $y = f(x)$ is a continuous curve, then the surface area A obtained by rotating $f(x)$ around the x-axis from $x = a$ to $x = b$ is:

$$A = 2\pi \int_a^b f(x) \cdot ds.$$

Example 3.23 Find the surface area of rotation of $y = \sqrt{x}$ about the x-axis from $x = 0$ to $x = 4$.

Solution:

Since $y = x^{1/2}$, $y' = \frac{1}{2}x^{-1/2} = \frac{1}{2\sqrt{x}}$. So: $ds = \sqrt{\frac{1}{4x} + 1} \cdot dx$ Using the formula from Knowledge Box 3.6 we obtain the surface area integral,

$$A = 2\pi \int_0^4 \sqrt{x} \cdot \sqrt{\frac{1}{4x} + 1} \cdot dx$$

$$= 2\pi \int_0^4 \sqrt{x \cdot \left(\frac{1}{4x} + 1\right)} \dot{d}x$$

$$= 2\pi \int_0^4 \sqrt{\frac{1 + 4x}{4}} \cdot dx$$

$$= 2\pi \int_0^4 \frac{1}{2}\sqrt{1 + 4x} \cdot dx$$

$$= \pi \int_0^4 \sqrt{1 + 4x} \cdot dx$$

$$\text{Let } u = 4x + 1, \ \frac{1}{4}du = dx$$

$$= \pi \int_1^{17} u^{1/2} \cdot \frac{1}{4}du$$

$$= \frac{\pi}{4} \int_1^{17} u^{1/2} \cdot du$$

$$= \frac{\pi}{4}\frac{2}{3}\, u^{3/2}\Big|_1^{17}$$

$$= \frac{\pi}{6}\left(17^{3/2} - 1\right) \cong 36.18 \text{ units}^2$$

◊

Example 3.24 Find the surface area of rotation for $y = e^x$ from $x = 0$ to $x = 2$.

Solution:

Noting that $y' = e^x$, we see that

$$ds = \sqrt{e^{2x} + 1} \cdot dx.$$

This means the surface area is:

$$\begin{aligned}
A &= 2\pi \int_0^2 e^x \sqrt{e^{2x} + 1} \cdot dx \\
&= 2\pi \int_0^2 \sqrt{e^{2x} + 1}\,(e^x \cdot dx)
\end{aligned}$$

Let $u = e^x$, then $du = e^x \cdot dx$

$$\begin{aligned}
&= 2\pi \int_?^? \sqrt{u^2 + 1} \cdot du \qquad \text{(known integral)} \\
&= 2\pi \left(u\sqrt{u^2 + 1} + \ln|u + \sqrt{u^2 + 1}| \right)\Big|_?^? \\
&= 2\pi \left(e^x\sqrt{e^{2x} + 1} + \ln|e^x + \sqrt{e^{2x} + 1}| \right)\Big|_0^2 \\
&= 2\pi \left(e^2\sqrt{e^4 + 1} + \ln(e^2 + \sqrt{e^4 + 1}) - \sqrt{2} - \ln|1 + \sqrt{2}| \right) \text{ units}^2
\end{aligned}$$

◊

The types of integrals that arise from arc length and rotational surface area problems are often quite challenging. This justifies the large number of integration techniques we learned when we studied methods of integration, several of which came up in this section.

If you study multivariate calculus more deeply, the differential of arc length will appear again for tasks like computing the work done moving a particle along a path through a field. This section is a bare introduction to the power and applications of the differential of arc length.

PROBLEMS

Problem 3.25 For each of the following functions, compute ds, the differential of arc length.

1. $f(x) = x^3$
2. $g(x) = \sin(x)$
3. $h(x) = \tan^{-1}(x)$
4. $r(x) = \dfrac{x}{x+1}$
5. $s(x) = e^{-x}$
6. $q(x) = x^{3/4}$
7. $a(x) = \dfrac{1}{3}$
8. $b(x) = 2^x$

Problem 3.26 For each of the following functions, compute the arc length of the graph of the function on the given interval.

1. $f(x) = 9x^{2/3};\ [0, 1]$
2. $g(x) = 2x + 1;\ [0, 4]$
3. $h(x) = 2x^{3/2};\ [2, 5]$
4. $r(x) = x^2 + 4x + 4;\ [0, 6]$
5. $s(x) = \sqrt{(x-2)^3};\ [4, 5]$
6. $q(x) = (x+1)^{2/3};\ [-1, 1]$
7. $a(x) = 4x^{3/2};\ [2, 4]$
8. $b(x) = x^2 + x + 1/4;\ [0, 8]$

Problem 3.27 For each of the following functions, compute the surface area of rotation of the function for the given interval.

1. $f(x) = x;\ [0, 3]$
2. $g(x) = \sqrt{x};\ [4, 9]$
3. $h(x) = e^x;\ [0, 1]$
4. $r(x) = \sin(x);\ [0, \pi]$
5. $s(x) = \cos(3x);\ [0, \pi/4]$
6. $q(x) = \sin(x)\cos(x);\ [0, \pi/2]$

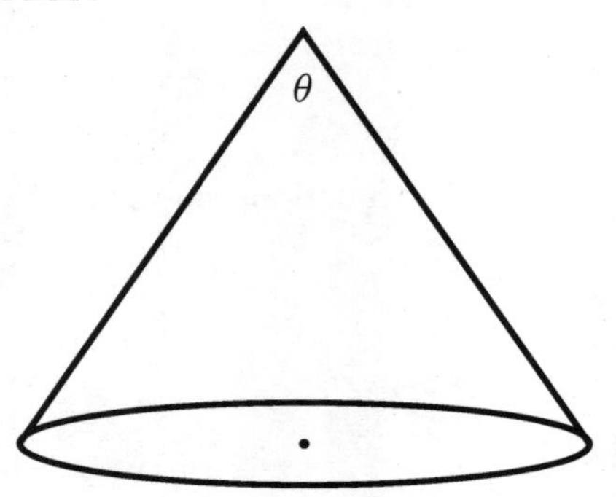

Problem 3.28 Using the techniques for surface area of revolution, find the formula for the surface area of a cone with apex angle θ, as shown above. Don't forget the area of the bottom.

Problem 3.29 If

$$y = x^{2/3},$$

find the formula for the arc length of the graph of this function on the interval $[a, b]$.

Problem 3.30 Based on the material in this section, if

$$(f(t), g(t))$$

is a parametric curve, what would the differential of arc length, ds, be?

Problem 3.31 Derive the polar differential of arc length.

Problem 3.32 Derive the parametric differential of arc length for

$$(x(t), y(t))$$

Problem 3.33 Find but do not evaluate the integral for computing the arc length of

$$y = \sin(x).$$

Discuss: what techniques might work for this integral.

Problem 3.34 Which is harder, finding the arc-length of

$$y = \sqrt{x}$$

or its surface area of rotation? Why?

Problem 3.35 We already know the area under $y = 1/x$ on the interval $[1, \infty)$ is infinite but that the enclosed volume of rotation is finite. Using comparison and cleverness, demonstrate the surface area of this shape is infinite.

3.3 MULTIPLE INTEGRALS

In our earlier study of integration, we learned to use integrals to find the area under a curve. The analogous task in three dimensions is to find the volume under a surface over some domain in the x-y plane. As with partial derivatives we will find the idea of a *currently active* variable useful.

When we were integrating a single-variable function to obtain an area, we integrated over an interval on the x-axis. When we are finding volumes under surfaces, we will integrate over *regions* or subsets of the plane. An example of a relatively simple region in the plane is shown in Figure 3.7.

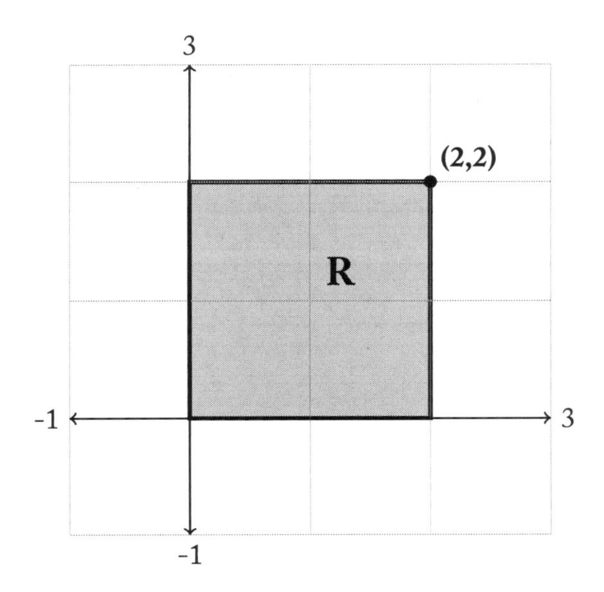

Figure 3.7: A rectangular region R, $0 \leq x, y \leq 2$ in the plane.

So far in this chapter we have found volumes by using integrals to add up slices. To perform volume integrals we will have to integrate slices that are, themselves, the result of integrating. This means we will use **multiple or iterated integrals**. Fortunately, these are done one at a time. We use differentials, like dy and dx, to cue which integral is performed next. Like forks at a fancy dinner party, the next integral is the one for the differential farthest to the left.

Example 3.36 Find the volume under

$$f(x, y) = x^2 + y^2$$

that is above R, $0 \leq x, y \leq 2$, as shown in Figure 3.7.

Solution:

The y and x variables are both in intervals of the form $[0, 2]$, so the limits of integration are 0 and 2 for both the integrals we must perform. The volume under the surface is:

$$\begin{aligned}
V &= \int_0^2 \int_0^2 (x^2 + y^2) \cdot dx\, dy && \text{The variable } x \text{ is active.} \\
&= \int_0^2 \left(\frac{1}{3}x^3 + xy^2 \right)\Big|_0^2 \cdot dy \\
&= \int_0^2 \left(\frac{8}{3} + 2y^2 - 0 - 0 \right) \cdot dy \\
&= \int_0^2 \left(2y^2 + \frac{8}{3} \right) \cdot dy && \text{Now } y \text{ is active.} \\
&= \frac{2}{3}y^3 + \frac{8}{3}y \Big|_0^2 \\
&= \left(\frac{16}{3} + \frac{16}{3} - 0 - 0 \right) = \frac{32}{3} \text{ units}^3
\end{aligned}$$

◊

Example 3.36 contained two integrals. During the first, x was the active variable, and y acted like a constant; when the limits of integration were substituted in they were substituted in for x

not y. The second integral took place in an environment where x was gone and y was the active variable.

Knowledge Box 3.7

Integration with respect to a variable

When we are computing an integral

$$V = \int_a^b \int_c^d f(x, y) \cdot dx\ dy,$$

the first integral treats x as a variable and is said to be an integral **with respect to** x*; the second integral treats y as a variable and is said to be an integral* **with respect to** y.

When we are integrating with respect to one variable, other variables act as constants and so can pass through integral signs. This permits us to simplify some integrals.

Example 3.37 Use the constant status of variables to compute

$$\int_0^1 \int_0^1 (x^2 y^2) \cdot dx\ dy$$

in a simple way.

Solution:

$$\int_0^1 \int_0^1 (x^2 y^2) \cdot dx\ dy = \left(\int_0^1 y^2 \cdot dy \int_0^1 x^2 \cdot dx \right) \qquad \text{Pass } y \text{ through the } x \text{ integral}$$

$$= \left(\int_0^1 y^2 \cdot dy \right) \cdot \left(\int_0^1 x^2 \cdot dx \right)$$

$$= \left(\int_0^1 x^2 \cdot dx \right)^2 \qquad \text{Since the integrals are equal}$$

$$= \left(\frac{1}{3} x^3 \Big|_0^1 \right)^2$$

$$= \left(\frac{1}{3} - 0 \right)^2 = \frac{1}{9} \text{ units}^3$$

◊

This sort of integral – that can be split up into two different integrals – is called a *decomposable integral.*

Knowledge Box 3.8

Decomposable Integrals

The multiple integral of the product of a function of one variable by a function of the other variable can be factored into two single-variable integrals.

$$\int \int f(x)g(y) \cdot dx\, dy = \left(\int f(x)\, dx \right) \left(\int g(y)\, dy \right).$$

Example 3.38 Use the decomposition of integrals to perform the following:

$$\int_0^{\pi/2} \int_0^2 (x \cdot \cos(y)) \cdot dx\, dy$$

Solution:

$$\int_0^{\pi/2} \int_0^2 (x \cdot \cos(y)) \cdot dx\, dy = \left(\int_0^2 x\, dx \right) \times \left(\int_0^{\pi/2} \cos(y)\, dy \right)$$

$$= \left(\frac{x^2}{2} \Big|_0^2 \right) \times \left(\sin(y) \Big|_0^{\pi/2} \right)$$

$$= (4/2 - 0/2) \times (\sin(\pi/2) - \sin(0)) = 2 \cdot 1 = 2$$

◊

Volume integration becomes more difficult when the region R is not a rectangle. Over a rectangular region, the limits of integration are constants. If a region is not rectangular, then the curves that describe the boundaries of the region become involved in the limits of integration.

Example 3.39 Integrate the function $f(x, y) = 2x - y + 4$ over the region R given in Figure 3.8.

Solution:

The function is simple, but the limits of integration are tricky. In this case, $0 \le x \le 2$ and, *for a given value of* x, $0 \le y \le x$. This is because the upper edge of the region of integration is the line $y = x$.

Now, we can set up the integral, choosing the order of integration to agree with the limits.

$$= \int_0^2 \int_0^x (2x - y + 4)\, dy\, dx$$

$$= \int_0^2 \left(2xy - \frac{1}{2}y^2 + 4y\right)\Big|_0^x dx$$

$$= \int_0^2 \left(2x^2 - \frac{1}{2}x^2 + 4x - 0 - 0 - 0\right) \cdot dx$$

$$= \int_0^2 \left(\frac{3}{2}x^2 + 4x\right) dx$$

$$= \frac{1}{2}x^3 + 2x^2\Big|_0^2$$

$$= \frac{1}{2}8 + 8 - 0 - 0$$

$$= 12 \text{ units}^3$$

◊

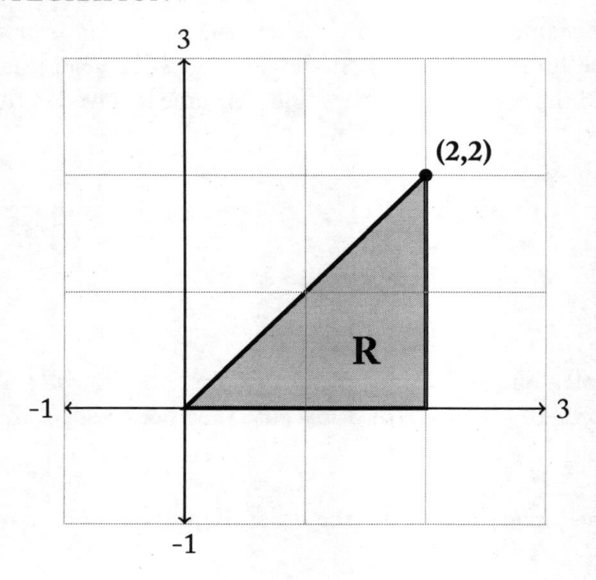

Figure 3.8: A non-rectangular region bounded by the x-axis, the line $x = 2$, and the line $y = x$.

Example 3.40 Find the volume underneath

$$f(x, y) = 4 - x^2$$

over the region bounded by

$$y = x$$

$$y = 2x$$

and

$$x = 2$$

Solution:

Start by drawing the region of integration.

This region has x bounds $0 \leq x \leq 2$.

For a given value of x the region goes from the line $y = x$ to the line $y = 2x$.

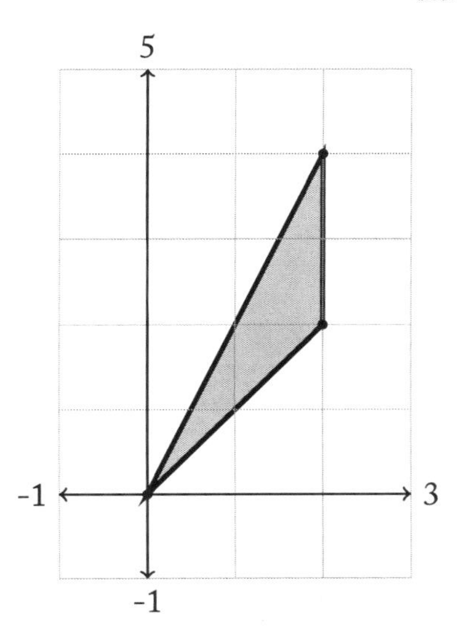

This gives us sufficient information to set up the integral.

$$V = \int_0^2 \int_x^{2x} (4 - x^2)\, dy\, dx$$

$$= \int_0^2 (4y - x^2 y)\Big|_x^{2x} \cdot dx$$

$$= \int_0^2 (8x - 2x^3 - 4x + x^3)\, dx$$

$$= \int_0^2 (4x - x^3)\, dx$$

$$= 2x^2 - \frac{1}{4}x^4 \Big|_0^2$$

$$= 8 - \frac{1}{4} \cdot 16 - 0 + 0$$

$$= 4 \text{ units}^3$$

◇

The regions we have used thus far have been based on functions that are easy to work with using Cartesian coordinates.

To deal with other sorts of regions, we need to first develop a broader point of view.

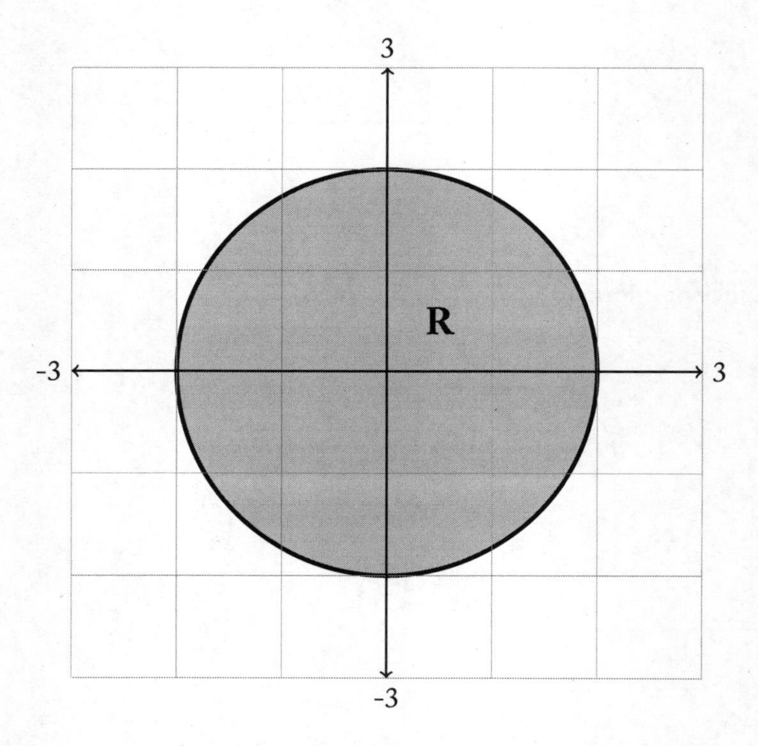

Figure 3.9: Another sort of region – a disk of radius 2 centered at the origin.

Knowledge Box 3.9

The differential of area

The **change of area** *is*

$$dA = dx\, dy = dy\, dx.$$

This permits us to change our integral notation to the following for the integral of $f(x, y)$ over a region R

$$V = \int\int_R f(x, y)\, dA.$$

In polar coordinates:

$$dA = r\, dr\, d\theta.$$

Example 3.41 Find the area enclosed below the curve $f(x, y) = x^2 + y^2$ over a disk of radius 2 centered at the origin.

Solution:

This volume is below the function $f(x, y)$ for points no farther from the origin than 2. This means the region R is

$$x^2 + y^2 \leq 4.$$

In polar coordinates, this region is those points (r, θ) for which $0 \leq r \leq 2$ and $0 \leq \theta \leq 2\pi$. Since the polar/rectangular conversion equations tell us $r^2 = x^2 + y^2$, $f(r, \theta) = r^2$.

Using polar coordinates the integral is:

$$\begin{aligned} V &= \int\int_R f(r, \theta)\, dA \\ &= \int_0^{2\pi} \int_0^2 r^2 \cdot r \cdot dr\, d\theta \end{aligned}$$

$$= \int_0^{2\pi} \int_0^2 r^3 \cdot dr\, d\theta$$

$$= \int_0^{2\pi} \frac{1}{4} r^4 \Big|_0^2 \, d\theta$$

$$= \int_0^{2\pi} (16/4 - 0)\, d\theta$$

$$= \int_0^{2\pi} 4 d\theta$$

$$= 4\theta \Big|_0^{2\pi}$$

$$= 8\pi - 0 = 8\pi \text{ units}^3$$

◊

The next example is a very important one for the theory of statistics. As you know if you have studied statistics, the normal distribution has a probability distribution function of:

$$\frac{1}{\sqrt{2\pi}} \mathrm{e}^{-x^2/2}$$

The area under the curve of a probability distribution function must be equal to one. Thus, if you have a function with an area greater than one, you must multiply it by a normalizing constant equal to one over the area.

The following example shows where the normalizing constant $\frac{1}{\sqrt{2\pi}}$ in the normal distribution probability distribution function comes from.

The integral relies on a trick: squaring the integral and then shifting the squared integral to polar coordinates. This changes an impossible integral into one that can be done without difficulty by u-substitution. Sadly, this only permits the evaluation of the integral on the interval $[-\infty, \infty]$; the coordinate change is intractable *except* on the full interval where the function exists.

Example 3.42 Find $A = \int_{-\infty}^{\infty} \mathrm{e}^{-x^2/2} \cdot dx$.

Solution:

The solution works if you compute the square of the integral.

$$
\begin{aligned}
A^2 &= \left(\int_{-\infty}^{\infty} \mathrm{e}^{-x^2/2} \cdot dx\right)^2 \\
&= \left(\int_{-\infty}^{\infty} \mathrm{e}^{-x^2/2} \cdot dx\right)\left(\int_{-\infty}^{\infty} \mathrm{e}^{-x^2/2} \cdot dx\right) \\
&= \left(\int_{-\infty}^{\infty} \mathrm{e}^{-x^2/2} \cdot dx\right)\left(\int_{-\infty}^{\infty} \mathrm{e}^{-y^2/2} \cdot dy\right) && \text{Rename} \\
&= \int_{-\infty}^{\infty} \int_{-\infty}^{\infty} \mathrm{e}^{-x^2/2} \mathrm{e}^{-y^2/2} \cdot dy\ dx \\
&= \int_{-\infty}^{\infty} \int_{-\infty}^{\infty} \mathrm{e}^{-1/2(x^2+y^2)} \cdot dA
\end{aligned}
$$

Change to polar coordinates

$$
\begin{aligned}
&= \int \int_R \mathrm{e}^{-1/2(r^2)} \cdot dA \\
&= \int \int_R \mathrm{e}^{-1/2(r^2)} \cdot r \cdot dr\ d\theta
\end{aligned}
$$

The polar region in question is $0 \le r < \infty$ and $0 \le \theta < 2\pi$. Rebuild the integral with these limits and we get:

$$
\begin{aligned}
A^2 &= \int_0^{2\pi} \int_0^{\infty} r \cdot \mathrm{e}^{-r^2/2} \cdot dr\ d\theta \\
&= \left(\int_0^{2\pi} d\theta\right) \cdot \left(\int_0^{\infty} r \cdot \mathrm{e}^{-r^2/2} \cdot dr\right) \\
&= \theta\Big|_0^{2\pi} \cdot \left(\int_0^{\infty} r \cdot \mathrm{e}^{-r^2/2} \cdot dr\right)
\end{aligned}
$$

$$= 2\pi \int_0^\infty r \cdot e^{-r^2/2} \cdot dr$$

$$= 2\pi \lim_{a\to\infty} \int_0^a r \cdot e^{-r^2/2} \cdot dr$$

Let $u = -r^2/2$, then $-du = r \cdot dr$

$$= 2\pi \lim_{a\to\infty} \int_?^? e^u \cdot -du$$

$$= -2\pi \lim_{a\to\infty} \int_?^? e^u \cdot du$$

$$= -2\pi \lim_{a\to\infty} e^{-r^2/2}\Big|_0^a$$

$$= -2\pi \lim_{a\to\infty} \left(e^{-a^2/2} - 1\right)$$

$$= -2\pi(0 - 1) = 2\pi$$

If $A^2 = 2\pi$ then $A = \sqrt{2\pi}$, which is the correct normalizing constant.

◊

3.3.1 MASS AND CENTER OF MASS

The center of mass for an object is the average position of all the mass in an object. This section demonstrates techniques for computing the center of mass of flat plates with a density function $\rho(x, y)$. Density is the rate at which mass changes as you move through an object, which, in turn, means that the mass of an object is the integral of its density.

Knowledge Box 3.10

Mass of a plate

Suppose that a flat plate occupies a region R with a density function $\rho(x, y)$ defined on R. Then the mass of the plate is

$$M = \int\int_R \rho(x, y) \cdot dA.$$

Remember that the function $\rho(x, y)$ is usually constant, or close enough to constant that we assume it to be constant, when we have a mass made of a relatively uniform material. The fairly high variation in the mass functions in the examples and homework problems is intended to give your integration skills a workout – not as a representation of situations encountered in physical reality.

Example 3.43 If a plate fills the triangular region R from Figure 3.8 with a density function $\rho(x, y) = x + 1$ grams/unit^2, find the mass of the plate.

Solution:

Using the mass formula, the integral is

$$\text{Mass} = \int_0^2 \int_0^x (x + 1)\, dy\, dx$$

$$= \int_0^2 (xy + y)\Big|_0^x \cdot dx$$

$$= \int_0^2 (x^2 + x) \cdot dx$$

$$= \frac{x^3}{3} + \frac{x^2}{2}\Big|_0^2$$

$$= 8/3 + 2 - 0 - 0 = 14/3 \text{ g}$$

◊

Once we have the ability to compute the mass of a plate from its dimensions and density, we can compute the coordinates of the center of mass of the plate using a type of averaging integral.

Knowledge Box 3.11

Center of mass

Suppose that a flat plate occupies a region R with a density function $\rho(x, y)$ defined on R. Then if

$$M_x = \int\int_R y\rho(x, y) \cdot dA$$

and

$$M_y = \int\int_R x\rho(x, y) \cdot dA$$

the center of mass of the plate is

$$(\overline{x}, \overline{y}) = \left(\frac{M_y}{M}, \frac{M_x}{M}\right).$$

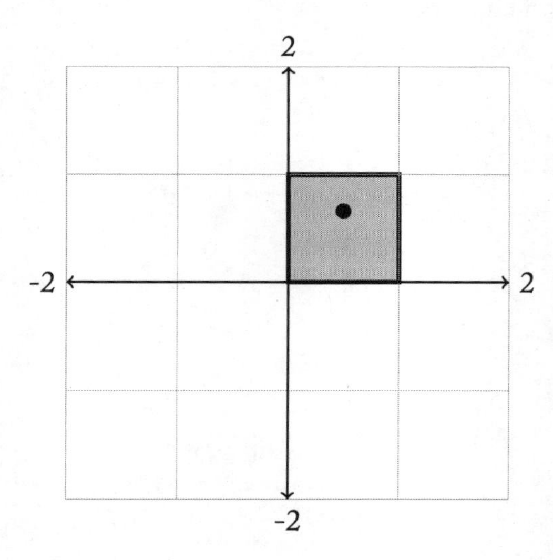

Figure 3.10: Center of mass of square region when $\rho(x, y) = 2y$ g/unit^2.

Example 3.44 Suppose that R is the square region

$$0 \le x,\ y \le 1$$

and that $\rho(x, y) = 2y$ g/unit^2 as shown in Figure 3.10.

Find the center of mass.

Solution:

This problem requires three integrals.

$$M = \int_0^1 \int_0^1 2y \cdot dy\, dx \qquad M_x = \int_0^1 \int_0^1 y \cdot 2y \cdot dy\, dx \qquad M_y = \int_0^1 \int_0^1 x \cdot 2y \cdot dy\, dx$$

$$M = \int_0^1 y^2\Big|_0^1 dx = \int_0^1 (1-0)dx = \int_0^1 dx = x\Big|_0^1 = (1-0) = 1 \text{ gram}$$

$$M_x = \int_0^1 \int_0^1 2y^2 \cdot dy\, dx = \int_0^1 \frac{2}{3}y^3\Big|_0^1 \cdot dx = \int_0^1 \left(\frac{2}{3} - 0\right) \cdot dx = \int_0^1 \frac{2}{3} \cdot dx = \frac{2}{3}x\Big|_0^1 = \frac{2}{3}$$

$$M_y = \int_0^1 xy^2\Big|_0^1 \cdot dx = \int_0^1 (x-0) \cdot dx = \frac{1}{2}x^2\Big|_0^1 = \frac{1}{2}$$

Now that we have the pieces we can use the formula for center of mass:

$$(\overline{x}, \overline{y}) = \left(\frac{1/2}{1}, \frac{2/3}{1}\right) = \left(\frac{1}{2}, \frac{2}{3}\right)$$

◊

PROBLEMS

Problem 3.45 Find the integral of each of the following functions over the specified region.

1. The function $f(x, y) = x + y^2$ on the strip

$$0 \leq x \leq 4 \quad 0 \leq y \leq 1.$$

2. The function $g(x, y) = xy$ on the rectangle

$$1 \leq x \leq 3 \quad 1 \leq y \leq 2.$$

3. The function $h(x, y) = x^2y + xy^2$ on the square

$$0 \leq x \leq 2 \quad 0 \leq y \leq 2.$$

4. The function $r(x, y) = 2x + 3y + 1$ on the region bounded by $x = 0$, $y = 1$, and $y = x$.
5. The function $s(x, y) = x^2 + y^2 + 1$ on the region bounded by the x axis and the function $y = 4 - x^2$.
6. The function $q(x, y) = x + y$ on the region bounded by the curves $y = \sqrt{x}$ and $y = x^2$.
7. The function $a(x, y) = x^2$ on the region bounded by the curves $y = 2x$ and $y = x^2$.
8. The function $b(x, y) = y^2$ on the region bounded by the curves $y = \sqrt[3]{x}$ and $y = x$ for $x \geq 0$.

Problem 3.46 Sketch the regions from Problem 3.45.

Problem 3.47 Explain why a density function $\rho(x, y)$ can never be negative.

Problem 3.48 Find a region R so that the integral over R of $f(x) = x^2 + y^2$ is 6 units^3.

Problem 3.49 Find the square region $0 \leq x,\ y \leq a$ so that

$$\int\int_R (x^3 + y) \cdot dA$$

is 12 units^3.

Problem 3.50 Find the mass of the plate $0 \leq x, y \leq 3$ if

$$\rho(x, y) = y^2 + 1.$$

Problem 3.51 Find the center of mass of the region bounded by the x-axis, the y-axis, and the line $x + y = 4$ if the density function is

$$\rho(x, y) = y + 1.$$

Problem 3.52 Find the center of mass of the region bounded by $y = \sqrt{x}$ and $y = x^2$ if the density function is

$$\rho(x, y) = x + 2.$$

Problem 3.53 Find the center of mass of the region $0 \leq x, y \leq 1$ if the density function is

$$\rho(x, y) = (x + y)/2.$$

Problem 3.54 Find the volume under

$$f(x, y) = \sqrt{x^2 + y^2}$$

above the region $x^2 + y^2 \leq 16$.

Problem 3.55 Find the volume under

$$f(x, y) = x^2 + y^2$$

above the region bounded by the petal curve $r = 2\cos(3\theta)$.

Problem 3.56 Find the volume under

$$f(x, y) = (x^2 + y^2)^{3/2}$$

above the region bounded by the petal curve $r = \cos(2\theta)$.

Problem 3.57 Find a plane $f(x, y)$ so that the area under the plane but over a circle of radius 2 centered at the origin is 16 units^3.

Problem 3.58 Derive the general formula for the volume over a rectangle and under a plane in a region where the plane has a positive z value.

CHAPTER 4

Sequences, Series, and Function Approximation

This chapter is quite different from the rest of the material in the *Fast Start* series. Calculus is mostly part of the mathematics of continuous functions, while the sequences and series we study in this chapter are a part of **discrete mathematics** – math that is broken up into individual pieces. Discrete math is about things you can count, rather than things you can measure. Since this is a calculus book, we will apply what we learn to understanding and increasing the power of our calculus. The ultimate goal of the chapter is a much deeper understanding of transcendental (non-polynomial) functions like e^x or $\cos(x)$. We start, however, at the beginning.

4.1 SEQUENCES AND THE GEOMETRIC SERIES

A sequence is an infinite list of numbers. Sometimes we give a sequence by listing an obvious pattern:

$$S = 1, \frac{1}{2}, \frac{1}{3}, \frac{1}{4}, \cdots$$

with the ellipsis meaning "and so on." We also can use more formal set notation to specify a sequence with a formula:

$$S = \left\{ \frac{1}{n} : n = 1, 2, \ldots \right\}$$

Like a function having a limit at infinity, there is a notion of a sequence **converging**.

Knowledge Box 4.1

Definition of the convergence of a sequence to a limit

We call L *the* **limit of a sequence** $\{x_n : n = 1, 2, \ldots\}$ *if, for each* $\epsilon > 0$, *there is a whole number* N *so that, whenever* $n > N$, *we have:*

$$|x_n - L| < \epsilon.$$

A sequence that has a limit is said to **converge**.

Example 4.1 Prove that the sequence

$$S = \left\{ \frac{1}{n} : n = 1, 2, \ldots \right\}$$

converges to zero.

Solution:

For $\epsilon > 0$ choose N to be the smallest whole number greater than $\frac{1}{\epsilon}$. This makes $\frac{1}{N} < \epsilon$. Then if $n > N$ we have:

$$n > N$$

$$\frac{1}{n} < \frac{1}{N} \qquad \text{Reciprocals reverse inequalities}$$

$$\frac{1}{n} < \epsilon \qquad \text{Since } 1/N < \epsilon$$

$$\left| \frac{1}{n} - 0 \right| < \epsilon \qquad \text{Value on the left did not change}$$

Which satisfies the definition of the limit of the sequence being $L = 0$.

As always there are shortcuts that mean we only need to rely on the definition of the limits of sequences occasionally.

Knowledge Box 4.2

Sequences drawn from functions

Suppose that we have a sequence

$$S = \{x_n\} = \{f(n) : n = 1, 2, \ldots\}$$

and that $\lim_{x \to \infty} f(x) = L$. *Then we may conclude that* $\lim_{n \to \infty} x_n = L$.

Example 4.2 Suppose that we have a sequence

$$\left\{x_n = \frac{1}{1+n^2} : n = 0, 1, 2, \ldots\right\}$$

Find $\lim_{n \to \infty} x_n$.

Solution:

We already know $\lim_{x \to \infty} \frac{1}{1+x^2} = 0$. So, using Knowledge Box 4.2, we have that

$$\lim_{n \to \infty} x_n = 0$$

◊

The rule in Knowledge Box 4.2 is *not reversible.*

Example 4.3 Determine if the sequence

$$S = \{\cos(2\pi n) : n = 0, 1, 2, \ldots\}$$

converges.

Solution:

Since $\lim_{x \to \infty} \cos(2\pi x)$ jumps back and forth in the range $-1 \leq y \leq 1$, the function that we drew the sequence from does not have a limit.

Leaving that aside, examine a listing of the first several terms of the sequence:

$$\{1, 1, 1, 1, 1, 1, \ldots\}$$

This sequence obviously converges to $L = 1$.

Again: if the function has a limit, then so does the sequence. The reverse need not be true.

The next sequence resolution technique requires that we define several terms.

Definition 4.1 *If a sequence* $\{x_n\}$ *has the property that* $x_n \leq x_{n+1}$, *then we say that the sequence is* **monotone increasing**.

Definition 4.2 *If a sequence* $\{x_n\}$ *has the property that* $x_n \geq x_{n+1}$, *then we say that the sequence is* **monotone decreasing**.

Definition 4.3 *If a sequence* $\{x_n\}$ *is either monotone increasing or monotone decreasing, then we say the sequence is* **monotone**.

Definition 4.4 *If a sequence* $\{x_n\}$ *has the property that*

$$x_n \leq C$$

for all n *and for some constant* C, *then we say the sequence is* **bounded above**.

Definition 4.5 *If a sequence* $\{x_n\}$ *has the property that*

$$x_n \geq C$$

for all n *and for some constant* C, *then we say the sequence is* **bounded below**.

Definition 4.6 *If a sequence* $\{x_n\}$ *is both bounded above and bounded below, then we say the sequence is* **bounded**.

Knowledge Box 4.3

Bounded monotone sequences

The following types of sequences all converge.

- *A monotone increasing sequence that is bounded above.*
- *A monotone decreasing sequence that is bounded below.*
- *Any monotone bounded sequence.*

The reason that the sequences listed in Knowledge Box 4.3 converge is that they are required to move toward a bound of some sort without passing it. This is another useful shortcut for determining if a sequence converges, although this shortcut does not tell us the value of the limit.

Example 4.4 Demonstrate that the sequence

$$\left\{x_n = \frac{n}{n+1} : n = 0, 1, 2, 3, \ldots\right\}$$

has a limit.

Solution:

While we could use Knowledge Box 4.2, let's use this example as a chance to demonstrate the technique from Knowledge Box 4.3.

First notice that this sequence is monotone increasing. To see this, notice that:

$$n^2 + 2n < n^2 + 2n + 1$$

$$n(n+2) < (n+1)^2$$

$$\frac{n}{n+1} < \frac{n+1}{n+2}$$

$$x_n < x_{n+1}$$

Since $n < n+1$, we can deduce that $x_n = \dfrac{n}{n+1} < 1$.

So the sequence is bounded above. This permits us to use Knowledge Box 4.3 to deduce that the sequence has a limit.

◊

It would have been easier to do Example 4.4 with Knowledge Box 4.2. But there will be times when the sequence is not drawn from a function when we have to use monotone sequence theory. The next Knowledge Box extends the reach of our ability to check sequences for convergence.

Knowledge Box 4.4

Arithmetic combinations of sequences

If $\{x_n\}$ has a limit of L, and $\{y_n\}$ has a limit of M, and if a, b are constants, then:

- $\lim_{n\to\infty} ax_n \pm by_n = aL \pm bM$
- $\lim_{n\to\infty} x_n \cdot y_n = L \cdot M$
- $\lim_{n\to\infty} \frac{x_n}{y_n} = \frac{L}{M}$ *if* $M \neq 0$.
- $\lim_{n\to\infty} x_n^k = L^k$

Example 4.5 Find the limit of

$$S = \left\{ x_n = \frac{1}{n} + 3 \cdot \frac{n}{n+1} : n = 1, 2, 3, \ldots \right\}$$

Solution:

We already know the limit of $1/n$, as a series, is zero. Using Knowledge Box 4.2 it is easy to see that:

$$\lim_{n\to\infty} \frac{n}{n+1} = 1$$

Combining these results using the information in Knowledge Box 4.4 we get that the limit is:

$$L = 0 + 3 \cdot 1 = 3$$

◊

This concludes our direct investigation of sequences. We now turn to using sequences as a tool to explore **series**. Where a sequence is an infinite list of numbers, a series is an infinite list of numbers *that you add up*. This may or may not result in a finite sum – and resolving that question requires sequence theory.

Knowledge Box 4.5

Definition of series

If $\{x_n : n = 0, 1, \ldots\}$ *is a sequence, then*

$$\sum_{n=0}^{\infty} x_n = x_0 + x_1 + \cdots + x_k + \cdots$$

is the corresponding **infinite series**. *If we sum only finitely many terms we have a* **finite series.**

Example 4.6 Show that the following infinite series sums to 1.

$$\sum_{n=1}^{\infty} \frac{1}{2}^n = \frac{1}{2} + \frac{1}{4} + \frac{1}{8} + \cdots$$

Solution:

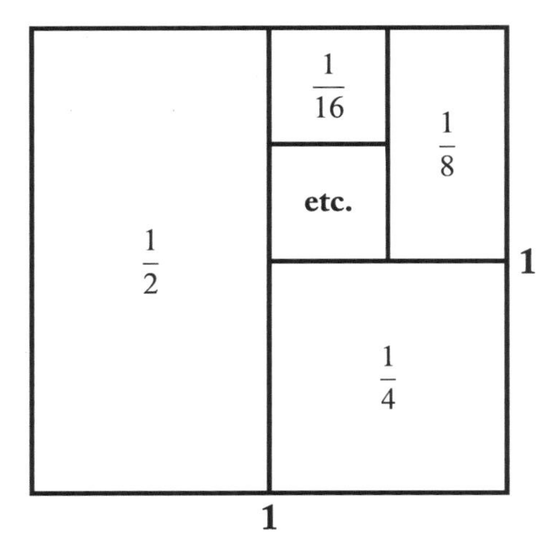

Figure 4.1: A unit square cut into pieces 1/2, 1/4, 1/8, ...

Examine Figure 4.1. The figure divides a unit square (with area 1) into rectangles of size 1/2, 1/4, 1/8, and every other number in the series we are trying to sum. This constitutes a geometric demonstration that the series sums to 1.

◊

Clearly, finding a cool picture is not a general technique for demonstrating that a series has a sum. Proving no such picture exists when the series fails to have a sum is even more impossible – we need a more general theory.

Knowledge Box 4.6

The sequence of partial sums of a series

Suppose that $S = \sum_{n=1}^{\infty} x_n$ *is a series. If we set*

$$p_n = \sum_{k=1}^{n} x_k,$$

then $\{p_n\}$ *is called the* **sequence of partial sums of** S. *We say that a series* **converges to a sum** L *if and only if its sequence of partial sums has L as a limit.*

Example 4.7 Find the sequence of partial sums of the series in Example 4.6 and compute its limit.

Solution:

Compute the first few members of the sequence of partial sums:

$$p_1 = 1/2$$

$$p_2 = 1/2 + 1/4 = 3/4$$

$$p_3 = 1/2 + 1/4 + 1/8 = 7/8$$

$$p_4 = 1/2 + 1/4 + 1/8 + 1/16 = 15/16$$

Which is a clear pattern, and it is easy to see

$$p_n = 1 - \frac{1}{2^n}$$

This gives us the sequence of partial sums. Computing the limit we get:

$$\lim_{n\to\infty} p_n = \lim_{n\to\infty} 1 - \frac{1}{2^n} = 1 - 0 = 1$$

So we get the same sum using this more formal approach.

◊

At this point we need to call forward an identity from *Fast Start Integral Calculus*:

$$1 + x + x^2 + \cdots x^n = \frac{x^{n+1} - 1}{x - 1} = \frac{1 - x^{n+1}}{1 - x}$$

This identity is one that is true for polynomials, but it also applies to summing a finite series. Additionally, this formula can be used as the partial sum of a particular type of infinite series – at least when its limit exists.

Knowledge Box 4.7

Finite and infinite geometric series

The polynomial identity in the text tells us, for a constant $a \neq 1$, *that:*

$$\sum_{k=0}^{n} a^k = \frac{a^{n+1} - 1}{a - 1}.$$

This is the **finite geometric series formula.**

If $|a| < 1$, *then the limit of the finite series gives us the* **infinite geometric series formula***:*

$$\sum_{n=0}^{\infty} a^n = \frac{1}{1-a}.$$

Applying Knowledge Box 4.4 we also get that:

$$\sum_{n=0}^{\infty} c \cdot a^n = \frac{c}{1-a}$$

for a constant c*. The number* a *is called the* **ratio** *of the geometric series.*

Example 4.8 Compute

$$\sum_{n=0}^{\infty} \frac{3}{5^n}$$

Solution:

This is a geometric series with ratio $a = 1/5$. It has a leading constant $c = 3$. Applying the appropriate formula we see that:

$$\sum_{n=0}^{\infty} \frac{3}{5^n} = \frac{3}{1 - 1/5} = \frac{3}{4/5} = \frac{15}{4}$$

PROBLEMS

Problem 4.9 Prove formally, using the definition of the limit of a sequence, that

$$\{\cos(2\pi n) : n = 0, 1, 2, \ldots\}$$

converges to 1.

Problem 4.10 Prove formally, using the definition of the limit of a sequence, that

$$\left\{\frac{1}{n^2} : n = 1, 2, 3, \ldots\right\}$$

converges to 0.

Problem 4.11 Prove formally, using the definition of the limit of a sequence, that

$$\left\{\frac{n}{n+1} : n = 0, 1, 2, \ldots\right\}$$

converges to 1.

Problem 4.12 Compute the limit of each of the following sequences or give a reason why the limit does not exist. Assume $n = 1, 2, \ldots$

1. $\{x_n = \tan^{-1}(n)\}$
2. $\left\{y_n = \sin\left(\frac{\pi}{2}n\right)\right\}$
3. $\left\{z_n = \frac{n^2}{n+1}\right\}$
4. $\{y_n = \sin(\pi n)\}$
5. $\left\{z_n = \frac{3n^2}{n^2+1}\right\}$
6. $\left\{z_n = \frac{\cos(n)}{n+1}\right\}$

Problem 4.13 Do the calculation to prove the infinite geometric series formula from the finite one: see Knowledge Box 4.7.

Problem 4.14 Compute the following sums or give a reason they fail to exist.

1. $\sum_{k=0}^{20} 1.2^k$
2. $\sum_{n=0}^{\infty} \left(\frac{1}{3}\right)^n$
3. $\sum_{n=0}^{\infty} 2 \cdot \left(\frac{1}{7}\right)^n$
4. $\sum_{n=0}^{\infty} \left(\frac{-1}{4}\right)^n$
5. $\sum_{n=0}^{\infty} 3\left(\frac{3}{2}\right)^n$
6. $\sum_{n=0}^{\infty} 0.05^n$
7. $\sum_{n=0}^{\infty} 112\,(0.065)^n$
8. $\sum_{n=0}^{\infty} 2 \cdot (-1)^n$

Problem 4.15 Compute the following sums or give a reason they fail to exist.

1. $\sum_{k=12}^{24} 3^n$
2. $\sum_{k=5}^{15} 2^n$
3. $\sum_{k=3}^{30} 1.2^n$
4. $\sum_{k=10}^{20} 4.5^n$
5. $\sum_{k=90}^{100} 1.1^n$
6. $\sum_{k=1}^{22} 7^n$

Problem 4.16 A swinging pendulum is losing energy. Its first swing is 2 m long, and each after that is 0.9985 times as long as the one before it. Estimate the total distance traveled by the pendulum.

Problem 4.17 A ball is dropped from a height of 8 m. If each bounce is 3/4 the height of the one before it, estimate the total vertical distance traveled by the ball.

Problem 4.18 A rod is initially displaced 2.1 mm from equilibrium and undergoes damped vibration with a decay in the length of each subsequent swing of 0.937. Find the total vertical distance traveled by the end of the rod.

Problem 4.19 A ball is dropped from a height of 4m. If each bounce is 0.86 times the height of the one before it, estimate the total vertical distance traveled by the ball.

Problem 4.20 A rod is initially displaced 3cm from equilibrium and undergoes damped vibration with a decay in the length of each subsequent swing of 0.987. Find the total vertical distance traveled by the end of the rod.

Problem 4.21 A very orderly and goes first north then east over an over. If the distances it travels before turning go 1, 1/3, 1/9, 1/27, and so on, what the is distance from its starting point that it approaches as it travels farther and farther?

4.2 SERIES CONVERGENCE TESTS

In this section we will develop a number of tests to determine if a series converges. At present, we know that an infinite geometric series with a ratio of a with $|a| < 1$ will converge, but not much else. We begin with a motivating example.

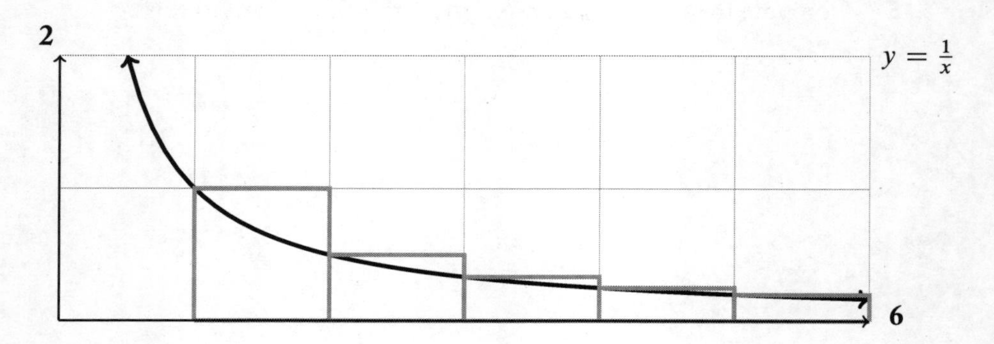

Figure 4.2: Shown is a portion of the graph of $y = 1/x$ and a sequence of rectangles of width 1 and height $1/n$ for $n = 1, 2, 3, \ldots$.

Example 4.22 Determine if the series $\sum_{n=1}^{\infty} \frac{1}{n}$ has a finite or infinite sum.

Solution:

Examine Figure 4.2. This shows that a series of rectangles with area $1, 1/2, 1/3, \ldots$ have a larger area than $\int_1^{\infty} \frac{dx}{x}$ because the area under the curve is strictly smaller than the sum of the areas of the rectangles. Computing the integral:

$$\begin{aligned} \int_1^{\infty} \frac{dx}{x} &= \lim_{a \to \infty} \int_1^{a} \frac{dx}{x} \\ &= \lim_{a \to \infty} \ln(x) \Big|_1^a \\ &= \lim_{a \to \infty} \ln(a) - \ln(1) \\ &= \infty \end{aligned}$$

shows us that: $\sum_{n=1}^{\infty} \frac{1}{n} \geq \infty$, and we conclude the sum is infinite.

◊

The series

$$\sum_{n=1}^{\infty} \frac{1}{n}$$

is sufficiently important that it has its own name: the **harmonic series**.

Definition 4.7 *When considering*

$$\sum_{n=1}^{\infty} x_n$$

we call x_n the **general term** *of the series.*

Example 4.22 is another example of proving something about a series sum by drawing a clever picture. An interesting technique, but, as before, we need more general tools. One such tool involves taking the limit of the sequence generating the series. If it does not approach zero, there is no hope that the infinite sum converges.

Knowledge Box 4.8

The divergence test

If $\lim_{n\to\infty} x_n \neq 0$ *then*

$$\sum_{n=0}^{\infty} x_n$$

does not have a finite value.

It is important to note that the divergence test is uni-directional. If the general term of a series *does* go to zero, that tells you exactly nothing about the behavior of the sum of the series.

Example 4.23 Show that

$$\sum_{n=0}^{\infty} \frac{n}{n+1}$$

diverges (has an infinite sum).

Solution:

Since

$$\lim_{n\to\infty} \frac{n}{n+1} = 1,$$

the series in question diverges by the divergence test. Colloquially, we are adding up an infinite number of terms that are approaching one – so the resulting sum is infinite.

◊

When you can use it, the divergence test is often short and sweet. The next test is the formal version of the test we used in Example 4.22.

Knowledge Box 4.9

The integral test

Suppose that $f(x)$ is a positive, decreasing function on $[0, \infty)$ and that $x_n = f(n)$. Then,

$$\sum_{n=1}^{\infty} x_n \text{ and } \int_a^{\infty} f(x)dx$$

both converge or both diverge for any finite $a \geq 1$.

Example 4.24 Show that

$$\sum_{n=1}^{\infty} \frac{1}{n^2}$$

is finite.

Solution:

Use the integral test.

$$\begin{aligned}\int_1^{\infty} \frac{dx}{x^2} &= \lim_{a\to\infty} \int_1^a \frac{dx}{x^2} \\ &= \lim_{a\to\infty} \left.\frac{-1}{x}\right|_1^a \\ &= \lim_{a\to\infty} \frac{-1}{a} - (-1) \\ &= 0 + 1 = 1\end{aligned}$$

Since the improper integral is finite, so is the sum. Note that this *does not* tell us the value of the sum.

◊

The fact that the series $\sum_{n=1}^{\infty} \frac{1}{n}$ diverges but $\sum_{n=1}^{\infty} \frac{1}{n^2}$ converges motivates our next test.

Definition 4.8 *A* p**-series** *is a series of the form*

$$\sum_{n=1}^{\infty} \frac{1}{n^p}$$

where p *is a constant.*

Knowledge Box 4.10

The p-series test

The p*-series*

$$\sum_{n=1}^{\infty} \frac{1}{n^p}$$

converges if $p > 1$ *and diverges if* $p \leq 1$.

The series convergence test in Knowledge Box 4.10 follows directly from the integral test – something you are asked to verify in the homework.

Example 4.25 Determine if

$$\sum_{n=1}^{\infty} \frac{1}{n^{\pi}}$$

converges to a finite number or diverges.

Solution:

This series has the form of a p-series with $p = \pi$. Since $\pi > 1$, we conclude the series converges.

◊

Example 4.26 Determine if

$$\sum_{n=1}^{\infty} \frac{1}{\sqrt{n}}$$

converges to a finite number or diverges.

Solution:

This series has the form of a p-series with $p = 1/2$. Since $1/2 \leq 1$, we conclude the series diverges.

$\diamond$

The next two tests leverage series we already understand to resolve even more series. The tests depend on various forms of comparison of a series under test to a series with known convergence or divergence behavior.

Knowledge Box 4.11

The comparison tests

- *If* $\sum_{n=0}^{\infty} x_n$ *converges and* $0 \leq y_n \leq x_n$ *for all* n, *we have that* $\sum_{n=0}^{\infty} y_n$ *also converges.*
- *If* $\sum_{n=0}^{\infty} x_n$ *diverges and* $y_n \geq x_n \geq 0$ *for all* n, *we have that* $\sum_{n=0}^{\infty} y_n$ *also diverges.*

The colloquial versions of the comparison tests are pretty easy to believe. The first says, of a series with positive general terms, that if it is smaller, term by term, than another series with a finite sum, then it has a finite sum. The second reverses that: saying that, if a series is larger, term by term, than another series with an infinite sum, then it has an infinite sum.

Example 4.27 Determine if

$$\sum_{n=1}^{\infty} \frac{1}{n/3}$$

converges or diverges.

Solution:

Notice that, for $n \geq 1$,

$$n \geq n/3$$

$$\frac{1}{n} \leq \frac{1}{n/3}$$

So $0 \leq \frac{1}{n} \leq \frac{1}{n/3}$, which permits us to deduce that the series in the example diverges by comparison to the harmonic series, which is known to diverge.

◊

Example 4.28 Determine if

$$\sum_{n=0}^{\infty} \frac{1}{2^n + 5}$$

converges or diverges.

Solution:

The key to using the comparison test is to find a good known series to compare to. In this case the geometric series with ratio 1/2 is natural.

$$2^n \leq 2^n + 5$$

$$\frac{1}{2^n} \geq \frac{1}{2^n + 5}$$

$$\left(\frac{1}{2}\right)^n \geq \frac{1}{2^n + 5}$$

Which means that the general term of the convergent geometric series

$$\sum_{n=0}^{\infty} \left(\frac{1}{2}\right)^n$$

is greater than the general term of our target series. Since all terms are positive, this tells us that we may conclude the target series converges, by comparison.

◊

The next test is one of the most all-around useful tests. It permits us to resolve any series drawn from a rational function, for example, by checking to see which p-series a rational function is most like.

Knowledge Box 4.12

The limit comparison test

Suppose that

$$\sum_{n=0}^{\infty} x_n \textit{ and } \sum_{n=0}^{\infty} y_n$$

are series, and that

$$\lim_{n\to\infty} \frac{x_n}{y_n} = C$$

where $0 < |C| < \infty$ *is a constant. Then, either both series converge or both series diverge.*

Example 4.29 Determine if

$$\sum_{n=1}^{\infty} \frac{n+5}{n^3+1}$$

converges or diverges.

Solution:

Perform limit comparison to

$$\sum_{n=1}^{\infty} \frac{1}{n^2}$$

$$\lim \frac{x_n}{y_n} = \lim_{n\to\infty} \frac{\frac{n+5}{n^3+1}}{\frac{1}{n^2}} = \lim_{n\to\infty} \frac{n+5}{n^3+1} \cdot \frac{n^2}{1} = \lim_{n\to\infty} \frac{n^3+5n^2}{n^3+1} = 1$$

Since $0 < 1 < \infty$, we can conclude that $\sum_{n=1}^{\infty} \frac{n+5}{n^3+1}$ converges, because $\sum_{n=1}^{\infty} \frac{1}{n^2}$ is a convergent p-series.

Notice that choosing a p-series with $p = 2$ was exactly what was needed to make the ratio of the general terms of the series be something that had a finite, non-zero limit. The limit comparison test is useful for putting rigour into the intuition that one series is "like" another.

The next test uses the fact that if you have a sum of positive terms that is finite, making some or all of those terms negative cannot take the sum farther from zero than the original finite sum – leaving the sum finite.

Knowledge Box 4.13

The absolute convergence test

If $\sum_{n=0}^{\infty} |x_n|$ *converges, then so does* $\sum_{n=0}^{\infty} x_n$.

Example 4.30 Prove that

$$1 - \frac{1}{4} + \frac{1}{9} - \frac{1}{16} + \frac{1}{25} - \cdots$$

converges.

Solution:

The series in question is given as an obvious pattern – begin by pulling it into summation notation:

$$\sum_{n=1}^{\infty} \frac{(-1)^{n+1}}{n^2}$$

If we take the absolute value of the general terms of this series, we obtain the series:

$$\sum_{n=1}^{\infty} \frac{1}{n^2}$$

We know this to be a convergent p-series. We may deduce that the series converges.

◊

Definition 4.9 *If* $\sum_{n=0}^{\infty} |x_n|$ *converges, then we say the series* $\sum_{n=0}^{\infty} x_n$ **converges in absolute value.**

This means we can restate Knowledge Box 4.13 as: "A series that converges in absolute value, converges."

The next test uses the fact that if you go up, then down by less, then up by even less, and so on, you end up a finite distance from your starting point.

Knowledge Box 4.14

The alternating series test

Suppose that x_n *is a series such that* $x_n > x_{n+1} \geq 0$, *and suppose that* $\lim_{n \to \infty} x_n = 0$. *Then the series*

$$\sum_{n=0}^{\infty} (-1)^n \cdot x_n$$

converges.

Example 4.31 Show that $\sum_{n=1}^{\infty} \frac{(-1)^n}{\sqrt{n}}$ converges.

Solution:

We already know that $\lim_{n \to \infty} \frac{1}{\sqrt{n}} = 0$, and the terms of the series clearly get smaller as n increases. So, we may conclude this series converges by the alternating series test.

◊

The next two tests both check to see if a series is like a geometric series and deduce its convergence or divergence from that similarity.

Knowledge Box 4.15

The ratio test

Suppose that $\sum_{n=0}^{\infty} x_n$ *is a series. Compute*

$$r = \lim_{n \to \infty} \left| \frac{x_{n+1}}{x_n} \right|.$$

Then:

- *if* $r < 1$, *then the series converges,*
- *if* $r > 1$, *then the series diverges,*
- *if* $r = 1$, *then the test is inconclusive.*

Definition 4.10 *The quantity* $n! = n \cdot (n-1) \cdot (n-2) \cdots 2 \cdot 1$ *is called* n **factorial**.

We define $0! = 1$.

Example 4.32 Determine if $\sum_{n=0}^{\infty} \frac{1}{n!}$ converges or diverges.

Solution:

Use the ratio test.

$$\lim_{n \to \infty} \left| \frac{1/(n+1)!}{1/n!} \right| = \lim_{n \to \infty} \left| \frac{n!}{(n+1)!} \right|$$

$$= \lim_{n \to \infty} \frac{n(n-1) \cdots 2 \cdot 1}{(n+1)n(n-1) \cdots 2 \cdot 1}$$

$$= \lim_{n \to \infty} \frac{1}{n+1} = 0$$

Since $0 < 1$, we can deduce that the series converges by the ratio test.

◊

Knowledge Box 4.16

The root test

Suppose that $\sum_{n=0}^{\infty} x_n$ *is a series. Compute*

$$s = \lim_{n\to\infty} \sqrt[n]{|x_n|}.$$

Then:

- *if* $s < 1$, *then the series converges,*
- *if* $s > 1$, *then the series diverges,*
- *if* $s = 1$, *then the test is inconclusive.*

Example 4.33 Determine if $\sum_{n=0}^{\infty} \frac{1}{n^n}$ converges or diverges.

Solution:

Use the root test.

$$\lim_{n\to\infty} \sqrt[n]{\left|\frac{1}{n^n}\right|} = \lim_{n\to\infty} \frac{1}{n} = 0$$

Since $0 < 1$, we can deduce that the series converges by the root test.

The root test and, especially, the ratio test will get a big workout in the next section. This section contains nine tests for series convergence – many of which depend on knowing the convergence or divergence behavior of other series. This creates a mental space very similar to the "which integration method do I use?" issue that arose in *Fast Start Integral Calculus*. The method for dealing with this is the same here as it was there: practice, practice, practice.

It's also good to keep in mind, when you are searching for series to compare to, that the examples in this section may be used as examples for comparison. A list of series with known behaviors is an excellent resource and, given different learning styles, somewhat personal: you may want to maintain your own annotated list.

4.2.1 TAILS OF SEQUENCES

You may have noticed that we are a little careless with where we start the index of summation on our infinite series. This is because, while the value of a convergent series depends on every term, the convergence or divergence behavior does not.

Definition 4.11 *If we take a sequence and make a new sequence by discarding a finite number of initial terms, the new sequence is a* **tail** *of the old sequence.*

Knowledge Box 4.17

Tail convergence

A sequence converges if and only if all its tails converge.

The practical effect of Knowledge Box 4.17 is that, if a few initial terms of a sequence are causing trouble, you may discard them and test the remainder of the sequence to determine convergence or divergence.

Example 4.34 Determine the convergence of the series:

$$\sum_{n=0}^{\infty} \frac{(-1)^n}{n^2 - 6n + 10}$$

Solution:

If we look at the first several terms of this series we get:

$$\frac{1}{10} - \frac{1}{5} + \frac{1}{2} - 1 + \frac{1}{2} - \frac{1}{5} + \frac{1}{10} - \frac{1}{17} + \cdots$$

The series alternates signs, getting larger in absolute value for the first four terms, but then getting smaller in absolute value for the remaining terms. This means that the alternating series test works *for the tail of the sequence starting at the fourth term.* Remember that the alternating series test requires that the terms shrink in absolute value.

We conclude that the series converges by the alternating series test applied to a tail of the sequence.

◊

The next example shows a very special sort of series for which we can compute the exact value. Once you understand how these series work, you can construct many examples of them. You may want to review *partial fractions* from *Fast Start Integral Calculus*.

Example 4.35 Prove that the sequence

$$\sum_{n=0}^{\infty} \frac{1}{n^2 + 3n + 2}$$

converges to exactly 1.

Solution:

We can see, by limit comparison to $\sum_{n=1}^{\infty} \frac{1}{n^2}$, a convergent p-series, that this series converges. Knowing the exact value of the sum is another matter. That will require a little algebra.

$$\begin{aligned}
\sum_{n=0}^{\infty} \frac{1}{n^2 + 3n + 2} &= \sum_{n=0}^{\infty} \frac{1}{(n+1)(n+2)} \\
&= \sum_{n=0}^{\infty} \left(\frac{A}{n+1} + \frac{B}{n+2} \right) && \text{-partial fractions} \\
&= \sum_{n=0}^{\infty} \left(\frac{1}{n+1} - \frac{1}{n+2} \right) \\
&= 1 - 1/2 + 1/2 - 1/3 + 1/3 - 1/4 + 1/4 - 1/5 + \cdots \\
&= 1 + 0 + 0 + 0 + 0 + \cdots \\
&= 1
\end{aligned}$$

◊

The series in Example 4.35 is called a *telescoping series*, in analogy to the way a small telescope or spyglass collapses for storage. In essence, the series is composed of a positive and a negative series that cancel out all but one term.

Knowledge Box 4.18

Telescoping series

If $b_n = a_n - a_{n+1}$ *then*

$$\sum_{n=0}^{\infty} b_n = a_0.$$

When expanded in terms of the a_k *values, everything except* a_0 *cancels out.*

Trace out the information in Knowledge Box 4.18 for Example 4.35.

The next Knowledge Box gives a guide for choosing a convergence test. As with choosing the best method for integration, the only real way to get better at choosing the correct convergence test is to work examples. Lots of examples.

Knowledge Box 4.19

Choosing a convergence test

1. *If the general term of the series fails to converge to zero, then it diverges by the divergence test. Remember that this does not work in reverse – if the general term converges to zero, anything might happen.*
2. *If the series* is *a geometric series or a* p*–series, use the tests for those series.*
3. *If the general term of the series is* $a_n = f(n)$ *for a function* $f(x)$ *that you can integrate, try the integral test.*
4. *A positive series that is, term-by-term, no larger than a series that converges, converges (comparison test) – look for this.*

continued

Knowledge Box 4.20

Choosing a convergence test—Continued

5. *Similarly, a positive series that is, term-by-term, no smaller than a divergent series diverges, again by the comparison test.*

6. *If you have a series that looks like a series you know how to deal with, try the limit comparison test. This is useful for things like slightly modified p series or letting you use simpler integral tests.*

7. *You can often set up a comparison test or limit comparison test by doing algebra or arithmetic to the general term of a series.*

8. *Remember that if the term-by-term absolute value of a series converges, then the series converges. This is the absolute convergence test.*

9. *If the terms of a series alternate in sign, look at the alternating series test.*

10. *If you can take the limit of adjacent terms of a series in a reasonable way, then the ratio test is a possibility.*

11. *If you can take the nth root of the general term of a series in a reasonable way, the root test is a possibility.*

12. *If some finite number of initial terms of a series are preventing you from using a test, tail convergence says you can ignore them and then do your test.*

13. *Unless you're taking a test or quiz, asking for advise and suggestions is not the worst possible option.*

PROBLEMS

Problem 4.36 For each of the following series, determine if the series converges or diverges. State the name of the test you are using.

1. $\sum_{n=1}^{\infty} \frac{1}{n^{e}}$
2. $\sum_{n=0}^{\infty} \frac{n+1}{n^2+1}$
3. $\sum_{n=0}^{\infty} \frac{n+1}{5n+4}$
4. $\sum_{n=1}^{\infty} \ln\left(\frac{1}{n^2}\right)$
5. $\sum_{n=0}^{\infty} \frac{2^n+1}{3^n+1}$
6. $\sum_{n=0}^{\infty} \frac{\sqrt{n}}{n^2+1}$

Problem 4.37 For each of the following series, determine if the series converges or diverges. State the name of the test you are using.

1. $\sum_{n=0}^{\infty} \frac{n^2}{2^n+1}$
2. $\sum_{n=0}^{\infty} 0.0462^n$
3. $\sum_{n=2}^{\infty} \frac{\sin(n)}{n\sqrt{2}}$
4. $\sum_{n=1}^{\infty} \frac{1}{\sqrt[3]{n}}$
5. $\sum_{n=1}^{\infty} \frac{(-1)^n}{\sqrt[3]{n}}$
6. $\sum_{n=0}^{\infty} \frac{1}{(2n)!}$

Problem 4.38 For each of the following series, determine if the series converges or diverges. State the name of the test you are using.

1. $\sum_{n=1}^{\infty} \frac{1}{n^{n/2}}$
2. $\sum_{n=1}^{\infty} \frac{5^n}{n^{n/2}}$
3. $\sum_{n=1}^{\infty} e^{-n}$
4. $\sum_{n=1}^{\infty} \frac{e^n}{\pi^n}$
5. $\sum_{n=1}^{\infty} \frac{e^{2n}}{\pi^n}$
6. $\sum_{n=1}^{\infty} e^{3+n-n^2}$

Problem 4.39 Give an example to demonstrate that the divergence test does not work in reverse, i.e., a sequence whose general term goes to zero but whose sum is infinite.

Problem 4.40 Use the integral test to prove that the p-test works.

Problem 4.41 Suppose that $p(x)$ is a polynomial. Use the integral test to demonstrate that

$$\sum_{n=0}^{\infty} p(n)e^{-n}$$

converges.

Problem 4.42 Suppose that $q(x)$ is a polynomial with exactly three roots, all of which are negative real numbers. Demonstrate that

$$\sum_{n=0}^{\infty} \frac{1}{q(n)}$$

converges.

Problem 4.43 If x_n and y_n are the general terms of a convergent series, then $x_n + y_n$ are as well. This requires only simple algebra. What is startling is that the reverse is not true. Find an example of $a_n = x_n + y_n$ so that

$$\sum_{n=1}^{\infty} a_n$$

converges, but neither of

$$\sum_{n=1}^{\infty} x_n \text{ or } \sum_{n=1}^{\infty} y_n$$

converge.

Problem 4.44 Show that, when you apply the ratio test to a geometric series, the limit that appears in the test is the ratio of the series.

Problem 4.45 Show that, when you apply the root test to a geometric series, the limit that appears in the test is the ratio of the series.

Problem 4.46 Suppose $r > 1$. Prove that

$$\sum_{n=0}^{\infty} \frac{n^k}{r^n}$$

converges when k is an integer ≥ 1.

Problem 4.47 Suppose that we are testing the convergence of a series

$$\sum_{n=0}^{\infty} \frac{p(n)}{q(n)}$$

where $p(x)$ and $q(x)$ are polynomials. If we add the assumption that $q(x)$ has no roots at any of the values of n involved in the sum, explain in terms of the degrees of the polynomials when the series converges.

Problem 4.48 For each of the following series, determine if the series converges or diverges. State the name of the test you are using.

1. $\displaystyle\sum_{n=0}^{\infty} \frac{n^3}{(1+n)(2+n)(3+n)(4+n)}$
2. $\displaystyle\sum_{n=0}^{\infty} \frac{n^2}{(1+n)(2+n)(3+n)(4+n)}$
3. $\displaystyle\sum_{n=0}^{\infty} \frac{n^5}{e^n}$
4. $\displaystyle\sum_{n=0}^{\infty} \frac{\sqrt{n}}{n^2+1}$
5. $\displaystyle\sum_{n=0}^{\infty} 1.25^{-n}$
6. $\displaystyle\sum_{n=0}^{\infty} e^{\pi-n}$

Problem 4.49 Compute exactly: $\displaystyle\sum_{n=1}^{\infty} \frac{1}{n^2+n}$

Problem 4.50 Compute exactly: $\displaystyle\sum_{n=1}^{\infty} \frac{1}{n^2+2n}$

Problem 4.51 Compute exactly: $\displaystyle\sum_{n=1}^{\infty} \frac{1}{n^2+5n}$

Problem 4.52 Demonstrate convergence of the series $\displaystyle\sum_{n=1}^{\infty} \frac{\sin(n)}{n^2+n}$

Problem 4.53 Compute exactly: $\displaystyle\sum_{n=0}^{\infty} \frac{1}{4n^2+8n+3}$

4.3 POWER SERIES

In this section we study series again. The good news is that we do not have any additional convergence tests. The bad news is that these series will have variables in them.

Definition 4.12 *A* **power series** *is a series of the form:*

$$\sum_{n=0}^{\infty} a_n x^n$$

In a way, a power series is actually an infinite number of different ordinary series, one for each value of x you could substitute into it. The goal of this section will be: given a power series, find values of x which cause it to converge.

Knowledge Box 4.21

The radius of convergence of a power series

The power series

$$\sum_{n=0}^{\infty} a_n x^n$$

converges in one of three ways:

1. *Only at* $x = 0$.
2. *For all* $|x| < r$ *and possibly at* $x = \pm r$.
3. *For all* x.

The number r *is the* **radius of convergence** *of the power series. In the first case above, we say the radius of convergence is zero; in the third, we say the radius of convergence is infinite.*

Definition 4.13 *The* **interval of convergence** *of a power series is the set of all* x *where it converges.*

Knowledge Box 4.21 implies that the interval of convergence of a power series is one of $[0, 0]$, $(-r, r)$, $[-r, r)$, $(-r, r]$, $[-r, r]$, or $(-\infty, \infty)$. The results with an r in them occur in the case where $0 < r < \infty$. Once we have the radius of convergence, in the case where r is positive and finite, we determine the interval of convergence by checking the behavior of the series when we set $x = \pm r$.

Example 4.54 Find the radius and interval of convergence of:

$$\sum_{n=0}^{\infty} x^n$$

Solution:

We start by trying to determine the radius of convergence.

Use the ratio test:

$$\lim_{n \to \infty} \left| \frac{x^{n+1}}{x^n} \right| = \lim_{n \to \infty} |x| = |x|$$

This is true because x does not depend on n.

The series thus converges when $|x| < 1$, meaning we have convergence for sure when $-1 < x < 1$.

This also means the radius of convergence is $r = 1$.

We now need to check $x = \pm 1$ to determine the interval of convergence.

These values both yield non-converging geometric series:

$$\sum_{n=0}^{\infty} (-1)^n \text{ and } \sum_{n=0}^{\infty} 1$$

So the potential endpoints of the interval of convergence are *not* part of the interval of convergence. This means that the interval of convergence is $(-1, 1)$.

◊

Example 4.55 Find the radius of convergence of:

$$\sum_{n=0}^{\infty} \frac{x^n}{2^n}$$

Solution:

Again, use the ratio test.

$$\begin{aligned}\lim_{n\to\infty}\left|\frac{a_{n+1}}{a_n}\right| &= \lim_{n\to\infty}\left|\frac{x^{n+1}/2^{n+1}}{x^n/2^n}\right| \\ &= \lim_{n\to\infty}\left|\frac{x^{n+1}}{2^{n+1}}\cdot\frac{2^n}{x^n}\right| \\ &= \lim_{n\to\infty}\left|\frac{x}{2}\right| = \frac{|x|}{2}\end{aligned}$$

So the series converges when:

$$-1 < \frac{x}{2} < 1$$

$$-2 < x < 2$$

The radius of convergence is $r = 2$.

Example 4.56 Find the radius of convergence of the series:

$$\sum_{n=0}^{\infty} \frac{x^n}{n!}$$

Solution:

Like before:

$$\begin{aligned}\lim_{n\to\infty}\left|\frac{a_{n+1}}{a_n}\right| &= \lim_{n\to\infty}\left|\frac{x^{n+1}/(n+1)!}{x^n/n!}\right| \\ &= \lim_{n\to\infty}\left|\frac{x}{n+1}\right| \\ &= |x|\lim_{n\to\infty}\left|\frac{1}{n+1}\right| \\ &= |x|\cdot 0 \\ &= 0\end{aligned}$$

Since $0 < 1$ for all values of x, this power series converges everywhere and the radius of convergence is infinite.

◊

Example 4.57 Find the radius of convergence of the series:

$$\sum_{n=0}^{\infty}\frac{x^n}{n^n}$$

Solution:

This time use the root test.

$$\begin{aligned}\lim_{n\to\infty}\sqrt[n]{\left|\frac{x^n}{n^n}\right|} &= \lim_{n\to\infty}\left|\frac{x}{n}\right| \\ &= |x|\cdot\lim_{n\to\infty}\frac{1}{n} = 0\end{aligned}$$

Which tells us that, as $0 < 1$ for all x, that this sequence converges everywhere.

◊

Example 4.58 Find the interval of convergence of the series:

$$\sum_{n=1}^{\infty} \frac{x^n}{n^2}$$

Solution:

Another natural job for the ratio test.

$$\begin{aligned}
\lim_{n\to\infty} \left| \frac{a_{n+1}}{a_n} \right| &= \lim_{n\to\infty} \left| \frac{x^{n+1}/(n+1)^2}{x^n/n^2} \right| \\
&= \lim_{n\to\infty} \left| \frac{n^2}{(n+1)^2} \cdot x \right| \\
&= |x| \cdot \lim_{n\to\infty} \frac{n^2}{n^2+2n+1} \\
&= |x| \cdot 1 = |x|
\end{aligned}$$

So $-1 < x < 1$, and the radius of convergence is $r = 1$. To find the interval of convergence we need to check the ends of the interval. These are of the form

$$\sum_{n=1}^{\infty} \frac{(\pm 1)^n}{n^2}$$

The absolute value of each of these is a p-series with $p = 2$. Both endpoints correspond to series that converge in absolute value – meaning they both converge. This makes the interval of convergence $[-1, 1]$.

4.3.1 USING CALCULUS TO FIND SERIES

It is sometimes useful to represent functions as power series. The geometric series formula (Knowledge Box 4.7) does this, for example, for the function $f(x) = \dfrac{1}{1-x}$. We can use calculus to derive power series for other functions.

Example 4.59 If $|x| < 1$, then the geometric series formula tells us that:

$$\sum_{n=0}^{\infty} x^n = \frac{1}{1-x}$$

If we plug $-x$ into this identity we find that:

$$\sum_{n=0}^{\infty} (-x)^n = \frac{1}{1-(-x)}$$

$$\sum_{n=0}^{\infty} (-1)^n x^n = \frac{1}{1+x}$$

or

$$\frac{1}{x+1} = 1 - x + x^2 - x^3 + x^4 - \cdots$$

Integrate both sides and we get:

$$\ln(x+1) + C = x - \frac{1}{2}x^2 + \frac{1}{3}x^3 - \frac{1}{4}x^4 + \frac{1}{5}x^5 - \cdots$$

Plug in $x = 0$

$$\ln(1) + C = 0$$

$$C = 0$$

Which means, at least when $|x| < 1$,

$$\ln(x+1) = \sum_{n=1}^{\infty} \frac{(-1)^{n+1} x^n}{n}$$

◊

This shows that we can find power series that, when they converge, are equal to familiar transcendental functions. The next section gives another technique for doing this, to be used when the sequences you already know don't give you enough power. Let's encode this as a Knowledge Box.

Knowledge Box 4.22

Using calculus to modify power series

Suppose that $f(x) = \sum_{n=0}^{\infty} a_n x^n$. *Then:*

$$\int f(x) \cdot dx = \sum_{n=0}^{\infty} \frac{a_n}{n+1} x^{n+1} \text{ and } f'(x) = \sum_{n=1}^{\infty} n \cdot a_n \, x^{n-1}.$$

Example 4.60 Find a power series for $\tan^{-1}(x)$.

Solution:

We already have seen that:

$$\frac{1}{u+1} = 1 - u + u^2 - u^3 + u^4 - \cdots$$

If we substitute $u = x^2$ we obtain:

$$\frac{1}{x^2+1} = 1 - x^2 + x^4 - x^6 + x^8 - \cdots$$

Integrate and we get:

$$\tan^{-1}(x) + C = x - \frac{1}{3}x^3 + \frac{1}{5}x^5 - \frac{1}{7}x^7 + \frac{1}{9}x^9 - \cdots$$

Substitute in $x = 0$ and we get $C = 0$. So we obtain the power series:

$$\tan^{-1}(x) = \sum_{n=0}^{\infty} \frac{(-1)^n x^{2n+1}}{2n+1}$$

◊

Example 4.61 What is the interval of convergence for the series for

$$f(x) = \tan^{-1}(x)$$

found in Example 4.60?

Solution:

Apply the ratio test:

$$\begin{aligned}
\lim_{n\to\infty}\left|\frac{a_{n+1}}{a_n}\right| &= \lim_{n\to\infty}\left|\frac{x^{2n+3}/(2n+3)}{x^{2n+1}/(2n+1)}\right| \\
&= \lim_{n\to\infty}\left|x^2\frac{2n+1}{2n+3}\right| \\
&= |x^2| \lim_{n\to\infty}\frac{2n+1}{2n+3} \\
&= |x^2| \cdot 1 = x^2
\end{aligned}$$

So $x^2 < 1$ when $-1 < x < 1$, making the radius of convergence $r = 1$.

Now check the endpoints $x = \pm 1$.

If $x = -1$ the resulting series is:

$$\sum_{n=0}^{n} \frac{(-1)^n(-1)^{2n+1}}{2n+1} = \sum_{n=0}^{n} \frac{(-1)^{n+1}}{2n+1}$$

which converges by the alternating series test.

If $x = 1$ we get:

$$\sum_{n=0}^{n} \frac{(-1)^n}{2n+1}$$

which also converges by the alternating series test.

The interval of convergence is thus $[-1, 1]$.

It is possible to find a power series by just using algebra on a known series.

Example 4.62 Find a power series for:

$$f(x) = \frac{x^2}{1-x^2}$$

Solution:

We start with the known form:

$$\frac{1}{1-u} = 1 + u + u^2 + u^3 + u^4 + \cdots$$

Substitute in $u = x^2$ and we get:

$$\frac{1}{1-x^2} = 1 + x^2 + x^4 + x^6 + x^8 + \cdots$$

Now multiply both sides by x^2 and we get:

$$\frac{x^2}{1-x^2} = x^2 + x^4 + x^6 + x^8 + x^{10} + \cdots$$

So

$$f(x) = \sum_{n=0}^{\infty} x^{2n+2}$$

$\diamond$

Example 4.63 Find a power series for:

$$g(x) = \ln(x^2 + 1)$$

Solution:

Start with the known result for $\frac{1}{1+x^2}$.

$$\frac{1}{1+x^2} = 1 - x^2 + x^4 - x^6 + x^8 - \cdots$$

$$\frac{2x}{1+x^2} = 2x - 2x^3 + 2x^5 - 2x^7 + 2x^9 - \cdots$$

$$\int \frac{2x}{1+x^2} \cdot dx = \int \left(2x - 2x^3 + 2x^5 - 2x^7 + 2x^9 - \cdots\right) dx$$

$$\ln(x^2+1) + C = 2\left(\frac{1}{2}x^2 - \frac{1}{4}x^4 + \frac{1}{6}x^6 - \frac{1}{8}x^8 + \frac{1}{10}x^{10} - \cdots\right)$$

Set $x = 0$ and we get $C = 0$.

$$\text{So: } \ln(x^2+1) = \sum_{n=0}^{\infty} \frac{2 \cdot (-1)^n x^{2n+2}}{2n+2}$$

$\diamond$

An integral that is both important in statistics, because it is related to the normal distribution, and famous for not having a closed form solution is:

$$\int e^{-x^2/2} \cdot dx$$

The next example shows why power series are useful – it is easy to get a power series for this integral.

Example 4.64 Find a power series for

$$F(x) = \int e^{-x^2/2} \cdot dx$$

Solution:

$$e^x = \sum_{n=0}^{\infty} \frac{x^n}{n!}$$

This will be shown in Section 4.4

$$e^{-x^2/2} = \sum_{n=0}^{\infty} \frac{(-x^2/2)^n}{n!}$$

$$e^{-x^2/2} = \sum_{n=0}^{\infty} \frac{(-1)^n x^{2n}}{2^n \cdot n!}$$

$$\int e^{-x^2/2} \cdot dx = \int \sum_{n=0}^{\infty} \frac{(-1)^n x^{2n}}{2^n \cdot n!} \cdot dx$$

$$\int e^{-x^2/2} \cdot dx = \sum_{n=0}^{\infty} \frac{(-1)^n x^{2n+1}}{(2n+1) \cdot 2^n \cdot n!}$$

This power series is useful for computing probabilities connected with the normal distribution.

◊

PROBLEMS

Problem 4.65 Find the radius of convergence for each of the following power series.

1. $\sum_{n=1}^{\infty} \frac{x^n}{n}$
2. $\sum_{n=0}^{\infty} \frac{x^n}{5^n}$
3. $\sum_{n=0}^{\infty} \frac{x^n}{(2n+1)!}$
4. $\sum_{n=0}^{\infty} \frac{x^{2n+1}}{n!}$
5. $\sum_{n=1}^{\infty} nx^n$
6. $\sum_{n=0}^{\infty} \frac{nx^n}{3n+1}$
7. $\sum_{n=0}^{\infty} \frac{x^n}{2n^2+4}$
8. $\sum_{n=0}^{\infty} \frac{x^n}{n^3+1}$

Problem 4.66 Demonstrate that the radius of convergence of

$$\sum_{n=0}^{\infty} a_n x^n \text{ and } \sum_{n=0}^{\infty} c a_n x^n$$

are the same for any constant c.

Problem 4.67 Compute the radius of convergence of

$$\sum_{n=1}^{\infty} \frac{x^n}{c^n}$$

where $c > 0$ is a constant.

Problem 4.68 If

$$f(x) = \frac{1}{1-x} + \frac{1}{1-2x} + \frac{1}{1-3x}$$

find and simplify a power series for $f(x)$.

Problem 4.69 If

$$g(x) = \frac{x}{2-x} + \frac{x}{3-x}$$

find and simplify a power series for $g(x)$.

Problem 4.70 If

$$h(x) = \frac{x^2}{2 - x} + \frac{x^2}{1 - 2x}$$

find and simplify a power series for $h(x)$.

Problem 4.71 Find the interval of convergence for each of the following power series.

1. $\displaystyle\sum_{n=0}^{\infty} \frac{x^n}{4^n}$
2. $\displaystyle\sum_{n=1}^{\infty} \frac{x^n}{(2n+1)!}$
3. $\displaystyle\sum_{n=0}^{\infty} \frac{nx^n}{2n^3 + 7}$
4. $\displaystyle\sum_{n=2}^{\infty} \frac{n^2 x^n}{n^4 - 1}$
5. $\displaystyle\sum_{n=0}^{\infty} \frac{x^n}{n^n}$
6. $\displaystyle\sum_{n=0}^{\infty} \frac{(-2x)^n}{n}$

Problem 4.72 Using calculus, find a power series for:

$$g(x) = \ln\left(\frac{1 + x}{1 - x}\right)$$

Problem 4.73 Using only algebra, and known series, find a power series for:

$$h(x) = \frac{3x^2}{1 + x^3}$$

Problem 4.74 Find a power series for:

$$q(x) = \ln\left(1 + x^3\right)$$

Problem 4.75 Find a power series for:

$$q(x) = \frac{1}{x^2 - 9}$$

Problem 4.76 Find a power series for:

$$q(x) = \frac{1}{x^2 - 3x + 2}$$

Problem 4.77 Find a power series for:

$$q(x) = x^2 \cdot \ln\left(1 + x^3\right)$$

Problem 4.78 Find a power series for:

$$q(x) = \frac{x^2 + 9}{x^2 - 9}$$

4.4 TAYLOR SERIES

In the last section, we managed to create power series for several of the standard transcendental functions. Notably absent were $\sin(x)$, $\cos(x)$, and e^x. The key to these is **Taylor series**. We need a little added notation to build Taylor series. We will denote the nth derivative of $f(x)$ by $f^{(n)}(x)$. Notice that this means that $f^{(0)}(x) = f(x)$.

Knowledge Box 4.23

Taylor series

If $f(x)$ is a function that can be differentiated any number of times,

$$f(x) = \sum_{n=0}^{\infty} \frac{f^{(n)}(c)(x-c)^n}{n!}.$$

This formula is called the **Taylor series expansion of $f(x)$ at c**. *The constant c is called the* **center** *of the expansion.*

Example 4.79 Use Taylor's formula to find a power series centered at $c = 0$ for $f(x) = e^x$ and find its radius of convergence.

Solution:

The function $f(x) = \mathrm{e}^x$ is a very good choice for a first demonstration of the Taylor expansion. This is because *every* derivative of of e^x is e^x. In other words,

$$f^{(n)}(x) = \mathrm{e}^x \text{ and so } f^{(n)}(0) = 1$$

Applying the formula we get:

$$\mathrm{e}^x = \sum_{n=0}^{\infty} \frac{f^{(n)}(0)(x-0)^n}{n!} = \sum_{n=0}^{\infty} \frac{1 \cdot x^n}{n!} = \sum_{n=0}^{\infty} \frac{x^n}{n!}$$

The radius of convergence of this series was computed in Example 4.56 – it is $r = \infty$. This means that the expansion of e^x converges everywhere.

◊

Here is an interesting calculation. Recall Euler's identity from *Fast Start Differential Calculus*: $\mathrm{e}^{ix} = i\sin(x) + \cos(x)$. Let's look at the Taylor series for e^{ix}:

$$\begin{aligned}
\mathrm{e}^{ix} &= \sum_{n=0}^{\infty} \frac{(ix)^n}{n!} \\
&= \sum_{n=0}^{\infty} \frac{i^n x^n}{n!} \\
&= 1 + ix - \frac{1}{2}x^2 - i\frac{1}{3!}x^3 + \frac{1}{4!}x^4 + i\frac{1}{5!}x^5 - \frac{1}{6!}x^6 - i\frac{1}{7!}x^7 + \cdots \\
&= \sum_{n=0}^{\infty} \frac{(-1)^n x^{2n}}{(2n)!} + i\sum_{n=0}^{\infty} \frac{(-1)^n x^{2n+1}}{(2n+1)!}
\end{aligned}$$

From Euler's identity, we get that the real part of the expression above is cosine, and the imaginary part is sine. This gives us power series for $\sin(x)$ and $\cos(x)$.

Knowledge Box 4.24

Taylor series for $\sin(x)$ **and** $\cos(x)$ **and** e^x

$$e^x = \sum_{n=0}^{\infty} \frac{x^n}{n!}$$

$$\sin(x) = \sum_{n=0}^{\infty} \frac{(-1)^n x^{2n+1}}{(2n+1)!}$$

$$\cos(x) = \sum_{n=0}^{\infty} \frac{(-1)^n x^{2n}}{(2n)!}$$

with all three expansions having an interval of convergence of $(-\infty, \infty)$.

The calculations above show that we can use algebraic manipulation to create power series based on the power series we get from Taylor expansions. Let's do a couple more examples.

Example 4.80 Find a power series for $h(x) = e^{2x}$.

Solution:

$$\begin{aligned} e^x &= \sum_{n=0}^{\infty} \frac{x^n}{n!} \\ e^{2x} &= \sum_{n=0}^{\infty} \frac{(2x)^n}{n!} \\ &= \sum_{n=0}^{\infty} \frac{2^n x^n}{n!} \\ &= \sum_{n=0}^{\infty} \frac{2^n}{n!} x^n \end{aligned}$$

◊

Example 4.81 Find the Taylor expansion for $s(x) = \cos(x)$ using $c = \frac{\pi}{2}$.

Solution:

The Taylor formula needs the nth derivatives at $\frac{\pi}{2}$. Let's start by computing these numbers.

n	$s^{(n)}(x)$	$s^{(n)}\left(\frac{\pi}{2}\right)$
0	$\cos(x)$	0
1	$-\sin(x)$	-1
2	$-\cos(x)$	0
3	$\sin(x)$	1
4	$\cos(x)$	0
5	$-\sin(x)$	-1

Which is enough to notice the values repeat every four steps.

Plug these values into the Taylor expansion formula with $c = \pi/2$ and we get:

$$\cos(x) = -\frac{(x - \pi/2)}{1!} + \frac{(x - \pi/2)^3}{3!} - \frac{(x - \pi/2)^5}{5!} + \frac{(x - \pi/2)^7}{7!}$$

$$\cos(x) = \sum_{n=0}^{\infty} \frac{(-1)^{n+1}(x - \pi/2)^{2n+1}}{(2n + 1)!}$$

◇

Notice that the expansion of $\cos(x)$ with a center of $c = \frac{\pi}{2}$ has the same form as the negative of the Taylor expansion of $-\sin(x)$ at $c = 0$. This is a fairly extreme way of proving the identity

$$\cos(x) = -\sin(x - \pi/2)$$

4.4.1 TAYLOR POLYNOMIALS

If we take the first n terms of a Taylor series for $f(x)$ the result is called the *Taylor polynomial of degree n* for $f(x)$. Taylor polynomials are approximations to the function they are derived from and, as we already know, polynomials are among the very nicest functions to work with.

Example 4.82 Find a fifth-degree Taylor polynomial for $f(x) = \sin(x)$ centered at $c = 0$.

Solution:

Since the Taylor series for $\sin(x)$ centered at $c = 0$ is

$$\sum_{n=0}^{\infty} \frac{(-1)^n \, x^{2n+1}}{(2n+1)!}$$

we need only extract the terms of degree five or less from the infinite series. This gives us a solution of

$$p(x) = x - \frac{x^3}{6} + \frac{x^5}{120}$$

◊

A natural question is "how good is this polynomial as an approximation to $\sin(x)$? Let's graph both functions on the same set of axes.

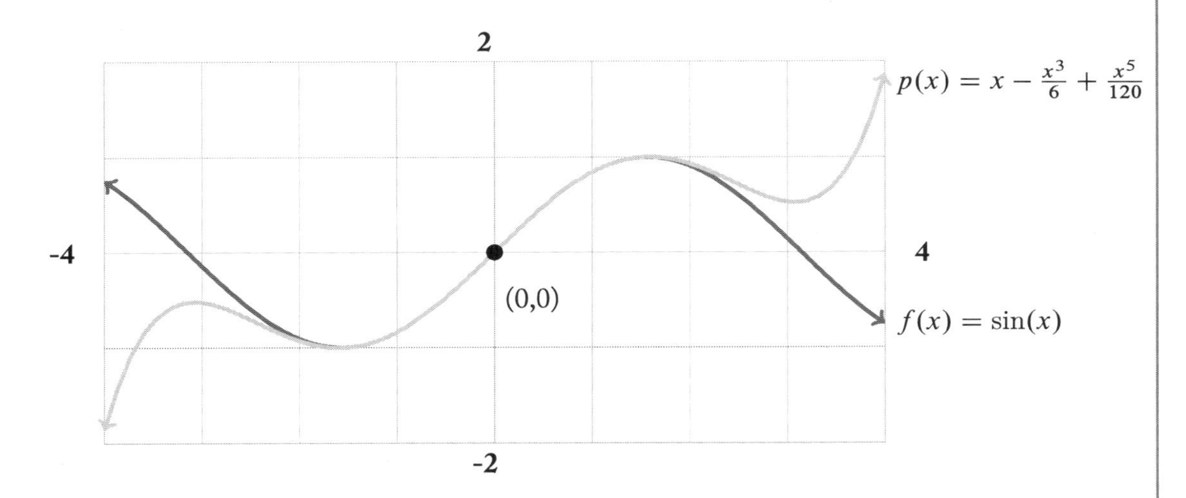

In the range $-2 \leq x \leq 2$ the polynomial and the sine function agree really well – after that they diverge from one another and the polynomial shoots off to positive and negative infinity. Not too surprisingly, the Taylor polynomial does a good job of approximating the function it was drawn from near the center c for expansion. How do we figure out where the polynomial is a good enough approximation?

If $T(x)$ is the Taylor series for a function $f(x)$ centered at c and $T_n(x)$ is the nth degree Taylor polynomial for that function, then we set $R_n(x) = T(x) - T_n(x)$ to get the *remainder* for the polynomial. Algebraically rearranging the terms:

$$T(x) = T_n(x) + R_n(x)$$

With this definition we can give Taylor's inequality for the remainder of a power series.

Knowledge Box 4.25

Taylor's Inequality

Suppose we are looking at a Taylor polynomial for the function $f(x)$ in the interval $|x - a| \leq d$ and that $|f^{(n+1)}(x)| \leq M$ everywhere in this interval. Then for values of x in the interval we have that

$$|R_n(x)| \leq \frac{M}{(n+1)!}|x - a|^{n+1}.$$

It is not immediately obvious how to use this result, so let's do a couple examples.

Example 4.83 Find a bound on the error of approximation of $T_5(x)$ for $f(x) = \sin(x)$ on the interval $-2 \leq x \leq 2$. Note that this is the interval that looked good in the graph associated with Example 4.82.

Solution:

We proceed by applying Taylor's inequality. The derivative of $\sin(x)$ are all sine and cosine functions. This means that $f^{(n+1)}(x)$ are always at most 1. So, we may set $M = 1$ for Taylor's inequality. The value of a is zero so $|x - a| \leq 2$ on the interval we are using.

Plugging these values into the formula for Taylor's inequality we get:

$$|R_5(x)| \leq \frac{1}{6!} \cdot 2^6 = \frac{64}{720} \cong 0.089$$

The error of $T_5(x)$ on $-2 \leq x \leq 2$ is *at most* 0.089. Not bad.

Notice that Taylor's inequality has $(n+1)!$ in the denominator. That means, if we use

$$T_7(x) = x - \frac{x^3}{6} + \frac{x^5}{120} - \frac{x^7}{5040}$$

for $f(x) = \sin(x)$ on the interval $-2 \leq x \leq 2$, then the estimate of maximum error drops to

$$|R_7(x)| \leq \frac{1}{8!} \cdot 2^8 = \frac{256}{40320} \cong 0.0063$$

The factorial in the denominator lets the error drop really quickly.

◊

The hard part of using Taylor's inequality is finding the constant M. Standard practice is to simply find the largest value of $f^{n+1}(x)$ in the interval and live with it. The fact that both sine and cosine are bounded in absolute value by 1 make them especially easy functions to work with. Let's do an example with an exponential function.

Example 4.84 Suppose we are approximating $f(x) = e^x$ in the interval $-3 \leq x \leq 3$ with $T_n(x)$. What value of n makes the Taylor's inequality estimate of error at most 0.1?

Solution:

If $f(x) = e^x$ then $f^{(n+1)} = e^x$, which is very convenient. We again have that $a = 0$. So, it is pretty easy to see that

$$|f^{(n+1)}(x)| \leq e^3$$

everywhere on the interval. So, we set $M = e^3$. The largest value of $|x - a|$ on the interval is 3, so we need to find the smallest value of n that makes

$$\frac{e^3}{(n+1)!} 3^{n+1} < 0.1$$

The simplest way to do this is to tabulate.

n	$\|R_n(x)\| \leq \|\frac{e^3}{(n+1)!} \cdot 3^{n+1}\|$	n	$\|R_n(x)\| \leq \|\frac{e^3}{(n+1)!} \cdot 3^{n+1}\|$
1	90.3849161543	6	8.7156883435
2	90.3849161543	7	3.2683831288
3	67.7886871158	8	1.0894610429
4	40.6732122695	9	0.3268383129
5	20.3366061347	10	0.0891377217

So the first value of n with an acceptable error is $n = 10$, and $T_{10}(x)$ is good enough to approximate $f(x) = e^x$ in the range $-3 \leq x \leq 3$.

◊

The error estimates given by Taylor's inequality are not the best possible – they are actually fairly conservative. They usually over-estimate the error. If you take a course in numerical analysis later in your career you may study methods for building better error estimates. There is also a lot of room to be clever with how you use Taylor polynomials. The sine and cosine function are periodic, and so if you know their values even on a very small interval, like $[0, \pi/2]$, you can use those values to deduce any other value.

PROBLEMS

Problem 4.85 Using the Taylor series formula, verify the formula for $y = \sin(x)$.

Problem 4.86 Using the Taylor series formula, verify the formula for $y = \cos(x)$.

Problem 4.87 Find the Taylor expansion of $f(x) = \sin(x)$ using a center of $c = \pi$. Use the formula for Taylor expansion.

Problem 4.88 Find the Taylor expansion of $f(x) = \sin(x)$ using a center of $c = \pi/2$. Use the formula for the Taylor expansion.

Problem 4.89 Using any method, find power series for the following functions.

1. $f(x) = e^{-x}$
2. $g(x) = xe^{2x}$
3. $h(x) = \cos 2x$
4. $r(x) = \sin(x^2)$
5. $s(x) = \ln(x^4)$
6. $q(x) = (e^x + e^{-x})/2$
7. $a(x) = \tan^{-1}(3x)$
8. $b(x) = \dfrac{3x^2}{1 - x^4}$

Problem 4.90 Find the Taylor expansion of $f(x) = \cos(x)$ using a center of $c = \pi/4$. Use the formula for the Taylor expansion. Warning: this is a little messy.

Problem 4.91 Using the Taylor series for $\sin(x)$, $\cos(x)$ and e^x, prove Euler's identity:

$$e^{i\theta} = i\sin(x) + \cos(x)$$

Problem 4.92 For each of the series you found in Problem 4.89, find the radius and interval of convergence.

Problem 4.93 Prove that the Taylor series for a polynomial function $p(x)$ is just the polynomial itself.

Problem 4.94 Find a power series expansion for

$$f(x) = \frac{1}{x^2 - 3x + 2}$$

Problem 4.95 Find a power series expansion for

$$f(x) = \frac{1}{4 - 4x + x^2}$$

Problem 4.96 Find a power series expansion for

$$f(x) = \frac{1}{x^3 - 6x^2 + 11x + 6}$$

Problem 4.97 Find a power series expansion for

$$f(x) = \frac{1}{x^3 + x}$$

Problem 4.98 If

$$f(x) = p(x)e^x$$

where $p(x)$ is a polynomial, demonstrate that $f(x)$ has a power series expansion with radius of convergence $r = \infty$.

Problem 4.99 Find the Taylor polynomial of degree n for the given function with the given center c.

1. $f(x) = \cos(x)$ for $n = 6$ at $c = 0$,
2. $g(x) = \sin(2x)$ for $n = 7$ at $c = 0$,
3. $h(x) = e^x$ for $n = 5$ at $c = 0$,
4. $r(x) = \log(x)$ for $n = 3$ at $c = 1$,
5. $s(x) = \tan^{-1}(x)$ for $n = 8$ at $c = 0$,
6. $q(x) = x^2 + 3x + 5$ for $n = 2$ at $c = 1$.

Problem 4.100 Suppose we have $T_5(x)$ for $f(x) = e^x$ at $c = 0$. Compute a bound on the size of $R_5(x)$ with Taylor's inequality.

Problem 4.101 Find the smallest n for which $T_n(x)$ on $f(x) = \cos(x)$ has $|R_n(x)| < 0.01$ on $-3 \le x \le 3$ with $c = 0$.

APPENDIX A

Useful Formulas

A.1 POWERS, LOGS, AND EXPONENTIALS

RULES FOR POWERS

- $a^{-n} = \frac{1}{a^n}$
- $a^n \times a^m = a^{n+m}$
- $\frac{a^n}{a^m} = a^{n-m}$
- $(a^n)^m = a^{n \times m}$

LOG AND EXPONENTIAL ALGEBRA

- $b^{\log_b(c)} = c$
- $\log_b(b^a) = a$
- $\log_b(xy) = \log_b(x) + \log_b(y)$
- $\log_b\left(\frac{x}{y}\right) = \log_b(x) - \log_b(y)$
- $\log_b(x^y) = y \cdot \log_b(x)$
- $\log_c(x) = \frac{\log_b(x)}{\log_b(c)}$
- If $\log_b(c) = a$, then $c = b^a$

A.2 TRIGONOMETRIC IDENTITIES

TRIG FUNCTION DEFINITIONS FROM SINE AND COSINE

- $\tan(\theta) = \frac{\sin(\theta)}{\cos(\theta)}$
- $\cot(\theta) = \frac{\cos(\theta)}{\sin(\theta)}$
- $\tan(\theta) = \frac{1}{\cot(\theta)}$
- $\sec(\theta) = \frac{1}{\cos(\theta)}$
- $\csc(\theta) = \frac{1}{\sin(\theta)}$

PERIODICITY IDENTITIES

- $\sin(x + 2\pi) = \sin(x)$
- $\cos(x + 2\pi) = \cos(x)$
- $\sin(x) = \cos\left(x - \frac{\pi}{2}\right)$
- $\tan(x) = -\cot\left(x - \frac{\pi}{2}\right)$
- $\sec(x) = \csc\left(x + \frac{\pi}{2}\right)$
- $\cos(-x) = \cos(x)$
- $\sin(-x) = -\sin(x)$
- $\tan(x) = -\tan(x)$
- $\sin(x + \pi) = -\sin(x)$
- $\cos(x + \pi) = -\cos(x)$
- $\tan(x + \pi) = \tan(x)$

THE PYTHAGOREAN IDENTITIES

- $\sin^2(\theta) + \cos^2(\theta) = 1$
- $\tan^2(\theta) + 1 = \sec^2(\theta)$
- $1 + \cot^2(\theta) = \csc^2(\theta)$

THE LAW OF SINES, THE LAW OF COSINES

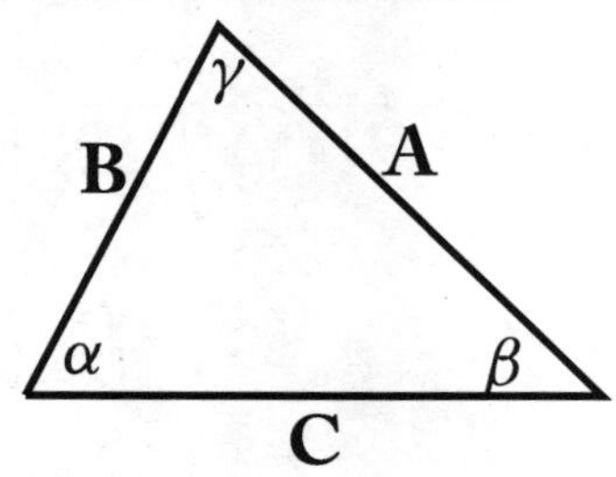

The Law of Sines

$$\frac{A}{\sin(\alpha)} = \frac{B}{\sin(\beta)} = \frac{C}{\sin(\gamma)}$$

The Law of Cosines

$$C^2 = A^2 + B^2 + 2AB \cdot \cos(\gamma)$$

The laws refer to the diagram.

SUM, DIFFERENCE, AND DOUBLE ANGLE

- $\sin(\alpha + \beta) = \sin(\alpha)\cos(\beta) + \sin(\beta)\cos(\alpha)$
- $\cos(\alpha + \beta) = \cos(\alpha)\cos(\beta) - \sin(\alpha)\sin(\beta)$
- $\sin(\alpha - \beta) = \sin(\alpha)\cos(\beta) - \sin(\beta)\cos(\alpha)$
- $\cos(\alpha - \beta) = \sin(\alpha)\sin(\beta) + \cos(\alpha)\cos(\beta)$
- $\sin(2\theta) = 2\sin(\theta)\cos(\theta)$
- $\cos(2\theta) = \cos^2(\theta) - \sin^2(\theta)$
- $\cos^2(\theta/2) = \dfrac{1 + \cos(\theta)}{2}$
- $\sin^2(\theta/2) = \dfrac{1 - \cos(\theta)}{2}$

A.3 SPEED OF FUNCTION GROWTH

- Logarithms grow faster than constants.
- Positive powers of x grow faster than logarithms.
- Larger positive powers of x grow faster than smaller positive powers of x.
- Exponentials (with positive exponents) grow faster than positive powers of x.
- Exponentials with larger exponents grow faster than those with smaller exponents.

A.4 DERIVATIVE RULES

- If $f(x) = x^n$ then
$$f'(x) = nx^{n-1}$$
- $(f(x) + g(x))' = f'(x) + g'(x)$
- $(a \cdot f(x))' = a \cdot f'(x)$
- If $f(x) = \ln(x)$, then $f'(x) = \frac{1}{x}$
- If $f(x) = \log_b(x)$, then $f'(x) = \frac{1}{x \ln(b)}$
- If $f(x) = e^x$, then $f'(x) = e^x$
- If $f(x) = a^x$, then $f'(x) = \ln(a) \cdot a^x$
- $(\sin(x))' = \cos(x)$
- $(\cos(x))' = -\sin(x)$
- $(\tan(x))' = \sec^2(x)$
- $(\cot(x))' = -\csc^2(x)$
- $(\sec(x))' = \sec(x)\tan(x)$
- $(\csc(x))' = -\csc(x)\cot(x)$
- $\left(\sin^{-1}(x)\right)' = \frac{1}{\sqrt{1-x^2}}$
- $\left(\cos^{-1}(x)\right)' = \frac{-1}{\sqrt{1-x^2}}$
- $\left(\tan^{-1}(x)\right)' = \frac{1}{1+x^2}$
- $\left(\cot^{-1}(x)\right)' = \frac{-1}{1+x^2}$
- $\left(\sec^{-1}(x)\right)' = \frac{1}{|x|\sqrt{x^2-1}}$
- $\left(\csc^{-1}(x)\right)' = \frac{-1}{|x|\sqrt{x^2-1}}$

The product rule

$$(f(x) \cdot g(x))' = f(x)g'(x) + f'(x)g(x)$$

The quotient rule

$$\left(\frac{f(x)}{g(x)}\right)' = \frac{g(x)f'(x) - f(x)g'(x)}{g^2(x)}$$

The reciprocal rule

$$\left(\frac{1}{f(x)}\right)' = \frac{-f'(x)}{f^2(x)}$$

The chain rule

$$(f(g(x)))' = f'(g(x)) \cdot g'(x)$$

A.5 SUMS AND FACTORIZATION RULES

FACTORIZATIONS OF POLYNOMIALS

- $x^2 - a^2 = (x - a)(x + a)$
- $x^3 - a^3 = (x - a)(x^2 + ax + a^2)$
- $x^3 + a^3 = (x + a)(x^2 - ax + a^2)$
- $x^n - a^n = (x - a)(x^{n-1} + ax^{n-2} + \cdots a^{n-2}x + a^{n-1})$

ALGEBRA OF SUMMATION

- $\sum_{i=a}^{b} f(i) + g(i) = \sum_{i=a}^{b} f(i) + \sum_{i=a}^{b} g(i)$
- $\sum_{i=a}^{b} c \cdot f(i) = c \cdot \sum_{i=a}^{b} f(i)$

CLOSED SUMMATION FORMULAS

- $\sum_{i=1}^{n} 1 = n$
- $\sum_{i=1}^{n} i = \frac{n(n+1)}{2}$
- $\sum_{i=1}^{n} i^2 = \frac{n(n+1)(2n+1)}{6}$
- $\sum_{i=1}^{n} i^3 = \frac{n^2(n+1)^2}{4}$

A.5.1 GEOMETRIC SERIES

- $\sum_{k=0}^{n} a^k = \frac{a^{n+1} - 1}{a - 1}$
- $\sum_{n=0}^{\infty} a^n = \frac{1}{1 - a}$ if $|a| < 1$
- $\sum_{n=0}^{\infty} qa^n = \frac{q}{1 - a}$ if $|a| < 1$

A.6 VECTOR ARITHMETIC

VECTOR ARITHMETIC AND ALGEBRA

- $c \cdot \vec{v} = (cv_1, cv_2, \ldots, cv_n)$
- $\vec{v} + \vec{w} = (v_1 + w_1, v_2 + w_2, \ldots, v_n + w_n)$
- $\vec{v} - \vec{w} = (v_1 - w_1, v_2 - w_2, \ldots, v_n - w_n)$
- $\vec{v} \cdot \vec{w} = v_1 w_1 + v_2 w_2 + \ldots + v_n w_n$
- $c \cdot (\vec{v} + \vec{w}) = c \cdot \vec{v} + c \cdot \vec{w}$
- $c \cdot (d \cdot \vec{v}) = (cd) \cdot \vec{v}$
- $\vec{v} + \vec{w} = \vec{w} + \vec{v}$
- $\vec{u} \cdot (\vec{v} + \vec{w}) = \vec{u} \cdot \vec{v} + \vec{u} \cdot \vec{w}$

CROSS PRODUCT OF VECTORS

- $\vec{v} \times \vec{w} = (\, v_2 w_3 - v_3 v_2,\ v_3 w_1 - v_1 w_3,\ v_1 w_2 - v_2 w_1 \,)$

Formula for the angle between vectors

$$\cos(\theta) = \frac{\vec{v} \cdot \vec{w}}{|v||w|}$$

A.7 POLAR AND RECTANGULAR CONVERSION

- $x = r \cdot \cos(\theta)$
- $y = r \cdot \sin(\theta)$
- $r = \sqrt{x^2 + y^2}$
- $\theta = \tan^{-1}(y/x)$

A.8 INTEGRAL RULES

BASIC INTEGRATION RULES

- $\int x^n \cdot dx = \frac{1}{n+1} x^{n+1} + C$
- $\int a \cdot f(x) \cdot dx = a \cdot \int f(x) \cdot dx$
- $\int (f(x) + g(x)) \cdot dx = \int f(x) \cdot dx + \int g(x) \cdot dx$

LOG AND EXPONENT

- $\int \frac{1}{x} \cdot dx = \ln(x) + C$
- $\int \mathrm{e}^x \cdot dx = \mathrm{e}^x + C$

TRIG AND INVERSE TRIG

- $\int \sin(x) \cdot dx = -\cos(x) + C$
- $\int \cos(x) \cdot dx = \sin(x) + C$
- $\int \sec^2(x) \cdot dx = \tan(x) + C$
- $\int \csc^2(x) \cdot dx = -\cot(x) + C$
- $\int \sec(x)\tan(x) \cdot dx = \sec(x) + C$
- $\int \csc(x)\cot(x) \cdot dx = -\csc(x) + C$
- $\int \frac{1}{\sqrt{1-x^2}} \cdot dx = \sin^{-1}(x) + C$
- $\int \frac{1}{1+x^2} \cdot dx = \tan^{-1}(x) + C$
- $\int \frac{1}{x\sqrt{x^2-1}} \cdot dx = \sec^{-1}(|x|) + C$
- $\int \tan(x) \cdot dx = \ln|\sec(x)| + C$
- $\int \sec(x) \cdot dx = \ln|\sec(x) + \tan(x)| + C$

INTEGRATION BY PARTS

$$\int U \cdot dV = UV - \int V \cdot dU$$

EXPONENTIAL/POLYNOMIAL SHORTCUT

$$\int p(x)e^x \cdot dx = \left(p(x) - p'(x) + p''(x) - p'''(x) + \cdots\right) e^x + C$$

VOLUME, SURFACE, ARC LENGTH

Volume of Rotation, Disks

$$V = \pi \int_a^b f(x)^2 \cdot dx$$

Volume of Rotation, Cylindrical Shells

$$V = 2\pi \int_{x=a}^{x=b} x \cdot f(x) \cdot dx$$

Volume of Rotation with Washers

$$V = \pi \int_a^b \left(f_1(x)^2 - f_2(x)^2\right) \cdot dx$$

Differential of Arc Length

$$ds = \sqrt{(y')^2} \cdot dx = \sqrt{(f'(x))^2} \cdot dx$$

Arc Length

$$S = \int ds$$

Surface Area of Rotation

$$A = 2\pi \int_a^b f(x) \cdot ds$$

A.9 SERIES CONVERGENCE TESTS

DIVERGENCE TEST

If $\lim_{n\to\infty} x_n \neq 0$ then $\sum_{n=0}^{\infty} x_n$ does not have a finite value.

INTEGRAL TEST

Suppose that $f(x)$ is a positive, decreasing function on $[0, \infty)$ and that $x_n = f(n)$. Then $\sum_{n=1}^{\infty} x_n$ and $\int_a^{\infty} f(x)dx$ both converge or both diverge for any finite $a \geq 1$.

P-SERIES TEST

The p-series $\sum_{n=1}^{\infty} \frac{1}{n^p}$ converges if $p > 1$ and diverges if $p \leq 1$.

COMPARISON TESTS

- If $\sum_{n=0}^{\infty}$ converges and $0 \leq y_n \leq x_n$ for all n, then $\sum_{n=0}^{\infty} y_n$ also converges.
- If $\sum_{n=0}^{\infty} x_n$ diverges and $y_n \geq x_n \geq 0$ for all n, then $\sum_{n=0}^{\infty} y_n$ also diverges.

LIMIT COMPARISON TEST

Suppose that $\sum_{n=0}^{\infty} x_n$ and $\sum_{n=0}^{\infty} y_n$ are series, and $\lim_{n\to\infty} \frac{x_n}{y_n} = C$ where $0 < |C| < \infty$ is a constant. Then, either both series converge or both series diverge.

ABSOLUTE CONVERGENCE TEST

If $\sum_{n=0}^{\infty} |x_n|$ converges, then so does $\sum_{n=0}^{\infty} x_n$.

ALTERNATING SERIES TEST

Suppose that x_n is a series such that $x_n > x_{n+1} \geq 0$, and suppose that $\lim_{n\to\infty} x_n = 0$. Then the series $\sum_{n=0}^{\infty} (-1)^n \cdot x_n$ converges.

RATIO TEST

Suppose that $\sum_{n=0}^{\infty} x_n$ is a series and $r = \lim_{n\to\infty} \left|\frac{x_{n+1}}{x_n}\right|$. Then

- if $r < 1$, then the series converges
- if $r > 1$, then the series diverges
- if $r = 1$, then the test is inconclusive

ROOT TEST

Suppose that $\sum_{n=0}^{\infty} x_n$ is a series and $s = \lim_{n\to\infty} \sqrt[n]{|x_n|}$. Then

- if $s < 1$, then the series converges
- if $s > 1$, then the series diverges
- if $s = 1$, then the test is inconclusive

A.10 TAYLOR SERIES

If $f(x)$ is a function that can be differentiated any number of times,

$$f(x) = \sum_{n=0}^{\infty} \frac{f^{(n)}(c)(x-c)^n}{n!}$$

SPECIAL TAYLOR SERIES

- $e^x = \sum_{n=0}^{\infty} \frac{x^n}{n!}$
- $\sin(x) = \sum_{n=0}^{\infty} \frac{(-1)^n x^{2n+1}}{(2n+1)!}$
- $\cos(x) = \sum_{n=0}^{\infty} \frac{(-1)^n x^{2n}}{(2n)!}$

TAYLOR'S INEQUALITY

Suppose we are looking at a Taylor polynomial for the function $f(x)$ in the interval $|x-a| \le d$ and that $|f^{(n+1)}(x)| \le M$ everywhere in this interval. Then for values of x in the interval we have that

$$|R_n(x)| \le \frac{M}{(n+1)!}|x-a|^{n+1}$$

Author's Biography

DANIEL ASHLOCK

Daniel Ashlock is a Professor of Mathematics at the University of Guelph. He has a Ph.D. in Mathematics from the California Institute of Technology, 1990, and holds degrees in Computer Science and Mathematics from the University of Kansas, 1984. Dr. Ashlock has taught mathematics at levels from 7th grade through graduate school for four decades, starting at the age of 17. Over this time Dr. Ashlock has developed a number of ideas about how to help students overcome both fear and deficient preparation. This text, covering the mathematics portion of an integrated mathematics and physics course, has proven to be one of the more effective methods of helping students learn mathematics with physics serving as an ongoing anchor and example.

Index

◎

编辑手记

世界著名数学家 R. Courant 曾说过:

> 对许多人来说,战争极大地刺激他们致力于发展数学与技术、科学之间的相互作用,我希望将来在没有战争时,那在知识界与工程界,在社会科学、在经济学、在生物学等方面处处皆有的巨大挑战能够被清楚地理解,能够对成长着的年轻一代具有说服力,也能够形成对数学家的挑战.

这段时间,全世界都在关注"俄乌战争",但我们的数学著作出版工作还要继续.本书是一部英文版的数学教程,中文书名可译为《高等微积分快速入门》.

本书的作者为丹尼尔·阿什洛克(Daniel Ashlock),他是美国人,圭尔夫大学的数学教授.他于1990年获得加州理工学院数学博士学位,并于1984年获得堪萨斯大学计算机科学和数学学位.丹尼尔·阿什洛克博士已经有40年的教学经验了,在这段时间里,阿什洛克博士提出了许多关于如何帮助学生克服恐惧和准备不充分的想法.本书涵盖了综合数学和物理课程的数学部分,证明了物理作为持续的支柱和示例,是帮助学生有效学习数学的方法之一.

正如作者在前言中所述:

> 本书涵盖了多变量积分和微积分的相关内容,并假设读者已熟悉前两本教材《微分学快速入门》和《积分学快速入门》中的单变量方法.这些教材是

为圭尔夫大学物理系课程而开发的介绍微积分的课程,其涵盖了学生在一年后会用到的内容.为了解决这个问题,一个多学科团队创建了一个两学期的综合微积分和物理课程.本书涵盖了该课程中的积分学主题以及有关多项式函数形式的材料.该课程的理念是,将微积分的相关内容提前教给学生,并且为了巩固学生对数学的理解,我们将列举大量有关物理学方面的激励性示例.

该课程在本书撰写之前已经进行了 3 次讲授,并以草稿形式用于该课程的第 4 次讲授,然后又使用了两年.本书背后有大量的课堂经验和测试内容,还有足够的信息来证实我们的假设,即该课程将对学生学习微积分有所帮助.这门课程的综合成绩下降率和不及格率始终低于 3%,而大学一年级的微积分课程中,不及格率通常为 20%.这是微积分和物理学的联合教学工作.值得注意的是,我们并没有通过淡化数学来实现这些结果.两个学期涵盖的内容大约是标准的第一年微积分课程所涵盖的内容的一半.这是一个大的惊喜:更快地覆盖更多内容提高了平均成绩并降低了失败率.使用物理学作为知识锚的效果比我们希望的还要好.

本书以及它的两个姊妹篇《微分学快速入门》和《积分学快速入门》都做出了一些创新,却引起了数学同仁的反对.在数学中,这是传统的,甚至是教条主义的,数学的教学顺序是在相关知识所依赖的概念已经被掌握之前,关于该知识的任何内容都不会出现.这对数学学生来说是正确的、有用的教条式教学.本书既不强调也不忽视理论,但它确实是将理论从课程的开头移开,以承认该材料在哲学上是困难的,同时在智力上具有挑战性的事实.该课程还尽可能展示了广泛的综合图景,本书也强调了聪慧与计算有效性,请记住"数学是避免计算的艺术".

我们需要说明,为了使这门课程和本书以原有的方式工作,我们做了哪些牺牲.对于数学专业的学生来说,这不是一本好书,除非他们能在之后的实分析课程中获得微积分的理论知识.本书相对来说是非正式的,内容几乎全部由示例与应用程序引领.本书作者是加州理工学院的数学教授,并取得了该校的博士学位,他有丰富的数学教学经验,指导了十几个成功的博士生.作者教授的微积分学包括数学和工程微积分、生物学微积分、商业微积分、多元以及矢量微积分.

本书的目录为:

1.2 梯度和方向导数

1.3 切面

2. 多变量与约束优化

2.1 带偏导数的优化

2.2 极值定理 Redux

2.3 拉格朗日乘数

3. 高等积分

3.1 旋转的体积

3.2 弧长与表面积

3.3 多重积分

4. 序列、级数与函数逼近

4.1 序列与几何级数

4.2 级数收敛试验

4.3 幂级数

4.4 泰勒级数

本书大致讲了两部分内容,一部分是级数论,讲的相当简单,相比笔者四十年前学习时的习题略显浅显,找到一道当年的习题:

问题 通过正弦与余弦级数相除,试决定级数

$$\tan x=\frac{\sin x}{\cos x}=a_0+a_1x+a_2x^2+\cdots$$

的开始4个异于零的系数 a_n.

问这个级数的收敛半径如何?

解 因为不仅正弦级数,而且余弦级数也都有正的收敛半径(它们甚至对一切 x 都收敛),此外,余弦级数的绝对项 $b_0=1$,就是说不等于0,所以级数相除是允许的. 我们得到一个收敛半径 $R>0$ 的幂级数.

令

$$\tan x=\sum_{n=0}^{\infty}a_nx^n$$

因为 $\tan x$ 是一个奇函数,所以每一个 $a_{2n}=0$

$$\tan x=\sum_{n=0}^{\infty}a_{2n+1}x^{2n+1}$$

这个级数当 $|x|\leqslant\gamma<R$ 时是绝对收敛的. 所以,做级数 $\tan x$ 与 $\cos x$ 的乘积,它还等于 $\sin x$

$$\left(\sum_{n=0}^{\infty}a_{2n+1}x^{2n+1}\right)\left(\sum_{n=0}^{\infty}b_{2n}x^{2n}\right)=\left(\sum_{n=0}^{\infty}c_{2n+1}x^{2n+1}\right)$$

左端可以逐项相乘. 按照柯西所做,我们得到

$$c_{2n+1}=\sum_{v=0}^{n}a_{2v+1}b_{2(n-v)}$$

对此,未知数 a_{2n+1} 可以依次算出来

$$a_{2n+1}=\frac{1}{b_0}\left[c_{2n+1}-\sum_{v=0}^{n-1}a_{2v+1}b_{2(n-v)}\right]$$

其中,$b_{2n}=\frac{(-1)^n}{(2n)!}$ 及 $c_{2n+1}=\frac{(-1)^n}{(2n+1)!}$.

起始几个系数的计算

$$n=0:\quad a_1=\frac{c_1}{b_0}=1$$

$$n=1:\quad a_3=\frac{1}{b_0}\{c_3-a_1b_2\}=\frac{(-1)^1}{3!}-1\cdot\frac{(-1)^1}{2!}=\frac{1}{3}$$

$$n=2:\quad a_5=\frac{1}{b_0}\{c_5-a_1b_4-a_3b_2\}$$

$$=\frac{(-1)^2}{5!}-\frac{(-1)^2}{4!}-\frac{1}{3}\frac{(-1)^1}{2!}=\frac{2}{15}$$

$$n=3:\quad a_7=\frac{1}{b_0}\{c_7-a_1b_6-a_3b_4-a_5b_2\}$$

$$=\frac{(-1)^3}{7!}-\frac{(-1)^3}{6!}-\frac{1}{3}\frac{(-1)^2}{4!}-\frac{2}{15}\frac{(-1)^1}{2!}$$

$$=\frac{17}{315}$$

等等.

$$\tan x=x+\frac{1}{3}x^3+\frac{2}{15}x^5+\frac{17}{315}x^7+\cdots$$

这个级数的收敛半径肯定是 $R>0$,按照一般理论,必须有 $|b_0|>\sum_{v=1}^{\infty}|b_vx^v|$. 这就是说

$$|1|=1>\sum_{v=1}^{\infty}\frac{|x|^{2v}}{(2v)!}$$

所以

$$2>1+\sum_{v=1}^{\infty}\frac{|x|^{2v}}{(2v)!}=\cosh|x|$$

上面所述的级数,对于一切满足 $\cosh|x|<2$ 的 x,即一切 $|x|=AR\cosh 2$,肯定是收敛的.

这道题现在做起来也有一定难度,好像是一本法国习题集中的题.

第二部分是多元微积分,特别是梯度导数之类,但是没有介绍散度,这些都是物理学特别是电磁学中的常用工具.

笔者注意到有两篇关于麦克斯韦方程组的文章在网络上广为流传. 一篇是清华大学物理系的王青教授在网络上发表的《深入理解"拓展的麦克斯韦方程组"》的文章,他介绍说[①]:

> 近十年来,为了提升学生对这两门课的兴趣,我在清华大学物理系开设的以电磁学为主的《费曼物理学 Ⅱ》课上从 2014 年起,《电动力学》课上从 2018 年起,实行在课堂上教师不再按传统方式授课,而完全改为讨论的问题驱动式的互动教学. 在这两门课中,最核心的内容是麦克斯韦方程组,而它所讨论的电场和磁场是物理学的基础. 在我所开展的针对它们的课堂讨论中,场的变换和场理论的协变性及麦克斯韦方程组等都是难点,从图像上不太容易理解. 为此,我几年前曾以《理解经典电磁学理论》为题撰写了专门的教学文章[②]. 前些日子,结合近期关于"拓展的麦克斯韦方程组"[③] 的争议,撰写了题为《理解王中林院士"拓展的麦克斯韦方程组"》的文章[④]. 本以为事情到此为止, 因为看到不少读者评价这篇文章[④]"把事情说清楚了". 后来又看到中国科学院理论物理所的同行们放到 arXiv 网站上,被人们误解读为支持王院士工作的两篇文章及其所引用的相关文献,促使我更深度地思考这篇文章[④] 中的理解,发现确有值得深度挖掘之处,而且这些进一步的理解对前面提到的场的变换和场理论的协变性及麦克斯韦方程组的观念和图像是一种冲击,有利于我们在电磁场的教学中更好地理解以及看待、定义和表达它们. 本文即在这些背景考虑之下,在这篇文章[④] 的讨论基础上,作为对这些后续发展的回应,结合关于"拓展的麦克斯韦方程组" 的争议,进一步开展一些更深入的学术讨论.
>
> 在这篇文章[④] 中,我指出对运动的电磁场可以给出两种描述方式:一种是理论物理中标准的参考系变换方式(这时场和时空坐标都要参与参考系的变换),从实验室参考系变换到运动者(介质) 静止的系中;另一种是运动的场线方式,通过使坐标含时间来反映运动(和质点运动学的讨论

① 摘编自微信公众号"物理与工程".

② 王青. 理解经典电磁学理论[J]. 物理与工程,2018,28(3):10-22.

③ WANG Z L. On the expanded Maxwell's equations for moving charged media system—General theory, mathematical solutions and applications.

④ 王青. 理解王中林院士"拓展的麦克斯韦方程组"[J]. 物理与工程,2022:网络首发.

模式相仿),这时场对时间的求导需要考虑进坐标含时间的额外贡献(类似于流体力学的随体导数).针对电磁场,我指出两种描述是不等价的,因为电磁场的场强不是参考系变换下的标量,形象地说它们在变换下不再是刚性不变的.进一步我分两种情形进行了简单分析:一种是参考系变换代表真实的物理,另一种是参考系变换不代表真实的物理.鉴于后者过于奇葩,不被学界所接受,本文只限于讨论前者,很多做理论物理的人认为只要承认标准的变换方式,也就是前者就足够了,不再有兴趣深入探究其他另类的描述.在这篇文章①中我们也同样地,对这种情况只讨论到运动的场线不再代表在运动的场点上所真实感受到的电磁场(那它究竟代表什么没再探究),就此打住,没继续深入.本文从这里出发,接着往下进行详细的分析和讨论.

之所以要如此咬文嚼字、追根究底地深入研究,其原因是王院士的拓展及最新的讨论②③,还有他所引用的早期文献目标最后都是这个运动的场线框架.很多人询问这些新出现的文章是否意味这篇文章① 中的讨论错了?目前情形下,我没有办法回避文献中业已存在的这一类做法,需要正面面对、理解和解读它们的含义,特别是要回答当我们认为参考系变换得到的是运动的场点上所真实感受到的电磁场时,与它不同的运动的场线所代表的究竟是个什么样的场?这些是这篇文章① 和本文的题目所特别指明,而在这篇文章① 中尚未完成的内容.

在这篇文章① 中,我们曾引用费曼的说法,说明对电磁场,运动的场线是不存在的.这是不是就把运动场线的做法一锤打死了呢?果真如此,如何理解一直还有这么多人在沿着这个路线持续在做讨论呢?

注意到费曼的讨论是在假定运动的场线描述和参考系变换描述等价的前提下进行的.他是用变换到运动者静止的参考系的结果来审视和评判所谓运动场线的正确性的.这个等价细化起来具体就是:在观察者静止的实验室参考系看到的运动的场完全等效于在观察者运动(实际是站在运动的介质上)的参考系看到的静止的场,这实际是一种相对性原理的表达形式.从这个意义上,目前所要坚持讨论的内容,即运动的场线描述不再与参考系变换等价的情形,已经破坏了费曼所依据的前提,或者说明显地破坏了相对性原理.目前的状态变成,如果承认相对性原理,按费曼

---

① 王青.理解王中林院士“拓展的麦克斯韦方程组”[J].物理与工程,2022:网络首发.

② WANG F, YANG J M. Relativistic origin of Hertz and extended Hertz equations for Maxwell theory of electromagnetism[J]. arXiv: 2201.10856v2 [physics.class-ph] 27 Jan 2022.

③ LI C, PEI J L, LI T J. Comments on the expanded Maxwell's equations for woving charged media system[J]. arXiv: 2201.11520vl[physics. class-ph] 27 Jan 2022.

的讨论,运动的场线根本不该存在,因此沿这种思路的讨论是错误的;不承认相对性原理,虽然运动的场线不再受费曼讨论结果的约束,可确实破坏了更加基本的相对性原理. 初看起来,怎么讨论似乎都有问题,由此,运动的场线的讨论是不是就真的完全错了呢?

在物理理论的表达中,相对性原理明显地显示出来的描述被称作理论的协变表达,这时理论的数学表达式在任何一个惯性系里都是完全形式上一样的,这是理论物理学家十分喜欢的一种优美的理论表达方式,此次争论异议方所参考比较多的文章[①]就是以这种方式来进行讨论的. 同时在理论描述中,也存在着大量非协变的理论表达. 在这类描述中,在不同的惯性参考系中,理论的数学表达式并不一样,相对性原理在这里正确与否需要额外地去讨论和证明,这时即使最后相对性原理被证明仍是在理论中正确的也是以隐性的方式. 对像麦克斯韦方程组这样的基本理论,物理学家们相信相对性原理一定是存在的,因此除了像费曼那样追求明显地实现相对性原理外,即使理论的表达表面上看似不显式地满足相对性原理,我们也必须证明它是以隐性的方式被内在地满足的. 如果达不到这一点,也就是理论怎么样都无法满足相对性原理,这个理论可能就真的彻底地错了.

从坚持相对性原理的角度,只要证明运动场线的描述可以通过方程的变形和物理量的重新定义转化为参考系变换的结果,就说明运动的场线描述不过是针对目前问题的一种相对性原理变形的"非协变"表达而已. 事实上这些文章[②③]都是从标准的参考系的洛伦兹变换出发,通过重新定义场量"推导出"王院士所给出的所谓的赫兹方程. 在这个意义上,麦克斯韦方程组的运动的场线描述只不过是换种等价的表达方式而已,不应被看作是拓展.

特别地,在其中一篇文章[②]中,法拉第电磁感应定律在实验室系中(用不带撇的变量描述),和固定在介质上随介质运动参考系中(用带撇的变量描述)都是满足相对性原理的标准形式

$$\nabla\times \boldsymbol{E}(\boldsymbol{r},t)=-\frac{\partial \boldsymbol{B}(\boldsymbol{r},t)}{\partial t}$$

$$\nabla'\times \boldsymbol{E}'(\boldsymbol{r}',t')=-\frac{\partial \boldsymbol{B}'(\boldsymbol{r}',t')}{\partial t'}$$

---

① 王雯宇,许洋. 运动介质洛伦兹协变电磁理论[J]. 物理与工程,2018,28(2):13-25.

② WANG F, YANG J M. Relativistic origin of Hertz and extended Hertz equations for Maxwell theory of electromagnetism[J]. arXiv: 2201.10856v2 [physics.class-ph] 27 Jan 2022.

③ LI C, PEI J L, LI T J. Comments on the expanded Maxwell's equations for woving charged media system[J]. arXiv: 2201.11520v1[physics. class-ph] 27 Jan 2022.

$$\boldsymbol{r}' = \boldsymbol{r} - \boldsymbol{v}t, t' = t \tag{1}$$

而运动的场线的电场强度 $\boldsymbol{\mathcal{E}}(\boldsymbol{r},t)$ 和磁感应强度 $\mathscr{B}(\boldsymbol{r},t)$ 被定义为一个由逆变换坐标给出的场(由此我们称其为非常规变换所给出的定义),由文章[①]中的式(2.9)给出,为

$$\begin{aligned}\boldsymbol{\mathcal{E}}(\boldsymbol{r}',t') &\equiv \boldsymbol{E}(\boldsymbol{r},t) = \boldsymbol{E}(\boldsymbol{r}' + \boldsymbol{v}t',t')\\ \mathscr{B}(\boldsymbol{r}',t') &\equiv \boldsymbol{B}(\boldsymbol{r},t) = \boldsymbol{B}(\boldsymbol{r}' + \boldsymbol{v}t',t')\end{aligned} \tag{2}$$

或

$$\boldsymbol{\mathcal{E}}(\boldsymbol{r},t) = \boldsymbol{E}(\boldsymbol{r} + \boldsymbol{v}t,t), \mathscr{B}(\boldsymbol{r},t) = \boldsymbol{B}(\boldsymbol{r} + \boldsymbol{v}t,t) \tag{3}$$

从这样定义的运动电磁场出发,由式(1)给出的法拉第电磁感应定律,很容易就导出赫兹方程

$$\begin{aligned}\nabla\times\boldsymbol{\mathcal{E}}(\boldsymbol{r},t) = \nabla\times\boldsymbol{E}(\boldsymbol{r} + \boldsymbol{v}t,t) &= -\frac{\mathrm{d}\boldsymbol{B}(\boldsymbol{r} + \boldsymbol{v}t,t)}{\mathrm{d}t}\\ &= -\frac{\partial\boldsymbol{B}(\boldsymbol{r} + \boldsymbol{v}t,t)}{\partial t} - \boldsymbol{v}\cdot\nabla\boldsymbol{B}(\boldsymbol{r} + \boldsymbol{v}t,t)\\ &= -\frac{\partial\mathscr{B}(\boldsymbol{r},t)}{\partial t} - \boldsymbol{v}\cdot\nabla\mathscr{B}(\boldsymbol{r},t)\\ &= -\frac{\partial\mathscr{B}(\boldsymbol{r},t)}{\partial t} + \nabla\times[\boldsymbol{v}\times\mathscr{B}(\boldsymbol{r},t)]\end{aligned} \tag{4}$$

实际上在电磁学里由式(4)等号右边的第一项,即传统的磁感应强度随时间的变化项所产生的电场叫感生电场,而由式(4)等号右边的第二项所产生的电场叫动生电场,它们都是电磁学里被广泛讲授和讨论的内容.

式(4)是在取式(1)中的坐标变换为伽利略变换的结果,如果把伽利略变换改为洛伦兹变换,方程就不再只能限制在法拉第电磁感应定律里讨论了.因为会出现电场的时间微商和磁感应强度的空间微商项,需要考虑进其他几个和介质性质有关的麦克斯韦方程组,由此介质的性质就进入了最后的结果.再准确到速度的一阶项的水平上,只需把式(1)中的时间变换式子右边加上一个项 $-\frac{\boldsymbol{r}\cdot\boldsymbol{v}}{c^2}$.在王飞等人的文章[①]中,经过详细的计算给出的最低阶相对论修正结果是在式(4)右边第二项上要乘个修正因子 α,其中参数 $\alpha = 1 - \frac{1}{\varepsilon\mu c^2}$,它和早年文献所给的结果一致.

强调一下如王飞等人的文章[①]所引的文献指出的,赫兹方程并不与实

① WANG F, YANG J M. Relativistic origin of Hertz and extended Hertz equations for Maxwell theory of electromagnetism[J]. arXiv: 2201.10856v2 [physics. class-ph] 27 Jan 2022.

验数据相符合,因此最简单的运动场线方程,也就是拓展的麦克斯韦方程组是不受实验支持的!只有考虑了含因子 α 的修正的方程才与实验相符合.从这里我们看到,即使是介质的低速运动,相对论效应也是非常重要的!

在李闯等人的文章①中,明确指出运动场线的方程不对应于我们习惯使用的实验室系或固定在运动介质身上的参考系里所使用的方程组.因为在这两个参考系里,由于相对性原理,方程组的数学形式是完全一样的,都是标准的麦克斯韦方程组.因此要强行制造出这个拓展了的方程组,必须打破习惯了的符合相对性原理的选择来重新定义我们的电磁场.王飞等人的文章②是通过改变场强所依赖的坐标,把它们变为逆变换之后的坐标来实现这种重新定义的.如果我们在王飞等人的文章② 给出的定义方式上再增做一次正变换(注意整个场和坐标都做变换,由相对性原理,是保麦克斯韦方程组形式不变的),结果是场要做正变换,而场所依赖的坐标要做一次正变换、一次逆变换,相互抵消,相当于坐标保持不变,因此独立于王飞等人的文章② 的选择的另一种是改用变换的场但坐标不变来定义运动的场线(它也是一种非常规性的定义).这就是李闯等人的文章① 的选择,具体地:按照李闯等人的文章① 所使用的符号,运动的场线的电场强度 $\boldsymbol{E}^*(\boldsymbol{r}',t')$ 和磁感应强度 $\boldsymbol{B}^*(\boldsymbol{r},t)$ 被定义为

$$\begin{aligned}\boldsymbol{E}^*(\boldsymbol{r}',t') &\equiv \boldsymbol{E}(\boldsymbol{r},t)=\boldsymbol{E}'(\boldsymbol{r}',t')-\boldsymbol{v}\times\boldsymbol{B}'(\boldsymbol{r}',t')\\ \boldsymbol{B}^*(\boldsymbol{r}',t') &\equiv \boldsymbol{B}(\boldsymbol{r},t)=\boldsymbol{B}'(\boldsymbol{r}',t')\end{aligned}\tag{5}$$

则我们得到赫兹方程

$$\begin{aligned}\nabla'\times\boldsymbol{E}^*(\boldsymbol{r}',t') &= \nabla'\times\boldsymbol{E}'(\boldsymbol{r}',t')-\nabla'\times[\boldsymbol{v}\times\boldsymbol{B}'(\boldsymbol{r}',t')]\\ &= -\frac{\partial\boldsymbol{B}'(\boldsymbol{r}',t')}{\partial t'}-\nabla'\times[\boldsymbol{v}\times\boldsymbol{B}'(\boldsymbol{r}',t')]\\ &= -\frac{\partial\boldsymbol{B}^*(\boldsymbol{r}',t')}{\partial t'}-\nabla'\times[\boldsymbol{v}\times\boldsymbol{B}^*(\boldsymbol{r}',t')]\end{aligned}\tag{6}$$

类似地,把式(5) 的伽利略变换替换为洛伦兹变换,重新做推导.准到速度的一阶项,可以得到式(6) 等号右方第二项前多乘了一个因子 α 的修正.

到此为止,我们给出了两种场强的以参考系变换来重新定义的方式.当变换是伽利略变换时,新定义出的电磁场满足不与实验相符合的赫兹方程,也即目前业界所争论的拓展的麦克斯韦方程组.而当变换是洛伦兹

---

① LI C, PEI J L, LI T J. Comments on the expanded Maxwell's equations for woving charged media system[J]. arXiv: 2201.11520vl[physics. class-ph] 27 Jan 2022.

② WANG F, YANG J M. Relativistic origin of Hertz and extended Hertz equations for Maxwell theory of electromagnetism[J]. arXiv: 2201.10856v2 [physics. class-ph] 27 Jan 2022.

变换时,相对论修正使得所导出的方程改变为与实验相符的扩展了的赫兹方程.这些方程与标准的麦克斯韦方程组完全等价,只是场量重新定义所导致的变换.

另一篇文章则是顾险峰教授写的《Maxell 方程与纤维丛》,用到的数学工具与前一篇比则要现代的多.顾先生指出①:

在大学的工科专业方向,Maxell 方程是重点内容,虽然在工业界中,电磁场设计极其普遍,几乎不太涉及现代理论,但是依随摩尔定律的终结,拓扑绝缘体的兴起,现代纤维丛理论必将日渐成为工科学生的必备知识.

经典电动力学认为电磁现象的本质描述的是电磁场强度,磁矢量势不是物理实在,只是为了数学的方便而引入;在量子力学中,磁矢量势更加本质,比磁场强度具有更加重要的地位.这涉及了纤维丛的联络概念.在下面的讨论中,我们采用更为简洁的微分形式的语言.

活动标架法

我们考察一个简单的纤维丛例子,以此解释相关的概念.假设 S 是嵌入在三维欧氏空间 $\mathbf{R}^3$ 中的光滑曲面,S 的所有单位切向量构成一个 3 维流形被称为曲面的单位切丛.我们取曲面 S 上的一个开集 U,其局部 C^2 参数为 (u,v).令 $\boldsymbol{p}\in U$ 为曲面上的点,其位置向量记为 $\boldsymbol{p}(u,v)$,其全微分记为

$$\mathrm{d}\boldsymbol{p}(u,v)=\frac{\partial \boldsymbol{p}(u,v)}{\partial u}\mathrm{d}u+\frac{\partial \boldsymbol{p}(u,v)}{\partial v}\mathrm{d}v=\boldsymbol{p}_u\mathrm{d}u+\boldsymbol{p}_v\mathrm{d}v$$

这里偏导数 $\boldsymbol{p}_u,\boldsymbol{p}_v$ 都是 $\mathbf{R}^3$ 中的向量.我们取曲面的单位正交标架场

$$\{\boldsymbol{e}_1(u,v),\boldsymbol{e}_2(u,v),\boldsymbol{e}_3(u,v)\}$$

这里 $\boldsymbol{e}_3(u,v)$ 是曲面的法向量场

$$\langle \boldsymbol{e}_i(u,v),\boldsymbol{e}_j(u,v)\rangle_{\mathbf{R}^3}=\delta_{ij}$$

这样我们可以将 $\boldsymbol{p}_u$ 和 $\boldsymbol{p}_v$ 在活动标架 $\{\boldsymbol{p}:\boldsymbol{e}_1,\boldsymbol{e}_2,\boldsymbol{e}_3\}$ 中表示为

$$\mathrm{d}\boldsymbol{p}=\boldsymbol{p}_u\mathrm{d}u+\boldsymbol{p}_v\mathrm{d}v=\omega_1 e_1+\omega_2 e_2$$

这里微分形式

$$\omega_1=\langle \boldsymbol{p}_u,\boldsymbol{e}_1\rangle\mathrm{d}u+\langle \boldsymbol{p}_u,\boldsymbol{e}_2\rangle\mathrm{d}v$$
$$\omega_2=\langle \boldsymbol{p}_v,\boldsymbol{e}_1\rangle\mathrm{d}u+\langle \boldsymbol{p}_v,\boldsymbol{e}_2\rangle\mathrm{d}v$$

---

① 摘编自微信公众号"老顾谈几何".

同样的

$$\mathrm{d}\begin{pmatrix}\boldsymbol{e}_1\\ \boldsymbol{e}_2\\ \boldsymbol{e}_3\end{pmatrix}=\begin{pmatrix}0 & \omega_{12} & \omega_{13}\\ -\omega_{12} & 0 & \omega_{23}\\ -\omega_{13} & -\omega_{23} & 0\end{pmatrix}\begin{pmatrix}\boldsymbol{e}_1\\ \boldsymbol{e}_2\\ \boldsymbol{e}_3\end{pmatrix}$$

这里微分形式 ω_{12} 满足一个非常关键的关系

$$\mathrm{d}\omega_{12}=-K\omega_1\wedge\omega_2$$

K 是曲面的高斯曲率,即联络的外微分等于曲率形式. 同时微分形式 ω_{12} 也定义了联络,而联络定义了协变微分算子 D,从而定义了平行移动,即

$$D\boldsymbol{e}_1=\omega_{12}\boldsymbol{e}_2, D\boldsymbol{e}_2=-\omega_{12}\boldsymbol{e}_1$$

假设 γ 是曲面上的一条曲线,其弧长参数为 s,$\boldsymbol{v}$ 是沿着 γ 的切向量场,我们说 $\boldsymbol{v}$ 沿着 γ 平行,如果

$$D_{\dot{\gamma}(s)}\boldsymbol{v}(s)=0,\quad \forall s$$

假设 $\boldsymbol{v}=f_1\boldsymbol{e}_1+f_2\boldsymbol{e}_2$,那么

$$\begin{aligned}D\boldsymbol{v}&=\mathrm{d}f_1\boldsymbol{e}_1+f_1D\boldsymbol{e}_1+\mathrm{d}f_2\boldsymbol{e}_2+f_2D\boldsymbol{e}_2\\&=\mathrm{d}f_1\boldsymbol{e}_1+f_1\omega_{12}\boldsymbol{e}_2+\mathrm{d}f_2\boldsymbol{e}_2-f_2\omega_{12}\boldsymbol{e}_1\end{aligned}$$

进一步展开得到

$$D_{\dot{\gamma}}\boldsymbol{v}=\langle D\boldsymbol{v},\dot{\gamma}\rangle=\langle \mathrm{d}f_1,\dot{\gamma}\rangle\boldsymbol{e}_1+f_1\langle\omega_{12},\dot{\gamma}\rangle\boldsymbol{e}_2+\langle\mathrm{d}f_2,\dot{\gamma}\rangle\boldsymbol{e}_2-f_2\langle\omega_{12},\dot{\gamma}\rangle\boldsymbol{e}_1$$

我们得到关于 $\boldsymbol{v}=f_1\boldsymbol{e}_1+f_2\boldsymbol{e}_2$ 的常微分方程

$$\begin{cases}\dot{f}_1-f_2\langle\omega_{12},\dot{\gamma}\rangle=0\\ \dot{f}_2-f_1\langle\omega_{12},\dot{\gamma}\rangle=0\end{cases}$$

给定初始条件,解存在并且唯一,即我们定义了切矢量沿着 γ 的平行移动. 我们可以证明平行移动保度量,即任给沿着 γ 的切矢量场 $\boldsymbol{v}$ 和 $\boldsymbol{w}$,有

$$\mathrm{d}\langle\boldsymbol{v},\boldsymbol{w}\rangle_{\mathbf{E}^3}=\langle D\boldsymbol{v},\boldsymbol{w}\rangle+\langle\boldsymbol{v},D\boldsymbol{w}\rangle$$

若 $\boldsymbol{v}$ 和 $\boldsymbol{w}$ 都沿着 γ 平行,则沿着 γ 它们的内积保持不变. 这时,我们说 ω_{12} 是黎曼度量的 Leve-Civita 联络. 一条曲线 γ 被称为是测地线,如果它的速度向量场 $\dot{\gamma}$ 沿着 γ 是自平行的,即

$$D_{\dot{\gamma}}\dot{\gamma}\equiv 0$$

假设 Ω 是曲面上的单连通区域,其边界为 $\partial\Omega$. 一个切向量沿着边界平行移动,跑了一圈回到原点之后此切向量会旋转一个角度,这个转角等于

$$\int_{\partial\Omega}\omega_{12}=\int_{\Omega}\mathrm{d}\omega_{12}=-\int_{\Omega}k\omega_1\wedge\omega_2$$

即围绕 Ω 边界平行移动一圈之后得到的转角(即和乐(holonomy))等于 Ω 的总高斯曲率.

示性类

纤维丛局部看具有直积结构,整体具有内在的扭曲. 我们希望在纤维

丛中找到一张曲面，它和每根纤维只相交于一点，即一个整体的截面. 但是由于内在的扭曲，这种截面未见得存在，这种截面整体存在性的障碍就是示性类.

我们考虑曲面所有的单位切向量构成的流形，曲面的单位切丛 $UTM(S)$，即

$$UTM(S):=\{(\boldsymbol{p},\boldsymbol{e})\mid \boldsymbol{p}\in S,\boldsymbol{e}\in T_pS,\mid\boldsymbol{e}\mid=1\}$$

固定点 $\boldsymbol{p}\in S$，所有的单位切向量 $\boldsymbol{e}\in T_pS$ 构成一个圆圈 $\mathbf{S}^2$，被称为是点 $\boldsymbol{p}$ 的纤维. 开集 U 上所有的单位切向量构成了一个直积 $U\times\mathbf{S}^1$. 但是整个曲面的单位切丛未见得是整体直积结构. 我们分析一下球面的单位切丛.

如图 1 所示，我们将球面沿着赤道分成上半球面和下半球面，每个半球面的单位切丛都具有直积结构，可以视为实心轮胎，每一点的纤维对应环绕实心轮胎的纬线. 我们一刀砍向实心轮胎，砍断所有纤维，得到一个截面，这个截面对应着半球面上的一个光滑单位切矢量场.

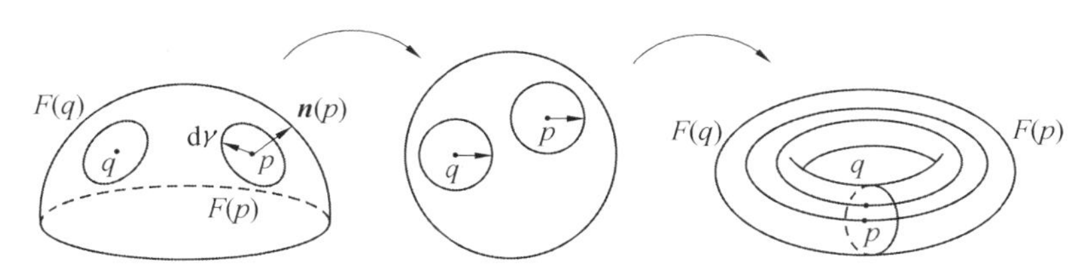

图 1　半球面的单位切丛具有直积结构

通过粘贴两个半球面的单位切丛，我们可以得到球面的单位切丛. 如图 2 所示，半球面单位切丛的边界是空心轮胎面 T^2，粘贴映射记为 $\varphi: T^2\to T^2$. 仔细考察，我们看到粘贴映射将纤维映成纤维，即纬线 a（红色）映成纬线（红线），但是经线 b（蓝色）映成复杂的圈 $a^{-2}b^{-1}$，即

$$\varphi_{\#}:[a]\mapsto[a],[b]\mapsto[a^{-2}b^{-1}]$$

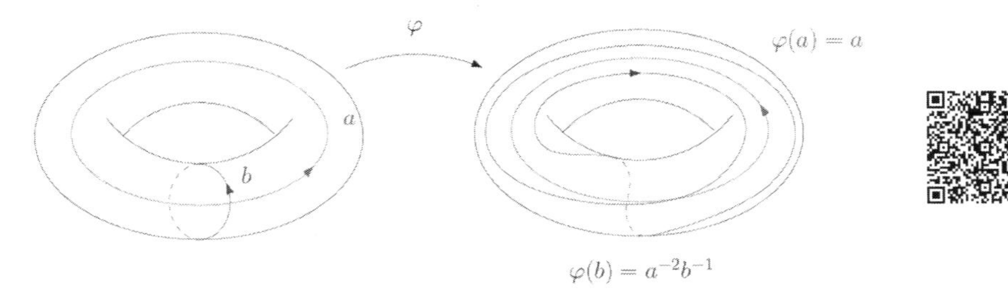

图 2　球面的单位切丛

粘贴之后得到的球面单位切丛是一个封闭的 3 维流形. 我们希望在单位切丛中找到一个曲面，曲面和每根纤维都相交于一点. 如果我们向左侧的

半球面单位切丛一刀砍去,斩断所有纤维,截面边界切痕为 b,截面为一个圆盘;切痕 b 映到右侧的 $\varphi(b)$,我们看到右侧半球面单位切丛内部不存在一个曲面以 $\varphi(b)$ 为边界. 这意味着 $\mathbf{S}^2$ 的单位切丛不存在全局截面. 纤维丛全局截面存在性的拓扑障碍由其示性类来表示. 曲率形式 $K\omega_1 \wedge \omega_2$ 可以被视为球面 de Rham 上同调群 $H^2(S,\mathbf{R})$ 的一个上同调类,被称为示性类. 积分

$$\frac{1}{2\pi}\int_S K\omega_1 \wedge \omega_2$$

为纤维丛的示性数. 这定量地描绘了纤维丛整体的扭曲程度.

在图 2 中,如果我们改变粘贴映射,只要满足特定条件,会得到不同的纤维丛. 不同纤维丛的内在扭曲也不同,可以用示性类来区分. 一般情形下,求取示性类的直观方法是逐步构造全局截面,中途遇到的障碍直接给出了示性类. 比如给定一张曲面上的圆丛,我们先计算曲面的三角剖分,然后在每个顶点的纤维上任取一点,得到顶点处的截面;圆丛限制在每条边上是一个圆柱面,其边界上确定了两点,我们在柱面上选取一条路径联结边界上的两点,这条路径是限制在边上的截面;圆丛限制在每个三角形面上是一个实心轮胎,其边界是空心轮胎曲面,曲面上已经确定了一条封闭曲线作为截面,如果这条封闭曲线在实心轮胎的一维同调群中为零,那么实心轮胎内部存在一张曲面以此曲线为边界,这张曲面就是此三角形面对应的截面;反之,如果这条封闭曲线在实心轮胎的一维同调群中非零而是一个整数,那么实心轮胎中不存在任何曲面以此曲线为边界,我们遇到了拓扑障碍. 由此,我们为每个三角形面赋予一个整数,即得到一个 2 - 形式,将其视为曲面 2 维上同调群中的一个类,这就是纤维丛的示性类.

我们也可以用几何上的曲率形式来定义示性类. 我们在每根纤维上定义一个内积,从而得到整体光滑的黎曼度量,与活动标架法类似得到唯一的 Levi-civita 联络,其外微分为曲率形式,曲率形式即为纤维丛的示性类. 这种方法将拓扑问题用几何来解决.

Maxwell 方程和 A - B 效应

经典的 Maxwell 方程组比较直接了当地描述电场强度 $\boldsymbol{E}$ 和磁感应强度 $\boldsymbol{B}$,核心是说它们的散度和旋量

$$\nabla\cdot \boldsymbol{E} = \frac{\rho}{\varepsilon_0}$$

$$\nabla\cdot \boldsymbol{B} = 0$$

$$\nabla\times \boldsymbol{E} = -\frac{\partial \boldsymbol{B}}{\partial t}$$

$$\nabla\times \boldsymbol{B} = \mu_0\left(\boldsymbol{J} + \varepsilon_0\frac{\partial \boldsymbol{E}}{\partial t}\right)$$

这里,ρ 是电荷密度,$\boldsymbol{J}$ 是电流密度,ε_0 是电常数,μ_0 是磁常数. 第一个方程是高斯定律,第二个方程是高斯磁定律,第三个方程是法拉第电磁感应定律,第四个方程是 Maxwell – Ampere 定律. 在自由空间中,ρ 和 $\boldsymbol{J}$ 为零,我们得到电场与磁场的波动方程

$$\frac{1}{c^2}\frac{\partial^2 \boldsymbol{E}}{\partial t^2} - \Delta \boldsymbol{E} = 0$$

$$\frac{1}{c^2}\frac{\partial^2 \boldsymbol{B}}{\partial t^2} - \Delta \boldsymbol{B} = 0$$

这意味着电场和磁场是平面行进正弦横波,电场和磁场彼此垂直,垂直于行进方向. 同时

$$c = \frac{1}{\sqrt{\mu_0\varepsilon_0}}$$

为光速,并且光速与观察者坐标系无关,因此 Maxwell 方程组本质上指向了相对论. 引入电势 ϕ 和磁矢量势 $\boldsymbol{A}$,高斯电场和磁场定律改写成

$$\boldsymbol{E} = -\nabla\phi - \frac{\partial \boldsymbol{A}}{\partial t}$$

$$\boldsymbol{B} = \nabla\times \boldsymbol{A}$$

在经典电动力学层次,磁矢量势 $\boldsymbol{A}$ 无法测量,因此人们一直以为它是为了数学的方便而非物理实在. 但在量子力学层次,它却比磁场强度具有更重要的地位. Aharonov 和 Bohm 于 1959 年提出 A – B 效应很好地阐述了这一点.

如图 3 所示,一束电子通过双缝,在背后的屏幕上形成干涉条纹. 我们在双缝后面放置一个无限长的螺线管,管内磁场强度为 $\boldsymbol{B}$,外部磁场、电场强度为零. 螺线管通电之后,干涉条纹在屏幕上发生了移动. 根据物理的局域原则,电子行为的改变一定是它和某种场相接触,但是电子的路径上电场强度和磁场强度均为零,电子不受洛伦兹力的作用. 历经数十年的争论,人们终于认识到干涉条纹的移动是由磁矢量势所引起. 通电的螺线管改变了时空的整体性质,改变了曲率的分布,从而改变了电子的相位,干涉条纹移动.

假设一带电粒子在 $\boldsymbol{B} = \boldsymbol{0}$ 而 $\boldsymbol{A} \neq \boldsymbol{0}$ 的区域内运动,哈密顿量

$$\hat{H} = \frac{1}{m}(-i\hbar\nabla - q\boldsymbol{A})^2 + V, \nabla\times \boldsymbol{A} = \boldsymbol{0}$$

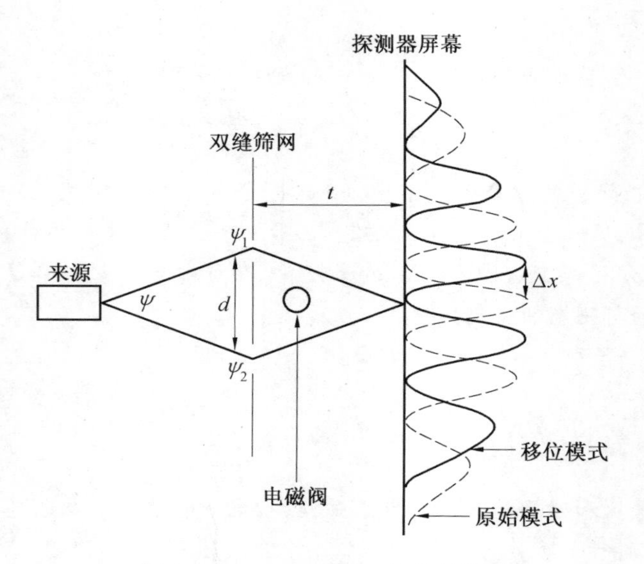

图3 Aharonov - Bohm 效应(互联网)

含时薛定谔方程

$$i\hbar \frac{\partial}{\partial t}\psi = \hat{H}\psi$$

可设波函数 $\psi = \phi e^{ig(\boldsymbol{r})}$,相位

$$g(\boldsymbol{r}) = \frac{q}{\hbar}\int_0^{r}\boldsymbol{A}(\boldsymbol{r}')\,d\boldsymbol{r}'$$

代入含时薛定谔方程,可得 ϕ 的方程

$$-\frac{\hbar^2}{2m}\Delta\phi + V\phi = i\hbar\frac{\partial}{\partial t}\phi$$

我们看到波函数的相因子体现了磁矢量势的物理效应.假设螺线管内部均匀磁场强度为 B,半径为 a,螺线管外足够远处磁场强度为零,磁矢量势非零.电子从左侧狭缝出发,到达屏幕某点,再返回右侧狭缝,假设两个狭缝足够接近,电子路径为封闭的圈 γ,那么相位的变化为

$$\frac{q}{\hbar}\oint_{\gamma}\boldsymbol{A}\cdot d\gamma = \frac{q\Phi}{\hbar}$$

这里磁通量 $\Phi = \pi a^2 B$.这个相位差导致可观测的物理效应(干涉条纹移动),即 A - B 效应.

纤维丛观点

我们将时空视为四维洛伦兹流形,洛伦兹度量为

$$ds^2 = -dt^2 + dx^2 + dy^2 + dz^2$$

在静磁场情形，$\frac{\partial \boldsymbol{B}}{\partial t}=0$，这时磁场强度无散，电场强度无旋，即

$$\nabla\cdot \boldsymbol{B}=0, \nabla\times \boldsymbol{E}=\boldsymbol{0}$$

令电场强度 $\boldsymbol{E}$ 表示为微分 1－形式

$$\boldsymbol{E}=E_x\mathrm{d}x+E_y\mathrm{d}y+E_z\mathrm{d}z$$

磁场强度 $\boldsymbol{B}$ 为

$$\boldsymbol{B}=B_x\mathrm{d}y\wedge\mathrm{d}z+B_y\mathrm{d}z\wedge\mathrm{d}x+B_z\mathrm{d}x\wedge\mathrm{d}y$$

由此我们得到外微分方程

$$\mathrm{d}\boldsymbol{B}=0, \mathrm{d}\boldsymbol{E}=0$$

电场和磁场强度表示成法拉第张量

$$\boldsymbol{F}=B_x\mathrm{d}y\wedge\mathrm{d}z+B_y\mathrm{d}z\wedge\mathrm{d}x+B_z\mathrm{d}x\wedge\mathrm{d}y+E_x\mathrm{d}x\wedge\mathrm{d}t+E_y\mathrm{d}y\wedge\mathrm{d}t+E_z\mathrm{d}z\wedge\mathrm{d}t$$

由此我们得到磁场无散、电场无旋的微分方程可以写成 $\mathrm{d}\boldsymbol{F}=0$. 更进一步，电磁势 1－形式是电势 φ 和磁矢量势 $\boldsymbol{A}=(A_x,A_y,A_z)$ 的结合

$$A=-\varphi\mathrm{d}t+A_x\mathrm{d}x+A_y\mathrm{d}y+A_z\mathrm{d}z$$

我们将电流密度 $\boldsymbol{J}=(J_x,J_y,J_z)$ 和电荷密度 ρ 结合成 1－形式

$$J=-\rho\mathrm{d}t+J_x\mathrm{d}x+J_y\mathrm{d}y+J_z\mathrm{d}z$$

这时，Maxwell 方程组具有更加简洁的形式

$$\boldsymbol{F}=\mathrm{d}A$$

$$\delta\boldsymbol{F}=J$$

这里 $\delta=*\mathrm{d}*$.

这里四维洛伦兹流形被视为是纤维丛的底空间，每一点 $(t,\boldsymbol{r})$ 处的纤维是复平面，电磁势 A 是联络，法拉第张量 $\boldsymbol{F}$ 是曲率形式，$\boldsymbol{F}=\mathrm{d}A$. 电子沿着封闭路径 γ 平行移动一圈，相位的变化为和乐，等于以 γ 为边界曲面的总曲率，即磁通量. 在螺线管没有通电的情形，时空曲率处处为零. 通电后，螺线管内部曲率非零，平行移动后电子相位发生变化. 我们用平面进行类比，开始我们有一个平面单位圆盘，沿着单位圆周平行移动一个向量，绕回初始位置后向量转角为零；如果我们将平面圆盘替换成圆锥面，圆锥顶点处曲率非零，其他各处曲率为零. 这时，我们沿着圆锥底面的边界平行移动一个切向量，绕回初始位置后切向量转角非零（和乐非零），这个圆锥面上的顶点为曲率奇点，就是 A－B 效应中的螺线管.

这里联络 A 是微分 1－形式，如果我们有一个函数 f，那么我们将 A 替换成 $A+\mathrm{d}f$，即所谓的规范变换，电磁场 $\boldsymbol{F}=\mathrm{d}A=\mathrm{d}(A+\mathrm{d}f)$ 没有变化. 这种性质被称为是规范对称. 在物理中有个规范原则：任何物理自然物理定律在规范变换下不变. 由此我们看到，从纤维丛的观点来解释 Maxwell 方程组，电势和磁矢量势成为联络，电场和磁场成为曲率形式，磁通量成为总

曲率,电子相位的偏转成为和乐.这些观点和数学工具简化了思维,揭示了更为深刻的物理,同时推动了理论的发展来进一步理解自然结构.

回顾

Maxwell 时代没有电气工业,他的理论超越于他所处的时代,因此在他的生命中,他的贡献一直没有得到社会承认.今天,由于电气工业、通信工业成为现代社会的基础,经典 Maxwell 方程组的电磁场、电磁波的理论成为大学工程科学系的必修内容,但是现代磁矢量势的理论依然被工科课程排斥在外.如果未来,拓扑绝缘体大规模进入工业,真正能够突破摩尔定律,那么纤维丛理论必将成为工科的核心课程内容.

最近元宇宙的浪潮高涨,这需要大量的数字几何算法,其背后的基础需要用到艰深的拓扑几何理论,其中也包括纤维丛示性类的理论.我们认识到这一点也是经历了非常漫长而曲折的过程,目前依然在继续探索中.

目前出版行业也开始进行品牌化建设,在这个众生喧嚣、注意力稀缺的时代,“无品牌不出版”越来越成为共识,不管是大而美还是小而美,必须以品牌带动方能突出重围.品牌力决定着出版力.

我们也试图进行这方面的尝试,当然过程会很漫长,但已经见到曙光!

刘培杰

2024 年 9 月 18 日

于哈工大

国外优秀数学著作
原版丛书（第三十辑）

Aspects of Differential Geometry(IV)

微分几何的各个方面

（第四卷）
（英文）

[西]埃斯特万·卡尔维尼奥－洛扎（Esteban Calviño-Louzao）
[西]爱德华多·加西亚－里奥（Eduardo García-Río）
[美]彼得·吉尔基（Peter Gilkey）
[韩]朴正阳（JeongHyeong Park）
[西]拉蒙·瓦茨克斯－洛伦佐（Ramón Vázquez-Lorenzo） 著

HITP
哈爾濱工業大學出版社
HARBIN INSTITUTE OF TECHNOLOGY PRESS

黑版贸登字 08-2021-051 号

图书在版编目(CIP)数据

微分几何的各个方面. 第四卷 = Aspects of Differerntial Geometry Ⅳ:英文/(西)埃斯特万·卡尔维尼奥-洛扎等著. —哈尔滨:哈尔滨工业大学出版社,2024.10
(国外优秀数学著作原版丛书. 第三十辑)
ISBN 978-7-5767-1340-4

Ⅰ. ①微… Ⅱ. ①埃… Ⅲ. ①微分几何-英文 Ⅳ. ①O186.1

中国国家版本馆 CIP 数据核字(2024)第 073692 号

WEIFEN JIHE DE GEGE FANGMIAN. DI SI JUAN

策划编辑 刘培杰 杜莹雪
责任编辑 刘家琳 张嘉芮
封面设计 孙茵艾
出版发行 哈尔滨工业大学出版社
社　　址 哈尔滨市南岗区复华四道街 10 号 邮编 150006
传　　真 0451-86414749
网　　址 http://hitpress. hit. edu. cn
印　　刷 哈尔滨圣铂印刷有限公司
开　　本 889 mm×1 194 mm 1/16 印张 45.5 字数 641 千字
版　　次 2024 年 10 月第 1 版 2024 年 10 月第 1 次印刷
书　　号 ISBN 978-7-5767-1340-4
定　　价 252.00 元(全 4 册)

Contents

Preface

This four-volume series arose out of work by the authors over a number of years both in teaching various courses and in their research endeavors. For technical reasons, the material is divided into four books and each book is largely self-sufficient. To facilitate cross references between the books, we have numbered the chapters of Book I from 1–3, the chapters of Book II from 4–8, the chapters of Book III from 9–11, and the chapters of the present Book IV on affine surfaces from 12–15. A final book in the series dealing with elliptic operator theory and its applications to Differential Geometry is proposed.

Up to isomorphism, there are two simply connected Lie groups of dimension 2. The translation group $\mathbb{R}^2$ is Abelian and acts on $\mathbb{R}^2$ by translation; the group structure is given by $(a,b)+(a',b')=(a+a',b+b')$. The $ax+b$ group $\mathbb{R}^+\times\mathbb{R}$ acts on $\mathbb{R}^+\times\mathbb{R}$ by

$$(x^1,x^2)\to(ax^1,ax^2+b)\quad\text{for}\quad a>0\quad\text{and}\quad b\in\mathbb{R}\,;$$

the group structure is given by composition and is non-Abelian;

$$(a,b)*(a',b')=(aa',ab'+b)\,.$$

An affine surface $\mathcal{M}$ is a pair (M,∇) where M is a smooth surface and ∇ is a torsion-free connection on the tangent bundle of M. One says $\mathcal{M}=(M,\nabla)$ is locally homogeneous if given any two points of M, there is the germ of a diffeomorphism mapping one point to the other point which preserves the connection ∇. Opozda [53] showed that any locally homogeneous affine surface geometry is *modeled* on one of the following three geometries:

- **Type $\mathcal{A}$.** $\mathcal{M}=(\mathbb{R}^2,\nabla)$ where ∇ has constant Christoffel symbols $\Gamma_{ij}{}^k=\Gamma_{ji}{}^k$. This geometry is homogeneous; the Type $\mathcal{A}$ connections are the left-invariant connections on the Lie group $\mathbb{R}^2$. An affine surface is modeled on such a geometry if and only if there exists a coordinate atlas so that the Christoffel symbols ${}^\alpha\Gamma_{ij}{}^k=\Gamma_{ij}{}^k$ are constant in each chart of the atlas.
- **Type $\mathcal{B}$.** $\mathcal{M}=(\mathbb{R}^+\times\mathbb{R},\nabla)$ where ∇ has Christoffel symbols $\Gamma_{ij}{}^k=(x^1)^{-1}A_{ij}{}^k$ where $A_{ij}{}^k=A_{ji}{}^k$ is constant. This geometry is homogeneous; the action of the $ax+b$ group sending $(x^1,x^2)\to(ax^1,ax^2+b)$ acts transitively on $\mathbb{R}^+\times\mathbb{R}$. If we identify the $ax+b$ group with $\mathbb{R}^+\times\mathbb{R}$, then the Type $\mathcal{B}$ connections are the left-invariant connections. An affine surface is modeled on such a geometry if there is a coordinate atlas so the Christoffel symbols ${}^\alpha\Gamma_{ij}{}^k=(x_\alpha^1)^{-1}A_{ij}{}^k$ in each chart of the atlas.
- **Type $\mathcal{C}$.** $\mathcal{M}=(M,\nabla)$ where ∇ is the Levi–Civita connection of a metric of constant non-zero sectional curvature.

This present volume is organized around this observation. There is a non-trivial intersection between the Type $\mathcal{A}$ and the Type $\mathcal{B}$ geometries. There is no geometry which is both Type $\mathcal{A}$ and $\mathcal{C}$. And the only Type $\mathcal{C}$ geometry which is not also Type $\mathcal{B}$ is modeled on the round sphere S^2 in $\mathbb{R}^3$. Chapter 12 of the book deals with preliminary material. We introduce the basics of affine geometry, discuss the affine quasi-Einstein equation, and establish its basic properties. We discuss affine gradient Ricci solitons and other preliminary matters. For surface geometries, the Ricci tensor

$$\rho(x, y) := \operatorname{Tr}\{z \to R(z, x)y\}$$

carries the geometry; an affine connection on a surface is flat if and only if $\rho = 0$.

Chapter 13 is devoted to a discussion of the geometry of Type $\mathcal{A}$ surfaces. Any Type $\mathcal{A}$ surface is strongly projectively flat. The solution space to the quasi-Einstein equation for the critical eigenvalue $\mu = -1$ will play a central role in our discussion as it is a complete invariant of strongly projectively flat surfaces. By identifying the Christoffel symbols $\{\Gamma_{ij}{}^k\}$ with a point of $\mathbb{R}^6$, we parameterize such surfaces. The Ricci tensor of any Type $\mathcal{A}$ surface is symmetric and any such surface is strongly projectively flat. The set of flat Type $\mathcal{A}$ surfaces where Γ does not vanish identically is a smooth 4-dimensional manifold which may be identified with a $\mathbb{Z}_2$ quotient of $S^1 \times S^2 \times \mathbb{R}$. The set of Type $\mathcal{A}$ surfaces where the Ricci tensor has rank 1 and is positive or negative semi-definite is a 5-dimensional manifold which may be identified with $S^1 \times S^1 \times \mathbb{R}^3$. It is natural to identify Type $\mathcal{A}$ geometries which differ by a change of coordinates or, equivalently, by the action of the general linear group $\mathrm{GL}(2, \mathbb{R})$. The resulting moduli spaces of flat Type $\mathcal{A}$ surfaces, of Type $\mathcal{A}$ surfaces where the Ricci tensor has rank 1, and of Type $\mathcal{A}$ surfaces where the Ricci tensor is non-degenerate and has signature (p, q) is determined quite explicitly. The surfaces which are geodesically complete are described up to linear equivalence. We discuss affine Killing vector fields and affine gradient Ricci solitons for such geometries.

In Chapter 14, we present an analogous discussion for the Type $\mathcal{B}$ surfaces. These surfaces are, in general, not strongly projectively flat, and thus the solution space to the quasi-Einstein equation is of less utility. The structure group here is the $ax + b$ group rather than the general linear group where the action this time is $(x^1, x^2) \to (x^1, bx^1 + ax^2)$. Let $\ker_{\mathcal{B}}\{\rho\} - \Gamma_0$ be the the space of flat connections other than the trivial connection where all the Christoffel symbols vanish and let $\ker_{\mathcal{B}}\{\rho_s\} - \ker_{\mathcal{B}}\{\rho\}$ be the space of all connections where the Ricci tensor is purely alternating but does not vanish identically. In contrast to the situation for Type $\mathcal{A}$ geometries, these two spaces are not smooth. The set $\ker_{\mathcal{B}}\{\rho\} - \Gamma_0$ (resp. $\ker_{\mathcal{B}}\{\rho_s\} - \ker_{\mathcal{B}}\{\rho\}$) is an immersed 3-dimensional (resp. 2-dimensional) manifold with transversal intersections. We also discuss affine Killing vector fields and affine gradient Ricci solitons in this context. We determine the locally symmetric Type $\mathcal{B}$ surfaces.

In Chapter 15, we present some applications of affine surface theory. If $\mathcal{M} = (M, \nabla)$ is an affine surface, the modified Riemannian extension gives rise to a neutral signature metric $g_{\nabla,\Phi,T,S,X}$ on the cotangent bundle of M where X is a tangent vector field on M, where Φ is a symmetric 2-tensor on M, and where T and S are endomorphisms of the tangent bundle of M.

There is an intimate relation between the geometry of the affine surface $\mathcal{M}$ and the geometry of $\mathcal{N} := (T^*M, g_{\nabla,\Phi,T,S,X})$. We relate solutions to the affine quasi-Einstein equation on $\mathcal{M}$ and the Riemannian quasi-Einstein equation. We also construct Bach flat signature $(2, 2)$ metrics using the Riemannian extension and construct vanishing scalar invariant (VSI) manifolds.

Esteban Calviño-Louzao, Eduardo García-Río, Peter Gilkey, JeongHyeong Park, and Ramón Vázquez-Lorenzo
April 2019

Acknowledgments

We report in this book, among other matters, on the research of the authors and our coauthors and would like, in particular, to note the contributions of D. D'Ascanio, M. Brozos-Vázquez, S. Gavino-Fernández, I. Gutiérrez-Rodríguez, D. Kupeli, S. Nikčević, P. Pisani, X. Valle-Regueiro, and M. E. Vázquez-Abal to this research.

D. D'Ascanio

M. Brozos-Vázquez

S. Gavino-Fernández

I. Gutiérrez-Rodríguez

S. Nikčević

P. Pisani

X. Valle-Regueiro

M. E. Vázquez-Abal

The research of the authors was supported by the Basic Science Research Program through the National Research Foundation of Korea (NRF) funded by the Ministry of Education (NRF-2016R1D1A1B03930449) and by Projects EM2014/009, MTM2013-41335-P, and MTM2016-75897-P (AEI/FEDER, UE). The assistance of Ekaterina Puffini of the Krill Institute of Technology has been invaluable.

Esteban Calviño-Louzao, Eduardo García-Río, Peter Gilkey, JeongHyeong Park, and Ramón Vázquez-Lorenzo
April 2019

CHAPTER 12

An Introduction to Affine Geometry

In Section 12.1, we summarize some basic definitions and facts. We discuss the curvature operator, the Ricci tensor, affine structures, and geodesics. We give a brief introduction to the theory of affine symmetric spaces and projective equivalence. We recall the definition of the Lie derivative and the Lie algebra of affine Killing vector fields. We define the Hessian, affine gradient Yamabe solitons, the affine quasi-Einstein operator, and affine gradient Ricci solitons. In Section 12.2, we present the classification of Wong of recurrent surfaces. In Section 12.3, we discuss the affine quasi-Einstein equation in the context of affine geometry. We prove the basic facts that we shall be using; strong projective equivalence plays an important role in our discussion. We exhibit some inhomogeneous examples. In Section 12.4 we discuss results of Opozda concerning locally affine homogeneous surfaces. In Section 12.5, we show that any locally affine homogeneous surface has a natural real analytic structure. We also discuss the structure of various spaces of smooth functions and other natural tensors as modules over the two 2-dimensional Lie algebras.

12.1 BASIC DEFINITIONS

In this section, we introduce the basic concepts we will be examining throughout the book. Section 12.1.1 discusses curvature, Section 12.1.2 introduces the theory of affine Killing vector fields, Section 12.1.3 treats geodesics, Section 12.1.4 treats geodesic completeness, and Section 12.1.5 defines affine maps, symmetric spaces, and models. Projective equivalence is introduced in Section 12.1.6. The Hessian and the affine quasi-Einstein operator (see Section 12.1.7) will play a central role in our development. Gradient Ricci and Yamabe solitons (see Section 12.1.8) and affine gradient Ricci and Yamabe solitons (see Section 12.1.9) will be examined in more detail in subsequent chapters. Classical results in ellipticity are given in Section 12.1.10.

Let $\vec{x} = (x^1, \dots, x^m)$ be a system of local coordinates on a smooth manifold M. We adopt the *Einstein convention* and sum over repeated indices henceforth. Let $\partial_{x^i} := \frac{\partial}{\partial x^i}$ and let $\theta^{(a_1,\dots,a_m)} := (\partial_{x^1})^{a_1} \dots (\partial_{x^m})^{a_m} \theta$. Let ∇ be a connection on the tangent bundle of M. Expand $\nabla_{\partial_{x^i}} \partial_{x^j} = \Gamma_{ij}{}^k \partial_{x^k}$ where $\Gamma = (\Gamma_{ij}{}^k)$ are the *Christoffel symbols* of the connection. We say that ∇ is *torsion-free* and that $\mathcal{M}$ is an *affine manifold* if $\nabla_X Y - \nabla_Y X - [X, Y] = 0$ or, equivalently, if $\Gamma_{ij}{}^k = \Gamma_{ji}{}^k$. The importance of the torsion-free condition lies in the observation that ∇ is torsion-free if and only if for every point P of M, there exist coordinates centered at P so that $\Gamma_{ij}{}^k(P) = 0$ (see Lemma 3.5 of Book I). We shall assume ∇ is torsion-free in this volume and refer to Arias-Marco and Kowalski [2], Gilkey [31, 32], and forthcoming work by D'Ascanio, Gilkey, and Pisani [23] for some results concerning the geometry of surfaces if the connection is permitted to have torsion.

12.1.1 CURVATURE. The *curvature operator* and the *Ricci tensor* are defined, respectively, by setting

$$R(X,Y) := \nabla_X \nabla_Y - \nabla_Y \nabla_X - \nabla_{[X,Y]},$$
$$\rho(X,Y) := \mathrm{Tr}\{Z \to R(Z,X)Y\}.$$

The components of these tensors may be expressed locally in the form:

$$\begin{aligned} R_{ijk}{}^\ell &= \partial_{x^i}\Gamma_{jk}{}^\ell - \partial_{x^j}\Gamma_{ik}{}^\ell + \Gamma_{in}{}^\ell\Gamma_{jk}{}^n - \Gamma_{jn}{}^\ell\Gamma_{ik}{}^n, \\ \rho_{jk} &= \partial_{x^i}\Gamma_{jk}{}^i - \partial_{x^j}\Gamma_{ik}{}^i + \Gamma_{in}{}^i\Gamma_{jk}{}^n - \Gamma_{jn}{}^i\Gamma_{ik}{}^n. \end{aligned} \tag{12.1.a}$$

In the affine setting, unlike in the pseudo-Riemannian context, the Ricci tensor need not be symmetric. Thus we introduce the *symmetric Ricci tensor* ρ_s and the *alternating Ricci tensor* ρ_a by defining:

$$\rho_s(X,Y) := \tfrac{1}{2}\{\rho(X,Y) + \rho(Y,X)\} \quad \text{and} \quad \rho_a(X,Y) := \tfrac{1}{2}\{\rho(X,Y) - \rho(Y,X)\}. \tag{12.1.b}$$

The following result shows that Ricci tensor carries the geometry for affine surfaces.

Lemma 12.1 *If $\mathcal{M} = (M, \nabla)$ is an affine surface, then $\mathcal{M}$ is flat if and only if $\rho = 0$.*

Proof. Let $\mathcal{M}$ be an affine surface. We have

$$\rho_{11} = R_{211}{}^2, \ \rho_{12} = R_{212}{}^2, \ \rho_{21} = R_{121}{}^1, \ \rho_{22} = R_{122}{}^1. \tag{12.1.c}$$

Thus the Ricci tensor determines the curvature operator; $\rho = 0$ if and only if $R = 0$ in dimension 2. □

12.1.2 AFFINE KILLING VECTOR FIELDS. If X is a smooth vector field on M, let Φ_t^X be the local flow defined by X. The *Lie derivative* $\mathcal{L}_X(\Xi)$ is defined by

$$\mathcal{L}_X(\Xi) := \lim_{t\to 0} \frac{\Phi_{-t}^X \circ \Xi \circ \Phi_t^X - \Xi}{t}.$$

If f is a function, then $\mathcal{L}_X(f) = X(f)$. If Y is a vector field, then $\mathcal{L}_X(Y) = [X,Y]$. If ω is a 1-form, then $\mathcal{L}_X\omega$ is characterized by the identity $\langle \mathcal{L}_X(Y), \omega\rangle + \langle Y, \mathcal{L}_X\omega\rangle = X\langle Y, \omega\rangle$. We refer to Kobayashi and Nomizu [39, Chapter VI] for the proof of the following result.

Lemma 12.2 *Let $\mathcal{M} = (M, \nabla)$ be an affine manifold of dimension m, possibly with torsion.*

1. *The following three conditions are equivalent and if any is satisfied, then X is said to be an* affine Killing vector field*:*
   (a) $(\Phi_t^X)_* \circ \nabla = \nabla \circ (\Phi_t^X)_*$ on the appropriate domain.
   (b) The Lie derivative $\mathcal{L}_X(\nabla)$ of ∇ vanishes.
   (c) $[X, \nabla_Y Z] - \nabla_Y[X, Z] - \nabla_{[X,Y]}Z = 0$ for all $Y, Z \in C^\infty(TM)$.
2. *Let $\mathfrak{K}(\mathcal{M})$ be the set of affine Killing vector fields. The Lie bracket gives $\mathfrak{K}(\mathcal{M})$ the structure of a real Lie algebra.*

The equation $[X, \nabla_Y Z] - \nabla_Y[X, Z] - \nabla_{[X,Y]}Z = 0$ is tensorial in $\{Y, Z\}$ so we may take $Y = \partial_{x^i}$ and $Z = \partial_{x^j}$. Expand $X = a^n \partial_{x^n}$. We compute:

$$\begin{aligned}
[X, \nabla_Y Z] &= [a^\ell \partial_{x^\ell}, \Gamma_{ij}{}^n \partial_{x^n}] = a^\ell \partial_{x^\ell} \Gamma_{ij}{}^n \partial_{x^n} - \Gamma_{ij}{}^n \partial_{x^n} a^\ell \partial_{x^\ell} \\
&= \{a^\ell \partial_{x^\ell} \Gamma_{ij}{}^k - \Gamma_{ij}{}^\ell \partial_{x^\ell} a^k\} \partial_{x^k}, \\
-\nabla_Y[X, Z] &= -\nabla_{\partial_{x^i}} [a^\ell \partial_{x^\ell}, \partial_{x^j}] = \nabla_{\partial_{x^i}} (\partial_{x^j} a^\ell \partial_{x^\ell}) \\
&= \partial_{x^i} \partial_{x^j} a^k \partial_{x^k} + \Gamma_{i\ell}{}^k \partial_{x^j} a^\ell \partial_{x^k}, \\
-\nabla_{[X,Y]}Z &= -\nabla_{[a^\ell \partial_{x^\ell}, \partial_{x^i}]} \partial_{x^j} = \Gamma_{\ell j}{}^k \partial_{x^i} a^\ell \partial_{x^k} .
\end{aligned}$$

We now obtain m^3 affine Killing equations for $1 \le i, j, k \le m$ from Assertion 1-c:

$$K_{ij}{}^k : \quad 0 = \frac{\partial^2 a^k}{\partial x^i \partial x^j} + \sum_\ell \left\{ a^\ell \frac{\partial \Gamma_{ij}{}^k}{\partial x^\ell} - \Gamma_{ij}{}^\ell \frac{\partial a^k}{\partial x^\ell} + \Gamma_{i\ell}{}^k \frac{\partial a^\ell}{\partial x^j} + \Gamma_{\ell j}{}^k \frac{\partial a^\ell}{\partial x^i} \right\} . \tag{12.1.d}$$

Let $\mathcal{M}$ be an affine manifold. We say that $\mathcal{M}$ is *affine Killing complete* if every affine Killing vector field of $\mathcal{M}$ is a complete vector field, i.e., the integral curves exist for all time. We examine this notion further in Section 13.2.7 and in Section 14.2.5. Let $\text{Aff}(\mathcal{M})$ be the group of all affine diffeomorphisms of an affine manifold $\mathcal{M}$.

Lemma 12.3 *Let $\mathcal{M}$ be a connected affine manifold $\mathcal{M}$.*

1. $\text{Aff}(\mathcal{M})$ *is a Lie group.*
2. *The Lie algebra $\mathfrak{a}(\mathcal{M})$ of* $\text{Aff}(\mathcal{M})$ *is the space of complete affine Killing vector fields.*
3. *$\mathcal{M}$ is affine Killing complete if and only if $\mathfrak{a}(\mathcal{M}) = \mathfrak{K}(\mathcal{M})$.*

Proof. We sketch the proof briefly and refer to Kobayashi and Nomizu [39] for a more complete explanation. Let $\mathcal{F}(\mathcal{M})$ be the frame bundle of the tangent bundle; this is a principal $\text{GL}(m, \mathbb{R})$ bundle over M. Fix a positive definite metric on $\mathbb{R}^m$ and a left-invariant Riemannian metric on $\text{GL}(m, \mathbb{R})$, i.e., a positive definite inner product on the Lie algebra $\mathfrak{gl}(m, \mathbb{R})$ of $\text{GL}(m, \mathbb{R})$. The connection ∇ (which may be permitted to have torsion) defines a canonical splitting of the tangent bundle of $\mathcal{F}(\mathcal{M})$ into vertical and horizontal subspaces. At a point F of $\mathcal{F}(\mathcal{M})$, we may use the frame to identify the horizontal subspace with $\mathbb{R}^m$ and the vertical subspace with $\mathfrak{gl}(m)$. We use the metrics on $\mathbb{R}^m$ and $\mathfrak{gl}(m)$ to define a Riemannian metric on $\mathcal{F}(\mathcal{M})$. If Φ is a (local) affine map, then the natural lift of Φ to the frame bundle is a (local) isometry of $\mathcal{F}(\mathcal{M})$. This defines a natural embedding of Aff($\mathcal{M}$) as a closed subgroup of the group of isometries of $\mathcal{F}(\mathcal{M})$. The group of isometries of a Riemannian manifold is a Lie group (see Theorem 7.6 of Book II). A closed subgroup of a Lie group is again a Lie group (see Theorem 6.10 of Book II). It now follows that Aff($\mathcal{M}$) is a Lie group. The remaining two Assertions are now immediate. □

The following is a useful observation that will play an important role in the classification of locally homogeneous surfaces in Section 12.4.

Lemma 12.4 *Let $\mathcal{M} = (M, \nabla)$ be an affine surface, possibly with torsion. Let X be an affine Killing vector field with $X(P) \neq 0$ at a point P of M. We can choose local coordinates centered at P so that*

$$X = \partial_{x^2}, \quad \Gamma_{ij}{}^k(x^1, x^2) = \Gamma_{ij}{}^k(x^1), \quad \Gamma_{11}{}^1(x^1) = 0, \quad \Gamma_{11}{}^2(x^1) = 0.$$

Proof. Choose initial coordinates (y^1, y^2) centered at P so that $X = \partial_{y^2}$. Since X is an affine Killing vector field, the Christoffel symbols do not depend on y^2, i.e., we may express $\Gamma_{ij}{}^k(y^1, y^2) = \Gamma_{ij}{}^k(y^1)$; the map $(y^1, y^2) \to (y^1, y^2 + t)$ is then an affine map. Let $\sigma(s) = (y^1(s), y^2(s))$ be an affine geodesic with $\sigma(0) = (0, 0)$ and with $\dot{\sigma}(0)$ and $X(0)$ linearly independent. Let $T(x^1, x^2) := (y^1(x^1), y^2(x^1) + x^2)$ define new coordinates with $\partial_{x^2} = \partial_{y^2}$. Since the curves $x^1 \to T(x^1, x^2)$ are affine geodesics for x^2 fixed and since ∂_{x^2} is an affine Killing vector field, the normalizations of the result hold. □

We present a technical lemma, which we shall need subsequently in Section 12.4, illustrating how the affine Killing equations are used. Let

$$\mathfrak{K}_{\mathbb{C}}(\mathcal{M}) := \mathfrak{K}(\mathcal{M}) \otimes_{\mathbb{R}} \mathbb{C}$$

be the algebra of complex affine Killing vector fields.

Lemma 12.5 *Let $\mathcal{M} = (M, \nabla)$ be an affine surface, possibly with torsion. Normalize the coordinates as in Lemma 12.4 to assume*

$$\Gamma_{ij}{}^k(x^1, x^2) = \Gamma_{ij}{}^k(x^1), \quad \Gamma_{11}{}^1(x^1) = 0, \quad \Gamma_{11}{}^2(x^1) = 0.$$

1. *If $0 \neq v(x^1)\partial_{x^2} \in \mathfrak{K}(\mathcal{M})$, with $v'(x^1) \neq 0$, then*

$$\begin{aligned} &\Gamma_{11}{}^1 = 0, \quad \Gamma_{11}{}^2 = 0, \quad \Gamma_{12}{}^1 = 0, \quad \Gamma_{21}{}^1 = 0, \\ &\Gamma_{22}{}^1 = 0, \quad \Gamma_{22}{}^2 = 0, \quad (\Gamma_{12}{}^2 + \Gamma_{21}{}^2)v' + v'' = 0. \end{aligned} \tag{12.1.e}$$

If $\{u_1(x^1)\cos(x^2) + u_2(x^1)\sin(x^2)\}\partial_{x^1} + w(x^1, x^2)\partial_{x^2}$ belongs to $\mathfrak{K}(\mathcal{M})$ and if Γ satisfies the relations of Equation (12.1.e), then $u_1 = u_2 = 0$.

2. *Assume* $0 \neq \alpha \in \mathbb{C}$. *If* $0 \neq e^{\alpha x^2} v(x^1)\partial_{x^2} \in \mathfrak{K}_{\mathbb{C}}(\mathcal{M})$, *then*

$$\begin{array}{llll} \Gamma_{11}{}^1 = 0, & \Gamma_{11}{}^2 = 0, & \Gamma_{12}{}^1 = 0, & \Gamma_{21}{}^1 = 0, \\ \Gamma_{22}{}^1 = 0, & \Gamma_{22}{}^2 = -\alpha. \end{array} \tag{12.1.f}$$

If $X := e^{\alpha x^2} u(x^1)\partial_{x^1} + w(x^1, x^2)\partial_{x^2}$ *belongs to* $\mathfrak{K}_{\mathbb{C}}(\mathcal{M})$ *and if* Γ *satisfies the relations of Equation (12.1.f), then* $u(x^1) = 0$.

Proof. The relations of Equation (12.1.e) follow from the affine Killing equations for $v(x^1)\partial_{x^2}$:

$$\begin{array}{ll} K_{11}{}^1 : \ 0 = (\Gamma_{12}{}^1 + \Gamma_{21}{}^1)v', & K_{12}{}^1 : \ 0 = \Gamma_{22}{}^1 v', \\ K_{12}{}^2 : \ 0 = (\Gamma_{22}{}^2 - \Gamma_{12}{}^1)v', & K_{21}{}^2 : \ 0 = (-\Gamma_{21}{}^1 + \Gamma_{22}{}^2)v', \\ K_{11}{}^2 : \ 0 = (\Gamma_{12}{}^2 + \Gamma_{21}{}^2)v' + v''. & \end{array}$$

Suppose that $\{u_1(x^1)\cos(x^2) + u_2(x^1)\sin(x^2)\}\partial_{x^1} + w(x^1, x^2)\partial_{x^2}$ is an affine Killing vector field and the relations of Equation (12.1.e) hold. We then have

$$K_{22}{}^1 : \ 0 = -u_1(x^1)\cos(x^2) - u_2(x^1)\sin(x^2)\,.$$

Consequently, $u_1 = u_2 = 0$. This proves Assertion 1

Assume $0 \neq e^{\alpha x^2} v(x^1)\partial_{x^2} \in \mathfrak{K}_{\mathbb{C}}(\mathcal{M})$ for $0 \neq \alpha \in \mathbb{C}$. We have $\Gamma_{11}{}^1 = 0$, $\Gamma_{11}{}^2 = 0$, $\Gamma_{ij}{}^k = \Gamma_{ij}{}^k(x^1)$, and $\partial_{x^2} \in \mathfrak{K}(\mathcal{M})$. We evaluate all affine Killing equations at $x^2 = 0$ to eliminate the exponential. The relations of Equation (12.1.f) follow from the affine Killing equations

$$\begin{array}{lll} K_{22}{}^1 : \ 0 = 2\alpha\Gamma_{22}{}^1 v, & \text{so} & \Gamma_{22}{}^1 = 0, \\ K_{12}{}^1 : \ 0 = \alpha\Gamma_{12}{}^1 v + \Gamma_{22}{}^1 v', & \text{so} & \Gamma_{12}{}^1 = 0, \\ K_{21}{}^1 : \ 0 = \alpha\Gamma_{21}{}^1 v + \Gamma_{22}{}^1 v', & \text{so} & \Gamma_{21}{}^1 = 0, \\ K_{22}{}^2 : \ 0 = \alpha v(\alpha + \Gamma_{22}{}^2) - \Gamma_{22}{}^1 v', & \text{so} & \Gamma_{22}{}^2 = -\alpha\,. \end{array}$$

Assume that $e^{\alpha x^2} u(x^1)\partial_{x^1} + w(x^1, x^2)\partial_{x^2} \in \mathfrak{K}_{\mathbb{C}}(\mathcal{M})$ and that the relations of Equation (12.1.f) hold. The associated affine Killing equation $K_{22}{}^1 : \ 0 = 2\alpha^2 u(x^1)$ implies that $u(x^1) = 0$ as desired and establishes Assertion 2. □

12.1.3 GEODESICS. We say that a curve γ in an affine manifold is an *affine geodesic* if $\ddot{\gamma} = 0$ or, in other words, $\nabla_{\dot{\gamma}}\dot{\gamma} = 0$. Express $\gamma(t) = (x^1(t), \dots, x^m(t))$ in a system of local coordinates. Then γ is an affine geodesic if and only if the *geodesic equation* is satisfied:

$$\ddot{x}^i(t) + \Gamma_{jk}{}^i(x(t))\dot{x}^j(t)\dot{x}^k(t) = 0\,.$$

Fix a point P of M and a tangent vector $\xi \in T_P M$. The Fundamental Theorem of Ordinary Differential Equations shows that the geodesic equation has a unique solution with initial conditions $x(0) = P$ and $\dot{x}(0) = \xi$ for $|t| < \varepsilon$. We say that $\mathcal{M}$ is *geodesically complete* if $\varepsilon(P, \xi) = \infty$ for every (P, ξ), or, in other words, if every affine geodesic extends for infinite time. We say that a submanifold N of M is *totally geodesic* if any affine geodesic in M which is tangent to N at a single point is entirely contained in N. We will discuss these notions further in the context of affine homogeneous surfaces. The *exponential map* $\exp_P$ (see Section 3.4.1 of Book I in the Riemannian setting) is a local diffeomorphism from a neighborhood of the origin in $T_P M$ to a neighborhood of P in M which is characterized by the fact that the curves $t \to \exp_P(t\xi)$ are affine geodesics starting at P with initial direction ξ. A bit of caution is needed as there are significant differences between the affine setting and the Riemannian setting.

Suppose given two affine manifolds $\mathcal{M}_1 = (M_1, \nabla_1)$ and $\mathcal{M}_2 = (M_2, \nabla_2)$. We say that a diffeomorphism Ψ from M_1 to M_2 is an *affine map* if Ψ intertwines the two connections, i.e., if $\Psi^*\nabla_2 = \nabla_1$.

Lemma 12.6 *If Φ is an affine diffeomorphism of an affine manifold $\mathcal{M}$, then the fixed point set of Φ is the union of smooth totally geodesic submanifolds.*

Proof. Let F be the fixed point set of Φ. If $P \in F$, let $V_0(P) := \{\xi \in T_P M : \Phi_*\xi = \xi\}$. Choose a small neighborhood $\mathcal{U}$ of the origin in $T_P M$ so that $\exp_P$ is a diffeomorphism from $\mathcal{U}$ to a neighborhood $\mathcal{O}$ of P in M and so that there is a unique affine geodesic joining any two points of $\mathcal{O}$. If $\xi \in T_P M$, then $\Phi_*\xi = \xi$ if and only if Φ fixes the affine geodesic through P with initial direction ξ. Furthermore, if $Q \in F \cap \mathcal{O}$, then there is a unique affine geodesic σ between P and Q lying in $\mathcal{O}$. Consequently, if ε is sufficiently small and if $Q \in \exp_P(B_\varepsilon(0))$, then $\Phi\sigma = \sigma$. This shows that $\exp_P(V_0 \cap B_\varepsilon(0)) = \mathcal{O} \cap F$. □

12.1.4 GEODESIC COMPLETENESS. If $\mathcal{M}$ is an affine manifold, we say that $\mathcal{M}$ is *geodesically complete* if every geodesic extends for infinite time; if $\mathcal{M}$ is not geodesically complete, we say $\mathcal{M}$ is *geodesically incomplete.*

Let $\mathcal{M}$ be the circle with the usual periodic parameter θ defined modulo 2π and with the connection given by ${\Gamma_{11}}^1 = 1$. The geodesic equation becomes $\ddot{\theta} + \dot{\theta}\dot{\theta} = 0$. Set $\theta(t) = \log(t)$ mod 2π for $t \in (0, \infty)$. Then $\dot{\theta} = t^{-1}$ and $\ddot{\theta} = -t^{-2}$. Thus this satisfies the geodesic equation and the maximal parameter range is $t \in (0, \infty)$. This provides an example of a compact affine geometry which is geodesically incomplete.

Let $\mathcal{M}_1$ be $S^1 \times [1, 4]$ with coordinates (θ, r) where θ is defined modulo 2π, where r belongs to the interval $[1, 4]$, and where ${\Gamma_{11}}^1 = 1$ is the only non-zero Christoffel symbol. Use polar coordinates to embed $S^1 \times [1, 4]$ as an annulus in the plane and use a partition of unity to extend the affine structure to the plane keeping the original geometry on $S^1 \times [1, 4]$ to obtain a flat geometry near infinity and near the origin. Again, we have an incomplete affine geodesic; this affine geodesic remains bounded, but the velocity increases without bound. It simply goes around the origin faster and faster on the circle of radius 2.

Sometimes there is an apparent singularity which can be removed. Let the non-zero Christoffel symbols of $\mathcal{M}$ be ${\Gamma_{11}}^1 = 1$, ${\Gamma_{12}}^2 = 1$, and ${\Gamma_{22}}^1 = -1$. Let $\arctan(\cdot)$ map $\mathbb{R}$ to

$(-\frac{\pi}{2}, \frac{\pi}{2})$. One verifies that the two curves

$$\begin{aligned}\sigma_1(t) &:= (\log(1+t), 0),\\ \sigma_2(t) &:= \left(\frac{1}{2}\log((1+t)^2+t^2), \arctan\left(\frac{t}{1+t}\right)\right) \quad \text{for} \quad t \in (-1, \infty)\end{aligned}$$

are affine geodesics. The first curve has a genuine singularity at $t = -1$. However, the apparent singularity at $t = -1$ in the second curve can be removed by taking a different branch of the arctangent function and setting

$$\tilde{\sigma}_2(t) := \left\{ \begin{array}{ll} (\frac{1}{2}\log((1+t)^2+t^2), \arctan(\frac{t}{1+t})) & \text{for } t \in (-1, \infty) \\ (0, -\frac{\pi}{2}) & \text{for } t = -1 \\ (\frac{1}{2}\log((1+t)^2+t^2), \arctan(\frac{t}{1+t}) - \pi) & \text{for } t \in (-\infty, -1) \end{array} \right\}.$$

One has the following picture where the upper curve is σ_2 and the lower curve is $\tilde{\sigma}_2$.

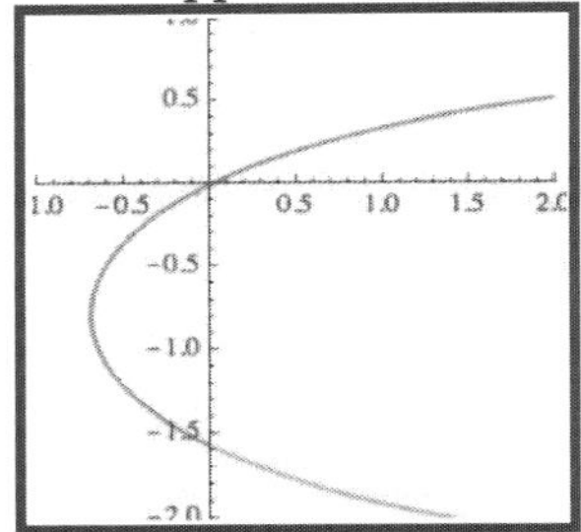

The crucial point in this example is, of course, that $\lim_{t\to -1^+} \sigma_2(t)$ exists. We have:

Lemma 12.7 *Let P be a point of an affine manifold $\mathcal{M}$. Let $\sigma : [0, T) \to M$ be an affine geodesic. Suppose $\lim_{t\to T} \sigma(t) = P$ exists. Then there exists $\varepsilon > 0$ so that σ can be extended to the parameter range $[0, T+\varepsilon)$ as an affine geodesic.*

Proof. Put a positive definite inner product $\langle\cdot,\cdot\rangle$ on T_PM to act as a reference metric. Let B_r be the ball of radius r about the origin in T_PM. Since the exponential map is a local diffeomorphism, we can use $\exp_P$ to identify B_ε with a neighborhood of P in M for some small ε. We use this identification to define a flat Riemannian metric near P on M so that $\exp_P$ is an isometry from B_ε to M. Let $d(\cdot,\cdot)$ be the associated distance function on M. Let $B_r(P) := \exp_P(B_r) = \{Q : d(P, Q) \le r\}$ for $r \le \varepsilon$. Choose linear coordinates on T_PM to put coordinates on $B_\varepsilon(P)$. This identifies T_QM with T_PM and extends $\langle\cdot,\cdot\rangle$ to T_QM. Compactness shows that there exists $0 < \tau < \frac{1}{2}\varepsilon$ so that if $Q \in B_{\frac{\varepsilon}{2}}(P)$ and if $\xi \in T_QM$ satisfies $\|\xi\| = 1$, then the affine geodesic $\sigma_{Q,\xi}(t) := \exp_Q(t\xi)$ exists for $t \in [0, \tau]$ and belongs to $B_\varepsilon(P)$. By continuity, we can choose $0 < \delta < \frac{1}{4}\tau$ so that if $Q \in B_\delta(P)$ and $\|\xi\| = 1$, then $d(\sigma_{Q,\xi}(\tau), \sigma_{P,\xi}(\tau)) < \frac{\tau}{2}$. Since $d(P, \sigma_{P,\xi}(\tau)) = \tau$, this implies $d(P, \sigma_{Q,\xi}(\tau)) \ge \frac{1}{2}\tau$. We conclude from these estimates that any non-trivial affine geodesic which begins in $B_\delta(P)$ continues to exist at least until it exits from $B_{\frac{1}{2}\tau}(P)$ and that it does in fact exit from $B_{\frac{1}{2}\tau}(P)$.

We assumed $\lim_{t\to T}\sigma(t) = P$. Choose $T_0 < T$ so $\sigma(T_0, T) \subset B_\delta(P)$. Then σ continues to exist until σ exits from $B_{\frac{1}{2}\tau}(P)$ and in particular extends to an affine geodesic defined on $(T_0, T+\varepsilon)$ for some ε. □

Suppose that $\mathcal{M}$ is *locally affine homogeneous*. We say that another affine manifold $\tilde{\mathcal{M}}$ is *modeled* on $\mathcal{M}$ if every point $\tilde{P}$ of $\tilde{\mathcal{M}}$ admits a neighborhood which is affine diffeomorphic to some small open subset of $\mathcal{M}$; the precise open subset being irrelevant as $\mathcal{M}$ was assumed locally affine homogeneous. We say that $\mathcal{M}$ is *essentially geodesically incomplete* if there is no locally affine homogeneous surface $\tilde{\mathcal{M}}$ which is modeled on $\mathcal{M}$ which is geodesically complete. The following provides useful criteria.

Lemma 12.8 *Let $\mathcal{M}$ be an affine homogeneous surface.*

1. *Suppose there exists an affine geodesic σ in $\mathcal{M}$ defined for $t \in (0, T)$ for $T < \infty$ such that* $\lim_{t\to T}\rho(\dot\sigma(t), \dot\sigma(t)) = \pm\infty$. *Then $\mathcal{M}$ is essentially geodesically incomplete.*
2. *Suppose there exists an affine Killing vector field X on $\mathcal{M}$ and an affine geodesic σ in $\mathcal{M}$ defined for $t \in (0, T)$ for $T < \infty$ such that* $\lim_{t\to T}\rho(X(t), \dot\sigma(t)) = \pm\infty$. *Then $\mathcal{M}$ is essentially geodesically incomplete.*

Proof. Let $f_\sigma(t) := \rho(\dot\sigma(t), \dot\sigma(t))$ such that $\lim_{t\to T} f_\sigma(t) = \pm\infty$ for some finite value of T. This clearly shows $\mathcal{M}$ is geodesically incomplete. Suppose, however, that there exists a geodesically complete geometry $\tilde{\mathcal{M}}$ which is modeled on $\mathcal{M}$. We argue for a contradiction. We will show in Theorem 12.52 that the coordinate atlas for $\tilde{\mathcal{M}}$ is real analytic. Copy a small piece of σ into $\tilde{\mathcal{M}}$ to define an affine geodesic $\tilde\sigma$. The function $\tilde f(t) := \tilde\rho(\dot{\tilde\sigma}, \dot{\tilde\sigma})$ would then agree with $f(t)$ on some parameter range. Since we are in the real analytic setting, we then have $\tilde f$ agrees with f on $(0, T)$ and thus $\lim_{t\to T}\tilde f(t) = \pm\infty$. This contradicts the assumption that $\tilde{\mathcal{M}}$ was geodesically complete. Consequently, $\mathcal{M}$ is essentially geodesically incomplete; this contradiction establishes Assertion 1. The proof of Assertion 2 is the same since, as we shall show in Lemma 12.14, any affine Killing vector field is real analytic. □

12.1.5 AFFINE MAPS, SYMMETRIC SPACES, AND MODELS. An affine manifold $\mathcal{M}$ is said to be *affine homogeneous* if given any two points P and $\tilde P$ of M, there is an affine map from M to M taking P to $\tilde P$. Correspondingly, $\mathcal{M}$ is said to be *locally affine homogeneous* if Ψ is only defined from some neighborhood of P to some neighborhood of $\tilde P$. Suppose that $\mathcal{M}_0$ is affine homogeneous. We say that $\mathcal{M}$ is *modeled* on $\mathcal{M}_0$ if there is an open neighborhood $\mathcal{O}_P$ of any point P of M and an affine diffeomorphism Φ from $\mathcal{O}_P$ to some open set of $\mathcal{M}_0$. We say that $\mathcal{M}$ is *locally affine symmetric* if $\nabla R = 0$.

Theorem 12.9 *Let $\mathcal{M}$ be a connected locally affine symmetric manifold.*

1. *$\mathcal{M}$ is locally affine homogeneous.*
2. *If ρ_s has maximal rank, then ∇ is the Levi-Civita connection of the locally affine symmetric pseudo-Riemannian manifold (M, ρ_s).*

3. *If $\mathcal{M}$ is 2-dimensional, then the Ricci tensor of $\mathcal{M}$ is symmetric.*

Proof. We establish Assertion 1 as follows. The exponential map is a diffeomorphism from an open neighborhood $\mathcal{O}$ of 0 in $T_P M$ to an open neighborhood $\tilde{\mathcal{O}}$ of P in M. We may assume that $-\mathcal{O} = \mathcal{O}$ without loss of generality. The *geodesic symmetry* is defined by setting

$$\sigma_P(Q) := \exp_P(-\exp_P^{-1}(Q)) \text{ for } Q \in \tilde{\mathcal{O}}.$$

Work of Nomizu [47] (see Theorem 17.1) shows that σ_P is an affine map. One can compose geodesic symmetries around various points to show that $\mathcal{M}$ is locally affine homogeneous. We refer to Koh [41] for subsequent related work. We also note that if $\mathcal{M}$ is locally affine symmetric, then $\mathcal{M}$ is k-affine curvature homogeneous for all k and this result follows from the work of Pecastaing [57] on the "Singer number" in a quite general context. Finally, Theorem 12.9 follows from work of Opozda [52] in the real analytic setting.

If (M, g) is a pseudo-Riemannian manifold, then there exists a unique affine connection so $\nabla^g g = 0$ (see Theorem 3.7 of Book I). Suppose ρ_s has rank m. Then (M, ρ_s) is a pseudo-Riemannian manifold. Furthermore, $\nabla = \nabla^g$. Since $\nabla R = 0$, $\nabla \rho_s = 0$ and (M, g) is a locally symmetric pseudo-Riemannian manifold. Assertion 2 follows.

Let R be the curvature operator; $R_{ij}(\partial_{x^k}) = R_{ijk}{}^{\ell}\partial_{x^\ell}$. We have a dual action on the cotangent bundle given by $R_{ij}(dx^\ell) = -R_{ijk}{}^{\ell}dx^k$. We extend this action to the whole exterior algebra by the Leibnitz rule. Let $\rho_a = (\rho_{12} - \rho_{21})dx^1 \otimes dx^2$ be the alternating Ricci tensor defined in Equation (12.1.b). Expand $\nabla^2 \rho_a = \rho_{a;ij}(dx^1 \wedge dx^2) \otimes dx^i \otimes dx^j$, where

$$\rho_{a;ij} = \nabla_{\partial_{x^j}} \nabla_{\partial_{x^i}} \rho_a .$$

The commutator of covariant differentiation is given by curvature. We compute

$$\begin{aligned} \rho_{a;21} - \rho_{a;12} &= (\rho_{12} - \rho_{21}) R_{12}(dx^1 \wedge dx^2) \\ &= (\rho_{12} - \rho_{21})(-R_{121}{}^1 - R_{122}{}^2)(dx^1 \wedge dx^2) \\ &= (\rho_{12} - \rho_{21})(-\rho_{21} + \rho_{12})(dx^1 \wedge dx^2) . \end{aligned}$$

If $\nabla \rho = 0$, then $\nabla \rho_a = 0$ and thus $(\rho_{12} - \rho_{21})^2 = 0$. This proves Assertion 3. We remark that Assertion 3 was known to Opozda [50]; the proof we have given is different from the proof appearing there and is due to D'Ascanio, Gilkey, and Pisani [23]. □

Remark 12.10 If the connection on the tangent bundle is not assumed to be torsion-free, then the situation is very different.

1. Let the non-zero Christoffel symbols of $\mathcal{M} = (\mathbb{R}^2, \nabla)$ be given by

$$\Gamma_{12}{}^2 = -\frac{1 - e^{2x^1}}{1 + e^{2x^1}} \quad \text{and} \quad \Gamma_{21}{}^2 = \frac{1 - e^{2x^1}}{1 + e^{2x^1}} ;$$

then a direct computation shows $\rho = dx^1 \otimes dx^1$ and $\nabla\rho = 0$. The space of affine Killing vector fields (see Section 12.1.2) takes the form $\text{span}\{\partial_{x^2}, x^1\partial_{x^2}, x^2\partial_{x^2}\}$ so this geometry is not locally affine homogeneous (although it is *cohomogeneity 1*, i.e., the orbits of the affine group have codimension 1) and thus Assertion 1 fails in this context.

2. Let the non-zero Christoffel symbols be given by

$$\Gamma_{11}{}^1 = -\frac{1}{x^1}, \quad \Gamma_{12}{}^2 = -\frac{1}{x^1}, \quad \Gamma_{21}{}^2 = -\frac{3}{x^1}, \quad \Gamma_{22}{}^1 = \frac{1}{x^1}.$$

This is a Type $\mathcal{B}$ affine surface with torsion (see Definition 12.42). We then have

$$\rho = (x^1)^{-2}\{-3dx^1 \otimes dx^1 - dx^2 \otimes dx^2\} \text{ and } \nabla\rho = 0.$$

Since the connection in question has torsion, it is not the Levi-Civita connection of any pseudo-Riemannian metric. Thus Assertion 2 also fails in this context. Note that in contrast to the previous example, this structure is homogeneous since $x^1\partial_{x^1} + x^2\partial_{x^2}$ and ∂_{x^2} are affine Killing vector fields for this geometry.

3. The proof we gave of Assertion 3 remains valid even if torsion is present; if $\mathcal{M}$ is an affine surface with torsion with $\nabla\rho = 0$, then ρ is symmetric. We refer to D'Ascanio, Gilkey, and Pisani [23] for further details.

12.1.6 PROJECTIVE EQUIVALENCE. Two connections ∇ and $\tilde{\nabla}$ on the tangent bundle of M are said to be *projectively equivalent* if there exists a smooth 1-form ω so

$$\nabla_X Y = \tilde{\nabla}_X Y + \omega(X)Y + \omega(Y)X \quad \text{for all} \quad X, Y;$$

ω is said to provide a projective equivalence from $\tilde{\nabla}$ to ∇. If, in addition, ω is closed, then ∇ and $\tilde{\nabla}$ are said to be *strongly projectively equivalent*. If ∇ is projectively equivalent (resp. strongly projectively equivalent) to a flat connection (i.e., to a connection which has zero Christoffel symbols in some coordinate atlas), then ∇ is said to be *projectively flat* (resp. *strongly projectively flat*). See, for example, the discussion in Opozda [51].

Theorem 12.11

1. *If ω provides a projective equivalence between ∇ and $\tilde{\nabla}$, then $d\omega = 0$ if and only if $\rho_{a,\nabla} = \rho_{a,\tilde{\nabla}}$.*
2. *If two projectively equivalent connections have symmetric Ricci tensors, then they are strongly projectively equivalent.*
3. *If $\mathcal{M}$ is a simply connected surface, then a connection ∇ is strongly projectively flat if and only if ρ and $\nabla\rho$ are totally symmetric.*
4. *If ∇ is a connection, then a curve with $\dot{C} \neq 0$ is an unparameterized geodesic if and only if $\nabla_{\dot{C}}\dot{C} = \alpha(t)\dot{C}$ for some smooth function α.*

5. *Two connections ∇ and $\tilde{\nabla}$ are projectively equivalent if and only if their unparameterized affine geodesics coincide.*

Proof. Let ω provide a projective equivalence between ∇ and $\tilde{\nabla}$. Let $j \neq k$. We use the identity $\Gamma_{ij}{}^k = \tilde{\Gamma}_{ij}{}^k + \delta_i^k \omega_j + \delta_j^k \omega_i$ and Equation (12.1.a) to compute

$$\begin{aligned}
\rho_{jk} - \rho_{kj} &= \partial_{x^k}\Gamma_{ij}{}^i - \partial_{x^j}\Gamma_{ik}{}^i, \quad \tilde{\rho}_{jk} - \tilde{\rho}_{kj} = \partial_{x^k}\tilde{\Gamma}_{ij}{}^i - \partial_{x^j}\tilde{\Gamma}_{ik}{}^i, \\
\rho_{jk} - \rho_{kj} - \tilde{\rho}_{jk} + \tilde{\rho}_{kj} &= \partial_{x^k}(\Gamma_{ij}{}^i - \tilde{\Gamma}_{ij}{}^i) - \partial_{x^j}(\Gamma_{ik}{}^i - \tilde{\Gamma}_{ik}{}^i) \\
&= (m+1)\{\partial_{x^k}\omega_j - \partial_{x^j}\omega_k\}.
\end{aligned}$$

It now follows that $\rho_{a,\nabla} = \rho_{a,\tilde{\nabla}}$ if and only if $d\omega = 0$ and Assertion 1 follows (see also the discussion in Steglich [58]). If ∇ and $\tilde{\nabla}$ have symmetric Ricci tensors, then $\rho_{a,\nabla} = \rho_{a,\tilde{\nabla}} = 0$, so Assertion 2 follows from Assertion 1. We refer to Eisenhart [29] (pages 88–96) for the proof of Assertion 3 (see also Theorem 3.3 in Kobayashi and Nomizu [39]).

We prove Assertion 4 as follows. Let $\sigma(s) = C(t(s))$ reparameterize a curve C. Then $\partial_s\sigma^i = (\partial_s t)\dot{C}^i$ and $\partial_s^2\sigma^i = (\partial_s^2 t)\dot{C}^i + (\partial_s t)^2\ddot{C}^i$. Thus:

$$\{\nabla_{\partial_s\sigma}\partial_s\sigma\}^k = \partial_s^2\sigma^k + (\partial_s\sigma^i)(\partial_s\sigma^j)\Gamma_{ij}{}^k = (\partial_s t)^2\{\ddot{C}^k + \Gamma_{ij}{}^k\dot{C}^i\dot{C}^j\} + (\partial_s^2 t)\dot{C}^k. \tag{12.1.g}$$

Suppose first that σ is an unparameterized affine geodesic. Then there exists $t(s)$ so that C is an affine geodesic, i.e., $\ddot{C}^k + \Gamma_{ij}{}^k\dot{C}^i\dot{C}^j = 0$ for all k. We use Equation (12.1.g) to see

$$\nabla_{\partial_s\sigma}\partial_s\sigma = (\partial_s^2 t)\dot{C} = (\partial_s t)^{-1}(\partial_s^2 t)\partial_s\sigma.$$

Set $\alpha(s) := (\partial_s t)^{-1}(\partial_s^2 t)$. Then $\nabla_{\partial_s\sigma}\partial_s\sigma = \alpha(s)\partial_s\sigma$ as desired. Conversely, if we suppose $\nabla_{\partial_s\sigma}\partial_s\sigma = \alpha(s)\partial_s\sigma$, define the function $t = t(s)$ by solving the ordinary differential equation (ODE) $\partial_s^2 t = \alpha(s)\partial_s t$. Then $\ddot{C}^k + \Gamma_{ij}{}^k\dot{C}^i\dot{C}^j = 0$ so $C(t)$ is an affine geodesic. This proves Assertion 4.

Let $\tilde{\nabla}$ be a second affine connection on M. Let $\Theta_{ij}{}^k := \tilde{\Gamma}_{ij}{}^k - \Gamma_{ij}{}^k$; Θ is tensorial and ω provides a projective equivalence from ∇ to $\tilde{\nabla}$ if and only if $\Theta_{ij}{}^k = \delta_i^k\omega_j + \delta_j^k\omega_i$. We compute:

$$\{\tilde{\nabla}_{\dot{C}}\dot{C} - \nabla_{\dot{C}}\dot{C}\}^k = \Theta_{ij}{}^k\dot{C}^i\dot{C}^j. \tag{12.1.h}$$

Suppose that ω provides a projective equivalence from $\tilde{\nabla}$ to ∇. Let C be a ∇ affine geodesic and let $\alpha := 2\omega_i\dot{C}^i$. We then have

$$\{\tilde{\nabla}_{\dot{C}}\dot{C}\}^k = \Theta_{ij}{}^k\dot{C}^i\dot{C}^j = (\delta_i^k\omega_j + \delta_j^k\omega_i)\dot{C}^i\dot{C}^j = \omega_j\dot{C}^k\dot{C}^j + \omega_i\dot{C}^i\dot{C}^k = \alpha\dot{C}^k$$

and C is an unparameterized $\tilde{\nabla}$ affine geodesic.

Conversely, suppose that every affine geodesic for ∇ is an unparameterized affine geodesic for $\tilde{\nabla}$. Let $\{e_1, \dots, e_m\}$ be a local frame for the tangent bundle. Fix a point P of the manifold in question. Let C be the ∇ affine geodesic with $C(0) = P$ and $\dot{C}(0) = e_1$. By Assertion 4, $\tilde{\nabla}_{\dot{C}}\dot{C}$

is a multiple of $\dot{C}$. Since $\dot{C}^k(P) = 0$ for $k > 1$, we use Equation (12.1.h) to see $\Theta_{11}{}^k = 0$ for $k > 1$. Of course, the use of the index "1" is illustrative only; it holds for any index and any local basis. We use polarization to establish this identity. Consider the shear

$$\begin{array}{llll} e_1(\varepsilon) := e_1 + \varepsilon e_2, & e_2(\varepsilon) = e_2, & e_k(\varepsilon) = e_k & \text{for} \quad k > 2, \\ e^1(\varepsilon) = e^1, & e^2(\varepsilon) = e^2 - \varepsilon e^1, & e^k(\varepsilon) = e^k & \text{for} \quad k > 2. \end{array}$$

If $k > 2$, we have

$$\begin{array}{l} 0 = \Theta_{11}{}^2(\varepsilon) = \Theta_{11}{}^2 + \varepsilon(2\Theta_{12}{}^2 - \Theta_{11}{}^1) + O(\varepsilon^2), \\ 0 = \Theta_{11}{}^k(\varepsilon) = \Theta_{11}{}^k + 2\varepsilon\Theta_{12}{}^k + O(\varepsilon^2). \end{array}$$

Thus $\Theta_{12}{}^k = 0$ and $\Theta_{11}{}^1 = 2\Theta_{12}{}^2$. Permuting the indices yields

$$\Theta_{ij}{}^k = \left\{ \begin{array}{ll} 0 & \text{if } \{i, j, k\} \text{ are distinct} \\ \frac{1}{2}\Theta_{ii}{}^i & \text{if } j = k \text{ and } k \neq i \end{array} \right\}.$$

Let $\omega_1 = \frac{1}{2}\Theta_{11}{}^1, \ldots, \omega_m := \frac{1}{2}\Theta_{mm}{}^m$. It now follows that $\Theta_{ij}{}^k = \delta_j^k \omega_i + \delta_i^k \omega_j$ and Assertion 5 follows. □

12.1.7 THE HESSIAN AND THE AFFINE QUASI-EINSTEIN OPERATOR.

If f belongs to $C^\infty(M)$, then the *Hessian* $\mathcal{H}f = \nabla^2 f = \nabla df$ is the symmetric $(0, 2)$-tensor

$$\mathcal{H}f := \nabla^2 f = (\partial_{x^i}\partial_{x^j} f - \Gamma_{ij}{}^k \partial_{x^k} f)\, dx^i \otimes dx^j .$$

Let $\mathfrak{Q}_{\mu,\nabla} f := \mathcal{H}f - \mu f \rho_s$ be the *affine quasi-Einstein operator*:

$$\mathfrak{Q}_{\mu,\nabla} f = \left\{ \frac{\partial^2 f}{\partial x^i \partial x^j} - \sum_{k=1}^m \Gamma_{ij}{}^k \frac{\partial f}{\partial x^k} - \mu f (\rho_s)_{ij} \right\} dx^i \otimes dx^j . \tag{12.1.i}$$

The eigenvalue μ is a parameter of the theory. Let

$$E(\mu, \nabla) := \ker\{\mathfrak{Q}_{\mu,\nabla}\} = \{f \in C^2(M) : \mathcal{H}f = \mu f \rho_s\}.$$

Note that if $\rho_s = 0$, then $E(\mu, \nabla) = E(0, \nabla)$ for any μ. Also observe that $E(0, \nabla) = \ker\{\mathcal{H}\}$ is the space of *affine gradient Yamabe solitons*, which we also denote by $\mathcal{Y}$. The equation

$$\mathfrak{Q}_{\mu,\nabla} f = 0$$

is called the *affine quasi-Einstein equation*. We will motivate this equation subsequently when we use the modified Riemannian extension to relate this equation to the quasi-Einstein equation on the tangent bundle in Section 15.1.5 (see Theorem 15.9). We shall study this equation in more detail in Section 12.3. The eigenvalue $\mu = -\frac{1}{m-1}$ is distinguished. If $\mathcal{M} = (M, \nabla)$, we set

$$\mathcal{Q}(\mathcal{M}) := E(-\tfrac{1}{m-1}, \nabla) = \{f \in C^2(M) : \mathcal{H}f = -\tfrac{1}{m-1} f \rho_s\}.$$

This solution space will play an important role in the study of strongly projectively flat geometries.

In Theorem 15.8, we will relate the affine quasi-Einstein equation on an affine surface to the corresponding quasi-Einstein equation for an associated pseudo-Riemannian manifold of neutral signature $(2,2)$; it is this relationship which motivates our notation for surfaces.

12.1.8 GRADIENT RICCI AND YAMABE SOLITONS. If (M,g) is a pseudo-Riemannian manifold, and if f is a smooth function on M, one says that (M,g,f) is a *pseudo-Riemannian gradient Yamabe soliton* if there exists a real constant λ so that $\mathcal{H}f = (\tau-\lambda)g$ where τ is the scalar curvature (see Chapter 11 of Book III). One says that (M,g,f) is a *pseudo-Riemannian gradient Ricci soliton* if there exists a real constant so $\mathcal{H}f + \rho = \lambda g$. We refer to Book III for further details concerning these matters.

12.1.9 AFFINE GRADIENT RICCI AND YAMABE SOLITONS. Many, but not all, notions of Riemannian geometry generalize to the affine category. However, in the affine setting, there is no metric. Consequently, it is not possible to define the scalar curvature, and one proceeds a bit differently. Let $\mathcal{M} = (M,\nabla)$ be an affine manifold. One says that (M,∇,f) is an *affine gradient Yamabe soliton* if $\mathcal{H}f = 0$, i.e., if f is a solution of the affine quasi-Einstein equation with eigenvalue $\mu = 0$. More generally, one says that (M,∇,f) is an *affine gradient Ricci soliton* if one has that $\mathcal{H}f + \rho_s = 0$. In local coordinates, this is equivalent to the following:

$$\frac{\partial^2 f}{\partial x^i \partial x^j}(\vec{x}) - \sum_{k=1}^{m} \Gamma_{ij}{}^k(\vec{x})\frac{\partial f}{\partial x^k}(\vec{x}) = -(\rho_s)_{ij}(\vec{x}) \quad \text{for all} \quad i,j\,. \tag{12.1.j}$$

Let $\mathcal{Y}(\mathcal{M})$ be the vector space of affine gradient Yamabe solitons for $\mathcal{M}$ and let $\mathfrak{A}(\mathcal{M})$ be the set of affine gradient Ricci solitons. If $\mathfrak{A}(\mathcal{M})$ is non-empty, then $\mathfrak{A}(\mathcal{M})$ is an affine space modeled on $\mathcal{Y}(\mathcal{M})$, i.e., if $f \in \mathfrak{A}(\mathcal{M})$ and if $\xi \in \mathcal{Y}(\mathcal{M})$, then $f + \xi \in \mathfrak{A}(\mathcal{M})$. We say that a 3-tensor field $a_{ijk}dx^i \otimes dx^j \otimes dx^k$ is *totally symmetric* if we have the symmetries $a_{ijk} = a_{jik} = a_{ikj}$. Recall that the action of the curvature on the cotangent bundle is given by $R_{ij}(dx^\ell) = -R_{ijk}{}^\ell dx^k$.

Lemma 12.12 *Let $\mathcal{M} = (M,\nabla)$ be an affine manifold.*

1. *If $f \in \mathfrak{A}(\mathcal{M})$ and if $X \in \mathfrak{K}(\mathcal{M})$, then $X(f) \in \mathcal{Y}(M)$.*
2. *If $h \in \ker\{\mathcal{H}\}$, then $R_{ij}(dh) = 0$ for $1 \le i < j \le m$.*
3. *If $f \in \mathfrak{A}(\mathcal{M})$ and if $\nabla\rho_s$ is totally symmetric, then $R_{ij}(df) = 0$ for $1 \le i < j \le m$.*

Proof. Let f be an affine gradient Ricci soliton and let X be an affine Killing vector field. We have by naturality that $(\Phi_t^X)^* f$ is again an affine gradient Ricci soliton. Since the difference of two affine gradient Ricci solitons belongs to $\ker\{\mathcal{H}\}$, $(\Phi_t^X)^* f - f \in \ker\{\mathcal{H}\}$. Differentiating this relation with respect to t and setting $t = 0$ yields Assertion 1. Assertion 2 follows from the identity $h_{;ijk} - h_{;ikj} = \{R_{kj}(dh)\}_i$. Assertion 3 follows similarly if we assume $\nabla\rho_s$ to be totally symmetric (see Brozos-Vázquez and García-Río [5]). □

12.1.10 ELLIPTICITY. We now state a simple result dealing with hypoellipticity and analytic hypoellipticity which follows from more general results in partial differential equations. We refer, for example, to Gilkey [30] for the proof of the first assertion and to Christ [17] or Treves [59] for the proof of the second assertion in the following result.

Lemma 12.13 *Let $\vec{f} = (f_1, \ldots, f_\ell)$ be a C^2 function which is defined on an open subset $\mathcal{O}$ of $\mathbb{R}^m$. Suppose that $\vec{f}$ satisfies a partial differential equation of the form*

$$\sum_{i=1}^{m} \frac{\partial^2 f_k}{\partial x^i \partial x^i}(\vec{x}) + \sum_{i,j=1}^{m} a_{ijk}(\vec{x}) \frac{\partial f_i}{\partial x^j}(\vec{x}) + \sum_{i=1}^{m} b_{ik}(\vec{x}) f_i(\vec{x}) = c_k(\vec{x}) \quad \text{for} \quad 1 \le k \le \ell .$$

1. *If $\{a_{ijk}, b_{ik}, c_k\}$ are smooth, then $\vec{f}$ is smooth.*
2. *If $\{a_{ijk}, b_{ik}, c_k\}$ are real analytic, then $\vec{f}$ is real analytic.*

So far, we have worked with global objects. We let $\mathfrak{K}_P(\mathcal{M})$ be the Lie algebra of *germs* of affine Killing vector fields based at P. The spaces $E_P(\mu, \nabla)$, $\mathcal{Y}_P(\mathcal{M})$, and $\mathfrak{A}_P(\mathcal{M})$ are defined similarly. Let $X \in \mathfrak{K}(\mathcal{M})$. If $F \in \mathfrak{K}(\mathcal{M})$, let $X(F) = [X, F]$. If $f \in E(\mu, \mathcal{M})$, we may apply X to define Xf. We use Lemma 12.13 to establish the following result which we have already used in the proof of Lemma 12.8.

Lemma 12.14 *Let $\mathcal{M}$ be a connected real analytic affine manifold.*

1. *Let $\mathfrak{V} \in \{\mathfrak{K}(\mathcal{M}),\ \mathfrak{A}(\mathcal{M}),\ E(\mu, \mathcal{M})\}$.*
   (a) *The elements of $\mathfrak{V}$ are real analytic.*
   (b) *If $\mathcal{M}$ is simply connected and if $\dim(\mathfrak{V}_P)$ is constant on M, then every element of $\mathfrak{V}_P$ extends uniquely to $\mathfrak{V}$.*
   (c) *If $X \in \mathfrak{V}$ satisfies $X(P) = 0$ and $\nabla X(P) = 0$, then X vanishes identically.*
2. $\dim(\mathfrak{K}(\mathcal{M})) \le m(m+1)$, $\dim(\mathfrak{A}(\mathcal{M})) \le m+1$, *and* $\dim(E(\preceq, \mathcal{M})) \le m+1$.
3. *Let $\mathfrak{V} \in \{\mathfrak{K}(\mathcal{M}),\ E(\mu, \mathcal{M})\}$. If $X \in \mathfrak{K}(\mathcal{M})$ and if $\xi \in \mathfrak{V}$, then $X(\xi) \in \mathfrak{V}$.*

Proof. Let $X = a^i \partial_{x^i} \in \mathfrak{K}(\mathcal{M})$. Fix k, set $i = j$ and sum over i in Equation (12.1.d) to see

$$0 = \sum_{i=1}^{m} \left\{ \frac{\partial^2 a^k}{\partial x^i \partial x^i}(\vec{x}) + \sum_{\ell} \left\{ a^\ell(\vec{x}) \frac{\partial \Gamma_{ii}{}^k}{\partial x^\ell}(\vec{x}) - \Gamma_{ii}{}^\ell(\vec{x}) \frac{\partial a^k}{\partial x^\ell}(\vec{x}) + \Gamma_{i\ell}{}^k(\vec{x}) \frac{\partial a^\ell}{\partial x^i}(\vec{x}) + \Gamma_{\ell i}{}^k(\vec{x}) \frac{\partial a^\ell}{\partial x^i}(\vec{x}) \right\} \right\}$$

for all k. We may now use Lemma 12.13 to see X is real analytic. If $f \in E(\mu, \mathcal{M})$, set $i = j$ and sum over i in Equation (12.1.i):

$$0 = \sum_{i=1}^{m} \left\{ \frac{\partial^2 f}{\partial x^i \partial x^i}(\vec{x}) - \mu f(\vec{x}) (\rho_s)_{ii}(\vec{x}) - \sum_{k=1}^{m} \Gamma_{ii}{}^k(\vec{x}) \frac{\partial f}{\partial x^k}(\vec{x}) \right\} .$$

Again, we may use Lemma 12.13 to see f is real analytic. If $f \in \mathfrak{A}(\mathcal{M})$, we set $i = j$ and sum over i in Equation (12.1.j) to see

$$\sum_{i=1}^{m} \left\{ \frac{\partial^2 f}{\partial x^i \partial x^i}(\vec{x}) - \sum_{k=1}^{m} \Gamma_{ii}{}^{k}(\vec{x}) \frac{\partial f}{\partial x^k}(\vec{x}) \right\} = -\sum_{i=1}^{m} (\rho_s)_{ii}(\vec{x})$$

and conclude f is real analytic. This proves Assertion 1-a. Assertion 1-b follows by analytic continuation. The defining equations permit us to solve recursively for the second and higher derivatives in terms of the value at P and the values of the derivatives at P. Assertion 1-c now follows as we are in the real analytic category. Assertion 2 follows from Assertion 1-c. Assertion 3 is immediate since the 1-parameter flow defined by an affine Killing vector field preserves the connection ∇ and since the partial differential equation (PDE) in question is homogeneous. □

Remark 12.15 In fact, all the assertions except 1-a continue to hold in the smooth category; see, for example, Brozos-Vázquez et al. [12] for a discussion of $E(\mu, \mathcal{M})$ or Kobayashi and Nomizu [39, Chapter VI] for a discussion of $\mathfrak{K}(\mathcal{M})$.

12.2 SURFACES WITH RECURRENT RICCI TENSOR

In Section 12.2.1 we define recurrence and in Section 12.2.2 we give Wong's classification of recurrent surfaces. In Section 12.2.3 we define the α invariant; this is a local invariant of recurrent surfaces whose Ricci tensor has rank 1 that will play a central role in our discussion of Type $\mathcal{A}$ models subsequently. In Section 12.2.4, we give various examples. Invariant and parallel tensors are defined in Section 12.2.5.

12.2.1 RECURRENCE. The following result is due to Wong [62].

Theorem 12.16 *Let $\mathcal{M}$ be an affine surface. We say ρ is recurrent if there exists a 1-form ω so $\nabla\rho = \omega \otimes \rho$. Let P be a point of $\mathcal{M}$ such that $\rho(P) \neq 0$. Then any of the two following conditions implies the third:*

1. *There exists a parallel non-trivial vector field near P.*
2. *There exists a parallel non-trivial 1-form near P.*
3. *ρ is recurrent near P.*

In this setting, the recurrence 1-form ω is closed if and only if $\rho_{a,\nabla} = 0$.

Remark 12.17 By Equation (12.1.c), we have $\rho_{11} = R_{211}{}^{2}$, $\rho_{12} = R_{212}{}^{2}$, $\rho_{21} = R_{121}{}^{1}$, and $\rho_{22} = R_{122}{}^{1}$. Let $\rho_{ij;k} = (\nabla_{\partial_{x^k}} \rho)_{ij}$ and $R_{ijk}{}^{\ell}{}_{;n} = (\nabla_{\partial_{x^n}} R)_{ijk}{}^{\ell}$ be the components of $\nabla\rho$ and ∇R. Lemma 3.5 of Book I shows that given $P \in M$, there exist local coordinates defined near

P so that $\Gamma_{ij}{}^k(P) = 0$. In those coordinates, covariant differentiation and partial differentiation coincide so we may covariantly differentiate these relations to see

$$\rho_{11;i} = R_{211}{}^2{}_{;i}, \ \rho_{12;i} = R_{212}{}^2{}_{;i}, \ \rho_{21;i} = R_{121}{}^1{}_{;i}, \ \rho_{22;i} = R_{122}{}^1{}_{;i} .$$

This shows that ρ is recurrent if and only if R is recurrent.

12.2.2 CLASSIFICATION.

A particularly simple example occurs when $\rho_s = 0$ so ρ is a 2-form.

Definition 12.18 Let $\theta = \theta(x^1, x^2)$. Let $\mathcal{R}_0(\theta)$ be the germ of an affine surface where the (possibly) non-zero Christoffel symbols are given by:

$$\mathcal{R}_0(\theta): \quad \Gamma_{11}{}^1 = -\theta^{(1,0)} \quad \text{and} \quad \Gamma_{22}{}^2 = \theta^{(0,1)} .$$

Let $A_0 := \begin{pmatrix} 0 & -1 \\ 1 & 0 \end{pmatrix}$. We compute that

$$\rho = \theta^{(1,1)} A_0, \quad \rho_{;1} = \{\theta^{(1,0)}\theta^{(1,1)} + \theta^{(2,1)}\} A_0, \quad \rho_{;2} = \{\theta^{(1,2)} - \theta^{(0,1)}\theta^{(1,1)}\} A_0 .$$

Thus $\rho_s = 0$. Furthermore, if $\theta^{(1,1)} \neq 0$, then the surface is recurrent and the recurrence tensor is given by $\omega = \frac{1}{\theta^{(1,1)}}\{\theta^{(1,0)}\theta^{(1,1)} + \theta^{(2,1)}\}dx^1 + \frac{1}{\theta^{(1,1)}}\{\theta^{(1,2)} - \theta^{(0,1)}\theta^{(1,1)}\}dx^2$.

We have the following result of Derdzinski [25] (see Theorem 6.1) which simplified and extended a previous result of Wong [62] (see Theorem 4.2).

Theorem 12.19 *If P is a point of an affine surface $\mathcal{M}$, then $\rho_s = 0$ near P if and only if $\mathcal{M}$ is locally affine isomorphic near P to the structure $\mathcal{R}_0(\theta)$ for some θ.*

There is a similar classification result if $\rho_s \neq 0$. We first establish some notation. Let $\theta = \theta(x^1, x^2)$. Let $\Delta\theta := \theta^{(2,0)} + \theta^{(0,2)}$. Let $\mathcal{R}_i(\theta)$ be the germ of an affine surface where the (possibly) non-zero Christoffel symbols are given by:

$\mathcal{R}_1(\theta)$: $\Gamma_{11}{}^1 = \Gamma_{12}{}^2 = \theta^{(0,1)}$ and $\Gamma_{11}{}^2 = \theta^{(1,0)} - \theta^{(0,1)}$ for $\theta^{(0,2)} \neq 0$.

$\mathcal{R}_2(\theta)$: $\Gamma_{11}{}^2 = \theta$ for $\theta^{(0,1)} \neq 0$.

$\mathcal{R}_3(\theta)$: $\Gamma_{22}{}^2 = \theta$ for $\theta^{(1,0)} \neq 0$.

$\mathcal{R}_4(\theta)$: $\Gamma_{11}{}^1 = (1+c)\theta^{(1,0)}$, $\Gamma_{22}{}^2 = (1-c)\theta^{(0,1)}$ for $c \notin \{0, \pm 1\}$ and $\theta^{(1,1)} \neq 0$.

$\mathcal{R}_5(\theta)$: $\Gamma_{11}{}^1 = \theta^{(1,0)}$ and $\Gamma_{22}{}^2 = \theta^{(0,1)}$ for $\theta^{(1,1)} \neq 0$.

$\mathcal{R}_6(\theta)$: $\Gamma_{11}{}^1 = \Gamma_{12}{}^2 = -\Gamma_{22}{}^1 = \theta^{(1,0)} + c\theta^{(0,1)}$
and $\Gamma_{22}{}^2 = \Gamma_{12}{}^1 = -\Gamma_{11}{}^2 = -c\theta^{(1,0)} + \theta^{(0,1)}$ for $c \neq 0$ and $\Delta\theta \neq 0$.

$\mathcal{R}_7(\theta)$: $\Gamma_{11}{}^1 = \Gamma_{12}{}^2 = -\Gamma_{22}{}^1 = \theta^{(1,0)}$ and $\Gamma_{22}{}^2 = \Gamma_{12}{}^1 = -\Gamma_{11}{}^2 = \theta^{(0,1)}$ for $\Delta\theta \neq 0$.

Lemma 12.20 *The structures $\mathcal{R}_i(\cdot)$ given above are recurrent.*

Proof. Set

$$
\begin{array}{llllll}
A_1 := \begin{pmatrix} 1 & -1 \\ 1 & 0 \end{pmatrix}, & A_2 := \begin{pmatrix} 1 & 0 \\ 0 & 0 \end{pmatrix}, & A_3 := \begin{pmatrix} 0 & 1 \\ 0 & 0 \end{pmatrix}, \\
A_4(c) := \begin{pmatrix} 0 & 1-c \\ 1+c & 0 \end{pmatrix}, & A_5 := \begin{pmatrix} 0 & 1 \\ 1 & 0 \end{pmatrix}, & A_6(c) := \begin{pmatrix} 1 & -c \\ c & 1 \end{pmatrix}, \\
A_7 := \begin{pmatrix} 1 & 0 \\ 0 & 1 \end{pmatrix}, & &
\end{array}
$$

$$
\begin{aligned}
\xi_{1,1} &:= 2\theta^{(0,1)}\theta^{(0,2)} - \theta^{(1,2)}, \\
\xi_{3,2} &:= \theta\theta^{(1,0)} - \theta^{(1,1)}, \\
\xi_{4,1} &:= (c+1)\theta^{(1,0)}\theta^{(1,1)} - \theta^{(2,1)}, \\
\xi_{4,2} &:= (1-c)\theta^{(0,1)}\theta^{(1,1)} - \theta^{(1,2)}, \\
\xi_{5,1} &:= \theta^{(1,0)}\theta^{(1,1)} - \theta^{(2,1)}, \\
\xi_{5,2} &:= \theta^{(0,1)}\theta^{(1,1)} - \theta^{(1,2)}, \\
\xi_{6,1} &:= 2c\theta^{(0,1)}\Delta\theta + 2\theta^{(0,2)}\theta^{(1,0)} + 2\theta^{(2,0)}\theta^{(1,0)} - \theta^{(1,2)} - \theta^{(3,0)}, \\
\xi_{6,2} &:= -2c\theta^{(0,2)}\theta^{(1,0)} - 2c\theta^{(1,0)}\theta^{(2,0)} - \theta^{(0,3)} + 2\theta^{(0,1)}\Delta\theta - \theta^{(2,1)}, \\
\xi_{7,1} &:= -\theta^{(1,2)} + 2\theta^{(1,0)}\Delta\theta - \theta^{(3,0)}, \\
\xi_{7,2} &:= -\theta^{(0,3)} + 2\theta^{(0,1)}\Delta\theta - \theta^{(2,1)}.
\end{aligned}
$$

Let $\rho_{;1} := \nabla_{\partial_{x^1}}\rho$ and $\rho_{;2} := \nabla_{\partial_{x^2}}\rho$. We compute:

$$
\begin{array}{llll}
\mathcal{R}_1(\theta): & \rho = -\theta^{(0,2)} A_1, & \rho_{;1} = \xi_{1,1} A_1, & \rho_{;2} = -\theta^{(0,3)} A_1, \\
\mathcal{R}_2(\theta): & \rho = \theta^{(0,1)} A_2, & \rho_{;1} = \theta^{(1,1)} A_2, & \rho_{;2} = \theta^{(0,2)} A_2, \\
\mathcal{R}_3(\theta): & \rho = -\theta^{(1,0)} A_3, & \rho_{;1} = -\theta^{(2,0)} A_3, & \rho_{;2} = \xi_{3,2} A_3, \\
\mathcal{R}_4(\theta): & \rho = -\theta^{(1,1)} A_4, & \rho_{;1} = \xi_{4,1} A_4, & \rho_{;2} = \xi_{4,2} A_4, \\
\mathcal{R}_5(\theta): & \rho = -\theta^{(1,1)} A_5, & \rho_{;1} = \xi_{5,1} A_5, & \rho_{;2} = \xi_{5,2} A_5, \\
\mathcal{R}_6(\theta): & \rho = -\Delta\theta\, A_6, & \rho_{;1} = \xi_{6,1} A_6, & \rho_{;2} = \xi_{6,2} A_6, \\
\mathcal{R}_7(\theta): & \rho = -\Delta\theta\, A_7, & \rho_{;1} = \xi_{7,1} A_7, & \rho_{;2} = \xi_{7,2} A_7.
\end{array}
$$

The result now follows. □

The following result is due to Wong [62] (see Theorem 4.1).

Theorem 12.21 *Let P be a point of an affine surface $\mathcal{M}$ with recurrent Ricci tensor and with $\rho_s(P) \neq 0$. Then $\mathcal{M}$ is locally affine isomorphic near P to one of the structures $\mathcal{R}_i(\cdot)$ given above.*

12.2.3 LOCAL INVARIANTS. In general, it can be difficult to define local invariants in the affine setting since we do not have a metric tensor to raise and lower indices. The following local invariants are not of Weyl type; they do not arise from contracting indices. Other local invariants will be constructed subsequently in Section 15.2 in our discussion of VSI manifolds. Let ρ_s be the symmetric Ricci tensor as defined in Equation (12.1.b).

Lemma 12.22 *Let P be a point of an affine surface $\mathcal{M}$ with recurrent Ricci tensor. Assume that* $\operatorname{Rank}(\rho_s) = 1$ *and that $\nabla \rho_s$ is totally symmetric. Choose a vector field X so that $\rho_s(X,X) \neq 0$. Let $\alpha_X := \rho_s(X,X)^{-3}\nabla\rho_s(X,X;X)^2$. Then α_X is independent of the particular X chosen and setting $\alpha := \alpha_X$ defines a smooth local invariant of $\mathcal{M}$.*

Proof. Let $\ker\{\rho_s\} := \{\xi : \rho_s(\xi,\eta) = 0 \text{ for all } \eta\}$. By assumption, $\dim(\ker\{\rho_s\}) = 1$. We work locally and find a smooth section ξ to this 1-dimensional distribution. Choosing local coordinates so $\partial_{x^2} = \xi$, we have

$$\begin{array}{lll} (\rho_s)_{11} = \theta_1, & (\rho_s)_{12} = 0, & (\rho_s)_{22} = 0, \\ (\rho_s)_{11;1} = \omega_1(\rho_s)_{11}, & (\rho_s)_{12;1} = \omega_1(\rho_s)_{12} = 0, & (\rho_s)_{22;1} = \omega_1(\rho_s)_{22} = 0, \\ (\rho_s)_{11;2} = \omega_2(\rho_s)_{11}, & (\rho_s)_{12;2} = \omega_2(\rho_s)_{12} = 0, & (\rho_s)_{22;2} = \omega_2(\rho_s)_{22} = 0\,. \end{array}$$

Since $\nabla\rho_s$ is totally symmetric, $(\rho_s)_{11;2} = (\rho_s)_{12;1} = 0$. This shows that $\rho_s = \theta_1 dx^1 \otimes dx^1$ and that $\nabla\rho_s = \omega_1\theta_1 dx^1 \otimes dx^1 \otimes dx^1$. Express $X = a_1\partial_{x^1} + a_2\partial_{x^2}$. Then

$$\alpha_X = \frac{\nabla\rho_s(X,X;X)^2}{\rho_s(X,X)^3} = \frac{a_1^6\omega_1^2\theta_1^2}{a_1^6\theta_1^3} = \frac{\omega_1^2}{\theta_1}$$

is independent of X. The desired result now follows. □

12.2.4 EXAMPLES. The proof of Lemma 12.20 exhibits the Ricci tensors and the covariant derivatives of the structures $\mathcal{R}_i(\theta)$; only $\mathcal{R}_1(\theta)$ and $\mathcal{R}_2(\theta)$ give rise Ricci tensors where $\operatorname{Rank}(\rho_s) = 1$.

Example 12.23 Suppose $\mathcal{M} = \mathcal{R}_1(\theta)$. Then $\rho = -\theta^{(0,2)}A_1$ so ∂_{x^2} spans $\ker\{\rho_s\}$. We have $\rho_{;2} = -\theta^{(0,3)}A_1$ so $\nabla\rho_s$ is totally symmetric if and only if $\theta^{(0,3)} = 0$. This means that we can express θ in the form $\theta = a_0(x^1) + a_1(x^1)x^2 + a_2(x^1)(x^2)^2$. We impose this condition and assume a_2 does not vanish identically. We compute

$$\alpha = -\frac{\left(4a_2(x^1)(a_1(x^1) + 2x^2a_2(x^1)) - 2a_2'(x^1)\right)^2}{8a_2(x^1)^3}\,.$$

Since α exhibits non-trivial x^2 dependence and since α is an affine invariant, this example is not homogeneous. If a_1 and a_2 are constant, and $a_0''(x^1) = 0$, then ∂_{x^1} is an affine Killing vector field so there are *cohomogeneity 1* examples in this family.

Example 12.24 Suppose $\mathcal{M} = \mathcal{R}_2(\theta)$ so ∂_{x^2} spans $\ker\{\rho_s\}$. We have $\rho = \theta^{(0,1)}A_2$. We have $\rho_{;1} = \theta^{(1,1)}A_2$, and $\rho_{;2} = \theta^{(0,2)}A_2$. Thus $\nabla\rho_s$ is totally symmetric if and only if $\theta^{(0,2)} = 0$. This means that we can express θ in the form $\theta = a_0(x^1) + a_1(x^1)x^2$. We compute

$$\alpha = \frac{a_1'(x^1)^2}{a_1(x^1)^3}\,.$$

This is constant if and only if $a_1(x^1) = c_0(x^1 - c_1)^{-2}$.

12.2.5 INVARIANT AND PARALLEL TENSORS OF TYPE (1,1).

Let $\mathcal{M}$ be a locally affine homogeneous manifold and let T be a smooth endomorphism of TM, i.e., a tensor of Type (1,1). We expand

$$T = T^i{}_j\partial_{x^i} \otimes dx^j \in C^\infty(TM \otimes T^*M) \quad \text{where} \quad T\partial_{x^j} = \sum_i T^i{}_j\partial_{x^i}\,.$$

We say that T is an *invariant tensor of Type (1,1)* if $\mathcal{L}_X T = 0$ for all $X \in \mathfrak{K}(\mathcal{M})$. If T is invariant, then $\operatorname{Tr}\{T\}$ is constant. Since Id is invariant, we may assume T is trace-free. We will investigate this subsequently in Section 13.2.8 and in Section 14.2.6 in the context of affine homogeneous surfaces. Let $X = a^k\partial_{x^k}$. We have $\mathcal{L}_X(\partial_{x^i}) = [X, \partial_{x^i}] = -(\partial_{x^i}a^k)\partial_{x^k}$ and dually we obtain $\mathcal{L}_X(dx^j) = \partial_{x^k}a^j dx^k$. Thus

$$\begin{aligned}\mathcal{L}_X T &= (a^k\partial_{x^k}T^i{}_j)\partial_{x^i} \otimes x^j - T^i{}_j(\partial_{x^i}a^k)\partial_{x^k} \otimes dx^j + T^i{}_j(\partial_{x^k}a^j)\partial_{x^i} \otimes dx^k \\ &= \{a^k\partial_{x^k}T^i{}_j - T^k{}_j(\partial_{x^k}a^i) + T^i{}_k(\partial_{x^j}a^k)\}\partial_{x^i} \otimes dx^j\,.\end{aligned}$$

Thus $\mathcal{L}_X T = 0$ if and only if we have the equations

$$a^k\partial_{x^k}T^i{}_j - T^k{}_j(\partial_{x^k}a^i) + T^i{}_k(\partial_{x^j}a^k) \quad \text{for all} \quad i, j\,. \tag{12.2.a}$$

This is studied further in Section 13.2.8 and Section 14.2.6.

We refer to Calviño-Louzao et al. [13] for the proof of the remaining results in this section. We say that a tensor field T of Type $(1,1)$ on $\mathcal{M}$ is *parallel* if $\nabla T = 0$. Let $\mathcal{P}(\mathcal{M})$ be the set of parallel tensors of Type $(1,1)$ on $\mathcal{M}$:

$$\mathcal{P}(\mathcal{M}) = \{T^i{}_j : \partial_{x^k}T^i{}_j + \Gamma_{k\ell}{}^i T^\ell{}_j - \Gamma_{kj}{}^\ell T^i{}_\ell = 0,\ \forall\, i, j, k\}\,.$$

The following is a basic observation.

Lemma 12.25 *If $\mathcal{M} = (M, \nabla)$ is a connected affine surface, then $\mathcal{P}(\mathcal{M})$ is a unital algebra with* $\dim(\mathcal{P}(\mathcal{M})) \le 4$. *Let $T \in \mathcal{P}(\mathcal{M})$. The eigenvalues of T are constant on M. If T vanishes at any point of M, then T vanishes identically.*

Let $\mathcal{P}^0(\mathcal{M}) := \{T \in \mathcal{P}(\mathcal{M}) : \operatorname{Tr}\{T\} = 0\}$ be the space of trace-free parallel tensors of Type (1,1), where $\operatorname{Tr}\{T\} := T^i{}_i$ is the trace of T. If $T \in \mathcal{P}(\mathcal{M})$, $\operatorname{Tr}\{T\}$ is constant and expressing $T = \frac{1}{2}\operatorname{Tr}\{T\}\operatorname{Id} + (T - \frac{1}{2}\operatorname{Tr}\{T\}\operatorname{Id})$ decomposes

$$\mathcal{P}(\mathcal{M}) = \operatorname{Id}\cdot\mathbb{R} \oplus \mathcal{P}^0(\mathcal{M})\,.$$

If $0 \neq T \in \mathcal{P}^0(\mathcal{M})$, then the eigenvalues of T are $\{\pm\lambda\}$ so $\operatorname{Tr}\{T^2\} = 2\lambda^2$. If $2\lambda^2 < 0$ (resp. $2\lambda^2 > 0$), we can rescale T so $T^2 = -\operatorname{Id}$ (resp. $T^2 = \operatorname{Id}$) and T defines a *Kähler* (resp. *para-Kähler*) structure on M; the almost complex (resp. almost para-complex) structure being integrable as M is a surface (see Cortés et al. [19] and Newlander and Nirenberg [46]). Finally, if $\lambda = 0$, then T is nilpotent and defines what we will call a *nilpotent Kähler structure*. The symmetric Ricci tensor plays a crucial role.

Theorem 12.26 *Let $\mathcal{M} = (M, \nabla)$ be a simply connected affine surface.*

1. *If* $\dim(\mathcal{P}^0(\mathcal{M})) = 1$, *then exactly one of the following possibilities holds:*
   (a) *$\mathcal{M}$ admits a Kähler structure and* $\operatorname{Rank}(\rho_s) = 2$.
   (b) *$\mathcal{M}$ admits a para-Kähler structure and* $\operatorname{Rank}(\rho_s) = 2$.
   (c) *$\mathcal{M}$ admits a nilpotent Kähler structure and* $\operatorname{Rank}(\rho_s) = 1$.
2. $\dim(\mathcal{P}^0(\mathcal{M})) \neq 2$.
3. $\dim(\mathcal{P}^0(\mathcal{M})) = 3$ *if and only if* $\rho_s = 0$. *This implies $\mathcal{M}$ admits Kähler, para-Kähler, and nilpotent Kähler structures.*

Generically, of course, $\dim(\mathcal{P}^0(\mathcal{M})) = 0$. Furthermore, there exist examples with $\operatorname{Rank}(\rho_s) = 1$ or $\operatorname{Rank}(\rho_s) = 2$ where $\dim(\mathcal{P}^0(\mathcal{M})) = 0$. What is somewhat surprising is that the existence of parallel $(1,1)$ tensor fields is completely characterized by the geometry of the symmetric part of the Ricci tensor ρ_s.

Theorem 12.27 *Let $\mathcal{M} = (M, \nabla)$ be a simply connected affine surface with $\rho_s \neq 0$.*

1. *$\mathcal{M}$ admits a Kähler structure if and only if* $\det(\rho_s) > 0$ *and ρ_s is recurrent.*
2. *$\mathcal{M}$ admits a para-Kähler structure if and only if* $\det(\rho_s) < 0$ *and ρ_s is recurrent.*
3. *$\mathcal{M}$ admits a nilpotent Kähler structure if and only if ρ_s is of rank 1 and recurrent.*

12.3 THE AFFINE QUASI-EINSTEIN EQUATION

Lemma 12.14 provides foundational results concerning $E_P(\mu, \nabla)$ that we will use repeatedly. We first use polarization to establish some results relating to the affine quasi-Einstein equation under strong projective equivalence. In what follows, it will be convenient to work with just one component of the affine quasi-Einstein operator. Suppose that Φ is a symmetric $(0,2)$-tensor defined on some vector space V and suppose that one could show that $\Phi_{11} = 0$ relative to any basis. It then follows that $\Phi = 0$; this process is called *polarization*.

Section 12.3.1 presents foundational results concerning strong projective equivalence. Section 12.3.2 shows that the solution space $\mathcal{Q}$ of the affine quasi-Einstein equation corresponding to the distinguished eigenvalue $\mu = -\frac{1}{m-1}$ transforms conformally under strong projective equivalence. Section 12.3.3 discusses the case when $\mathcal{Q}$ has maximal dimension $m+1$. In Section 12.3.4, the solution space $\mathcal{Q}$ is used to give an algorithm for constructing affine geodesics for strongly projectively flat manifolds. The role of the alternating Ricci tensor in the study of $\mathcal{Q}$ is treated in Section 12.3.5. Yamabe solitons are examined in Section 12.3.6.

12.3.1 STRONG PROJECTIVE EQUIVALENCE.

Let $\omega = dg$ provide a strong projective equivalence from ∇ to ${}^g\nabla$, i.e., ${}^g\nabla_X Y = \nabla_X Y + X(g)Y + Y(g)X$.

Lemma 12.28 *Adopt the notation established above.*

1. $\rho_{s,{}^g\nabla} = \rho_{s,\nabla} - (m-1)\{\mathcal{H}_\nabla g - dg \otimes dg\}$.
2. *If* $\mu = -\frac{1}{m-1}$ *or if* $\mathcal{H}_\nabla g - dg \otimes dg = 0$, *then* $e^g \mathfrak{Q}_{\mu,\nabla} e^{-g} = \mathfrak{Q}_{\mu,{}^g\nabla}$.

Proof. By definition, ${}^g\Gamma_{ij}{}^k = \Gamma_{ij}{}^k + \delta_i^k \partial_{x^j} g + \delta_j^k \partial_{x^i} g$. Fix a point P of M. Since we are working in the category of torsion-free connections, we can choose a coordinate system so $\Gamma(P) = 0$. We compute at the point P and set $\Gamma_{ij}{}^k(P) = 0$. By Equation (12.1.a),

$$\begin{aligned}
(\rho_{{}^g\nabla})_{11}(P) &= \{\partial_{x^i} {}^g\Gamma_{11}{}^i - \partial_{x^1} {}^g\Gamma_{i1}{}^i + {}^g\Gamma_{in}{}^i\, {}^g\Gamma_{11}{}^n - {}^g\Gamma_{1n}{}^i\, {}^g\Gamma_{i1}{}^n\}(P) \\
&= \{\partial_{x^i}\Gamma_{11}{}^i - \partial_{x^1}\Gamma_{i1}{}^i + (1-m)\partial_{x^1x^1} g \\
&\qquad + 2(m+1)(\partial_{x^1} g)^2 - (m+3)(\partial_{x^1} g)^2\}(P) \\
&= \{(\rho_\nabla)_{11} - (m-1)(\partial_{x^1x^1} g - (\partial_{x^1} g)^2)\}(P) \\
&= \{\rho_\nabla - (m-1)(\mathcal{H}_\nabla g - dg \otimes dg)\}_{11}(P)\,.
\end{aligned}$$

We use polarization to establish Assertion 1. To prove Assertion 2, we examine $\mathfrak{Q}_{\mu,\nabla,11}$ and $\{e^{-g}\,\mathfrak{Q}_{\mu,{}^g\nabla}\, e^g\}_{11}$ at P. We compute:

$$\begin{aligned}
&\{e^{-g}\mathcal{H}_{{}^g\nabla,11} e^g f\}(P) = \{e^{-g}\partial_{x^1x^1}(fe^g) - {}^g\Gamma_{11}{}^k\, e^{-g}\partial_{x^k}(fe^g)\}(P) \\
&\quad = \{\partial_{x^1x^1} f + 2\partial_{x^1} f\, \partial_{x^1} g + f\partial_{x^1x^1} g + f(\partial_{x^1} g)^2 - 2\partial_{x^1} g(\partial_{x^1} f + f\partial_{x^1} g)\}(P) \\
&\quad = \{\mathcal{H}_{\nabla,11} f + f(\partial_{x^1x^1} g - (\partial_{x^1} g)^2)\}(P)\,.
\end{aligned}$$

We use Assertion 1 to obtain:

$$\begin{aligned}
&\{e^{-g}\mathfrak{Q}_{\mu,{}^g\nabla}\, e^g f - \mathfrak{Q}_{\mu,\nabla} f\}_{11}(P) \\
&\quad = \{e^{-g}(\mathcal{H}_{{}^g\nabla} e^g f - \mu\rho_{s,{}^g\nabla} e^g f)_{11} - (\mathcal{H}_\nabla f - \mu\rho_{s,\nabla} f)_{11}\}(P) \\
&\quad = \{f(1 + (m-1)\mu)(\partial_{x^1x^1} g - (\partial_{x^1} g)^2)\}(P)\,.
\end{aligned}$$

We use polarization to establish the following identity from which Assertion 2 follows:

$$e^{-g}\mathfrak{Q}_{\mu,{}^g\nabla} e^g f - \mathfrak{Q}_{\mu,\nabla} f = (1 + (m-1)\mu) f(\mathcal{H}_\nabla g - dg \otimes dg)\,. \qquad \square$$

12.3.2 CONFORMAL EQUIVALENCE. Recall that

$$\mathcal{Q}(\mathcal{M}) := \left\{ f \in C^\infty(M) : \mathcal{H} f + \tfrac{1}{m-1} f \rho_s = 0 \right\} .$$

Theorem 12.29 *If $g \in C^\infty(M)$, let ${}^g\mathcal{M} = (M, {}^g\nabla)$.*

1. *$\mathcal{Q}({}^g\mathcal{M}) = e^g \mathcal{Q}(\mathcal{M})$.*
2. *The following assertions are equivalent:*
   (a) $\rho_{s,{}^g\nabla} = \rho_{s,\nabla}$. *(b) $\mathcal{H} g - dg \otimes dg = 0$.* *(c) $e^{-g} \in E(0, \nabla)$.*
3. *If any of the assertions in (2) hold, then $E(\mu, {}^g\nabla) = e^g E(\mu, \nabla)$ for any μ.*

Proof. Assertion 1 is immediate from the intertwining relation of Lemma 12.28. The equivalence of Assertion 2-a and Assertion 2-b again follows from Lemma 12.28. The equivalence of Assertion 2-b and Assertion 2-c follows by noting

$$\mathfrak{Q}_{0,\nabla}(e^{-g}) = \mathcal{H}(e^{-g}) = -e^{-g}\{\mathcal{H}(g) - dg \otimes dg\} .$$

Assertion 3 now follows from Assertion 2-b and Lemma 12.28. □

Remark 12.30 Theorem 12.11 and Theorem 12.29 deal with strong projective equivalence. If $\mathcal{M}$ and $\tilde{\mathcal{M}}$ are strongly projectively equivalent, then

$$\rho_{a,\mathcal{M}} = \rho_{a,\tilde{\mathcal{M}}} \quad \text{and} \quad \dim(\mathcal{Q}(\mathcal{M})) = \dim(\mathcal{Q}(\tilde{\mathcal{M}})) .$$

However, if $\mathcal{M}$ and $\tilde{\mathcal{M}}$ are only projectively equivalent, then these two identities can fail. Let $\mathcal{M}$ be the usual flat structure on $\mathbb{R}^2$. Let $\omega = x^2 dx^1$ define a projective equivalence from $\mathcal{M}$ to $\tilde{\mathcal{M}}$. Let $P = 0$. A direct computation shows that

$$\dim(\mathcal{Q}_P(\mathcal{M})) = 3, \quad \dim(\mathcal{Q}_P(\tilde{\mathcal{M}})) = 0,$$
$$\rho_{\mathcal{M}} = \begin{pmatrix} 0 & 0 \\ 0 & 0 \end{pmatrix}, \quad \rho_{\tilde{\mathcal{M}}} = \begin{pmatrix} (x^2)^2 & 1 \\ -2 & 0 \end{pmatrix}.$$

We can draw the following conclusion from Theorem 12.29.

Lemma 12.31 *Let $\mathcal{M}$ be an affine manifold of dimension m.*

1. *If $\mu \neq 0$, then $\rho_{s,\mathcal{M}} = 0$ if and only if $\mathbb{1} \in E(\mu, \mathcal{M})$.*
2. *$\mathcal{M}$ is strongly projectively equivalent to an affine surface $\mathcal{M}_1$ with $\rho_{s,\mathcal{M}_1} = 0$ if and only if there exists $f \in \mathcal{Q}_P(\mathcal{M})$ with $f(P) \neq 0$.*

Proof. We have $f \in E_P(\mu, \mathcal{M})$ if and only if $\mathcal{H} f = \mu f \rho_s$. Since $\mathcal{H}\mathbb{1} = 0$, $\mathbb{1} \in E_P(\mu, \mathcal{M})$ if and only if $\mu\rho_s = 0$. Assertion 1 follows since $\mu \neq 0$. Suppose dg provides a local strong projective equivalence from $\mathcal{M}$ to $\mathcal{M}_1$ with $\rho_{s,\mathcal{M}_1} = 0$. By Theorem 12.29, $\mathcal{Q}(\mathcal{M}) = e^{-g}\mathcal{Q}(\mathcal{M}_1)$. By Assertion 1, $\mathbb{1} \in \mathcal{Q}(\mathcal{M}_1)$. Thus $f := e^{-g} \in \mathcal{Q}(\mathcal{M})$ and we obtain as desired that $f(P) \neq 0$.

Suppose there exists $f \in \mathcal{Q}_P(\mathcal{M})$ so $f(P) \neq 0$. By replacing f by $-f$ if necessary, we may assume $f(P) > 0$ and set $g = \log(f)$. Let $\mathcal{M}_1 = e^{-g}\mathcal{M}$; $\mathbb{1} = e^{-g}e^{g} \in \mathcal{Q}(\mathcal{M}_1)$. Assertion 1 then yields $\rho_{s,\mathcal{M}_1} = 0$. This completes the proof of Assertion 2. □

12.3.3 DIM($\mathcal{Q}(\mathcal{M})$) = DIM(M) + 1. The solution space $\mathcal{Q}$ of the affine quasi-Einstein equation corresponding to the eigenvalue $\mu = -\frac{1}{m+1}$ provides a complete system of invariants if the manifold in question is strongly projectively flat. We establish the results in this section that will play a central role in our discussion of Type $\mathcal{A}$ geometries in Chapter 13. We work locally for the most part. We begin with a useful technical result.

Lemma 12.32 *Let $\mathcal{M}$ be an affine manifold of dimension m with $\dim(E_P(\mu, \nabla)) = m + 1$ for some μ.*

1. *There is a basis $\mathcal{B} = \{\phi_0, \phi_1, \dots, \phi_m\}$ for $E_P(\mu, \nabla)$ so*

$$\phi_0(P) = 1,\ d\phi_0(P) = 0,\ \textit{and for } 1 \le i \le m,\ \phi_i(P) = 0,\ d\phi_i(P) = dx^i\,. \qquad (12.3.a)$$

   Let $g = \log(\phi_0)$ and let $x^i := e^{-g}\phi_i$ for $i \ge 1$. Then $(x^1, \dots, x^m)$ is a system of local coordinates on M centered at P such that $E_P(\mu, \nabla) = e^{g}\operatorname{span}\{x^1, \dots, x^m\}$.

2. *$\mathcal{M}$ is locally strongly projectively flat. If $\mu \neq -\frac{1}{m+1}$, then $\rho_s = 0$.*

Proof. Suppose $\dim(E_P(\mu, \nabla)) = m + 1$. Let $\Theta_P(\phi) := (\phi, \partial_{x^1}\phi, \dots, \partial_{x^m}\phi)(P) \in \mathbb{R}^{m+1}$. If $\Theta_P(\phi) = 0$, then $\phi(P) = 0$ and $\nabla\phi(P) = 0$ so $\phi \equiv 0$ by Lemma 12.14. Consequently, Θ_P is an injective map from $E_P(\mu, \nabla)$ to $\mathbb{R}^{m+1}$. Thus, for dimensional reasons, ϕ_P is bijective and we can choose a basis satisfying the normalizations of Equation (12.3.a). The remaining assertions of Assertion 1 now follow. We compute:

$$\begin{aligned} 0 &= e^{-g}\left\{\mathfrak{Q}_{\mu,\nabla}(x^k e^g) - x^k\mathfrak{Q}_{\mu,\nabla}(e^g)\right\}_{ij} = e^{-g}\{\mathcal{H}(x^k e^g) - x^k\mathcal{H}(e^g)\}_{ij} \\ &= \delta_j^k\,\partial_{x^i} g + \delta_i^k\,\partial_{x^j} g - \Gamma_{ij}{}^k\,. \end{aligned}$$

Let $\tilde{\Gamma}_{ij}{}^k \equiv 0$ define $\tilde{\nabla}$. We have $\Gamma_{ij}{}^k = \tilde{\Gamma}_{ij}{}^k + \delta_j^k\,\partial_{x^i} g + \delta_i^k\,\partial_{x^j} g$ so dg is a strong projective equivalence from $\tilde{\nabla}$ to ∇. Consequently, ∇ is strongly projectively flat. By Theorem 12.29, $\mathcal{Q}_P(\mathcal{M}) = e^{g}\operatorname{span}\{\mathbb{1}, x^1, \dots, x^m\}$ so

$$0 = \left\{\mathfrak{Q}_{\nu,\nabla} - \mathfrak{Q}_{-\frac{1}{m-1},\nabla}\right\} e^g = \left(\nu + \frac{1}{m-1}\right) e^g \rho_s\,.$$

Thus if $\nu \neq -\frac{1}{m-1}$, $\rho_s = 0$ and Assertion 2 follows. □

We use essentially the same argument to prove the following result; it will play an important role in our analysis of Type $\mathcal{A}$ models in Section 13.3.

Lemma 12.33 *If $\mathcal{M}$ is a strongly projectively flat surface, then*

$$\mathcal{Q}(\mathcal{M}) \neq e^{g(x^1,x^2)} \operatorname{span}\{f_1(x^1), f_2(x^1), f_3(x^1)\}.$$

Proof. Suppose to the contrary that $\mathcal{Q}(\mathcal{M}) = e^{g(x^1,x^2)} \operatorname{span}\{f_1(x^1), f_2(x^1), f_3(x^1)\}$. We argue for a contradiction. Let $\tilde{\nabla} := {}^{-g}\nabla$; by Theorem 12.29,

$$\mathcal{Q}(M, \tilde{\nabla}) = \operatorname{span}\{f_1(x^1), f_2(x^1), f_3(x^1)\}.$$

Define $f = c_1 f_1 + c_2 f_2 + c_3 f_3$. Since $f = f(x^1)$, we may choose $(c_1, c_2, c_3) \neq (0,0,0)$ so that $f(P) = 0$ and $df(P) = 0$; thus $f \equiv 0$ and the functions $\{f_1, f_2, f_3\}$ are not linearly independent which is false. □

The following result is due to Gilkey and Valle-Regueiro [35].

Theorem 12.34 *Let P be a point of an affine manifold $\mathcal{M}$ of dimension m.*

1. *$\mathcal{M}$ is locally strongly projectively flat if and only if $\dim(\mathcal{Q}_P(\mathcal{M})) = m + 1$.*
2. *If $\dim(\mathcal{Q}_P(\mathcal{M})) = m + 1$ and if $\mathbb{1} \in \mathcal{Q}_P(\mathcal{M})$, then $\mathcal{M}$ is flat near P.*
3. *Let $\mathcal{M}_i$ be affine manifolds of dimension m with $\dim(\mathcal{Q}(\mathcal{M}_i)) = m + 1$. Let Φ be a diffeomorphism from M_1 to M_2. If $\mathcal{Q}(\mathcal{M}_1) = \Phi^* \mathcal{Q}(\mathcal{M}_2)$, then $\mathcal{M}_1 = \Phi^* \mathcal{M}_2$.*

Proof. By Lemma 12.14, $\dim(\mathcal{Q}_P(\mathcal{M})) \leq m + 1$ for any affine manifold of dimension m. By Lemma 12.32, if equality holds, then $\mathcal{M}$ is locally strongly projectively flat. Suppose dg provides a local strong projective equivalence between ∇ and a flat connection ∇_f. We can choose local coordinates $(x^1, \ldots, x^m)$ near P so that the Christoffel symbols of ∇_f vanish. It is then immediate that $\{\mathbb{1}, x^1, \ldots, x^m\} \subset \mathcal{Q}_P(\mathcal{M}_f)$ so $\dim(\mathcal{Q}_P(\mathcal{M}_f)) = m + 1$. By Theorem 12.29, $\mathcal{Q}_P(\mathcal{M}) = e^g \mathcal{Q}_P(\mathcal{M}_f)$ so $\dim(\mathcal{Q}_P(\mathcal{M})) = m + 1$. This proves Assertion 1.

Suppose $\mathbb{1} \in \mathcal{Q}_P(\mathcal{M})$. Let $\mathcal{B}$ be the basis of Lemma 12.32. Since $\{\phi_0 - \mathbb{1}\}(P) = 1$ and $d\{\phi_0 - \mathbb{1}\}(P) = 0$, $\phi_0 \equiv \mathbb{1}$ by Lemma 12.14. Thus $g = 0$ and $\mathcal{Q}_P(\mathcal{M}) = \operatorname{span}\{\mathbb{1}, x^1, \ldots, x^m\}$ in the new system of coordinates. Since $\mathcal{H}\mathbb{1} = 0$ and $\mathbb{1} \in \mathcal{Q}_P(\mathcal{M})$, $\rho_s = 0$. As $x^i \in \mathcal{Q}_P(\mathcal{M})$ and $\rho_s = 0$, $0 = \mathcal{H}x^i = \{\partial_j \partial_k x^i - \Gamma_{jk}{}^{\ell} \partial_{x^\ell} x^i\} dx^j \otimes dx^k = -\Gamma_{jk}{}^i dx^j \otimes dx^k$. Consequently, $\Gamma = 0$ so $\mathcal{M}$ is locally flat which establishes Assertion 2.

By replacing $\mathcal{M}_2$ by $\Phi^* \mathcal{M}_2$, it suffices to prove Assertion 3 under the additional hypothesis that $M_1 = M_2$ and $\mathcal{Q}(\mathcal{M}_1) = \mathcal{Q}(\mathcal{M}_2)$. We adopt the notation of Lemma 12.32. Let $\mathcal{Q}(\mathcal{M}_1) = \mathcal{Q}(\mathcal{M}_2) = e^g \operatorname{span}\{\mathbb{1}, x^1, \ldots, x^m\}$. Let $\tilde{\nabla}_1 := {}^{-g}\nabla_1$ and $\tilde{\nabla}_2 := {}^{-g}\nabla_2$. By Theorem 12.29, $\mathcal{Q}(\tilde{\mathcal{M}}_1) = \mathcal{Q}(\tilde{\mathcal{M}}_2) = \operatorname{span}\{\mathbb{1}, x^1, \ldots, x^m\}$. The argument given to prove Assertion 2 then shows $\tilde{\mathcal{M}}_1 = \tilde{\mathcal{M}}_2$ is defined by the flat connection where all the Christoffel symbols vanish in the new coordinate system. Since $\mathcal{M}_1 = {}^g\tilde{\mathcal{M}}_2$ and $\mathcal{M}_2 = {}^g\tilde{\mathcal{M}}_2$, we obtain $\mathcal{M}_1 = \mathcal{M}_2$ and Assertion 3 follows. □

By Theorem 12.34, $\mathcal{M}$ is strongly projectively flat if and only if $\dim(\mathcal{Q}(\mathcal{M})) = m + 1$. In the category of strongly projective flat affine manifolds of dimension $m + 1$, Theorem 12.34 shows that the transformation $\mathcal{M} \rightsquigarrow \mathcal{Q}(\mathcal{M})$ is a natural embedding of contravariant functors from the category of strongly projective flat affine manifolds of dimension $m + 1$ (where the morphisms are affine diffeomorphisms) into the category of $(m + 1)$-dimensional vector spaces of functions on smooth manifolds of dimension $m + 1$ (where the morphisms are given by pull-backs induced by diffeomorphisms). This latter category is much more tractable.

12.3.4 GEODESICS IN STRONGLY PROJECTIVELY FLAT MANIFOLDS. Let $\mathcal{M}$ be an affine manifold of dimension m. Choose a system of local coordinates centered at a point P of M. Let $\sigma_{\vec{a}}(t)$ be the affine geodesic on $\mathcal{M}$ with $\sigma_{\vec{a}}(0) = P$ and $\dot{\sigma}_{\vec{a}}(0) = \vec{a}$ in $\mathbb{R}^m$. In general, the ordinary differential equation (ODE) defining $\sigma_{\vec{a}}$ is a quadratic system of ODEs in m functions of t. In the strongly projectively flat category, the following result reduces the geodesic equation to an ODE in a single function of t. This will inform our discussion of affine geodesics in Chapter 13 subsequently.

Theorem 12.35 *Let $\mathcal{M}$ be an affine manifold of dimension m with $\dim(\mathcal{Q}_P(\mathcal{M})) = m + 1$. Adopt the notation of Lemma 12.32 and express $\mathcal{Q}_P(\mathcal{M}) = \operatorname{span}\{\phi_0, \dots, \phi_m\}$ where the ϕ_i satisfy Equation (12.3.a). Let $\Psi = (\phi_1/\phi_0, \dots, \phi_m/\phi_0)$ define a local diffeomorphism from (M, P) to $(\mathbb{R}^m, 0)$. There exists the germ of a smooth function $\psi_{\vec{a}}(t)$ with $\psi_{a,b}(0) = 0$ and $\psi'_{a,b}(0) = 1$ so that $\sigma_{a,b}(t) = \Phi^{-1}(\psi_{\vec{a}}(t)\vec{a})$.*

Proof. Let $\mathcal{M}_0$ be the flat affine structure on $\mathbb{R}^m$. The affine geodesics in $\mathcal{M}_0$ are straight lines; they take the form $\sigma(t) := t\vec{a}$. By Theorem 12.11, two projectively equivalent affine manifolds have the same unparameterized affine geodesics. The desired result now follows since e^g provides a strong projective equivalence between $\Psi^*\mathcal{M}_0$ and $\mathcal{M}$. □

We will be working in the real analytic category (see Theorem 12.52) in our treatment of locally affine homogeneous surface geometries. Thus we can apply analytic continuation to pass from germs of affine geodesics to the affine geodesics themselves. In Lemma 13.6, we will show any Type $\mathcal{A}$ model is strongly projectively flat and thus this analysis pertains by Theorem 12.34. This ansatz will inform our discussion of geodesic completeness in the Type $\mathcal{A}$ setting.

12.3.5 THE ALTERNATING RICCI TENSOR. We use Theorem 12.19 to establish the following result.

Lemma 12.36 *Assume ρ_s vanishes identically near a point P of an affine surface $\mathcal{M}$.*

1. *If ρ_a does not vanish identically near P, then $E_P(\mu, \mathcal{M}) = \operatorname{span}\{\mathbb{1}\}$.*
2. *If ρ_a vanishes identically near P, then $\dim(E_P(\mu, \mathcal{M})) = 3$.*

Proof. Since ρ_s vanishes identically near P, we may use Lemma 12.31 to see that $\mathbb{1}$ belongs to $E(\mu, \mathcal{M})$. Suppose $\rho_a(P) \neq 0$. Use Theorem 12.19 to choose local coordinates near P so that $\Gamma_{11}{}^1 = -\partial_{x^1}\phi$ and $\Gamma_{22}{}^2 = \partial_{x^2}\phi$. Since $\rho = -\partial_{x^1}\partial_{x^2}\phi dx^1 \wedge dx^2$, $\partial_{x^1}\partial_{x^2}\phi \neq 0$. We have $E_P(\mu, \mathcal{M}) = \ker_P\{\mathcal{H}\}$. If $\mathcal{H}f = 0$, then $0 = \mathcal{H}_{12}f = \partial_{x^1}\partial_{x^2}f$ and, consequently, $f(x^1, x^2) = a_1(x^1) + a_2(x^2)$. Setting $\mathcal{H}f = 0$ then yields the relations

$$\begin{aligned} 0 &= \mathcal{H}_{11}f = a_1''(x^1) + a_1'(x^1)\partial_{x^1}\phi(x^1, x^2), \\ 0 &= \mathcal{H}_{22}f = a_2''(x^2) - a_2'(x^2)\partial_{x^2}\phi(x^1, x^2). \end{aligned}$$

Differentiating $\mathcal{H}_{11}f$ with respect to x^2 and $\mathcal{H}_{22}f$ with respect to x^1 yields

$$a_1'(x^1)\partial_{x^1}\partial_{x^2}\phi(x^1, x^2) = 0 \quad \text{and} \quad a_2'(x^2)\partial_{x^1}\partial_{x^2}\phi(x^1, x^2) = 0.$$

Since $\rho_a(P) \neq 0$, we have $\partial_{x^1}\partial_{x^2}\phi(x^1, x^2)(P) \neq 0$. Consequently, $a_1' = 0$ and $a_2' = 0$. This shows that f is a multiple of $\mathbb{1}$ and $E_P(\mu, \mathcal{M}) = \text{span}\{\mathbb{1}\}$.

Suppose next that $\rho_a(P) = 0$ but there is a sequence of points Q_n which converge to P so that $\rho_a(Q_n) \neq 0$. We then have that $E_{Q_n}(\mu, \mathcal{M}) = \text{span}\{\mathbb{1}\}$. Suppose that

$$\dim(E_P(\mu, \mathcal{M})) \neq 1.$$

Since $\mathbb{1} \in E_P(\mu, \mathcal{M})$, we may choose $0 \neq f \in E_P(\mu, \mathcal{M})$ so that $f(P) = 0$; thus by Lemma 12.14, $df(P) \neq 0$. As $f \in E_{Q_n}(\mu, \mathcal{M})$ for n large satisfies $df(Q_n) \neq 0$, this shows $\dim(E_{Q_n}(\mu, \mathcal{M})) \geq 2$, which is false. Thus $\dim(E_P(\mu, \mathcal{M})) = 1$ and Assertion 1 follows.

If $\rho_s = 0$ and $\rho_a = 0$ near P, then $\mathcal{M}$ is flat near P. Consequently, by Theorem 12.34, $\dim(\mathcal{Q}_P(\mathcal{M})) = 3$. Since $\rho_s = 0$, $\mathcal{Q}_P(\mathcal{M}) = \ker_P\{\mathcal{H}\} = E_P(\mu, \mathcal{M})$ for any μ. This proves Assertion 2. □

We now establish a result of Brozos-Vázquez et al. [12].

Theorem 12.37 *If $\mathcal{M}$ is an affine surface, then* $\dim(\mathcal{Q}_P(\mathcal{M})) \neq 2$.

Proof. We suppose to the contrary that $\dim(\mathcal{Q}_P(\mathcal{M})) = 2$ and argue for a contradiction. Suppose first that $f(P) \neq 0$ for some $f \in \mathcal{Q}_P(\mathcal{M})$. Then $\mathcal{M}$ is locally strongly projectively equivalent to a surface $\mathcal{M}_1$ with $\rho_{s,\mathcal{M}_1} = 0$ by Lemma 12.31. By Lemma 12.36, $\dim(\mathcal{Q}_P(\mathcal{M}_1)) \neq 2$. By Theorem 12.29, $\dim(\mathcal{Q}_P(\mathcal{M}_1)) = \dim(\mathcal{Q}_P(\mathcal{M})) = 2$. This provides the desired contradiction.

Next we suppose $f(P) = 0$ for all $f \in \mathcal{Q}_P(\mathcal{M})$. By Lemma 12.14, $df(P) \neq 0$. Since $\dim(\mathcal{Q}_P(\mathcal{M})) = 2$, we can choose $\{\phi_1, \phi_2\} \subset \mathcal{Q}_P(\mathcal{M})$ so that $d\phi_i(P) = dx^i$. Thus we can change coordinates to ensure $\phi_i = x^i$. If $\tilde{P} \neq P$, then $\dim(\mathcal{Q}_{\tilde{P}}(\mathcal{M})) \geq 2$. Since there is an element of $\mathcal{Q}_{\tilde{P}}(\mathcal{M})$ which is non-zero, $\dim(\mathcal{Q}_{\tilde{P}}(\mathcal{M})) \neq 2$. Consequently, $\dim(\mathcal{Q}_{\tilde{P}}(\mathcal{M})) = 3$. Theorem 12.34 then shows that $\mathcal{M}$ is strongly projectively flat near $\tilde{P}$. By Theorem 12.11, we conclude ρ and $\nabla\rho$ are totally symmetric at $\tilde{P}$. Continuity then implies ρ and $\nabla\rho$ are totally symmetric at P as well. Hence, by Theorem 12.11, $\mathcal{M}$ is strongly projectively flat near P. This implies $\dim(\mathcal{Q}_P(\mathcal{M})) = 3$, which is a contradiction. □

Remark 12.38 Theorem 12.37 fails if we replace $\dim(\mathcal{Q}_P(\mathcal{M}))$ by $\dim(\mathcal{Q}(\mathcal{M}))$ and if $\mathcal{M}$ is not simply connected. Let $\mathcal{M} = \mathbb{R}^2$ with the standard flat structure, let $\mathcal{M}_2$ be the cylinder with the standard flat structure where we identify

$$(x^1, x^2) \text{ with } (x^1 + 2n_1\pi, x^2)$$

for $n_1 \in \mathbb{Z}$, and let $\mathcal{M}_1$ be the torus with the standard flat structure where we identify (x^1, x^2) with $(x^1 + 2n_1\pi, x^2 + 2n_2\pi)$ for $(n_1, n_2) \in \mathbb{Z}^2$. The holonomy plays a crucial role as

$$\mathcal{Q}(\mathcal{M}) = \operatorname{span}\{\mathbb{1}, x^1, x^2\}, \quad \mathcal{Q}(\mathcal{M}_1) = \operatorname{span}\{\mathbb{1}, x^2\}, \quad \mathcal{Q}(\mathcal{M}_2) = \operatorname{span}\{\mathbb{1}\}.$$

12.3.6 YAMABE SOLITONS. The space of Yamabe solitons plays an important role in the study of flat connections. Let Γ_0 be the flat connection all of whose Christoffel symbols vanish identically. The following observation is an immediate consequence of Theorem 12.34 since $\mathcal{Q}(\mathcal{M}) = \mathcal{Y}(\mathcal{M})$ in the flat setting.

Lemma 12.39 *Let Γ define a flat connection on $\mathbb{R}^2$. Then $\dim(\mathcal{Y}(\Gamma)) = 3$. Let $F = (f^1, f^2)$ be the germ of a diffeomorphism. We have*

1. *$F^*\Gamma_0 = \Gamma$ if and only if $\{1, f^1, f^2\}$ is a basis for $\mathcal{Y}(\Gamma)$.*
2. *$F^*\Gamma = \Gamma$ if and only if $F^*\{\mathcal{Y}(\Gamma)\} = \mathcal{Y}(\Gamma)$.*

The following will be a useful observation.

Lemma 12.40 *Let $\mathcal{M}$ be an affine manifold of dimension m with $\dim(\mathcal{Y}(\mathcal{M})) \geq m + 1$. Then $\mathcal{M}$ is flat and $\dim(\mathcal{Y}(\mathcal{M})) = m + 1$.*

Proof. We apply Lemma 12.14 to see $\dim(\mathcal{Y}(\mathcal{M})) \leq m + 1$, and hence equality holds. We have $\mathbb{1} \in \mathcal{Y}(\mathcal{M})$. Consequently, we may apply Lemma 12.32 with $\mu = 0$ to choose local coordinates so $\mathcal{Y}(\mathcal{M}) = \operatorname{span}\{1, x^1, \dots, x^m\}$. We have $0 = \mathcal{H}(x^i) = \Gamma_{jk}{}^i dx^j \otimes dx^j$. Consequently, all the Christoffel symbols vanish in this coordinate system. This implies the connection in question is flat. □

12.4 THE CLASSIFICATION OF LOCALLY HOMOGENEOUS AFFINE SURFACES WITH TORSION

We permit the surface in question to have torsion throughout this section as the analysis is no more difficult in this setting. The approach we will take will be quite different in flavor from the proof of the classification result given by Opozda [53] in the torsion-free setting; her work rested upon a detailed analysis of the curvature tensor. A subsequent proof was given by Kowalski, Opozda, and Vlášek [44] which was group theoretic in nature. The classification was completed by Arias-Marco and Kowalski [2] by permitting the surface in question to have torsion. Their work involved a careful examination of the classification of the Lie algebras of vector fields due to Olver [49]. By contrast, the proof we shall give mixes Lie theoretic methods, specifically an analysis of root systems, and a detailed examination of the eight affine Killing Equations (12.1.d). We also refer to related work of Guillot and Sánchez-Godinez [37], Kowalski, Opozda, and Vlášek [42], and Opozda [54].

T. Arias-Marco

O. Kowalski

P. Olver[1]

B. Opozda

There is always a question of the local versus the global geometry of an object in Differential Geometry. Let $\mathcal{M}$ be a locally affine homogeneous surface. By Lemma 12.14, for the objects under consideration, questions of passing from the local to the global involve the holonomy action of the fundamental group; there is no obstruction if $\mathcal{M}$ is assumed simply connected. We shall not belabor the point and ignore the question of passing from local to global henceforth for the most part.

Choose a reference point $P \in M$; any other point will do as well since we have assumed that $\mathcal{M}$ is locally affine homogeneous and connected. Thus we can assume without loss of generality that M is an arbitrarily small neighborhood of P. Consequently, for example, if f is a smooth function with $f(P) \neq 0$, we may assume f never vanishes on M. Similarly, if X_1 and X_2 are two vector fields which are linearly independent at P, we may assume that X_1 and X_2 are linearly independent on all of M.

Here is a brief outline to this section. In Section 12.4.1, we present the four families of locally homogeneous pseudo-Riemannian surfaces. In Section 12.4.2, we discuss the two simply connected 2-dimensional Lie groups. These are the additive group $\mathbb{R}^2$, which is Abelian, and the $ax + b$ group, which is non-Abelian. Denote the associated Lie algebras of these groups by $\mathfrak{K}_{\mathcal{A}}$ and $\mathfrak{K}_{\mathcal{B}}$, respectively. Let $\mathfrak{so}(3)$ be the Lie algebra of the 3-dimensional special orthogonal group SO(3). A Lie subalgebra of the Lie algebra of affine Killing vector fields is said to be *effective* if it spans the tangent space of M at some point. In Section 12.4.3, we discuss affine surface geometries with torsion where $\mathfrak{K}(\mathcal{M})$ contains an effective Lie subalgebra isomorphic to $\mathfrak{K}_{\mathcal{A}}$, $\mathfrak{K}_{\mathcal{B}}$, or $\mathfrak{so}(3)$. In Section 12.4.4, we discuss the exceptional Lie algebra $A_{3,7}^{a}$.

In Section 12.4.5, we use the adjoint action of $\mathfrak{K}(\mathcal{M})$ on itself to show that $\mathfrak{K}(\mathcal{M})$ must contain an effective Lie subalgebra isomorphic to $\mathfrak{K}_{\mathcal{A}}$, to $\mathfrak{K}_{\mathcal{B}}$ or to $\mathfrak{so}(3)$. These results are then

[1]Photo credit University of Minnesota

used in Section 12.4.6 to complete the classification of locally homogeneous surfaces with torsion in Theorem 12.49. Metrizability is treated briefly in Section 12.4.7.

12.4.1 LOCALLY HOMOGENEOUS PSEUDO-RIEMANNIAN SURFACES.

A pseudo-Riemannian manifold is said to be *locally homogeneous* if given any two points of the manifold, there is a local isometry taking one point to the other. An affine manifold is said to be *locally affine homogeneous* if given any two points of the manifold there is a local affine map taking one point to the other. We delete the word "local" if the maps in question are global diffeomorphisms.

A pseudo-Riemannian surface is locally homogeneous if and only if it has constant scalar curvature. Up to a non-necessarily positive rescaling and isometry, there are four simply connected geodesically complete homogeneous pseudo-Riemannian geometries:

1. Let $\mathbb{E}^2$ be $\mathbb{R}^2$ with the flat metric $ds^2 = (dx^1)^2 + (dx^2)^2$.
2. Let $\mathfrak{H}$ be the unit sphere in $\mathbb{R}^3$ with the round metric.
3. Let $\mathbb{H}$ be the upper half-plane $\mathbb{R}^+ \times \mathbb{R}$ with the metric $ds^2 = (x^1)^{-2}((dx^1)^2 + (dx^2)^2)$.
4. Let $\mathfrak{L}$ be the pseudo-sphere $(x^1)^2 + (x^2)^2 - (x^3)^2 = -1$ for $x^3 > 0$ in $\mathbb{R}^3$ with the Lorentzian metric induced by the Minkowski metric $(dx^1)^2 + (dx^2)^2 - (dx^3)^2$.

The manifolds $\{\mathbb{E}^2, \mathfrak{H}, \mathbb{H}\}$ are Riemannian with sectional curvature $\{0, +1, -1\}$, respectively. The manifold $\mathfrak{L}$ is Lorentzian. There is another useful model for $\mathfrak{L}$. Let $\mathbb{L}$ be the upper half-plane $\mathbb{R}^+ \times \mathbb{R}$ with the metric $ds^2 = (x^1)^{-2}((dx^1)^2 - (dx^2)^2)$. This should be thought of as the Lorentzian hyperbolic plane. The manifolds $\{\mathbb{E}^2, \mathfrak{H}, \mathbb{H}, \mathfrak{L}\}$ are all geodesically complete. By contrast $\mathbb{L}$ is not geodesically complete. However, it is homogeneous and it embeds isometrically in $\mathfrak{L}$. We refer to D'Ascanio, Gilkey, and Pisani [22] for further details concerning the relationship between $\mathfrak{L}$ and $\mathbb{L}$.

In the pseudo-Riemannian category there are two levels of homogeneity. A locally homogeneous pseudo-Riemannian manifold is necessarily locally affine homogeneous, but there are pseudo-Riemannian manifolds which are not locally homogeneous but whose Levi-Civita connection is locally affine homogeneous (see Kowalski, Opozda, and Vlášek [43]).

12.4.2 2-DIMENSIONAL LIE GROUPS.

Give $\mathbb{R}^2$ the structure of an Abelian Lie group by using vector addition $(a^1, a^2) + (b^1, b^2) := (a^1 + b^1, a^2 + b^2)$. If we identify $\mathbb{R}^2$ with the translation group $T_{(a^1,a^2)} : (x^1, x^2) \to (x^1 + a^1, x^2 + a^2)$, then

$$T_{(a^1,a^2)+(b^1,b^2)} = T_{(a^1,a^2)} \circ T_{(b^1,b^2)} .$$

This Lie group can be identified with the additive group

$$\left\{ \begin{pmatrix} a_1 & 0 \\ 0 & a_2 \end{pmatrix} \text{ for } a^1 \in \mathbb{R},\ a^2 \in \mathbb{R} \right\} \subset M_2(\mathbb{R}) .$$

The multiplication $(a,b) * (c,d) := (ac, ad + b)$ makes $\mathbb{R}^+ \times \mathbb{R}$ into a non-Abelian Lie group. If we identify (a,b) with the affine map $A_{(a,b)} : x \to ax + b$, then $A_{(a,b)} \circ A_{(c,d)} = A_{(a,b)*(c,d)}$. Consequently, this non-Abelian Lie group is called the $ax + b$ group. We refer to the discussion in Section 6.7.2 of Book II for further details. It can be identified with the multiplicative group

$$\left\{ \begin{pmatrix} a & b \\ 0 & 1 \end{pmatrix} \text{ for } a > 0,\ b \in \mathbb{R} \right\} \subset \mathrm{GL}(2, \mathbb{R}) .$$

Lemma 12.41

1. *$\mathfrak{K}_{\mathcal{A}} := \operatorname{span}\{\partial_{x^1}, \partial_{x^2}\}$ is the Lie algebra of $\mathbb{R}^2$.*
2. *$\mathfrak{K}_{\mathcal{B}} := \operatorname{span}\{x^1\partial_{x^1} + x^2\partial_{x^2}, \partial_{x^2}\}$ is the Lie algebra of the $ax + b$ group.*
3. *A connection on $\mathbb{R}^2$ is left-invariant if and only if $\Gamma_{ij}{}^k \in \mathbb{R}$.*
4. *A connection on the $ax + b$ group is left-invariant if and only if $x^1\Gamma_{ij}{}^k \in \mathbb{R}$.*
5. *A simply connected 2-dimensional Lie group is isomorphic to one of these two groups.*

Proof. A direct computation establishes Assertions 1 and 2. If g belongs to a Lie group G, let L_g denote the action of G on itself by left-multiplication; an affine connection on G is *left-invariant* if and only if $L_g \circ \nabla = \nabla \circ L_g$, i.e., if left-multiplication is an affine map. Suppose $G = \mathbb{R}^2$. Let $(a_1, a_2) \in \mathbb{R}^2$. As $L_{(a_1,a_2)}$ preserves ∂_{x^1} and ∂_{x^2}, $L_{(a^1,a^2)}$ preserves ∇ if and only if

$$\Gamma_{ij}{}^k(x^1 + a^1, x^2 + a^2) = \Gamma_{ij}{}^k(x^1, x^2) \quad \text{for all} \quad (x^1, a^1, x^2, a^2) \in \mathbb{R}^4 .$$

This implies that $\Gamma_{ij}{}^k \in \mathbb{R}$ and establishes Assertion 3. Let G be the $ax + b$ group. Then $L_{(a,b)}(x^1, x^2) = (ax^1, ax^2 + b)$ so $L_{(a,b)}\partial_{x^1} = a\partial_{x^1}$ and $L_{(a,b)}\partial_{x^2} = a\partial_{x^2}$. Thus ∇ is left-invariant if and only if $\Gamma_{ij}{}^k$ is independent of x^2 and $\Gamma_{ij}{}^k(ax^1) = a^{-1}\Gamma_{ij}{}^k(x^1)$; Assertion 4 follows. Let $\mathfrak{K}$ be the Lie algebra of a simply connected Lie group. If $\mathfrak{K}$ is Abelian, then $\mathfrak{K}$ is isomorphic to $\mathfrak{K}_{\mathcal{A}}$ and if $\mathfrak{K}$ is non-Abelian, then $\mathfrak{K}$ is isomorphic to $\mathfrak{K}_{\mathcal{B}}$; Assertion 5 now follows from this analysis. □

Definition 12.42 Let $\mathcal{M} = (M, \nabla)$ be an affine surface, possibly with torsion.

1. We say that $\mathcal{M}$ is *Type* $\mathcal{A}$ if there exists a coordinate atlas for $\mathcal{M}$ so that $\Gamma_{ij}{}^k \in \mathbb{R}$.
2. We say that $\mathcal{M}$ is *Type* $\mathcal{B}$ if there exists a coordinate atlas for $\mathcal{M}$ so that $x^1\Gamma_{ij}{}^k \in \mathbb{R}$.
3. We say that $\mathcal{M}$ is *Type* $\mathcal{C}$ if there exists a coordinate atlas for $\mathcal{M}$ so that ∇ is isomorphic to the Levi-Civita connection defined by the metric of the round sphere.

These structures have a profound affect on the geometry of the underlying affine structure. Let $\mathfrak{so}(3) := \operatorname{span}\{X, Y, Z\}$ be the 3-dimensional Lie algebra of the special orthogonal group. The bracket is given by $[X, Y] = Z$, $[Y, Z] = X$, and $[Z, X] = Y$. We refer to the discussion in Chapter 6 of Book II for further details.

Lemma 12.43 *Let $\mathcal{M}$ be a simply connected affine surface, possibly with torsion.*

1. *If $\mathcal{M}$ is Type $\mathcal{A}$, Type $\mathcal{B}$, or Type $\mathcal{C}$, then $\mathcal{M}$ is locally affine homogeneous.*
2. *If $\mathcal{M}$ is Type $\mathcal{A}$, there is an effective Lie subalgebra of $\mathfrak{K}(\mathcal{M})$ isomorphic to $\mathfrak{K}_{\mathcal{A}}$.*
3. *If $\mathcal{M}$ is Type $\mathcal{B}$, there is an effective Lie subalgebra of $\mathfrak{K}(\mathcal{M})$ isomorphic to $\mathfrak{K}_{\mathcal{B}}$.*
4. *If $\mathcal{M}$ is Type $\mathcal{C}$, there is an effective Lie subalgebra of $\mathfrak{K}(\mathcal{M})$ isomorphic to $\mathfrak{so}(3)$.*

Proof. It is clear from the discussion of Section 12.4.2 that any such geometry is locally affine homogeneous. Since $\mathcal{M}$ is simply connected, any local affine Killing vector field extends to a global affine Killing vector field. Assertion 1 and Assertion 2 follow from the proof of Lemma 12.41; the proof of Assertion 3 follows similar lines. □

12.4.3 THE LIE ALGEBRAS $\mathfrak{K}_{\mathcal{A}}$, $\mathfrak{K}_{\mathcal{B}}$, AND $\mathfrak{so}(3)$. We have the following converse to Lemma 12.43.

Lemma 12.44 *Let $\mathcal{M} = (M, \nabla)$ be an affine surface, possibly with torsion. Let $\tilde{\mathfrak{K}}$ be an effective Lie subalgebra of $\mathfrak{K}(\mathcal{M})$.*

1. *If $\tilde{\mathfrak{K}}$ is isomorphic to $\mathfrak{K}_{\mathcal{A}}$, then there exists a coordinate atlas for $\mathcal{M}$ so $\Gamma_{ij}{}^k \in \mathbb{R}$.*
2. *If $\tilde{\mathfrak{K}}$ is isomorphic to $\mathfrak{K}_{\mathcal{B}}$, then there exists a coordinate atlas for $\mathcal{M}$ so $x^1\Gamma_{ij}{}^k \in \mathbb{R}$.*
3. *If $\tilde{\mathfrak{K}}$ is isomorphic to $\mathfrak{so}(3)$, then there exists a coordinate atlas for $\mathcal{M}$ so ∇ is the Levi-Civita connection defined by the metric of the round sphere.*

Proof. Since $\mathcal{M}$ is assumed to be locally affine homogeneous, the analysis is local and we can work in an arbitrarily small neighborhood of the distinguished point P. We distinguish cases.

Suppose first that there exists an effective Lie subalgebra of $\mathfrak{K}(\mathcal{M})$ which is isomorphic to $\mathfrak{K}_{\mathcal{A}}$. Choose affine Killing vector fields X and Y so that $X(P)$ and $Y(P)$ are linearly independent and so that $[X, Y] = 0$. By the Frobenius Theorem, there are local coordinates (x^1, x^2) centered at P on M so that $X = \partial_{x^1}$ and $Y = \partial_{x^2}$. It then follows that $\Gamma_{ij}{}^k \in \mathbb{R}$, which establishes Assertion 1.

Suppose next that there exists an effective Lie subalgebra of $\mathfrak{K}(\mathcal{M})$ which is isomorphic to $\mathfrak{K}_{\mathcal{B}}$. Choose affine Killing vector fields X and Y so $X(P)$ and $Y(P)$ are linearly independent and so that $[X, Y] = aY$ for $a \neq 0$. We replace X by $a^{-1}X$ to assume that $[X, Y] = Y$. Choose local coordinates such that $Y = \partial_{x^2}$ and expand

$$X = u(x^1, x^2)\partial_{x^1} + v(x^1, x^2)\partial_{x^2}$$

for some functions u, v. The bracket relation $[X, Y] = Y$ shows $\partial_{x^2}u = 0$ and $\partial_{x^2}v = -1$. Consequently, $X = u(x^1)\partial_{x^1} + (v_0(x^1) - x^2)\partial_{x^2}$.

We change coordinates setting $\tilde{x}^1 = x^1$ and $\tilde{x}^2 = x^2 + \varepsilon(x^1)$. Then

$$\begin{aligned} d\tilde{x}^1 &= dx^1, & d\tilde{x}^2 &= dx^2 + \varepsilon'(x^1)dx^1, \\ \partial_{\tilde{x}^1} &= \partial_{x^1} - \varepsilon'(x^1)\partial_{x^2}, & \partial_{\tilde{x}^2} &= \partial_{x^2}. \end{aligned}$$

We then have $X = u(\tilde{x}^1)\partial_{\tilde{x}^1} + \{-\tilde{x}^2 + \varepsilon(\tilde{x}^1) + v_0(\tilde{x}^1) + u(\tilde{x}^1)\varepsilon'(\tilde{x}^1)\}\partial_{\tilde{x}^2}$. We may then solve the ODE $\varepsilon(x^1) + v_0(x^1) + u(x^1)\varepsilon'(x^1) = 0$ to express $X = u(\tilde{x}^1)\partial_{\tilde{x}^1} - \tilde{x}^2\partial_{\tilde{x}^2}$. Hence one may assume $X = u(x^1)\partial_{x^1} - x^2\partial_{x^2}$ without changing $Y = \partial_{x^2}$. Finally replace x^1 by $\hat{x}^1$ to ensure $u(x^1)\partial_{x^1} = -\hat{x}^1\partial_{\hat{x}^1}$ and $X = -\hat{x}^1\partial_{\hat{x}^1} - \hat{x}^2\partial_{\hat{x}^2}$. It now follows that the Christoffel symbols have the form given in Assertion 2.

Suppose finally that there is is an effective Lie subalgebra of $\mathfrak{K}(\mathcal{M})$ which is isomorphic to $\mathfrak{K}_{\mathcal{B}}$. Choose affine Killing vector fields so $\{X, Y, Z\}$ is effective and so

$$[X, Y] = Z, \quad [Y, Z] = X, \quad [Z, X] = Y .$$

Since Z does not vanish identically, we may assume that $Z(P) \neq 0$ and choose local coordinates (x^1, x^2) which are defined near P so that $Z = \partial_{x^2}$. Decompose

$$X = u_1(x^1, x^2)\partial_{x^1} + \star\partial_{x^2}$$

where $\star$ indicates a coefficient which is not of interest. We then have $\partial^2_{x^2}u_1 = -u_1$ so $u_1(x^1, x^2) = r(x^1)\cos(x^2 + \theta(x^1))$. Since $\{X, Y, Z\}$ is a linearly independent set, $r(x^1) \neq 0$.

Change coordinates $(\tilde{x}^1, \tilde{x}^2) = (x^1, x^2 + \theta(x^1))$ so that $\partial_{\tilde{x}^2} = \partial_{x^2}$ and rewrite X in the form $X = r_1(\tilde{x}^1)\cos(\tilde{x}^2)\partial_{\tilde{x}^1} + \star\partial_{\tilde{x}^2}$ without changing ∂_{x^2}. We now choose coordinates $(z^1, z^2) = (f(\tilde{x}^1), \tilde{x}^2)$ so that $\partial_{z^1} = r_1(\tilde{x}^1)\partial_{\tilde{x}^1}$ and $\partial_{z^2} = \partial_{\tilde{x}^2}$. Consequently, $Z = \partial_{z^2}$. We use the Lie algebra relations defining $\mathfrak{so}(3)$ to expand X and Y in the form

$$\begin{aligned} X &= \cos(z^2)\partial_{z^1} + \{v_c(z^1)\cos(z^2) + v_s(z^1)\sin(z^2)\}\partial_{z^2}, \\ Y &= -\sin(z^2)\partial_{z^1} + \{-v_c(z^1)\sin(z^2) + v_s(z^1)\cos(z^2)\}\partial_{z^2} . \end{aligned}$$

To simplify the notation we replace (z^1, z^2) by (x^1, x^2). The bracket relation $[X, Y] = Z$ then yields the relations $-v_c(x^1) = 0$ and $-v_c(x^1)^2 - v_s(x^1)^2 + v_s'(x^1) = 1$. We solve this to obtain $v_c(x^1) = 0$ and $v_s(x^1) = \tan(x^1 + c)$. We replace $x^1 + c$ by x^1 to assume $c = 0$. We still have $Z = \partial_{x^2}$, but we now may express X and Y in a somewhat simpler form:

$$\begin{aligned} X &= \cos(x^2)\partial_{x^1} + \tan(x^1)\sin(x^2)\partial_{x^2}, \\ Y &= -\sin(x^2)\partial_{x^1} + \tan(x^1)\cos(x^2)\partial_{x^2} . \end{aligned}$$

Since $Z = \partial_{x^2}$ is an affine Killing vector field, the Christoffel symbols depend only on x^1. The affine Killing equations in general are quite complicated. For example, we have

$$\begin{aligned} K_{11}{}^1 : 0 \;=\;& \cos(x^2)(\Gamma_{11}{}^1)'(x^1) + \Gamma_{11}{}^2(x^1)\sin(x^2) \\ & +\Gamma_{12}{}^1(x^1)\sec^2(x^1)\sin(x^2) + \Gamma_{21}{}^1(x^1)\sec^2(x^1)\sin(x^2) . \end{aligned}$$

To simplify the equations, we set $x^2 = 0$ to obtain

$$\begin{aligned} &K_{11}{}^1 : 0 = (\Gamma_{11}{}^1)', \\ &K_{11}{}^2 : 0 = -\Gamma_{11}{}^2\tan(x^1) + (\Gamma_{11}{}^2)', \\ &K_{12}{}^1 : 0 = \Gamma_{12}{}^1\tan(x^1) + (\Gamma_{12}{}^1)', \\ &K_{12}{}^2 : 0 = \sec^2(x^1) + (\Gamma_{12}{}^2)', \end{aligned} \tag{12.4.a}$$

and

$$\begin{aligned}
K_{21}{}^1 &: 0 = \Gamma_{21}{}^1 \tan(x^1) + (\Gamma_{21}{}^1)', \\
K_{21}{}^2 &: 0 = \sec^2(x^1) + (\Gamma_{21}{}^2)', \\
K_{22}{}^1 &: 0 = -1 + 2\Gamma_{22}{}^1 \tan(x^1) + (\Gamma_{22}{}^1)', \\
K_{22}{}^2 &: 0 = \Gamma_{22}{}^2 \tan(x^1) + (\Gamma_{22}{}^2)'.
\end{aligned} \tag{12.4.b}$$

Let $a_{ij}{}^k$ be constants to be determined. We solve the ODEs given in Equation (12.4.a) and in Equation (12.4.b) to see:

$$\begin{aligned}
\Gamma_{11}{}^1 &= a_{11}{}^1, & \Gamma_{11}{}^2 &= a_{11}{}^2 \sec(x^1), \\
\Gamma_{12}{}^1 &= a_{12}{}^1 \cos(x^1), & \Gamma_{12}{}^2 &= a_{12}{}^2 - \tan(x^1), \\
\Gamma_{21}{}^1 &= a_{21}{}^1 \cos(x^1), & \Gamma_{21}{}^2 &= a_{21}{}^2 - \tan(x^1), \\
\Gamma_{22}{}^1 &= a_{22}{}^1 \cos(x^1)^2 + \cos(x^1)\sin(x^1), & \Gamma_{22}{}^2 &= a_{22}{}^2 \cos(x^1).
\end{aligned}$$

Equation (12.4.a) and Equation (12.4.b) were obtained by specializing the affine Killing equations at $x^2 = 0$. We now specialize the affine Killing equations at $x^2 = \frac{\pi}{2}$ to determine the constants. Equation (12.4.a) becomes:

$$\begin{aligned}
K_{11}{}^1 &: 0 = (a_{11}{}^2 + a_{12}{}^1 + a_{21}{}^1)\sec(x^1), \\
K_{11}{}^2 &: 0 = (-a_{11}{}^1 + a_{12}{}^2 + a_{21}{}^2)\sec(x^1)^2, \\
K_{12}{}^1 &: 0 = -a_{11}{}^1 + a_{12}{}^2 + a_{22}{}^1, \\
K_{12}{}^2 &: 0 = -(a_{11}{}^2 + a_{12}{}^1 - a_{22}{}^2)\sec(x^1)
\end{aligned}$$

which leads to the relations.

$$\begin{aligned}
a_{21}{}^1 &= -a_{11}{}^2 - a_{12}{}^1, & a_{21}{}^2 &= a_{11}{}^1 - a_{12}{}^2, \\
a_{22}{}^1 &= a_{11}{}^1 - a_{12}{}^2, & a_{22}{}^2 &= a_{11}{}^2 + a_{12}{}^1.
\end{aligned}$$

We specialize the affine Killing equations $K_{21}{}^1$ and $K_{22}{}^2$ at $x^2 = \frac{\pi}{2}$ to obtain

$$K_{21}{}^1 : 0 = a_{11}{}^1 - 2a_{12}{}^2, \quad K_{22}{}^2 : 0 = -2a_{11}{}^1 + a_{12}{}^2.$$

This yields that $a_{11}{}^1 = 0$ and $a_{12}{}^2 = 0$. The remaining affine Killing equations without specialization yield

$$(2a_{11}{}^2 + a_{12}{}^1)\cos(x^1)\sin(x^2) = 0 \quad \text{and} \quad (a_{11}{}^2 + 2a_{12}{}^1)\sec(x^1)\sin(x^2) = 0$$

so $a_{11}{}^2 = 0$ and $a_{12}{}^1 = 0$. This yields finally

$$\begin{aligned}
\Gamma_{11}{}^1 &= 0, & \Gamma_{11}{}^2 &= 0, & \Gamma_{12}{}^1 &= 0, & \Gamma_{12}{}^2 &= -\tan(x^1), \\
\Gamma_{21}{}^1 &= 0, & \Gamma_{21}{}^2 &= -\tan(x^1), & \Gamma_{22}{}^1 &= \cos(x^1)\sin(x^1), & \Gamma_{22}{}^2 &= 0.
\end{aligned}$$

We summarize matters. If we assume $\mathfrak{so}(3)$ is an effective Lie subalgebra of $\mathfrak{K}(\mathcal{M})$, then the Christoffel symbols are uniquely determined. Thus any two such geometries are locally isomorphic. Since $\mathfrak{so}(3)$ is the Lie algebra of SO(3) and since SO(3) is the Lie group of orientation-preserving isometries of S^3, we obtain that any such geometry is locally modeled on the geometry of the round sphere. □

Remark 12.45 One may show directly that there are no additional affine Killing vector fields for the round sphere and thus $\mathfrak{so}(3) = \mathfrak{K}(\mathcal{M})$. Consequently, no Type $\mathcal{C}$ geometry is either Type $\mathcal{A}$ or Type $\mathcal{B}$. Any flat geometry is trivially both Type $\mathcal{A}$ and Type $\mathcal{B}$. However, as we shall see in Section 14.2.4, there are locally affine homogeneous surfaces which are not flat and which are both Type $\mathcal{A}$ and Type $\mathcal{B}$; they admit both an effective Lie subalgebra isomorphic to $\mathfrak{K}_{\mathcal{A}}$ and an effective Lie subalgebra isomorphic to $\mathfrak{K}_{\mathcal{B}}$.

12.4.4 THE LIE ALGEBRA $A^a_{3,7}$. We shall adopt the notation of Patera et al. [55] and let $A^a_{3,7} = \text{span}\{X, Y\}$ be the 3-dimensional Lie algebra defined by the bracket relations $[X, Y] = 0$, $[X, Z] = aX + Y$, and $[Y, Z] = aY - X$. Note that $A^a_{3,7}$ is a semi-direct product $\mathbb{R} \ltimes_\varphi \mathbb{R}^2$ where $\mathbb{R}Z$ acts on $\mathbb{R}^2 = \text{span}\{X, Y\}$ by the derivation

$$\varphi = \begin{pmatrix} -a & 1 \\ -1 & -a \end{pmatrix}.$$

The Lie algebra $A^a_{3,7}$ is a Lie subalgebra of $\mathfrak{K}(\mathcal{M})$ for the Type $\mathcal{A}$ structures $\mathcal{M}^4_5(c)$ (see Definition 13.7). The following result shows that $A^a_{3,7}$ does not affect any classification results.

Lemma 12.46 *Suppose that there is an effective Lie subalgebra of $\mathfrak{K}(\mathcal{M})$ which is isomorphic to $A^a_{3,7}$. Then there exists an effective Lie subalgebra of $\mathfrak{K}(\mathcal{M})$ isomorphic to $\mathfrak{K}_{\mathcal{A}}$.*

Proof. The result is immediate if $\{X, Y\}$ is effective. Consequently, we assume that Y is a multiple of X and $\{X, Z\}$ is effective. We use Lemma 12.4 to normalize the coordinate system so that

$$X = \partial_{x^2} \text{ and } Y = v(x^1, x^2)\partial_{x^2}\,.$$

Since $[X, Y] = 0$, $\partial_{x^2} v = 0$. Thus, $v = v(x^1)$. Since Y is not a constant multiple of X, we may assume that $v'(x^1) \neq 0$. Therefore, we have that $X = \partial_{x^2}$ and $Y = v(x^1)\partial_{x^2}$. Consequently, by Assertion 1 of Lemma 12.5, v solves the ODE $(\Gamma_{12}{}^2 + \Gamma_{21}{}^2)v' + v'' = 0$. Expand

$$\begin{aligned} Z &= u(x^1, x^2)\partial_{x^1} + w(x^1, x^2)\partial_{x^2}, \\ [X, Z] &= \partial_{x^2} u(x^1, x^2)\partial_{x^1} + \partial_{x^2} w(x^1, x^2)\partial_{x^2} \\ &= aX + Y = (a + v(x^1))\partial_{x^2}\,. \end{aligned}$$

Thus $u = u(x^1)$ and $w = (a + v(x^1))x^2 + v_0(x^1)$; as $\{X, Z\}$ is effective, $u \neq 0$ and

$$X = \partial_{x^2}, \quad Y = v(x^1)\partial_{x^2}, \quad Z = u(x^1)\partial_{x^1} + \{(a + v(x^1))x^2 + v_0(x^1)\}\partial_{x^2}\,.$$

The affine Killing equation $K_{11}{}^2$ for Z yields $0 = (\Gamma_{12}{}^2 + \Gamma_{21}{}^2)v_0' + v_0''$. Thus v and v_0 are solutions of the same linear homogeneous ODE. Since $\{\mathbb{1}, v\}$ are linearly independent solutions, they form a basis for the solution space. Thus after subtracting a suitable linear combination of ∂_{x^2} and $v(x^1)\partial_{x^2}$, we may assume $v_0 = 0$ so $Z = u(x^1)\partial_{x^1} + (a + v(x^1))x^2\partial_{x^2}$. The relations

$$\begin{array}{llll} \Gamma_{11}{}^1 = 0, & \Gamma_{11}{}^2 = 0, & \Gamma_{12}{}^1 = 0, & \Gamma_{21}{}^1 = 0, \\ \Gamma_{22}{}^1 = 0, & \Gamma_{22}{}^2 = 0, & (\Gamma_{12}{}^2 + \Gamma_{21}{}^2)v' + v'' = 0 & \end{array}$$

in Lemma 12.5 1 show that $x^2\partial_{x^2}$ is an affine Killing vector field. Since $x^2\partial_{x^2}$ commutes with Z, we obtain an effective Lie subalgebra of $\mathfrak{K}(\mathcal{M})$ isomorphic to $\mathfrak{K}_A$. □

12.4.5 THE ADJOINT ACTION ON $\mathfrak{K}(\mathcal{M})$. Let X be an affine Killing vector field with $X(P) \neq 0$. We impose the normalizations of Lemma 12.4 to assume

$$X = \partial_{x^2}, \quad \Gamma_{ij}{}^k(x^1, x^2) = \Gamma_{ij}{}^k(x^1), \quad \Gamma_{11}{}^1(x^1) = 0, \quad \Gamma_{11}{}^2(x^1) = 0\,.$$

Let $\operatorname{ad}(\partial_{x^2})(Y) := [X, Y]$ denote the adjoint action of ∂_{x^2} on $\mathfrak{K}_{\mathbb{C}}(\mathcal{M}) := \mathfrak{K}(\mathcal{M}) \otimes_{\mathbb{R}} \mathbb{C}$. Denote the generalized eigenspaces of this action by

$$\mathfrak{E}(\alpha) := \{X_\alpha \in \mathfrak{K}_{\mathbb{C}}(\mathcal{M}) : (\operatorname{ad}(\partial_{x^2}) - \alpha)^6 X_\alpha = 0\}\,.$$

Choose $X \in \mathfrak{E}(\alpha)$ for some α so $\{X, \partial_{x^2}\}$ is effective. Expand

$$X = e^{\alpha x^2}\sum_{i=0}^{i_0} u_i(x^1)(x^2)^i\partial_{x^1} + e^{\alpha x^2}\sum_{j=0}^{j_0} v_j(x^1)(x^2)^j\partial_{x^2}\,.$$

Since $\{X, \partial_{x^2}\}$ is effective, $u_i \neq 0$ for some i. Choose i_0 maximal so $u_{i_0} \neq 0$. By applying $(\operatorname{ad}(\partial_{x^2}) - \alpha)^{i_0-1}$ to X, we may assume that $i_0 = 0$ so

$$X = e^{\alpha x^2}\{u(x^1)\partial_{x^1} + \sum_{j=0}^{j_0} v_j(x^1)(x^2)^j\partial_{x^2}\} \quad \text{for} \quad u \neq 0\,. \tag{12.4.c}$$

We first examine $\mathfrak{E}(\alpha)$ for $\alpha \neq 0$. The case $\alpha = 0$ will be considered in Lemma 12.48.

Lemma 12.47 *If $\alpha \neq 0$, then there exists an effective Lie subalgebra of $\mathfrak{K}(\mathcal{M})$ isomorphic to $\mathfrak{K}_A$, $\mathfrak{K}_B$, or $\mathfrak{so}(3)$.*

Proof. Adopt the notation established above. We wish to show $j_0 = 0$. Suppose to the contrary that $v_j \neq 0$ for some $j > 0$. Choose v_{j_0} maximal so $v_{j_0} \neq 0$. Hence

$$0 \neq (\operatorname{ad}(\partial_{x^2}) - \alpha)^{j_0} X = j_0! e^{\alpha x^2} v_{j_0}(x^1)\partial_{x^2} \in \mathfrak{K}_\alpha(\mathcal{M})\,.$$

Lemma 12.5 2 now implies $u(x^1) = 0$ contrary to our assumption. Thus $j_0 = 0$. We consider subsequently the different possibilities for α to be real or complex.

Case 1. Suppose $\alpha \in \mathbb{R}$. Since $X = e^{\alpha x^2}\{u(x^1)\partial_{x^1} + v(x^1)\partial_{x^2}\}$, one has $[\partial_{x^2}, X] = \alpha X$ so $[\alpha^{-1}\partial_{x^2}, X] = X$. Since $\{X, \partial_{x^2}\}$ is effective, we have an effective Lie subalgebra isomorphic to $\mathfrak{K}_{\mathcal{B}}$.

Case 2. Suppose $\alpha \in \mathbb{C} \setminus \mathbb{R}$. By rescaling x^2, we may suppose $\alpha = a + \sqrt{-1}$ for $a \geq 0$. Now we consider the following possibilities.

Case 2.1. Assume that the real part $a \neq 0$. Choose a maximal so that there exists X in $\mathfrak{E}(a + \sqrt{-1})$ so $\{X, \partial_{x^2}\}$ is effective. Observe that $[\mathfrak{E}(\alpha), \mathfrak{E}(\beta)] \subset \mathfrak{E}(\alpha + \beta)$ for arbitrary complex numbers α and β. Indeed, $X_\alpha \in \mathfrak{E}(\alpha)$ and $X_\beta \in \mathfrak{E}(\beta)$ if and only if

$$X_\alpha = e^{\alpha x^2} \sum_i u_i^\alpha(x^1)(x^2)^i \partial_{x^1} + e^{\alpha x^2} \sum_j v_j^\alpha(x^1)(x^2)^j \partial_{x^2},$$
$$X_\beta = e^{\beta x^2} \sum_k \tilde{u}_k^\beta(x^1)(x^2)^k \partial_{x^1} + e^{\beta x^2} \sum_\ell \tilde{v}_\ell^\beta(x^1)(x^2)^\ell \partial_{x^2}.$$

This leads to an expansion for $[X_\alpha, X_\beta]$ where the relevant exponential is $e^{(\alpha+\beta)x^2}$ that shows $[X_\alpha, X_\beta] \in \mathfrak{E}(\alpha + \beta)$.

Expand $X \in \mathfrak{E}(a + \sqrt{-1})$ as $X = e^{ax^2} e^{\sqrt{-1}x^2}\{u(x^1)\partial_{x^1} + v(x^1)\partial_{x^2}\}$. We have now that $\bar{X} \in \mathfrak{E}(\bar{\alpha})$. Let $Y_1 := \sqrt{-1}[X, \bar{X}]$. We showed previously that

$$Y_1 \in \mathfrak{E}(\alpha + \bar{\alpha}) = \mathfrak{E}(2a).$$

Since $\bar{Y}_1 = Y_1$, Y_1 is real. Decompose

$$Y_1 = e^{2ax^2}\{u_1(x^1)\partial_{x^1} + v_1(x^1)\partial_{x^2}\}.$$

Case 2.1.1. If $u_1 \neq 0$, then we may apply Case 1 to Y_1.

Case 2.1.2. If $u_1 = 0$ and if $v_1 \neq 0$, then ${\Gamma_{22}}^2 = -2a$ by Lemma 12.5 2. Set

$$Y_2 := [X, Y_1] = e^{(3a+\sqrt{-1})x^2}\{u_2(x^1)\partial_{x^1} + v_2(x^1)\partial_{x^2}\} \in \mathfrak{E}(3a + \sqrt{-1}).$$

Then one of the following three possibilities hold.

Case 2.1.2.a. If $u_2 \neq 0$, this contradicts the maximality of a.

Case 2.1.2.b. If $u_2 = 0$ and $v_2 \neq 0$, then by Lemma 12.5 2 with ${\Gamma_{22}}^2(x^1) = -(3a + \sqrt{-1})$. This contradicts the fact that ${\Gamma_{22}}^2(x^1) = -2a$.

Case 2.1.2.c. Finally, if $u_2 = 0$ and $v_2 = 0$, then X and Y_1 commute. Since Y_1 is real, $[\Re(X), Y_1] = 0$ and $[\Im(X), Y_1] = 0$. Either $\{\Re(X), Y_1\}$ or $\{\Im(X), Y_1\}$ generates an effective 2-dimensional Lie subalgebra of $\mathfrak{K}(\mathcal{M})$ which is isomorphic to $\mathfrak{K}_{\mathcal{A}}$.

Case 2.1.3. If $u_1 = 0$ and $v_1 = 0$, then $[X, \bar{X}] = 0$ and $\{\Re(X), \Im(X), \partial_{x^2}\}$ span a Lie algebra

$$[\Im(X), \Re(X)] = 0, \quad [\Im(X), -\partial_{x^2}] = a\Im(X) + \Re(X), \quad [\Re(X), -\partial_{x^2}] = a\Re(X) - \Im(X)$$

isomorphic to $A_{3,7}^a$. Hence there exists an effective Lie subalgebra of $\mathfrak{K}(\mathcal{M})$ isomorphic to $\mathfrak{K}_{\mathcal{A}}$ by Lemma 12.46.

Case 2.2. Assume that α is purely imaginary $\alpha = \sqrt{-1}$. We have X_i in $\mathfrak{K}(\mathcal{M})$ with $\{X_i, \partial_{x^2}\}$ effective where

$$\begin{aligned} &X_1 = u(x^1, x^2)\partial_{x^1} + v(x^1, x^2)\partial_{x^2}, \quad X_2 = \partial_{x^2} X_1, \\ &u(x^1, x^2) = u_1(x^1)\cos(x^2) + u_2(x^1)\sin(x^2), \\ &v(x^1, x^2) = v_1(x^1)\cos(x^2) + v_2(x^1)\sin(x^2). \end{aligned}$$

Let $X_3 := [X_1, X_2] \in \mathfrak{E}(0)$. There are no polynomial terms in X_1 or X_2. Consequently,

$$X_3 = u_3(x^1)\partial_{x^1} + v_3(x^1)\partial_{x^2}$$

and one of the following possibilities pertain.

Case 2.2.1. If $u_3 \neq 0$, then $\{X_3, \partial_{x^2}\}$ is an effective Lie algebra isomorphic to $\mathfrak{K}_{\mathcal{A}}$.

Case 2.2.2. If $u_3 = 0$ but $v_3 \neq 0$, then $X_3 = v_3(x^1)\partial_{x^2}$ and one of the following two possibilities occurs.

Case 2.2.2.a. If $v_3' \neq 0$, then Assertion 2 in Lemma 12.5 1 now gives $u_1 = u_2 = 0$, which is false.

Case 2.2.2.b. Suppose $v_3' = 0$ so $[X_1, X_2]$ is a constant non-zero multiple of ∂_{x^2}. This gives the Lie algebra $\mathfrak{so}(3)$.

Case 2.2.3. If $X_3 = 0$, we have $[X_1, X_2] = 0$. Then $\{X = X_1, Y = -X_2, Z = \partial_{x^2}\}$ span the Lie algebra $A^0_{3,7}$ and we can apply Lemma 12.46. □

We finally examine $\mathfrak{E}(0)$.

Lemma 12.48 *Assume that $\alpha = 0$ and that there exists $X \in \mathfrak{E}(0)$ such that $\{X, \partial_{x^2}\}$ is effective. Then there exists an effective Lie subalgebra of $\mathfrak{K}(\mathcal{M})$ isomorphic to $\mathfrak{K}_{\mathcal{A}}$, $\mathfrak{K}_{\mathcal{B}}$, or $\mathfrak{so}(3)$.*

Proof. Choose $X \in \mathfrak{E}(0)$ of the form given in Equation (12.4.c), i.e.,

$$X = e^{\alpha x^2}\{u(x^1)\partial_{x^1} + \sum_{j=0}^{j_0} v_j(x^1)(x^2)^j \partial_{x^2}\} \quad \text{for} \quad u \neq 0.$$

If $j_0 = 0$, then $\{X, \partial_{x^2}\}$ is an effective algebra isomorphic to $\mathfrak{K}_{\mathcal{A}}$. We may therefore assume that $j_0 \geq 1$. We suppose $j_0 \geq 2$ and argue for a contradiction. Since $j_0 - 1 \leq 2j_0 - 3$, $u(x^1)\,\partial_{x^2}$ contributes lower-order terms and plays no role. Set:

$$\begin{aligned} Y_1 &:= [\partial_{x^2}, X] = \{c_1 v_{j_0}(x^1)(x^2)^{j_0-1} + O((x^2)^{j_0-2})\}\partial_{x^2}, \\ Y_2 &:= [X, Y_1] = \{c_2 v_{j_0}^2(x^1)(x^2)^{2(j_0-1)} + O((x^2)^{2(j_0-1)-1})\}\partial_{x^2}, \\ &\dots \qquad \dots \\ Y_n &:= [X, Y_{n-1}] = \{c_n v_{j_0}^n(x^1)(x^2)^{n(j_0-1)} + O((x^2)^{n(j_0-1)-1})\}\partial_{x^2} \end{aligned}$$

where the constants $c_n \neq 0$. This creates an infinite string of linearly independent elements of $\mathfrak{K}(\mathcal{M})$ which is not possible. We therefore suppose $j_0 = 1$ henceforth so

$$X = u(x^1)\partial_1 + (v_1(x^1)x^2 + v_0(x^1))\partial_{x^2} \quad \text{for} \quad v_1 \neq 0\,.$$

If $v_1' = 0$, then $[X, \partial_{x^2}] = v_1\partial_{x^2}$ and we obtain a subalgebra isomorphic to $\mathfrak{K}_\mathcal{A}$ or $\mathfrak{K}_\mathcal{B}$. We therefore suppose $v' \neq 0$ and apply Lemma 12.5 1 to obtain the relations of Assertion 1. If $w(x^1)\partial_{x^1} \in \mathfrak{K}(\mathcal{M})$, we obtain an affine Killing equation

$$K_{11}{}^2:\ 0 = (\Gamma_{12}{}^2 + \Gamma_{21}{}^2)w' + w'' = 0\,.$$

This is a linear homogeneous second-order ODE. Since both $v_1(x^1)$ and $\mathbb{1}$ satisfy this ODE, w is a linear combination of v_1 and $\mathbb{1}$. Let $\mathfrak{K}_0 := \operatorname{span}\{v(x^1)\partial_{x^2}, \partial_{x^2}\}$; $\operatorname{ad}(X)$ preserves this space. We now change our perspective and decompose $\mathfrak{K}_0 \otimes \mathbb{C}$ into generalized eigenspaces under the action of $\operatorname{ad}(X)$. If there is an eigenvalue β with $\beta \neq 0$, the analysis of Lemma 12.47 pertains since $\{X, \xi\}$ is an effective set for any $0 \neq \xi \in \mathfrak{K}_0$. On the other hand if 0 is an eigenvalue, there exists $\xi \in \mathfrak{K}_0$ so that $[X, \xi] = 0$ and there exists an effective Lie subalgebra which is isomorphic to $\mathfrak{K}_\mathcal{A}$. □

12.4.6 CLASSIFICATION OF AFFINE HOMOGENEOUS SURFACES.

The classification of affine homogeneous surfaces by Opozda [53] (see also Arias-Marco and Kowalski [2]) can now be stated in terms of the existence of distinguished coordinates as follows.

Theorem 12.49 *Let M be a smooth connected manifold and let ∇ be a connection, possibly with torsion, on the tangent bundle of M.*

1. *If $\mathcal{M} = (M, \nabla)$ is locally affine homogeneous, then there exists an effective Lie subalgebra $\tilde{\mathfrak{K}}$ of $\mathfrak{K}_P(\mathcal{M})$ which is isomorphic to $\mathfrak{K}_\mathcal{A}$, to $\mathfrak{K}_\mathcal{B}$, or to $\mathfrak{so}(3)$.*
   - *(a) If $\tilde{\mathfrak{K}} \approx \mathfrak{K}_\mathcal{A}$, then there exists a coordinate atlas so $\Gamma_{ij}{}^k \in \mathbb{R}$.*
   - *(b) If $\tilde{\mathfrak{K}} \approx \mathfrak{K}_\mathcal{B}$, then there exists a coordinate atlas so $\Gamma_{ij}{}^k = (x^1)^{-1}A_{ij}{}^k$ for $A_{ij}{}^k \in \mathbb{R}$.*
   - *(c) If $\tilde{\mathfrak{K}} \approx \mathfrak{so}(3)$, then there exists a coordinate atlas so ∇ is the Levi-Civita connection defined by the metric of the round sphere.*
2. *If there exists an effective Lie subalgebra of $\mathfrak{K}(\mathcal{M})$ which is isomorphic to $\mathfrak{K}_\mathcal{A}$, $\mathfrak{K}_\mathcal{B}$, or $\mathfrak{so}(3)$, then $\mathcal{M}$ is locally affine homogeneous.*
3. *If there exists a coordinate atlas for $\mathcal{M}$ normalized as in Assertions 1-a, 1-b or 1-c, then $\mathcal{M}$ is locally affine homogeneous.*

Proof. The proof of Theorem 12.49 now follows from the previous lemmas. Lemma 12.47 and Lemma 12.48 show the existence of an effective Lie subalgebra $\mathfrak{K}_0 \subset \mathfrak{K}(\mathcal{M})$ isomorphic to $\mathfrak{K}_\mathcal{A}$, $\mathfrak{K}_\mathcal{B}$ or $\mathfrak{so}(3)$. The existence of distinguished coordinates then follows from Lemma 12.44. The remaining assertions follow from Lemma 12.41 and Lemma 12.43. □

We note, in passing, that recent work of D'Ascanio et al. [24] completes the analysis of Assertion 1 by determining exactly which Lie algebras arise, up to isomorphism, as the Lie algebras of affine Killing vector fields of locally affine homogeneous simply connected manifolds; this is implicit, of course, in the work of Arias-Marco and Kowalski [2].

12.4.7 METRIZABILITY. If (M, g) is a pseudo-Riemannian manifold, then we can obtain an affine manifold by taking ∇ to be the Levi-Civita connection; such an affine structure is said to be *metrizable*. There are, however, many affine manifolds which are not metrizable. For example, the Ricci tensor of any pseudo-Riemannian manifold is symmetric; in Chapter 14, we will discuss structures where this fails and which therefore are not metrizable.

The situation is more tractable in the 2-dimensional case since the Ricci tensor of (M, g) satisfies $\rho = \frac{1}{2}\,{}^{g}\mathrm{Sc}\, g$, where ${}^{g}\mathrm{Sc}$ denotes the scalar curvature of (M, g). Hence the Ricci tensor is either parallel or recurrent since $\nabla\rho = d\,{}^{g}\mathrm{Sc} \otimes \rho$ (see Vanžurová [60]).

Locally homogeneous metrizable affine connections on surfaces were determined by Kowalski, Opozda, and Vlášek [43] as follows.

Theorem 12.50 *The only 2-dimensional locally non-homogeneous pseudo-Riemannian metrics with locally homogeneous Levi-Civita connection are those which are modeled on* $\mathcal{N}_\varepsilon := (\mathbb{R}^+ \times \mathbb{R}, g_\varepsilon^k)$, *where* $g_\varepsilon^k = (x^1)^{2k}(dx^1 \otimes dx^1 + \varepsilon\, dx^2 \otimes dx^2)$ *for* $\varepsilon = \pm 1$ *and* $k \neq 0, -1$.

The corresponding connections, which are of Type $\mathcal{B}$, are completely determined by ${}^{\varepsilon}\Gamma_{ij}{}^{k} = (x^1)^{-1}\,{}^{\varepsilon}A_{ij}{}^{k}$, where

$$
{}^{\varepsilon}A_{11}{}^{1} = k,\ {}^{\varepsilon}A_{11}{}^{2} = 0,\ {}^{\varepsilon}A_{12}{}^{1} = 0,\ {}^{\varepsilon}A_{12}{}^{2} = k,\ {}^{\varepsilon}A_{22}{}^{1} = -\varepsilon k,\ {}^{\varepsilon}A_{22}{}^{2} = 0\,.
$$

It is now a routine calculation to show that $\rho_\varepsilon = \frac{k}{(x^1)^2}(dx^1 \otimes dx^1 + \varepsilon\, dx^2 \otimes dx^2)$ is recurrent, symmetric, and of rank 2. Furthermore, $\dim(\mathfrak{K}(\mathcal{N}_\varepsilon)) = 2$ and moreover $f(x^1) = (x^1)^{1+2k}$ is a solution of the affine quasi-Einstein equation for $\mu = 1 + 2k$.

12.5 ANALYTIC STRUCTURE FOR HOMOGENEOUS AFFINE SURFACES

In Section 12.5.1, we examine the structure group of a Type $\mathcal{A}$ or Type $\mathcal{B}$ coordinate atlas. We use this analysis in Section 12.5.2 to show such an atlas is real analytic and thereby, in light of Theorem 12.49, give a natural real analytic structure to any locally affine homogeneous surface. In Section 12.5.3, we examine the module structure of various tensors under the action of the Lie algebras $\mathfrak{K}_\mathcal{A}$ and $\mathfrak{K}_\mathcal{B}$ of the two 2-dimensional Lie groups.

12.5.1 COORDINATE ATLAS. The general linear group $\mathrm{GL}(2, \mathbb{R})$ acts by matrix multiplication on $\mathbb{R}^2$; we say that two connections on $\mathbb{R}^2$ are linearly equivalent if they are intertwined by an element of $\mathrm{GL}(2, \mathbb{R})$. We say that two connections on $\mathbb{R}^+ \times \mathbb{R}$ are linearly equivalent if they are intertwined by a linear transformation $(x^1, x^2) \to (x^1, ux^1 + vx^2)$ for $u \in \mathbb{R}$ and $v \neq 0$.

Theorem 12.51 *Let $\mathcal{M}$ be a locally affine homogeneous surface with* $\dim(\mathfrak{K}_P(\mathcal{M})) = 2$ *for all points P in M. If $\mathcal{M}$ admits a Type $\mathcal{A}$ atlas, then the coordinate transformations have the form $\vec{x} \to A\vec{x} + \vec{b}$ for $A \in \mathrm{GL}(2, \mathbb{R})$ and $\vec{b} \in \mathbb{R}^2$. If $\mathcal{M}$ admits a Type $\mathcal{B}$ atlas, then coordinate transformations have the form $(x^1, x^2) \to (ax^1, bx^1 + cx^2 + d)$ for $a > 0$ and $c \neq 0$.*

Proof. Let $(\mathcal{O}_\alpha, \phi_\alpha)$ be a Type $\mathcal{A}$ coordinate atlas. Let ∇_α be the associated Type $\mathcal{A}$ connections on $\mathcal{O}_\alpha$ and let $\mathfrak{K}_\alpha$ be the associated Lie algebras of affine Killing vector fields. Then the transition functions $\phi_{\alpha\beta}$ intertwine ∇_α and ∇_β and, consequently, intertwine the Lie algebras $\mathfrak{K}_\alpha$ and $\mathfrak{K}_\beta$. Suppose that $\dim(\mathfrak{K}_\alpha) = 2$. We then have $\mathfrak{K}_\alpha = \mathrm{span}\{\partial_{x_\alpha^1}, \partial_{x_\alpha^2}\}$. Consequently,

$$(\phi_{\alpha\beta})_* \partial_{x_\alpha^i} = a_i^j \partial_{x_\beta^j} \text{ for } (a_i^j) \in \mathrm{GL}(2, \mathbb{R}) .$$

If we express

$$\phi_{\alpha\beta}(x^1, x^2) = (\phi_{\alpha\beta}^1(x^1, x^2), \phi_{\alpha\beta}^2(x^1, x^2)) ,$$

then $\partial_{x^i} \phi_{\alpha\beta}^j = a_i^j$ is constant so $\phi_{\alpha\beta}$ has the required form.

Suppose the atlas is a Type $\mathcal{B}$ atlas. Let $X := x^1\partial_{x^1} + x^2\partial_{x^2}$. Since the left-action of the $ax + b$ group defines a transitive action on $\mathbb{R}^+ \times \mathbb{R}$, we may assume without loss of generality that the transition functions $\phi_{\alpha\beta}$ satisfy $\phi_{\alpha\beta}(1, 0) = (1, 0)$. Suppose $\dim(\mathfrak{K}(\mathcal{M})) = 2$. Expand $\phi_{\alpha\beta} = (\phi_{\alpha\beta}^1, \phi_{\alpha\beta}^2)$. We must show $\phi_{\alpha\beta}^1 = x^1$ and $\phi_{\alpha\beta}^2 = bx^1 + cx^2$ for $c \neq 0$. Because $\mathfrak{K}_\alpha = \mathrm{span}\{X, \partial_{x^2}\}$, $(\phi_{\alpha\beta})_* X = rX + s\partial_{x^2}$ and $(\phi_{\alpha\beta})_* \partial_{x^2} = tX + v\partial_{x^2}$ for suitably chosen constants. We have that $\partial_{x^2} = [\partial_{x^2}, X]$ and thus

$$tX + v\partial_{x^2} = [tX + v\partial_{x^2}, rX + s\partial_{x^2}] = (vr - ts)\partial_{x^2} .$$

This implies that $t = 0$ and $r = 1$ so $(\phi_{\alpha\beta})_* X = X + s\partial_{x^2}$ and $(\phi_{\alpha\beta})_* \partial_{x^2} = v\partial_{x^2}$. Let $\tilde{\phi}(x^1, x^2) = (x^1, \alpha x^1 + \beta x^2 - \alpha)$. Then $\tilde{\phi}(1, 0) = (1, 0)$, and

$$\begin{aligned}
&\tilde{\phi}^*(dx^1) = dx^1, && \tilde{\phi}^*(dx^2) = \alpha dx^1 + \beta dx^2, \\
&\tilde{\phi}_*(\partial_{x^1}) = \partial_{x^1} - \alpha\beta^{-1}\partial_{x^2}, && \tilde{\phi}_*(\partial_{x^2}) = \beta^{-1}\partial_{x^2}, \\
&\tilde{\phi}_*(x^1\partial_{x^1} + x^2\partial_{x^2}) = x^1(\partial_{x^1} - \alpha\beta^{-1}\partial_{x^2}) + (\alpha x^1 + \beta x^2 - \alpha)(\beta^{-1}\partial_{x^2}) \\
&\qquad = (x^1\partial_{x^1} + x^2\partial_{x^2}) - \alpha\beta^{-1}\partial_{x^2} .
\end{aligned}$$

To ensure $(\tilde{\phi} - \phi_{\alpha\beta})_* = 0$, we take $-\alpha\beta^{-1} = s$ and $\beta^{-1} = v$. We then have $\tilde{\phi}_* - \phi_{\alpha\beta}$ is constant. Since $(\tilde{\phi} - \phi_{\alpha\beta})(1, 0) = (0, 0)$, we conclude $\tilde{\phi} = \phi_{\alpha\beta}$ as desired. □

12.5.2 REAL ANALYTIC STRUCTURES.

Theorem 12.52 *If $\mathcal{M}$ is a Type $\mathcal{A}$ (resp. Type $\mathcal{B}$) surface, then the Type $\mathcal{A}$ (resp. Type $\mathcal{B}$) coordinate atlas is real analytic.*

Proof. Suppose first that $\mathcal{M}$ has a Type $\mathcal{A}$ model. Let $\{\mathcal{U}_\alpha, \vec{x}_\alpha\}$ be the associated Type $\mathcal{A}$ coordinate atlas. The transition map $\Phi_{\alpha\beta}$ from an open subset of $\mathcal{U}_\alpha$ to an open subset of $\mathcal{U}_\beta$ is an affine map. We must show $\Phi_{\alpha\beta}$ is real analytic. The vector fields $\{\partial_{x_\beta^1}, \partial_{x_\beta^2}\}$ are affine Killing vector fields. The dual frame for the cotangent bundle takes the form $\{dx_\beta^1, dx_\beta^2\}$. Since $\Phi_{\alpha\beta}$ is an affine morphism, $\{\Phi_{\alpha\beta}^*\partial_{x_\beta^1}, \Phi_{\alpha\beta}^*\partial_{x_\beta^2}\}$ are affine Killing vector fields on $\mathcal{U}_\alpha$. Consequently, by Lemma 12.14, this is a real analytic frame. Thus, the dual frame for the cotangent bundle $\{\Phi_{\alpha\beta}^* dx_\beta^1, \Phi_{\alpha\beta}^* dx_\beta^2\}$ is real analytic as well. This implies that the coordinate functions $\Phi_{\alpha\beta}^* x_\beta^1$ and $\Phi_{\alpha\beta}^* x_\beta^2$ are real analytic. Consequently, $\Phi_{\alpha\beta}$ is real analytic. This shows that the Type $\mathcal{A}$ coordinate atlas is real analytic.

Suppose next that $\mathcal{M}$ has a Type $\mathcal{B}$ model. If $\dim(\mathfrak{K}(\mathcal{M})) = 2$, we may apply Theorem 12.51 to see that $\mathcal{M}$ is real analytic. We will show subsequently in Theorem 14.17 that if $\dim(\mathfrak{K}(\mathcal{M})) = 4$, then $\mathcal{M}$ is also a Type $\mathcal{A}$ geometry and the analysis performed above pertains. The only remaining possibility is that $\dim(\mathfrak{K}(\mathcal{M})) = 3$. The Riemannian and Lorentzian hyperbolic planes are real analytic. If $\mathcal{M}$ is not modeled on the Riemannian or Lorentzian hyperbolic plane, then Lemma 14.15 shows that

$$\mathfrak{K}(\mathcal{M}) = \operatorname{span}\{e_1 := x^1\partial_{x^1} + x^2\partial_{x^2},\ e_2 := \partial_{x^2},\ e_3 := 2x^1x^2\partial_{x^1} + (x^2)^2\partial_{x^2}\}.$$

Thus the elements of $\mathfrak{K}(\mathcal{M})$ are real analytic. Note that $\{e_1, e_2\}$ is a frame for the tangent bundle. Let $e^1 := (x^1)^{-1}dx^1$ and $e^2 := (x^1)^{-1}x^2dx^1 - dx^2$ be the corresponding dual frame. Since $\Phi^*(e_1)$ and $\Phi^*(e_2)$ are real analytic, we conclude $\Phi^*(e^1)$ and $\Phi^*(e^2)$ are real analytic. We have $\Phi^*(e^1)\{\Phi^*(e_3)\} = 2\Phi^*(x^2)$ so $\Phi^*(x^2)$ is real analytic. Furthermore $\Phi^*(e^1) = d\log(\Phi^*(x^1))$, and thus $\Phi^*(x^1)$ is real analytic as well. This implies Φ is real analytic as desired. □

12.5.3 $\mathfrak{K}_\mathcal{A}$ AND $\mathfrak{K}_\mathcal{B}$-MODULES.

Recall that $\mathfrak{K}_\mathcal{A} = \operatorname{span}\{\partial_{x^1}, \partial_{x^2}\}$ is the Lie algebra of $\mathbb{R}^2$ and that $\mathfrak{K}_\mathcal{B} = \operatorname{span}\{x^1\partial_{x^1} + x^2\partial_{x^2}, \partial_{x^2}\}$ is the Lie algebra of the $ax+b$ group. We define the associated polynomial algebras of functions by setting

$$\mathfrak{P}^\mathcal{A} := \mathbb{C}[e^{\alpha_1x^1+\alpha_2x^2}, x^1, x^2]_{\alpha_i\in\mathbb{C}} \quad\text{and}\quad \mathfrak{P}^\mathcal{B} := \mathbb{C}[(x^1)^\alpha, \log(x^1), x^2]_{\alpha\in\mathbb{C}}.$$

Lemma 12.53 *Let $\mathcal{E}$ be a k-dimensional space of smooth complex functions on $\mathbb{R}^2$ (resp. $\mathbb{R}^+\times\mathbb{R}$). If $\mathcal{E}$ is invariant under the action of $\mathfrak{K}_\mathcal{A}$ (resp. $\mathfrak{K}_\mathcal{B}$), then $\mathcal{E}\subset\mathfrak{P}^\mathcal{A}$ (resp. $\mathcal{E}\subset\mathfrak{P}^\mathcal{B}$).*

Proof. Suppose $\mathcal{E}$ is invariant under the action of $\mathfrak{K}_\mathcal{A}$. Decompose $\mathcal{E} = \oplus_{\alpha_1,\alpha_2}E_{\alpha_1,\alpha_2}$ where

$$E_{\alpha_1,\alpha_2} := \{f\in\mathcal{E} : (\partial_{x^1}-\alpha_1)^k f = 0 \text{ and } (\partial_{x^2}-\alpha_2)^k f = 0\}$$

are the simultaneous generalized eigenspaces of ∂_{x^1} and ∂_{x^2}. Let

$$f(x^1,x^2) = e^{\alpha_1x^1+\alpha_2x^2}\tilde f(x^1,x^2)\in E_{\alpha_1,\alpha_2}.$$

We have $0 = (\partial_{x^1}-\alpha_1)^k f = e^{\alpha_1x^1+\alpha_2x^2}\partial_{x^1}^k\tilde f$ and $0 = (\partial_{x^2}-\alpha_2)^k f = e^{\alpha_1x^1+\alpha_2x^2}\partial_{x^2}^k\tilde f$. Consequently, $\partial_{x^1}^k\tilde f = 0$ and $\partial_{x^2}^k\tilde f = 0$. This implies $\tilde f$ is polynomial.

Suppose $\mathcal{E}$ is invariant under the action of $\mathfrak{K}_{\mathcal{B}}$. We may decompose $\mathcal{E} = \oplus_\alpha E_\alpha$ into the generalized eigenspaces of $X := x^1\partial_{x^1} + x^2\partial_{x^2}$ where

$$E_\alpha := \{f \in \mathcal{E} : (X - \alpha)^k f = 0\}.$$

If $(X - \alpha)^k f = 0$, then $f = (x^1)^\alpha \sum_{i \leq k} \log(x^1)^i f_i(x^2)$. Because $[X, \partial_{x^2}] = -\partial_{x^2}$,

$$(X - \alpha - 1)^k \partial_{x^2} f = \partial_{x^2}(X - \alpha)^k f = 0 \quad \text{so} \quad \partial_{x^2} : E_\alpha \to E_{\alpha-1}.$$

Since $\dim(\mathcal{E}) = k$, $\partial_{x^2}^k f = 0$. Consequently, each of the $f_i(x^2)$ is polynomial in x^2. □

We complexify and set $E_{\mathbb{C}}(\mu, \mathcal{M}) := E(\mu, \mathcal{M}) \otimes_{\mathbb{R}} \mathbb{C}$ and $\mathfrak{K}_{\mathbb{C}}(\mathcal{M}) := \mathfrak{K}(\mathcal{M}) \otimes_{\mathbb{R}} \mathbb{C}$. We can take the real and imaginary parts to obtain corresponding real bases. It follows from Lemma 12.14 that the elements of $\mathfrak{K}(\mathcal{M})$ and $E(\mu, \mathcal{M})$ are real analytic if $\mathcal{M}$ determines either a Type $\mathcal{A}$ or a Type $\mathcal{B}$ geometry. We use Lemma 12.14 and Lemma 12.53 to establish the following result, which makes this observation much more specific; it will inform our discussion of $E(\mu, \mathcal{M})$ and $\mathfrak{K}(\mathcal{M})$ subsequently in Chapter 13 and Chapter 14.

Lemma 12.54 *Let $f \in E_{\mathbb{C}}(\mu, \mathcal{M})$. Let $\xi \in \mathfrak{K}_{\mathbb{C}}(\mathcal{M})$. Decompose $\xi = \xi_1\partial_{x^1} + \xi_2\partial_{x^2}$. If $\mathcal{M}$ is a Type $\mathcal{A}$ (resp. Type $\mathcal{B}$) model, then $\{f, \xi_1, \xi_2\} \subset \mathfrak{P}^{\mathcal{A}}$ (resp. $\{f, \xi_1, \xi_2\} \subset \mathfrak{P}^{\mathcal{B}}$).*

Proof. By Lemma 12.14, $\mathfrak{K}(\mathcal{M})$ and $E(\mu, \mathcal{M})$ are finite-dimensional $\mathfrak{K}(\mathcal{M})$-modules. If $\mathcal{M}$ is a Type $\mathcal{A}$ model, then $\mathfrak{K}_{\mathcal{A}}$ is a Lie subalgebra of $\mathfrak{K}(\mathcal{M})$; if $\mathcal{M}$ is a Type $\mathcal{B}$ model, then $\mathfrak{K}_{\mathcal{B}}$ is a Lie subalgebra of $\mathfrak{K}(\mathcal{M})$. Let $f \in E_{\mathbb{C}}(\mu, \mathcal{M})$. We apply Lemma 12.53 to see $f \in \mathfrak{P}^{\mathcal{A}}$ if $\mathcal{M}$ is a Type $\mathcal{A}$ model and $f \in \mathfrak{P}^{\mathcal{B}}$ if $\mathcal{M}$ is a Type $\mathcal{B}$ model. Suppose that

$$\xi = \xi_1\partial_{x^1} + \xi_2\partial_{x^2} \in \mathfrak{K}_{\mathbb{C}}(\mathcal{M})$$

and that $\mathcal{M}$ is a Type $\mathcal{A}$ model. Then $[\partial_{x^i}, \xi] = \partial_{x^i}\xi_1\partial_{x^1} + \partial_{x^i}\xi_2\partial_{x^2}$ and, consequently, ξ_i belongs to $\mathfrak{P}^{\mathcal{A}}$. Let $X = x^1\partial_{x^1} + x^2\partial_{x^2}$. If $\mathcal{M}$ is a Type $\mathcal{B}$ model, then

$$[X, \xi] = (X(\xi_1) - \xi_1)\partial_{x^1} + (X(\xi_2) - \xi_2)\partial_{x^2} \quad \text{and} \quad [\partial_{x^2}, \xi] = \partial_{x^2}(\xi_1)\partial_{x^1} + \partial_{x^2}(\xi_2)\partial_{x^2}.$$

Thus although the module action of $\mathfrak{K}_{\mathcal{B}}$ on $\mathfrak{K}(\mathcal{M})$ is a bit different than the action on C^∞, the analysis of Lemma 12.53 again yields $\xi_i \in \mathfrak{P}^{\mathcal{B}}$. □

CHAPTER 13

The Geometry of Type $\mathcal{A}$ Models

In Chapter 13, we shall report on work of Brozos-Vázquez et al. [9–12] and Gilkey and Valle-Regueiro [35] that deals with the affine quasi-Einstein equation, work of Brozos-Vázquez, García-Río, and Gilkey [7] that deals with affine Killing vector fields and gradient Ricci solitons, work of Brozos-Vázquez, García-Río, and Gilkey [6] that deals with moduli spaces, and work of other authors as cited. We shall assume that $\mathcal{M} = (\mathbb{R}^2, \nabla)$ is a Type $\mathcal{A}$ model unless otherwise noted; this means that the Christoffel symbols of ∇ are constant. We begin by recalling some results from Chapter 12 that will play a crucial role in what follows.

Observation 13.1 We have defined $\mathcal{Q}(\mathcal{M}) = \{f \in C^\infty(\mathbb{R}^2) : \mathcal{H}f + f\rho_s = 0\}$ for a surface $\mathcal{M}$. Let g be a smooth function which defines a strong projective equivalence between two affine manifolds $\mathcal{M}$ and ${}^g\mathcal{M}$, i.e.,

$$\begin{array}{ll} {}^g\Gamma_{11}{}^1 = 2\partial_{x^1}g + \Gamma_{11}{}^1, & {}^g\Gamma_{11}{}^2 = \Gamma_{11}{}^2, \\ {}^g\Gamma_{12}{}^1 = \partial_{x^2}g + \Gamma_{12}{}^1, & {}^g\Gamma_{12}{}^2 = \partial_{x^1}g + \Gamma_{12}{}^2, \\ {}^g\Gamma_{22}{}^1 = \Gamma_{22}{}^1, & {}^g\Gamma_{22}{}^2 = 2\partial_{x^2}g + \Gamma_{22}{}^2 . \end{array}$$

We then have by Theorem 12.29 that $\mathcal{Q}({}^g\mathcal{M}) = e^g\mathcal{Q}(\mathcal{M})$. Suppose that $\mathcal{M}$ is a Type $\mathcal{A}$ model. We will show in Lemma 13.6 that $\mathcal{M}$ is strongly projectively flat. Consequently, by Theorem 12.34, $\dim(\mathcal{Q}(\mathcal{M})) = 3$. The space of functions $\mathcal{Q}(\mathcal{M})$ is a complete invariant of Type $\mathcal{A}$ models; by Theorem 12.34, if $\mathcal{M}_i$ are Type $\mathcal{A}$ models, then $\mathcal{M}_1 = \mathcal{M}_2$ if and only if $\mathcal{Q}(\mathcal{M}_1) = \mathcal{Q}(\mathcal{M}_2)$. Let Ψ be a diffeomorphism from M_1 to M_2. Then Ψ intertwines $\mathcal{M}_1$ and $\mathcal{M}_2$ if and only if $\Psi^*\mathcal{Q}(\mathcal{M}_2) = \mathcal{Q}(\mathcal{M}_1)$. By Lemma 12.53, if $\mathcal{M}$ is a Type $\mathcal{A}$ model, then $\mathcal{Q}(\mathcal{M})$ is a 3-dimensional space which is spanned by products of linear exponentials and polynomials and which is invariant under the action of ∂_{x^i}. This will inform our discussion.

In Section 13.1, we present some foundational results. In Section 13.2, see Definition 13.7, we will present some basic examples of Type $\mathcal{A}$ models. We will determine the Ricci tensor, the algebra of affine Killing vector fields, and the spaces $\mathcal{Q}$ for these examples. In Section 13.3, we will use the analytic facts outlined in Observation 13.1 to show that any Type $\mathcal{A}$ model is linearly equivalent to one of the examples discussed in Section 13.2. The α invariant of Lemma 12.22 is determined when the Ricci tensor has rank 1. Theorem 12.35 gives an ansatz for determining the geodesic structure of a strongly projectively flat surface. We will use this ansatz subsequently to determine which Type $\mathcal{A}$ models are geodesically complete. In Section 13.4, we discuss moduli spaces of Type $\mathcal{A}$ models.

13.1 TYPE $\mathcal{A}$: FOUNDATIONAL RESULTS AND BASIC EXAMPLES

In Section 13.1.1, we discuss various results concerning the Ricci tensor for Type $\mathcal{A}$ geometries. In Section 13.1.2, we give a useful criteria to ensure the Ricci tensor has rank 1 and show such a tensor is recurrent. In Section 13.1.3, we relate the rank of the space of affine Killing vector fields to the Ricci tensor (see Lemma 13.4); we will subsequently improve this result in Corollary 13.25, but this weaker form will be useful in our analysis at this stage. In Section 13.1.4, we show linear equivalence and affine equivalence are equivalent concepts if the Ricci tensor has rank 2; this fails if the Ricci tensor is degenerate (see Remark 13.13). In Section 13.1.5, we will give a direct combinatorial proof that any Type $\mathcal{A}$ geometry is linearly strongly projectively flat. This will permit us to use various results concerning the solution space to the affine quasi-Einstein equation $\mathcal{Q}$ which were established in Chapter 12.

13.1.1 THE RICCI TENSOR.

Lemma 13.2 *Let $\mathcal{M} = (\mathbb{R}^2, \nabla)$ be a Type $\mathcal{A}$ model.*

1. *ρ is symmetric and $\nabla\rho$ is totally symmetric.*
2. *Let $f \in E(\mu, \mathcal{M})$ for $\mu \neq -1$. Then $R_{12}(df) = 0$.*
3. *Let f be an affine gradient Ricci soliton. Then $R_{12}(df) = 0$.*

Proof. Let $\mathcal{M}$ be a Type $\mathcal{A}$ model. We show that ρ is symmetric by computing:

$$\begin{aligned}
\rho_{11} &= (\Gamma_{11}{}^1 - \Gamma_{12}{}^2)\Gamma_{12}{}^2 + \Gamma_{11}{}^2(\Gamma_{22}{}^2 - \Gamma_{12}{}^1),\\
\rho_{12} &= \rho_{21} = \Gamma_{12}{}^1\Gamma_{12}{}^2 - \Gamma_{11}{}^2\Gamma_{22}{}^1,\\
\rho_{22} &= -(\Gamma_{12}{}^1)^2 + \Gamma_{22}{}^2\Gamma_{12}{}^1 + (\Gamma_{11}{}^1 - \Gamma_{12}{}^2)\Gamma_{22}{}^1\,.
\end{aligned}\tag{13.1.a}$$

We show that $\nabla\rho$ is totally symmetric and establish Assertion 1 by computing:

$$\begin{aligned}
\rho_{11;1} &= 2\{-(\Gamma_{11}{}^1)^2\Gamma_{12}{}^2 + \Gamma_{11}{}^1(\Gamma_{11}{}^2(\Gamma_{12}{}^1 - \Gamma_{22}{}^2) + (\Gamma_{12}{}^2)^2)\\
&\qquad +\Gamma_{11}{}^2(\Gamma_{11}{}^2\Gamma_{22}{}^1 - \Gamma_{12}{}^1\Gamma_{12}{}^2)\},\\
\rho_{12;1} &= \rho_{21;1} = \rho_{11;2} = 2\left(\Gamma_{11}{}^2\left((\Gamma_{12}{}^1)^2 - \Gamma_{12}{}^1\Gamma_{22}{}^2 + \Gamma_{12}{}^2\Gamma_{22}{}^1\right) - \Gamma_{11}{}^1\Gamma_{12}{}^1\Gamma_{12}{}^2\right),\\
\rho_{12;2} &= \rho_{21;2} = \rho_{22;1} = 2\left(\Gamma_{12}{}^2(-\Gamma_{11}{}^1\Gamma_{22}{}^1 - \Gamma_{12}{}^1\Gamma_{22}{}^2 + \Gamma_{12}{}^2\Gamma_{22}{}^1) + \Gamma_{11}{}^2\Gamma_{12}{}^1\Gamma_{22}{}^1\right),\\
\rho_{22;2} &= 2\{\Gamma_{22}{}^1(\Gamma_{22}{}^2(\Gamma_{12}{}^2 - \Gamma_{11}{}^1) + \Gamma_{11}{}^2\Gamma_{22}{}^1) + (\Gamma_{12}{}^1)^2\Gamma_{22}{}^2\\
&\qquad -\Gamma_{12}{}^1(\Gamma_{12}{}^2\Gamma_{22}{}^1 + (\Gamma_{22}{}^2)^2)\}\,.
\end{aligned}$$

Let $f \in E(\mu, \mathcal{M})$. We covariantly differentiate the relation $\mathcal{H}f = \mu f \rho$ to conclude that $f_{;jki} = \mu\{f_{;i}\rho_{jk} + f\rho_{jk;i}\}$. Since $\nabla\rho$ is totally symmetric, anti-symmetrize in i and k to see

$$R_{kij}{}^{\ell} f_{;\ell} = f_{;jki} - f_{;jik} = \mu\{f_{;i}\rho_{jk} - f_{;k}\rho_{ij}\}\,.$$

We take $i = 2$ and $k = 1$ and unpack this expression after setting $\rho_{12} = \rho_{21}$:

$$\begin{aligned}
(j = 1) \qquad & R_{121}{}^{1} f_{;1} + R_{121}{}^{2} f_{;2} = \mu(f_{;2}\rho_{11} - f_{;1}\rho_{12}) \\
& \quad = \mu\{f_{;2}\rho_{11} - f_{;1}\rho_{21}\} = \mu(-R_{121}{}^{1} f_{;1} + R_{211}{}^{2} f_{;2}), \\
(j = 2) \qquad & R_{122}{}^{1} f_{;1} + R_{122}{}^{2} f_{;2} = \mu(f_{;2}\rho_{12} - f_{;1}\rho_{22}) \\
& \quad = \mu(f_{;2}\rho_{21} - f_{;1}\rho_{22}) = \mu(-R_{122}{}^{1} f_{;1} + R_{212}{}^{2} f_{;2})\,.
\end{aligned}$$

This shows $(\mu + 1)R_{12j}{}^{\ell} f_{;\ell} = 0$ for $j = 1, 2$ or, equivalently, $R_{12}(df) = 0$ since $\mu \neq -1$. This establishes Assertion 2; the proof of Assertion 3 follows by applying the same argument to the identity $f_{;jki} = -\rho_{jk;i}$. □

13.1.2 RECURRENT TENSORS. The following is a useful observation.

Lemma 13.3 *Let $\mathcal{M}$ be a Type $\mathcal{A}$ model which is not flat.*

1. *The following conditions are equivalent:*
   (a) $\rho = \rho_{22} dx^2 \otimes dx^2$.
   (b) $\Gamma_{11}{}^{2} = 0$ *and* $\Gamma_{12}{}^{2} = 0$.
   (c) $\rho = \{\Gamma_{12}{}^{1}(\Gamma_{22}{}^{2} - \Gamma_{12}{}^{1}) + \Gamma_{11}{}^{1}\Gamma_{22}{}^{1}\} dx^2 \otimes dx^2$.
2. *$\mathcal{M}$ is recurrent if and only if* $\operatorname{Rank}(\rho) = 1$.

Proof. We use Equation (13.1.a) to compute ρ. We first show that Assertion 1-a implies Assertion 1-b. We distinguish cases.

Case 1. Suppose that $\Gamma_{22}{}^{1}$ is non-zero. By rescaling, we may suppose $\Gamma_{22}{}^{1} = 1$. To ensure $\rho_{12} = 0$, we set $\Gamma_{11}{}^{2} = \Gamma_{12}{}^{1}\Gamma_{12}{}^{2}$ and obtain $\rho_{11} = \Gamma_{12}{}^{2}\rho_{22}$. Since $\rho_{22} \neq 0$, $\Gamma_{12}{}^{2} = 0$, and hence $\Gamma_{11}{}^{2} = 0$ as well as desired.

Case 2. Suppose that $\Gamma_{22}{}^{1} = 0$. Then $0 = \rho_{12} = \Gamma_{12}{}^{1}\Gamma_{12}{}^{2}$. Since $\rho_{22} = \Gamma_{12}{}^{1}(\Gamma_{22}{}^{2} - \Gamma_{12}{}^{1})$, $\Gamma_{12}{}^{1} \neq 0$. Thus $\Gamma_{12}{}^{2} = 0$, $\rho_{11} = \Gamma_{11}{}^{2}(\Gamma_{22}{}^{2} - \Gamma_{12}{}^{1})$, and $\rho_{22} = \Gamma_{12}{}^{1}(\Gamma_{22}{}^{2} - \Gamma_{12}{}^{1})$. Consequently, $\Gamma_{11}{}^{2} = 0$ as desired.

We have shown that Assertion 1-a implies Assertion 1-b. We make a direct computation to show that Assertion 1-b implies Assertion 1-c. It is immediate that Assertion 1-c implies Assertion 1-a. This completes the proof of Assertion 1. We now establish Assertion 2. If $\operatorname{Rank}(\rho) = 1$, we can change coordinates to assume $\rho = \rho_{22} dx^2 \otimes dx^2$. We then use Assertion 1 to see that $\Gamma_{11}{}^{2} = \Gamma_{12}{}^{2} = 0$. This yields

$$\rho = \rho_{22} dx^2 \otimes dx^2 \quad \text{and} \quad \nabla\rho = -2\Gamma_{22}{}^{2}\rho_{22} dx^2 \otimes dx^2 \otimes dx^2\,.$$

Consequently, ρ is recurrent. Conversely, suppose ρ is recurrent. Make a linear change of coordinates to diagonalize ρ. Since $\mathcal{M}$ is not flat, ρ does not vanish identically and thus we may assume the notation is chosen so $\rho_{22} \neq 0$. Again we distinguish cases setting $\rho_{12} = 0$.

Case 1. Suppose that $\Gamma_{22}{}^1 \neq 0$. We rescale to assume $\Gamma_{22}{}^1 = 1$ and set $\Gamma_{11}{}^2 = \Gamma_{12}{}^1\Gamma_{12}{}^2$ to ensure $\rho_{12} = 0$. We obtain $\rho_{11} = \Gamma_{12}{}^2\rho_{22}$ and $\rho_{12;2} = -2\Gamma_{12}{}^2\rho_{22}$. Since $\rho_{22} \neq 0$ and since $\rho_{12;2} = \omega_2\rho_{12} = 0$, we obtain $\Gamma_{12}{}^2 = 0$. Thus $\rho_{11} = 0$ and $\text{Rank}(\rho) = 1$ as desired.

Case 2. Suppose that $\Gamma_{22}{}^1 = 0$. Setting $\rho_{12} = 0$ and $\rho_{22} \neq 0$ then yields $\Gamma_{12}{}^2 = 0$. We have $0 = \omega_1\rho_{12} = \rho_{12;1} = -2\Gamma_{11}{}^2\rho_{22}$ and thus $\Gamma_{11}{}^2 = 0$. We apply Assertion 1 to see $\text{Rank}(\rho) = 1$ as desired. □

13.1.3 AFFINE KILLING VECTOR FIELDS.

For Type $\mathcal{A}$ models, there is a close link between the rank of the Ricci tensor and the dimension of the space of affine Killing vector fields. This is not the case for the Type $\mathcal{B}$ models.

Lemma 13.4 *Let $\mathcal{M}$ be a Type $\mathcal{A}$ model.*

1. *If* $\text{Rank}(\rho) = 0$, *then* $\dim(\mathfrak{K}(\mathcal{M})) = 6$.
2. *If* $\text{Rank}(\rho) = 1$, *then* $\dim(\mathfrak{K}(\mathcal{M})) \leq 4$.
3. *If* $\text{Rank}(\rho) = 2$, *then* $\dim(\mathfrak{K}(\mathcal{M})) = 2$.

We shall show in Corollary 13.25 that equality holds in Assertion 2 of Lemma 13.4; $\dim(\mathfrak{K}(\mathcal{M})) = 4$ if and only if the Ricci tensor of $\mathcal{M}$ has rank 1.

Proof. By Lemma 12.14, $\dim(\mathfrak{K}(\mathcal{M})) \leq 6$. Let $\mathbb{A} = (\mathbb{R}^2, \Gamma_0)$ be the affine plane where all the Christoffel symbols vanish identically. Let (a_i^j) in $M_2(\mathbb{R})$ be a constant matrix and let (b^1, b^2) in $\mathbb{R}^2$. One verifies $a_i^j x^i \partial_{x^j} + b^k \partial_{x^k} \in \mathfrak{K}(\mathbb{A})$ and thus $\dim(\mathfrak{K}(\mathbb{A})) = 6$. Let $\mathcal{M}$ be a Type $\mathcal{A}$ model with $\text{Rank}(\rho) = 0$. Then $\rho = 0$ so by Lemma 12.1, $\mathcal{M}$ is flat. Thus $\dim(\mathfrak{K}_P(\mathcal{M})) = 6$ for every point $P \in \mathbb{R}^2$. Since $\mathbb{R}^2$ is simply connected, $\dim(\mathfrak{K}(\mathcal{M})) = 6$ which establishes Assertion 1.

Suppose $\text{Rank}(\rho) = 1$. By Lemma 13.3, we can make a linear change of coordinates to assume ρ is a non-zero multiple of $dx^2 \otimes dx^2$. Let $X = a^1(x^1, x^2)\partial_{x^1} + a^2(x^1, x^2)\partial_{x^2}$ be an affine Killing vector field. By subtracting appropriate multiples of ∂_{x^1} and ∂_{x^2}, we may assume $X(0) = 0$. Then

$$\begin{aligned}
0 &= (\mathcal{L}_X\rho)(\partial_{x^1}, \partial_{x^2}) = X(\rho(\partial_{x^1}, \partial_{x^2})) - \rho([X, \partial_{x^1}], \partial_{x^2}) - \rho(\partial_{x^1}, [X, \partial_{x^2}]) \\
&= 0 + (\partial_{x^1}a^2)\rho(\partial_{x^2}, \partial_{x^2}) + 0 \Rightarrow \partial_{x^1}a^2 = 0, \\
0 &= (\mathcal{L}_X\rho)(\partial_{x^2}, \partial_{x^2}) = X(\rho(\partial_{x^2}, \partial_{x^2})) - 2\rho([X, \partial_{x^2}], \partial_{x^2}) \\
&= 0 + 2(\partial_{x^2}a^2)\rho(\partial_{x^2}, \partial_{x^2}) \Rightarrow \partial_{x^2}a^2 = 0\,.
\end{aligned}$$

Thus a^2 is constant. Since $X(0) = 0$, there is no ∂_{x^2} dependence and $X = a^1(x^1, x^2)\partial_{x^1}$. Since $X(0) = 0$ and X is determined by $X(0)$ and $dX(0)$, the set of such X forms at most a 2-dimensional space so $\dim(\mathfrak{K}(\mathcal{M})) \leq 4$. This proves Assertion 2.

Suppose that ρ has rank 2 so ρ is non-degenerate. Let X be an affine Killing vector field with $X(0) = 0$. We wish to show X vanishes identically and thus $\mathfrak{K}(\mathcal{M}) = \operatorname{span}\{\partial_{x^1}, \partial_{x^2}\}$. Suppose to the contrary that there exists an affine Killing vector field which is non-trivial with $X(0) = 0$. Let Φ_t^X be the (local) flow of X. The relations $(\Phi_t^X)^*\rho = \rho$ and $\Phi_t^X(0) = \mathrm{Id}$ imply that $\Phi_t^X \in \mathrm{SO}(\rho)$. If ρ is definite, we can choose coordinates to assume that

$$\rho = \pm\{(dx^1)^2 + (dx^2)^2\} \quad \text{and} \quad \mathrm{SO}(\rho) = \mathrm{SO}(2)\,.$$

Consequently, after rescaling X if necessary, we may identify Φ_t^X with a rotation through an angle t. No non-zero constant Christoffel symbol Γ is preserved by the action of SO(2) (two indices are down and one is up), so this provides the desired contradiction. The argument is analogous if ρ is Lorentzian if we replace SO(2) by $\mathrm{SO}(1,1)$. Assertion 3 follows. □

13.1.4 LINEAR EQUIVALENCE. In the context of Type $\mathcal{A}$ models with non-degenerate Ricci tensor, linear equivalence and affine equivalence are the same concept. This vastly simplifies the analysis.

Theorem 13.5 *Let $\mathcal{M}_i$ be Type $\mathcal{A}$ models such that $\rho_{\mathcal{M}_i}$ are non-degenerate. Then $\mathcal{M}_1$ is linearly equivalent to $\mathcal{M}_2$ if and only if $\mathcal{M}_1$ is affinely equivalent to $\mathcal{M}_2$.*

Proof. Although this follows from work of Brozos-Vázquez, García-Río, and Gilkey [7], we give a slightly different derivation to keep our present treatment as self-contained as possible. Let $\mathcal{M}_i$ be Type $\mathcal{A}$ models with non-degenerate Ricci tensors. By Lemma 13.4, $\dim(\mathfrak{K}(\mathcal{M}_i)) = 2$. If ϕ is an affine transformation from $\mathcal{M}_1$ to $\mathcal{M}_2$, then $\phi_*\mathfrak{K}(\mathcal{M}_1) = \mathfrak{K}(\mathcal{M}_2)$. Thus $\phi_*(\partial_{x^i}) = a_i^j\,\partial_{x^j}$ for some $(a_i^j) \in \mathrm{GL}(2,\mathbb{R})$. By subtracting an appropriate translation, we may assume $\phi(0) = 0$. It now follows that $\phi \in \mathrm{GL}(2,\mathbb{R})$. □

In Remark 13.13, we will show that Theorem 13.5 fails if the Ricci tensor is not assumed to be non-degenerate.

13.1.5 LINEARLY STRONGLY PROJECTIVELY FLAT. We say that a Type $\mathcal{A}$ model $\mathcal{M}$ is *linearly strongly projectively flat* if there exists a linear function g which provides projective equivalence from $\mathcal{M}$ to a flat geometry $\mathcal{M}_f$; note that $\mathcal{M}_f$ is again a Type $\mathcal{A}$ model in this setting.

Lemma 13.6 *Let $\mathcal{M} = (\mathbb{R}^2, \nabla)$ be a Type $\mathcal{A}$ model. There exists a linear function of the form $g(x^1, x^2) = a_1x^1 + a_2x^2$ which provides a strong projective equivalence from $\mathcal{M}$ to a flat Type $\mathcal{A}$ model and which satisfies $e^{-g} \in \mathcal{Q}(\mathcal{M})$. Thus, in particular, $\mathcal{M}$ is linearly strongly projectively flat.*

Proof. Although Lemma 13.2 and Theorem 12.11 imply that $\mathcal{M}$ is strongly projectively flat, we do not obtain that this can be done by choosing $\omega = a_1dx^1 + a_2dx^2$ to be constant. We follow the discussion in Gilkey and Valle-Regueiro [35]. We work modulo linear equivalence.

We use Equation (13.1.a) to study the Ricci tensor ρ of $\mathcal{M}$. Let $g(x^1, x^2) = a_1 x^1 + a_2 x^2$ for $(a_1, a_2) \in \mathbb{R}^2$.

Case 1. Suppose $\Gamma_{11}{}^2 \neq 0$. Rescale x^2 to ensure $\Gamma_{11}{}^2 = 1$. We have

$$(\rho_{^g\mathcal{M}})_{11} = a_1^2 + a_1\Gamma_{11}{}^1 - \Gamma_{12}{}^1 + \Gamma_{11}{}^1\Gamma_{12}{}^2 - (\Gamma_{12}{}^2)^2 + a_2 + \Gamma_{22}{}^2 .$$

We set $a_2 := -a_1^2 - a_1\Gamma_{11}{}^1 + \Gamma_{12}{}^1 - \Gamma_{11}{}^1\Gamma_{12}{}^2 + (\Gamma_{12}{}^2)^2 - \Gamma_{22}{}^2$ to ensure $(\rho_{^g\mathcal{M}})_{11} = 0$. Then

$$\begin{aligned} (\rho_{^g\mathcal{M}})_{12} &= -\Gamma_{22}{}^1 - (a_1 + \Gamma_{12}{}^2)(a_1^2 + a_1\Gamma_{11}{}^1 - 2\Gamma_{12}{}^1 + \Gamma_{11}{}^1\Gamma_{12}{}^2 - (\Gamma_{12}{}^2)^2 + \Gamma_{22}{}^2), \\ (\rho_{^g\mathcal{M}})_{22} &= (-a_1 - \Gamma_{11}{}^1 + \Gamma_{12}{}^2)(\rho_{^g\mathcal{M}})_{12} . \end{aligned}$$

Since $(\rho_{^g\mathcal{M}})_{12}$ is cubic in a_1 with non-zero leading coefficient, we can find a_1 so $(\rho_{^g\mathcal{M}})_{12} = 0$. Since $(\rho_{^g\mathcal{M}})_{12}$ divides $(\rho_{^g\mathcal{M}})_{22}$, $(\rho_{^g\mathcal{M}})_{22} = 0$ as well so $\rho_{^g\mathcal{M}} = 0$.

Case 2. Suppose $\Gamma_{11}{}^2 = 0$. Since the argument is the same as that given in Case 1 if $\Gamma_{22}{}^1 \neq 0$, we may assume $\Gamma_{22}{}^1 = 0$ as well. We make a direct computation to see that taking $a_1 = -\Gamma_{12}{}^2$ and $a_2 = -\Gamma_{12}{}^1$ yields $\rho_{^g\mathcal{M}} = 0$.

We have chosen a linear function g so that $\rho_{^g\mathcal{M}} = 0$. This implies $^g\mathcal{M}$ is flat. Consequently, $\mathcal{M} = {}^{-g}\{^g\mathcal{M}\}$ is linearly strongly projectively flat. Since $\mathcal{Q}(\mathcal{M}) = e^{-g}\mathcal{Q}(^g\mathcal{M})$ and since $\mathbb{1} \in \mathcal{Q}(^g\mathcal{M})$, we conclude $e^{-g} \in \mathcal{Q}(\mathcal{M})$. □

13.2 TYPE $\mathcal{A}$: DISTINGUISHED GEOMETRIES

In Section 13.2.1 (see Definition 13.7), we define 15 families of Type $\mathcal{A}$ models that will form the focus of our investigations in Chapter 13. We will show subsequently in Section 13.3 (see Theorem 13.22) that any Type $\mathcal{A}$ model is linearly equivalent to one of these examples, and thus establishing their basic properties is crucial. Section 13.2.2 gives the Ricci tensor, Section 13.2.3 gives the solution space $\mathcal{Q}$ of the affine quasi-Einstein equation for $\mu = -1$, and Section 13.2.4 gives the spaces $E(\mu)$ for $\mu \neq -1$. Some useful local affine embeddings, immersions, and equivalences are given in Section 13.2.5. Section 13.2.6 treats affine gradient Ricci solitons. The space of affine Killing vector fields is given in Section 13.2.7. Section 13.2.8 treats invariant and parallel tensors of Type (1,1) for these examples. We thought it best to gather all the basic information concerning these examples in one section for the convenience of the reader; we will use these computations throughout the rest of this section.

13.2.1 DEFINING TYPE $\mathcal{A}$ MODELS.

We introduce the following notational conventions.

Definition 13.7 For $\vec{\xi} := (\xi_1, \xi_2, \xi_3, \xi_4, \xi_5, \xi_6) \in \mathbb{R}^6$, let

$$\begin{aligned} \Gamma(\vec{\xi}) &:= \Gamma(\xi_1, \xi_2, \xi_3, \xi_4, \xi_5, \xi_6) \\ &= \{\Gamma_{11}{}^1 = \xi_1,\ \Gamma_{11}{}^2 = \xi_2,\ \Gamma_{12}{}^1 = \xi_3, \\ &\qquad \Gamma_{12}{}^2 = \xi_4,\ \Gamma_{22}{}^1 = \xi_5,\ \Gamma_{22}{}^2 = \xi_6\}, \\ \mathcal{M}(\vec{\xi}) &:= \mathcal{M}(\xi_1, \xi_2, \xi_3, \xi_4, \xi_5, \xi_6) = (\mathbb{R}^2, \Gamma(\vec{\xi}))\,. \end{aligned} \tag{13.2.a}$$

Define the following Type $\mathcal{A}$ affine models.

$\mathcal{M}_i^6$: $\mathcal{M}_0^6 := \mathcal{M}(0,0,0,0,0,0)$, $\mathcal{M}_1^6 := \mathcal{M}(1,0,0,1,0,0)$,
$\mathcal{M}_2^6 := \mathcal{M}(-1,0,0,0,0,1)$, $\mathcal{M}_3^6 := \mathcal{M}(0,0,0,0,0,1)$,
$\mathcal{M}_4^6 := \mathcal{M}(0,0,0,0,1,0)$, $\mathcal{M}_5^6 := \mathcal{M}(1,0,0,1,-1,0)$.

$\mathcal{M}_i^4$: $\mathcal{M}_1^4 := \mathcal{M}(-1,0,1,0,0,2)$, $\mathcal{M}_2^4(c_1) := \mathcal{M}(-1,0,c_1,0,0,1+2c_1)$,
$\mathcal{M}_3^4(c_1) := \mathcal{M}(0,0,c_1,0,0,1+2c_1)$, $\mathcal{M}_4^4(c) := \mathcal{M}(0,0,1,0,c,2)$,
$\mathcal{M}_5^4(c) := \mathcal{M}(1,0,0,0,1+c^2,2c)$,
where $c \in \mathbb{R}$ and $c_1 \notin \{0,-1\}$.

$\mathcal{M}_i^2$: $\mathcal{M}_1^2(a_1,a_2) := \mathcal{M}\left(\frac{a_1^2+a_2-1, a_1^2-a_1, a_1a_2, a_1a_2, a_2^2-a_2, a_1+a_2^2-1}{a_1+a_2-1}\right)$,

$\mathcal{M}_2^2(b_1,b_2) := \mathcal{M}\left(1+b_1, 0, b_2, 1, \frac{1+b_2^2}{b_1-1}, 0\right)$,

$\mathcal{M}_3^2(c_2) := \mathcal{M}(2,0,0,1,c_2,1)$, $\mathcal{M}_4^2(\pm1) := \mathcal{M}(2,0,0,1,\pm1,0)$,
where $a_1a_2 \neq 0$, $a_1+a_2 \neq 1$, $b_1 \neq 1$, $(b_1,b_2) \neq (0,0)$, and $c_2 \neq 0$.

The notation is chosen (see Lemma 13.8 and the computations of Section 13.2.7) so that

$$\dim(\mathfrak{K}(\mathcal{M}_i^\nu(\cdot))) = \nu \quad \text{and} \quad \operatorname{Rank}(\rho_{\mathcal{M}_i^\nu(\cdot)}) = \left\{ \begin{array}{ll} 0 & \text{if } \nu = 6 \\ 1 & \text{if } \nu = 4 \\ 2 & \text{if } \nu = 2 \end{array} \right\}.$$

13.2.2 THE RICCI TENSOR.

Lemma 13.8 *Adopt the notation of Definition 13.7.*

1. *The Ricci tensor of the geometries $\mathcal{M}_i^6$ vanishes. These geometries are flat.*
2. *The Ricci tensor of the geometries $\mathcal{M}_i^4$ has rank* 1.

$$\begin{aligned} &\rho_{\mathcal{M}_1^4} = dx^2 \otimes dx^2, && \rho_{\mathcal{M}_2^4(c_1)} = (c_1 + c_1^2) dx^2 \otimes dx^2, \\ &\rho_{\mathcal{M}_3^4(c_1)} = (c_1 + c_1^2) dx^2 \otimes dx^2, && \rho_{\mathcal{M}_4^4(c)} = dx^2 \otimes dx^2, \\ &\rho_{\mathcal{M}_5^4(c)} = (1 + c^2) dx^2 \otimes dx^2\,. \end{aligned}$$

These geometries are recurrent. Let $\Theta := dx^2 \otimes dx^2 \otimes dx^2$.

$$\begin{aligned} &\nabla\rho_{\mathcal{M}_1^4} = -4\Theta, && \nabla\rho_{\mathcal{M}_2^4(c_1)} = -2(1+2c_1)(c_1+c_1^2)\Theta, \\ &\nabla\rho_{\mathcal{M}_3^4(c_1)} = -2(1+2c_1)(c_1+c_1^2)\Theta, && \nabla\rho_{\mathcal{M}_4^4(c)} = -4\Theta, \\ &\nabla\rho_{\mathcal{M}_5^4(c)} = -4c(1+c^2)\Theta\,. \end{aligned}$$

Since $c_1 + c_1^2 \neq 0$, *only* $\mathcal{M}_2^4(-\frac{1}{2})$, $\mathcal{M}_3^4(-\frac{1}{2})$, *and* $\mathcal{M}_5^4(0)$ *are affine symmetric spaces.*

3. *The Ricci tensor of the geometries* $\mathcal{M}_i^2$ *has rank* 2.

$$\rho_{\mathcal{M}_1^2(a_1,a_2)} = \begin{pmatrix} \frac{(a_1-1)a_1}{a_1+a_2-1} & \frac{a_1a_2}{a_1+a_2-1} \\ \frac{a_1a_2}{a_1+a_2-1} & \frac{(a_2-1)a_2}{a_1+a_2-1} \end{pmatrix}, \quad \det(\rho_{\mathcal{M}_1^2(a_1,a_2)}) = -\frac{a_1a_2}{a_1+a_2-1},$$

$$\rho_{\mathcal{M}_2^2(b_1,b_2)} = \begin{pmatrix} b_1 & b_2 \\ b_2 & \frac{b_2^2+b_1}{b_1-1} \end{pmatrix}, \quad \det(\rho_{\mathcal{M}_2^2(b_1,b_2)}) = \frac{b_1^2+b_2^2}{b_1-1},$$

$$\rho_{\mathcal{M}_3^2(c_2)} = dx^1 \otimes dx^1 + c_2 dx^2 \otimes dx^2, \quad \rho_{\mathcal{M}_4^2}(\pm 1) = dx^1 \otimes dx^1 \pm dx^2 \otimes dx^2.$$

$$\begin{aligned} &(\rho_{\mathcal{M}_1^2(a_1,a_2)})_{11;1} = -\frac{2a_1(a_1^2-1)}{a_1+a_2-1}, && (\rho_{\mathcal{M}_1^2(a_1,a_2)})_{11;2} = -\frac{2a_1^2a_2}{a_1+a_2-1}, \\ &(\rho_{\mathcal{M}_1^2(a_1,a_2)})_{12;2} = -\frac{2a_1a_2^2}{a_1+a_2-1}, && (\rho_{\mathcal{M}_1^2(a_1,a_2)})_{22;2} = -\frac{2a_2(a_2^2-1)}{a_1+a_2-1}, \\ &(\rho_{\mathcal{M}_2^2(b_1,b_2)})_{11;1} = -2b_1(b_1+1), && (\rho_{\mathcal{M}_2^2(b_1,b_2)})_{11;2} = -2(b_1+1)b_2, \\ &(\rho_{\mathcal{M}_2^2(b_1,b_2)})_{12;2} = -\frac{2b_1(b_2^2+1)}{b_1-1}, && (\rho_{\mathcal{M}_2^2(b_1,b_2)})_{22;2} = -\frac{2b_2(b_2^2+1)}{b_1-1}, \\ &(\rho_{\mathcal{M}_3^2(c_2)})_{11;1} = -4, && (\rho_{\mathcal{M}_3^2(c_2)})_{11;2} = 0, \\ &(\rho_{\mathcal{M}_3^2(c_2)})_{12;2} = -2c_2, && (\rho_{\mathcal{M}_3^2(c_2)})_{22;2} = -2c_2, \\ &(\rho_{\mathcal{M}_4^2(\pm 1)})_{11;1} = -4, && (\rho_{\mathcal{M}_4^2(\pm 1)})_{11;2} = 0, \\ &(\rho_{\mathcal{M}_4^2(\pm 1)})_{12;2} = \mp 2, && (\rho_{\mathcal{M}_4^2(\pm 1)})_{22;2} = 0\,. \end{aligned}$$

None of the geometries where $\text{Rank}(\rho) = 2$ *are affine symmetric spaces.*

Proof. The desired result follows by a direct computation; since $c_1 \notin \{0, -1\}$, $a_1a_2 \neq 0$, and $c_2 \neq 0$, ρ has the correct rank.

13.2.3 THE SOLUTION SPACE $\mathcal{Q}$.

Lemma 13.9 *Adopt the notation of Definition 13.7.*

$$\begin{aligned} \mathcal{Q}(\mathcal{M}_i^6)\colon\ &\mathcal{Q}(\mathcal{M}_0^6) = \operatorname{span}\{\mathbb{1}, x^1, x^2\}, && \mathcal{Q}(\mathcal{M}_1^6) = \operatorname{span}\{\mathbb{1}, e^{x^1}, x^2e^{x^1}\}, \\ &\mathcal{Q}(\mathcal{M}_2^6) = \operatorname{span}\{\mathbb{1}, e^{x^2}, e^{-x^1}\}, && \mathcal{Q}(\mathcal{M}_3^6) = \operatorname{span}\{\mathbb{1}, x^1, e^{x^2}\}, \\ &\mathcal{Q}(\mathcal{M}_4^6) = \operatorname{span}\{\mathbb{1}, x^2, (x^2)^2 + 2x^1\}, && \mathcal{Q}(\mathcal{M}_5^6) = \operatorname{span}\{\mathbb{1}, e^{x^1}\cos(x^2), e^{x^1}\sin(x^2)\}\,. \end{aligned}$$

$\mathcal{Q}(\mathcal{M}_i^4)$: $\mathcal{Q}(\mathcal{M}_1^4) = \operatorname{span}\{e^{x^2}, x^2 e^{x^2}, e^{-x^1+x^2}\}$, $\mathcal{Q}(\mathcal{M}_2^4(c_1)) = e^{c_1 x^2}\operatorname{span}\{\mathbb{1}, e^{x^2}, e^{-x^1}\}$,
$\mathcal{Q}(\mathcal{M}_3^4(c_1)) = e^{c_1 x^2}\operatorname{span}\{\mathbb{1}, e^{x^2}, x^1\}$, $\mathcal{Q}(\mathcal{M}_4^4(c)) = e^{x^2}\operatorname{span}\{\mathbb{1}, x^2, c(x^2)^2 + 2x^1\}$,
$\mathcal{Q}(\mathcal{M}_5^4(c)) = \operatorname{span}\{e^{cx^2}\cos(x^2), e^{cx^2}\sin(x^2), e^{x^1}\}$.

$\mathcal{Q}(\mathcal{M}_i^2)$: $\mathcal{Q}(\mathcal{M}_1^2(a_1, a_2)) = \operatorname{span}\{e^{x^1}, e^{x^2}, e^{a_1 x^1 + a_2 x^2}\}$,
$\mathcal{Q}(\mathcal{M}_2^2(b_1, b_2)) = \operatorname{span}\{e^{x^1}\cos(x^2), e^{x^1}\sin(x^2), e^{b_1 x^1 + b_2 x^2}\}$,
$\mathcal{Q}(\mathcal{M}_3^2(c_2)) = e^{x^1}\operatorname{span}\{\mathbb{1}, x^1 - c_2 x^2, e^{x^2}\}$,
$\mathcal{Q}(\mathcal{M}_4^2(\pm 1)) = \operatorname{span}\{e^{x^1}, x^2 e^{x^1}, (2x^1 \pm (x^2)^2)e^{x^1}\}$.

Proof. Our initial investigation was informed by Lemma 12.53, which shows that $\mathcal{Q}(\cdot)$ was generated by exponentials and polynomials; it was not simply routine exercise using mathematica. But once we had determined the solution space, one could then perform a direct computation to see that the functions given belong to $\mathcal{Q}(\cdot)$. As $\mathcal{M}$ is strongly projectively flat, $\dim(\mathcal{Q}(\cdot)) = 3$. Consequently, for dimensional reasons, these functions span. □

Corollary 13.10 *If $\mathcal{M}_i^j(\cdot)$ is linearly equivalent to $\mathcal{M}_k^\ell(\cdot)$, then $i = k$ and $j = \ell$.*

Proof. It is immediate by inspection that $\mathcal{Q}(\mathcal{M}_i^j(\cdot))$ is not linearly isomorphic to $\mathcal{Q}(\mathcal{M}_k^\ell(\cdot))$ for $(i, j) \neq (k, \ell)$. □

We note that there can be linear equivalences within these classes. For example, interchanging the roles of x^1 and x^2 interchanges $\mathcal{Q}(\mathcal{M}_1^2(a_1, a_2))$ and $\mathcal{Q}(\mathcal{M}_1^2(a_2, a_1))$, and thus $\mathcal{M}_1^2(a_1, a_2)$ and $\mathcal{M}_1^2(a_2, a_1)$ are linearly equivalent. We will study this point in more detail in Section 13.4.

13.2.4 OTHER EIGENSPACES OF THE AFFINE QUASI-EINSTEIN EQUATION.

We complexify to define $E_{\mathbb{C}}(\mu, \mathcal{M}_i^j(\cdot))$; real solutions can then be obtained by taking the real and imaginary parts. Our computations were informed by Lemma 12.53; the eigenspaces $E(\mu, \mathcal{M})$ are spanned by the product of linear exponentials and polynomials.

Theorem 13.11 *Let $\mu \neq -1$.*

1. *$E(\mu, \mathcal{M}_i^6) = \mathcal{Q}(\mathcal{M}_i^6)$ for $0 \le i \le 5$.*
2. *$E_{\mathbb{C}}(\mu, \mathcal{M}_1^4) = \operatorname{span}\{e^{(1+\sqrt{1+\mu})x^2}, e^{(1-\sqrt{1+\mu})x^2}\}$.*
3. *Let $D := 1 + 4c_1 + 4c_1^2 + 4c_1\mu + 4c_1^2\mu$. Let $\lambda_\pm := \frac{1}{2}(1 + 2c_1 \pm \sqrt{D})$.*
   (a) If $D \neq 0$, $E_{\mathbb{C}}(\mu, \mathcal{M}_2^4(c_1)) = E_{\mathbb{C}}(\mu, \mathcal{M}_3^4(c_1)) = \operatorname{span}\{e^{\lambda_+ x^2}, e^{\lambda_- x^2}\}$.
   (b) If $D = 0$, $E_{\mathbb{C}}(\mu, \mathcal{M}_2^4(c_1)) = E_{\mathbb{C}}(\mu, \mathcal{M}_3^4(c_1)) = \operatorname{span}\{e^{\lambda_+ x^2}, x^2 e^{\lambda_+ x^2}\}$.
4. *$E_{\mathbb{C}}(\mu, \mathcal{M}_4^4(c)) = \operatorname{span}\{e^{(1+\sqrt{1+\mu})x^2}, e^{(1-\sqrt{1+\mu})x^2}\}$.*
5. *Let $D = c^2 + \mu + c^2\mu$. Let $\lambda_\pm := c \pm \sqrt{D}$.*

(a) *If* $D \neq 0$, $E_{\mathbb{C}}(\mu, \mathcal{M}_5^4(c)) = \text{span}\{e^{\lambda + x^2}, e^{\lambda - x^2}\}$.

(b) *If* $D = 0$, $E_{\mathbb{C}}(\mu, \mathcal{M}_5^4(c)) = \text{span}\{e^{cx^2}, x^2 e^{cx^2}\}$.

6. $E(0, \mathcal{M}_i^2) = \text{span}\{\mathbb{1}\}$. *If* $\mu \neq 0$, *then* $E(\mu, \mathcal{M}_i^2) = \{0\}$.

Proof. Assertion 1 is a direct consequence of Definition 13.7 since $\rho = 0$ for $\mathcal{M}_i^6$. Let $\mathcal{M} = \mathcal{M}_i^j(\cdot)$ for $j = 2, 4$. Since $\mathcal{M}$ is not flat, $\dim(E(\mu, \mathcal{M})) \neq 3$ for $\mu \neq -1$. Thus $\dim(E(\mu, \mathcal{M})) \leq 2$. We may then verify Assertions 2–5 by a direct computation. Suppose the Ricci tensor has rank 2. Let $f \in E(\mu, \mathcal{M})$ for $\mu \neq -1$. We apply Lemma 13.2 to see $R_{12}(df) = 0$. Since R_{12} is injective, this implies $df = 0$ and f is constant. Assertion 6 follows. □

13.2.5 LOCAL AFFINE EQUIVALENCES.

Lemma 13.12

1. $\Phi_1^6(x^1, x^2) := (e^{x^1}, x^2 e^{x^1})$ *is an affine embedding of* $\mathcal{M}_1^6$ *in* $\mathcal{M}_0^6$.
2. $\Phi_2^6(x^1, x^2) := (e^{x^2}, e^{-x^1})$ *is an affine embedding of* $\mathcal{M}_2^6$ *in* $\mathcal{M}_0^6$.
3. $\Phi_3^6(x^1, x^2) := (x^1, e^{x^2})$ *is an affine embedding of* $\mathcal{M}_3^6$ *in* $\mathcal{M}_0^6$.
4. $\Phi_4^6(x^1, x^2) := (x^2, (x^2)^2 + 2x^1)$ *is an affine isomorphism from* $\mathcal{M}_4^6$ *to* $\mathcal{M}_0^6$.
5. $\Phi_5^6(x^1, x^2) := (e^{x^1}\cos(x^2), e^{x^1}\sin(x^2))$ *is an affine immersion of* $\mathcal{M}_5^6$ *in* $\mathcal{M}_0^6$.
6. $\Phi_1^4(x^1, x^2) := (e^{-x^1}, x^2)$ *is an affine embedding of* $\mathcal{M}_1^4$ *in* $\mathcal{M}_4^4(0)$.
7. $\Phi_2^4(x^1, x^2) := (e^{-x^1}, x^2)$ *is an affine embedding of* $\mathcal{M}_2^4(c_1)$ *in* $\mathcal{M}_3^4(c_1)$.
8. $\Phi_3^4(x^1, x^2) := (x^1 e^{-x^2}, -x^2)$ *is an affine isomorphism from* $\mathcal{M}_3^4(c_1)$ *to* $\mathcal{M}_3^4(-c_1 - 1)$.
9. $\Phi_4^4(c)(x^1, x^2) := (x^1 + \frac{1}{2}c(x^2)^2, x^2)$ *is an affine isomorphism from* $\mathcal{M}_4^4(c)$ *to* $\mathcal{M}_4^4(0)$.
10. $\Phi_5^4(x^1, x^2) := (x^1, -x^2)$ *is an affine isomorphism from* $\mathcal{M}_5^4(c)$ *to* $\mathcal{M}_5^4(-c)$.

Proof. We use Lemma 13.9 to show that the Φ_i^j intertwine the solution spaces $\mathcal{Q}$; these maps are therefore affine maps by Theorem 12.34. □

Remark 13.13 We use Lemma 13.9 to obtain that

$$\begin{aligned}
\mathcal{Q}(\mathcal{M}_4^6) &= \text{span}\{\mathbb{1}, x^2, (x^2)^2 + 2x^1\}, \\
\mathcal{Q}(\mathcal{M}_0^6) &= \text{span}\{\mathbb{1}, x^1, x^2\}, \\
\mathcal{Q}(\mathcal{M}_4^4(c)) &= e^{x^2}\,\text{span}\{\mathbb{1}, x^2, c(x^2)^2 + 2x^1\}.
\end{aligned}$$

The solution spaces $\mathcal{Q}(\mathcal{M}_0^6)$ and $\mathcal{Q}(\mathcal{M}_4^6)$ are not linearly equivalent. Consequently, by Theorem 12.34, $\mathcal{M}_0^6$ is not linearly equivalent to $\mathcal{M}_4^6$. However, these two spaces are affine equivalent by Lemma 13.12. Similarly, $\mathcal{Q}(\mathcal{M}_4^4(1))$ is not linearly equivalent to $\mathcal{Q}(\mathcal{M}_4^4(0))$, and thus

$\mathcal{M}_4^4(1)$ is not linearly equivalent to $\mathcal{M}_4^4(0)$. However, again by Lemma 13.12, there is an affine isomorphism between $\mathcal{M}_4^4(c)$ and $\mathcal{M}_4^4(0)$ for any c. Thus Theorem 13.5 fails if the Ricci tensor is not assumed to be non-degenerate.

13.2.6 AFFINE GRADIENT RICCI SOLITONS. We say that a smooth 1-form ω is an *affine Ricci soliton* if $\nabla\omega + \rho = 0$ and that a smooth function f is an *affine gradient Ricci soliton* if df is an affine Ricci soliton or, equivalently, if $\mathcal{H}f + \rho_s = 0$. In the Type $\mathcal{A}$ setting, the Ricci tensor is symmetric. Thus $\nabla\omega = -\rho$ implies $\omega_{i;j} = \omega_{j;i} = -\rho_{ij}$. This symmetry implies that ω is closed. Consequently, since $\mathbb{R}^2$ is simply connected, ω is exact, and thus these two notions coincide in the setting at hand.

Let $\mathfrak{A}(\mathcal{M})$ be the set of all affine gradient Ricci solitons. If $f_0 \in \mathfrak{A}(\mathcal{M})$ is an affine gradient Ricci soliton and if $f_1 \in E(0,\mathcal{M})$ is a Yamabe soliton, then $f_0 + f_1 \in \mathfrak{A}(\mathcal{M})$ is again an affine gradient Ricci soliton. Thus $\mathfrak{A}(\mathcal{M})$ is an affine space. The spaces $E(0,\mathcal{M})$ of Yamabe solitons are given in Theorem 13.11. Consequently, to describe $\mathfrak{A}(\mathcal{M})$, we must either show $\mathfrak{A}(\mathcal{M})$ is empty or exhibit a single element.

Theorem 13.14 *Adopt the notation of Definition 13.7.*

1. *Let* $\mathcal{M} = \mathcal{M}_i^6$ *for* $0 \le i \le 5$. *Then* $\mathbb{1} \in \mathfrak{A}(\mathcal{M})$.
2. *Let* $\mathcal{M} = \mathcal{M}_1^4$ *or* $\mathcal{M} = \mathcal{M}_4^4(c)$. *Then* $\frac{x^2}{2} \in \mathfrak{A}(\mathcal{M})$.
3. *Let* $\mathcal{M} = \mathcal{M}_2^4(c_1)$ *or* $\mathcal{M} = \mathcal{M}_3^4(c_1)$. *If* $c_1 \neq -\frac{1}{2}$, *then* $\frac{c_1(1+c_1)}{1+2c_1}x^2 \in \mathfrak{A}(\mathcal{M})$. *If* $c_1 = -\frac{1}{2}$, *then* $\frac{1}{8}(x^2)^2 \in \mathfrak{A}(\mathcal{M})$.
4. *Let* $\mathcal{M} = \mathcal{M}_5^4(c)$. *If* $c \neq 0$, *then* $\frac{1+c^2}{2c}x^2 \in \mathfrak{A}(\mathcal{M})$. *If* $c = 0$, *then* $-\frac{1}{2}(x^2)^2 \in \mathfrak{A}(\mathcal{M})$.
5. *If* $\mathcal{M} = \mathcal{M}_i^2(\cdot)$, $1 \le i \le 4$, *then* $\mathfrak{A}(\mathcal{M})$ *is empty.*

Proof. A direct calculation establishes the first four assertions. Suppose that $\mathcal{M}$ is a Type $\mathcal{A}$ model where the Ricci tensor is non-degenerate. By Lemma 13.2, $R_{12}(df) = 0$. Since ρ is non-degenerate, this implies $df = 0$. This shows that $\mathcal{H}f = 0$. Thus it is not possible that one has $\mathcal{H}f = -\rho_s$. This proves Assertion 5. □

13.2.7 AFFINE KILLING VECTOR FIELDS. Recall that $\mathcal{M}$ is said to be *affine Killing complete* if every affine Killing vector field of $\mathcal{M}$ is a complete vector field, i.e., the integral curves exist for all time or, equivalently by Lemma 12.3, that $\mathfrak{K}(\mathcal{M})$ is the Lie algebra of the Lie group of affine diffeomorphisms $\text{Aff}(\mathcal{M})$. We refer to Nomizu [48]. We divide our analysis into three cases depending upon the rank of the Ricci tensor. Our initial investigation was informed by Lemma 12.53, which shows that $\mathfrak{K}(\cdot)$ is generated by exponentials and polynomials; the results of this section are not simply routine exercise using mathematica. But once we had determined a sufficient number of affine Killing vector fields, we then proved they spanned for dimensional reasons by Lemma 13.4. We first study the flat geometries; the affine plane is affine Killing complete and is modeled on any of these geometries.

Lemma 13.15 *The geometries $\mathcal{M}_i^6$ are all flat. Consequently,* $\dim(\mathfrak{K}(\mathcal{M}_i^6)) = 6$.

1. $\mathfrak{K}(\mathcal{M}_0^6) = \operatorname{span}\{\partial_{x^1}, x^2\partial_{x^1}, x^1\partial_{x^1}, \partial_{x^2}, x^1\partial_{x^2}, x^2\partial_{x^2}\}$.
2. $\mathfrak{K}(\mathcal{M}_1^6) = \operatorname{span}\{\partial_{x^1}, \partial_{x^2}, e^{-x^1}(\partial_{x^1} - x^2\partial_{x^2}), x^2\partial_{x^1} - (x^2)^2\partial_{x^2}, x^2\partial_{x^2}, e^{-x^1}\partial_{x^2}\}$.
3. $\mathfrak{K}(\mathcal{M}_2^6) = \operatorname{span}\{\partial_{x^1}, \partial_{x^2}, e^{x^1}\partial_{x^1}, e^{x^1+x^2}\partial_{x^1}, e^{-x^2}\partial_{x^2}, e^{-x^1-x^2}\partial_{x^2}\}$.
4. $\mathfrak{K}(\mathcal{M}_3^6) = \operatorname{span}\{\partial_{x^1}, \partial_{x^2}, x^1\partial_{x^1}, e^{x^2}\partial_{x^1}, e^{-x^2}\partial_{x^2}, x^1 e^{-x^2}\partial_{x^2}\}$.
5. $\mathfrak{K}(\mathcal{M}_4^6) = \operatorname{span}\{\partial_{x^1}, \partial_{x^2}, x^2\partial_{x^1}, (-x^1x^2 - \frac{1}{2}(x^2)^3)\partial_{x^1} + (x^1 + \frac{1}{2}(x^2)^2)\partial_{x^2},$
   $(x^1 + \frac{1}{2}(x^2)^2)\partial_{x^1}, -(x^2)^2\partial_{x^1} + x^2\partial_{x^2}\}$.
6. $\mathfrak{K}(\mathcal{M}_5^6) = \operatorname{span}\{\partial_{x^1}, \partial_{x^2}, \cos(2x^2)\partial_{x^1} - \sin(2x^2)\partial_{x^2}, \sin(2x^2)\partial_{x^1} + \cos(2x^2)\partial_{x^2},$
   $e^{-x^1}(\cos(x^2)\partial_{x^1} - \sin(x^2)\partial_{x^2}), e^{-x^1}(\sin(x^2)\partial_{x^1} + \cos(x^2)\partial_{x^2})\}$.
7. *$\mathcal{M}_0^6$ and $\mathcal{M}_4^6$ are affine Killing complete; $\mathcal{M}_1^6$, $\mathcal{M}_2^6$, $\mathcal{M}_3^6$, and $\mathcal{M}_5^6$ are affine Killing incomplete. All these Lie algebras are isomorphic to the full Lie algebra of the 6-dimensional affine group.*

Proof. Assertions 1–6 follow by a direct computation. We have $\operatorname{Aff}(\mathcal{M}_0^6)$ is the 6-dimensional affine group. Thus, for dimensional reasons, the Lie algebra of $\operatorname{Aff}(\mathcal{M}_0^6)$ is the full Lie algebra of affine Killing vector fields and $\mathcal{M}_0^6$ is affine Killing complete. By Lemma 13.12, $\mathcal{M}_4^6$ is affine diffeomorphic to $\mathcal{M}_0^6$. This shows that $\mathcal{M}_4^6$ is affine Killing complete. The remaining geometries $\mathcal{M}_i^6$ all have proper affine embeddings or immersions into $\mathcal{M}_0^6$. It then follows that these must be affine Killing incomplete. It is instructive, however, to give a direct argument.

Case 1. $\mathcal{M}_1^6$. Let $X = x^2\partial_{x^1} - (x^2)^2\partial_{x^2}$. The flow is determined by the ODE $\dot{x}^1 = x^2$ and $\dot{x}^2 = -(x^2)^2$. We solve this ODE to see $x^2(t) = \frac{1}{t+a}$ so the integral curve is not defined for all t and $\mathcal{M}_1^6$ is affine Killing incomplete.

Case 2. $\mathcal{M}_2^6$. Let $X = e^{x^1}\partial_{x^1}$. The flow is determined by the ODE $\dot{x}^1 = e^{x^1}$ and $\dot{x}^2 = 0$. The first ODE can be solved by setting $x^1(t) = -\log(-t + a)$ so the integral curve is not defined for all t and $\mathcal{M}_2^6$ is affine Killing incomplete.

Case 3. $\mathcal{M}_3^6$. Let $X = e^{-x^2}\partial_{x^2}$. The flow is determined by the ODE $\dot{x}^1 = 0$ and $\dot{x}^2 = e^{-x^2}$. We obtain $x^2(t) = \log(t + a)$ so the integral curve is not defined for all t and $\mathcal{M}_3^6$ is affine Killing incomplete.

Case 4. $\mathcal{M}_5^6$. Let $X = e^{-x^1}\cos(x^2)\partial_{x^1} - e^{-x^1}\sin(x^2)\partial_{x^2}$. The flow is determined by the ODE $\dot{x}^1 = e^{-x^1}\cos(x^2)$ and $\dot{x}^2 = -e^{-x^1}\sin(x^2)$. We set $x^2(t) = 0$ to solve the second ODE. The first ODE then becomes $\dot{x}^1 = e^{-x^1}$, which yields $x^1(t) = \log(t + a)$ so the integral curve is not defined for all t and $\mathcal{M}_5^6$ is affine Killing incomplete. □

Lemma 13.16 *The geometries* $\mathcal{M}_i^4(\cdot)$ *satisfy* $\dim(\mathfrak{K}(\mathcal{M}_i^4(\cdot))) = 4$.

1. $\mathfrak{K}(\mathcal{M}_1^4) = \operatorname{span}\{\partial_{x^1}, \partial_{x^2}, e^{x^1}\partial_{x^1}, x^2 e^{x^1}\partial_{x^1}\}$.
2. $\mathfrak{K}(\mathcal{M}_2^4(c_1)) = \operatorname{span}\{\partial_{x^1}, \partial_{x^2}, e^{x^1}\partial_{x^1}, e^{x^1+x^2}\partial_{x^1}\}$.
3. $\mathfrak{K}(\mathcal{M}_3^4(c_1)) = \operatorname{span}\{\partial_{x^1}, \partial_{x^2}, e^{x^2}\partial_{x^1}, x^1\partial_{x^1}\}$.
4. $\mathfrak{K}(\mathcal{M}_4^4(c)) = \operatorname{span}\{\partial_{x^1}, \partial_{x^2}, (x^1 + \frac{1}{2}c(x^2)^2)\partial_{x^1}, x^2\partial_{x^1}\}$.
5. $\mathfrak{K}(\mathcal{M}_5^4(c)) = \operatorname{span}\{\partial_{x^1}, \partial_{x^2}, e^{-x^1+cx^2}\cos(x^2)\partial_{x^1}, e^{-x^1+cx^2}\sin(x^2)\partial_{x^1}\}$.
6. *The geometries* $\mathcal{M}_3^4(c_1)$ *and* $\mathcal{M}_4^4(c)$ *are affine Killing complete.*
7. *The geometries* $\mathcal{M}_1^4$, $\mathcal{M}_2^4(c_1)$,*and* $\mathcal{M}_5^4(c)$ *are affine Killing incomplete.*

Proof. By Lemma 13.8, the Ricci tensor has rank 1. Lemma 13.4 yields $\dim(\mathfrak{K}(\cdot)) \le 4$. We perform a direct computation to verify that the vector fields given are all affine Killing vector fields and thus for dimensional reasons span $\mathfrak{K}(\cdot)$. Assertions 1–5 now follow. We break the proof of Assertion 6 into two cases.

Case 1. $\mathcal{M} = \mathcal{M}_3^4(c_1)$. We have $\mathcal{Q}(\mathcal{M}) = \operatorname{span}\{e^{c_1 x^2}, e^{(1+c_1)x^2}, x^1 e^{c_1 x^2}\}$. This is not a particularly convenient form of this surface to work with. We set $u^1 := x^1 e^{c_1 x^2}$ and $u^2 := x^2$ to express $\mathcal{Q}(\mathcal{M}_3^4(c_1)) = \operatorname{span}\{e^{c_1 u^2}, e^{(1+c_1)u^2}, u^1\}$. We define

$$T(\alpha,\beta,\gamma,\delta)(u^1,u^2) = (e^{\alpha}u^1 + \beta e^{c_1 u^2} + \gamma e^{(1+c_1)u^2}, u^2 + \delta)\,.$$

Because $T(\alpha,\beta,\gamma,\delta)$ preserves $\mathcal{Q}(\mathcal{M})$, $T(\alpha,\beta,\gamma,\delta) \in \operatorname{Aff}(\mathcal{M})$. The set of these elements is closed under composition and inverse:

$$\begin{aligned} T(\alpha,\beta,\gamma,\delta) \circ T(\tilde\alpha,\tilde\beta,\tilde\gamma,\tilde\delta) &= T(\alpha+\tilde\alpha, \tilde\beta e^{\alpha} + \beta e^{c\tilde\delta}, \tilde\gamma e^{\alpha} + \gamma e^{(1+c)\tilde\delta}, \delta+\tilde\delta)\,,\\ T(\alpha,\beta,\gamma,\delta)^{-1} &= T(-\alpha, -\beta e^{-\alpha-c\delta}, -\gamma e^{-\alpha+(-1-c)\delta}, -\delta)\,. \end{aligned}$$

This gives a group structure to $\mathbb{R}^4$ and identifies $\mathbb{R}^4$ with the connected component of the identity of the group $\operatorname{Aff}(\mathcal{M})$. The Lie algebra of this group is 4-dimensional and consists of the full Lie algebra of Killing vector fields. This shows that $\mathcal{M}$ is affine Killing complete. This action is transitive on $\mathbb{R}^2$ so this is a homogeneous geometry.

Case 2. By Lemma 13.12, $\mathcal{M}_4^4(c)$ is affine equivalent to $\mathcal{M}_4^4(0)$ for any c so it is only necessary to consider $\mathcal{M} = \mathcal{M}_4^4(0)$. We have $\mathcal{Q}(\mathcal{M}) = \operatorname{span}\{e^{x^2}, x^2 e^{x^2}, x^1 e^{x^2}\}$. We clear the previous notation and set $T(\alpha,\beta,\gamma,\delta)(x^1,x^2) := (e^{\alpha}x^1 + \beta x^2 + \gamma, x^2 + \delta)$. As in Case 1, $T(\alpha,\beta,\gamma,\delta)$ belongs to $\operatorname{Aff}(\mathcal{M})$. We complete the analysis by checking

$$\begin{aligned} T(\alpha,\beta,\gamma,\delta) \circ T(\tilde\alpha,\tilde\beta,\tilde\gamma,\tilde\delta) &= T(\alpha+\tilde\alpha, \beta + \tilde\beta e^{\alpha}, \gamma + \beta\tilde\delta + \tilde\gamma e^{\alpha}, \delta+\tilde\delta),\\ T(\alpha,\beta,\gamma,\delta)^{-1} &= T(-\alpha, -\beta e^{-\alpha}, e^{-\alpha}(\beta\delta-\gamma), -\delta)\,. \end{aligned}$$

We perform a direct computation to show $\mathcal{M}_1^4$, $\mathcal{M}_2^4(c_1)$, and $\mathcal{M}_5^4(c)$ are affine Killing incomplete, although this also follows from Lemma 13.12. If $\mathcal{M} = \mathcal{M}_1^4$ or $\mathcal{M} = \mathcal{M}_2^4(c_1)$, then

$e^{x^1}\partial_{x^1}$ is an affine Killing vector field. We noted in the proof of Lemma 13.15 (see Case 2) that this vector field was incomplete. If $\mathcal{M} = \mathcal{M}_5^4(c)$, we take $X = e^{-x^1+cx^2}\cos(x^2)\partial_{x^1}$. We take $x^2 = 0$ to obtain $X = e^{-x^1}\partial_{x^1}$. We showed in the proof of Lemma 13.15 (see Case 3 and interchange the roles of x^1 and x^2) that this vector field was incomplete. □

Patera et al. [55] have classified the low-dimensional Lie algebras. Let $\{e_1, e_2, e_3, e_4\}$ be a basis of $\mathbb{R}^4$. We define the following solvable Lie algebras by specifying their bracket relations.

1. $A_2 \oplus A_2$: the relations of the bracket are given by $[e_1, e_2] = e_2$ and $[e_3, e_4] = e_4$.
2. $A_{4,9}^b$: the relations of the bracket for $-1 \le b \le 1$ are given by $[e_2, e_3] = e_1$, $[e_1, e_4] = (1 + b)e_1$, $[e_2, e_4] = e_2$, and $[e_3, e_4] = be_3$.
3. $A_{4,12}$: the relations of the bracket are given by $[e_1, e_3] = e_1$, $[e_2, e_3] = e_2$, $[e_1, e_4] = -e_2$, $[e_2, e_4] = e_1$.

Lemma 13.17

1. *$\mathfrak{K}(\mathcal{M}_1^4) \approx \mathfrak{K}(\mathcal{M}_4^4(c)) \approx A_{4,9}^0$.*
2. *$\mathfrak{K}(\mathcal{M}_2^4(c_1)) \approx \mathfrak{K}(\mathcal{M}_3^4(c_1)) \approx A_2 \oplus A_2$.*
3. *$\mathfrak{K}(\mathcal{M}_5^4(c)) \approx A_{4,12}$.*

Proof. In view of the embeddings and isomorphisms of Lemma 13.12, we need only consider three cases. We follow the discussion of Brozos-Vázquez, García-Río, and Gilkey [7].

Case 1. $\mathcal{M} = \mathcal{M}_4^4(0)$. Set $e_1 := \partial_{x^1}$, $e_2 := x^2\partial_{x^1}$, $e_3 := -\partial_{x^2}$, $e_4 := x^1\partial_{x^1}$ to obtain the bracket relations of the Lie algebra $A_{4,9}^0$: $[e_2, e_3] = e_1$, $[e_1, e_4] = e_1$, and $[e_2, e_4] = e_2$.

Case 2. $\mathcal{M} = \mathcal{M}_3^4(c_1)$. Set $e_1 := -x^1\partial_{x^1} - \partial_{x^2}$, $e_2 := -\partial_{x^1}$, $e_3 := \partial_{x^2}$, and $e_4 := e^{x^2}\partial_{x^1}$ to obtain the bracket relations of the Lie algebra $A_2 \oplus A_2$: $[e_1, e_2] = e_2$ and $[e_3, e_4] = e_4$.

Case 3. $\mathcal{M} = \mathcal{M}_5^4(c)$. Set $e_1 := e^{x^1}\cos(x^2)\partial_{x^1}$, $e_2 := e^{x^1}\sin(x^2)\partial_{x^1}$, $e_3 := -\partial_{x^1}$, and $e_4 := -\partial_{x^2}$, to obtain the bracket relations of $A_{4,12}$: $[e_1, e_3] = e_1$, $[e_2, e_3] = e_2$, $[e_1, e_4] = -e_2$, and $[e_2, e_4] = e_1$. □

By Lemma 13.12, the map $(x^1, x^2) \to (e^{-x^1}, x^2)$ is a proper affine embedding of $\mathcal{M}_1^4$ in $\mathcal{M}_4^4(0)$ and of $\mathcal{M}_2^4(c_1)$ in $\mathcal{M}_3^4(c_1)$. Thus these geometries can be affine Killing completed. We now introduce geometries $\tilde{\mathcal{M}}_5^4(c)$ which contain $\mathcal{M}_5^4(c)$ as a proper submanifold and which can be regarded as a completion of the geometries $\mathcal{M}_5^4(c)$.

Lemma 13.18 *Let $\tilde{\mathcal{M}}_5^4(c) := (\mathbb{R}^2, \tilde{\nabla})$ where the only (possibly) non-zero Christoffel symbols of $\tilde{\nabla}$ are $\tilde{\Gamma}_{22}{}^1 = (1 + c^2)x^1$ and $\tilde{\Gamma}_{22}{}^2 = 2c$.*

1. *$\mathcal{Q}(\tilde{\mathcal{M}}_5^4(c)) = \operatorname{span}\{e^{cx^2}\cos(x^2), e^{cx^2}\sin(x^2), x^1\}$.*
2. *$\tilde{\mathcal{M}}_5^4(c)$ is a homogeneous geometry which is affine Killing complete.*
3. *The map $\Theta_5^4(x^1, x^2) := (e^{x^1}, x^2)$ is an affine embedding of $\mathcal{M}_5^4(c)$ in $\tilde{\mathcal{M}}_5^4(c)$.*

4. *$\tilde{\mathcal{M}}_5^4(c)$ is an affine symmetric space if and only if $c = 0$.*

Proof. We make a direct computation to see

$$\{e^{cx^2}\cos(x^2), e^{cx^2}\sin(x^2), x^1\} \subset \mathcal{Q}(\tilde{\mathcal{M}}_5^4(c)).$$

Since $\dim(\mathcal{Q}(\tilde{\mathcal{M}}_5^4(c)) \le 3$, Assertion 1 holds for dimensional reasons. Let

$$T(\alpha,\beta,\gamma,\delta)(x^1,x^2) := (e^\alpha x^1 + \beta e^{cx^2}\cos(x^2) + \gamma e^{cx^2}\sin(x^2), x^2+\delta).$$

Since $T(\alpha,\beta,\gamma,\delta)$ preserves $\mathcal{Q}(\tilde{\mathcal{M}}_5^4(c))$, $T(\alpha,\beta,\gamma,\delta) \in \text{Aff}(\tilde{\mathcal{M}}_5^4(c))$. The set of these elements is closed under composition and inverse:

$$\begin{aligned}
&T(\alpha,\beta,\gamma,\delta)\circ T(\tilde\alpha,\tilde\beta,\tilde\gamma,\tilde\delta)\\
&\quad = T(\alpha+\tilde\alpha, e^\alpha\tilde\beta + \beta e^{c\tilde\delta}\cos(\tilde\delta) + \gamma e^{c\tilde\delta}\sin(\tilde\delta), e^\alpha\tilde\gamma - \beta e^{c\tilde\delta}\sin(\tilde\delta) + \gamma e^{c\tilde\delta}\cos(\tilde\delta), \delta+\tilde\delta),\\
&T(\alpha,\beta,\gamma,\delta)^{-1}\\
&\quad = T(-\alpha, -e^{-\alpha-c\delta}(\beta\cos(\delta)-\gamma\sin(\delta)), -e^{-\alpha-c\delta}(\beta\sin(\delta)+\gamma\cos(\delta)), -\delta).
\end{aligned}$$

We argue as in the proof of Lemma 13.16 to see that $\tilde{\mathcal{M}}_5^4(c))$ is affine Killing complete. This action is transitive on $\mathbb{R}^2$ so this is a homogeneous geometry. This establishes Assertion 2. Assertion 3 follows since Θ_5^4 intertwines $\mathcal{Q}(\mathcal{M}_5^4(c))$ and $\mathcal{Q}(\tilde{\mathcal{M}}_5^4(c))$. Assertion 4 follows from the corresponding assertion for $\mathcal{M}_5^4(c)$ given in Lemma 13.8. □

Lemma 13.19 *Let $\mathcal{M} = \mathcal{M}_i^2(\cdot)$ for $1 \le i \le 4$. Then $\mathfrak{K}(\mathcal{M}) = \text{span}\{\partial_{x^1}, \partial_{x^2}\}$ and $\mathcal{M}$ is affine Killing complete.*

Proof. By Lemma 13.8, $\text{Rank}(\rho_{\mathcal{M}_i^2(\cdot)}) = 2$ and thus by Lemma 13.4, $\dim(\mathfrak{K}(\mathcal{M}_i^2(\cdot))) = 2$. Thus $\mathfrak{K}(\mathcal{M}_i^2(\cdot)) = \text{span}\{\partial_{x^1}, \partial_{x^2}\}$. Since the coordinate vector fields are complete, $\mathcal{M}_i^2(\cdot)$ is affine Killing complete. □

13.2.8 TYPE (1,1) INVARIANT AND PARALLEL TENSORS ON TYPE $\mathcal{A}$ SURFACES. Let $\mathcal{M} = (\mathbb{R}^2, \nabla)$ be a Type $\mathcal{A}$ structure which is not flat. Suppose given a tensor of Type (1,1) $T = T^i{}_j\partial_{x^i}\otimes dx^j$, i.e., an endomorphism of the tangent bundle. We assume T is an *invariant* tensor of Type (1,1), i.e., $\mathcal{L}_X T = 0$ for all $X \in \mathfrak{K}(\mathcal{M})$.

Theorem 13.20 *Let $T = T^i{}_j(x^1, x^2)$ be a tensor of Type (1,1).*

1. *If $\mathcal{M} = \mathcal{M}_i^2$, then T is invariant if and only if the component functions $T^i{}_j$ are constant.*
2. *If $\mathcal{M} = \mathcal{M}_i^6$ or if $\mathcal{M} = \mathcal{M}_i^4$, then T is invariant if and only if T is a constant multiple of the identity.*

Proof. We have ∂_{x^1} and ∂_{x^2} are affine Killing vector fields. Since $\mathcal{L}_{\partial_{x^i}}\partial_{x^j} = 0$, we have dually that $\mathcal{L}_{\partial_{x^i}} dx^j = 0$. Consequently, $\mathcal{L}_{\partial_{x^i}}(T^a{}_b \partial_{x^a} \otimes dx^b) = (\partial_{x^i} T^a{}_b)\partial_{x^a} \otimes dx^b$. Thus T is invariant implies $T \in M_2(\mathbb{R})$ is constant. If $\dim(\mathfrak{K}(\mathcal{M})) = 2$, this implies T is invariant; Assertion 1 now follows. More generally, let $X = a^1\partial_{x^1} + a^2\partial_{x^2}$. By Equation (12.2.a), $\mathcal{L}_X T = 0$ is equivalent to the conditions

$$0 = -T^k{}_j(\partial_{x^k} a^i) + T^i{}_k(\partial_{x^j} a^k) \quad \text{for all} \quad i, j\,.$$

Clearly if T is a constant multiple of the identity, then $\mathcal{L}_X T = 0$. Thus we may assume T is trace-free henceforth, i.e., $T^1{}_1 + T^2{}_2 = 0$. We impose this condition. Then $\mathcal{L}_X T = 0$ is equivalent to the conditions

$$\begin{aligned}
0 &= \{\mathcal{L}_X T\}^1{}_1 = T^1{}_2(a^2)^{(1,0)} - T^2{}_1(a^1)^{(0,1)},\\
0 &= \{\mathcal{L}_X T\}^1{}_2 = T^1{}_2\{(a^2)^{(0,1)} - (a^1)^{(1,0)}\} + 2T^1{}_1(a^1)^{(0,1)},\\
0 &= \{\mathcal{L}_X T\}^2{}_1 = T^2{}_1(a^1)^{(1,0)} - 2T^1{}_1(a^2)^{(1,0)} - T^2{}_1(a^2)^{(0,1)},\\
0 &= \{\mathcal{L}_X T\}^2{}_2 = T^2{}_1(a^1)^{(0,1)} - T^1{}_2(a^2)^{(1,0)}\,.
\end{aligned}$$

We apply Lemma 13.12 to see that we need only consider the geometries $\mathcal{M}_0^6$, $\mathcal{M}_3^4(c_1)$, $\mathcal{M}_4^4(0)$, and $\mathcal{M}_5^4(c)$. We consider cases. We apply the results of Section 13.2.7 to determine the Lie algebra of affine Killing vector fields.

Case 1. $\mathcal{M} = \mathcal{M}_0^6$. We have $X = (b_{11}x^1 + b_{12}x^2)\partial_{x^1} + (b_{21}x^1 + b_{22}x^2)\partial_{x^2}$ is an affine Killing vector field. We set $T^2{}_2 = -T^1{}_1$ and compute:

$$\mathcal{L}_X T = \begin{pmatrix} b_{21}T^1{}_2 - b_{12}T^2{}_1 & 2b_{12}T^1{}_1 + (b_{22} - b_{11})T^1{}_2 \\ (b_{11} - b_{22})T^2{}_1 - 2b_{21}T^1{}_1 & b_{12}T^2{}_1 - b_{21}T^1{}_2 \end{pmatrix}.$$

Since this must vanish identically for all values of the parameters b_{ij}, we conclude $T = 0$.

Case 2. $\mathcal{M} = \mathcal{M}_3^4(c_1)$. We have $X = (b_1 e^{x^2} + b_2 x^1)\partial_{x^1}$ is an affine Killing vector field. We compute

$$\mathcal{L}_X T = \begin{pmatrix} -b_1 e^{x^2} T^2{}_1 & 2b_1 e^{x^2} T^1{}_1 - b_2 T^1{}_2 \\ b_2 T^2{}_1 & b_1 e^{x^2} T^2{}_1 \end{pmatrix}.$$

This vanishes identically for all (b_1, b_2) if and only if $T = 0$.

Case 3. $\mathcal{M} = \mathcal{M}_4^4(c)$. We have $X = \{b_1(x^1 + \frac{1}{2}c(x^2)^2) + b_2 x^2\}\partial_{x^1}$ is an affine Killing vector field. We compute

$$\mathcal{L}_X T = \begin{pmatrix} -T^2{}_1(b_2 + b_1 c x^2) & 2T^1{}_1(b_2 + b_1 c x^2) - b_1 T^1{}_2 \\ b_1 T^2{}_1 & T^2{}_1(b_2 + b_1 c x^2) \end{pmatrix}.$$

This vanishes identically for all (b_1, b_2) if and only if $T = 0$.

Case 4. $\mathcal{M} = \mathcal{M}_5^4(c)$. We have $X = e^{-x^1+cx^2}\{b_1\cos(x^2) + b_2\sin(x^2)\}\partial_{x^1}$ is an affine Killing vector field. We compute

$$(\mathcal{L}_X T)_{x^1=0,x^2=0} = \begin{pmatrix} -(b_2+b_1c)T^2{}_1 & 2(b_2+b_1c)T^1{}_1 + b_1T^1{}_2 \\ -b_1T^2{}_1 & (b_2+b_1c)T^2{}_1 \end{pmatrix}.$$

This vanishes identically for all (b_1, b_2) if and only if $T = 0$. □

The Ricci tensor of any Type $\mathcal{A}$ homogeneous model is symmetric. Furthermore, the Ricci tensor is recurrent if and only if it is of rank 1. Therefore Theorem 12.27 3 shows that a Type $\mathcal{A}$ homogeneous surface admits a parallel tensor field if and only if the Ricci tensor is of rank 1, in which case it is a nilpotent Kähler surface.

Theorem 13.21 *Let $\mathcal{M} = (\mathbb{R}^2, \nabla)$ be a Type $\mathcal{A}$ model. The following assertions are equivalent.*
(1) $\operatorname{Rank}(\rho) = 1$. (2) $\mathcal{P}^0(\mathcal{M}) \neq \{0\}$. (3) $\dim(\mathcal{P}^0(\mathcal{M})) = 1$. (4) $\dim(\mathfrak{K}(\mathcal{M})) = 4$.

13.3 TYPE $\mathcal{A}$: PARAMETERIZATION

In Section 13.3.1, we parameterize the Type $\mathcal{A}$ models (see Theorem 13.22). In the remainder of this section, we exploit this classification. Section 13.3.2 relates the rank of the Ricci tensor and the dimension of the space of affine Killing vector fields. Section 13.3.3 treats strong projective equivalence and linear equivalence. The α invariant is computed for the examples where the Ricci tensor has rank 1 in Section 13.3.4. The geodesic equation is explicitly solved for the examples where the Ricci tensor has rank 0 or rank 1 and geodesic completeness is discussed in Section 13.3.5.

13.3.1 CLASSIFICATION. Gilkey and Valle-Regueiro [35] classified the Type $\mathcal{A}$ structures up to linear equivalence using the space of solutions of the affine quasi-Einstein equation $\mathcal{Q}$; a similar classification had been given previously by Brozos-Vázquez, García-Río, and Gilkey [6] using different methods. The following is the main result of this section.

Theorem 13.22 *If $\mathcal{M} = (\mathbb{R}^2, \nabla)$ is a Type $\mathcal{A}$ model, then $\mathcal{M}$ is linearly equivalent to one of the structures given in Definition 13.7.*

We say that a function $L(x^1, x^2)$ is linear if $L(x^1, x^2) = a_1x^1 + a_2x^2$; as we permit $(a^1, a^2) = (0, 0)$, the zero function is permitted as a degenerate case. We say that a function $Q(x^1, x^2)$ is quadratic if $Q(x^1, x^2) = a_{11}(x^1)^2 + a_{12}x^1x^2 + a_{22}(x^2)^2 + a_1x^1 + a_2x^2$; as we permit $(a_{11}, a_{12}, a_{22}) = (0, 0, 0)$, linear functions are permitted as a degenerate case.

Lemma 13.23 *Let $\mathcal{M} = (\mathbb{R}^2, \nabla)$ be a Type $\mathcal{A}$ model. There exists a basis $\mathcal{B}$ for $\mathfrak{Q}(\mathcal{M})$ which has one of the following forms.*

1. *$\mathcal{B} = \{e^{L_1}\cos(L_2), e^{L_1}\sin(L_2), e^{L_3}\}$ where L_i are linear functions.*
2. *$\mathcal{B} = \{e^{L_1}, e^{L_2}, e^{L_3}\}$ where L_i are linear functions.*
3. *$\mathcal{B} = \{e^{L_1}, L_2 e^{L_1}, e^{L_3}\}$ where L_i are linear functions.*
4. *$\mathcal{B} = \{e^{L_1}, L_2 e^{L_1}, Q e^{L_1}\}$ where L_i are linear functions and Q is quadratic.*

Proof. We complexify and use Lemma 12.53 to see that $\mathfrak{Q}_{\mathbb{C}}(\mathcal{M}) := \mathfrak{Q}(\mathcal{M}) \otimes_{\mathbb{R}} \mathbb{C}$ is spanned by functions of the form $p(x^1, x^2)e^{a_1x^1+a_2x^2}$ for $(a_1, a_2) \in \mathbb{C}^2$ and p polynomial. We have that $\mathfrak{Q}_{\mathbb{C}}(\mathcal{M})$ is a $\mathfrak{K}(\mathcal{M})$-module. By applying $(\partial_{x^i} - a_i)$, we see that $\partial_{x^i} p(x^1, x^2)e^{a_1x^1+a_2x^2}$ belongs to $\mathfrak{Q}_{\mathbb{C}}(\mathcal{M})$ as well and thus inductively $e^{a_1x^1+a_2x^2} \in \mathfrak{Q}_{\mathbb{C}}(\mathcal{M})$. We consider cases.

Case 1. Suppose that $e^{a_1x^1+a_2x^2} \in \mathfrak{Q}_{\mathbb{C}}(\mathcal{M})$ where $(a_1, a_2) \notin \mathbb{R}^2$. Define real functions by setting $L_1 = \Re(a_1x^1 + a_2x^2)$ and $L_2 = \Im(a_1x^1 + a_2x^2)$. As the affine quasi-Einstein operator $\mathfrak{Q}$ is real, we may take the real and imaginary parts of $e^{a_1x^1+a_2x^2}$ to see that $e^{L_1}\cos(L_2) \in \mathfrak{Q}(\mathcal{M})$ and $e^{L_1}\sin(L_2) \in \mathfrak{Q}(\mathcal{M})$. We now search for the remaining generator of $\mathfrak{Q}(\mathcal{M})$. Suppose that there is a non-constant polynomial $p(x^1, x^2)$ so that $p(x^1, x^2)e^{a_1x^1+a_2x^2} \in \mathfrak{Q}_{\mathbb{C}}(\mathcal{M})$. Then

$$\{\Re(pe^{a_1x^1+a_2x^2}), \Im(pe^{a_1x^1+a_2x^2}), \Re(e^{a_1x^1+a_2x^2}), \Im(pe^{a_1x^1+a_2x^2})\} \subset \mathfrak{Q}(\mathcal{M}).$$

This is impossible as $\dim(\mathfrak{Q}(\mathcal{M})) = 3$. Suppose next that the missing generator takes the form $p(x^1, x^2)e^{b_1x^1+b_2x^2}$ for $(b_1, b_2) \neq (a_1, a_2)$. Since $e^{b_1x^1+b_2x^2} \in \mathfrak{Q}_{\mathbb{C}}(\mathcal{M})$, p must be a polynomial of degree 0 and $b_1x^1 + b_2x^2$ is a real linear function. This is the possibility of Assertion 1.

Case 2. We suppose that all the L_i are real and that there are no polynomial factors. This is the possibility of Assertion 2.

Case 3. We suppose that all the L_i are real and there is a single polynomial factor. There must be two distinct exponentials defined by linear functions and the polynomial factor must be linear for dimensional reasons and the possibility of Assertion 3 holds.

Case 4. There is a single exponential defined by a linear function which is multiplied by polynomial factors. Two of the polynomial factors can be linear. Or one can be linear and the other quadratic; the possibility of Assertion 4 then holds. □

We complete the proof of Theorem 13.22 by examining the possibilities given by Lemma 13.23.

Lemma 13.24 *Let $\mathcal{M}$ be a Type $\mathcal{A}$ model.*

1. *Suppose that $\mathfrak{Q}(\mathcal{M}) = \operatorname{span}\{e^{L_1}\cos(L_2), e^{L_1}\sin(L_2), e^{L_3}\}$. Then $\mathcal{M}$ is linearly equivalent to $\mathcal{M}_5^6$, to $\mathcal{M}_5^4(c)$, or to $\mathcal{M}_2^2(b_1, b_2)$.*
2. *Suppose that $\mathfrak{Q}(\mathcal{M}) = \operatorname{span}\{e^{L_1}, e^{L_2}, e^{L_3}\}$. Then $\mathcal{M}$ is linearly equivalent to $\mathcal{M}_2^6$, to $\mathcal{M}_2^4(c_1)$, or to $\mathcal{M}_1^2(a_1, a_2)$.*

3. *Suppose that* $\mathcal{Q}(\mathcal{M}) = \text{span}\{e^{L_1}, L_2 e^{L_1}, e^{L_3}\}$. *Then* $\mathcal{M}$ *is linearly equivalent to* $\mathcal{M}_3^2(c_2)$, *to* $\mathcal{M}_1^4$, *to* $\mathcal{M}_3^4(c_1)$, *to* $\mathcal{M}_1^6$, *or to* $\mathcal{M}_3^6$.
4. *Suppose that* $\mathcal{Q}(\mathcal{M}) = \text{span}\{e^{L_1}, L_2 e^{L_1}, Q e^{L_1}\}$. *Then* $\mathcal{M}$ *is linearly equivalent to* $\mathcal{M}_4^2(\pm 1)$, *to* $\mathcal{M}_4^4(c)$, *to* $\mathcal{M}_0^6$, *or to* $\mathcal{M}_4^6$.

Proof. We apply Theorem 12.34 to see that $\mathcal{Q}(\mathcal{M})$ determines $\mathcal{M}$ throughout the proof.

Case 1. Suppose that $\mathcal{Q}(\mathcal{M}) = \text{span}\{e^{L_1}\cos(L_2), e^{L_1}\sin(L_2), e^{L_3}\}$. Since L_2 must be non-trivial, we can make a linear change of coordinates to assume $L_2 = x^2$.

Case 1.a. Suppose first that L_1 is not a multiple of L_2. Change coordinates to set $L_1 = x^1$ and $L_3 = b_1 x^1 + b_2 x^2$. Then $\mathcal{Q}(\mathcal{M}) = \text{span}\{e^{x^1}\cos(x^2), e^{x^1}\sin(x^2), e^{b_1 x^1 + b_2 x^2}\}$. By Lemma 12.33, $b_1 \neq 1$.

1.1. If $(b_1, b_2) \neq (0, 0)$, then we obtain $\mathcal{M}_2^2(b_1, b_2)$.

1.2. If $(b_1, b_2) = (0, 0)$, then we obtain $\mathcal{M}_5^6$.

Case 1.b. Suppose that L_2 is a multiple of L_1, i.e., that $L_2 = cL_1$. In this case we may express $\mathcal{Q}(\mathcal{M}) = \text{span}\{e^{cx^2}\cos(x^2), e^{cx^2}\sin(x^2), e^{L_3}\}$. By Lemma 12.33, L_3 is not independent of x^1.

1.3. Make a linear change of coordinates to assume $L_3 = x^1$ and obtain $\mathcal{M}_5^4(c)$.

Case 2. Suppose that $\mathcal{Q}(\mathcal{M}) = \text{span}\{e^{L_1}, e^{L_2}, e^{L_3}\}$. By Lemma 12.14, $\text{span}\{dL_1, dL_2, dL_3\}$ is 2-dimensional. Choose the notation so that dL_1 and dL_2 are linearly independent. Change coordinates to assume that $x^1 = L_1$ and $x^2 = L_2$. We then have

$$\mathcal{Q}(\mathcal{M}) = \text{span}\{e^{x^1}, e^{x^2}, e^{a_1 x^1 + a_2 x^2}\}.$$

If $a_1 + a_2 = 1$, then $a_1 - 1 = -a_2$ and $\mathcal{Q}(\mathcal{M}) = e^{x^1}\text{span}\{\mathbb{1}, e^{x^2 - x^1}, e^{a_2(x^2 - x^1)}\}$. This contradicts Lemma 12.33. Thus, $a_1 + a_2 \neq 1$. A brief calculation yields:

$$\mathcal{M} = \mathcal{M}\left(\frac{a_1^2 + a_2 - 1, a_1^2 - a_1, a_1 a_2, a_1 a_2, a_2^2 - a_2, a_1 + a_2^2 - 1}{a_1 + a_2 - 1}\right),$$

$$\rho_{\mathcal{M}} = \begin{pmatrix} \frac{(a_1-1)a_1}{a_1+a_2-1} & \frac{a_1 a_2}{a_1+a_2-1} \\ \frac{a_1 a_2}{a_1+a_2-1} & \frac{(a_2-1)a_2}{a_1+a_2-1} \end{pmatrix} \quad \text{and} \quad \det(\rho_{\mathcal{M}}) = -\frac{a_1 a_2}{a_1 + a_2 - 1}.$$

2.1. If $a_1 a_2 \neq 0$, then $\text{Rank}(\rho) = 2$ and $\mathcal{M}$ is linearly equivalent to $\mathcal{M}_1^2(a_1, a_2)$.

2.2. If $a_1 = a_2 = 0$, then $\text{Rank}(\rho) = 0$ and we replace x^1 by $-x^1$ to see that $\mathcal{M}$ is linearly equivalent to $\mathcal{M}_2^6$.

2.3. If $a_1 \neq 0$ and $a_2 = 0$ (the case $a_1 = 0$ and $a_2 \neq 0$ is analogous), then

$$\mathcal{Q}(\mathcal{M}) = \text{span}\{e^{x^1}, e^{x^2}, e^{a_1 x^1}\} \quad \text{for} \quad a_1 \neq 0, 1.$$

Change variables to replace x^1 by $c_1 x^1$ and obtain $\mathcal{Q}(\mathcal{M}) = \text{span}\{e^{c_1 x^1}, e^{x^2}, e^{a_1 c_1 x^1}\}$. Setting $a_1 = \frac{c_1+1}{c_1}$ then yields $\mathcal{Q}(\mathcal{M}) = \text{span}\{e^{c_1 x^1}, e^{x^2}, e^{(c_1+1)x^1}\}$. Since $a_1 \neq 0$, we

have that $c_1 \neq 0, -1$. Interchanging the roles of x^1 and x^2 yields
$$\mathcal{Q}(\mathcal{M}) = \operatorname{span}\{e^{c_1 x^2}, e^{(c_1+1)x^2}, e^{x^1}\}.$$
Replacing x^1 by $c_1 x^2 - x^1$ then yields $\mathcal{M}_2^4(c_1)$.

Case 3. Suppose that $\mathcal{Q}(\mathcal{M}) = \operatorname{span}\{e^{L_1}, L_2 e^{L_1}, e^{L_3}\}$.

Case 3.a. Suppose that $L_1 \neq 0$ and that L_1 and L_3 are linearly independent. Change coordinates to ensure $L_1 = x^1$, $L_2 = a_1 x^1 + a_2 x^2$ and $L_3 = x^1 + x^2$ so

$$\mathcal{Q}(\mathcal{M}) = e^{x^1} \operatorname{span}\{\mathbb{1}, a_1 x^1 + a_2 x^2, e^{x^2}\}.$$

By Lemma 12.33, $a_1 \neq 0$. We rescale $a_1 = 1$ to obtain $\mathcal{Q}(\mathcal{M}) = e^{x^1} \operatorname{span}\{\mathbb{1}, x^1 + a_2 x^2, e^{x^2}\}$.

3.1. If $a_2 \neq 0$, we obtain $\mathcal{M}_3^2(-a_2)$.

3.2. If $a_2 = 0$, we obtain $\mathcal{Q}(\mathcal{M}) = \operatorname{span}\{e^{x^1}, x^1 e^{x^1}, e^{x^1+x^2}\}$. We make a linear change of coordinates to assume $\mathcal{Q}(\mathcal{M}) = \operatorname{span}\{e^{x^2}, x^2 e^{x^2}, e^{x^2-x^1}\}$ and obtain $\mathcal{M}_1^4$.

Case 3.b. Suppose that $L_1 \neq 0$ and that $L_3 = aL_1$ for $a \neq 1$. Change coordinates to ensure that $L_1 = x^1$ and $L_2 = a_1 x^1 + a_2 x^2$ so $\mathcal{Q}(\mathcal{M}) = \operatorname{span}\{e^{x^1}, e^{ax^1}, (a_1 x^1 + a_2 x^2)e^{x^1}\}$. By Lemma 12.33, $a_2 \neq 0$ so after replacing $a_1 x^1 + a_2 x^2$ by x^2, we obtain

$$\mathcal{Q}(\mathcal{M}) = \operatorname{span}\{e^{x^1}, e^{ax^1}, x^2 e^{x^1}\}.$$

Replace x^1 by cx^1 to obtain $\mathcal{Q}(\mathcal{M}) = \operatorname{span}\{e^{cx^1}, e^{acx^1}, x^2 e^{cx^1}\}$. Set $a = \frac{c+1}{c}$; we then have that $c \neq 0$. We then have $\mathcal{Q}(\mathcal{M}) = \operatorname{span}\{e^{cx^1}, e^{(c+1)x^1}, x^2 e^{cx^1}\}$.

3.3. If $c \neq -1$, we obtain $\mathcal{M}_3^4(c)$.

3.4. If $c = -1$, then $\mathcal{Q}(\mathcal{M}) = \operatorname{span}\{e^{-x^2}, x^1 e^{-x^2}, \mathbb{1}\}$ which is linearly equivalent to $\mathcal{M}_1^6$.

Case 3.c. Suppose $L_1 = 0$. Since $L_3 \neq 0$, we may change coordinates to assume $L_3 = x^2$. We then have L_2 is not a multiple of L_3, so we may assume $L_2 = x^1$.

3.5. $\mathrm{Q}(\mathcal{M}) = \operatorname{span}\{\mathbb{1}, x^1, e^{x^2}\}$ and we obtain $\mathcal{M}_3^6$.

Case 4. Suppose $\mathcal{Q}(\mathcal{M}) = e^{L_1} \operatorname{span}\{\mathbb{1}, L_2, Q\}$. Set $\tilde{\mathcal{M}} = w^{-L_1}\mathcal{M}$; $\mathcal{Q}(\tilde{\mathcal{M}}) = \operatorname{span}\{\mathbb{1}, L_2, Q\}$.

Case 4.a. If $Q = L_3$ is linear, then $\mathcal{Q}(\tilde{\mathcal{M}}) = \operatorname{span}\{\mathbb{1}, L_2, L_3\}$. Since L_2 and L_3 are linearly independent, $\mathcal{Q}(\tilde{\mathcal{M}}) = \operatorname{span}\{\mathbb{1}, x^1, x^2\}$.

4.1. If $L_1 = 0$, then we obtain $\mathcal{M}_0^6$.

4.2. If $L_1 \neq 0$, choose coordinates to assume $L_1 = x^2$. We then obtain $\mathcal{M}_4^4(0)$.

Case 4.b. If Q is genuinely quadratic, change coordinates to assume $L_2 = x^2$. Since $\partial_{x^1} Q$ belongs to $\mathcal{Q}(\tilde{\mathcal{M}})$ and since $\partial_{x^1} Q$ is a multiple of x^2, $(x^1)^2$ does not appear in Q. Since $\partial_{x^2} Q$ is a multiple of x^2, $x^1 x^2$ does not appear in Q so $Q = (x^2)^2 + a_1 x^1 + a_2 x^2$. Subtract a multiple of x^2 to assume $a_2 = 0$ so $\mathcal{Q}(\tilde{\mathcal{M}}) = \operatorname{span}\{\mathbb{1}, x^2, (x^2)^2 + a_1 x^1\}$. Lemma 12.33 ensures $a_1 \neq 0$, so we rescale x^1 to get $\mathcal{Q}(\tilde{\mathcal{M}}) = \operatorname{span}\{\mathbb{1}, x^2, (x^2)^2 + 2x^1\}$.

4.3. If $L_1 = 0$, then we obtain $\mathcal{M}_4^6$.

Suppose that $L_1 \neq 0$ and $\mathcal{Q}(\mathcal{M}) = e^{b_1 x^1 + b_2 x^2} \operatorname{span}\{\mathbb{1}, x^2, (x^2)^2 + 2x^1\}$.

4.4. Suppose $b_1 = 0$. Set $\tilde{x}^2 := b_2 x^2$ so $\mathcal{Q}(\mathcal{M}) = e^{\tilde{x}^2} \operatorname{span}\{\mathbb{1}, \tilde{x}^2, (2x^1 + b_2^{-2}(\tilde{x}^2)^2)\}$. Set $c = b_2^{-2} \neq 0$ to see that we have $\mathcal{M}_4^4(c)$; we obtained $\mathcal{M}_4^4(0)$ previously in 4.2.

4.5. Suppose $b_1 \neq 0$. Let $b_1 = \pm c^2$ and $\tilde{x}^2 = cx^2$ setting $\tilde{x}^1 = b_1 x^1 + b_2 x^2$. We have

$$\begin{aligned} \mathcal{Q}(\mathcal{M}) &= e^{\tilde{x}^1} \operatorname{span}\{\mathbb{1}, x^2, (x^2)^2 + 2b_1^{-1}(\tilde{x}^1 - b_2 x^2)\} \\ &= e^{\tilde{x}^1} \operatorname{span}\{\mathbb{1}, x^2, b_1(x^2)^2 + 2\tilde{x}^1\} = e^{\tilde{x}^1} \operatorname{span}\{\mathbb{1}, \tilde{x}^2, \pm(\tilde{x}^2)^2 + 2\tilde{x}^1\}. \end{aligned}$$

We obtain $\mathcal{M}_4^2(\pm 1)$.

This proves Lemma 13.24 and thereby completes the proof of Theorem 13.22. □

We now draw some consequences of Theorem 13.22.

13.3.2 RANK(ρ) AND DIM($\mathfrak{K}(\mathcal{M})$).

We first relate the rank of the Ricci tensor and the dimension of the space of Killing vector fields for the Type $\mathcal{A}$ geometries by extending Lemma 13.4.

Corollary 13.25 *If $\mathcal{M} = (\mathbb{R}^2, \nabla)$ is a Type $\mathcal{A}$ model, then exactly one of the following possibilities holds.*

1. *$\mathcal{M}$ is flat,* $\operatorname{Rank}(\rho) = 0$, *and* $\dim(\mathfrak{K}(\mathcal{M})) = 6$.
2. $\operatorname{Rank}(\rho) = 1$ *and* $\dim(\mathfrak{K}(\mathcal{M})) = 4$.
3. $\operatorname{Rank}(\rho) = 2$ *and* $\dim(\mathfrak{K}(\mathcal{M})) = 2$.

Proof. By Theorem 13.22, $\mathcal{M}$ is linearly isomorphic to one of the geometries $\mathcal{M}_i^j(\cdot)$. If $j = 6$, Assertion 1 holds. If $j = 4$, Assertion 2 holds. If $j = 2$, Assertion 3 holds. □

13.3.3 STRONG PROJECTIVE AND LINEAR EQUIVALENCE.

We showed in Lemma 13.6 that every Type $\mathcal{A}$ affine model $\mathcal{M}$ is strongly linearly projectively equivalent to a flat Type $\mathcal{A}$ affine model $\tilde{\mathcal{M}}$. The following result now follows by inspection from the definitions given and from Theorem 12.34; it describes the extent to which $\tilde{\mathcal{M}}$ is not unique.

Theorem 13.26 *The Type $\mathcal{A}$ models $\mathcal{M}_1^6$ and $\mathcal{M}_3^6$ are strongly linearly projectively equivalent; otherwise $\mathcal{M}_i^6$ and $\mathcal{M}_j^6$ are not strongly linearly projectively equivalent for $i \neq j$.*

13.3.4 THE α INVARIANT.

By Lemma 13.3, a Type $\mathcal{A}$ model $\mathcal{M}$ is recurrent if and only if $\operatorname{Rank}(\rho) = 1$; these are the models $\mathcal{M}_i^4(\cdot)$ given in Definition 13.7. Let α be the invariant of Lemma 12.22. We use Lemma 13.8 to see:

Lemma 13.27 $\alpha\{\mathcal{M}_1^4\} = 16, \quad \alpha\{\mathcal{M}_2^4(c_1)\} = 4\frac{(1+2c_1)^2}{c_1+c_1^2}, \quad \alpha\{\mathcal{M}_3^4(c_1)\} = 4\frac{(1+2c_1)^2}{c_1+c_1^2},$
$\alpha\{\mathcal{M}_4^4(c)\} = 16, \quad \alpha\{\mathcal{M}_5^4(c)\} = \frac{16c^2}{1+c^2}.$

This invariant together with the sign of ρ is a complete invariant in this setting.

Theorem 13.28 *Let $\mathcal{M}_1$ and $\mathcal{M}_2$ be two Type $\mathcal{A}$ models whose Ricci tensor has rank 1. Assume that $\alpha(\mathcal{M}_1) = \alpha(\mathcal{M}_2)$. If $\alpha(\mathcal{M}_1) = 0$, also assume either $\rho_{\mathcal{M}_1} \geq 0$ and $\rho_{\mathcal{M}_2} \geq 0$ or that $\rho_{\mathcal{M}_1} \leq 0$ and $\rho_{\mathcal{M}_2} \leq 0$. Then $\mathcal{M}_1$ and $\mathcal{M}_2$ are locally isomorphic.*

Proof. By Theorem 13.22, it suffices to prove this result where the models in question belong to the families $\mathcal{M}_i^4(\cdot)$. By Lemma 13.12, $\mathcal{M}_1^4$ is locally affine isomorphic to $\mathcal{M}_4^4(0)$, $\mathcal{M}_4^4(c)$ is locally affine isomorphic to $\mathcal{M}_4^4(0)$, and $\mathcal{M}_2^4(c_1)$ is locally affine isomorphic to $\mathcal{M}_3^4(c_1)$. Thus we must only show that these invariants distinguish the spaces $\mathcal{M}_3^4(c_1)$, $\mathcal{M}_4^4(0)$, and $\mathcal{M}_5^4(c)$. We have $\mathcal{M}_3^4(c_1)$ is symmetric around the axis $c_1 = -\frac{1}{2}$ and $\mathcal{M}_5^4(c)$ is symmetric around the axis $c = 0$. By Lemma 13.27,

$$\alpha\{\mathcal{M}_3^4(c_1)\} = 4\frac{(1+2c_1)^2}{c_1+c_1^2},\ \alpha\{\mathcal{M}_4^4(0)\} = 16,\ \alpha\{\mathcal{M}_5^4(c)\} = \frac{16c^2}{1+c^2}\,.$$

We graph below these functions. The graph of $\alpha(\mathcal{M}_3^4(t))$ is given in red; it is the upper and the lower curve and has a vertical asymptote at $t \in \{-1, 0\}$; it is symmetric around the line $t = -\frac{1}{2}$. The graph of $\alpha(\mathcal{M}_4^4(0))$ is given in brown; it takes the constant value 16. The graph of $\alpha(\mathcal{M}_5^4(t))$ is given in blue. It lies in the middle and is symmetric around the line $t = 0$. The vertical scale is compressed by a factor of $\frac{1}{16}$.

Figure 13.1: The graph of α

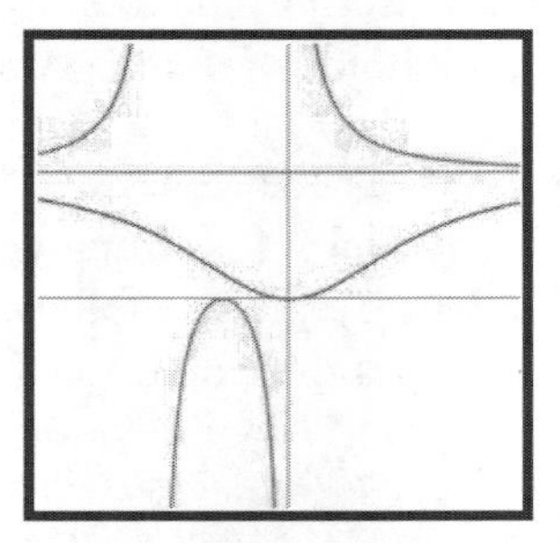

The α invariant of these three families is distinct except for the values

$$\alpha\{\mathcal{M}_5^4(0)\} = 0 \quad \text{and} \quad \alpha\{\mathcal{M}_3^4(-\tfrac{1}{2})\} = 0\,;$$

these two manifolds are affine symmetric spaces. We have $\rho_{\mathcal{M}_3^4(-\frac{1}{2})} \leq 0$ and $\rho_{\mathcal{M}_5^4(0)} \geq 0$ so the sign of the Ricci tensor distinguishes these two spaces. The function $4\frac{(1+2t)^2}{t+t^2}$ is 1-1 on $(-\infty, -\frac{1}{2}) \cup (-\frac{1}{2}, 0)$ and the function $\frac{16t^2}{1+t^2}$ is 1-1 on $(-\infty, 0)$. □

The affine symmetric spaces $\mathcal{M}_5^4(0)$ and $\mathcal{M}_3^4(-\frac{1}{2})$ were already considered by Opozda [50], who provided a parameterization of $\tilde{\mathcal{M}}_5^4(0)$.

13.3.5 GEODESIC COMPLETENESS. Let $\mathcal{M}$ be a Type $\mathcal{A}$ model. Let $\sigma_{a,b}(t)$ be the affine geodesic in $\mathcal{M}$ so that $\sigma_{a,b}(0) = (0,0)$ and $\dot{\sigma}_{a,b}(0) = (a,b)$. Our investigation was informed by Theorem 12.35 and Lemma 13.9; it was not a routine computation. When the Ricci tensor had rank 2, in the generic case, we could not in general solve the resulting ODE for the reparameterizing function and we contented ourselves with obtaining enough information for later use. We shall follow the treatment in Gilkey, Park, and Valle-Regueiro [34] and refer to D'Ascanio, Gilkey, and Pisani [21] for a different approach. We first deal with the flat geometries:

Lemma 13.29 *Let $\mathcal{M}$ be a Type $\mathcal{A}$ model with* $\dim(\mathfrak{K}(\mathcal{M})) = 6$ *and let $\sigma(a,b)(t)$ be an affine geodesic in $\mathcal{M}$ with $\sigma(a,b)(0) = (0,0)$ and $\dot{\sigma}(a,b)(0) = (a,b)$. Then:*

1. *Let $\mathcal{M} = \mathcal{M}_0^6$. Then $\sigma(a,b)(t) = t(a,b)$. $\mathcal{M}_0^6$ is geodesically complete.*
2. *Let $\mathcal{M} = \mathcal{M}_1^6$. Then $\sigma(a,b)(t) = (\log(1+at), (1+at)^{-1}bt)$. $\mathcal{M}_1^6$ is geodesically incomplete.*
3. *Let $\mathcal{M} = \mathcal{M}_2^6$. Then $\sigma(a,b)(t) = (-\log(1-at), \log(1+bt))$. $\mathcal{M}_2^6$ is geodesically incomplete.*
4. *Let $\mathcal{M} = \mathcal{M}_3^6$. Then $\sigma(a,b)(t) = (at, \log(1+bt))$. $\mathcal{M}_3^6$ is geodesically incomplete.*
5. *Let $\mathcal{M} = \mathcal{M}_4^6$. Then $\sigma(a,b)(t) = (at - \frac{1}{2}b^2t^2, bt)$. $\mathcal{M}_4^6$ is geodesically complete.*
6. *Let $\mathcal{M} = \mathcal{M}_5^6$. Then*
$$\sigma(a,b)(t) = \left(\tfrac{1}{2}\log(1+2at+a^2t^2+b^2t^2), \arctan\left(\tfrac{1+(a+b)t}{1+at-bt}\right) - \tfrac{\pi}{4}\right).$$
$\mathcal{M}_5^6$ is geodesically incomplete.

Proof. A direct computation verifies that the curves given satisfy the geodesic equation for the structures given with initial position $\sigma(0) = (0,0)$ and initial velocity $\dot{\sigma}(0) = (a,b)$. The curves are defined for all time if $\mathcal{M} = \mathcal{M}_0^6$ or $\mathcal{M} = \mathcal{M}_4^6$. This shows that these structures are geodesically complete. In the remaining structures, some of the affine geodesics blow up at finite time for some value of the parameters so these structures are geodesically incomplete. By Lemma 13.12, all these structures have an affine embedding or immersion in $\mathcal{M}_0^6$, so they can all be geodesically completed. □

We now turn to the geometries where the Ricci tensor has rank 1. By Lemma 12.8, if there exists an affine geodesic so $\lim_{t\to T}\rho(\dot{\sigma}(t),\dot{\sigma}(t)) = \pm\infty$, the geometry is essentially geodesically incomplete. In all these geometries, $\rho = \lambda dx^2 \otimes dx^2$ where $\lambda \neq 0$. Thus we shall examine $\dot{x}^2$ with particular care. Since we can solve the geodesic equations explicitly, the following result is now immediate.

Lemma 13.30 *Let $\mathcal{M}$ be a Type $\mathcal{A}$ model with* $\dim(\mathfrak{K}(\mathcal{M})) = 4$ *and let $\sigma(a,b)(t)$ be an affine geodesic in $\mathcal{M}$ with $\sigma(a,b)(0) = (0,0)$ and $\dot{\sigma}(a,b)(0) = (a,b)$. Then:*

1. *Let* $\mathcal{M} = \mathcal{M}_1^4$.
   (a) *If* $b \neq 0$, *then* $\sigma(a,b)(t) = \left(-\log\left(1 - \frac{a\log(1+2bt)}{2b}\right), \frac{1}{2}\log(1+2bt)\right)$.
   (b) *If* $b = 0$, *then* $\sigma(a,0)(t) = (-\log(1-at), 0)$.
   (c) $\mathcal{M}$ *is essentially geodesically incomplete.*

2.1. *Let* $\mathcal{M} = \mathcal{M}_2^4(c_1)$, $c_1 \neq -\frac{1}{2}$.
   (a) *If* $b \neq 0$ *and* $a + b \neq 0$, *then*
$$\sigma(a,b)(t) = \left(\log(\tfrac{b}{a+b}) - \log(1 - \tfrac{a(2bc_1t+bt+1)^{\frac{1}{2c_1+1}}}{a+b}), \tfrac{\log(2bc_1t+bt+1)}{2c_1+1}\right).$$
   (b) *If* $b \neq 0$ *and* $a + b = 0$, *then* $\sigma(-b,b)(t) = \left(-\frac{\log(1+bt+2bc_1t)}{1+2c_1}, \frac{\log(1+bt+2bc_1t)}{1+2c_1}\right)$.
   (c) *If* $b = 0$, *then* $\sigma(a,0)(t) = (-\log(1-at), 0)$.
   (d) $\mathcal{M}$ *is essentially geodesically incomplete.*

2.2. *Let* $\mathcal{M} = \mathcal{M}_2^4(-\frac{1}{2})$.
   (a) *If* $b \neq 0$ *and* $a \neq 0$, *then* $\sigma(a,b)(t) = \left(-\log\left(\frac{a+b-ae^{bt}}{b}\right), bt\right)$.
   (b) *If* $b \neq 0$ *and* $a = 0$, *then* $\sigma(0,b)(t) = (0, bt)$.
   (c) *If* $b = 0$ *and* $a \neq 0$, *then* $\sigma(a,0)(t) = (-\log(1-at), 0)$.
   (d) $\mathcal{M}$ *is geodesically incomplete. However,* $\mathcal{M}$ *affine embeds in* $\mathcal{M}_3^4(-\frac{1}{2})$ *which is geodesically complete.*

3.1. *Let* $\mathcal{M} = \mathcal{M}_3^4(c_1)$, $c_1 \neq -\frac{1}{2}$.
   (a) *If* $b \neq 0$, *then*
$$\sigma(a,b)(t) = \left(\tfrac{a}{b}(-1 + (1 + b(t + 2c_1t))^{\frac{1}{1+2c_1}}), \tfrac{1}{1+2c_1}\log(1 + bt + 2bc_1t)\right).$$
   (b) *If* $b = 0$, *then* $\sigma(a,0)(t) = (at, 0)$.
   (c) $\mathcal{M}$ *is essentially geodesically incomplete.*

3.2. *Let* $\mathcal{M} = \mathcal{M}_3^4(-\frac{1}{2})$.
   (a) *If* $b \neq 0$, *then* $\sigma(a,b)(t) = (\frac{a}{b}(e^{bt} - 1), bt)$.
   (b) *If* $b = 0$, *then* $\sigma(a,0)(t) = (at, 0)$.
   (c) $\mathcal{M}$ *is geodesically complete.*

4. *Let* $\mathcal{M} = \mathcal{M}_4^4(c)$ *and* $\tau(t) := 1 + 2bt$.
   (a) *If* $b \neq 0$, *then* $\sigma(a,b)(t) = \left(-\frac{\log(\tau(t))(bc\log(\tau(t))-4a)}{8b}, \frac{1}{2}\log(\tau(t))\right)$.
   (b) *If* $b = 0$, *then* $\sigma(a,0)(t) = (at, 0)$.
   (c) $\mathcal{M}$ *is essentially geodesically incomplete.*

5.1. *Let* $\mathcal{M} = \mathcal{M}_5^4(c)$, $c \neq 0$ *and* $\tau(t) := \log(1 + 2bct)$.

(a) *If* $b \neq 0$, *then* $\sigma(a,b)(t) = \left(\log\left(\frac{(a-bc)\sin\left(\frac{\tau(t)}{2c}\right)}{b} + \cos\left(\frac{\tau(t)}{2c}\right)\right) + \frac{1}{2}\tau(t), \frac{\tau(t)}{2c}\right)$.

(b) *If* $b = 0$, *then* $\sigma(a,0)(t) = (\log(1+at), 0)$.

(c) $\mathcal{M}$ *is essentially geodesically incomplete.*

5.2. *Let* $\mathcal{M} = \mathcal{M}_5^4(0)$.

(a) *If* $b \neq 0$, *then* $\sigma(a,b)(t) = \left(\frac{1}{2}\log\left(\frac{a^2}{b^2}+1\right) + \log\left(\cos\left(bt - \arctan\left(\frac{a}{b}\right)\right)\right), bt\right)$.

(b) *If* $b = 0$, *then* $\sigma(a,0)(t) = (\log(1+at), 0)$.

(c) $\mathcal{M}$ *is geodesically incomplete. However,* $\mathcal{M}$ *affine embeds in the geodesically complete manifold* $\tilde{\mathcal{M}}_5^4(0)$.

5.3. *Let* $\mathcal{M} = \tilde{\mathcal{M}}_5^4(c)$, $c \neq 0$.

(a) *If* $b \neq 0$, *then* $\sigma(a,b)(t) = \left(\frac{a}{b}(1+2bct)^{1/2}\sin\left(\frac{\log(1+2bct)}{2c}\right), \frac{\log(1+2bct)}{2c}\right)$.

(b) *If* $b = 0$, *then* $\sigma(a,0)(t) = (at, 0)$.

(c) $\mathcal{M}$ *is essentially geodesically incomplete.*

5.4. *Let* $\mathcal{M} = \tilde{\mathcal{M}}_5^4(0)$.

(a) *If* $b \neq 0$, *then* $\sigma(a,b)(t) = \left(\frac{a}{b}\sin(bt), bt\right)$.

(b) *If* $b = 0$, *then* $\sigma(a,0)(t) = (at, 0)$.

(c) $\mathcal{M}$ *is geodesically complete*

Let $\mathcal{M}$ be a Type $\mathcal{A}$ model with $\text{Rank}(\rho) = 2$. By Lemma 12.8, if there exists an affine geodesic and an affine Killing vector field X so $\lim_{t\to T}\rho(\dot\sigma(t), X(\sigma(t))) = \pm\infty$ for some finite time T, then $\mathcal{M}$ is essentially geodesically incomplete. Since ∂_{x^1} and ∂_{x^2} are affine Killing vector fields and since ρ is non-degenerate, $\mathcal{M}$ is essentially geodesically incomplete if

$$\lim_{t\to T}\dot x^1(t) = \pm\infty \text{ or if } \lim_{t\to T}\dot x^2(t) = \pm\infty\,.$$

We were unable to solve the ODE for the geodesic equation in closed form for the geometries $\mathcal{M}_i^2$ where the Ricci tensor has rank 2. However, we were able to obtain sufficient information to establish the following result.

Lemma 13.31 *Let* $\mathcal{M}$ *be a Type* $\mathcal{A}$ *model with* $\dim(\mathfrak{K}(\mathcal{M})) = 2$ *and let* $\sigma(a,b)(t)$ *be an affine geodesic in* $\mathcal{M}$ *with* $\sigma(a,b)(0) = (0,0)$ *and* $\dot\sigma(a,b)(0) = (a,b)$. *Then:*

1. $\mathcal{M}_1^2(a_1,a_2)$ *is essentially geodesically incomplete.*
2. $\mathcal{M}_3^2(c_2)$ *and* $\mathcal{M}_4^2(\pm1)$ *are essentially geodesically incomplete.*
3. *If* $b_1 \neq -1$, *then* $\mathcal{M}_2^2(b_1,b_2)$ *is essentially geodesically incomplete.*
4. $\mathcal{M}_2^2(-1,b_2)$ *is geodesically complete.*

Proof. We apply the criteria of Lemma 12.8 to prove the first three assertions; the proof of the final assertion requires a more delicate argument.

Assertion 1. $\mathcal{M} = \mathcal{M}_1^2(a_1, a_2)$. We obtain three possible affine geodesics which have the form $\sigma_i(t) = \log(t)\vec{\alpha}_i$ where

$$\vec{\alpha}_1 = \tfrac{1}{1+a_1+a_2}(1,1), \quad \vec{\alpha}_2 = \tfrac{1}{1+a_1-a_2}(1-a_2, a_1), \quad \vec{\alpha}_3 = \tfrac{1}{1-a_1+a_2}(a_2, 1-a_1).$$

The first affine geodesic is defined for $a_1 + a_2 + 1 \neq 0$, the second for $a_1 - a_2 + 1 \neq 0$, and the third for $-a_1 + a_2 + 1 \neq 0$. At least two affine geodesics are defined for any given geometry. These geometries are essentially geodesically incomplete.

Assertion 2. $\mathcal{M} = \mathcal{M}_3^2(c_2)$ or $\mathcal{M} = \mathcal{M}_4^2(\pm 1)$. These geometries are essentially geodesically incomplete since we have an affine geodesic of the form $\sigma(t) = (\frac{1}{2}\log(1+t), 0)$.

Assertion 3. $\mathcal{M} = \mathcal{M}_2^2(b_1, b_2)$ for $b_1 \neq -1$. These geometries are essentially geodesically incomplete since we have an affine geodesic of the form $\sigma(t) = \log(t)(\frac{1}{1+b_1}, 0)$.

Assertion 4. $\mathcal{M} = \mathcal{M}_2^2(-1, b_2)$. Suppose, to the contrary, that $\mathcal{M}$ is geodesically incomplete. Let σ be an affine geodesic in $\mathcal{M}$ which is defined on a parameter range (t_0, t_1) where $t_1 < \infty$ (resp. $-\infty < t_0$) which cannot be extended to a parameter range $(t_0, t_1 + \varepsilon)$ (resp. $(t_0 - \varepsilon, t_1)$) for any $\varepsilon > 0$. By Lemma 12.7, this implies that $\lim_{t \downarrow t_0^+} \sigma(t)$ (resp. $\lim_{t \uparrow t_1^-} \sigma(t)$) does not exist. We argue for a contradiction; our proof is motivated by the paper of Bromberg and Medina [4]. The non-zero Christoffel symbols of $\mathcal{M}$ are $\Gamma_{12}{}^1 = b_2$, $\Gamma_{12}{}^2 = 1$, and $\Gamma_{22}{}^1 = -\frac{1}{2}(1 + b_2^2)$. We work in the tangent bundle and introduce variables $u^1(t) := \dot{x}^1(t)$ and $u^2(t) := \dot{x}^2(t)$. This yields the geodesic equations

$$\dot{u}^1 + 2b_2 u^1 u^2 - \tfrac{1}{2}(1 + b_2^2) u^2 u^2 = 0 \quad \text{and} \quad \dot{u}^2 + 2u^1 u^2 = 0. \tag{13.3.a}$$

If $u^2(t) = 0$ for any $s \in (t_0, t_1)$, then $\dot{u}^1(s) = 0$ and $\dot{u}^2(s) = 0$. Consequently, $u^1(t) = u^1(s)$ and $u^2(t) = u^2(s)$ solves this ODE and $\vec{u} = (u^1, u^2)$ is constant on the interval (t_0, t_1). We may therefore assume u^2 does not change on the interval (t_0, t_1). Set

$$\begin{aligned} u^1(t) &:= e^{-b_2\tau(t)} \left(\tfrac{1}{2}\left(-2ab_2 + bb_2^2 + b\right)\sin(\tau(t)) + a\cos(\tau(t))\right), \\ u^2(t) &:= e^{-b_2\tau(t)}((bb_2 - 2a)\sin(\tau(t)) + b\cos(\tau(t))) \end{aligned}$$

where τ is a function which needs to be determined. Equation (13.3.a) then gives rise to a single ODE to be satisfied: $(2a - bb_2)\sin(\tau(t)) - b\cos(\tau(t)) + e^{b_2\tau(t)}\tau'(t)$, i.e., we have $\dot{\tau}(t) = u^2(\tau(t))$. Since u^2 does not change sign, $\tau(t)$ is restricted to a parameter interval of length at most π. If u^2 is positive (resp. negative), then $\dot{\tau}(t)$ is positive (resp. negative) and bounded so $\tau(t)$ is monotonically increasing (resp. decreasing) and bounded on the interval (t_0, t_1). Thus $\lim_{t \downarrow t_0} \tau(t)$ and $\lim_{t \uparrow t^1} \tau(t)$ exist. This shows that $\lim_{t \downarrow t_0} \dot{\sigma}(t)$ and $\lim_{t \uparrow t_1} \dot{\sigma}(t)$ exist. We integrate to conclude $\lim_{t \downarrow t_0} \dot{\sigma}(t)$ and $\lim_{t \uparrow t_1} \dot{\sigma}(t)$ exist. This provides the desired contradiction. □

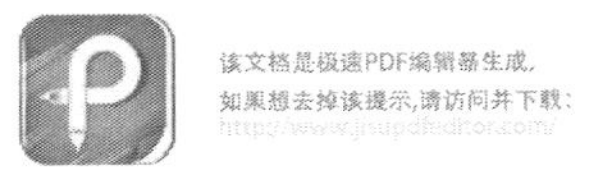

The next result follows from the previous analysis and from Theorem 13.22.

Theorem 13.32 *Let $\mathcal{M}$ be a Type $\mathcal{A}$ model.*

1. *If the Ricci tensor of $\mathcal{M}$ is zero, then $\mathcal{M}$ is geodesically complete if and only if $\mathcal{M}$ is linearly equivalent to $\mathcal{M}_0^6$ or to $\mathcal{M}_4^6$.*
2. *If the Ricci tensor of $\mathcal{M}$ has rank 1, then $\mathcal{M}$ is geodesically complete if and only if $\mathcal{M}$ is linearly equivalent to $\mathcal{M}_3^4(-\frac{1}{2})$.*
3. *If the Ricci tensor of $\mathcal{M}$ has rank 2, then $\mathcal{M}$ is geodesically complete if and only if $\mathcal{M}$ is linearly equivalent to $\mathcal{M}_2^2(-1, b)$ for some b.*
4. *If $\tilde{\mathcal{M}}$ is a geodesically complete homogeneous surface which is modeled on $\mathcal{M}$, then $\mathcal{M}$ can be taken to be $\mathcal{M}_0^6$, $\mathcal{M}_3^4(-\frac{1}{2})$, $\mathcal{M}_5^4(0)$, or $\mathcal{M}_2^2(-1, b)$ for some b.*
5. *The models $\{\mathcal{M}_i^6, \mathcal{M}_2^4(-\frac{1}{2}), \mathcal{M}_5^4(0)\}$, for $i \notin \{0, 4\}$, are geodesically incomplete. These models can be geodesically completed. All the remaining models $\mathcal{M}_i^j(\cdot)$ which are geodesically incomplete are essentially geodesically incomplete.*

We give below pictures of the geodesic structures for the models $\mathcal{M}_0^6$, $\mathcal{M}_3^4(-\frac{1}{2})$, $\tilde{\mathcal{M}}_5^4(0)$, and $\mathcal{M}_2^2(-1, b)$ for $b = 0, 1, 2$. The pictures for $\mathcal{M}_0^6$, $\mathcal{M}_3^4(-\frac{1}{2})$, and $\tilde{\mathcal{M}}_5^4(0)$ arise from the parameterizations obtained in this section; the pictures for $\mathcal{M}_2^2(\cdot)$ were obtained numerically.

Figure 13.2: Geodesic structure of $\mathcal{M}_0^6$, $\mathcal{M}_4^6$, $\mathcal{M}_3^4(-\frac{1}{2})$, $\tilde{\mathcal{M}}_5^4(0)$

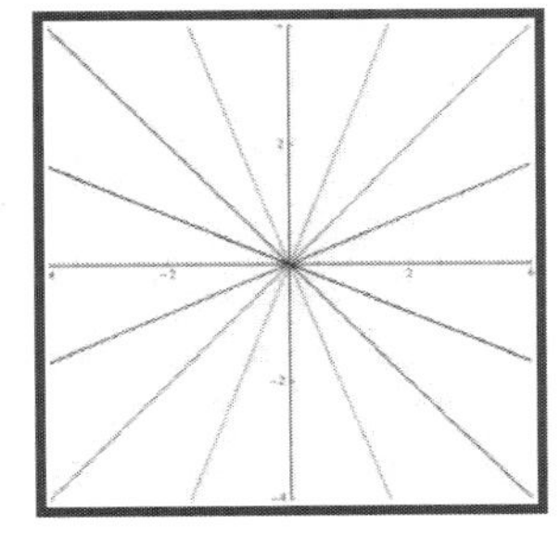
,
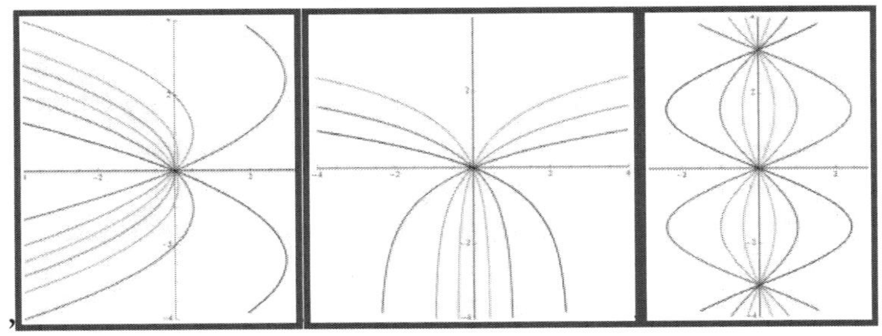

Figure 13.3: Geodesics for $\mathcal{M}_2^2(-1,0)$, $\mathcal{M}_2^2(-1,1)$, $\mathcal{M}_2^2(-1,2)$

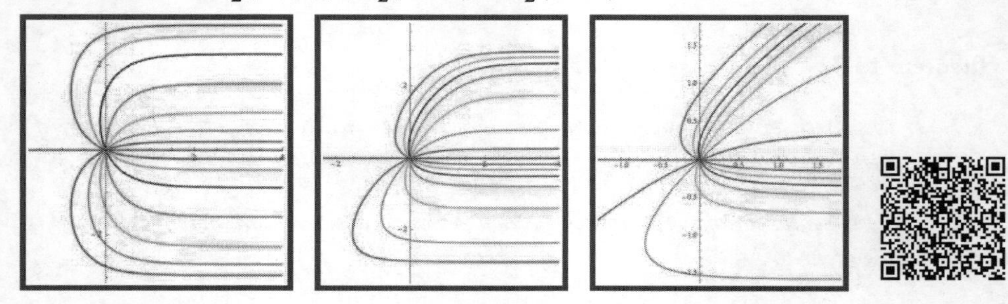

13.4 TYPE $\mathcal{A}$: MODULI SPACES

If $\mathcal{M}$ is a Type $\mathcal{A}$ model, let $\mathcal{G}(\mathcal{M}) := \{T \in \mathrm{GL}(2,\mathbb{R}) : T^*\mathcal{M} = \mathcal{M}\}$ be the associated symmetry group; by Lemma 12.39,

$$\mathcal{G}(\mathcal{M}) = \{T \in \mathrm{GL}(2,\mathbb{R}) : T^*\mathcal{Q}(\mathcal{M}) = \mathcal{Q}(\mathcal{M})\}\,.$$

Let $\mathcal{S}(\mathcal{M})$ be the set of all Type $\mathcal{A}$ models which are linearly equivalent to $\mathcal{M}$. Since $\mathrm{GL}(2,\mathbb{R})$ acts smoothly and transitively on $\mathcal{S}(\mathcal{M})$, $\mathcal{S}(\mathcal{M})$ is a homogeneous space and

$$\mathcal{S}(\mathcal{M}) = \mathrm{GL}(2,\mathbb{R})/\mathcal{G}(\mathcal{M})\,.$$

In Section 13.4.1, we study flat Type $\mathcal{A}$ connections. In Theorem 13.33, we determine the symmetry groups $\mathcal{G}(\mathcal{M}_i^6)$ and thereby determine the spaces $\mathcal{S}(\mathcal{M}_i^6)$. Let

$$\mathcal{S}_{\mathcal{A}}^6 = \cup_{1\le i\le 5}\mathcal{S}(\mathcal{M}_i^6)$$

be the space of all flat Type $\mathcal{A}$ models other than the trivial model $\mathcal{M}_0^6$ which otherwise would be a cone point. Let $\mathbb{L}$ be the Möbius strip over the circle and let $\mathbb{1}$ be the trivial real line bundle over the circle. In Theorem 13.34, we show that $\mathcal{S}_{\mathcal{A}}^6$ is diffeomorphic to $\mathbb{L}\oplus\mathbb{1}\oplus\mathbb{1}$ minus the zero section. In Section 13.4.2 (see Theorem 13.35), we perform a similar analysis for the set of all $\mathcal{A}$ models with rank 1 Ricci tensor; this space naturally decomposes into two components depending on whether ρ is positive or negative semi-definite. Note that the α invariant together with the signature of ρ completely determines the affine equivalence classes by Theorem 13.28. In Section 13.4.3, we discuss the two invariants (ψ,Ψ) which, together with the signature of the Ricci tensor, completely detect the equivalence class of a Type $\mathcal{A}$ model with non-degenerate Ricci tensor up to linear equivalence or, in view of Theorem 13.5, up to affine equivalence. In Section 13.4.4, we discuss linear equivalence for the manifolds $\mathcal{M}_1^2(a_1,a_2)$, and in Section 13.4.5, we discuss linear equivalence for the manifolds $\mathcal{M}_2^2(b_1,b_2)$.

13.4.1 FLAT TYPE $\mathcal{A}$ CONNECTIONS. The following result is an immediate application of Theorem 12.34 and Lemma 13.9.

Theorem 13.33

1. $\mathcal{G}(\mathcal{M}_0^6) = \mathrm{GL}(2,\mathbb{R})$.
2. $\mathcal{G}(\mathcal{M}_1^6) = \left\{\begin{pmatrix} 1 & 0 \\ 0 & a \end{pmatrix} : a \neq 0\right\}$.
3. $\mathcal{G}(\mathcal{M}_2^6) = \left\{\begin{pmatrix} 1 & 0 \\ 0 & 1 \end{pmatrix}, \begin{pmatrix} 0 & -1 \\ -1 & 0 \end{pmatrix}\right\}$.
4. $\mathcal{G}(\mathcal{M}_3^6) = \left\{\begin{pmatrix} a & 0 \\ 0 & 1 \end{pmatrix} : a \neq 0\right\}$.
5. $\mathcal{G}(\mathcal{M}_4^6) = \left\{\begin{pmatrix} a^2 & b \\ 0 & a \end{pmatrix} : a \neq 0,\ b \in \mathbb{R}\right\}$.
6. $\mathcal{G}(\mathcal{M}_5^6) = \left\{\begin{pmatrix} 1 & 0 \\ 0 & \pm 1 \end{pmatrix}\right\}$.

Theorem 13.34 *$\mathcal{S}_{\mathcal{A}}^6$ is a smooth submanifold of $\mathbb{R}^6 - \{0\}$ which is diffeomorphic to the total space of the bundle $\mathbb{L} \oplus 1\!\!1 \oplus 1\!\!1$ over the circle minus the zero section.*

Proof. Let (Ξ_1, Ξ_2) be a point of the unit circle S^1 in $\mathbb{R}^2$ and let (x^1, x^2, x^3) denote a point of $\mathbb{R}^3 - (0,0,0)$. We identify $((\Xi_1, \Xi_2), (x^1, x^2, x^3))$ with $((-\Xi_1, -\Xi_2), (-x^1, x^2, x^3))$ to define the bundle $\mathbb{L} \oplus 1\!\!1 \oplus 1\!\!1$. Let $\mathcal{M}(\xi_1, \xi_2, \xi_3, \xi_4, \xi_5, \xi_6)$ be the Type $\mathcal{A}$ model given in Equation (13.2.a). We make a linear change of variables and set

$$\mathcal{M}_1(p,q,t,s,v,w) := \mathcal{M}(2q,\, p+t,\, w,\, q+s,\, v,\, p-t)\,.$$

We have:

$$\begin{aligned} &\rho_{11} = p^2 + q^2 - s^2 - t^2 - pw - tw, \quad \rho_{12} = \rho_{21} = -(p+t)v + (q+s)w, \\ &\rho_{22} = qv - sv + (p - t - w)w\,. \end{aligned} \tag{13.4.a}$$

The equations $\rho_{12} = 0$ and $\rho_{22} = 0$ can be written in matrix form

$$\begin{pmatrix} -v & w \\ w & v \end{pmatrix} \begin{pmatrix} p \\ q \end{pmatrix} = \begin{pmatrix} tv - sw \\ sv + (t+w)w \end{pmatrix}.$$

If $v^2 + w^2 \neq 0$, we solve these equations to express

$$p = \frac{2svw + t\left(w^2 - v^2\right) + w^3}{v^2 + w^2} \quad \text{and} \quad q = \frac{s\left(v^2 - w^2\right) + vw(2t + w)}{v^2 + w^2}\,. \tag{13.4.b}$$

Substituting these values of p and q into Equation (13.4.a) then yields $\rho_{11} = 0$ as well; somewhat surprisingly, perhaps, there is no additional relation. This parameterizes the portion of $\mathcal{S}_{\mathcal{A}}^6$

where $(v, w) \neq (0, 0)$ as a graph over the (s, t, v, w) coordinate 4-plane. If we set $v = w = 0$, the equation $\rho = 0$ becomes the single equation

$$p^2 + q^2 - s^2 - t^2 = 0 . \tag{13.4.c}$$

There is an apparent singularity in Equation (13.4.b) at $(v, w) = (0, 0)$. We introduce polar coordinates $v = r\cos(\theta)$ and $w = r\sin(\theta)$ to obtain

$$\begin{aligned} p &= p(\theta, r, s, t) := r\sin^3(\theta) + s\sin(2\theta) - t\cos(2\theta),\\ q &= q(\theta, r, s, t) := \cos(\theta)\sin^2(\theta) + s\cos(2\theta) + t\sin(2\theta) . \end{aligned}$$

This no longer has a singularity at $r = 0$; when $r = 0$, we obtain

$$\begin{pmatrix} \sin(2\theta) & -\cos(2\theta) \\ \cos(2\theta) & \sin(2\theta) \end{pmatrix} \begin{pmatrix} s \\ t \end{pmatrix} = \begin{pmatrix} p \\ q \end{pmatrix}$$

which parameterizes the singular locus given by Equation (13.4.c). To avoid the trivial connection $\Gamma_0 = \Gamma(0, 0, 0, 0, 0, 0)$, we need to assume $(r, s, t) \neq (0, 0, 0)$; the parameterization is singular at this point since θ no longer plays a role. Thus our parameters take the form $\theta \in \mathbb{R}/(2\pi\mathbb{Z})$ and $(r, s, t) \in \mathbb{R}^3 - \{0\}$; since we are permitting r to be negative in polar coordinates, we must identify (θ, r) with $(\theta + \pi, -r)$. These are exactly the identifications defining the total space of the bundle $\mathbb{L} \oplus 1\!\!1 \oplus 1\!\!1$ minus the zero section once we identify $\mathbb{R}/(2\pi\mathbb{Z})$ with S^1. □

13.4.2 TYPE $\mathcal{A}$ CONNECTIONS WITH RANK 1 RICCI TENSOR.

The set of all Type $\mathcal{A}$ connections where the Ricci tensor has rank 1 is the disjoint union $\mathcal{S}^4_+ \dot{\cup} \mathcal{S}^4_-$ of two spaces where

$$\begin{aligned} \mathcal{S}^4_+ &:= \{\Gamma \in \mathbb{R}^6 : \text{Rank}(\rho_\Gamma) = 1 \text{ and } \rho_\Gamma \geq 0\},\\ \mathcal{S}^4_- &:= \{\Gamma \in \mathbb{R}^6 : \text{Rank}(\rho_\Gamma) = 1 \text{ and } \rho_\Gamma \leq 0\} . \end{aligned}$$

Theorem 13.34 generalizes to this situation to become the following result.

Theorem 13.35 *$\mathcal{S}^4_\pm$ is a smooth submanifold of $\mathbb{R}^6 - \{0\}$ diffeomorphic to $S^1 \times S^1 \times \mathbb{R}^3$.*

Proof. Let $\mathcal{S}^4_{+,0}$ (resp. $\mathcal{S}^4_{-,0}$) be the space of all Type $\mathcal{A}$ models where the Ricci tensor is a positive (resp. negative) multiple of $dx^2 \otimes dx^2$. By Lemma 13.3, an element of $\mathcal{S}^4_{\pm,0}$ satisfies the conditions

$$\Gamma_{11}{}^2 = 0, \quad \Gamma_{12}{}^2 = 0, \quad \rho_{22} = \Gamma_{22}{}^2\Gamma_{12}{}^1 - (\Gamma_{12}{}^1)^2 + \Gamma_{11}{}^1\Gamma_{22}{}^1 \neq 0 .$$

We make a change of variables setting

$$\begin{aligned} &\Gamma_{11}{}^1(p, q, u, v) := q + v, && \Gamma_{11}{}^2(p, q, u, v) := 0, && \Gamma_{12}{}^1(p, q, u, v) := u + p,\\ &\Gamma_{12}{}^2(p, q, u, v) := 0, && \Gamma_{22}{}^1 := q - v, && \Gamma_{22}{}^2 := 2p . \end{aligned}$$

We then have $\rho_{22} = (p^2 + q^2 - u^2 - v^2)dx^2 \otimes dx^2$. Thus we have

$$\mathcal{S}^4_{+,0} = \{\Gamma(p,q,u,v) : p^2 + q^2 > u^2 + v^2\},$$
$$\mathcal{S}^4_{-,0} = \{\Gamma(p,q,u,v) : p^2 + q^2 < u^2 + v^2\}.$$

We examine $\mathcal{S}^4_{+,0}$ as the analysis of $\mathcal{S}^4_{-,0}$ is the same after interchanging the roles of (p,q) and (u,v). Let $\dot{D}^2 := \{(\tilde{u},\tilde{v}) \in \mathbb{R}^2 : \tilde{u}^2 + \tilde{v}^2 < 1\}$. We construct a diffeomorphism

$$\tilde{\Gamma}(\theta, r, \tilde{u}, \tilde{v}) := \Gamma(p(\cdot), q(\cdot), u(\cdot), v(\cdot)) : S^1 \times \mathbb{R}^+ \times \dot{D}^2 \to \mathcal{S}^4_{+,0}$$

by setting $p = r\cos\theta$, $q = r\sin\theta$, $u = r\tilde{u}$, $v = r\tilde{v}$. We then have

$$-\Gamma_+(\theta, r, \tilde{u}, \tilde{v}) = \Gamma_+(\theta + \pi, r, -\tilde{u}, -\tilde{v}).$$

Consider the coordinate rotation through an angle of ϕ

$$T_\phi(x^1, x^2) = (\cos(\phi)x^1 + \sin(\phi)x^2, -\sin(\phi)x^1 + \cos(\phi)x^2).$$

Let $\tilde{\mathcal{M}}$ be an arbitrary Type $\mathcal{A}$ model with $\operatorname{Rank}(\rho_{\tilde{\mathcal{M}}}) = 1$ and $\rho_{\tilde{\mathcal{M}}}$ positive semi-definite. There exists $\lambda > 0$ and ϕ so that

$$\rho_{\tilde{\mathcal{M}}} = \lambda(-\sin(\phi)dx^1 + \cos(\phi)dx^2) \otimes (-\sin(\phi)dx^1 + \cos(\phi)dx^2);$$

the angle ϕ is not uniquely determined but only defined modulo π and not the more usual 2π. Let $\mathcal{M} := (T_\phi)_*\tilde{\mathcal{M}}$. Then $\rho_{\mathcal{M}} = \lambda dx^2 \otimes dx^2$ and, consequently, $\mathcal{M} \in \mathcal{S}^4_{+,0}$. Let $-\mathcal{M}$ be defined by $-\Gamma$. If we replace ϕ by $\phi + \pi$, then we replace $\mathcal{M}$ by $-\mathcal{M}$ and thus we have that

$$\mathcal{S}^4_{+} = \{(\mathbb{R}/(2\pi\mathbb{Z})) \times \mathcal{S}^4_{+,0}\}/(\phi, \mathcal{M}) \sim (\phi + \pi, -\mathcal{M})$$

where the gluing reflects the fact that when $\phi = \pi$ we have replaced (x^1, x^2) by $(-x^1, -x^2)$ and thus replaced Γ by $-\Gamma$. Using our previous parameterization of $\mathcal{S}^4_{+,0}$, this yields

$$\mathcal{S}^4_{+} = (\mathbb{R}^2/(2\pi\mathbb{Z})^2) \times \mathbb{R}^+ \times \dot{D}^2/\{(\phi, \theta, r, \tilde{u}, \tilde{v}) \sim (\phi + \pi, \theta + \pi, r, -\tilde{u}, -\tilde{v})\}.$$

After setting $\tilde{\theta} = \theta + \phi$, we can rewrite this equivalence relation in the form

$$(\phi, \tilde{\theta}, r, \tilde{u}, \tilde{v}) \sim (\phi + \pi, \tilde{\theta}, r, -\tilde{u} - \tilde{v}).$$

The variable $\tilde{\theta}$ now no longer plays a role in the gluing. After replacing $\mathbb{R}^+$ by $\mathbb{R}$ and $\dot{D}^2$ by $\mathbb{R}^2$, we see $\mathcal{S}^4_{+}$ is diffeomorphic to $S^1 \times S^1 \times \mathbb{R}^3$ modulo the gluing relation

$$(\phi, \tilde{\theta}, x^1, x^2, x^3) \sim (\phi + \pi, \tilde{\theta}, x^1, -x^2, -x^3). \tag{13.4.d}$$

The gluing of Equation (13.4.d) defines the total space of the bundle $1\!\!1 \oplus \mathbb{L} \oplus \mathbb{L}$ over (S^1, ϕ). Since $\mathbb{L} \oplus \mathbb{L}$ is just the trivial 2-plane bundle $1\!\!1 \oplus 1\!\!1$, we obtain finally that $\mathcal{S}^4_{+,0}$ is diffeomorphic to $S^1 \times S^1 \times \mathbb{R}^3$. □

13.4.3 THE INVARIANTS ψ AND Ψ WHEN RANK$(\rho) = 2$. We suppose the Ricci tensor is non-degenerate for the remainder of this section and let ρ^{ij} be the inverse of the matrix ρ_{ij}. Linear equivalence and affine equivalence are the same in this setting by Theorem 13.5. Since contraction of an upper index against a lower index is invariant under the action of GL(2, ℝ), the following are GL(2, ℝ) invariants:

$$\rho_{v,ij} := \Gamma_{ik}{}^{\ell}\Gamma_{j\ell}{}^{k}, \quad \psi := \mathrm{Tr}_{\rho}\{\rho_v\} = \rho^{ij}\rho_{v,ij}, \quad \Psi := \det(\rho_v)/\det(\rho).$$

We refer to Brozos-Vázquez, García-Río, and Gilkey [6] for the proof of the following result as it is beyond the scope of this book.

Theorem 13.36 *Let $\mathcal{M}$ and $\tilde{\mathcal{M}}$ be two Type $\mathcal{A}$ connections such that ρ_∇ and $\rho_{\tilde\nabla}$ are non-degenerate and have the same signature. Then $\mathcal{M}$ and $\tilde{\mathcal{M}}$ are affine equivalent if and only if we have that $(\psi, \Psi)(\mathcal{M}) = (\psi, \Psi)(\tilde{\mathcal{M}})$.*

We show the image of (ψ, Ψ) below in Figure 13.4; the region on the far right is the moduli space for positive definite Ricci tensor, the central region is the moduli space for indefinite Ricci tensor, and the region on the left the moduli space for negative definite Ricci tensor. The left boundary curve between negative definite and indefinite Ricci tensors is σ_ℓ (given in red) and the right boundary curve between indefinite and positive definite Ricci tensors is $\sigma_r(t)$ (given in blue) where

$$\sigma_\ell(t) := (-4t^2 - t^{-2} + 2, 4t^4 - 4t^2 + 2),$$
$$\sigma_r(t) := (4t^2 + t^{-2} + 2, 4t^4 + 4t^2 + 2).$$

Figure 13.4: Moduli spaces of Type $\mathcal{A}$ models with $\det(\rho) \neq 0$.

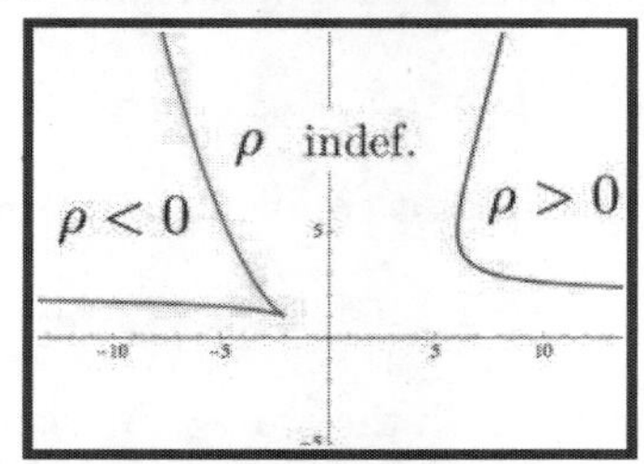

Note that although (ψ, Ψ) is 1-1 on each of the three cases separately, the images intersect along the smooth curves σ_ℓ and σ_r.

13.4.4 LINEAR EQUIVALENCE FOR THE MANIFOLDS $\mathcal{M}_1^2(\cdot)$. If we have that $\mathcal{M} = \mathcal{M}_1^2(a_1, a_2)$, then $\mathcal{Q}(\mathcal{M}) = \mathrm{span}\{e^{L_1}, e^{L_2}, e^{L_3}\}$. Suppose that $\{L_i, L_j\}$ are linearly independent for $i \neq j$. Let σ be a permutation of the integers $\{1, 2, 3\}$. Introduce new coordinates

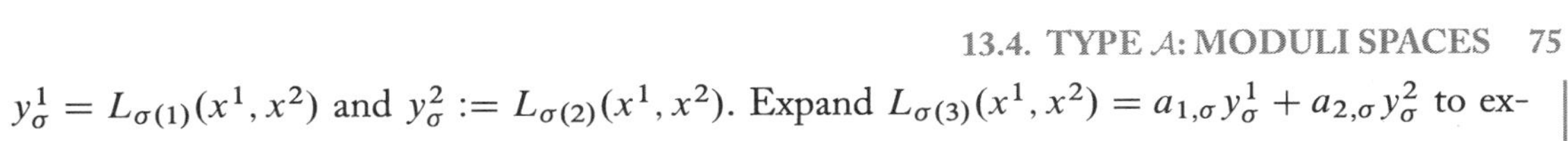

$y^1_\sigma = L_{\sigma(1)}(x^1, x^2)$ and $y^2_\sigma := L_{\sigma(2)}(x^1, x^2)$. Expand $L_{\sigma(3)}(x^1, x^2) = a_{1,\sigma} y^1_\sigma + a_{2,\sigma} y^2_\sigma$ to express

$$\mathcal{Q}(\mathcal{M}) = \operatorname{span}\{e^{y^1_\sigma}, e^{y^2_\sigma}, e^{a_{1,\sigma} y^1_\sigma + a_{2,\sigma} y^2_\sigma}\}.$$

This structure is defined by the pair $(a_{1,\sigma}, a_{2,\sigma})$; there are, generically, six such pairs that give rise to the same affine structure up to linear equivalence. If $\mathcal{M}^2_1(a_1, a_2)$ is linearly equivalent to $\mathcal{M}^2_1(\tilde{a}_1, \tilde{a}_2)$, then we shall write $(a_1, a_2) \sim (\tilde{a}_1, \tilde{a}_2)$. This means that there exists $T \in \mathbb{R}^2$ so $T \operatorname{span}\{e^{x^1}, e^{x^2}, e^{a_1 x^1 + a_2 x^2}\} = \operatorname{span}\{e^{\tilde{x}^1}, e^{\tilde{x}^2}, e^{\tilde{a}_1 \tilde{x}^1 + \tilde{a}_2 \tilde{x}^2}\}$. Suppose that $L_1 = x^1$, $L_2 = x^2$, and $L_3 = a_1 x^1 + a_2 x^2$. Let σ_{ijk} be the permutation $1 \to i, 2 \to j, 3 \to k$. We have

$$\begin{array}{lllll}
\sigma_{123}: & y^1 = L_1, & y^2 = L_2, & L_3 = a_1 y^1 + a_2 y^2, & (a_1, a_2) \sim (a_1, a_2). \\
\sigma_{213}: & y^1 = L_2, & y^2 = L_1, & L_3 = a_2 y^1 + a_1 y^2, & (a_1, a_2) \sim (a_2, a_1). \\
\sigma_{132}: & y^1 = L_1, & y^2 = L_3, & L_2 = -\frac{a_1}{a_2} y^1 + \frac{1}{a_2} y^2, & (a_1, a_2) \sim (-\frac{a_1}{a_2}, \frac{1}{a_2}). \\
\sigma_{321}: & y^1 = L_3, & y^2 = L_2, & L_1 = \frac{1}{a_1} y^1 - \frac{a_2}{a_1} y^2, & (a_1, a_2) \sim (\frac{1}{a_1}, -\frac{a_2}{a_1}). \\
\sigma_{231}: & y^1 = L_2, & y^2 = L_3, & L_1 = -\frac{a_2}{a_1} y^1 + \frac{1}{a_1} y^2, & (a_1, a_2) \sim (-\frac{a_2}{a_1}, \frac{1}{a_1}). \\
\sigma_{312}: & y^1 = L_3, & y^2 = L_1, & L_2 = \frac{1}{a_2} y^1 - \frac{a_1}{a_2} y^2, & (a_1, a_2) \sim (\frac{1}{a_2}, -\frac{a_1}{a_2}).
\end{array}$$

We observe that since ψ and Ψ are linear invariants, they are constant under the action of the group of permutations s_3. Although generically s_3 acts without fixed points, there are degenerate cases where the action is not fixed point free.

If $\det(\rho) > 0$ and $\operatorname{Tr}\{\rho\} < 0$, then ρ is negative definite; if $\det(\rho) > 0$ and $\operatorname{Tr}\{\rho\} > 0$, then ρ is positive definite; if $\det(\rho) < 0$, then ρ is indefinite. The six lines

$$\{x = 0,\ x = -1,\ y = 0,\ y = -1,\ x + y = 1,\ x = y\}$$

are given in black below; they further divide the regions where ρ is negative definite (light blue), ρ is positive definite (yellow), and ρ is indefinite (green); the three regions in different colors can be further divided into six regions under the action of s_3 under the picture given below.

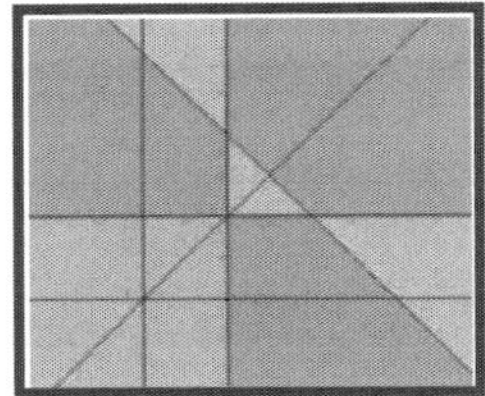

13.4.5 LINEAR EQUIVALENCE FOR THE MANIFOLDS $\mathcal{M}^2_2(\cdot)$.

We have that $\mathcal{Q}(\mathcal{M}) = \operatorname{span}\{e^{L_1}\cos(L_2), e^{L_1}\sin(L_2), e^{L_3}\}$. We set $\mathcal{M} = \mathcal{M}^2_2(b_1, b_2)$ where $b_1 \neq 1$ and $(b_1, b_2) \neq (0, 0)$. We have $b_1 > 1$ corresponds to ρ positive definite and $b_1 < 1$ corresponds to ρ indefinite; (b_1, b_2) and $(\tilde{b}_1, \tilde{b}_2)$ are linearly equivalent if and only if $b_1 = \tilde{b}_1$ and $b_2 = \pm\tilde{b}_2$.

CHAPTER 14

The Geometry of Type $\mathcal{B}$ Models

By Theorem 12.49, a locally affine homogeneous surface falls into one of three types. The Type $\mathcal{A}$ models are precisely those where there exists an effective Abelian 2-dimensional Lie subalgebra of the Lie algebra of affine Killing vector fields. They are modeled on the geometry of a left-invariant affine connection on $\mathbb{R}^2$ and the geometry of these surfaces is discussed at some length in Chapter 13. The Type $\mathcal{B}$ models are those where there exists an effective non-Abelian 2-dimensional Lie subalgebra of the Lie algebra of affine Killing vector fields. These are modeled on the geometry of a left-invariant affine connection on the $ax + b$ group. The Christoffel symbols take the form

$$\Gamma_{ij}{}^{k}(x^1, x^2) = \frac{1}{x^1} A_{ij}{}^{k} \quad \text{for} \quad A_{ij}{}^{k} \in \mathbb{R}\,.$$

Chapter 14 is devoted to the study of their geometry.

14.1 TYPE $\mathcal{B}$: DISTINGUISHED GEOMETRIES

Here is a brief outline to this section. We begin in Definition 14.1 by defining the fundamental structures of interest. In Lemma 14.2 of Section 14.1.1, we discuss how the Christoffel symbols transform under a shear

$$(x^1, x^2) \to (x^1, ax^2 + bx^1)\,.$$

In Lemma 14.3 of Section 14.1.2, we determine the Ricci tensor of the non-flat models of Definition 14.1; the spaces $\mathcal{Q}(\cdot)$ are treated in Lemma 14.4 of Section 14.1.3 and in Lemma 14.5 of Section 14.1.4. Eigenspaces of the affine quasi-Einstein equation for eigenvalues $\mu \notin \{0, -1\}$ are studied in Lemma 14.6 of Section 14.1.5. The topology of the space of flat non-trivial Type $\mathcal{B}$ models is presented in Theorem 14.7 in Section 14.1.6; by contrast with the Type $\mathcal{A}$ setting, it is not smooth but consists of three smooth closed surfaces which intersect transversally. In Theorem 14.8 of Section 14.1.7, we perform similar analysis for the Type $\mathcal{B}$ models with $\rho_s = 0$ but $\rho_a \neq 0$. In Theorem 14.9 of Section 14.1.8, we use the parameterizations of Theorem 14.7 and Theorem 14.8 to show that any flat geometry is linearly equivalent to one of the models $\mathcal{N}_i^6(\cdot)$ and that any model which is not flat but with $\rho_s = 0$ is linearly equivalent to one of the models $\mathcal{N}_3^2(0, c, \pm)$ for $c \neq 0$ or $\mathcal{N}_2^2(c)$. We also show that $\mathcal{N}$ is strongly projectively flat but not flat if it is linearly equivalent to one of the models $\mathcal{N}_i^4(\cdot)$, $\mathcal{N}_3^3$, $\mathcal{N}_4^3$, or $\mathcal{N}_1^2(\cdot)$. In Lemma 14.11 of Section 14.1.9, we establish some useful affine immersion and embedding results. In Lemma 14.12

of Section 14.1.10, we treat Yamabe solitons, and in Lemma 14.13 of Section 14.1.11, we examine affine gradient Ricci solitons.

In analogy to the Type $\mathcal{A}$ models which were defined in Equation (13.2.a), we let $\mathcal{N}(\xi_1, \xi_2, \xi_3, \xi_4, \xi_5, \xi_6)$ be the Type $\mathcal{B}$ model defined by the Christoffel symbols

$$\left\{\Gamma_{11}{}^1 = \frac{\xi_1}{x^1},\ \Gamma_{11}{}^2 = \frac{\xi_2}{x^1},\ \Gamma_{12}{}^1 = \frac{\xi_3}{x^1},\ \Gamma_{12}{}^2 = \frac{\xi_4}{x^1},\ \Gamma_{22}{}^1 = \frac{\xi_5}{x^1},\ \Gamma_{22}{}^2 = \frac{\xi_6}{x^1}\right\}.$$

Definition 14.1 We define the following Type $\mathcal{B}$ structures.

$\mathcal{N}_i^6$: $\mathcal{N}_0^6 := \mathcal{N}(0,0,0,0,0,0)$, $\mathcal{N}_1^6(\pm) := \mathcal{N}(1,0,0,0,\pm 1,0)$,
$\mathcal{N}_2^6(c_1) := \mathcal{N}(c_1 - 1, 0, 0, c_1, 0, 0)$, $\mathcal{N}_3^6 := \mathcal{N}(-2,1,0,-1,0,0)$,
$\mathcal{N}_4^6 := \mathcal{N}(0,1,0,0,0,0)$, $\mathcal{N}_5^6 := \mathcal{N}(-1,0,0,0,0,0)$,
$\mathcal{N}_6^6(c_2) := \mathcal{N}(c_2,0,0,0,0,0)$,
where $c_1 \neq 0$ and $c_2 \notin \{0,-1\}$.

$\mathcal{N}_i^4$: $\mathcal{N}_1^4(c_2) := \mathcal{N}(2c_2, 1, 0, c_2, 0, 0)$,
$\mathcal{N}_2^4(\kappa,\theta) := \mathcal{N}(2\kappa + \theta - 1, 0, 0, \kappa, 0, 0)$, $\mathcal{N}_3^4(c_1) := \mathcal{N}(2c_1 - 1, 0, 0, c_1, 0, 0)$,
where $c_1 \neq 0$, $c_2 \notin \{0,-1\}$, $\theta \neq 0$ and $\kappa(\kappa+\theta) \neq 0$.

$\mathcal{N}_i^3$: $\mathcal{N}_1^3(\pm) := \mathcal{N}(-\frac{3}{2}, 0, 0, -\frac{1}{2}, \mp\frac{1}{2}, 0)$, $\mathcal{N}_2^3(c) := \mathcal{N}(-\frac{3}{2}, 0, 1, -\frac{1}{2}, c, 2)$,
$\mathcal{N}_3^3 := \mathcal{N}(-1,0,0,-1,-1,0)$, $\mathcal{N}_4^3 := \mathcal{N}(-1,0,0,-1,1,0)$,
where $c \in \mathbb{R}$.

$\mathcal{N}_i^2$: $\mathcal{N}_1^2(v,\pm) := \mathcal{N}(1+2v, 0, 0, v, \pm 1, 0)$, $v \notin \{0,-1\}$,
$\mathcal{N}_2^2(c) := \mathcal{N}(0, c, 1, 0, 0, 1)$, $c \in \mathbb{R}$,
$\mathcal{N}_3^2(a,c,\pm) := \mathcal{N}(\frac{1}{2}\left(a^2 + 4a \mp 2c^2 + 2\right), c, 0, \frac{1}{2}\left(a^2 + 2a \mp 2c^2\right), \pm 1, \pm 2c)$,
where $(a,c) \neq (0,0)$ and $c \geq 0$.

$\mathcal{N}_3^3$ is the Lorentzian hyperbolic plane and $\mathcal{N}_4^3$ is the hyperbolic plane; the connections are the Levi-Civita connections of the metrics $ds^2 = (x^1)^{-2}\{(dx^1)^2 \mp (dx^2)^2\}$. The notation is chosen so that (see Theorem 14.16) $\dim(\mathfrak{K}(\mathcal{N}_i^\nu)) = \nu$ except for $\mathcal{N}_3^2(0, \frac{3}{\sqrt{2}}, +)$, which is equivalent to $\mathcal{N}_2^3(\frac{1}{2})$.

14.1.1 LINEAR EQUIVALENCE. The natural gauge group for the Type $\mathcal{A}$ geometries was the general linear group. The corresponding gauge group in the Type $\mathcal{B}$ setting is much smaller. If $T(x^1,x^2) = (ax^1, ax^2 + b)$ for $a \neq 0$ and if $\mathcal{N}$ is a Type $\mathcal{B}$ geometry, then $T^*\mathcal{N} = \mathcal{N}$. Thus these are homogeneous geometries. The $ax+b$ group has another action on $\mathbb{R}^+ \times \mathbb{R}$ by means of the shear transformation $T(x^1,x^2) = (x^1, ax^2 + bx^1)$ and is the appropriate gauge group in this setting. We say that two Type $\mathcal{B}$ geometries are *linearly equivalent* if there exists a shear intertwining them. This is a much smaller gauge group than $GL(2,\mathbb{R})$ and for that reason the Type $\mathcal{B}$ geometries are much richer. The following is a useful result which follows by a direct computation.

Lemma 14.2 *Let $(y^1, y^2) = (x^1, a^{-1}(x^2 - bx^1))$ be a change of variables which defines a shear. Then*

$$\begin{aligned}
&dy^1 = dx^1, \quad dy^2 = a^{-1}(dx^2 - bdx^1), \quad \partial_{y^1} = \partial_{x^1} + b\partial_{x^2}, \quad \partial_{y^2} = a\partial_{x^2},\\
&{}^yA_{11}{}^1 = {}^xA_{11}{}^1 + 2b\,{}^xA_{12}{}^1 + b^2\,{}^xA_{22}{}^1,\\
&{}^yA_{11}{}^2 = \tfrac{1}{a}\{{}^xA_{11}{}^2 + b(2\,{}^xA_{12}{}^2 - {}^xA_{11}{}^1) + b^2({}^xA_{22}{}^2 - 2\,{}^xA_{12}{}^1) - b^3\,{}^xA_{22}{}^1\},\\
&{}^yA_{12}{}^1 = a({}^xA_{12}{}^1 + b\,{}^xA_{22}{}^1),\\
&{}^yA_{12}{}^2 = {}^xA_{12}{}^2 + b\,{}^xA_{22}{}^2 - b({}^xA_{12}{}^1 + b\,{}^xA_{22}{}^1),\\
&{}^yA_{22}{}^1 = a^2\,{}^xA_{22}{}^1,\\
&{}^yA_{22}{}^2 = a({}^xA_{22}{}^2 - b\,{}^xA_{22}{}^1).
\end{aligned}$$

14.1.2 THE RICCI TENSOR. To clear denominators, we will let $\tilde{\rho} := (x^1)^2\rho$ throughout this section. A direct computation shows

$$\begin{aligned}
\tilde{\rho}_{11} &= (1 + A_{11}{}^1 - A_{12}{}^2)A_{12}{}^2 + A_{11}{}^2(-A_{12}{}^1 + A_{22}{}^2),\\
\tilde{\rho}_{12} &= A_{12}{}^1A_{12}{}^2 - A_{11}{}^2A_{22}{}^1 + A_{22}{}^2,\\
\tilde{\rho}_{21} &= A_{12}{}^1(-1 + A_{12}{}^2) - A_{11}{}^2A_{22}{}^1,\\
\tilde{\rho}_{22} &= -(A_{12}{}^1)^2 + (-1 + A_{11}{}^1 - A_{12}{}^2)A_{22}{}^1 + A_{12}{}^1A_{22}{}^2.
\end{aligned} \tag{14.1.a}$$

In particular, note that the Ricci tensor is not in general symmetric. We make a direct computation to establish the following result.

Lemma 14.3 *The Ricci tensor of the geometries $\mathcal{N}_i^6(\cdot)$ vanishes. These geometries are flat. Moreover:*

$\mathcal{N}_i^4$: $\tilde{\rho}_{\mathcal{N}_1^4(c_2)} = c_2(1 + c_2)dx^1 \otimes dx^1$, $\quad\tilde{\rho}_{\mathcal{N}_2^4(\kappa,\theta)} = \kappa(\kappa + \theta)dx^1 \otimes dx^1$,

$\tilde{\rho}_{\mathcal{N}_3^4(c_1)} = c_1^2dx^1 \otimes dx^1$.

$\mathcal{N}_i^3$: $\tilde{\rho}_{\mathcal{N}_1^3(\pm)} = \pm dx^2 \otimes dx^2$, $\quad\tilde{\rho}_{\mathcal{N}_2^3(c)} = \begin{pmatrix} 0 & \frac{3}{2} \\ -\frac{3}{2} & 1 - 2c \end{pmatrix}$,

$\tilde{\rho}_{\mathcal{N}_3^3} = -(dx^1)^2 + (dx^2)^2$, $\quad\tilde{\rho}_{\mathcal{N}_4^3} = -(dx^1)^2 - (dx^2)^2$.

$\mathcal{N}_i^2$: $\tilde{\rho}_{\mathcal{N}_1^2(v,\pm)} = v(2 + v)(dx^1)^2 \pm v(dx^2)^2$, $\quad\tilde{\rho}_{\mathcal{N}_2^2(c)} = \begin{pmatrix} 0 & 1 \\ -1 & 0 \end{pmatrix}$,

$$\tilde{\rho}_{\mathcal{N}_3^2(a,c,\pm)} = \begin{pmatrix} \frac{1}{2}a\,(a^2 + 4a \mp 2c^2 + 4) & \pm c \\ \mp c & \pm a \end{pmatrix}.$$

14.1.3 THE AFFINE QUASI-EINSTEIN EQUATION.

All the geometries of Definition 14.1 with the exception of $\mathcal{N}_1^3(\pm)$, $\mathcal{N}_2^3(c)$, $\mathcal{N}_2^2(c)$, and $\mathcal{N}_3^2(a,c,\pm)$ are strongly projectively flat; we shall show in Theorem 14.9 that, up to linear equivalence, there are no other strongly projectively flat geometries.

Lemma 14.4 *We have:*

$\mathcal{Q}(\mathcal{N}_i^6)$: $\mathcal{Q}(\mathcal{N}_0^6) = \operatorname{span}\{\mathbb{1}, x^1, x^2\}$, $\quad \mathcal{Q}(\mathcal{N}_1^6(\pm)) = \operatorname{span}\{\mathbb{1}, x^2, (x^1)^2 \pm (x^2)^2\}$,
$\quad \mathcal{Q}(\mathcal{N}_2^6(c_1)) = \operatorname{span}\{\mathbb{1}, (x^1)^{c_1}, (x^1)^{c_1}x^2\}$, $\quad \mathcal{Q}(\mathcal{N}_3^6) = \operatorname{span}\{\mathbb{1}, \frac{1}{x^1}, \frac{x^2}{x^1} + \log(x^1)\}$,
$\quad \mathcal{Q}(\mathcal{N}_4^6) = \operatorname{span}\{\mathbb{1}, x^1, x^2 + x^1\log(x^1)\}$, $\quad \mathcal{Q}(\mathcal{N}_5^6) = \operatorname{span}\{\mathbb{1}, \log(x^1), x^2\}$,
$\quad \mathcal{Q}(\mathcal{N}_6^6(c_2)) = \operatorname{span}\{\mathbb{1}, (x^1)^{1+c_2}, x^2\}$.

$\mathcal{Q}(\mathcal{N}_i^4)$: $\mathcal{Q}(\mathcal{N}_1^4(c_2)) = (x^1)^{c_2}\mathcal{Q}(\mathcal{N}_4^6)$,
$\quad \mathcal{Q}(\mathcal{N}_2^4(\kappa,\theta)) = (x^1)^{\kappa}\operatorname{span}\{\mathbb{1}, x^2, (x^1)^{\theta}\}$, $\quad \mathcal{Q}(\mathcal{N}_3^4(c_1)) = (x^1)^{c_1}\mathcal{Q}(\mathcal{N}_5^6)$.

$\mathcal{Q}(\mathcal{N}_i^3)$: $\mathcal{Q}(\mathcal{N}_1^3(\pm)) = \mathcal{Q}(\mathcal{N}_2^3(c)) = \{0\}$, *for* $c \neq \frac{1}{2}$, $\mathcal{Q}(\mathcal{N}_2^3(\frac{1}{2})) = \mathbb{R}\cdot\mathbb{1}$,
$\quad \mathcal{Q}(\mathcal{N}_3^3) = \operatorname{span}\{\frac{1}{x^1}, \frac{x^2}{x^1}, \frac{(x^2)^2-(x^1)^2}{x^1}\}$, $\quad \mathcal{Q}(\mathcal{N}_4^3) = \operatorname{span}\{\frac{1}{x^1}, \frac{x^2}{x^1}, \frac{(x^2)^2+(x^1)^2}{x^1}\}$.

$\mathcal{Q}(\mathcal{N}_i^2)$: $\mathcal{Q}(\mathcal{N}_1^2(v,\pm)) = (x^1)^{v}\mathcal{Q}(\mathcal{N}_1^6(\pm))$, $\quad \mathcal{Q}(\mathcal{N}_2^2(c)) = \mathbb{R}\cdot\mathbb{1}$.
$\quad$ *If* $a \neq 0$, *then* $\mathcal{Q}(\mathcal{N}_3^2(a\neq 0, c, \pm)) = \{0\}$.
$\quad$ *If* $a = 0$, *then* $\mathcal{Q}(\mathcal{N}_3^2(0, c, \pm)) = \mathbb{R}\cdot\mathbb{1}$.

Proof. By Lemma 12.14, $\dim(\mathcal{Q}(\cdot)) \le 3$. With the exception of the geometries $\mathcal{N}_1^3(\pm)$, $\mathcal{N}_2^3(c)$, $\mathcal{N}_2^2(c)$, and $\mathcal{N}_3^2(a,c,\pm)$, we need only verify that the functions given belong to $\mathcal{Q}(\cdot)$; they span for dimensional reasons. Theorem 12.34 then implies all these geometries are strongly projectively flat. We apply Lemma 12.53 to see that if $\mathcal{Q}(\mathcal{N})$ is non-trivial, then $(x^1)^{\alpha} \in \mathcal{Q}_{\mathbb{C}}(\mathcal{N})$ for some $\alpha \in \mathbb{C}$. This fails for the geometries $\mathcal{N}_1^3(\pm)$ and $\mathcal{N}_2^3(c)$ with $c \neq \frac{1}{2}$. This shows that $\mathcal{Q}(\cdot) = \{0\}$ in these instances. In the geometry $\mathcal{N}_2^2(c)$ as well as $\mathcal{N}_2^3(\frac{1}{2})$, the Ricci tensor is alternating; we verify $\mathbb{1}$ spans $\mathcal{Q}(\cdot)$. In the geometry $\mathcal{N}_3^2(a,c,\pm)$, a similar computation pertains; the Ricci tensor is alternating if and only if $a = 0$. □

14.1.4 TYPE $\mathcal{B}$ GEOMETRIES WITH DIM($\mathcal{Q}$) = 1.

We use Lemma 12.32 to see that if $\dim(\mathcal{Q}(\mathcal{N})) = 3$, then $\mathcal{N}$ is strongly projectively flat; we will show in Theorem 14.9 that if $\mathcal{N}$ is strongly projectively flat, then it appears among the manifolds of Definition 14.1. We refer to Brozos-Vázquez et al. [9] for a slightly different discussion.

Lemma 14.5 *Suppose $\mathcal{N}$ is a Type $\mathcal{B}$ geometry with* $\dim(\mathcal{Q}(\mathcal{N})) = 1$ *and* $\rho_s \neq 0$. *Then one of the following possibilities pertains.*

1. *$\mathcal{N}$ is linearly equivalent to $\mathcal{N}(1 + 2d + c^2\varepsilon, c, 0, d, \varepsilon, 2c\varepsilon)$ for some (c,d) where we have that $c \neq 0$, $c^2\varepsilon + d \neq 0$, and $\varepsilon = \pm 1$. We have $\mathcal{Q}(\mathcal{N}) = \operatorname{span}\{(x^1)^{d+c^2\varepsilon}\}$ and*

$$\tilde{\rho} = \begin{pmatrix} (d+2)(\varepsilon c^2 + d) & c\varepsilon \\ -c\varepsilon & \varepsilon(\varepsilon c^2 + d) \end{pmatrix}.$$

2. *$\mathcal{N}$ is linearly equivalent to $\mathcal{N}(0, b, 1, a, 0, 1)$ for some (a, b) with $a \neq 0$. We then have that $\mathcal{Q}(\mathcal{N}) = \text{span}\{(x^1)^a\}$ and*

$$\tilde{\rho} = \begin{pmatrix} a - a^2 & a + 1 \\ a - 1 & 0 \end{pmatrix}.$$

Proof. Let $\Gamma_{ij}{}^k = (x^1)^{-1} A_{ij}{}^k$ be a Type $\mathcal{B}$ geometry. We set $\mu = -1$. We use Lemma 12.53 and distinguish cases.

Case 1. Suppose $A_{22}{}^1 \neq 0$. We can rescale x^2 to assume $A_{22}{}^1 = \varepsilon = \pm 1$. We use Lemma 14.2 to apply a shear to assume $A_{12}{}^1 = 0$. This fixes the gauge. We renormalize the affine quasi-Einstein equation by multiplying by $(x^1)^{2-\alpha}$ to simplify expressions. We set $f = (x^1)^\alpha$ to obtain

$$\begin{aligned} &\tilde{\mathfrak{Q}}_{11} : 0 = (1 + A_{11}{}^1) A_{12}{}^2 - (A_{12}{}^2)^2 + A_{22}{}^2 A_{11}{}^2 + \alpha(-1 - A_{11}{}^1 + \alpha), \\ &\tilde{\mathfrak{Q}}_{12} : 0 = \tfrac{1}{2}(A_{22}{}^2 - 2A_{11}{}^2 \varepsilon), \qquad \tilde{\mathfrak{Q}}_{22} : 0 = (-1 + A_{11}{}^1 - A_{12}{}^2 - \alpha)\varepsilon . \end{aligned}$$

We set $A_{11}{}^2 = c$ and $A_{12}{}^2 = d$ and solve these relations to obtain the defining relations of Assertion 1:

$$\begin{aligned} &A_{11}{}^1 = 1 + 2d + c^2\varepsilon, && A_{11}{}^2 = c, && A_{12}{}^1 = 0, \\ &A_{12}{}^2 = d, && A_{22}{}^1 = \varepsilon, && A_{22}{}^2 = 2c\varepsilon, \\ &\tilde{\rho} = \begin{pmatrix} (2 + d)(d + c^2\varepsilon) & c\varepsilon \\ -c\varepsilon & \varepsilon(d + c^2\varepsilon) \end{pmatrix}, && \alpha = d + c^2\varepsilon . \end{aligned} \tag{14.1.b}$$

To ensure $\rho_s \neq 0$, we assume $c^2\varepsilon + d \neq 0$. We must show there are no other solutions if Equation (14.1.b) holds. Since α is unique, any other solution must have the form

$$f = (x^1)^\alpha \log(x^1) \quad \text{or} \quad f = (x^1)^\alpha (a_1 x^1 + x^2) .$$

1. If $f = (x^1)^\alpha \log(x^1)$, then $\tilde{\mathfrak{Q}}_{22} : 0 = -\varepsilon$ which is impossible as $\varepsilon = \pm 1$.
2. If $f = (x^1)^\alpha (a_1 x^1 + x^2)$, then $\tilde{\mathfrak{Q}}_{12} : 0 = c^2 \varepsilon x^1$ so $c = 0$. Conversely, if $c = 0$, then $(x^1)^\alpha x^2 \in \mathcal{Q}(\mathcal{N})$. Consequently, the restriction that $c \neq 0$ is essential. This establishes the possibility discussed in Assertion 1.

Case 2. We assume $A_{22}{}^1 = 0$ but $A_{12}{}^1 \neq 0$. We can rescale x^2 to assume $A_{12}{}^1 = 1$ and then normalize the gauge by performing a shear to assume $A_{11}{}^1 = 0$. Set $f = (x^1)^\alpha$ and $\mu = -1$; we obtain $\tilde{\mathfrak{Q}}_{22}$: $0 = -1 + A_{22}{}^2$. We set $A_{22}{}^2 = 1$ and obtain that $\tilde{\mathfrak{Q}}_{12}$: $0 = A_{12}{}^2 - \alpha$. This implies that $\alpha = A_{12}{}^2$. This yields $\tilde{\mathfrak{Q}}_{11} = 0$. We set $A_{11}{}^2 = b$ and $A_{12}{}^2 = a$ to obtain the defining relations of Assertion 2:

$$\begin{aligned} &A_{11}{}^1 = 0, && A_{11}{}^2 = b, \quad A_{12}{}^1 = 1, \\ &A_{12}{}^2 = a, && A_{22}{}^1 = 0, \quad A_{22}{}^2 = 1, \\ &\tilde{\rho} = \begin{pmatrix} a - a^2 & a + 1 \\ a - 1 & 0 \end{pmatrix}, && \alpha = a . \end{aligned} \tag{14.1.c}$$

To ensure $\rho_s \neq 0$, we have $a \neq 0$. We impose the relations of Equation (14.1.c). We suppose there exists another solution f and argue for a contradiction. The exponent is unique; therefore the other possibilities are:

$$f = (x^1)^\alpha \log(x^1) \quad \text{or} \quad f = (x^1)^\alpha (a_1 x^1 + x^2).$$

1. If $f = (x^1)^\alpha \log(x^1)$, then $\tilde{\mathfrak{Q}}_{12} : 0 = -1$ so this is impossible.
2. If $f = (x^1)^\alpha (a_1 x^1 + x^2)$, then $\tilde{\mathfrak{Q}}_{22} : 0 = -x^1$ so this is impossible.

Case 3. We assume $A_{22}{}^1 = 0$ and $A_{12}{}^1 = 0$. We obtain $\tilde{\mathfrak{Q}}_{12} : 0 = \frac{1}{2} A_{22}{}^2$ so $A_{22}{}^2 = 0$ as well. We will show subsequently in Theorem 14.17 that this implies that $\mathcal{N}$ is linearly equivalent to $\mathcal{N}_i^4(\cdot)$ so $\dim(\mathcal{Q}(\mathcal{N})) = 3$ and this case does not occur. Nevertheless, it is worth giving a direct argument. We suppose $A_{12}{}^1 = A_{22}{}^1 = A_{22}{}^2 = 0$. The remaining equation becomes

$$\tilde{\mathfrak{Q}}_{11} : 0 = -(A_{12}{}^2 - \alpha)(-1 - A_{11} + A_{12}{}^2 + \alpha).$$

1. If $A_{11}{}^1 \neq 2A_{12}{}^2 - 1$, then there are two possible values for α. Thus, $\dim(\mathcal{Q}(\mathcal{N})) > 1$, and this case is impossible.
2. If $A_{11}{}^1 = 2A_{12}{}^2 - 1$, then $(x^1)^\alpha (a_0 + a_1 \log(x^1))$ is a solution for $\alpha = A_{12}{}^2$ and a_0 and a_1 arbitrary. Thus again $\dim(\mathcal{Q}(\mathcal{N})) > 1$ so this case is impossible as well. □

14.1.5 OTHER EIGENVALUES OF THE AFFINE QUASI-EINSTEIN EQUATION.

We shall assume $\mathcal{N}$ is not flat. If $\rho_s = 0$, then $\rho_a \neq 0$ and it follows from Theorem 14.9 that $\mathcal{N} = \mathcal{N}_2^2(\cdot)$ or $\mathcal{N} = \mathcal{N}_3^2(\cdot)$. We suppose $\mu \neq 0$ since we will study Yamabe solitons in Lemma 14.12. We also suppose $\mu \neq -1$ since we examined that setting in the previous section. By Lemma 12.32, $\dim(E(\mu, \mathcal{N})) = 3$ if and only if $\rho_s = 0$. Consequently, $\dim(E(\mu, \mathcal{N})) \leq 2$. By Lemma 12.31, $\mathbb{1} \notin E(\mu, \mathcal{N})$. Thus we will examine $(x^1)^\alpha$ for $\alpha \neq 0$. We shall follow the same argument used to prove Lemma 14.5. We refer to Brozos-Vázquez et al. [9] for a slightly different discussion.

Lemma 14.6 *Suppose $\mathcal{N}$ is a Type $\mathcal{B}$ model with $\rho_s \neq 0$, with $\dim(E(\mu, \mathcal{N})) \geq 1$, and with $\mu \neq 0, -1$. Then one of the following possibilities pertains.*

1. *$\mathcal{N}$ is linearly equivalent to $\mathcal{N} = \mathcal{N}(a, c, 0, b, \varepsilon, 2c\varepsilon)$ for $a - b - 1 \neq 0$,*

   $$\mu = \frac{-a^2+2ab-b^2+2b+2c^2\varepsilon+1}{(a-b-1)^2}, \ \alpha = \frac{a^2-2ab+b^2-2b-2c^2\varepsilon-1}{a-b-1}, \text{ and } E(\mu, \mathcal{N}) = \operatorname{span}\{(x^1)^\alpha\}.$$

2. *$\mathcal{N}$ is linearly equivalent to $\mathcal{N}_i^4(\cdot)$. We have $A_{12}{}^1 = A_{22}{}^1 = A_{22}{}^2 = 0$ and $\tilde{\rho} = (1 + A_{11}{}^1 - A_{12}{}^2) A_{12}{}^2 dx^1 \otimes dx^1 \neq 0$. Denote the discriminant by $D := (A_{11}{}^1)^2 + 4A_{11}{}^1 A_{12}{}^2 \mu + 2A_{11}{}^1 - 4(A_{12}{}^2)^2 \mu + 4A_{12}{}^2 \mu + 1$.*
   (a) *If $D \neq 0$, then $E_{\mathbb{C}}(\mu, \mathcal{N}) = \operatorname{span}\{(x^1)^{\frac{1}{2}(1+A_{11}{}^1+\sqrt{D})}, (x^1)^{\frac{1}{2}(1+A_{11}{}^1-\sqrt{D})}\}$. If $D < 0$, we must take real and imaginary parts to get real solutions.*

(b) *If* $D = 0$, *then* $\mu = \frac{(A_{11}{}^1)^2+2A_{11}{}^1+1}{4A_{12}{}^2(-A_{11}{}^1+A_{12}{}^2-1)}$ *and* $E(\mu,\mathcal{N}) = \operatorname{span}\{(x^1)^{\frac{1}{2}(1+A_{11}{}^1)}, (x^1)^{\frac{1}{2}(1+A_{11}{}^1)}\log(x^1)\}$.

Proof. We distinguish cases. We assume $\mu \neq 0, -1$ and $\alpha \neq 0$.

Case 1. Suppose $A_{22}{}^1 \neq 0$. We rescale x^2 to assume $A_{22}{}^1 = \varepsilon = \pm 1$. Use Lemma 14.2 to choose a shear to assume that $A_{12}{}^1 = 0$; this fixes the gauge. We obtain

$$\tilde{\mathfrak{Q}}_{\mu,12}: \ 0 = A_{11}{}^2\mu\varepsilon - \tfrac{A_{22}{}^2\mu}{2}, \quad \tilde{\mathfrak{Q}}_{\mu,22}: \ 0 = \varepsilon(\mu(-A_{11}{}^1 + A_{12}{}^2 + 1) - \alpha)\,.$$

We solve these equations for α and $A_{22}{}^2$. The remaining equation can then be solved for μ and we obtain the structure of Assertion 1; $-a + b + 1 \neq 0$ since $\alpha \neq 0$. Since α is unique, we need only test $(x^1)^\alpha \log(x^1)$ and $(x^1)^{\alpha-1}(a_1x^1 + x^2)$; these fail so $E(\mu,\mathcal{N})$ is 1-dimensional.

Case 2. We assume $A_{22}{}^1 = 0$ but $A_{12}{}^1 \neq 0$. We can rescale to assume $A_{12}{}^1 = 1$ and then fix the gauge by performing a shear to assume $A_{11}{}^1 = 0$. We obtain $\tilde{\mathfrak{Q}}_{\mu,22}$: $0 = \mu - A_{22}{}^2\mu$. Consequently, $A_{22}{}^2 = 1$. We then obtain $\tilde{\mathfrak{Q}}_{\mu,12}$: $0 = -\alpha - A_{12}{}^2\mu$ so $\alpha = -A_{12}{}^2\mu$. The final equation then becomes $(A_{12}{}^2)^2\mu(1+\mu) = 0$. Since $\mu \neq 0$ and $\mu \neq -1$, this forces $A_{12}{}^2 = 0$. This in turn forces $\alpha = 0$, which is false.

Case 3. $A_{22}{}^1 = A_{12}{}^1 = 0$. We obtain $\tilde{\mathfrak{Q}}_{\mu,22} = 0$ and $\tilde{\mathfrak{Q}}_{\mu,12}$: $0 = -\frac{1}{2}A_{22}{}^2\mu$ so $A_{22}{}^2 = 0$ as well. Thus, as we shall establish subsequently in Theorem 14.17, this is linearly equivalent to $\mathcal{N}_1^4(c_2)$, $\mathcal{N}_2^4(\kappa,\theta)$, or $\mathcal{N}_3^4(c_1)$. Since $\mu \neq -1$ and $\rho_s \neq 0$, Lemma 12.32 implies $\dim(E(\mu,\mathcal{N})) \leq 2$. The final equation becomes

$$\tilde{\mathfrak{Q}}_{\mu,11} = \alpha^2 - \alpha(A_{11}{}^1 + 1) + A_{12}{}^2\mu(-A_{11}{}^1 + A_{12}{}^2 - 1)\,.$$

We can take μ arbitrary and solve this for α. Generically, we will obtain two different values of α; if α is complex we must take real and imaginary part. The discriminant is

$$D = (A_{11}{}^1)^2 + 4A_{11}{}^1A_{12}{}^2\mu + 2A_{11}{}^1 - 4(A_{12}{}^2)^2\mu + 4A_{12}{}^2\mu + 1\,.$$

Since $\tilde{\rho} = A_{12}{}^2(A_{11}{}^1 - A_{12}{}^2 + 1)dx^1 \otimes dx^1 \neq 0$, we can solve the equation $D = 0$ for μ to obtain the exceptional value

$$\mu = \frac{(A_{11}{}^1)^2 + 2A_{11}{}^1 + 1}{4A_{12}{}^2(-A_{11}{}^1 + A_{12}{}^2 - 1)}\,.$$

At this value of μ, we obtain $E(\mu,\mathcal{N}) = \operatorname{span}\{(x^1)^{\frac{1}{2}(1+A_{11}{}^1)}, (x^1)^{\frac{1}{2}(1+A_{11}{}^1)}\log(x^1)\}$. □

It is striking that in the setting of Assertion 1, there is exactly one eigenvalue and the solution space is 1-dimensional. By contrast, in the setting of Assertion 2, the eigenvalue μ can be any real number other than 0 or -1 and the solution space is 2-dimensional.

14.1.6 FLAT TYPE $\mathcal{B}$ GEOMETRIES. Set

$$\begin{aligned} \mathcal{U}_1(r,s) &:= \mathcal{N}(1+rs^2, -s(1+rs^2), rs, -rs^2, r, -rs), & \mathcal{S}^6_{1,\mathcal{B}} &:= \text{range}\{\mathcal{U}_1\}, \\ \mathcal{U}_2(u,v) &:= \mathcal{N}(u,v,0,0,0,0), & \mathcal{S}^6_{2,\mathcal{B}} &:= \text{range}\{\mathcal{U}_2\}, \\ \mathcal{U}_3(u,v) &:= \mathcal{N}(u,v,0,1+u,0,0), & \mathcal{S}^6_{3,\mathcal{B}} &:= \text{range}\{\mathcal{U}_3\}. \end{aligned}$$

Let $\mathcal{S}^6_{\mathcal{B}} \subset \mathbb{R}^6$ be the space of flat Type $\mathcal{B}$ models other than the cone point $\mathcal{N}^6_0$. Unlike the Type $\mathcal{A}$ setting described in Theorem 13.34, $\mathcal{S}^6_{\mathcal{B}}$ is not a smooth manifold but consists of the union of three smooth submanifolds of $\mathbb{R}^6$ which intersect transversally along the union of three smooth curves in $\mathbb{R}^6$.

Theorem 14.7 *$\mathcal{S}^6_{\mathcal{B}} = \mathcal{S}^6_{1,\mathcal{B}} \cup \mathcal{S}^6_{2,\mathcal{B}} \cup \mathcal{S}^6_{3,\mathcal{B}}$. $\mathcal{S}^6_{2,\mathcal{B}}$ and $\mathcal{S}^6_{3,\mathcal{B}}$ are closed smooth surfaces in $\mathbb{R}^6$ which are diffeomorphic to $\mathbb{R}^2$ and which intersect transversally along the curve $\mathcal{N}(-1,v,0,0,0,0)$. $\mathcal{S}^6_{1,\mathcal{B}}$ can be completed to a smooth closed surface $\tilde{\mathcal{S}}^6_{1,\mathcal{B}}$ which intersects $\mathcal{S}^6_{2,\mathcal{B}}$ transversally along the curve $\mathcal{N}(1,v,0,0,0,0)$ and which intersects $\mathcal{S}^6_{3,\mathcal{B}}$ transversally along the curve $\mathcal{N}(0,v,0,1,0,0)$.*

Proof. We use Equation (14.1.a) to compute the Ricci tensor. Let $\Gamma_{ij}{}^k = (x^1)^{-1}A_{ij}{}^k$ where $A_{ij}{}^k \in \mathbb{R}$. Let $\tilde{\rho}_{ij} := (x^1)^2\rho_{ij}$. A direct computation shows the structures $\mathcal{U}_i(\cdot)$ are flat. We distinguish cases to establish the converse. Because $\tilde{\rho}_{12} - \tilde{\rho}_{21} = A_{12}{}^1 + A_{22}{}^2$, we have that $A_{22}{}^2 = -A_{12}{}^1$.

Case 1. Assume $A_{22}{}^1 \neq 0$. Set $A_{12}{}^1 = rs$, $A_{22}{}^1 = r$, and $A_{22}{}^2 = -rs$ for $r \neq 0$. Then

$$\tilde{\rho}_{22} = -r(1 - A_{11}{}^1 + A_{12}{}^2 + 2rs^2) \quad \text{and} \quad \tilde{\rho}_{21} = -r(A_{11}{}^2 - A_{12}{}^2 s + s).$$

We solve these equations to obtain $A_{11}{}^1 = 1 + A_{12}{}^2 + 2rs^2$ and $A_{11}{}^2 = (A_{12}{}^2 - 1)s$. We have $\tilde{\rho}_{11} = 2(A_{12}{}^2 + rs^2)$. Thus $A_{12}{}^2 = -rs^2$, which gives the parameterization $\mathcal{U}_1$.

Case 2. Suppose $A_{22}{}^1 = 0$. Set $A_{11}{}^1 = u$, $A_{11}{}^2 = v$, and $A_{22}{}^2 = -A_{12}{}^1$ to obtain

$$\tilde{\rho} = \begin{pmatrix} -(A_{12}{}^2)^2 + uA_{12}{}^2 + A_{12}{}^2 - 2A_{12}{}^1 v & A_{12}{}^1(A_{12}{}^2 - 1) \\ A_{12}{}^1(A_{12}{}^2 - 1) & -2(A_{12}{}^1)^2 \end{pmatrix}.$$

This yields $A_{12}{}^1 = 0$ and $A_{12}{}^2(1 + u - A_{12}{}^2) = 0$. If we set $A_{12}{}^2 = 0$, we obtain the parameterization $\mathcal{U}_2$; if we set $A_{12}{}^2 = 1 + u$, we obtain the parameterization $\mathcal{U}_3$. This establishes the first part of the result.

The parameterizations $\mathcal{U}_2$ and $\mathcal{U}_3$ intersect when $u = -1$; the intersection is transversal along the curve $\mathcal{N}(-1,v,0,0,0,0)$. We wish to extend the parameterization $\mathcal{U}_1$ to study the limiting behavior as $A_{22}{}^1 \to 0$. We distinguish cases.

Case A. Suppose $\lim_{n\to\infty} \mathcal{U}_1(r_n, s_n) \in \text{range}\{\mathcal{U}_2\}$. We have

$$\begin{aligned} &\lim\nolimits_{n\to\infty}(1 + r_n s_n^2) = u, && \lim\nolimits_{n\to\infty} -s_n(1 + r_n s_n^2) = v, && \lim\nolimits_{n\to\infty} r_n s_n = 0, \\ &\lim\nolimits_{n\to\infty} r_n s_n^2 = 0, && \lim\nolimits_{n\to\infty} r_n = 0, && \lim\nolimits_{n\to\infty} r_n s_n = 0. \end{aligned}$$

These equations imply $u = 1$, $\lim_{n\to\infty} r_n = 0$, $\lim_{n\to\infty} s_n = -v$. Thus we may simply set $r = 0$ to obtain a transversal intersection along the curve $\mathcal{N}(1, v, 0, 0, 0, 0)$.

Case B. Suppose $\lim_{n\to\infty} \mathcal{U}_1(r_n, s_n) \in \text{range}\{\mathcal{U}_3\}$. We have

$$\begin{array}{lll} \lim_{n\to\infty}(1 + r_n s_n^2) = u, & \lim_{n\to\infty} -s_n(1 + r_n s_n^2) = v, & \lim_{n\to\infty} r_n s_n = 0, \\ \lim_{n\to\infty} r_n s_n^2 = 1 + u, & \lim_{n\to\infty} r_n = 0, & \lim_{n\to\infty} -r_n s_n = 0\,. \end{array}$$

These equations imply $u = 0$, $\lim_{n\to\infty} r_n = 0$, and $\lim_{n\to\infty} r_n s_n^2 = -1$. We change variables setting $r = -t^2$ and $s = \frac{1}{t} + w$ to express

$$\begin{aligned} \mathcal{U}_1(-t^2, \tfrac{1}{t} + w) = \mathcal{N}(&-2tw - t^2w^2, 2w + 3tw^2 + t^2w^3, -t - t^2w, \\ &1 + 2tw + t^2w^2, -t^2, t + t^2w)\,. \end{aligned}$$

We may now safely set $t = 0$ to obtain the intersection with $\text{range}\{\mathcal{U}_3\}$ along the curve $\mathcal{N}(0, 2w, 0, 1, 0, 0)$. □

14.1.7 THE ALTERNATING RICCI TENSOR. In the Type $\mathcal{B}$ setting, it is possible for the symmetric Ricci tensor to vanish without the geometry being flat; this is not possible in the Type $\mathcal{A}$ setting. For $r \neq 0$ and $u \neq 0$, set:

$$\begin{aligned} \mathcal{V}_1(r, s, t) &:= \mathcal{N}(s, t, r, 0, 0, r), \\ \mathcal{V}_2(u, v, w) &:= \mathcal{N}(1 - 2uw + vw^2, w(1 - uw + vw^2), u - vw, -vw^2, v, u + vw)\,. \end{aligned}$$

Theorem 14.8

1. *The sets* $\text{range}\{\mathcal{V}_i\}$ *are smooth 3-dimensional submanifolds of* $\mathbb{R}^3$ *for* $r \neq 0$ *and* $u \neq 0$. *They intersect transversally along a smooth 2-dimensional submanifold.*
2. *The set of all Type* $\mathcal{B}$ *structures where* $\rho_s = 0$ *but* $\rho_a \neq 0$ *is the union of* $\text{range}\{\mathcal{V}_1\}$ *and* $\text{range}\{\mathcal{V}_2\}$.
3. *Let* $\mathfrak{V}_1 := \{\mathcal{N}_2^2(c) : c \in \mathbb{R}\}$ *and* $\mathfrak{V}_2 := \{\mathcal{N}_3^2(0, c, \pm) : c \neq 0\}$. *Let* G *be the* $ax + b$ *group acting as in Lemma 14.2. The map* $(\mathcal{N}, T) \to T^*\mathcal{N}$ *is a diffeomorphism from* $\mathfrak{V}_i \times G$ *to* $\mathcal{V}_i$. *Thus* $\mathcal{V}_i$ *can be regarded as a principal* $ax + b$ *bundle over the curve* $\mathfrak{V}_i$.

Proof. It is clear $\text{range}\{\mathcal{V}_1\}$ is a smooth 3-dimensional submanifold. To see $\mathcal{V}_2$ is smooth, we note $u = \frac{1}{2}\{A_{12}{}^1 + A_{22}{}^2\}$ and $v = A_{22}{}^1$. If $v \neq 0$, then $w = \frac{1}{v}(A_{22}{}^2 - u)$, while if $v = 0$, then $w = \frac{1}{2u}(1 - A_{11}{}^1)$. Thus $\mathcal{V}_2$ is 1-1; it is not difficult to verify the Jacobian determinant is non-zero. Thus $\text{range}\{\mathcal{V}_2\}$ is smooth as well. We set $v = 0$ and $u = r$ to see they intersect along the surface $v = 0$, $u = r$, $s = 1 - 2uw$ and $t = w(1 - uw)$. Assertion 1 follows.

A direct computation shows

$$\tilde{\rho}_{\mathcal{V}_1} = r\begin{pmatrix} 0 & 1 \\ -1 & 0 \end{pmatrix} \quad \text{and} \quad \tilde{\rho}_{\mathcal{V}_2} = u\begin{pmatrix} 0 & 1 \\ -1 & 0 \end{pmatrix}.$$

So they belong to the set of Type $\mathcal{B}$ models where $\rho_s = 0$ and $\rho_a \neq 0$.

Let $\mathcal{N}$ be a Type $\mathcal{B}$ model with $\rho_s = 0$ and $(\tilde{\rho}_a)_{12} = \frac{1}{2}(A_{12}{}^1 + A_{22}{}^2) \neq 0$. We distinguish cases to prove Assertion 2.

Case 1. Suppose $A_{22}{}^1 = 0$. Set $A_{12}{}^1 = 2r - A_{22}{}^2$ for $r \neq 0$. Setting $\rho_s = 0$ yields

$$
\begin{aligned}
(\tilde{\rho}_s)_{11} &: \ 0 = A_{12}{}^2 + A_{11}{}^1 A_{12}{}^2 - (A_{12}{}^2)^2 + 2A_{11}{}^2 A_{22}{}^2 - 2A_{11}{}^2 r, \\
(\tilde{\rho}_s)_{12} &: \ 0 = (1 - A_{12}{}^2) A_{22}{}^2 - r + 2A_{12}{}^2 r, \\
(\tilde{\rho}_s)_{22} &: \ 0 = 2(3A_{22}{}^2 r - (A_{22}{}^2)^2 - 2r^2) \,.
\end{aligned}
$$

We solve the equation $-2((A_{22}{}^2)^2 - 3A_{22}{}^2 r + 2r^2) = 0$ to obtain $A_{22}{}^2 = r$ or $A_{22}{}^2 = 2r$. Setting $A_{22}{}^2 = 2r$ yields $(\tilde{\rho}_s)_{12}$: $0 = r$ which is false. Consequently, $A_{22}{}^2 = r$. We obtain $(\tilde{\rho}_s)_{12} = A_{12}{}^2 r$ so $A_{12}{}^2 = 0$. Set $A_{11}{}^1 = s$ and $A_{11}{}^2 = t$ to obtain the parameterization $\mathcal{V}_1$.

Case 2. Set $A_{12}{}^1 = 2u - A_{22}{}^2$ and $A_{22}{}^1 = v$ for $u \neq 0$ and $v \neq 0$. We obtain

$$
\begin{aligned}
(\tilde{\rho}_s)_{11} &: \ 0 = A_{12}{}^2 + A_{11}{}^1 A_{12}{}^2 - (A_{12}{}^2)^2 + 2A_{11}{}^2 A_{22}{}^2 - 2A_{11}{}^2 u, \\
(\tilde{\rho}_s)_{12} &: \ 0 = (1 - A_{12}{}^2) A_{22}{}^2 - u + 2A_{12}{}^2 u - A_{11}{}^2 v, \\
(\tilde{\rho}_s)_{22} &: \ 0 = -2(A_{22}{}^2)^2 + 6A_{22}{}^2 u - 4u^2 - (1 - A_{11}{}^1 + A_{12}{}^2) v \,.
\end{aligned}
$$

Setting $(\tilde{\rho}_s)_{12} = 0$ and $(\tilde{\rho}_s)_{22} = 0$ yields

$$
\begin{aligned}
A_{11}{}^2 &= \tfrac{1}{v}(A_{22}{}^2 - A_{12}{}^2 A_{22}{}^2 - u + 2A_{12}{}^2 u), \\
A_{11}{}^1 &= \tfrac{1}{v}(2(A_{22}{}^2)^2 - 6A_{22}{}^2 u + 4u^2 + v + A_{12}{}^2 v) \,.
\end{aligned}
$$

We obtain $(\tilde{\rho}_s)_{11} = \frac{2}{v}((A_{22}{}^2)^2 - 2A_{22}{}^2 u + u^2 + A_{12}{}^2 v)$. This implies

$$
A_{12}{}^2 = \tfrac{1}{v}(-(A_{22}{}^2)^2 + 2A_{22}{}^2 u - u^2) \,.
$$

Setting $A_{22}{}^2 = vw + u$ yields the parameterization $\mathcal{V}_2$. This parameterization can be extended safely to $v = 0$. Assertion 2 follows.

We apply Lemma 14.2 to compare the parameterizations in order to prove Assertion 3. Again, we distinguish cases.

Case A. Suppose $\mathcal{N} = \mathcal{N}(s, t, r, 0, 0, r)$ is given by the parameterization $\mathcal{V}_1$. Consider a shear $(x^1, x^2) \to (x^1, ax^2 + bx^1)$ where $a = \frac{1}{r}$ and $b = -\frac{s}{2r}$. By Lemma 14.2 this yields the structure $\mathcal{N}(0, \frac{s^2}{4} + rt, 1, 0, 0, 1)$ which is $\mathcal{N}_2^2(c = \frac{s^2}{4} + rt)$. Thus every element of range$\{\mathcal{V}_1\}$ is linearly equivalent to $\mathcal{N}_2^2(c)$ for some suitably chosen, and unique, c. Conversely, if we apply a shea to $\mathcal{N}_2^2(c) = \mathcal{N}(0, c, 1, 0, 0, 1)$, we obtain $\mathcal{N}(2b, \frac{c-b^2}{a}, a, 0, 0, a)$. To ensure this has the form $\mathcal{N}_2^2(\bar{c})$ for some $\bar{c}$, we need $a = 1$ and $b = 0$; this implies $\bar{c} = c$ as desired. Thus the $ax + b$ group acts without fixed points on $\mathfrak{V}_1$ and the quotient of $\mathfrak{V}_1$ by this action is the curve $\mathcal{N}_2^2(c)$.

Case B. Suppose $\mathcal{N} = \mathcal{V}_2(u, v, w)$. Let $\varepsilon = v$, and $c = u/\varepsilon$. A similar computation taking $a = 1$ and $b = w - c$ yields $\mathcal{N}(1 - c^2\varepsilon, c, 0, -c^2\varepsilon, \varepsilon, 2c\varepsilon)$. We can then rescale x^2 to ensure $\varepsilon = \pm 1$

and obtain $\mathcal{N}_3^2(0,c,\varepsilon)$ for $c \neq 0$. Thus every element of range$\{\mathcal{V}_2(u,v,w)\}$ is represented by some point of the curve $\mathcal{N}_3^2(0,c,\pm)$. Conversely, suppose we apply a shear to $\mathcal{N}_3^2(0,c,\varepsilon)$ and obtain $\mathcal{N}_3^2(0,\bar{c},\bar{\varepsilon})$. Examining $A_{12}{}^1 = ab\varepsilon$ so $b = 0$. Examining $A_{22}{}^1$ yields $a^2\varepsilon = \bar{\varepsilon}$. Consequently, $\varepsilon = \bar{\varepsilon}$ and $a = \pm 1$. If $a = -1$, we replace c by $-c$ and assume $c > 0$. □

14.1.8 PARAMETERIZATIONS. The $ax+b$ group acts on $\mathcal{S}_{\mathcal{B}}^6$ sending (x^1,x^2) to (x^1, bx^1+ax^2); we permit $a \neq 0$ here to define the notion of linear equivalence; the orbits of this action are the linear equivalence classes in this setting. The analogue of Theorem 13.22 in this case becomes:

Theorem 14.9 *Let $\mathcal{N}$ be a Type $\mathcal{B}$ model.*

1. *$\mathcal{N}$ is flat if and only if $\mathcal{N}$ is linearly equivalent to $\mathcal{N}_i^6(\cdot)$ for some i.*
2. *$\mathcal{N}$ is not flat, but the symmetric Ricci tensor vanishes if and only if $\mathcal{N}$ is linearly equivalent to $\mathcal{N}_3^2(a,c,\pm)$ for $a = 0$ and $c \neq 0$ or to $\mathcal{N}_2^2(c)$.*
3. *$\mathcal{N}$ is strongly projectively flat but not flat if and only if $\mathcal{N}$ is linearly equivalent to $\mathcal{N}_i^4(\cdot)$, $\mathcal{N}_3^3$, $\mathcal{N}_4^3$, or $\mathcal{N}_1^2(\cdot)$.*

Proof. We first prove Assertion 1. By Lemma 14.3, the geometries $\mathcal{N}_i^6(\cdot)$ are all flat. Conversely, suppose $\mathcal{N}$ is flat. By Theorem 12.34, $\dim(\mathcal{Q}(\mathcal{N})) = 3$ and $\mathbb{1} \in \mathcal{Q}(\mathcal{N})$. Furthermore, the geometry is determined by $\mathcal{Q}(\mathcal{N})$. We use the parameterization of Theorem 14.7.

Case 1.1. Consider $\mathcal{U}_1(r,s) = \mathcal{N}(1+rs^2, -s(1+rs^2), rs, -rs^2, r, -rs)$.

Case 1.1.1. Suppose $r = 0$. We verify $\{\mathbb{1}, x^2+sx^1, (x^1)^2\} \subset \mathcal{Q}(\mathcal{N})$. We can make a linear change of variables to replace x^2+sx^1 by x^2 and obtain $\mathcal{N}_6^6(1)$.

Case 1.1.2. Suppose $r \neq 0$. We verify $\{\mathbb{1}, sx^1+x^2, (x^1)^2 + r(sx^1+x^2)^2\} \subset \mathcal{Q}(\mathcal{N})$. We can make a linear change of variables to replace $|r|^{\frac{1}{2}}(sx^1+x^2)$ by x^2 and obtain $\mathcal{N}_1^6(\pm)$.

Case 1.2. Consider $\mathcal{U}_2(u,v) = \mathcal{N}(u,v,0,0,0,0)$.

Case 1.2.1. Suppose $u = -1$. We verify $\{\mathbb{1}, \log(x^1), x^2+vx^1\} \subset \mathcal{Q}(\mathcal{N})$. We make a linear change of variables to replace x^2+vx^1 by x^2 and obtain $\mathcal{N}_5^6$.

Case 1.2.2. Suppose $u = 0$. We verify $\{\mathbb{1}, x^1, -vx^1+x^2+vx^1\log(x^1)\} \subset \mathcal{Q}(\mathcal{N})$. If $v = 0$, we obtain $\mathcal{N}_0^6$. If $v \neq 0$, we can make a linear change of coordinates to obtain $\mathcal{N}_4^6$.

Case 1.2.3. Suppose $u \notin \{-1,0\}$. We verify $\{\mathbb{1}, (x^1)^{1+u}, x^2 - \frac{v}{u}x^1\} \subset \mathcal{Q}(\mathcal{N})$. We can make a linear change of variables to replace $x^2 - \frac{v}{u}x^1$ by x^2 and obtain $\mathcal{N}_6^6(c_2)$ for $c_2 = u$. Note that we also obtained $\mathcal{N}_6^6(1)$ previously in Case 1.1.1.

Case 1.3. Consider $\mathcal{U}_3(u,v) = \mathcal{N}(u,v,0,1+u,0,0)$.

Case 1.3.1. Suppose $u = -1$. We verify $\{\mathbb{1}, \log(x^1), x^2+vx^1\} \subset \mathcal{Q}(\mathcal{N})$. We obtain once again $\mathcal{N}_5^6$ since Case 1.3.1 and Case 1.2.1 coincide.

Case 1.3.2. Suppose $u = -2$. We verify $\{\mathbb{1}, \frac{1}{x^1}, \frac{x^2}{x^1} + v\log(x^1)\} \subset \mathcal{Q}(\mathcal{N})$. If $v \neq 0$, we can rescale x^2 to obtain $\mathcal{N}_3^6$. If $v = 0$, we obtain $\mathcal{N}_2^6(-1)$.

Case 1.3.3. Suppose $u \neq -1$ and $u \neq -2$. We verify $\{\mathbb{1}, (x^1)^{1+u}, (x^1)^{1+u}(x^2 + \frac{vx^1}{2+u})\} \subset \mathcal{Q}(\mathcal{N})$. We rescale x^2 to obtain $\mathcal{N}_2^6(1+u)$; the value $u = -2$ was obtained previously in Case 1.3.2.

Assertion 2 follows from Theorem 14.8. Finally, we establish Assertion 3. Suppose $\mathcal{N}$ is strongly projectively flat but not flat. A direct computation shows that any strongly projectively flat surface has both ρ and $\nabla\rho$ totally symmetric; this also follows from Theorem 12.11 of course, but it is not necessary to invoke this result. Set $\tilde{\rho} := (x^1)^2\rho$ and $\bar{\rho}_{ij;k} := (x^1)^3\rho_{ij;k}$ to eliminate denominators. We consider cases.

Case 3.1. Suppose $A_{22}{}^1 = \varepsilon \neq 0$. Let $(\tilde{x}^1, \tilde{x}^2) = (x^1, x^2 + bx^1)$. It follows from Lemma 14.2 that we may choose b so that $A_{12}{}^1 = 0$. Since $\tilde{\rho}_{12} - \tilde{\rho}_{21} = A_{22}{}^2$, $A_{22}{}^2 = 0$. This implies

$$\bar{\rho}_{11;2} - \bar{\rho}_{21;1} = -3A_{11}{}^2\varepsilon \text{ so } A_{11}{}^2 = 0\,.$$

Finally, $\bar{\rho}_{12;2} - \bar{\rho}_{22;1} = -2\varepsilon(1 - A_{11}{}^1 + 2A_{12}{}^2)$ so $A_{11}{}^1 = 1 + 2A_{12}{}^2$. We set $A_{12}{}^2 = v$ to obtain

$$\begin{aligned}
&A_{11}{}^1 = 1 + 2v, A_{11}{}^2 = 0, A_{12}{}^1 = 0, A_{12}{}^2 = v, A_{22}{}^1 = \varepsilon, A_{22}{}^2 = 0,\\
&\tilde{\rho} = v(2+v)(dx^1)^2 + v\varepsilon(dx^2)^2\,.
\end{aligned}$$

We obtain $\{(x^1)^v, x^2(x^1)^v, ((x^1)^2 + \varepsilon(x^2)^2)(x^1)^v\} \subset \mathcal{Q}(\mathcal{N})$. Since $v \neq 0$, we obtain $\mathcal{N}_1^2(v, \pm)$ if $v \neq -1$; if $v = -1$, we obtain $\mathcal{N}_3^3$ or $\mathcal{N}_4^3$.

Case 3.2. Suppose $A_{22}{}^1 = 0$. Assuming ρ and $\nabla\rho$ are symmetric then yields $A_{12}{}^1 = 0$ and $A_{22}{}^2 = 0$ as well. We then have $\tilde{\rho} = (1 + A_{11}{}^1 - A_{12}{}^2)A_{12}{}^2 dx^1 \otimes dx^1$ so $A_{12}{}^2 \neq 0$ and $A_{12}{}^2 \neq 1 + A_{11}{}^1$.

Case 3.2.1. Suppose $A_{12}{}^2 \neq 1 + A_{11}{}^1 - A_{12}{}^2$ and $A_{11}{}^1 \neq 2A_{12}{}^2$. Let $s = -\frac{A_{11}{}^2}{A_{11}{}^1 - 2A_{12}{}^2}$. We then have $\{(x^1)^{A_{12}{}^2}, (x^2 + sx^1)(x^1)^{A_{12}{}^2}, (x^1)^{1+A_{11}{}^1-A_{12}{}^2}\} \subset \mathcal{Q}(\mathcal{N})$. Change variables to obtain $\mathcal{N}_2^4(\kappa = A_{12}{}^2, \theta = 1 + A_{11}{}^1 - 2A_{12}{}^2 \neq 1)$ with $\kappa \neq 0$, $\theta \neq 0$, $\kappa + \theta \neq 0$.

Case 3.2.2. Suppose $A_{12}{}^2 \neq 1 + A_{11}{}^1 - A_{12}{}^2$ and $A_{11}{}^1 = 2A_{12}{}^2$.

Case 3.2.2.a. If $A_{11}{}^2 = 0$, then $\{(x^1)^{A_{12}{}^2}, (x^1)^{A_{12}{}^2+1}, (x^1)^{A_{12}{}^2}x^2\} \subset \mathcal{Q}(\mathcal{N})$ and we obtain $\mathcal{N}_2^4(\kappa = A_{12}{}^2, \theta = 1)$ with $\kappa \neq 0$ and $\kappa + \theta \neq 0$.

Case 3.2.2.b. If $A_{11}{}^2 \neq 0$, then

$$\{(x^1)^{A_{12}{}^2}, (x^1)^{1+A_{12}{}^2}, (x^2 + A_{11}{}^2 x^1 \log(x^1))(x^1)^{A_{12}{}^2}\} \subset \mathcal{Q}(\mathcal{N})$$

and we obtain $\mathcal{N}_1^4(c_2 = A_{12}{}^2 \neq 0)$ after rescaling x^2 appropriately.

Case 3.2.3. Suppose finally that $A_{12}{}^2 = 1 + A_{11}{}^1 - A_{12}{}^2$. In this case we have

$$(x^1)^{A_{12}{}^2}\{\mathbb{1}, A_{11}{}^2 x^1 + x^2, \log(x^1)\} \subset \mathcal{Q}(\mathcal{N})$$

so after adjusting x^2, we obtain $\mathcal{N}_3^4(c_1 = A_{12}{}^2 \neq 0)$. □

14.1.9 AFFINE IMMERSIONS AND EMBEDDINGS.

Definition 14.10 Let

$$\Psi_i^6: \Psi_0^6(x^1,x^2)=(x^1,x^2), \qquad \Psi_1^6(\pm 1)(x^1,x^2) = (x^2,(x^1)^2 \pm (x^2)^2),$$
$$\Psi_2^6(c_1)(x^1,x^2)=((x^1)^{c_1},(x^1)^{c_1}x^2), \qquad \Psi_3^6(x^1,x^2)=(\tfrac{1}{x^1},\tfrac{x^2}{x^1}+\log(x^1)),$$
$$\Psi_4^6(x^1,x^2)=(x^1,x^2+x^1\log(x^1)) \qquad \Psi_5^6(x^1,x^2)=(\log(x^1),x^2),$$
$$\Psi_6^6(c_2)(x^1,x^2)=((x^1)^{1+c_2},x^2).$$
$$\Psi_i^4: \Psi_1^4(x^1,x^2)=(x^2+x^1\log(x^1),\log(x^1)), \quad \Psi_2^4(\kappa,\theta)(x^1,x^2)=(x^2,\theta\log(x^1)),$$
$$\Psi_3^4(c_1)(x^1,x^2)=(x^2,c_1\log(x^1)).$$

The following result will play the role in the Type $\mathcal{B}$ setting that Lemma 13.12 played in the Type $\mathcal{A}$ setting.

Lemma 14.11

1. *$\Psi_i^6(\cdot)$ is an affine embedding of $\mathcal{N}_i^6(\cdot)$ in $\mathcal{M}_0^6$ for any i.*
2. *Ψ_1^4 is an affine isomorphism from $\mathcal{N}_1^4(c_2)$ to $\mathcal{M}_3^4(c_2)$.*
3. *$\Psi_2^4(\kappa,\theta)$ is an affine isomorphism from $\mathcal{N}_2^4(\kappa,\theta)$ to $\mathcal{M}_3^4(\frac{\kappa}{\theta})$.*
4. *$\Psi_3^4(c_1)$ is an affine isomorphism from $\mathcal{N}_3^4(c_1)$ to $\mathcal{M}_4^4(0)$.*

Proof. By Lemma 13.6 and Theorem 14.9, the geometries in question are all strongly projectively flat. The spaces $\mathcal{Q}(\mathcal{M}_i^j)$ are given in Lemma 13.9 and the spaces $\mathcal{Q}(\mathcal{N}_i^j)$ are given in Lemma 14.4. The diffeomorphisms in question intertwine these solution spaces. The result now follows from Theorem 12.34. □

14.1.10 YAMABE SOLITONS. We compute the components of the Hessian in this setting:

$$\mathcal{H}_{11}(f)=-(x^1)^{-1}\{A_{11}{}^1 f^{(1,0)}+A_{11}{}^2 f^{(0,1)}-x^1 f^{(2,0)}\},$$
$$\mathcal{H}_{12}(f)=\mathcal{H}_{21}(f)=-(x^1)^{-1}\{A_{12}{}^1 f^{(1,0)}+A_{12}{}^2 f^{(0,1)}-x^1 f^{(1,1)}\},$$
$$\mathcal{H}_{22}(f)=-(x^1)^{-1}\{A_{22}{}^1 f^{(1,0)}+A_{22}{}^2 f^{(0,1)}-x^1 f^{(0,2)}\}.$$

Let $\mathcal{Y}(\mathcal{N})=E(0,\mathcal{N})=\ker\{\mathcal{H}\}$ be the space of Yamabe solitons. By Lemma 12.14, we have that $\dim(\mathcal{Y}(\mathcal{N}))\le 3$. By Lemma 12.40, $\dim(\mathcal{Y}(\mathcal{N}))=3$ implies $\mathcal{N}$ is flat. In that setting, $\mathcal{Y}(\mathcal{N})=\mathcal{Q}(\mathcal{N})$ and those structures are treated in Lemma 14.4. Since $\mathbb{1}\in\mathcal{Y}(\mathcal{N})$, we shall be searching for Type $\mathcal{B}$ models where $\dim(\mathcal{Y}(\mathcal{N}))=2$, i.e., where there is an extra Yamabe soliton.

Lemma 14.12 *Let $\mathcal{N}$ be a Type $\mathcal{B}$ model with* $\dim(\mathcal{Y}(\mathcal{N}))=2$. *Then exactly one of the following possibilities pertains.*

1. $A_{12}{}^1=0$, $A_{22}{}^1=0$, $A_{11}{}^1\ne -1$, $\mathcal{Y}(\mathcal{N})=\operatorname{span}\{\mathbb{1},(x^1)^{1+A_{11}{}^1}\}$, *and*

$$\tilde{\rho} = \begin{pmatrix} -(A_{12}{}^2)^2 + A_{11}{}^1 A_{12}{}^2 + A_{12}{}^2 + A_{11}{}^2 A_{22}{}^2 & A_{22}{}^2 \\ 0 & 0 \end{pmatrix} \neq 0\,.$$

2. $A_{12}{}^1 = 0$, $A_{22}{}^1 = 0$, $A_{11}{}^1 = -1$, $\mathcal{Y}(\mathcal{N}) = \operatorname{span}\{\mathbb{1}, \log(x^1)\}$, *and*

$$\tilde{\rho} = \begin{pmatrix} A_{11}{}^2 A_{22}{}^2 - (A_{12}{}^2)^2 & A_{22}{}^2 \\ 0 & 0 \end{pmatrix} \neq 0\,.$$

3. $\mathcal{N}$ *is linearly equivalent to* $\mathcal{N}(r,0,s,0,t,0)$, $\mathcal{Y}(\mathcal{N}) = \operatorname{span}\{\mathbb{1}, x^2\}$, *and*

$$\tilde{\rho} = \begin{pmatrix} 0 & 0 \\ -s & (r-1)t - s^2 \end{pmatrix} \neq 0\,.$$

Proof. We apply Lemma 12.53. Let $\mathcal{Y}_\alpha := \ker\{(x^1\partial_{x^1} + x^2\partial_{x^2} - \alpha)^2\}$. A priori, α could be complex, although in fact this does not arise. We suppose $\mathcal{Y}_\alpha \neq \{0\}$ and, if $\alpha = 0$, we suppose $\mathcal{Y}_\alpha \neq \operatorname{span}\{\mathbb{1}\}$.

Case 1. Suppose $A_{12}{}^1 = 0$, $A_{22}{}^1 = 0$, and $A_{11}{}^1 \neq -1$. We verify that $(x^1)^{1+A_{11}{}^1} \in \mathcal{Y}(\mathcal{N})$. This is the possibility of Assertion 1.

Case 2. Suppose $A_{12}{}^1 = 0$, $A_{22}{}^1 = 0$, and $A_{11}{}^1 = -1$. We verify that $\log(x^1) \in \mathcal{Y}(\mathcal{N})$. This is the possibility of Assertion 2.

Case 3. Suppose $(A_{12}{}^1, A_{22}{}^1) \neq (0,0)$. To clear denominators, we set $\tilde{\mathcal{H}} := (x^1)^{2-\alpha}\mathcal{H}$.

Case 3.1. Suppose $\alpha \neq 0, 1$. By applying $x^1\partial_{x^1} + x^2\partial_{x^2} - \alpha$, we may assume that there is no $\log(x^1)$ dependence. Express $f = f_\alpha(x^1) + f_{\alpha-1}(x^1)x^2$. Since

$$f_{\alpha-1} \in E(\alpha - 1, \mathcal{N}) \neq E(0, \mathcal{N})\,,$$

we may assume there is no x^2 dependence. Thus $f = (x^1)^\alpha$. The Hessian equations cannot be satisfied; $\tilde{\mathcal{H}}_{12}: \ 0 = -\alpha A_{12}{}^1$ and $\tilde{\mathcal{H}}_{22}: \ 0 = -\alpha A_{22}{}^1$.

Case 3.2. Suppose $\alpha = 0$. Express $f = f_0(x^1) + f_{-1}(x^1)x^2$. Since $\mathcal{Y}_{-1} = \{0\}$, $f_{-1}(x^1) = 0$. Since $\mathbb{1} \in \mathcal{Y}(\mathcal{N})$, we may suppose $f = \log(x^1)$. The Hessian equations cannot be satisfied; $\tilde{\mathcal{H}}_{12}$: $0 = -A_{12}{}^1$ and $\tilde{\mathcal{H}}_{22}$: $0 = -A_{22}{}^1$.

Case 3.3. Suppose $\alpha = 1$. We can eliminate log dependence and suppose $f = a_1x^1 + a_2x^2$. If $a_2 = 0$, we obtain $\tilde{\mathcal{H}}_{12}$: $0 = -a_1A_{12}{}^1$ and $\tilde{\mathcal{H}}_{22}$: $0 = -a_1A_{22}{}^1$. This is not possible since $(A_{12}{}^1, A_{22}{}^1) \neq (0,0)$. Since $a_2 \neq 0$, $(\hat{x}^1, \hat{x}^2) = (x^1, a_2x^2 + a_1x^1)$ defines new coordinates $(\hat{x}^1, \hat{x}^2)$ so that $f(\hat{x}^1, \hat{x}^2) = \hat{x}^2 \in \mathcal{Y}(\mathcal{N})$. The Yamabe soliton equations show that one has $\tilde{\mathcal{H}}_{11}$: $0 = -A_{11}{}^2$, $\tilde{\mathcal{H}}_{12}$: $0 = -A_{12}{}^2$ and $\tilde{\mathcal{H}}_{22}$: $0 = -A_{22}{}^2$. This is the possibility of Assertion 3. □

14.1.11 AFFINE GRADIENT RICCI SOLITONS. Recall that f is said to be an *affine gradient Ricci soliton* if $\mathcal{H}f + \rho_s = 0$. Let $\mathfrak{A}$ be the set of affine gradient Ricci solitons. If $\mathfrak{A}(\mathcal{N})$ is non-empty, then the most general possible affine gradient Ricci soliton can be written in the form $f + \mathcal{Y}(\mathcal{N})$ for any $f \in \mathfrak{A}(\mathcal{N})$. Thus it suffices to exhibit a single element as the space $\mathcal{Y}(\mathcal{N})$ was studied in Lemma 14.12. If $\mathcal{N}$ is flat, we can take $f = \mathbb{1}$, so we eliminate this possibility.

Lemma 14.13 *Let $\mathcal{N}$ be a Type $\mathcal{B}$ model which is not flat. If $\mathcal{N}$ admits an affine gradient Ricci soliton, then one of the following possibilities pertains.*

1. $A_{12}{}^1 = 0$, $A_{22}{}^1 = 0$, $A_{22}{}^2 = 0$, $A_{11}{}^1 \neq -1$; $\frac{1+A_{11}{}^1 - A_{12}{}^2}{1+A_{11}{}^1} A_{12}{}^2 \log(x^1) \in \mathfrak{A}(\mathcal{N})$.
2. $A_{12}{}^1 = 0$, $A_{22}{}^1 = 0$, $A_{22}{}^2 = 0$, $A_{11}{}^1 = -1$; $\frac{1}{2}(A_{12}{}^2)^2(\log(x^1))^2 \in \mathfrak{A}(\mathcal{N})$.
3. *$\mathcal{N}$ is linearly equivalent to $\mathcal{N}_3^2(a, c, \pm)$; $a \log(x^1) \in \mathfrak{A}(\mathcal{N})$.*
4. *$\mathcal{N}$ is linearly equivalent to $\mathcal{N}_2^2(c)$; $0 \in \mathfrak{A}(\mathcal{N})$.*

We shall show in Theorem 14.17 that the geometries discussed in Assertion 1 are also Type $\mathcal{A}$.

Proof. A direct computation shows that the structures of Assertions 1–4 admit the gradient Ricci solitons listed. To complete the proof of the lemma, we must establish the converse implication. To clear denominators, we normalize the soliton equations defining

$$\tilde{S} := (x^1)^2 S = (x^1)^2\{\mathcal{H}f + \rho_s\}.$$

Assume that $f \in \mathfrak{A}(\mathcal{N})$ and $\mathcal{N}$ is not flat. If $\rho_s = 0$, then $\rho_a \neq 0$ and Theorem 14.9 shows either $\mathcal{N}$ is linearly equivalent to $\mathcal{N}_2^2(c)$ or to $\mathcal{N}_3^2(0, c, \pm)$. Thus we shall suppose that $\rho_s \neq 0$ henceforth and, consequently, that $\mathfrak{A}(\mathcal{N})$ is disjoint from $\mathcal{Y}(\mathcal{N})$. If X is an affine Killing vector field, then $X(f) \in \mathcal{Y}(\mathcal{N})$. We will use this observation in our analysis.

Assertions 1–2. Suppose $A_{12}{}^1 = 0$ and $A_{22}{}^1$=0. By Lemma 14.12, Yamabe solitons are independent of x^2 . Since $\partial_{x^2} f \in \mathcal{Y}(\mathcal{N})$, $f = f_0(x^1) + f_1(x^1)x^2$. We have

$$\tilde{S}_{22}:\ 0 = -A_{22}{}^2 x^1 f_1(x^1).$$

If $f_1 \neq 0$, then $A_{22}{}^2 = 0$. If $f_1 = 0$, we obtain $\tilde{S}_{12}$: $0 = \frac{1}{2}A_{22}{}^2$ so $A_{22}{}^2 = 0$. Thus in either event, $A_{22}{}^2 = 0$ and the possibilities of Assertions 1 and 2 hold.

Assertions 3–4. We apply Lemma 14.12. Suppose $(A_{12}{}^1, A_{22}{}^1) \neq (0, 0)$. Then either

$$\mathcal{Y}(\mathcal{N}) = \operatorname{span}\{\mathbb{1}, x^2\} \text{ or } \mathcal{Y}(\mathcal{N}) = \operatorname{span}\{\mathbb{1}\}.$$

Suppose $\mathcal{Y}(\mathcal{N}) = \operatorname{span}\{\mathbb{1}, x^2\}$. Since $\partial_{x^2} f \in \mathcal{Y}(\mathcal{N})$, $f = f_0(x^1) + x^2(a_0 + a_1 x^2)$. We apply the affine Killing vector field $x^1\partial_{x^1} + x^2\partial_{x^2}$ to see

$$x^1 f_0'(x^1) + x^2(a_0 + 2a_1 x^2) = (x^1\partial_{x^1} + x^2\partial_{x^2}) f \in \mathcal{Y}(\mathcal{N}).$$

Consequently, $a_1 = 0$. Since $x^2 \in \mathcal{Y}(\mathcal{N})$, we can delete the $a_0 x^2$ term and assume $f = f_0(x^1)$. Since $x^1 f_0' = a_2$, we have $f_0 = a_2 \log(x^1)$ since we may remove the constant term. On the other hand, if $\mathcal{Y}(\mathcal{N}) = \operatorname{span}\{\mathbb{1}\}$, then $f = f_0(x^1) + a_0 x^2$. Because $x^1 f_0'(x^1) + a_0 x^2 \in \mathcal{Y}(\mathcal{N})$, we may assume that $a_0 = 0$. Again we have $x^1 f_0' = a_2$. Consequently, we shall suppose that $f = a \log(x^1)$.

Case 1. Suppose $A_{22}{}^1 \neq 0$. We can normalize $A_{22}{}^1 = \varepsilon = \pm 1$. We use Lemma 14.2 to choose a shear so that we may assume $A_{12}{}^1 = 0$. We set $A_{11}{}^2 = c$. We have

$$\tilde{S}_{12}: \ 0 = \tfrac{1}{2}(A_{22}{}^2 - 2c\varepsilon) \quad \text{and} \quad \tilde{S}_{22}: \ 0 = \varepsilon(A_{11}{}^1 - A_{12}{}^2 - 1 - a).$$

This yields $A_{22}{}^2 = 2c\varepsilon$ and $A_{11}{}^1 = 1 + a + A_{12}{}^2$. The remaining soliton equation is then

$$\tilde{S}_{11}: \ 0 = -2a - a^2 + 2A_{12}{}^2 + 2c^2\varepsilon.$$

We use this equation to determine $A_{12}{}^2$ and establish Assertion 3 by computing:

$$\mathcal{N} = \mathcal{N}(\tfrac{1}{2}(2 + 4a + a^2 - 2c^2\varepsilon), c, 0, a + \tfrac{1}{2}a^2 - c^2\varepsilon, \varepsilon, 2c\varepsilon) = \mathcal{N}_3^2(a, c, \pm).$$

Case 2. Suppose $A_{22}{}^1 = 0$ but $A_{12}{}^1 \neq 0$. Rescale to set $A_{12}{}^1 = 1$. Set $c := A_{11}{}^2$ and obtain

$$\tilde{S}_{12}: \ 0 = \tfrac{1}{2}(-1 - 2a + 2A_{12}{}^2 + A_{22}{}^2) \quad \text{and} \quad \tilde{S}_{22}: \ 0 = -1 + A_{22}{}^2.$$

We obtain $A_{22}{}^2 = 1$ and $A_{12}{}^2 = a$; $\tilde{S}_{11}$: $0 = -a^2$ yields $a = 0$; the Ricci tensor is alternating. By Theorem 14.9, $A_{22}{}^1 = 0$ and $\mathcal{N} = \mathcal{N}_2^2(c)$; Assertion 4 follows. □

14.2 TYPE $\mathcal{B}$: AFFINE KILLING VECTOR FIELDS

In this section, we present some results related to the Lie algebra of affine Killing vector fields. In Section 14.2.1, we study the space of affine Killing vector fields for flat geometries, and in Section 14.2.2, we perform a similar analysis for the non-flat geometries. In Section 14.2.3, we classify the models where $\dim(\mathfrak{K}) > 2$. We use these results in Section 14.2.4 to classify the non-flat models which are both Type $\mathcal{A}$ and Type $\mathcal{B}$ models. In Section 14.2.5, we determine which Type $\mathcal{B}$ models are affine Killing complete. In Section 14.2.6, we examine invariant and parallel tensors of Type (1,1).

14.2.1 AFFINE KILLING VECTOR FIELDS FOR FLAT GEOMETRIES.

Lemma 14.14 *The geometries $\mathcal{N}_i^6(\cdot)$ are all flat. Consequently,* $\dim(\mathfrak{K}(\mathcal{N}_i^6(\cdot))) = 6$. *Let a_i be arbitrary real constants.*

1. $\mathfrak{K}(\mathcal{N}_0^6) = \operatorname{span}\{(a_1 + a_2x^1 + a_3x^2)\partial_{x^1} + (a_4 + a_5x^1 + a_6x^2)\partial_{x^2}\}$.
2. $\mathfrak{K}(\mathcal{N}_1^6(\pm)) = \operatorname{span}\{Z_1\partial_{x^1} + Z_2\partial_{x^2}\}$ *where*
$$Z_1 = a_1x^1 + \tfrac{a_2}{x^1} + \tfrac{a_3x^2}{x^1} + \tfrac{a_4((x^1)^2 \pm (x^2)^2)}{x^1} - \tfrac{a_6x^2(\pm(x^1)^2 + (x^2)^2)}{x^1},$$
$$Z_2 = a_1x^2 + a_5 + a_6\left((x^1)^2 \pm (x^2)^2\right).$$
3. $\mathfrak{K}(\mathcal{N}_2^6(c_1)) = \operatorname{span}\{Z_1\partial_{x^1} + Z_2\partial_{x^2}\}$ *where*
$$Z_1 = x^1\left(a_1 + a_2x^2 + a_3(x^1)^{-c_1}\right),$$
$$Z_2 = -a_2c_1(x^2)^2 - a_3c_1x^2(x^1)^{-c_1} + a_4x^2 + a_5 + a_6(x^1)^{-c_1}.$$

4. $\mathfrak{K}(\mathcal{N}_3^6) = \operatorname{span}\{Z_1\partial_{x^1} + Z_2\partial_{x^2}\}$ *where for* $\tau(x^1,x^2) := x^2 + x^1\log(x^1)$ *we have*
$$Z_1 = a_1x^1 - a_2(x^1)^2 - a_3x^1\tau(x^1,x^2),$$
$$Z_2 = a_1x^2 + a_2\left((x^1)^2 - x^1x^2\right) + a_3(x^1-x^2)\tau(x^1,x^2) + a_4 + a_5x^1 + a_6\tau(x^1,x^2).$$
5. $\mathfrak{K}(\mathcal{N}_4^6) = \operatorname{span}\{Z_1\partial_{x^1} + Z_2\partial_{x^2}\}$ *where we have for* $\tau(x^1,x^2) := x^2 + x^1\log(x^1)$ *and for* $\tau_1(x^1) := \log(x^1) + 1$ *that*
$$Z_1 = a_1x^1 + a_2\tau(x^1,x^2) + a_3,$$
$$Z_2 = a_1x^2 - a_2\tau_1(x^1)\tau(x^1,x^2) - a_3\tau_1(x^1) + a_4 + a_5x^1 + a_6\tau(x^1,x^2).$$
6. $\mathfrak{K}(\mathcal{N}_5^6) = \operatorname{span}\left\{\left(x^1(a_1 + a_2x^2 + a_3\log(x^1))\right)\partial_{x^1} + \left(a_4 + a_5x^2 + a_6\log(x^1)\right)\partial_{x^2}\right\}$.
7. $\mathfrak{K}(\mathcal{N}_6^6(c_2)) = \operatorname{span}\{Z_1\partial_{x^1} + Z_2\partial_{x^2}\}$ *where*
$$Z_1 = a_1x^2(x^1)^{-c_2} + a_2(x^1)^{-c_2} + a_3x^1,$$
$$Z_2 = a_4 + a_5x^2 + a_6(x^1)^{c_2+1}.$$

Proof. Let $\mathcal{N}$ be a Type $\mathcal{B}$ geometry. By Lemma 12.53, the components of an affine Killing vector field are sums of elements of the form $(x^1)^\alpha p(x^2, \log(x^1))$ where p is polynomial. This fact informed our investigation. Suppose $\mathcal{N}$ is flat. Let $\mathcal{Q}(\mathcal{N}) = \operatorname{span}\{1\!\!1, \phi_1, \phi_2\}$. By Lemma 12.32, $\Phi := (\phi_1, \phi_2)$ defines an immersion. Since $\Phi^*\mathcal{Q}(\mathcal{M}_0^6) = \mathcal{Q}(\mathcal{N})$, Φ is an affine immersion by Theorem 12.34. Consequently, the affine Killing vector fields of $\mathcal{N}$ are the pull-back of the affine Killing vector fields of $\mathcal{M}_0^6$. Since the elements of $\mathfrak{K}(\mathcal{M}_0^6)$ take the form $a_i\partial_{x^i} + a_i^j x^i\partial_{x^j}$, this provides an algorithm for determining $\mathfrak{K}(\mathcal{N})$. We used the computation of $\mathcal{Q}(\mathcal{N})$ given in Lemma 14.4 to inform our investigation. One can now check directly that the vector fields given in the lemma are in fact affine Killing vector fields; they span for dimensional reasons. We emphasize, this was not a routine mathematica calculation; it was informed by the theory noted above. □

14.2.2 AFFINE KILLING VECTOR FIELDS FOR NON-FLAT MODELS.

Lemma 14.15 *For the non-flat geometries of Definition 14.1 we have that:*

1. $\mathfrak{K}(\mathcal{N}_1^4(c_2)) = \operatorname{span}\left\{\left(a_1x^1 + a_2x^1\right)\partial_{x^1} + \left(a_1x^2 - a_2x^1\log(x^1) + a_3 + a_4x^1\right)\partial_{x^2}\right\}$.
2. $\mathfrak{K}(\mathcal{N}_2^4(\kappa,\theta)) = \operatorname{span}\left\{a_1x^1\partial_{x^1} + \left(a_2x^2 + a_3 + a_4(x^1)^\theta\right)\partial_{x^2}\right\}$.
3. $\mathfrak{K}(\mathcal{N}_3^4(c_1)) = \operatorname{span}\left\{a_1x^1\partial_{x^1} + \left(a_2x^2 + a_3 + a_4\log(x^1)\right)\partial_{x^2}\right\}$.
4. $\mathfrak{K}(\mathcal{N}_1^3(\pm)) = \mathfrak{K}(\mathcal{N}_2^3(c)) = \operatorname{span}\left\{\left(a_1x^1 + 2a_2x^1x^2\right)\partial_{x^1} + \left(a_1x^2 + a_2(x^2)^2 + a_3\right)\partial_{x^2}\right\}$.
5. $\mathfrak{K}(\mathcal{N}_3^3) = \operatorname{span}\left\{\left(a_1x^1 + 2a_2x^1x^2\right)\partial_{x^1} + \left(a_1x^2 + a_2((x^2)^2 + (x^1)^2) + a_3\right)\partial_{x^2}\right\}$.
6. $\mathfrak{K}(\mathcal{N}_4^3) = \operatorname{span}\left\{\left(a_1x^1 + 2a_2x^1x^2\right)\partial_{x^1} + \left(a_1x^2 + a_2((x^2)^2 - (x^1)^2) + a_3\right)\partial_{x^2}\right\}$.
7. $\mathfrak{K}(\mathcal{N}_i^2(\cdot)) = \operatorname{span}\left\{a_1x^1\partial_{x^1} + \left(a_1x^2 + a_2\right)\partial_{x^2}\right\}$ *for all* $1 \le i \le 3$ *excepting the case* $\mathcal{N}_3^2(0, \frac{3}{\sqrt{2}}, +)$, *which is equivalent to* $\mathcal{N}_2^3\left(\frac{1}{2}\right)$.

Proof. We perform a direct computation to see the vector fields given in the lemma are actually affine Killing vector fields. We use Lemma 14.11 to verify that the structures $\mathcal{N}_i^4(\cdot)$ are isomorphic to Type $\mathcal{A}$ manifolds $\mathcal{M}_j^4(\cdot)$. Consequently, $\dim(\mathfrak{K}(\mathcal{N}_i^4(\cdot))) = 4$, which proves Assertions 1–3. Using the ansatz of Lemma 12.53, one verifies there are no additional affine Killing vector fields to establish the remaining assertions. □

14.2.3 CLASSIFICATION OF TYPE $\mathcal{B}$ MODELS WITH DIM($\mathfrak{K}$)> 2.

Theorem 14.16 *Let $\mathcal{N}$ be a Type $\mathcal{B}$ model.*

1. $\dim(\mathfrak{K}(\mathcal{N})) \in \{2, 3, 4, 6\}$.
2. $\dim(\mathfrak{K}(\mathcal{N})) = 3$ *if and only if* $\mathcal{N}$ *is linearly equivalent to* $\mathcal{N}_i^3(\cdot)$ *for some* i.
3. *The following assertions are equivalent.*
   (a) $\dim(\mathfrak{K}(\mathcal{N})) = 4$.
   (b) $\mathcal{N}$ *is linearly equivalent to* $\mathcal{N}_i^4(\cdot)$ *for some* i;
   (c) $\mathcal{N}$ *is also Type* $\mathcal{A}$ *and not flat.*
4. *The following assertions are equivalent.*
   (a) $\dim(\mathfrak{K}(\mathcal{N})) = 6$.
   (b) $\mathcal{N}$ *is linearly equivalent to* $\mathcal{N}_i^6(\cdot)$ *for some* i.
   (c) $\mathcal{N}$ *is flat.*

Proof. We distinguish cases. Assertion 4 follows from Theorem 14.9, so we shall assume $\mathcal{N}$ is not flat henceforth. Assertions 1–3 were first established by Brozos-Vázquez, García-Río, and Gilkey [7]. The variable $A_{22}{}^1$ plays a central role in our analysis as it is unchanged by any shear. Our discussion is informed by the the corresponding discussion which was used to prove Lemma 12.53. We will examine the generalized eigenspaces $E(\alpha)$ of the adjoint map $\operatorname{ad}(x^1\partial_{x^1} + x^2\partial_{x^2})$; we change notation slightly and shift the eigenvalue to define

$$E(\alpha) := \{X \in \mathfrak{K}_{\mathbb{C}}(\mathcal{N}) : (\operatorname{ad}(x^1\partial_{x^1} + x^2\partial_{x^2}) - (\alpha - 1))^6 X = 0\};$$

thus $E(\alpha)$ is generated by elements of the form $(x^1)^\beta (x^2)^{\alpha-\beta} \log(x^1)^j$, which are homogeneous of degree α in $\{x^1, x^2\}$ jointly; the variable $\log(x^1)$ contributes Jordan normal form but is homogeneous of degree 0. Clearly $\operatorname{ad}(\partial_{x^2}) : E(\alpha) \to E(\alpha - 1)$. Assume $\dim(\mathfrak{K}(\mathcal{N})) > 2$ so there is an element $X \in E(\alpha)$ for some α which is neither $\partial_{x^2} \in E(0)$ nor $x^1\partial_{x^1} + x^2\partial_{x^2} \in E(1)$. Let $K_{ij}{}^k$ be the affine Killing equations of Equation (12.1.d).

Case 1. Suppose $A_{22}{}^1 \neq 0$. Except when considering $E(2)$, we will apply Lemma 14.2 to see there exists a unique shear $(x^1, x^2) \to (x^1, x^2 + bx^1)$ so that $A_{12}{}^1 = 0$. We then rescale x^2 to ensure $A_{22}{}^1 = \varepsilon = \pm 1$. This fixes the gauge in these cases; when considering $E(2)$, a different gauge normalization will be convenient.

Case 1.1. Suppose $\Re(\alpha) < 0$ and $\alpha \neq -1$. Choose α so $\Re(\alpha)$ is minimal. As $\operatorname{ad}(\partial_{x^2})X$ belongs to $E(\alpha - 1)$ and since $\Re(\alpha)$ is minimal, $\operatorname{ad}(\partial_{x^2})X = 0$ so there is no x^2 dependence in X. By applying $\operatorname{ad}(x^1\partial_{x^1} + x^2\partial_{x^2}) - \alpha$, we may also assume there is no $\log(x^1)$ dependence. Consequently, $X = (x^1)^\alpha\{a_1\partial_{x^1} + a_2\partial_{x^2}\}$. We show $X = 0$ and thus this case does not happen by considering

$$K_{12}{}^1 : 0 = (x^1)^{\alpha-2}a_2\alpha\varepsilon \quad \text{and} \quad K_{22}{}^1 : 0 = -(x^1)^{\alpha-2}a_1(1+\alpha)\varepsilon . \tag{14.2.a}$$

Case 1.2. Suppose $\alpha = -1$. As in Case 1.1, we may assume $X = (x^1)^{-1}(a_1\partial_{x^1} + a_2\partial_{x^2})$; Equation (14.2.a) then implies $a_2 = 0$ so $X = (x^1)^{-1}\partial_{x^1}$. We obtain a flat geometry which is false:

$$\begin{aligned} &K_{11}{}^1 : 0 = 2(x^1)^{-3}(1 - A_{11}{}^1), && K_{11}{}^2 : 0 = -3(x^1)^{-3}A_{11}{}^2, \\ &K_{12}{}^2 : 0 = -2(x^1)^{-3}A_{12}{}^2, && K_{22}{}^2 : 0 = -(x^1)^{-3}A_{22}{}^2, \quad \rho = 0 . \end{aligned}$$

Case 1.3. Suppose $\alpha = 0$. Any non-trivial x^2 dependence would give rise to an element of $E(-1)$, which has been dealt with in Case 1.2. If there is a $\log(x^1)^k$ dependence where $k > 1$, we may apply $\operatorname{ad}(x^1\partial_{x^1} + x^2\partial_{x^2})^{k-1}$ to assume $k = 1$. We may therefore assume that

$$X = (a_1 + a_2\log(x^1))\partial_{x^1} + (b_1 + b_2\log(x^1))\partial_{x^2}.$$

Since $\partial_{x^2} \in \mathfrak{K}(\mathcal{N})$, we may assume without loss of generality that $b_1 = 0$ since we want an additional vector field. The following equations imply $a_1 = 0$, $a_2 = 0$, and $b_2 = 0$ which is false:

$$\begin{aligned} &K_{22}{}^1 : 0 = -\varepsilon(x^1)^{-2}(a_1 + a_2 + a_2\log(x^1)), \\ &K_{22}{}^2 : 0 = -(x^1)^{-2}(a_1A_{22}{}^2 + b_2\varepsilon + a_2A_{22}{}^2\log(x^1)) . \end{aligned}$$

Case 1.4. Suppose $\alpha = 1$. Any non-trivial x^2 dependence divided by a power of x^1 would give rise to an element of $E(\alpha)$ for $\alpha < 0$ which has already been eliminated. Similarly, any $\log(x^1)^2$ dependence can be eliminated. So we assume

$$\begin{aligned} X \;=\; &(a_1x^1 + a_2x^2 + a_3x^1\log(x^1) + a_4x^2\log(x^1))\partial_{x^1} \\ &+(b_1x^1 + b_2x^2 + b_3x^1\log(x^1) + b_4x^2\log(x^1))\partial_{x^2} . \end{aligned}$$

Since $x^1\partial_{x^1} + x^2\partial_{x^2}$ belongs to $\mathfrak{K}(\mathcal{N})$, we may assume $b_2 = 0$ to eliminate this term. We have

$$\operatorname{ad}(\partial_{x^2})X = (a_2 + a_4\log(x^1))\partial_{x^1} + (b_2 + b_4\log(x^1))\partial_{x^2} .$$

By Case 1.3, we have that $a_2 = a_4 = b_4 = 0$. We show that $a_1 = a_3 = b_1 = b_3 = 0$ and, consequently, there are no unexpected terms which are homogeneous of degree 1 by computing

$$\begin{aligned} &X = (a_1x^1 + a_3x^1\log(x^1))\partial_{x^1} + (b_1x^1 + b_3x^1\log(x^1))\partial_{x^2}, \\ &K_{12}{}^1 : 0 = (x^1)^{-1}\varepsilon(b_1 + b_3 + b_3\log(x^1)), \\ &K_{22}{}^1 : 0 = -(x^1)^{-1}\varepsilon(2a_1 + a_3 + 2a_3\log(x^1)) . \end{aligned}$$

Case 1.5. Suppose $\alpha = 2$. We recall that we use a different gauge normalization in this case; we will obtain all the $\mathcal{N}_i^3(\cdot)$ models except $\mathcal{N}_2^3(0)$. The arguments given previously show we may suppose

$$X = (a_{11}x^1x^1 + 2a_{12}x^1x^2 + a_{22}x^2x^2)\partial_{x^1} + (b_{11}x^1x^1 + 2b_{12}x^1x^2 + b_{22}x^2x^2)\partial_{x^2}\,.$$

We have $\operatorname{ad}(\partial_{x^2})^2X = a_{22}\partial_{x^1} + b_{22}\partial_{x^2}$; $a_{22} = 0$ by Case 1.1. We apply Case 1.4 to see $b_{12} = 0$ and $b_{22} = a_{12}$ since $\operatorname{ad}(\partial_{x^2})X = 2a_{12}x^1\partial_{x^1} + (2b_{12}x^1 + 2b_{22}x^2)\partial_{x^2}$. Consequently,

$$X = (a_{11}(x^1)^2 + 2a_{12}x^1x^2)\partial_{x^1} + (b_{11}(x^1)^2 + a_{12}(x^2)^2)\partial_{x^2}\,.$$

We set $A_{22}{}^1 = \varepsilon \neq 0$, but we do not normalize $A_{12}{}^2$. If $a_{12} = 0$, then $K_{22}{}^1$: $0 = -3a_{11}\varepsilon$ and $K_{22}{}^2 = -a_{11}A_{22}{}^2 - 2b_{11}\varepsilon$. This would imply $X = 0$, which is false. Therefore $a_{12} \neq 0$, so we may assume $a_{12} = 1$. Set $\tilde{x}^2 := x^2 + \frac{1}{2}a_{11}x^1$ to ensure $X = 2x^1\tilde{x}^2\partial_{x^1} + \star\partial_{x^2}$. This fixes the gauge up to a possible rescaling of x^2. We have

$$\begin{aligned} X &= 2x^1x^2\partial_{x^1} + (b_{11}(x^1)^2 + (x^2)^2)\partial_{x^2},\\ K_{12}{}^1 &: 0 = 2(A_{11}{}^1 - A_{12}{}^2 + b_{11}\varepsilon + 1),\\ K_{22}{}^1 &: 0 = 4A_{12}{}^1 - 2A_{22}{}^2\,. \end{aligned}$$

This determines b_{11} and shows $A_{22}{}^2 = 2A_{12}{}^1$. We determine $A_{11}{}^2$ from the affine Killing equation $K_{12}{}^2$: $0 = 2(A_{11}{}^2 + \frac{1}{\varepsilon}A_{12}{}^1(-1 - A_{11}{}^1 + A_{12}{}^2))$ and then determine $A_{11}{}^1$ from the affine Killing equation $K_{22}{}^2$: $0 = 2(2 + A_{11}{}^1 + A_{12}{}^2)$. The remaining two affine Killing equations become

$$K_{11}{}^1 : 0 = \tfrac{6}{\varepsilon}A_{12}{}^1(1 + 2A_{12}{}^2) \quad \text{and} \quad K_{11}{}^2 : 0 = \tfrac{6}{\varepsilon}(1 + 3A_{12}{}^2 + 2(A_{12}{}^2)^2)\,.$$

We solve the equation $1 + 3A_{12}{}^2 + 2(A_{12}{}^2)^2 = 0$ to see either $A_{12}{}^2 = -\frac{1}{2}$ or $A_{12}{}^2 = -1$.

Case 1.5.1. $A_{12}{}^2 = -\frac{1}{2}$. We have $X = 2x^1x^2\partial_{x^1} + (x^2)^2\partial_{x^2}$ satisfies the affine Killing equations with $\mathcal{N} = \mathcal{N}(-\frac{3}{2}, 0, A_{12}{}^1, -\frac{1}{2}, \varepsilon, 2A_{12}{}^1)$. If $A_{12}{}^1 = 0$, we obtain $\mathcal{N}(-\frac{3}{2}, 0, 0, -\frac{1}{2}, \varepsilon, 0)$. We can rescale x^2 to ensure $\varepsilon = \pm\frac{1}{2}$ and obtain $\mathcal{N}_1^3(\pm)$. If $A_{12}{}^1 \neq 0$, we can rescale x^2 to ensure $A_{12}{}^1 = 1$ and obtain $\mathcal{N}(-\frac{3}{2}, 0, 1, -\frac{1}{2}, \varepsilon, 2)$. This is $\mathcal{N}_2^3(c)$ for $c = \varepsilon \neq 0$.

Case 1.5.2. $A_{12}{}^2 = -1$. $K_{11}{}^1$: $0 = -\frac{6}{\varepsilon}A_{12}{}^1$. Set $A_{12}{}^1 = 0$ so $\mathcal{N} = \mathcal{N}(-1, 0, 0, -1, \varepsilon, 0)$ and $X = 2x^1x^2\partial_{x^1} + ((x^2)^2 - \frac{1}{\varepsilon}(x^1)^2)\partial_{x^2}$. Rescale $\varepsilon = \pm 1$ to obtain either $\mathcal{N}_3^3$ or $\mathcal{N}_4^3$.

Case 1.6. For $\alpha \geq 0$ we have dealt with the eigenspaces $E(0)$, $E(1)$, and $E(2)$. If $\alpha \notin \{0, 1, 2\}$, we could apply $\operatorname{ad}(\partial_{x^2})$ to eliminate any x^2 terms and $\operatorname{ad}(x^1\partial_{x^1} + x^2\partial_{x^2})$ to eliminate any $\log(x^1)$ terms to ensure $X = (x^1)^\alpha(a_1\partial_{x^1} + a_2\partial_{x^2})$; this is eliminated by Equation (14.2.a).

Case 2. Suppose $A_{22}{}^1 = 0$. We will obtain $\mathcal{N}_2^3(0)$ and $\mathcal{N}_i^4(\cdot)$.

Case 2.1. Suppose $X = (x^1)^\alpha(a_1\partial_{x^1} + a_2\partial_{x^2})$ where $\alpha \neq 0, 1$; if $\Re(\alpha) < 0$, the argument of Case 1.1 permits us to assume that X has this form whereas for other values of α, it is a separate assumption.

Case 2.1.1. If $a_1 \neq 0$, then $K_{12}{}^1$: $0 = -a_1 A_{12}{}^1 (x^1)^{\alpha-2}$ and $K_{22}{}^2$: $0 = -a_1 A_{22}{}^2 (x^1)^{\alpha-2}$. Thus $A_{12}{}^1 = A_{22}{}^2 = 0$. We obtain $K_{12}{}^2$: $0 = a_1 A_{12}{}^2 (x^1)^{\alpha-2}(\alpha-1)$, so $A_{12}{}^2 = 0$ and $\rho = 0$, which is false.

Case 2.1.2. Suppose $a_1 = 0$. Set $X = (x^1)^\alpha \partial_{x^2}$. We have that

$$K_{11}{}^1 : 0 = 2(x^1)^{\alpha-2}\alpha A_{12}{}^1, \qquad K_{11}{}^2 : 0 = (x^1)^{\alpha-2}\alpha(-1 - A_{11}{}^1 + 2A_{12}{}^2 + \alpha),$$
$$K_{12}{}^2 : 0 = (x^1)^{\alpha-2}\alpha(A_{22}{}^2 - A_{12}{}^1).$$

We set $\alpha = 1 + A_{11}{}^1 - 2A_{12}{}^2$, $A_{12}{}^1 = 0$, and $A_{22}{}^2 = 0$. Since $\alpha \neq 1$, $A_{11}{}^1 - 2A_{12}{}^2 \neq 0$ and we can apply Lemma 14.2 to perform a shear to ensure $A_{11}{}^2 = 0$. Set $A_{11}{}^1 := 2\kappa + \theta - 1$ and $A_{12}{}^2 = \kappa$. Then $(x^1)^\theta \partial_{x^2} \in \mathfrak{K}(\mathcal{N})$ for $\mathcal{N}(\theta + 2\kappa - 1, 0, 0, \kappa, 0, 0) = \mathcal{N}_2^4(\kappa, \theta)$. Because we have $\tilde{\rho} = \kappa(\kappa + \theta) dx^1 \otimes dx^1 \neq 0$, we obtain $\kappa(\theta + \kappa) \neq 0$. Because $\alpha \neq 0$, $\theta \neq 0$.

Case 2.2. Suppose $\alpha = 0$. We first assume there are no log terms and, since ∂_{x^2} is an affine Killing vector field, examine $X = \partial_{x^1}$. The affine Killing equations yield $\Gamma = 0$, which is a flat geometry. Thus ∂_{x^1} is not an affine Killing vector field. Suppose $X = \log(x^1)\partial_{x^2}$.

$$K_{11}{}^1 : 0 = (x^1)^{-2} 2A_{12}{}^1, \qquad K_{11}{}^2 : 0 = -(x^1)^{-2}(1 + A_{11}{}^1 - 2A_{12}{}^2),$$
$$K_{12}{}^2 : 0 = (x^1)^{-2}(-A_{12}{}^1 + A_{22}{}^2).$$

We impose the resulting relations; set $A_{12}{}^2 = t$ and $A_{11}{}^2 = s$ to obtain

$$\tilde{\rho} = t^2 dx^1 \otimes dx^1, \quad X = \log(x^1)\partial_{x^2} \in \mathfrak{K}(\mathcal{N}), \quad \mathcal{N}(t,s) = \mathcal{N}(2t-1, s, 0, t, 0, 0).$$

To ensure $\mathcal{N}(t,s)$ is not flat, we assume $t \neq 0$. We can apply Lemma 14.2 to assume $s = 0$ and obtain $\mathcal{N}_3^4(c_1)$ for $c_1 = t \neq 0$.

Case 2.3. Suppose $\alpha = 1$. We do not need to worry about x^2 divided by a power of x^1 since we have considered the case $\alpha < 0$ in Case 2.1. We first suppose there are no log terms so $X = (a_1 x^1 + a_2 x^2)\partial_{x^1} + (b_1 x^1 + b_2 x^2)\partial_{x^2}$. By subtracting an appropriate multiple of $x^1\partial_{x^1} + x^1\partial_{x^2}$, we may assume $b_2 = 0$. If $a_2 \neq 0$, we would obtain $\partial_{x^1} \in E(0)$, and we have already dealt with this possibility in Case 2.2. Thus we have $X = x^1(a_1\partial_{x^1} + b_1\partial_{x^2})$.

Case 2.3.1. Suppose $a_1 \neq 0$. We may assume $a_1 = 1$. We set $\tilde{x}^2 = x^2 - b_1 x^1$. We then have $\partial_{\tilde{x}^1} = \partial_{x^1} + b_1\partial_{x^2}$ and $\partial_{\tilde{x}^2} = \partial_{x^2}$. We may therefore assume $b_1 = 0$. Consequently, we have that $K_{12}{}^1$: $0 = -(x^1)^{-1}A_{12}{}^1$ and $K_{22}{}^2$: $0 = -(x^1)^{-1}A_{22}{}^2$. Set $A_{12}{}^1 = 0$ and $A_{22}{}^2 = 0$ and obtain the relation $K_{11}{}^2$: $0 = (x^1)^{-1}A_{11}{}^2$. We have solved all the affine Killing equations so $x^1\partial_{x^1} \in \mathfrak{K}(\mathcal{N})$. Set $A_{11}{}^1 = r$ and $A_{12}{}^2 = s$ to obtain

$$\tilde{\rho} = (1 + r - s)s\,dx^1 \otimes dx^1 \text{ and } \mathcal{N} = \mathcal{N}(r, 0, 0, s, 0, 0).$$

Depending on the values of $\{r, s\}$, this is either $\mathcal{N}_2^4(\kappa, \theta)$ for $\kappa = s$ and $\theta = r - (2s - 1) \neq 0$ or $\mathcal{N}_3^4(c_1)$ for $s = c_1$ and $r = 2s - 1$.

Case 2.3.2. Suppose $a_1 = 0$ so $X = x^1\partial_{x^2}$. We obtain

$$K_{11}{}^1 : 0 = (x^1)^{-1}2A_{12}{}^1, \qquad K_{11}{}^2 : 0 = (x^1)^{-1}(-A_{11}{}^1 + 2A_{12}{}^2),$$
$$K_{12}{}^2 : 0 = (x^1)^{-1}(-A_{12}{}^1 + A_{22}{}^2).$$

We impose these relations and set $A_{12}{}^2 = c_2$ to obtain a solution with

$$\tilde{\rho} = (c_2 + c_2^2)dx^1 \otimes dx^1, \quad X = x^1\partial_{x^2}, \quad \mathcal{N} = \mathcal{N}(2c_2, A_{11}{}^2, 0, c_2, 0, 0).$$

To ensure $\tilde{\rho} \neq 0$, we require $c_2 \notin \{0, -1\}$. If $A_{11}{}^2 \neq 0$, we can rescale x^2 to set $A_{11}{}^2 = 1$ and obtain $\mathcal{N}_1^4(c_2)$. If $A_{11}{}^2 = 0$, we obtain $\mathcal{N}_2^4(\kappa, \theta)$ for $\theta = 1$ and $\kappa = c_2$.

Case 2.3.3. Suppose we have $\log(x^1)$ terms. There cannot be negative powers of x^1. Any terms $x^2\log(x^1)$ give rise to a $\log(x^1)$ in $E(0)$ which is false. Any terms with $x^1\log(x^1)$ give rise to an x^1 term. This does not occur.

Case 2.4. Suppose $\alpha = 2$. Since there are no terms with $\Re(\alpha) < 0$, we need not worry about x^2 divided by a power of x^1. Nor need we worry about $\log(x^1)$ terms. Thus X is a pure quadratic polynomial. The analysis of Case 1.5 permits us to assume

$$X = (a_{11}(x^1)^2 + 2a_{12}x^1x^2)\partial_{x^1} + (b_{11}(x^1)^2 + a_{12}(x^2)^2)\partial_{x^2}.$$

Case 2.4.1. Suppose $a_{12} = 0$ but $a_{11} = 1$. We have

$$K_{12}{}^1 : 0 = -A_{12}{}^1, \; K_{12}{}^2 : 0 = A_{12}{}^2 - 2A_{12}{}^1b_{11} + 2A_{22}{}^2b_{11}, \; K_{22}{}^2 : 0 = -A_{22}{}^2.$$

If $A_{12}{}^1 = 0$, $A_{22}{}^2 = 0$, and $A_{22}{}^1 = 0$, then $\rho = 0$. This case does not appear.

Case 2.4.2. Suppose $a_{12} = 0$, $a_{11} = 0$, and $b_{11} = 1$. We have

$$K_{11}{}^1 : 0 = 4A_{12}{}^1, \quad K_{11}{}^2 : 0 = 2 - 2A_{11}{}^1 + 4A_{12}{}^2, \quad K_{12}{}^2 : 0 = -2A_{12}{}^1 + 2A_{22}{}^2.$$

We solve these equations and set $A_{12}{}^2 = s$ to obtain an affine Killing vector field $(x^1)^2\partial_{x^2}$ where $\tilde{\rho} = s(2+s)dx^1 \otimes dx^1$ and $\mathcal{N} = \mathcal{N}(1 + 2s, A_{11}{}^2, 0, s, 0, 0)$. We can use Lemma 14.2 and take a shear with $b = A_{12}{}^2$ and $a = 1$ to assume $A_{11}{}^2 = 0$ and obtain $\mathcal{N}_2^4(s, 2)$; this was also obtained previously in Case 2.1.2.

Case 2.4.3. Suppose $a_{12} \neq 0$. We can assume $a_{12} = 1$ and perform a gauge transformation to assume $a_{11} = 0$ and obtain $X = 2x^1x^2\partial_{x^1} + (b_{11}(x^1)^2 + (x^2)^2)\partial_{x^2}$,

$$K_{22}{}^1 : 0 = 4A_{12}{}^1 - 2A_{22}{}^2 \quad \text{and} \quad K_{22}{}^2 : 0 = 2 + 4A_{12}{}^2.$$

We set $A_{12}{}^2 = -\frac{1}{2}$ and $A_{22}{}^2 = 2A_{12}{}^1$. We obtain $K_{12}{}^1$: $0 = 3 + 2A_{11}{}^1$ so $A_{11}{}^1 = -\frac{3}{2}$. We obtain $K_{11}{}^1$: $0 = -2A_{11}{}^2 + 4A_{12}{}^1b_{11}$ and $K_{11}{}^2$: $0 = 3b_{11}$. Thus, $b_{11} = 0$ and $A_{11}{}^2 = 0$. We obtain a solution where $X = 2x^1x^2\partial_{x^1} + (x^2)^2\partial_{x^2}$, $\mathcal{N} = \mathcal{N}(-\frac{3}{2}, 0, A_{12}{}^1, -\frac{1}{2}, 0, 2A_{12}{}^1)$ and

$$\tilde{\rho} = \begin{pmatrix} 0 & \frac{3}{2}A_{12}{}^1 \\ -\frac{3}{2}A_{12}{}^1 & (A_{12}{}^1)^2 \end{pmatrix}.$$

To ensure the geometry is not flat, we have $A_{12}{}^1 \neq 0$. We can therefore rescale x^2 to set $A_{12}{}^1 = 1$ and obtain $\mathcal{N}_2^3(0)$; this was missing in Case 1.5.1.

Case 2.5. We suppose $\alpha \neq 0, 1, 2$. Since $E(\beta) = \{0\}$ for $\Re(\beta < 0)$, there are no terms with a power of x^2 divided by a power of x^1. Similarly, we can assume there are no $\log(x^1)$ terms. Thus the analysis of Case 2.1 pertains. □

14.2.4 RELATING TYPE $\mathcal{A}$ AND TYPE $\mathcal{B}$ MODELS. The following result describes the intersection between the Type $\mathcal{A}$ and the Type $\mathcal{B}$ models; it is due to Brozos-Vázquez, García-Río, and Gilkey [7].

Theorem 14.17

1. *Let $\mathcal{N}$ be a non-flat Type $\mathcal{B}$ model. The following assertions are equivalent.*
   (a) $\mathcal{N}$ is locally affine isomorphic to a Type $\mathcal{A}$ model.
   (b) $\dim(\mathfrak{K}(\mathcal{N})) = 4$.
   (c) $\mathcal{N}$ is linearly isomorphic to $\mathcal{N}_1^4(c_2)$, $\mathcal{N}_2^4(\kappa, \theta)$, or $\mathcal{N}_3^4(c_1)$.
   (d) $A_{12}{}^1 = A_{22}{}^1 = A_{22}{}^2 = 0$.
2. *Let $\mathcal{M}$ be a non-flat Type $\mathcal{A}$ model. The following assertions are equivalent.*
   (a) $\mathcal{M}$ is locally affine isomorphic to a Type $\mathcal{B}$ model.
   (b) $\mathcal{M}$ is locally affine isomorphic to $\mathcal{M}_3^4(c_1)$ or to $\mathcal{M}_4^4(0)$.

Proof. We establish Assertion 1 as follows. Let $\mathcal{N}$ be a non-flat Type $\mathcal{B}$ model. If $\mathcal{N}$ is also a Type $\mathcal{A}$ model, then $\mathfrak{K}(\mathcal{N})$ contains both a 2-dimensional Abelian Lie subalgebra and a 2-dimensional non-Abelian Lie subalgebra. Consequently, $\dim(\mathfrak{K}(\mathcal{N})) \geq 3$. Since there are no Type $\mathcal{A}$ models with $\dim(\mathfrak{K}(\mathcal{N})) = 3$ and since $\mathcal{N}$ is non-flat, $\dim(\mathfrak{K}(\mathcal{N})) = 4$. Next suppose $\dim(\mathfrak{K}(\mathcal{N})) = 4$. By Theorem 14.16, $\mathcal{N}$ is linearly isomorphic to $\mathcal{N}_i^4(\cdot)$. Suppose $\mathcal{N}$ is linearly isomorphic to $\mathcal{N}_i^4(\cdot)$. By Lemma 14.11, $\mathcal{N}$ is also a Type $\mathcal{A}$ model. Thus Assertions 1-a, 1-b, and 1-c are equivalent. Suppose $\mathcal{N}$ is linearly isomorphic to $\mathcal{N}_i^4(\cdot)$. These models satisfy $A_{12}{}^1 = A_{22}{}^1 = A_{22}{}^2 = 0$. By Lemma 14.2, this condition is preserved by a shear. Consequently, Assertion 1-c implies Assertion 1-d. Conversely, suppose Assertion 1-d holds. We compute $\tilde{\rho} = A_{12}{}^2(1 + A_{11}{}^1 - A_{12}{}^2)dx^1 \otimes dx^1$ so $A_{12}{}^2 \neq 0$ and $1 + A_{11}{}^1 - A_{12}{}^2 \neq 0$.

Case 1. If $2A_{12}{}^2 - A_{11}{}^1 \neq 0$, perform a shear to set $A_{11}{}^2 = 0$. Set $A_{11}{}^1 = 2\kappa + \theta - 1$ and $A_{12}{}^2 = \kappa$. We have $\kappa \neq 0$ and $1 + A_{11}{}^1 - A_{12}{}^2 = \theta + \kappa$. If $\theta = 0$, we obtain $\mathcal{N}_3^4(c_1)$ for $c_1 = \kappa$. If $\theta \neq 0$, we obtain $\mathcal{N}_2^4(\kappa, \theta)$.

Case 2. If $2A_{12}{}^2 - A_{11}{}^1 = 0$, then we can rescale x^2 to assume either $A_{11}{}^2 = 1$ and obtain $\mathcal{N}_1^4(c_2)$ or that $A_{11}{}^2 = 0$ and obtain $\mathcal{N}_2^4(\kappa, 1)$.

This shows that Assertion 1-d implies Assertion 1-c. Assertion 2 is immediate from Assertion 1. □

14.2.5 AFFINE KILLING COMPLETE.

Theorem 14.18 *Let $\mathcal{N}$ be a Type $\mathcal{B}$ model.*

1. *If $\dim(\mathfrak{K}(\mathcal{N})) = 2$ or if $\dim(\mathfrak{K}(\mathcal{N})) = 4$, then $\mathcal{N}$ is affine Killing complete.*
2. *If $\dim(\mathfrak{K}(\mathcal{N})) = 3$, then $\mathcal{N}$ is affine Killing complete if and only if $\mathcal{N}$ is linearly equivalent to the hyperbolic plane.*
3. *If $\dim(\mathfrak{K}(\mathcal{N})) = 6$, then $\mathcal{N}$ is affine Killing complete if and only if $\mathcal{N}$ is linearly equivalent to $\mathcal{N}_0^6$ or $\mathcal{N}_5^6$.*

Proof. If $\dim(\mathfrak{K}(\mathcal{N})) = 4$, then $\mathcal{N}$ is linearly equivalent to $\mathcal{N}_i^4(\cdot)$ for some i by Theorem 14.16. By Lemma 14.11, these manifolds are affine isomorphic to $\mathcal{M}_3^4(c_1)$, for some c_1, or to $\mathcal{M}_4^4(0)$; these manifolds are affine Killing complete by Lemma 13.16. If $\dim(\mathfrak{K}(\mathcal{N})) = 2$, then the vector fields in question are the Lie algebra of the $ax + b$ group and are complete. The Lie algebra of the hyperbolic plane $\mathcal{N}_4^3$ is the 3-dimensional group of isometries of $\mathcal{N}_4^3$; this is affine Killing complete. We will discuss this structure further in Section 14.3.

We complete the proof by examining the geometries $\mathcal{N}_i^3(\cdot)$ for $i = 1, 2, 3$.

Case 1. Suppose $\mathcal{N} = \mathcal{N}_1^3(\pm)$ or $\mathcal{N} = \mathcal{N}_2^3(c)$. Then $X = 2x^1x^2\partial_{x^1} + (x^2)^2\partial_{x^2}$ is an affine Killing vector field. We let $\sigma(t) = \left(\frac{1}{t^2}, -\frac{1}{t}\right)$. This is a flow curve for X, so this geometry is affine Killing incomplete.

Case 2. Suppose $\mathcal{N} = \mathcal{N}_3^3$. Then $X = 2x^1x^2\partial_{x^1} + ((x^1)^2 + (x^2)^2)\partial_{x^2}$ is an affine Killing vector field. We let $\sigma(t) = \left(\frac{1}{t^2-1}, -\frac{t}{t^2-1}\right)$. This is a flow curve for X, so this geometry is affine Killing incomplete. □

The structure $\mathcal{N}_3^3$ embeds in the pseudo-sphere, as we shall see in Section 14.3. Thus this geometry can be completed. We do not know if the remaining geometries $\mathcal{N}_1^3(\pm)$ and $\mathcal{N}_2^3(c)$ can be affine Killing completed.

14.2.6 INVARIANT AND PARALLEL TENSORS OF TYPE (1,1). If $\mathcal{N}$ is of Type $\mathcal{B}$, then ∂_{x^2} and $x^1\partial_{x^1} + x^2\partial_{x^2}$ are affine Killing vector fields. We have

$$\mathcal{L}_{\partial_{x^2}}(\partial_{x^i} \otimes \partial_{x^j}) = 0 \quad \text{and} \quad \mathcal{L}_{x^1\partial_{x^1}+x^2\partial_{x^2}}(\partial_{x^i} \otimes \partial_{x^j}) = 0\,.$$

Thus T is invariant implies $T = T^i{}_j\partial_{x^i} \otimes dx^j$ where the coefficients $T^i{}_j$ are constant. We suppose $\mathrm{Tr}\{T\} = 0$. We apply Theorem 14.16 and Theorem 14.17. If $\dim(\mathfrak{K}(\mathcal{N})) = 4$, then $\mathcal{N}$ is also Type $\mathcal{A}$ and the analysis in Section 13.2.8 pertains. However, it is worth doing an independent calculation. We have:

Example 14.19

1. $\mathfrak{K}(\mathcal{N}) = \operatorname{span}\{x^1\partial_{x^1} + x^2\partial_{x^2}, \partial_{x^2}, x^1\partial_{x^1} - x^1\log(x^1)\partial_{x^2}, x^1\partial_{x^2}\}$. We have:

$$\mathcal{L}_{x^1\partial_{x^1} - x^1\log(x^1)\partial_{x^2}} T = \begin{pmatrix} -T^1{}_2(1+\log(x^1)) & -T^1{}_2 \\ 2T^1{}_1 + T^2{}_1 + 2T^1{}_1\log(x^1) & T^1{}_2(1+\log(x^1)) \end{pmatrix},$$

$$\mathcal{L}_{x^1\partial_{x^2}} T = \begin{pmatrix} T^1{}_2 & 0 \\ -2T^1{}_1 & -T^1{}_2 \end{pmatrix}.$$

2. $\mathfrak{K}(\mathcal{N}) = \operatorname{span}\{x^1\partial_{x^1} + x^2\partial_{x^2}, \partial_{x^2}, x^1\partial_{x^1}, (x^1)^\alpha\partial_{x^2}\}$ for $\alpha \neq 0$. We have:

$$\mathcal{L}_{x^1\partial_{x^1}} T = \begin{pmatrix} 0 & -T^1{}_2 \\ T^2{}_1 & 0 \end{pmatrix} \quad \text{and} \quad \mathcal{L}_{(x^1)^\alpha\partial_{x^2}} T = \alpha(x^1)^{\alpha-1}\begin{pmatrix} T^1{}_2 & 0 \\ -2T^1{}_1 & -T^1{}_2 \end{pmatrix}.$$

3. $\mathfrak{K}(\mathcal{N}) = \operatorname{span}\{x^1\partial_{x^1} + x^2\partial_{x^2}, \partial_{x^2}, x^1\partial_{x^1}, \log(x^1)\partial_{x^2}\}$. We have:

$$\mathcal{L}_{x^1\partial_{x^1}} T = \begin{pmatrix} 0 & -T^1{}_2 \\ T^2{}_1 & 0 \end{pmatrix} \quad \text{and}$$

$$\mathcal{L}_{\log(x^1)\partial_{x^2}} T = (x^1)^{-1}\begin{pmatrix} T^1{}_2 & 0 \\ -2T^1{}_1 & -T^1{}_2 \end{pmatrix}.$$

Consequently, there is no non-trivial invariant trace-free tensor of Type (1,1) if $\dim(\mathfrak{K}(\mathcal{N})) = 4$. We now consider the situation when $\dim(\mathfrak{K}(\mathcal{N})) = 3$ and obtain a solution.

Example 14.20 $\mathfrak{K}(\mathcal{N}) = \operatorname{span}\{x^1\partial_{x^1} + x^2\partial_{x^2}, \partial_{x^2}, 2x^1x^2\partial_{x^1} + ((x^2)^2 + \sigma(x^1)^2)\partial_{x^2}\}$ where we have $\sigma \in \{-1, 0, 1\}$. We compute:

$$\mathcal{L}_{2x^1x^2\partial_{x^1} + ((x^2)^2+\sigma(x^1)^2)\partial_{x^2}} T = 2x^1\begin{pmatrix} T^1{}_2\sigma - T^2{}_1 & 2T^1{}_1 \\ -2T^1{}_1\sigma & T^2{}_1 - T^1{}_2\sigma \end{pmatrix}.$$

We set $T^1{}_1 = 0$ and $T^2{}_1 = T^1{}_2\sigma$. We have $\det(T) = -(T^1{}_2)^2\sigma$. There are three cases.

1. Suppose $\sigma = -1$. Taking $A_{11}{}^1 = -1$, $A_{11}{}^2 = 0$, $A_{12}{}^1 = 0$, $A_{12}{}^2 = -1$, $A_{22}{}^1 = 1$, and $A_{22}{}^2 = 0$ we have the Christoffel symbols of the metric of constant Gaussian curvature -1 on the hyperbolic plane. Take $T^1{}_2 = -1$ and $T^2{}_1 = 1$ to define the parallel complex structure on this Kähler surface.
2. Suppose $\sigma = 1$. For $A_{11}{}^1 = -1$, $A_{11}{}^2 = 0$, $A_{12}{}^1 = 0$, $A_{12}{}^2 = -1$, $A_{22}{}^1 = -1$, and $A_{22}{}^2 = 0$ we have the Christoffel symbols of the metric of constant Gaussian curvature on the Lorentzian hyperbolic plane. Take $T^1{}_2 = 1$ and $T^2{}_1 = 1$ to define the parallel para-complex structure; the hyperbolic plane is a para-Kähler surface.
3. Suppose $\sigma = 0$ so $T = \varepsilon\partial_{x^1} \otimes dx^2$. This gives an invariant nilpotent structure on the surfaces $\mathcal{N}_1^3(\pm)$ and $\mathcal{N}_2^3(c)$.

The following characterization of trace-free parallel $(1,1)$-tensor fields on Type $\mathcal{B}$ models was obtained by Calviño-Louzao et al. [13]. It follows from Lemma 12.53.

Lemma 14.21 *If ∇ is a Type $\mathcal{B}$ connection on $M = \mathbb{R}^+ \times \mathbb{R}$ and if $\mathcal{P}^0(\mathcal{M}) \neq \{0\}$, then there exists $\alpha \in \mathbb{C}$ and $0 \neq \mathfrak{t} \in M_2^0(\mathbb{C})$ so that $(x^1)^\alpha \mathfrak{t} \in \mathcal{P}^0_{\mathbb{C}}(\mathcal{M})$.*

Lemma 13.3 shows that ρ_s is recurrent if and only if it is of rank 1 in the Type $\mathcal{A}$ setting. This fact is no longer true for Type $\mathcal{B}$ geometries.

Example 14.22 Let $\mathcal{M}$ be the Type $\mathcal{B}$ surface defined by setting $A_{22}{}^2 = (3+2\sqrt{3})/3$ and $A_{ij}{}^k = 1$ otherwise. We compute

$$\rho_s = \frac{1}{(x^1)^2}\begin{pmatrix} 1+\frac{2}{\sqrt{3}} & \frac{1}{\sqrt{3}} \\ \frac{1}{\sqrt{3}} & \frac{2}{\sqrt{3}}-1 \end{pmatrix}$$

and, consequently, ρ_s has rank 1. Assume $\dim(\mathcal{P}^0(\mathcal{M})) \geq 1$. It follows from Lemma 14.21 that there exists an element in $\mathcal{P}^0_{\mathbb{C}}(\mathcal{M})$ of the form $T = (x^1)^\alpha(\mathfrak{t}^i{}_j)$ where $0 \neq (\mathfrak{t}^i{}_j) \in M_2^0(\mathbb{C})$. Setting $T^i{}_{j;2} = 0$ yields the relations:

$$(x^1)^{\alpha-1}\begin{pmatrix} \mathfrak{t}^2{}_1 - \mathfrak{t}^1{}_2 & -2\mathfrak{t}^1{}_1 - \frac{2}{\sqrt{3}}\mathfrak{t}^1{}_2 \\ 2\mathfrak{t}^1{}_1 + \frac{2}{\sqrt{3}}\mathfrak{t}^2{}_1 & \mathfrak{t}^1{}_2 - \mathfrak{t}^2{}_1 \end{pmatrix} = \begin{pmatrix} 0 & 0 \\ 0 & 0 \end{pmatrix}.$$

We solve this relation to see $\mathfrak{t}^2{}_1 = \mathfrak{t}^1{}_2$ and $\mathfrak{t}^1{}_1 = -\frac{1}{\sqrt{3}}\mathfrak{t}^1{}_2$. Substituting these relations and setting $T^i{}_{j;1} = 0$ then yields:

$$(x^1)^{\alpha-1}\begin{pmatrix} -\frac{\alpha}{\sqrt{3}}\mathfrak{t}^1{}_2 & \left(\alpha+\frac{2}{\sqrt{3}}\right)\mathfrak{t}^1{}_2 \\ \left(\alpha-\frac{2}{\sqrt{3}}\right)\mathfrak{t}^1{}_2 & \frac{\alpha}{\sqrt{3}}\mathfrak{t}^1{}_2 \end{pmatrix} = \begin{pmatrix} 0 & 0 \\ 0 & 0 \end{pmatrix}.$$

This shows $\mathfrak{t}^1{}_2 = 0$, so $T = 0$. This shows $\mathcal{P}^0(\mathcal{M})$ is trivial.

14.3 SYMMETRIC SPACES

In Section 14.3.1 (see Lemma 14.23), we will show that up to linear equivalence, the only non-flat Type $\mathcal{B}$ models which are affine symmetric spaces are given by

1. $\mathcal{N}_1^4(c_2 = -\frac{1}{2})$.
2. $\mathcal{N}_2^4(\kappa = c, \theta = -2c)$ for $c \neq 0$.
3. $\mathbb{L} := \mathcal{N}_3^3$ (the Lorentzian hyperbolic plane).
4. $\mathbb{H} := \mathcal{N}_4^3$ (the hyperbolic plane).

We have $\mathcal{N}_2^4(c,-2c)$ for $c \neq 0$, and $\mathcal{N}_3^4(-\frac{1}{2})$ are affine diffeomorphic to the affine symmetric space $\mathcal{M}_3^4(-\frac{1}{2})$; this Type $\mathcal{A}$ geometry was studied in Chapter 13 and up to linear equivalence, the only other Type $\mathcal{A}$ geometry which is an affine symmetric space and non-flat is $\mathcal{M}_5^4(0)$, which is not a Type $\mathcal{B}$ geometry. In Section 14.3.2 (see Theorem 14.24), we will show that the only Type $\mathcal{B}$ geometries with $\dim(\mathfrak{K}(\mathcal{N})) > 2$ which are geodesically complete are affine symmetric spaces; $\mathbb{L}$ is not geodesically complete while $\mathcal{N}_2^4(c,-2c)$, $\mathcal{N}_3^4(-\frac{1}{2})$, and $\mathbb{H}$ are geodesically complete.

In Section 14.3.3, we discuss the geometry of spaces of constant sectional curvature and generalize results concerning $\mathfrak{H}$ and $\mathfrak{L}$ to this setting, in Section 14.3.4, we will study the geometry of the hyperbolic plane $\mathbb{H}$, and in Section 14.3.5, we study the hyperbolic pseudo-sphere $\mathfrak{H}$, which is a different but equivalent model for hyperbolic geometry. In Section 14.3.6, we treat the geometry of the Lorentzian hyperbolic plane $\mathbb{L}$, and in Section 14.3.7, we present the pseudo-sphere $\mathfrak{L}$ and its universal cover $\tilde{\mathfrak{L}}$; $\mathfrak{L}$ and $\tilde{\mathfrak{L}}$ are geodesically complete. The Lorentzian hyperbolic plane is geodesically incomplete but admits a proper affine embedding in $\mathfrak{L}$.

14.3.1 CLASSIFICATION OF TYPE $\mathcal{B}$ SYMMETRIC SPACES.

It follows from Lemma 13.8 and Theorem 13.22 that any Type $\mathcal{A}$ model which is a non-flat affine symmetric space is linearly equivalent to $\mathcal{M}_2^4(-\frac{1}{2})$, $\mathcal{M}_3^4(-\frac{1}{2})$, or $\mathcal{M}_5^4(0)$. There is a similar classification here.

Lemma 14.23 *If $\mathcal{N}$ is a non-flat Type $\mathcal{B}$ model, then $\mathcal{N}$ is an affine symmetric space if and only if $\mathcal{N}$ is linearly equivalent to $\mathcal{N}_1^4(-\frac{1}{2})$, $\mathcal{N}_2^4(c,-2c)$ for $c \neq 0$, $\mathbb{L} = \mathcal{N}_3^3$, or $\mathbb{H} = \mathcal{N}_4^3$.*

Proof. Let $\mathcal{N}$ be a Type $\mathcal{B}$ model so $\nabla\rho = 0$. By Theorem 12.9, ρ is symmetric. Consequently, $A_{12}{}^1 + A_{22}{}^2 = 0$. Let $\tilde{\rho} := (x^1)^2\rho$ and $\bar{\rho}_{ij;k} := (x^1)^3\rho_{ij;k}$. We examine the relations which arise from setting $\bar{\rho}_{ij;k} = 0$.

Case 1. Suppose $A_{22}{}^1 \neq 0$. We rescale to set $A_{22}{}^1 = \varepsilon$ for $\varepsilon = \pm 1$ and then, by Lemma 14.2, perform a shear to ensure $A_{12}{}^1 = 0$. Since $A_{12}{}^1 + A_{22}{}^2 = 0$, we have $A_{22}{}^2 = 0$ as well. We have that $\bar{\rho}_{22;2}$: $0 = 2A_{11}{}^2$ and $\bar{\rho}_{21;2}$: $0 = 2A_{12}{}^2(A_{12}{}^2 - A_{11}{}^1)\varepsilon$. We set $A_{11}{}^2 = 0$ and consider subcases.

Case 1.1. Suppose $A_{12}{}^2 \neq 0$. Then $A_{11}{}^1 = A_{12}{}^2$. We obtain $\bar{\rho}_{22;1}$: $0 = 2(1 + A_{12}{}^2)\varepsilon$. This yields $A_{12}{}^2 = -1$. We obtain $\mathcal{N} = \mathcal{N}(-1,0,0,-1,\varepsilon,0)$, which is $\mathbb{L} = \mathcal{N}_3^3$ if $\varepsilon = -1$ and is $\mathbb{H} = \mathcal{N}_4^3$ if $\varepsilon = +1$.

Case 1.2. Suppose $A_{12}{}^2 = 0$. We obtain $\bar{\rho}_{22;1}$: $0 = 2(1 - A_{11}{}^1)\varepsilon$. This yields $A_{11}{}^1 = 1$. We then obtain $\rho = 0$ contrary to our assumption. This case does not occur.

Case 2. Suppose $A_{22}{}^1 = 0$ and $A_{12}{}^1 \neq 0$. We then obtain $\bar{\rho}_{22;2}$: $0 = -4(A_{12}{}^1)^3$, so this case is impossible.

Case 3. We assume $A_{22}{}^1 = 0$ and $A_{12}{}^1 = 0$ to obtain $A_{22}{}^2 = 0$ as well. The remaining non-zero component of $\nabla\rho$ is $\bar{\rho}_{11;1}$: $0 = -2(1 + A_{11}{}^1)(1 + A_{11}{}^1 - A_{12}{}^2)A_{12}{}^2$. Since $\mathcal{N}$ is not flat,

$\tilde{\rho} = (1 + A_{11}{}^1 - A_{12}{}^2)A_{12}{}^2 dx^1 \otimes dx^1 \neq 0$. Thus we have that $A_{11}{}^1 = -1$ and, consequently, we have $\mathcal{N} = \mathcal{N}(-1, A_{11}{}^2, 0, A_{12}{}^2, 0, 0)$. If $A_{12}{}^2 = -\frac{1}{2}$ and $A_{11}{}^2 \neq 0$, we can rescale x^2 to set $A_{11}{}^2 = 1$ and obtain $\mathcal{N}(-1, 1, 0, -\frac{1}{2}, 0, 0) = \mathcal{N}_1^4(-\frac{1}{2})$. If $c = A_{12}{}^2 \neq -\frac{1}{2}$, apply a shear to set $A_{11}{}^2 = 0$ and obtain $\mathcal{N}(-1, 0, 0, c, 0, 0) = \mathcal{N}_2^4(\kappa = c, \theta = -2c)$; we assume $c \neq 0$ to ensure that $\rho \neq 0$. □

A similar argument can be used to analyze the situation when the surface in question has torsion; we refer to D'Ascanio, Gilkey, and Pisani [23] as the analysis is beyond the scope of the current volume.

14.3.2 GEODESIC COMPLETENESS.

In Section 13.3.5, we determined which Type $\mathcal{A}$ models were geodesically complete. The situation in the Type $\mathcal{B}$ setting is considerably more difficult and the answer in general is not known. We apply Theorem 14.16 to see that any Type $\mathcal{B}$ model with $\dim(\mathfrak{K}(\mathcal{N})) > 2$ is linearly isomorphic to one of the models $\mathcal{N}_i^\nu(\cdot)$ of Definition 14.1 for $\nu = 3, 4, 6$. We examine the situation for these examples as follows.

Theorem 14.24 *Let $\mathcal{N}$ be a Type $\mathcal{B}$ model with* $\dim(\mathfrak{K}(\mathcal{N})) > 2$.

1. *$\mathcal{N}_5^6$ is geodesically complete and $\mathcal{N}_i^6(\cdot)$ is geodesically incomplete for $i \neq 5$.*
2. *$\mathcal{N}_1^4(-\frac{1}{2})$ and $\mathcal{N}_2^4(c, -2c)$ for $c \neq 0$ are geodesically complete. $\mathcal{N}_i^4(\cdot)$ is essentially geodesically incomplete otherwise.*
3. *$\mathcal{N}_1^3(\pm)$ and $\mathcal{N}_2^3(c)$ are essentially geodesically incomplete.*
4. *$\mathbb{L}$ is geodesically incomplete and $\mathbb{H}$ is geodesically complete.*

Remark 14.25 As in the Type $\mathcal{A}$ setting, all of the examples of Type $\mathcal{B}$ models which are geodesically complete, non-flat, and with $\dim(\mathfrak{K}(\mathcal{N})) > 2$ are affine symmetric spaces. The geodesically incomplete Type $\mathcal{A}$ models which are symmetric spaces can be completed; similarly we shall see in Section 14.3.7 that the Lorentzian hyperbolic plane $\mathbb{L}$ can be completed by embedding it in the pseudo-sphere $\mathfrak{L}$. As noted previously, $\mathcal{N}_1^4(-\frac{1}{2})$ and $\mathcal{N}_2^4(c, -2c)$ are affine equivalent to $\mathcal{M}_3^4(-\frac{1}{2})$; this geometry was studied in Chapter 13.

Proof. Suppose first $\mathcal{N}$ is flat. By Lemma 14.11, the maps Ψ_i^6 given in Definition 14.10 are affine embeddings of $\mathcal{N}_i^6(\cdot)$ in $\mathcal{M}_0^6$. They are not surjective for $i \neq 5$. Consequently, these geometries are geodesically incomplete. The map Ψ_5^6 is an affine diffeomorphism. Therefore, $\mathcal{N}_5^6$ is geodesically complete; Assertion 1 follows. Suppose next $\mathcal{N}$ is linearly isomorphic to $\mathcal{N}_i^4(\cdot)$. We have $\mathcal{N}_1^4(c_2)$ is affine isomorphic to $\mathcal{M}_3^4(c_2)$, $\mathcal{N}_2^4(\kappa, \theta)$ is linearly isomorphic to $\mathcal{M}_3^4(\frac{\kappa}{\theta})$, and $\mathcal{N}_3^4(c_1)$ is linearly isomorphic to $\mathcal{M}_4^4(0)$. By Lemma 13.30, $\mathcal{M}_3^4(c_1)$ is essentially geodesically incomplete for $c_1 \neq -\frac{1}{2}$, while $\mathcal{M}_3^4(-\frac{1}{2})$ is an affine symmetric space which is geodesically complete (see Lemma 13.8). Furthermore, $\mathcal{M}_4^4(0)$ is essentially geodesically incomplete. Assertion 2 follows. Finally, assume $\dim(\mathfrak{K}(\mathcal{N})) = 3$.

Case 1. Let $\mathcal{N} = \mathcal{N}_1^3(\pm) = \mathcal{N}(-\frac{3}{2}, 0, 0, -\frac{1}{2}, \mp\frac{1}{2}, 0)$ or $\mathcal{N} = \mathcal{N}_2^3(c) = \mathcal{N}(-\frac{3}{2}, 0, 1, -\frac{1}{2}, c, 2)$. Let $\sigma(t) = (t^{-2}, 0)$. Then $\dot{\sigma} = t^{-3}(-2, 0)$ and $\ddot{\sigma} = t^{-4}(6, 0)$. The only non-trivial geodesic equation is $\ddot{x}^1 - \frac{3}{2}(x^1)^{-1}\dot{x}^1\dot{x}^1 = 6t^{-4} - 4\frac{3}{2}t^2t^{-3}t^{-3} = 0$. Thus these geometries are geodesically incomplete. Suppose there is a geodesically complete geometry $\tilde{\mathcal{N}}$ modeled on $\mathcal{N}_1^3(\pm)$ or on $\mathcal{N}_2^3(c)$. We may suppose without loss of generality that $\tilde{\mathcal{N}}$ is simply connected. Copy a small part of σ into $\tilde{\mathcal{N}}$. Extend σ to a globally defined affine geodesic τ. By Lemma 14.15, $\mathfrak{K}(\mathcal{N}) = \operatorname{span}\{x^1\partial_{x^1} + x^2\partial_{x^2}, \partial_{x^2}, 2x^1x^2\partial_{x^1} + (x^2)^2\partial_{x^2}\}$. By Theorem 12.52, $\tilde{\mathcal{N}}$ is real analytic. Choose global affine Killing vector fields on $\tilde{\mathcal{N}}$ so that near σ we have

$$\xi_1 = x^1\partial_{x^1} + x^2\partial_{x^2}, \quad \xi_2 = \partial_{x^2}, \quad \xi_3 = 2x^1x^2\partial_{x^1} + (x^2)^2\partial_{x^2}.$$

Since ξ_3 vanishes on σ and since the structures are real analytic, ξ_3 vanishes identically on τ. Consequently, $\{\xi_1, \xi_2\}$ forms a frame for the tangent bundle along $\tilde{\sigma}$. Thus we can express $\dot{\tau} = \kappa^1(t)\xi_1(t) + \kappa^2(t)\xi_2(t)$ where the κ^i are now real analytic functions defined for $t \in \mathbb{R}$. Since $\kappa^1(t) = -2t^{-1/2}$, this is impossible. Thus the structures $\mathcal{N}_1^3(\pm)$ and $\mathcal{N}_2^3(c)$ are essentially geodesically incomplete.

Case 2. We show $\mathbb{L} = \mathcal{N}(-1, 0, 0, -1, -1, 0)$ is geodesically incomplete by verifying that the geodesic equations are satisfied by the curve $\sigma(t) = t^{-1}(1, 1)$: $x^1\ddot{x}^1 - \dot{x}^1\dot{x}^1 - \dot{x}^2\dot{x}^2 = 0$ and $x^1\ddot{x}^2 - 2\dot{x}^1\dot{x}^2 = 0$. We will show that $\mathbb{H}$ is geodesically complete in Section 14.3.4. □

We give below the geodesic structure of these geometries with basepoint $(1, 0)$. The geodesics of the hyperbolic plane $\mathcal{N}_4^3$ are circles perpendicular to the vertical axis that go through $(1, 0)$; the geodesic structure of the remaining models are more complicated. Note that none of the geodesics ever reach the asymptote $x^1 = 0$.

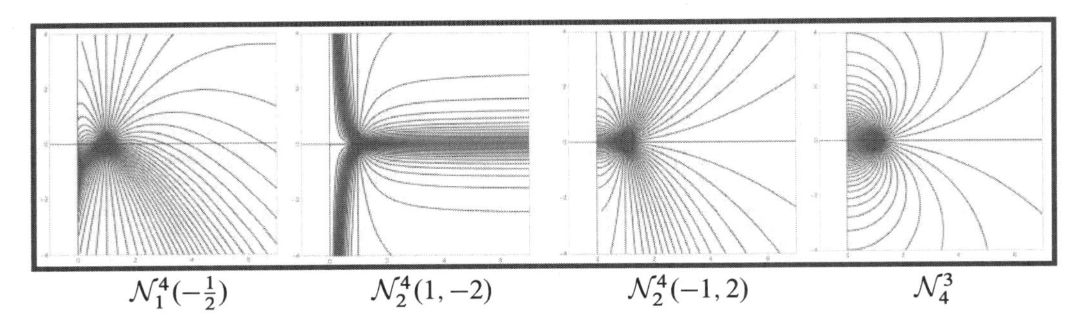

$\mathcal{N}_1^4(-\frac{1}{2})$ $\quad$ $\mathcal{N}_2^4(1, -2)$ $\quad$ $\mathcal{N}_2^4(-1, 2)$ $\quad$ $\mathcal{N}_4^3$

14.3.3 THE GEOMETRY OF PSEUDO-SPHERES. Before discussing the hyperbolic plane $\mathbb{H}$ and the Lorentzian hyperbolic plane $\mathbb{L}$, it is convenient to digress and discuss *pseudo-spheres* in complete generality. These provide models for spaces of constant sectional curvature. We adopt the following notational conventions. Let $\varepsilon := (\varepsilon_{ij})$ where $\varepsilon_{ij} = 0$ if $i \neq j$ and $\varepsilon_{ij} = \pm 1$ if $i = j$. We shall assume that $\varepsilon_{m+1,m+1} = +1$ to simplify certain sign conventions.

For example, we will take

$$\varepsilon_{\mathcal{S}} := \begin{pmatrix} 1 & 0 & 0 \\ 0 & 1 & 0 \\ 0 & 0 & 1 \end{pmatrix}, \ \varepsilon_{\mathfrak{L}} := \begin{pmatrix} -1 & 0 & 0 \\ 0 & 1 & 0 \\ 0 & 0 & 1 \end{pmatrix}, \ \varepsilon_{\mathfrak{H}} := \begin{pmatrix} -1 & 0 & 0 \\ 0 & -1 & 0 \\ 0 & 0 & 1 \end{pmatrix} \tag{14.3.a}$$

to discuss the round sphere, a model $\mathfrak{L}$ for the Lorentzian hyperbolic plane $\mathbb{L}$, and a model $\mathfrak{H}$ for the hyperbolic plane $\mathbb{H}$; we obtain a metric on $\mathfrak{H}$ of signature $(2, 0)$ rather than the usual Riemannian metric of signature $(0, 2)$, but this does not change the underlying affine structure.

Let signature$(\varepsilon) = (p, q)$ where p is the number of $+1$ and q the number of -1 entries. Set $\langle \cdot, \cdot \rangle_\varepsilon := \varepsilon_{ij} dx^i \otimes dx^j$. Let the *pseudo-sphere*

$$\mathcal{S}_\varepsilon := \{\vec{x} \in \mathbb{R}^{m+1} : \langle \vec{x}, \vec{x} \rangle_\varepsilon = +1\} \tag{14.3.b}$$

given the induced pseudo-Riemannian metric g_ε of signature $(q, p-1)$. Denote the associated *orthogonal group* by $\mathcal{O}_\varepsilon := \{T \in \mathrm{GL}(m+1, \mathbb{R}) : T^*\langle \cdot, \cdot \rangle_\varepsilon = \langle \cdot, \cdot \rangle_\varepsilon\}$. Let g_ε be the restriction of $\langle \cdot, \cdot \rangle_\varepsilon$ to the tangent bundle of $\mathcal{S}_\varepsilon$. We summarize as follows the basic facts concerning pseudo-spheres.

Lemma 14.26 *$(\mathcal{S}_\varepsilon, g_\varepsilon)$ is a pseudo-Riemannian manifold of signature $(p, q-1)$ on which $\mathcal{O}_\varepsilon$ acts transitively. If T is the germ of an isometry of $(\mathcal{S}_\varepsilon, g_\varepsilon)$, then $T \in \mathcal{O}_\varepsilon$. $(\mathcal{S}_\varepsilon, g_\varepsilon)$ is an affine symmetric space, $R(x, y, z, w) = \langle x, w \rangle_\varepsilon \langle y, z \rangle_\varepsilon - \langle x, z \rangle_\varepsilon \langle y, w \rangle_\varepsilon$, $\rho_\varepsilon = (m-1) g_\varepsilon$, and $(\mathcal{S}_\varepsilon, g_\varepsilon)$ has constant sectional curvature $+1$. $(\mathcal{S}_\varepsilon, g_\varepsilon)$ is geodesically complete and any pseudo-Riemannian manifold of constant sectional curvature $+1$ is locally isometric to $(\mathcal{S}_\varepsilon, g_\varepsilon)$ for some ε.*

By replacing ε by $-\varepsilon$ and considering $\{\vec{x} : \langle \vec{x}, \vec{x} \rangle = -1\}$, we can obtain models for pseudo-Riemannian manifolds with constant sectional curvature -1. As the underlying affine structure is unchanged, we need not deal with this setting.

Proof. If $\vec{x} \in \mathcal{S}_\varepsilon$, then $\vec{x}$ is spacelike and $T_{\vec{x}}(\mathcal{S}_\varepsilon) = \vec{x}^\perp$. Thus the restriction of $\langle \cdot, \cdot \rangle_\varepsilon$ to $T_{\vec{x}} \mathcal{S}_\varepsilon$ defines a pseudo-Riemannian metric g_ε on $\mathcal{S}_\varepsilon$ of signature $(p, q-1)$. The orthogonal group $\mathcal{O}_\varepsilon$ preserves the structures and acts isometrically on $(\mathcal{S}_\varepsilon, g_\varepsilon)$. Let $\xi_{m+1} \in \mathcal{S}_\varepsilon$. Choose an orthonormal basis $\{\xi_1, \ldots, \xi_m\}$ for $\xi_{m+1}^\perp$ so $\langle \xi_i, \xi_j \rangle = \varepsilon_{ij}$. Then the matrix $T = (\xi_1, \ldots, \xi_{m+1}) \in \mathcal{O}_\varepsilon$ and $Te_i = \xi_i$ where $\{e_i\}$ is the standard basis for $\mathbb{R}^{m+1}$. In particular, $\mathcal{O}_\varepsilon$ acts transitively on $\mathcal{S}_\varepsilon$ by isometries. Suppose T is an isometry of some component of $\mathcal{S}_\varepsilon$. Let $\{e_1, \ldots, e_{m+1}\}$ be the standard basis for $\mathbb{R}^{m+1}$. Then $\xi_{m+1} := Te_{m+1}$ is a space-like unit vector. Furthermore, $\xi_1 := T_*(e_1)$, ..., $\xi_m := T_*(e_m)$ is a basis for $T_{\xi_{m+1}} \mathcal{S}_\varepsilon = \xi_{m+1}^\perp$ with $\langle \xi_i, \xi_j \rangle_\varepsilon = \langle e_i, e_j \rangle_\varepsilon$. Let $\tilde{T}$ be the matrix with columns $(\xi_1, \ldots, \xi_{m+1})$. Then $\tilde{T} \in \mathcal{O}_\varepsilon$. We have $\tilde{T}^{-1} T e_{m+1} = e_{m+1}$ and $(\tilde{T}^{-1} T)_* e_i = e_i$ for $1 \le i \le m$. Thus $\tilde{T}^{-1} T$ preserves e_{m+1} and is the identity on $T_{e_{m+1}} \mathcal{S}_\varepsilon$. Since $\tilde{T}^{-1} T$ is the germ of an isometry of $\mathcal{S}_\varepsilon$ it follows $\tilde{T}^{-1} T$ is the identity. Therefore $T = \tilde{T} \in \mathcal{O}_\varepsilon$. This is, of course, only a local result. Global isometries may be represented by different elements of the orthogonal group on different arc components of $\mathcal{S}_\varepsilon$. As this will play no role in our analysis, we ignore this point.

Let $T(e_i) = -e_i$ for $1 \le i \le m$ and $Te_{m+1} = e_{m+1}$ define an element of $\mathcal{O}_\varepsilon$. Then $Te_{m+1} = e_{m+1}$ and $T_* = -\mathrm{Id}$ on $T_{e_{m+1}}\mathcal{S}_\varepsilon = e_{m+1}^\perp$. Since $(\mathcal{S}_\varepsilon, g_\varepsilon)$ is a homogeneous space, it follows that $(\mathcal{S}_\varepsilon, g_\varepsilon)$ is a global affine symmetric space. Since $(\mathcal{S}_\varepsilon, g_\varepsilon)$ is a homogeneous space, it suffices to compute the curvature tensor at a single point. We choose the point e_{m+1}. We have $\varepsilon_{11}(x^1)^2 + \cdots + \varepsilon_{mm}(x^m)^2 + (x^{m+1})^2 = 1$. Thus it is natural to introduce

$$\begin{aligned}&\Phi(u^1,\dots,u^m) = (u^1,\dots,u^m,(1-\varepsilon_{11}(u^1)^2 - \cdots - \varepsilon_{mm}(u^m)^2)^{1/2}),\\ &\Phi_*(\partial_{u^i}) = e_i - \varepsilon_{ii}u^i(1-\varepsilon_{11}(u^1)^2 - \cdots - \varepsilon_{mm}(u^m)^2)^{-1/2}e_{m+1}\,.\end{aligned} \tag{14.3.c}$$

We then have

$$g_{ij} = \varepsilon_{ij} + \varepsilon_{ii}\varepsilon_{jj}u^iu^j + O(\|u\|^3) \quad \text{for} \quad 1 \le i, j \le m\,. \tag{14.3.d}$$

In the proof of Lemma 3.8 of Book I, we showed that if all the derivatives of the metric vanish at the origin, then

$$R_{ijk\ell}(0) = \tfrac{1}{2}\{g_{j\ell/ik} + g_{ik/j\ell} - g_{jk/i\ell} - g_{i\ell/jk}\}(0)\,. \tag{14.3.e}$$

Combining Equation (14.3.d) and Equation (14.3.e) then determines R; the calculation of the Ricci tensor is then immediate.

A curve σ in a non-degenerate hypersurface $\mathcal{S}$ in $(\mathbb{R}^m, \langle\cdot,\cdot\rangle_\varepsilon)$ is an affine geodesic if and only if $\ddot\sigma \perp \mathcal{S}$. We use this observation in what follows. Since $\mathcal{S}_\varepsilon$ is a homogeneous space, it suffices to consider an affine geodesic σ that starts from e_{m+1}. Let $\xi := \dot\sigma(0)$. There are three cases to be considered. Suppose first ξ is timelike. We may assume $\langle\xi,\xi\rangle_\varepsilon = -1$ and assume the coordinates on $\mathbb{R}^{m+1}$ are chosen so that $\xi = e_1$ where

$$\langle\cdot,\cdot\rangle = -dx^1 \otimes dx^1 \pm dx^2 \otimes dx^2 \pm \cdots \pm dx^m \otimes dx^m + dx^{m+1} \otimes dx^{m+1}\,.$$

Let $\sigma(t) := \sinh(t)e_1 + \cosh(t)e_{m+1}$. We verify that σ is an affine geodesic by checking that $\langle\sigma(t),\sigma(t)\rangle_\varepsilon = 1$, $\langle\sigma(t),\dot\sigma(t)\rangle_\varepsilon = 0$, and $\langle\dot\sigma(t),\dot\sigma(t)\rangle_\varepsilon = -1$. This implies σ is a timelike affine geodesic starting from e_{m+1} with initial direction e_1 which exists for all time. If ξ is spacelike, we assume the coordinates on $\mathbb{R}^{m+1}$ are chosen so $\xi = e_1$ where

$$\langle\cdot,\cdot\rangle = dx^1 \otimes dx^1 \pm dx^2 \otimes dx^2 \pm \cdots \pm dx^m \otimes dx^m + dx^{m+1} \otimes dx^{m+1}\,.$$

The remainder of the analysis is the same if we set $\sigma(t) = \sin(t)e_1 + \cos(t)e_{m+1}$. Finally, if ξ is null, assume $\xi = e_1 + e_2$ where

$$\langle\cdot,\cdot\rangle = dx^1 \otimes dx^1 - dx^2 \otimes dx^2 \pm \cdots \pm dx^m \otimes dx^m + dx^{m+1} \otimes dx^{m+1}\,.$$

Set $\sigma(t) = t(e_1 + e_2) + e_{m+1}$. We verify that σ is an affine geodesic by checking that $\langle\sigma(t),\sigma(t)\rangle_\varepsilon = 1$, $\langle\sigma(t),\dot\sigma(t)\rangle_\varepsilon = 0$, and $\langle\dot\sigma(t),\dot\sigma(t)\rangle_\varepsilon = 0$. Consequently, σ is a null affine geodesic which is defined for all time.

Any pseudo-Riemannian manifold of constant sectional curvature is locally symmetric. We apply Lemma 7.15 of Book II, which showed that a pseudo-Riemannian symmetric space is characterized up to local isometry by the form of its curvature tensor and metric at a single point. □

Let ∇_ε be the Levi-Civita connection of the pseudo-Riemannian manifold. Denote the underlying affine structure by $\mathcal{M}_\varepsilon := (\mathcal{S}_\varepsilon, \nabla_\varepsilon)$. Since the Levi-Civita connection of g_ε is the same as the Levi-Civita connection of $-g_\varepsilon$, the sign of the sectional curvature is irrelevant. Let $X_{ij} := \varepsilon_{ii} x^i \partial_{x^j} - \varepsilon_{jj} x^j \partial_{x^i}$ where we do not sum over repeated indices. If $f(x) = \langle x, x\rangle$, then $X_{ij}(f) = 0$. Thus the vector fields X_{ij} are tangential to $\mathcal{S}_\varepsilon$.

Lemma 14.27 Let $\mathcal{M}_\varepsilon$ be the underlying affine manifold defined by the pseudo-sphere $\mathcal{S}_\varepsilon$ of Equation (14.3.b). Then $\mathfrak{A}(\mathcal{M}_\varepsilon) = \text{span}_{i<j}\{X_{ij}\}$, there are no affine gradient Ricci solitons on $\mathcal{M}_\varepsilon$, $\mathcal{M}_\varepsilon$ is strongly projectively flat, and

$$E(\mu, \mathcal{M}_\varepsilon) = \left\{ \begin{array}{ll} \text{span}\{x^1, \dots, x^{m+1}\} & \text{if } \mu = -\frac{1}{m-1} \\ \text{span}\{\mathbb{1}\} & \text{if } \mu = 0 \\ \{0\} & \text{otherwise} \end{array} \right\}.$$

Proof. We apply Lemma 14.26. Let X be an affine Killing vector field and let Φ_t^X be the 1-parameter family of local diffeomorphisms corresponding to X. Then Φ_t^X commutes with ∇_ε, and hence preserves the Ricci tensor. We have $\rho_\varepsilon = (m-1)g_\varepsilon$. Consequently, the Φ_t^X are isometries, and thus X is an affine Killing vector field. The $\{X_{ij}\}$ represent the action of the Lie algebra of $\mathcal{O}_\varepsilon$ on $\mathbb{R}^{m+1}$ and, consequently, $\mathfrak{A}(\mathcal{M}_\varepsilon) = \text{span}_{i<j}\{X_{ij}\}$.

Suppose ψ is an affine gradient Ricci soliton on $\mathcal{M}_\varepsilon$. Since $\nabla_\varepsilon \rho_\varepsilon = 0$, Lemma 12.12 shows $R_{ij}d\psi = 0$ for all i, j. We then have $R_{ij}e^k = \pm(\delta_{ik}e^j - \delta_{jk}e^i)$. Consequently, we have that $R_{ij}d\psi = 0$ for all i, j implies $d\psi = 0$ so $\psi = \mathbb{1}$. This implies $\rho_\varepsilon = 0$, which is false. Consequently, there are no affine gradient Ricci solitons on $\mathcal{M}_\varepsilon$.

Suppose $\mu = -\frac{1}{m-1}$. We wish to show $E(\mu, \mathcal{M}_\varepsilon) = \text{span}\{x^1, \dots, x^{m+1}\}$. Since $\mathcal{M}_\varepsilon$ is a homogeneous space, it suffices to show that, at the point e_{m+1}, the coordinate functions $\{x^1, \dots, x^{m+1}\}$ solve the affine quasi-Einstein equation. We choose the coordinate system of Equation (14.3.c). At the origin, all the first derivatives of the metric and the Christoffel symbols vanish. We have $x^i = u^i$ for $1 \le i \le m$. Consequently, $\mathcal{H}(x^i)(0) = 0$ and $x^i \rho_\varepsilon(0) = 0$ and the affine quasi-Einstein equation is satisfied. Because

$$x^{m+1} = (1 - \varepsilon_i (u^i)^2)^{\frac{1}{2}} \text{ and } \mathcal{H}(x^{m+1})(0) = \partial_{u^i}\partial_{u^j}(1 - \varepsilon_i(u^i)^2)^{\frac{1}{2}}(0),$$

we have that $\mathcal{H}_{ij}(x^{m+1})(0) = -\varepsilon_{ij} = -\frac{1}{m-1}(-(m-1))\varepsilon_{ij} = -\frac{1}{m-1}x^{m+1}(\rho_\varepsilon)_{ij}(0)$. Suppose next that μ is not the critical eigenvalue, i.e., that $\mu(m-1) \ne -1$. If $\{\mathcal{H}f\}_{ij} = \mu f(\rho_\varepsilon)_{ij}$, then we use the fact that $(f\rho_\varepsilon)_{ij;k} = f_{;k}(\rho_\varepsilon)_{ij}$ and the identity $h_{;ijk} - h_{;ikj} = \{R_{kj}(dh)\}_i$ to equate

$$\begin{aligned} \{R_{kj}(dh)\}_i &= R_{jki}{}^{\ell} f_{;\ell} = \varepsilon_{ik} f_{;j} - \varepsilon_{ij} f_{;k} \quad \text{and} \\ \mu\{f_{;k}(\rho_\varepsilon)_{ij} - f_{;j}(\rho_\varepsilon)_{ik}\} &= \mu(m-1)\{f_{;j}\varepsilon_{ij} - \varepsilon_{ik} f_{;k}\}. \end{aligned}$$

Since $\mu(m-1) \ne -1$, this implies $f_{;k} = 0$ for all k. We may rescale f to ensure that $f = \mathbb{1}$. Since $\mathcal{H}\mathbb{1} = 0$, this implies $\mu\rho_\varepsilon = 0$. This shows that $\mu = 0$. □

We have the following consequence.

Corollary 14.28 *Let $S^2 := \{(x, y, z) \in \mathbb{R}^3 : x^2 + y^2 + z^2 = 1\}$ be given the metric inherited from the flat Euclidean metric. Then S^2 has no affine gradient Ricci solitons,*

$$\mathcal{Y}(S^2) = \mathbb{1} \cdot \mathbb{R}, \ \mathcal{Q}(S^2) = \operatorname{span}\{x, y, z\}, \textit{ and } E(\mu, S^2) = \{0\} \textit{ for } \mu \neq 0 \textit{ and } \mu \neq -1\,.$$

Proof. Take $\varepsilon = \varepsilon_S$ as defined in Equation (14.3.a) and apply Lemma 14.26 and Lemma 14.27. □

14.3.4 THE HYPERBOLIC PLANE. Let $ds^2 = (x^1)^{-2}((dx^1)^2 + (dx^2)^2)$. A direct computation using the Koszul formula (see Theorem 3.7 of Book I) shows that the non-zero Christoffel symbols of the Levi-Civita connection are

$$\Gamma_{11}{}^1 = \Gamma_{12}{}^2 = -(x^1)^{-1} \quad \text{and} \quad \Gamma_{22}{}^1 = (x^1)^{-1}\,.$$

This is the geometry $\mathcal{N}_4^3$ of Definition 14.1. Thus, in particular, the Ricci tensor is given by $\rho = -(x^1)^{-2}((dx^1)^2 + (dx^2)^2)$. This surface is Einstein with Einstein constant -1. We summarize briefly well known facts concerning hyperbolic geometry as an introduction to our discussion of the Lorentzian analogue in Section 14.3.6. Set $z := x^2 + \sqrt{-1}x^1$ to identify $\mathbb{R}^+ \times \mathbb{R}$ with the upper half-plane $\mathbb{H} := \{z \in \mathbb{C} : \Im(z) > 0\}$. Let $\mathrm{SL}(2, \mathbb{R})$ be the Lie group of linear transformations of $\mathbb{R}^2$ with determinant 1. We refer to Lemma 6.25 of Book II for further information concerning this Lie group. The subgroup of diagonal matrices $\pm\,\mathrm{Id}$ is a normal subgroup of $\mathrm{SL}(2, \mathbb{R})$ and we set $\mathrm{PSL}(2, \mathbb{R}) := \mathrm{SL}(2, \mathbb{R})/\{\pm\,\mathrm{Id}\}$. We define a *linear fractional transformation* preserving $\mathbb{H}$ by setting:

$$T_A(z) := \frac{az + b}{cz + d} \quad \text{for} \quad A = \begin{pmatrix} a & b \\ c & d \end{pmatrix} \in \mathrm{SL}(2, \mathbb{R})\,.$$

The following is well known. Since it follows from a direct computation, we omit the proof.

Lemma 14.29 *Let A and B belong to $\mathrm{SL}(2, \mathbb{R})$. Then $T_A \circ T_B = T_{AB}$. We have T_A is an orientation-preserving isometry of $\mathbb{H}$, and the map $A \to T_A$ identifies $\mathrm{PSL}(2, \mathbb{R})$ with the group of orientation-preserving isometries of $\mathbb{H}$.*

The curve $\sigma(t) = (e^t, c)$ satisfies the geodesic equation $x^1\ddot{x}^1(t) - \dot{x}^1(t)\dot{x}^1(t) = 0$. Thus horizontal rays from the vertical axis are complete affine geodesics. Linear fractional transformations preserve angles and map straight lines/circles to straight lines/circles. It now follows that any circle whose center is on the vertical axis (i.e., the imaginary axis) is a complete affine geodesic and thus $\mathbb{H}$ is geodesically complete.

14.3.5 THE HYPERBOLIC PSEUDO-SPHERE $\mathfrak{H}$. Define a non-degenerate inner product of signature $(1,2)$ on $\mathbb{R}^3$ by setting $\langle x, y\rangle = x^1y^1 + x^2y^2 - x^3y^3$. Give the level set

$$\mathfrak{H} := \{\vec{x} : \langle\vec{x},\vec{x}\rangle = -1 \text{ and } x^3 > 0\}$$

the induced Riemannian inner product to define the hyperbolic pseudo-sphere $\mathfrak{H}$. Subsequently in Section 14.3.7, we will consider the level set $\langle\vec{x},\vec{x}\rangle = +1$ in studying the Lorentzian hyperbolic plane. We assume $x^3 > 0$ to ensure $\mathfrak{H}$ is connected as the equation $\langle\vec{x},\vec{x}\rangle = -1$ is a hyperboloid of two sheets. The x^3 axis is vertical.

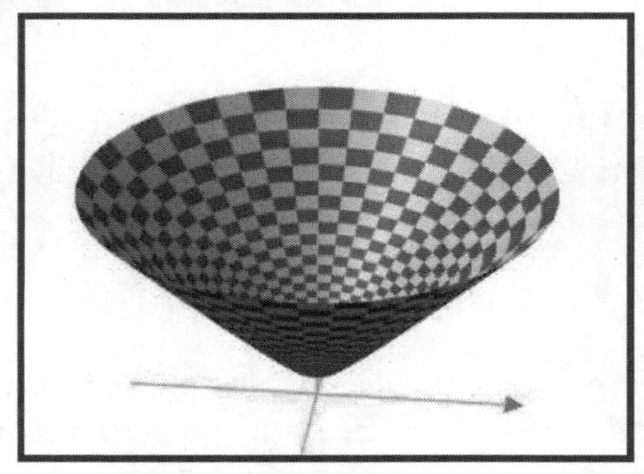

Lemma 14.30

1. *The Lorentz group $O(1,2)$ acts transitively on $\mathfrak{H}$ by isometries. The level set $\mathfrak{H}$ is geodesically complete, $\rho = -g$, $\nabla\rho = 0$, and $\mathcal{Q}(\mathfrak{H}) = \operatorname{span}\{x^1, x^2, x^3\}$.*
2. *The exponential map is surjective.*
3. *Let $\Phi(x^1,x^2) := \frac{1}{2x^1}((x^1)^2 + (x^2)^2 - 1, \quad 2x^2, \quad (x^1)^2 + (x^2)^2 + 1)$. Then Φ is an affine diffeomorphism between $\mathbb{H}$ and $\mathfrak{H}$.*

Proof. Adopt the notation of Equation (14.3.a) and set

$$\varepsilon_{\mathfrak{H}} := \begin{pmatrix} -1 & 0 & 0 \\ 0 & -1 & 0 \\ 0 & 0 & +1 \end{pmatrix}.$$

Then $\varepsilon_{\mathcal{H}} = -\langle\cdot,\cdot\rangle$ so $g = -g_{\varepsilon_{\mathcal{H}}}$ but allowing for the changes in sign, Assertion 1 follows from Lemma 14.26. Let $\sigma(t) := (u\sinh(t), v\sinh(t), \cosh(t))$. The analysis of Lemma 14.29 shows that these curves are affine geodesics. Consequently, the exponential map is surjective. Let $\Phi = (\Phi_1, \Phi_2, \Phi_3)$. We verify $\Phi_1^2 + \Phi_2^2 - \Phi_3^2 = \langle\Phi,\Phi\rangle = -1$ so Φ takes values in $\mathfrak{H}$. We can recover x^1 from $\Phi_1 + \Phi_3$, and once x^1 is known, we can recover x^2 from Φ_2. Thus Φ is a smooth embedding. We use Lemma 14.4 to see $\mathcal{Q}(\mathbb{H}) = \operatorname{span}\{\frac{1}{x^1}, \frac{x^2}{x^1}, \frac{(x^1)^2+(x^2)^2}{x^1}\}$. An algebraic computation shows $\mathcal{Q}(\mathbb{H}) = \operatorname{span}\{\Phi_1,\Phi_2,\Phi_3\}$ so $\mathcal{Q}(\mathbb{H}) = \Phi^*\operatorname{span}\{x^1,x^2,x^3\}$. By Assertion 1, $\mathcal{Q}(\mathfrak{H}) = \operatorname{span}\{x^1,x^2,x^3\}$. Theorem 12.34 then implies Φ is an affine map. As the exponential map is a diffeomorphism for both $\mathbb{H}$ and $\mathfrak{H}$, Φ is a diffeomorphism. □

Let Φ be the map of Lemma 14.30. Since Φ is an affine map, Φ intertwines the Ricci tensors. Since the metrics are essentially given by the Ricci tensors up to sign, Φ is an isometry. One can also check this directly by verifying that

$$\Phi^*(g_{\mathfrak{H}}) = d\Phi_1^2 + d\Phi_2^2 - d\Phi_3^2 = (x^1)^{-2}((dx^1)^2 + (dx^2)^2).$$

14.3.6 THE LORENTZIAN HYPERBOLIC PLANE. The Euclidean inner product is defined by the quadratic form $((x^1, x^2), (y^1, y^2)) = x^1y^1 + x^2y^2$. The Lorentzian inner product is defined by setting $\langle (x^1, x^2), (y^1, y^2)\rangle := -x^1y^1 + x^2y^2$. The null vectors are defined by the equation $x^1 = \pm x^2$. In analogy with the hyperbolic plane, let $\mathbb{L} = (\mathbb{R}^+ \times \mathbb{R})$ with the Lorentzian metric $ds^2 = (x^1)^{-2}\{-(dx^1)^2 + (dx^2)^2\}$. The Koszul formula now yields

$$\Gamma_{11}{}^1 = \Gamma_{12}{}^2 = \Gamma_{22}{}^1 = -(x^1)^{-1}\,.$$

This is the structure $\mathcal{N}_3^3$ of Definition 14.1 and thus by Lemma 14.3,

$$\rho = (x^1)^{-2}(-(dx^1)^2 + (dx^2)^2)\,,$$

so the Einstein constant is 1. We remark that we could equally well have taken

$$ds^2 = (x^1)^{-2}\{(dx^1)^2 - (dx^2)^2\}$$

since we are only interested in the underlying affine structure and the resulting Levi-Civita connection is the same; the Einstein constant here would be -1. Thus these two structures are not isometric, although the underlying affine structure is the same. As was the case for the hyperbolic plane, the linear fractional transformations play a central role. Let

$$\mathrm{Id} := \begin{pmatrix} 1 & 0 \\ 0 & 1 \end{pmatrix} \quad \text{and} \quad \iota := \begin{pmatrix} 0 & 1 \\ 1 & 0 \end{pmatrix}.$$

Let $\hat{\mathbb{C}} := \mathrm{span}_{\mathbb{R}}\{\mathrm{Id}, \iota\} \subset M_2(\mathbb{R})$ be the *para-complex* numbers; we refer to Cruceanu, Fortuny, and Gadea [20] for further details. Since $\iota^2 = \mathrm{Id}$, $\hat{\mathbb{C}}$ is a Abelian 2-dimensional unital algebra; unlike the complex numbers, $\hat{\mathbb{C}}$ is not a field as there are zero divisors. We set $\hat{z} = x^2\,\mathrm{Id} + x^1\iota$. Let $\hat{z}^* = x^2\,\mathrm{Id} - x^1\iota$. We compute $\hat{z}\,\hat{z}^* = (x^2)^2 - (x^1)^2$. Consequently, $\hat{z}$ is invertible if and only if $x^1 \neq \pm x^2$, i.e., $\hat{z}$ is not a null vector. Let

$$\hat{T}_A(\hat{z}) = \frac{a\hat{z} + b}{c\hat{z} + d} \quad \text{for} \quad A = \begin{pmatrix} a & b \\ c & d \end{pmatrix} \in \mathrm{SL}(2, \mathbb{R})\,.$$

The transformation $\hat{T}_A$ is a *para-complex linear fractional transformation*. Unlike the complex setting, $\hat{T}_A$ is not defined for $c\hat{z} + d = 0$, but only on the open dense subset where we have that $cx^2 + d \neq \pm cx^1$. The following result is due to Catoni et al. [16] and provides the appropriate generalization of Lemma 14.29 to this setting; as it follows by a direct computation, we shall omit the proof.

Lemma 14.31 *Let A and B belong to* $\mathrm{SL}(2,\mathbb{R})$. *Then* $\hat{T}_A \circ \hat{T}_B = \hat{T}_{AB}$. *We have* $\hat{T}_A$ *is an orientation-preserving isometry of* $\mathbb{L}$, *where defined, and the map* $A \to \hat{T}_A$ *identifies* $\mathrm{PSL}(2,\mathbb{R})$ *with the group of orientation-preserving isometries of* $\mathbb{L}$.

The Lorentzian hyperbolic plane is the only non-complete symmetric space of Type $\mathcal{B}$, and the exponential map is not surjective, although it is 1-1. Let $X = \xi_1\partial_{x^1} + \xi_2\partial_{x^2}$; X is a *null vector field* if $\xi_1 = \pm\xi_2$, X is a *timelike vector field* if $|\xi_1| > |\xi_2|$, and X is a *spacelike vector field* if $|\xi_1| < |\xi_2|$. The following result is due to D'Ascanio, Gilkey, and Pisani [22]; we shall omit the proof in the interests of brevity.

Theorem 14.32

1. *The affine geodesics of* $\mathbb{L}$ *have one of the following forms for some* $\alpha, \beta, c \in \mathbb{R}$, *up to rescaling:*
   - *(a)* $\sigma(t) = (e^t, \alpha)$ *for* $-\infty < t < \infty$. *This affine geodesic is complete.*
   - *(b)* $\sigma(t) = (t^{-1}, \pm t^{-1} + \alpha)$ *for* $0 < t < \infty$. *This affine geodesic is incomplete at one end and complete at the other end.*
   - *(c)* $\sigma(t) = (\frac{1}{c\sinh(t)}, \pm\frac{\coth(t)}{c} + \beta)$ *for* $t \in (0,\infty)$ *and* $c > 0$. *This tends asymptotically to the line* $x^1 = 0$ *as* $t \to \infty$ *and escapes to the right as* $t \to 0$. *These affine geodesics are incomplete at one end and complete at the other. These affine geodesics all have infinite (and negative) length.*
   - *(d)* $\sigma(t) = (\frac{1}{c\sin(t)}, \pm\frac{\cot(t)}{c} + \beta)$ *for* $t \in (0,\pi)$ *and* $c > 0$. *These affine geodesic escape upwards and to the right as* $t \to 0$ *and downwards and to the right as* $t \to \pi$. *The affine geodesic* σ *is incomplete at both ends and has total length* π.
   - *(e) The affine geodesics in* (c) *and* (d) *solve the equation* $(x^1)^2 - \frac{\lambda}{c^2} = (x^2+\beta)^2$; *they are hyperbolas; the affine geodesic is "vertical" if* $\lambda = +1$, *"horizontal" if* $\lambda = -1$, *and null if* $\lambda = 0$.
2. *The exponential map is an embedding of* T_PM *which omits a region of* $\mathbb{L}$.

We picture the geodesic structure below; the region omitted by the exponential map consists of the regions $x^2 \geq 1 + x^1$ and $x^2 \leq -1 - x^1$. The "vertical" affine geodesics (in red) point up (resp. below) and to the right; they lie above and below the two null half-lines with slope $\pm\frac{\pi}{4}$. The "horizontal" affine geodesics (in black) point to the right and the left and lie between the null half-lines with slope $\pm\frac{\pi}{4}$. The x^1 axis is horizontal and the x^2 axis is vertical.

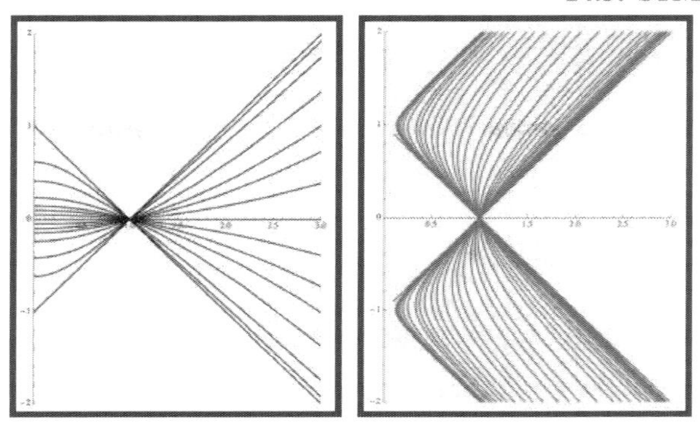

14.3.7 THE LORENTZIAN PSEUDO-SPHERE $\mathfrak{L}$. We continue our investigation of the Lorentzian hyperbolic plane using a different model following the discussion of D'Ascanio, Gilkey, and Pisani [22]. As before, we set $\langle x, y\rangle = x^1y^1 + x^2y^2 - x^3y^3$. We now consider the level set $\{\vec{x} : \langle\vec{x},\vec{x}\rangle = +1\}$; the induced metric is a Lorentzian inner product defining the Lorentzian pseudo-sphere $\mathfrak{L}$. The x^3 axis is vertical.

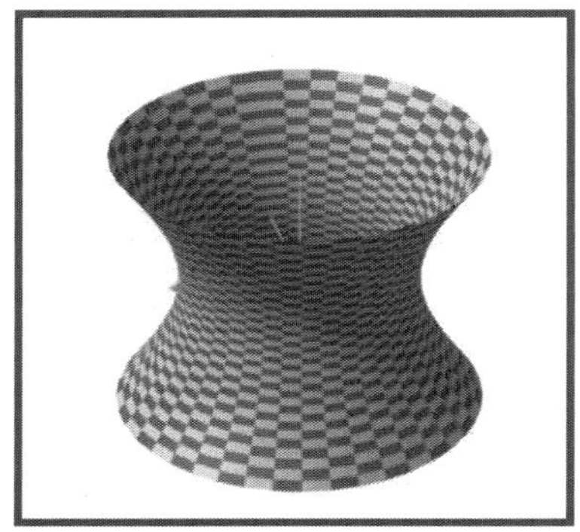

The pseudo-sphere $\mathfrak{L}$ is not simply connected since the hyperboloid of one sheet given by $(x^1)^2 + (x^2)^2 - (x^3)^2 = 1$ is diffeomorphic to $S^1 \times \mathbb{R}$. Set

$$T(x^1, x^2) := (\cosh(x^1)\cos(x^2), \cosh(x^1)\sin(x^2), \sinh(x^1))\,;$$

T exhibits $\mathbb{R}^2$ as the universal cover $\tilde{\mathfrak{L}}$ of $\mathfrak{L}$.

Lemma 14.33

1. *$\mathfrak{L}$ is geodesically complete and $\mathcal{Q}(\mathfrak{L}) = \operatorname{span}\{x^1, x^2, x^3\}$.*
2. *The exponential map in $\mathfrak{L}$ is not surjective.*
3. $\mathcal{Q}(\tilde{\mathfrak{L}}) = \operatorname{span}\{\cosh(x^1)\cos(x^2), \cosh(x^1)\sin(x^2), \sinh(x^1)\}$, $\Gamma_{22}{}^2 = \cosh(x^1)\sinh(x^1)$ *and* $\Gamma_{12}{}^2 = \frac{\sinh(x^1)}{\cosh(x^1)}$ *give the induced affine structure on* $\mathbb{R}^2$, $\rho_{\tilde{\mathfrak{L}}} = \operatorname{diag}(-1, \cosh^2(x^1))$, *and* $\nabla\rho_{\tilde{\mathfrak{L}}} = 0$.
4. *Let* $\Phi(x^1, x^2) := (2x^1)^{-1}(1 + (x^1)^2 - (x^2)^2, \quad 2x^2, \quad -1 + (x^1)^2 - (x^2)^2)$. *Then* Φ *is a*

proper affine embedding of $\mathbb{L}$ *in* $\mathfrak{L}$.

Proof. We have $\langle\cdot,\cdot\rangle$ is defined by the bilinear form $\varepsilon_{\mathfrak{L}}$ of Equation (14.3.a). We may therefore apply Lemma 14.29 to see $\mathfrak{L}$ is geodesically complete and that $O(2,1)$ acts transitively on

$$\mathfrak{L} \cup -\mathfrak{L}\,.$$

To see that the exponential map is not surjective, we must examine the geodesic structure in a bit more detail. Since $\mathfrak{L}$ is homogeneous, we may assume that $e_1 = (1,0,0)$ in proving the remaining assertions. We observe that $T_{e_1}\mathfrak{L} = e_1^\perp = \operatorname{span}\{e_2, e_3\}$. Let $\xi = ae_2 + be_3$.

Case 1. Assume $a^2 - b^2 > 0$ so ξ is spacelike. We rescale ξ to ensure that $a^2 - b^2 = 1$. Let $\sigma(\theta) = \cos(\theta)e_1 + \sin(\theta)\xi : \mathbb{R} \to \mathfrak{L}$. Since $\ddot\sigma = -\sigma$, $\ddot\sigma \perp T_\sigma\mathfrak{L}$. Consequently, $\ddot\sigma \perp \mathfrak{L}$. This implies σ is an affine geodesic which is defined for all time.

Case 2. Assume $a^2 - b^2 = 0$ so ξ is null. We can let $\sigma(t) = e_1 + t\xi$. Since $\ddot\sigma = 0$, this is an affine geodesic which extends for all time.

Case 3. Assume $a^2 - b^2 < 0$ so ξ is timelike.

We can rescale ξ to ensure that $b^2 - a^2 = 1$. Let $\sigma(t) = \cosh(t)e_1 + \sinh(t)\xi$. Again, $\ddot\sigma \perp \mathfrak{L}$ so this affine geodesic is defined for all time.

We note that affine geodesics can never reach $-e_1 + \xi$ for $0 \neq \xi$ null, and hence the exponential map is not surjective. We prove the third assertion by computing:

$$\begin{aligned}
\partial_{x^1} T &= (\cos(x^2)\sinh(x^1), \sin(x^2)\sinh(x^1), \cosh(x^1)),\\
\partial_{x^2} T &= (-\sin(x^2)\cosh(x^1), \cos(x^2)\cosh(x^1), 0),\\
g_{11} &= \langle \partial_{x^1}T, \partial_{x^1}T\rangle = -1, \qquad g_{12} = \langle \partial_{x^1}T, \partial_{x^2}T\rangle = 0,\\
g_{22} &= \langle \partial_{x^2}T, \partial_{x^2}T\rangle = \cosh^2(u)\,.
\end{aligned}$$

We use the Koszul formula (see Theorem 3.7 of Book I) to see $\Gamma_{122} = \cosh(x^1)\sinh(x^1)$ and $\Gamma_{221} = -\cosh(x^1)\sinh(x^2)$. We raise indices to determine $\Gamma_{12}{}^2$ and $\Gamma_{22}{}^1$. We perform a direct computation to find $\rho_{\tilde{\mathfrak{L}}}$ and $\nabla\rho_{\tilde{\mathfrak{L}}}$; we use Lemma 14.29 to see $\mathcal{Q}(\mathfrak{L}) = \operatorname{span}\{x^1, x^2, x^3\}$ and then pull-back to compute $\mathcal{Q}(\tilde{\mathfrak{L}})$.

We compute $\Phi_1^2 + \Phi_2^2 - \Phi_3^2 = 1$. Consequently, Φ takes values in $\mathfrak{L}$. We can recover x^1 from $\Phi_1 - \Phi_3$ and once x^1 is known, we can recover x^2 from Φ_2. Thus Φ is a smooth embedding; since $\mathfrak{L}$ is not simply connected and $\mathbb{L}$ is simply connected, the embedding is proper. By Lemma 14.4,

$$\mathcal{Q}(\mathbb{L}) = \operatorname{span}\{\frac{1}{x^1}, \frac{x^2}{x^1}, \frac{-(x^1)^2 + (x^2)^2}{x^1}\}\,.$$

By Assertion 1, the coordinate functions $\{x^1, x^2, x^3\}$ span $\mathcal{Q}(\mathfrak{L})$. Let $\Phi = (\Phi_1, \Phi_2, \Phi_3)$. We observe that $\mathcal{Q}(\mathbb{L}) = \operatorname{span}\{\Phi_1, \Phi_2, \Phi_3\}$. Since $\Phi^*(\mathcal{Q}(\mathfrak{L})) = \mathcal{Q}(\mathbb{L})$, Theorem 12.34 yields that Φ is an affine map. $\square$

We have the following picture of the geodesic structure in $\tilde{\mathfrak{L}}$; the corresponding geodesic structure in $\mathfrak{L}$ may be obtained by identifying (x^1, x^2) with $(x^1, x^2 + 2n\pi)$ – i.e., wrapping it up to form a cylinder. The x^1 axis is horizontal.

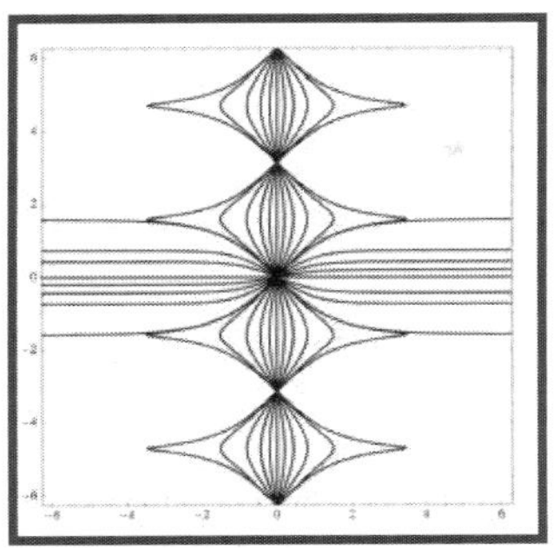

Let Φ be the map of Lemma 14.33. We note as before that since Φ is an affine map, Φ preserves the Ricci tensor, and hence the metric so Φ is an isometry. One can also compute directly $d\Phi_1^2 + d\Phi_2^2 - d\Phi_3^2 = (x^1)^{-2}(-(dx^1)^2 + (dx^2)^2)$.

CHAPTER 15

Applications of Affine Surface Theory

In Section 15.1, we present some basic material concerning the Riemannian extension and other notions which we shall need. In Section 15.2, we discuss a family of VSI (vanishing scalar invariants) manifolds. In Section 15.3, we present material of Calviño-Louzao et al. [14] concerning Bach flat manifolds.

15.1 PRELIMINARY MATTERS

In Section 15.1.1, we define the Bach, Cotton, and Weyl tensors and we introduce Walker coordinates. In Section 15.1.2, we discuss nilpotent endomorphisms of the tangent bundle of a surface. In Section 15.1.3, we discuss various modifications of the Riemannian extension which provide useful metrics on the cotangent bundle. In Section 15.1.4, we construct Bach flat structures on the cotangent bundle of an affine surface. We introduce quasi-Einstein geometry in Section 15.1.5.

15.1.1 BASIC DEFINITIONS. Let $\mathcal{N} = (N, g)$ be a pseudo-Riemannian manifold of dimension $n \geq 4$. Let ∇^g be the associated Levi-Civita connection, and let $R(\cdot, \cdot)$ be the curvature operator of ∇^g. Let

$$R_{ijk\ell} := g(R(\partial_{x^i}, \partial_{x^j})\partial_{x^k}, \partial_{x^\ell})$$

be the components of the curvature tensor. Let $\rho_{ij} = R_{kij}{}^k$ be the components of the Ricci tensor, and let $\tau := g^{ij}\rho_{ij}$ be the scalar curvature (see Section 3.5.2 of Book I). Let Δ be the scalar Laplacian. The *Weyl tensor* is conformally invariant. It is defined by setting

$$W_{ijk\ell} := R_{ijk\ell} + \frac{\rho_{i\ell}g_{jk} - \rho_{ik}g_{j\ell} + R_{jk}g_{i\ell} - R_{j\ell}g_{ik}}{n-2} + \tau\frac{g_{ik}g_{j\ell} - g_{i\ell}g_{jk}}{(n-1)(n-2)}\,.$$

We say that $\mathcal{N}$ is *conformally flat* if $W = 0$ or, equivalently, if there exists a coordinate atlas so $ds^2 = e^{\phi}((dx^1)^2 + \cdots + (dx^n)^2)$. Suppose that $n = 4$ and that N is oriented. We may decompose W under the eigenvalues of the Hodge $\star$ operator (see Section 5.2 of Book II) into components $W^{\pm}$. We say that $\mathcal{N}$ is self-dual (resp. anti-self-dual) if W^+ (resp. W^-) vanishes. One says that $\mathcal{N}$ is *half conformally flat* if either $W^+ = 0$ or $W^- = 0$ but $\mathcal{N}$ is not conformally flat; $\mathcal{N}$ is conformally flat if and only if $W = 0$, i.e., $W^+ = 0$ and $W^- = 0$.

The *Cotton tensor* is conformally invariant in dimension 3 and vanishes if $\mathcal{M}$ is conformally flat; it is defined in arbitrary dimension by setting

$$\mathfrak{C}(X,Y,Z) := (\nabla_X\rho)(Y,Z) - (\nabla_Y\rho)(X,Z) - \frac{d\tau(X)g(Y,Z) - d\tau(Y)g(X,Z)}{2(n-1)}.$$

The *Bach tensor* of $\mathcal{N}$ is defined by setting:

$$\mathfrak{B}_{ij} = \nabla^k\nabla^\ell W_{kij\ell} + \tfrac{1}{2}\rho^{k\ell}W_{kij\ell}.$$

It is a conformally invariant tensor. We say that $\mathcal{N}$ is *Bach flat* if $\mathfrak{B} = 0$. We will construct various examples of Bach flat manifolds in Theorem 15.4 which are not conformally flat, i.e., which have non-vanishing Weyl tensors.

We say that a subbundle $\mathcal{D}$ of the tangent bundle of a pseudo-Riemannian manifold $\mathcal{N} = (N, g)$ is *totally isotropic* if the restriction of the metric g to $\mathcal{D}$ is trivial. If $\dim(N) = 2\ell$ is even, or if $\dim(N) = 2\ell + 1$ is odd and if $\mathcal{D}$ is totally isotropic, then $\dim(\mathcal{D}) \le \ell$; we say that $\dim(\mathcal{D})$ is *maximal* if equality holds. The bundle is said to be *parallel* if any section ξ to $\mathcal{D}$ satisfies $\nabla_X\xi$ is again a section to $\mathcal{D}$ for any tangent vector field X. We say that $\mathcal{N} = (N, g)$ is a *Walker manifold* if there is a non-trivial totally isotropic parallel subbundle $\mathcal{D}$ of the tangent bundle of maximal dimension; $\mathcal{N}$ is said to be *strict* if $\mathcal{D}$ is generated by parallel vector fields. If $n = 4$ and if $\mathcal{N}$ is a Walker manifold, then $\mathcal{N}$ has neutral signature. The work of Walker [61](see also the discussion of Section 5.3 of Brozos-Vázquez et al. [8]) when restricted to $n = 4$ yields the following result:

Theorem 15.1 *A pseudo-Riemannian manifold $\mathcal{N}$ of dimension 4 is said to be a Walker manifold if and only if there is a coordinate atlas with local coordinates (x^1, x^2, y_1, y_2) and if there exist smooth functions $a(\cdot)$, $b(\cdot)$, and $c(\cdot)$ so that*

$$\begin{aligned} g \;=\;& dx^1\otimes dy_1 + dy_1\otimes dx^1 + dx^2\otimes dy_2 + dy_2\otimes dx^2\\ &+a(x^1,x^2,y_1,y_2)dx^1\otimes dx^1 + b(x^1,x^2,y_1,y_2)dx^2\otimes dx^2\\ &+c(x^1,x^2,y_1,y_2)(dx^1\otimes dx^2 + dx^2\otimes dx^1). \end{aligned}$$

In this setting, the distinguished distribution takes the form $\mathcal{D} = \operatorname{span}\{\partial_{y_1}, \partial_{y_2}\}$. Such a manifold N has a distinguished orientation given locally by $dx^1\wedge dx^2\wedge dy_1\wedge dy_2$; this ensures that $\mathcal{D}$ is self-dual. Thus self-duality and anti-self-duality are not equivalent concepts in this context. We refer to Brozos-Vázquez et al. [8] for further details.

15.1.2 NILPOTENT ENDOMORPHISMS. Let T be an endomorphism of the tangent bundle of a manifold M. Let $T\partial_{x^j} = T^i{}_j\partial_{x^i}$ define the components of T. We then have that $T = T^i{}_j\partial_{x^i}\otimes dx^j$; consequently, T is said to be of *Type* $(1,1)$. We say that T is *nilpotent* if $T^2 = 0$; nilpotent endomorphism will play a central role in our discussion of Bach flat manifolds in Section 15.3. The following is a useful technical observation.

Lemma 15.2 *Let $0 \neq T$ be a nilpotent tensor of Type (1,1) on a surface M. We can choose local coordinates near any point of M so that $T = \partial_{x^1} \otimes dx^2$, i.e., so $T^1{}_2$ is the only non-zero component of T.*

Proof. Let $0 \neq T$ be nilpotent. Fix a point P of M. We work locally in a small neighborhood of P. Choose a non-zero vector field W_2 defined near P so that $W_1 := TW_2 \neq 0$. Choose local coordinates (y^1, y^2) on M centered at P so that $W_1 = \partial_{y^1}$. Since T is nilpotent, $T\partial_{y^1} = T(TW_2) = 0$. Since T is nilpotent, range$\{T\}$ = ker$\{T\}$. Since ∂_{y^1} spans range$\{T\}$, we have that $T\partial_{y^2} = f\partial_{y^1}$ for some non-zero function f. Set $\tilde{W}_1 = f\partial_{y^1}$ and $\tilde{W}_2 = F\partial_{y^1} + \partial_{y^2}$ where F remains to be determined. We compute

$$[\tilde{W}_1, \tilde{W}_2] = [f\partial_{y^1}, F\partial_{y^1} + \partial_{y^2}] = \{f\partial_{y^1}F - F\partial_{y^1}f - \partial_{y^2}f\}\partial_{y^1} .$$

The equation $[\tilde{W}_1, \tilde{W}_2] = 0$ is equivalent to the ODE $\partial_{y^1}F = f^{-1}\{F\partial_{y^1}f + \partial_{y^2}f\}$. We solve this ODE with initial condition $F(0, y^2) = 0$ to ensure that $[\tilde{W}_1, \tilde{W}_2] = 0$. We then use the *Frobenius Theorem* to choose local coordinates $\{x^1, x^2\}$ so $\tilde{W}_1 = \partial_{x^1}$ and $\tilde{W}_2 = \partial_{x^2}$. We then have $T\partial_{x^1} = \partial_{x^2}$ and $T\partial_{x^2} = 0$ so $T = \partial_{x^1} \otimes dx^2$ as desired. □

Ferdinand Georg Frobenius (1849–1917)

15.1.3 METRICS ON THE COTANGENT BUNDLE. The subject arises first in the classic papers of Patterson and Walker [56] and Walker [61]. We also refer to related work of Afifi [1] and of Derdzinski and Roter [26]. Let $\mathcal{M} = (M, \nabla)$ be an affine manifold. Definition 11.39 of Book III gives various canonically defined metrics on the cotangent bundle of M. We recall these metrics for the convenience of the reader. Choose local coordinates (x^1, x^2) on M. We introduce *dual coordinates* (y_1, y_2) on the cotangent bundle T^*M by expressing any 1-form locally in the form $\omega = y_1 dx^1 + y_2 dx^2$. The classical *Riemannian extension* is defined by setting:

$$g_\nabla = dx^i \otimes dy_i + dy_i \otimes dx^i - 2y_k \Gamma_{ij}{}^k(\vec{x}) dx^i \otimes dx^j . \qquad (15.1.a)$$

If $\Phi = \Phi_{ij} dx^i \otimes dx^j$ is a smooth symmetric 2-tensor on M, then we can use Φ to deform the metric g_∇ of Equation (15.1.a) by setting:

$$g_{\nabla,\Phi} = g_\nabla + \Phi_{ij}(\vec{x}) dx^i \otimes dx^j . \qquad (15.1.b)$$

We omit the proof of the following result as it is a direct computation.

Lemma 15.3 *Let $g = g_{\nabla,\Phi}$ and let π be the canonical projection from the cotangent bundle to M. Then $\mathcal{H}_g(\pi^* f) = \pi^*\mathcal{H}(f)$, $\|d\pi^* f\|_g^2 = 0$, and $\rho_g = 2\pi^*\rho_{s,\nabla}$.*

Finally, suppose that $X = a^i \partial_{x^i}$ is a vector field on M and that $S = S^j{}_i$ and $T = T^j{}_i$ are tensors of Type (1,1). We modify $g_{\nabla,\Phi}$ to define

$$g_{\nabla,\Phi,T,S,X} = g_{\nabla,\Phi} + y_r y_s T^r{}_i(\vec{x}) S^s{}_j(\vec{x}) dx^i \otimes dx^j + a^i y_i y_j y_k dx^j \otimes dx^k . \qquad (15.1.c)$$

By Lemma 11.40 of Book III, these metrics are invariantly defined and independent of the particular local coordinate system (x^1, x^2) chosen; it is immediate from Theorem 15.1 that these all are Walker metrics. If one or more of these parameters is zero, we shall drop it from the notation.

15.1.4 BACH FLAT GEOMETRY. We take $X = 0$ and $S = T$ in Equation (15.1.c) to define the metric $g_{\nabla,\Phi,T}$. We use the metrics $g_{\nabla,\Phi,T}$ to provide examples of Bach flat neutral signature manifolds; we will consider these geometries in more detail in Section 15.3 and for the moment content ourselves with constructing some non-trivial examples.

Theorem 15.4 *Let $\mathcal{M} = (M, \nabla)$ be an affine surface, let T be a tensor of Type $(1,1)$, and let Φ be a symmetric 2-tensor. Let $\mathcal{N} := (T^*M, g_{\nabla,\Phi,T})$.*

1. *Suppose that $T(P) = \lambda(P)\,\mathrm{Id}$ is a scalar multiple of the identity for every point P of M. Then $\mathcal{N}$ is half conformally flat, and hence Bach flat.*
2. *Suppose that $T(P)$ is not a scalar multiple of the identity for any point of M. If $\mathcal{N}$ is Bach flat, then T is nilpotent.*
3. *Suppose that T is nilpotent and non-vanishing on M. Fix $P \in M$ and use Lemma 15.2 to choose local coordinates on an open neighborhood $\mathcal{O}$ of P so that $T = \partial_{x^1} \otimes dx^2$. The following assertions are equivalent on $\mathcal{O}$:*
   (a) $\mathcal{N}$ is Bach flat.
   (b) $\Gamma_{11}{}^2 = 0$ and $(\Gamma_{11}{}^1)^2 - \Gamma_{11}{}^1\Gamma_{12}{}^2 + \partial_{x^1}(\Gamma_{11}{}^1 - \Gamma_{12}{}^2) = 0$.

We note that the auxiliary tensor Φ plays no role in the analysis. We can express the conditions of Assertion 3-b in the form

$$\Gamma_{11}{}^2 = 0, \qquad \Gamma_{11}{}^1 = -\phi^{(1,0)}, \qquad \Gamma_{12}{}^2 = \Gamma_{11}{}^1 + c \cdot e^{\phi}$$

for smooth functions $c = c(x^2)$ and $\phi = \phi(x^1, x^2)$. Since we are interested in Bach flat manifolds which are not half conformally flat, we will concentrate on the case in which T is nilpotent in Section 15.3.

Proof. A direct computation shows that if $T = f\,\mathrm{Id}$ for $f \in C^\infty(M)$, then $\mathcal{N}$ is half conformally flat and that $\mathfrak{B} = 0$; this establishes Assertion 1 of Theorem 15.4.

We now turn to the proof of Assertion 2. Let $\Theta_{ijk\ell}$ be the coefficient of $y_i y_j$ in $\mathfrak{B}_{k\ell}$. A straightforward computation shows that $\Theta_{ijk\ell}$ is a polynomial which is homogeneous of degree 6 in the $T^u{}_v$ variables for $k, \ell \in \{1,2\}$ and zero otherwise; the Christoffel symbols and their derivatives, the auxiliary endomorphism Φ and its derivatives, and the derivatives of T do not appear in these terms. Consequently, $\Theta = \{\Theta_{ijk\ell}\}$ is tensorial. Assume that $\mathcal{N}$ is Bach flat. This implies $\Theta(T) = 0$. We suppose $T(P)$ is not a scalar multiple of the identity. Let $\{\lambda_1, \lambda_2\}$ be the (possibly complex) eigenvalues of $T(P)$. We can make a complex linear change of coordinates in the $\{x^1, x^2\}$ variables to put T in upper triangular form; this induces a corresponding dual complex linear change of coordinates in the $\{y_1, y_2\}$ variables. This is, of course, just Jordan normal form. Thus we may assume that:

$$T(P) := \begin{pmatrix} \lambda_1 & \varepsilon \\ 0 & \lambda_2 \end{pmatrix}.$$

Since $T(P)$ is not diagonal, we have either that $\lambda_1 \neq \lambda_2$ or $\lambda_1 = \lambda_2$ but $\varepsilon \neq 0$. We distinguish cases.

Case 1. Suppose $\lambda_1 \neq \lambda_2$. We compute that

$$\begin{aligned}\Theta_{1111}(T(P)) &= \tfrac{1}{6}\lambda_1^2(\lambda_1 - \lambda_2)^2(\lambda_1^2 + \lambda_1\lambda_2 - 5\lambda_2^2),\\ \Theta_{2222}(T(P)) &= \tfrac{1}{6}\lambda_2^2(\lambda_1 - \lambda_2)^2(-5\lambda_1^2 + \lambda_1\lambda_2 + \lambda_2^2).\end{aligned}$$

Note that the parameter ε does not appear; these two terms are not sensitive to the precise Jordan normal form but only to the eigenvalues. Since $\mathfrak{B} = 0$ and since $\lambda_1 - \lambda_2 \neq 0$, we obtain

$$\lambda_1^2(\lambda_1^2 + \lambda_1\lambda_2 - 5\lambda_2^2) = 0 \quad \text{and} \quad \lambda_2^2(-5\lambda_1^2 + \lambda_1\lambda_2 + \lambda_2^2) = 0. \tag{15.1.d}$$

If $\lambda_2 = 0$, then $\lambda_1 \neq 0$ and the first identity of Equation (15.1.d) fails. Similarly, if $\lambda_1 = 0$, then $\lambda_2 \neq 0$ and the second identity of Equation (15.1.d) fails. We may conclude therefore that $\lambda_1 \neq 0$ and $\lambda_2 \neq 0$. We now obtain

$$\lambda_1^2 + \lambda_1\lambda_2 - 5\lambda_2^2 = 0 \quad \text{and} \quad -5\lambda_1^2 + \lambda_1\lambda_2 + \lambda_2^2 = 0.$$

Subtracting these two identities yields $6\lambda_1^2 - 6\lambda_2^2 = 0$. Since $\lambda_1 \neq \lambda_2$, we have $\lambda_1 = -\lambda_2$, so $-5\lambda_1^2 = 0$ and again $\lambda_1 = \lambda_2 = 0$ which is false.

Case 2. Suppose that $\lambda_1 = \lambda_2$ but $\varepsilon \neq 0$. We obtain $\Theta_{1122}(T) = -3\varepsilon^2\lambda_1^4$; this term is sensitive to the Jordan normal form. Since $\varepsilon \neq 0$, $\lambda_1 = 0$ and $T(P)$ is nilpotent.

We now turn to the proof of Assertion 3. Suppose $\mathfrak{B} = 0$. Examining $\mathfrak{B}_{11}$ shows that ${\Gamma_{11}}^2 = 0$. Examining $\mathfrak{B}_{22}$ yields the remaining relation of Assertion 3-b. A direct computation shows that if the relations of Assertion 3-b are satisfied, then the Riemannian extension is Bach flat. □

A special case of Theorem 15.4 occurs if the tensor field T is parallel. Indeed, a nilpotent $(1,1)$-tensor field $T = \partial_{x^1} \otimes dx^2$ is parallel if and only if

$$\Gamma_{11}{}^1 = 0, \quad \Gamma_{11}{}^2 = 0, \quad \Gamma_{12}{}^1 = \Gamma_{22}{}^2, \quad \Gamma_{12}{}^2 = 0.$$

We apply Theorem 15.4 to derive the following result.

Corollary 15.5 *Let $\mathcal{M} = (M, \nabla)$ be an affine surface, let T be a parallel tensor field of Type $(1,1)$, and let Φ be a symmetric $(0,2)$-tensor field. Then $\mathcal{N} := (T^*M, g_{\nabla,\Phi,T})$ is Bach flat if and only if either $T = \mathrm{Id}$ or T is nilpotent.*

We have assumed in Theorem 15.4 either that T is a scalar multiple of the identity or that T is non-zero and not nilpotent. There are, however, examples where one can pass from one setting to the other. Let $M = \mathbb{R}^2$, let $\alpha(x^2)$ be a smooth real-valued function which vanishes to infinite-order at $x^2 = 0$ and which is positive for $x^2 \neq 0$. Assume that

$$\Gamma_{11}{}^2 = 0 \text{ and } (\Gamma_{11}{}^1)^2 - \Gamma_{11}{}^1\Gamma_{12}{}^2 + \partial_{x^1}(\Gamma_{11}{}^1 - \Gamma_{12}{}^2) = 0\,.$$

Let

$$T(x^1, x^2) = \left\{ \begin{array}{ll} \begin{pmatrix} \alpha(x^2) & 0 \\ 0 & \alpha(x^2) \end{pmatrix} & \text{if} \quad x^2 \leq 0 \\ \begin{pmatrix} 0 & \alpha(x^2) \\ 0 & 0 \end{pmatrix} & \text{if} \quad x^2 \geq 0. \end{array} \right\}.$$

One may then compute that $\mathfrak{B} = 0$ so this yields a Bach flat manifold where the Jordan normal form of T changes at $x^2 = 0$. Furthermore, if we only assume that α is C^k for $k \geq 2$, we still obtain a solution; thus there is no hypo-ellipticity present when considering the solutions to the equations $\mathfrak{B} = 0$ and Remark 12.15 fails in this context.

15.1.5 QUASI-EINSTEIN MANIFOLDS. Let $\mathcal{N} = (N, g)$ be a pseudo-Riemannian manifold. If the associated Ricci tensor ρ_g is a constant multiple of the metric, then $\mathcal{N}$ is said to be an *Einstein manifold*. If there is a smooth function h so that $e^h g$ is Einstein, then $\mathcal{N}$ is said to be a *conformally Einstein manifold*. More generally, let μ_N be a real constant and let F be a smooth function on N. Let $\mathcal{H}_g$ be the Hessian defined by the Levi-Civita connection. We say that $\mathcal{N} := (N, g, F, \mu_N)$ is a *quasi-Einstein manifold* if the following relation is satisfied:

$$\mathcal{H}_g F + \rho_g - \mu_N dF \otimes dF = \lambda\, g \quad \text{for some} \quad \lambda \in \mathbb{R}\,. \tag{15.1.e}$$

We say that $\mathcal{N}$ is *isotropic* if $\|dF\| = 0$. If λ is allowed to be a smooth function, then $\mathcal{N}$ is said to be a *generalized quasi-Einstein manifold*. We make the following observations:

Remark 15.6

1. If F is constant, then Equation (15.1.e) is the Einstein equation; quasi-Einstein manifolds are a generalization of Einstein manifolds.
2. If $n \geq 3$ and if $\mu_N = -\frac{1}{n-2}$ is the *critical eigenvalue*, then $\mathcal{N}$ is a generalized quasi-Einstein manifold if and only if $(N, e^{-\frac{2}{n-2}F} g)$ is Einstein. Thus any conformally Einstein manifold is quasi-Einstein with $\mu_N = -\frac{1}{n-2}$. For this reason we will often suppose $\mu_N \neq -\frac{1}{n-2}$. We refer to Brinkmann [3] or Gover and Nurowski [36] for further details concerning conformally Einstein manifolds.
3. For $\mu = 1$, the change of variable $h = e^{-F}$ yields $\mathcal{H}_g h - h\rho = -h\lambda g$. Let $\lambda = -\frac{\Delta h}{h}$. We then have $\mathcal{H}_g f - h\rho = (\Delta h)g$. This is the defining equation of the so-called *static manifolds* that arise in the study of static space-times as discussed in Kobayashi [38] and in Kobayashi and Obata [40].
4. If $\mu = 0$ and if λ is constant, one obtains the gradient Ricci soliton equation. This yields self-similar solutions of the Ricci flow. See, for example, the discussion in Cao and Chen [15] or in Munteanu and Sesum [45].

Let $\mathcal{M}$ be an affine surface. There is a useful change of variables.

Lemma 15.7 *Let $\mathcal{M} = (M, \nabla)$ be an affine manifold. Set $f = e^{-\mu h}$.*

1. $\mathcal{H}f - \mu f \rho_s = -\mu e^{-\mu h(P)}\{\mathcal{H}h + \rho_s - \mu dh \otimes dh\}$.
2. *If $\mu \neq 0$, then $f \in E(\mu, \mathcal{M})$ if and only if $\mathcal{H}h + \rho_s = \mu dh \otimes dh$.*

Proof. Fix a point P of M. Since ∇ is torsion-free, we may apply Lemma 9.1 of Book III to choose local coordinates so that $\Gamma_{ij}{}^k(P) = 0$. We then have

$$\begin{aligned}
\{\mathcal{H}f\}_{ij}(P) &= \partial_{x^i}\partial_{x^j}\{e^{-\mu h}\}(P) = e^{-\mu h(P)}\left\{-\mu\partial_{x^i}\partial_{x^j}h + \mu^2\partial_{x^i}h\partial_{x^j}h\right\}(P) \\
&= e^{-\mu h(P)}\left\{-\mu\mathcal{H}h + \mu^2 dh \otimes dh\right\}_{ij}(P), \\
\{\mathcal{H}f - \mu f\rho_s\}_{ij}(P) &= -\mu e^{-\mu h(P)}\{\mathcal{H}h + \rho_s - \mu dh \otimes dh\}_{ij}(P)\,.
\end{aligned}$$

Assertion 1 follows since P was arbitrary. Since μ was non-zero, Assertion 2 follows from Assertion 1. □

We can now use the metric $g_{\nabla,\Phi}$ of Equation (15.1.b) to construct quasi-Einstein manifolds; it is this construction which motivated the terminology *quasi-Einstein equation* in affine geometry in the first instance.

Theorem 15.8 *Let $\mathcal{M}$ be an affine surface, let $\mu \neq 0$, let $f = e^{-\mu h} \in E(\mu, \mathcal{M})$, and let Φ be an arbitrary symmetric $(0,2)$-tensor field on M. Then $\mathcal{N} := (T^*M, g_{\nabla,\Phi}, 2\pi^*h, \frac{1}{2}\mu)$ is a self-dual isotropic quasi-Einstein Walker manifold of signature $(2,2)$ with $\lambda = 0$.*

Proof. Since $f \in E(\mu, \mathcal{M}), \mathcal{H}f - \mu f \rho_s = 0$ so $\mathcal{H}h + \rho_s - \mu dh \otimes dh = 0$ by Lemma 15.7. We apply Lemma 15.3 to see that

$$\mathcal{H}_g 2\pi^* h + \rho_g - \tfrac{1}{2}\mu\pi^*(d2h \otimes d2h) = \pi^*\{2\mathcal{H}h + 2\rho_s - 2\mu dh \otimes dh\} = 0\,.$$

We conclude that $(2\pi^* h, \frac{1}{2}\mu, \lambda = 0)$ satisfies Equation (15.1.e). □

We note the factor of $\frac{1}{2}$ in passing from the eigenvalue μ on the base to the eigenvalue $\mu_N = \frac{1}{2}\mu$ on the cotangent bundle. The critical eigenvalue is $\mu = -\frac{1}{m-1}$ for the affine quasi-Einstein equation on an affine manifold as noted in Section 12.3. The corresponding critical eigenvalue is $\mu_N = -\frac{1}{2}\frac{1}{m-1} = -\frac{1}{n-2}$ on the cotangent space; this eigenvalue corresponds to the case $\mathcal{N}$ is conformally Einstein, as noted in Remark 15.6.

Theorem 15.8 gave an algorithm for constructing quasi-Einstein metrics on the cotangent bundle given a solution to the affine quasi-Einstein equation on the base. The following result of Brozos-Vázquez et al. [11] provides a partial converse to this result. We shall omit the proof as it is somewhat long and technical.

Theorem 15.9 *Let $\mathcal{N} := (N, g, F, \frac{1}{2}\mu)$ be a self-dual isotropic quasi-Einstein manifold with λ constant, $\mu \neq -1$, and $\|\nabla F\|^2 = 0$. Fix a point P of N. There exists an affine surface $\mathcal{M}$, there exists $f = e^{-\mu h} \in E(\mu, \mathcal{M})$, and there exists a symmetric $(0,2)$-tensor field on M so that near P, $\mathcal{N}$ is locally isomorphic to $(T^*M, g_{\nabla,\Phi}, 2\pi^* h, \frac{1}{2}\mu)$.*

Based on Corollary 15.5 and proceeding in a similar way as in Theorem 15.8, one may construct Bach flat quasi-Einstein metrics. We use the nilpotent tensor T to define the tensor $\hat{\Phi}(X,Y) = \Phi(TX, TY)$. If $T = \partial_{x^1} \otimes dx^2$, then $\hat{\Phi} = \Phi_{11} dx^1 \otimes dx^1$, and one has:

Theorem 15.10 *Let $\mathcal{M}$ be an affine surface and let T be a parallel nilpotent $(1,1)$-tensor field on $\mathcal{M}$. Let $h \in C^\infty(M)$ be a smooth function. Then $\mathcal{N} := (T^*M, g_{\nabla,\Phi}, 2\pi^* h, \frac{1}{2}\mu)$ is an isotropic Bach flat quasi-Einstein manifold if and only if $dh(\ker\{T\}) = 0$ and the deformation tensor field Φ satisfies $\hat{\Phi} = -\{\mathcal{H}h + \rho_s - \mu dh \otimes dh\}$.*

Theorem 15.10 can be further specialized so that the Bach flat structure is anti-self-dual. This shows that self-duality and anti-self-duality cannot be interchanged in Theorem 15.9. We refer to Brozos-Vázquez et al. [11] for the proof of the following result, which is an application of the distinguished coordinates considered in Section 12.2.

Corollary 15.11 *Let $\mathcal{M}$ be an affine surface with symmetric and recurrent Ricci tensor of rank 1. Let T be a parallel nilpotent $(1,1)$-tensor field on $\mathcal{M}$ and let $h \in C^\infty(M)$ be a smooth function with $dh(\ker\{T\}) = 0$ satisfying $\mathcal{H}h + \rho_s - \mu dh \otimes dh = 0$.*

1. *If the deformation tensor Φ vanishes, then $\mathcal{N} := (T^*M, g_{\nabla,\Phi}, 2\pi^* h, \frac{1}{2}\mu)$ is an isotropic Bach flat quasi-Einstein manifold.*

2. *If the recurrence one-form* ω *given by* $\nabla\rho = \omega \otimes \rho$ *satisfies* $\omega(\ker\{T\}) = 0$, *then we have that* $\mathcal{N} := (T^*M, g_{\nabla,\Phi}, 2\pi^*h, \frac{1}{2}\mu)$ *is anti-self-dual for* $\Phi = 0$.

15.2 SIGNATURE (2, 2) VSI MANIFOLDS

Let $R_{i_1,\dots,i_4;j_1,\dots,j_k}$ be the components of the kth covariant derivative of the Levi-Civita connection of a pseudo-Riemannian manifold $\mathcal{N}$. Scalar invariants of the metric can be formed by using the metric tensors g^{ij} and g_{ij} to *fully contract* all indices. For example, the scalar curvature τ, the *norm of the Ricci tensor* $\|\rho\|^2$, and the *norm of the full curvature tensor* $\|R\|^2$ are scalar invariants which are given by:

$$\tau := g^{ij} R_{kij}{}^{k}, \qquad \|\rho\|^2 := g^{i_1 j_1} g^{i_2 j_2} R_{k i_1 i_2}{}^{k} R_{\ell j_1 j_2}{}^{\ell}, \quad \text{and}$$
$$\|R\|^2 := g^{i_1 j_1} g^{i_2 j_2} g^{i_3 j_3} g_{i_4 j_4} R_{i_1 i_2 i_3}{}^{i_4} R_{j_1 j_2 j_3}{}^{j_4}.$$

Such invariants are called *Weyl scalar invariants*; if all possible such invariants vanish, then $\mathcal{M}$ is said to be *VSI* (vanishing scalar invariants). We refer, for example, to Coley et al. [18] for further details. In the Riemannian setting, the vanishing of $\|R\|^2$ implies $\mathcal{N}$ is flat; this is not the case for pseudo-Riemannian manifolds as we shall see subsequently.

We take $X = 0$ (i.e., $a^i = 0$ for all i) and $S = T$ in Equation (15.1.c) to define

$$g_{\nabla,\Phi,T} := 2\,dx^i \circ dy_i + \left\{\frac{1}{2} y_r y_s (T^r{}_i T^s{}_j + T^r{}_j T^s{}_i) - 2 y_k \Gamma_{ij}{}^k + \Phi_{ij}\right\} dx^i \circ dx^j \qquad (15.2.a)$$

where $\xi \circ \eta := \frac{1}{2}(\xi \otimes \eta + \eta \otimes \xi)$ denotes the symmetric tensor product. In Section 15.2.1, we show that if $T = 0$ or if $0 \neq T$ and T is nilpotent, then $g_{\nabla,\Phi,T}$ is VSI. In Section 15.2.2, we construct scalar invariants of such manifolds which are not of Weyl type. In Section 15.2.3, we consider various examples defined by Type $\mathcal{A}$ or Type $\mathcal{B}$ models.

15.2.1 T IS NILPOTENT. If the underlying manifold is flat, then the Riemannian extension takes the form $g_{\nabla,\Phi} = 2dx^i \otimes dy_i + \Phi_{ij}(\vec{x})dx^i \otimes dx^j$. This is called a *plane wave manifold*; such manifolds are VSI by results of Gilkey and Nikčević [33]. We will show in Theorem 15.14 that suitable generalizations of this structure are VSI; these structures will play an important role in our discussion of Bach flat manifolds subsequently in Section 15.3. If T is nilpotent, we apply Lemma 15.2 to choose coordinates so the only non-zero component of T is $T^1{}_2 = 1$. To allow for the possibility that T could vanish identically, we shall suppose $T^1{}_2$ is constant in the following result. To simplify the notation, we let $x^3 = y_1$ and $x^4 = y_2$ henceforth.

Lemma 15.12 *Let* $g = g_{\nabla,\Phi,T}$ *where* $T = c\,\partial_{x^1} \otimes dx^2$ *where* c *is a real constant. Then the variables* $\{g_{ij}, g^{ij}, {}^g\Gamma_{ij}{}^k, R_{abcd;e_1\dots e_k}\}$ *depend polynomially on the variables* $\{\Gamma_{ij}{}^k, \Phi_{ij}, y_i\}$ *and the derivatives of the variables* $\{\Gamma_{ij}{}^k, \Phi_{ij}\}$ *with respect to* x^1 *and* x^2.

1. $g_{11} = a_{11}(\vec{x}, \vec{y})$, $g_{12} = a_{12}(\vec{x}, \vec{y})$, $g_{22} = a_{22}(\vec{x}, \vec{y})$, $g_{13} = g_{24} = 1$, $g_{14} = g_{23} = g_{33} = g_{34} = g_{44} = 0$.
2. $g^{11} = g^{12} = g^{22} = g^{14} = g^{23} = 0$, $g^{13} = g^{24} = 1$, $g^{33} = -a_{11}(\vec{x}, \vec{y})$, $g^{34} = -a_{12}(\vec{x}, \vec{y})$, $g^{44} = -a_{22}(\vec{x}, \vec{y})$.
3. *The only possibly non-zero Christoffel symbols of the Levi-Civita connection of g are* ${}^g\Gamma_{11}{}^1$, ${}^g\Gamma_{11}{}^2$, ${}^g\Gamma_{11}{}^3$, ${}^g\Gamma_{11}{}^4$, ${}^g\Gamma_{12}{}^1$, ${}^g\Gamma_{12}{}^2$, ${}^g\Gamma_{12}{}^3$, ${}^g\Gamma_{12}{}^4$, ${}^g\Gamma_{13}{}^3$, ${}^g\Gamma_{13}{}^4$, ${}^g\Gamma_{14}{}^3$, ${}^g\Gamma_{14}{}^4$, ${}^g\Gamma_{22}{}^1$, ${}^g\Gamma_{22}{}^2$, ${}^g\Gamma_{22}{}^3$, ${}^g\Gamma_{22}{}^4$, ${}^g\Gamma_{23}{}^3$, ${}^g\Gamma_{23}{}^4$, ${}^g\Gamma_{24}{}^3$, ${}^g\Gamma_{24}{}^4$.
4. *The only possible non-zero components of the curvature tensor are* R_{1212}, R_{1213}, R_{1214}, R_{1223}, R_{1224}, *and* R_{2323}.

Proof. The first assertion is immediate from Equation (15.2.a). If we write, schematically,

$$g = \begin{pmatrix} a & I \\ I & 0 \end{pmatrix} \quad \text{then} \quad g^{-1} = \begin{pmatrix} 0 & I \\ I & -a \end{pmatrix},$$

then Assertion 2 follows. We use the Koszul formula (see Theorem 3.7 of Book I) to see that ${}^g\Gamma_{abc} = 0$ unless at least two indices belong to $\{1, 2\}$. Thus

$$\begin{aligned} 0 &= {}^g\Gamma_{133} = {}^g\Gamma_{331} = {}^g\Gamma_{134} = {}^g\Gamma_{143} = {}^g\Gamma_{341} = {}^g\Gamma_{144} = {}^g\Gamma_{441}, \\ 0 &= {}^g\Gamma_{233} = {}^g\Gamma_{332} = {}^g\Gamma_{234} = {}^g\Gamma_{243} = {}^g\Gamma_{342} = {}^g\Gamma_{244} = {}^g\Gamma_{442}, \\ 0 &= {}^g\Gamma_{333} = {}^g\Gamma_{334} = {}^g\Gamma_{343} = {}^g\Gamma_{444} = {}^g\Gamma_{344} = {}^g\Gamma_{443}\,. \end{aligned}$$

We raise indices using Assertion 2 to see

$$\begin{aligned} 0 &= {}^g\Gamma_{13}{}^1 = {}^g\Gamma_{13}{}^2 = {}^g\Gamma_{14}{}^1 = {}^g\Gamma_{14}{}^2 = {}^g\Gamma_{24}{}^1 = {}^g\Gamma_{24}{}^2 = {}^g\Gamma_{23}{}^1, \\ 0 &= {}^g\Gamma_{23}{}^2 = {}^g\Gamma_{33}{}^1 = {}^g\Gamma_{33}{}^2 = {}^g\Gamma_{33}{}^3 = {}^g\Gamma_{33}{}^4 = {}^g\Gamma_{34}{}^1 = {}^g\Gamma_{34}{}^2, \\ 0 &= {}^g\Gamma_{34}{}^3 = {}^g\Gamma_{34}{}^4 = {}^g\Gamma_{44}{}^1 = {}^g\Gamma_{44}{}^2 = {}^g\Gamma_{44}{}^3 = {}^g\Gamma_{44}{}^4. \end{aligned}$$

The Levi-Civita connection in dimension 4 has 40 components. We have listed 20; the remaining 20 appear in Assertion 3. A direct computation shows

$$\begin{aligned} 0 &= R_{1234} = R_{1313} = R_{1314} = R_{1323} = R_{1324} = R_{1334} = R_{1414}, \\ 0 &= R_{1423} = R_{1424} = R_{1434} = R_{2324} = R_{2334} = R_{2424} = R_{2434} = R_{3434}\,. \end{aligned}$$

The curvature tensor has 21 components in dimension 4. We have listed 15; the remaining 6 appear in Assertion 4. □

Lemma 15.12 controls which curvature tensors are potentially non-zero. We introduce some additional notation to control the higher covariant derivatives of the curvature tensor. Let $\mathfrak{o}(\cdot)$ be the maximal-order of an expression in the dual variables $\{y_1 = x^3, y_2 = x^4\}$. Thus if $\mathfrak{o}(\cdot) = 0$, these variables do not occur, if $\mathfrak{o}(\cdot) = 1$, the expression is linear in the variables $\{x_3, x_4\}$, and so forth. If $\mathfrak{o}(R_{ijk\ell}) = 2$, then $R_{ijk\ell}$ is at most quadratic in $\{x_3, x_4\}$; if $\mathfrak{o}(R_{ijk\ell}) = 1$, then $R_{ijk\ell}$ is at most linear in $\{x_3, x_4\}$; and if $\mathfrak{o}(R_{ijk\ell}) = 0$, then $R_{ijk\ell}$ does not involve $\{x_3, x_4\}$. A direct computation shows

$$\begin{array}{llll}
\mathfrak{o}({}^g\Gamma_{11}{}^1) = 0, & \mathfrak{o}({}^g\Gamma_{11}{}^2) = 0, & \mathfrak{o}({}^g\Gamma_{11}{}^3) = 1, & \mathfrak{o}({}^g\Gamma_{11}{}^4) = 2, \\
\mathfrak{o}({}^g\Gamma_{12}{}^1) = 0, & \mathfrak{o}({}^g\Gamma_{12}{}^2) = 0, & \mathfrak{o}({}^g\Gamma_{12}{}^3) = 1, & \mathfrak{o}({}^g\Gamma_{12}{}^4) = 2, \\
\mathfrak{o}({}^g\Gamma_{13}{}^3) = 0, & \mathfrak{o}({}^g\Gamma_{13}{}^4) = 0, & \mathfrak{o}({}^g\Gamma_{14}{}^3) = 0, & \mathfrak{o}({}^g\Gamma_{14}{}^4) = 0, \\
\mathfrak{o}({}^g\Gamma_{22}{}^1) = 1, & \mathfrak{o}({}^g\Gamma_{22}{}^2) = 0, & \mathfrak{o}({}^g\Gamma_{22}{}^3) = 2, & \mathfrak{o}({}^g\Gamma_{22}{}^4) = 2, \\
\mathfrak{o}({}^g\Gamma_{23}{}^3) = 0, & \mathfrak{o}({}^g\Gamma_{23}{}^4) = 1, & \mathfrak{o}({}^g\Gamma_{24}{}^3) = 0, & \mathfrak{o}({}^g\Gamma_{24}{}^4) = 0, \\
\mathfrak{o}(R_{1212}) = 2, & \mathfrak{o}(R_{1213}) = 1, & \mathfrak{o}(R_{1214}) = 0, & \mathfrak{o}(R_{1223}) = 1, \\
\mathfrak{o}(R_{1224}) = 1, & \mathfrak{o}(R_{2323}) = 0\,. & &
\end{array}$$

We define the *defect* by setting

$$\begin{aligned}
\mathfrak{d}({}^g\Gamma_{ij}{}^k) &= -\sum_{n=1}^{2}\{\delta_{i,n} + \delta_{j,n} - \delta_{k,n}\} + \sum_{n=3}^{4}\{\delta_{i,n} + \delta_{j,n} - \delta_{k,n}\}, \\
\mathfrak{d}(R_{i_1 i_2 i_3 i_4; i_5 \dots i_\nu}) &:= \sum_{n=1}^{\nu}\{\delta_{i_n,3} + \delta_{i_n,4} - \delta_{i_n,1} - \delta_{i_n,2}\}\,.
\end{aligned}$$

In brief, we count, with multiplicity, each lower index "1" or "2" with the sign -1 and "3" or "4" with the sign $+1$ and, dually, reverse the sign for upper indices. Set $\mathfrak{r} = \mathfrak{o} + \mathfrak{d}$ and compute:

$$\begin{array}{llll}
\mathfrak{r}({}^g\Gamma_{11}{}^1) = -1, & \mathfrak{r}({}^g\Gamma_{11}{}^2) = -1, & \mathfrak{r}({}^g\Gamma_{11}{}^3) = -2, & \mathfrak{r}({}^g\Gamma_{11}{}^4) = -1, \\
\mathfrak{r}({}^g\Gamma_{12}{}^1) = -1, & \mathfrak{r}({}^g\Gamma_{12}{}^2) = -1, & \mathfrak{r}({}^g\Gamma_{12}{}^3) = -2, & \mathfrak{r}({}^g\Gamma_{12}{}^4) = -1, \\
\mathfrak{r}({}^g\Gamma_{13}{}^3) = -1, & \mathfrak{r}({}^g\Gamma_{13}{}^4) = -1, & \mathfrak{r}({}^g\Gamma_{14}{}^3) = -1, & \mathfrak{r}({}^g\Gamma_{14}{}^4) = -1, \\
\mathfrak{r}({}^g\Gamma_{22}{}^1) = 0, & \mathfrak{r}({}^g\Gamma_{22}{}^2) = -1, & \mathfrak{r}({}^g\Gamma_{22}{}^3) = -1, & \mathfrak{r}({}^g\Gamma_{22}{}^4) = -1, \\
\mathfrak{r}({}^g\Gamma_{23}{}^3) = -1, & \mathfrak{r}({}^g\Gamma_{23}{}^4) = 0, & \mathfrak{r}({}^g\Gamma_{24}{}^3) = -1, & \mathfrak{r}({}^g\Gamma_{24}{}^4) = -1, \\
\mathfrak{r}(R_{1212}) = -2, & \mathfrak{r}(R_{1213}) = -1, & \mathfrak{r}(R_{1214}) = -2, & \mathfrak{r}(R_{1223}) = -1, \\
\mathfrak{r}(R_{1224}) = -1, & \mathfrak{r}(R_{2323}) = 0\,. & &
\end{array} \tag{15.2.b}$$

We can now control the non-zero higher covariant derivatives.

Lemma 15.13 *Suppose that* $R_{i_1 i_2 i_3 i_4; i_5 \dots i_\nu} \neq 0$. *Then* $\mathfrak{r}(R_{i_1 i_2 i_3 i_4; i_5 \dots i_\nu}) \leq 0$. *Furthermore,* $\mathfrak{r}(R_{i_1 i_2 i_3 i_4; i_5 \dots i_\nu}) = 0$ *if and only if* $R_{i_1 i_2 i_3 i_4; i_5 \dots i_\nu} = \pm R_{2323}$.

Proof. Let $R_{ijk\ell} \neq 0$. By Equation (15.2.b), we have $\mathfrak{r}(R_{ijk\ell}) \leq 0$ with equality if and only if $R_{ijk\ell} = \pm R_{2323}$. This establishes the result if $\nu = 4$. Next we suppose $\nu = 5$ and examine ${}^g\nabla R$. By Lemma 15.12, $R_{i_1 i_2 i_3 i_4; n}$ is polynomial in its arguments; we suppose this polynomial does not vanish identically and that $\mathfrak{x}(R_{i_1i_2i_3i_4;n}) \geq 0$. We argue for a contradiction. Expand

$$\begin{aligned} R_{i_1i_2i_3i_4;n} &= \partial_n R_{i_1i_2i_3i_4} - \sum_a {}^g\Gamma_{ni_1}{}^a R_{ai_2i_3i_4} - \sum_a {}^g\Gamma_{ni_2}{}^a R_{i_1ai_3i_4} \\ &\quad - \sum_a {}^g\Gamma_{ni_3}{}^a R_{i_1i_2ai_4} - \sum_a {}^g\Gamma_{ni_4}{}^a R_{i_1i_2i_3a}\,. \end{aligned}$$

Since $R_{i_1i_2i_3i_4;n}$ does not vanish identically, at least one of the terms in the previous display is non-zero. We distinguish cases.

Case 1. Suppose that $\partial_n R_{i_1i_2i_3i_4} \neq 0$. If $n \in \{1,2\}$, then

$$\begin{aligned} &\mathfrak{d}(R_{i_1i_2i_3i_4;n}) = \mathfrak{d}(R_{i_1i_2i_3i_4}) - 1 < 0, \quad \mathfrak{o}(\partial_n R_{i_1i_2i_3i_4}) \leq \mathfrak{o}(R_{i_1i_2i_3i_4}), \\ &\mathfrak{x}(\partial_n R_{i_1i_2i_3i_4}) \leq \mathfrak{x}(R_{i_1i_2i_3i_4}) - 1 < 0\,. \end{aligned}$$

If $n \in \{3,4\}$, then necessarily $\mathfrak{o}(R_{i_1i_2i_3i_4}) > 0$ to ensure $R_{i_1i_2i_3i_4}$ in fact depends on (x_3, x_4). Thus $R_{i_1i_2i_3i_4} \neq R_{2323}$. We have

$$\begin{aligned} &\mathfrak{d}(R_{i_1i_2i_3i_4;n}) = \mathfrak{d}(R_{i_1i_2i_3i_4}) + 1, \quad \mathfrak{o}(\partial_n R_{i_1i_2i_3i_4}) \leq \mathfrak{o}(R_{i_1i_2i_3i_4}) - 1, \\ &\mathfrak{x}(R_{i_1i_2i_3i_4;n}) \leq \mathfrak{x}(R_{i_1i_2i_3i_4}) < 0\,. \end{aligned}$$

Thus in any event, $\mathfrak{x}(R_{i_1i_2i_3i_4;n}) < 0$, which is false.

Case 2. Suppose that ${}^g\Gamma_{ni_1}{}^a R_{ai_2i_3i_4} \geq 0$ for some a; the remaining four cases involving ${}^g\Gamma_{ni_2}{}^a R_{i_1ai_3i_4}$, ${}^g\Gamma_{ni_3}{}^a R_{i_1i_2ai_4}$, and ${}^g\Gamma_{ni_4}{}^a R_{i_1i_2i_3a}$ are similar. Since $0 \geq \mathfrak{x}({}^g\Gamma_{ni_1}{}^a)$ and since $0 \geq \mathfrak{x}(R_{ai_2i_3i_4})$, we have that

$$0 \geq \mathfrak{x}({}^g\Gamma_{ni_1}{}^a) + \mathfrak{x}(R_{ai_2i_3i_4}) = \mathfrak{x}({}^g\Gamma_{ni_1}{}^a R_{ai_2i_3i_4}) \geq 0\,.$$

Thus $\mathfrak{x}({}^g\Gamma_{ni_1}{}^a) = 0$ and $\mathfrak{x}(R_{ai_2i_3i_4}) = 0$. By Equation (15.2.b), $R_{ai_2i_3i_4} = \pm R_{2323}$ so $a = 2$ or $a = 3$. Since $\mathfrak{x}({}^g\Gamma_{ni_1}{}^a) = 0$, Equation (15.2.b) then shows ${}^g\Gamma_{ni_1}{}^a$ is either ${}^g\Gamma_{22}{}^1$ or ${}^g\Gamma_{23}{}^4$. This shows that $a = 1$ or $a = 4$. This provides the desired contradiction and establishes the desired result for the components of ${}^g\nabla R$. The argument for ${}^g\nabla^\ell R$ for $\ell \geq 2$ now proceeds by induction; we do not have the additional complexity involved in considering variables $R_{i_1i_2i_3i_4;i_5\dots i_\nu}$ where $\mathfrak{x}(R_{i_1i_2i_3i_4;i_5\dots i_\nu}) = 0$. □

We can extend the results of Gilkey and Nikčević [33] from the setting of generalized plane wave manifolds to the setting at hand as follows.

Theorem 15.14 *Let $\mathcal{N} = (T^*M, g_{\nabla,\Phi,T})$. Suppose either T vanishes identically or T never vanishes. The following assertions are equivalent:*

(1) T is nilpotent. *(2) $\mathcal{N}$ is VSI.* *(3) $\|R\|^2 = \|\rho\|^2 = 0$.* *(4) $\|\rho\|^2 = \tau = 0$.*

Proof. Suppose T is nilpotent. Let $\mathcal{W}$ be a Weyl scalar invariant formed from the curvature tensor and its covariant derivatives. By Lemma 15.12, we can contract an index "1" against an index "3" and an index "2" against an index "4". We can also contract indices $\{3,4\}$ against $\{3,4\}$. We may not, however, contract indices $\{1,2\}$ against indices $\{1,2\}$. Consequently, if $A = R_{i_1 i_2 i_3 i_4; i_5 \dots i_\nu} \dots$ is a monomial, then $\deg_1(A) \le \deg_3(A)$ and $\deg_2(A) \le \deg_4(A)$. The inequality can, of course, be strict as we can also contract an index 3 or 4 against an index 3 or 4. This implies that $\mathfrak{d}(A) \ge 0$. Since $\mathfrak{o}(A) \ge 0$, this implies $\mathfrak{x}(A) \ge 0$. By Lemma 15.13, $\mathfrak{x}(A) \le 0$. Thus we have $\mathfrak{x}(A) = 0$. This implies A is a power of R_{2323}. We cannot contract an index "2" against an index "3". Consequently, $\mathcal{W} = 0$. Thus Assertion 1 implies Assertion 2. It is immediate that Assertion 2 implies Assertion 3 and Assertion 4 since if $\mathcal{N}$ is VSI, we have $\|R\|^2 = \|\rho\|^2 = \tau = 0$.

Fix P and $\{\lambda_1, \lambda_2\}$ be the (possibly complex) eigenvalues of $T(P)$. As in the proof of Theorem 15.4, we can make a complex linear change of coordinates in the $\{x^1, x^2\}$ variables to put T in Jordan normal form and express

$$T(P) := \begin{pmatrix} \lambda_1 & \varepsilon \\ 0 & \lambda_2 \end{pmatrix}.$$

This induces a corresponding dual complex linear change of coordinates in the $\{y_1, y_2\}$ variables. The parameter ε plays no role and we obtain at P that

$$\begin{aligned} \tau &= 2\left(\lambda_1^2 + \lambda_1\lambda_2 + \lambda_2^2\right), \qquad \|R\|^2 = 4(\lambda_1^4 + \lambda_1^2\lambda_2^2 + \lambda_2^4), \\ \|\rho\|^2 &= 2\lambda_1^4 + 2\lambda_1^3\lambda_2 + \lambda_1^2\lambda_2^2 + 2\lambda_1\lambda_2^3 + 2\lambda_2^4. \end{aligned}$$

Suppose that $\|R\|^2 = \|\rho\|^2 = 0$. If the eigenvalues are real, then the vanishing of $\|R\|^2$ implies $(\lambda_1, \lambda_2) = (0,0)$ so T is nilpotent. Thus we assume the eigenvalues are complex. Consequently, $\lambda_2 = \bar\lambda_1 \ne 0$. Set $\lambda_1 = re^{i\theta}$ and $\lambda_2 = re^{-i\theta}$ for $r \ne 0$. The equations in question are homogeneous so we may assume without loss of generality $r = 1$. We have

$$0 = \|\rho\|^2 - \tfrac{1}{2}\|R\|^2 = 2\lambda_1^3\lambda_2 - \lambda_1^2\lambda_2^2 + 2\lambda_1\lambda_2^3.$$

Dividing this equation by $\lambda_1\lambda_2$ yields $0 = 2\lambda_1^2 - \lambda_1\lambda_2 + 2\lambda_2^2$. Setting $\lambda_1 = e^{i\theta}$ and $\lambda_2 = e^{-i\theta}$, we have $0 = e^{4i\theta} + 1 + e^{-4i\theta}$ so $\cos(4\theta) = -\frac{1}{2}$ and $0 = 2e^{2i\theta} - 1 + 2e^{2i\theta}$. Thus $\cos(2\theta) = \frac{1}{4}$. The angle addition formulas yield $-\frac{1}{2} = \cos(4\theta) = 2\cos^2(2\theta) - 1 = \frac{1}{8} - 1$ so this case does not occur. Thus Assertion 3 implies Assertion 1.

Assume $\tau = 0$ and $\|\rho\|^2 = 0$. We compute

$$\|\rho\|^2 - \tfrac{1}{2}\tau^2 = -\lambda_1\lambda_2(2\lambda_1 + \lambda_2)(\lambda_1 + 2\lambda_2).$$

Setting $\tau = 0$ then yields $\lambda_1 = \lambda_2 = 0$. Thus T is nilpotent and Assertion 4 implies Assertion 1. □

The conditions of Assertion 3 and Assertion 4 are optimal in a certain sense. For θ constant and $r(x^1, x^2) \neq 0$, define an endomorphism of the tangent bundle which is not nilpotent by setting

$$T(x) := r(x^1, x^2) \begin{pmatrix} \cos(\theta) & \sin(\theta) \\ -\sin(\theta) & \cos(\theta) \end{pmatrix}.$$

Then $\tau = 2r^2(2\cos(2\theta) + 1)$ and $\|R\|^2 = 4r^4(2\cos(4\theta) + 1)$. Taking $\theta = \frac{\pi}{3}$ then yields $\|R\|^2 = \tau = 0$. Similarly,

$$\|\rho\|^2 = r^4(4\cos(4\theta) + 4\cos(2\theta) + 1) = 0 \text{ for } \theta = \frac{1}{2}\arctan\left(\frac{\sqrt{7}+1}{\sqrt{7}-1}\right).$$

Thus the conditions $\{\|R\|^2 = 0, \tau = 0\}$ or $\|\rho\|^2 = 0$ do not suffice to show T is nilpotent.

15.2.2 INVARIANTS OF VSI MANIFOLDS NOT OF WEYL TYPE. We continue the discussion of Section 15.2.1. Let $\mathcal{M} = (M, \nabla)$ be an affine surface, let $0 \neq T$ be nilpotent, and let $\mathcal{N} := (T^*M, g_{\nabla,\Phi,T})$ be the associated VSI Riemannian extension. Apply Lemma 15.2 to choose local coordinates so that $T = \partial_{x^1} \otimes dx^2$. Take dual coordinates $y_1 = x^3$ and $y_2 = x^4$. Let $\{R, \rho\}$ denote the curvature operator and Ricci tensor of $\mathcal{N}$ and let $\{R^\nabla, \rho^\nabla, \rho_a^\nabla, \rho_s^\nabla\}$ be the curvature operator, Ricci tensor, alternating Ricci tensor, and symmetric Ricci tensor of $\mathcal{M}$, respectively. Let $\mathcal{V} := \operatorname{span}\{\partial_{x_3}, \partial_{x_4}\}$ be the "vertical" space and let $\mathfrak{H} := \operatorname{span}\{\partial_{x^1}, \partial_{x^2}\}$ be the "horizontal" space. The vertical space $\mathcal{V}$ is the kernel of $\pi_* : T_*(TN) \to TM$; this is the Walker distribution and is a maximal parallel null 2-dimensional subspace. The horizontal space $\mathcal{H}$, however, is not invariantly defined. We may then decompose

$$R(X,Y) = \begin{pmatrix} R_{\mathfrak{H}}^{\mathfrak{H}} = \begin{pmatrix} R_{XY1}{}^1 & R_{XY2}{}^1 \\ R_{XY1}{}^2 & R_{XY2}{}^2 \end{pmatrix} & R_{\mathcal{V}}^{\mathfrak{H}} = \begin{pmatrix} R_{XY3}{}^1 & R_{XY4}{}^1 \\ R_{XY3}{}^2 & R_{XY4}{}^2 \end{pmatrix} \\ R_{\mathfrak{H}}^{\mathcal{V}} = \begin{pmatrix} R_{XY1}{}^3 & R_{XY2}{}^3 \\ R_{XY1}{}^4 & R_{XY2}{}^4 \end{pmatrix} & R_{\mathcal{V}}^{\mathcal{V}} = \begin{pmatrix} R_{XY3}{}^3 & R_{XY4}{}^3 \\ R_{XY3}{}^4 & R_{XY4}{}^4 \end{pmatrix} \end{pmatrix}.$$

The following result follows by a direct computation.

Lemma 15.15 *Let $\mathcal{N} := (T^*M, g_{\nabla,\Phi,T})$ where $T = \partial_{x^1} \otimes dx^2$.*

1. *$R_{\mathcal{V}}^{\mathfrak{H}}(X,Y) = 0$ for all (X,Y), i.e., $R_{abi}{}^j = 0$ for $3 \leq i \leq 4$, $1 \leq j \leq 2$.*
2. *$\{R_{\mathfrak{H}}^{\mathfrak{H}} + (R_{\mathcal{V}}^{\mathcal{V}})^t\}(X,Y) = 0$ for all (X,Y) i.e., $R_{ab1}{}^1 + R_{ab3}{}^3 = 0$, $R_{ab2}{}^2 + R_{ab4}{}^4 = 0$, $R_{ab1}{}^2 + R_{ab4}{}^3 = 0$, and $R_{ab2}{}^1 + R_{ab3}{}^4 = 0$.*
3. *$R_{\mathfrak{H}}^{\mathfrak{H}}(\partial_{x^i}, \partial_{x^j}) = 0$ for $i < j$ and $(i,j) \notin \{(1,2),(2,3)\}$.*
4. $\begin{pmatrix} R_{231}{}^1 & R_{232}{}^1 \\ R_{231}{}^2 & R_{232}{}^2 \end{pmatrix} = \begin{pmatrix} 0 & 1 \\ 0 & 0 \end{pmatrix}.$
5. $\begin{pmatrix} R_{121}{}^1 & R_{122}{}^1 \\ R_{121}{}^2 & R_{122}{}^2 \end{pmatrix} = \begin{pmatrix} R_{121}^\nabla{}^1 & R_{122}^\nabla{}^1 \\ R_{121}^\nabla{}^2 & R_{122}^\nabla{}^2 \end{pmatrix} - x_3 \begin{pmatrix} -\Gamma_{11}{}^2 & \Gamma_{11}{}^1 - \Gamma_{12}{}^2 \\ 0 & \Gamma_{11}{}^2 \end{pmatrix}.$

6. $\mathrm{Tr}\{R^{\tilde{\mathfrak{H}}}_{\tilde{\mathfrak{H}}}(X,Y)\} = -2(\pi^*\rho_a^\nabla)(X,Y)$.
7. $\rho_{ij} = 0$ *if* $i \geq 3$ *or* $j \geq 3$.
8. $\begin{pmatrix} \rho_{11} & \rho_{21} \\ \rho_{12} & \rho_{22} \end{pmatrix} = 2\rho_s^\nabla + \begin{pmatrix} 0 & 2x_3\Gamma_{11}{}^2 \\ 2x_3\Gamma_{11}{}^2 & -4x_3\Gamma_{11}{}^1 - 2x_4\Gamma_{11}{}^2 + 2x_3\Gamma_{12}{}^2 + \Phi_{11} \end{pmatrix}$.
9. ${}^g\nabla R(i,j,1,1;k) + {}^g\nabla R(i,j,2,2;k) = 0$ *unless* $\{i,j,k\} \in \{1,2\}$.

Generically, $\mathcal{V}$ is the only parallel null distribution. Consequently, it defines an additional piece of structure. The distribution $\mathrm{span}\{\partial_{x^1}, \partial_{x^2}\}$ is not invariantly defined. To obtain an invariant object, we set $\tilde{\mathfrak{H}} := TN/\mathcal{V}$ and let $\tilde{\pi} : TN \to \tilde{\mathfrak{H}}$ be the natural projection. By Lemma 15.15, $\tilde{\pi} R(X,Y)v = 0$ for $v \in \mathcal{V}$ and thus $\tilde{\pi} R(X,Y)$ descends to a well defined map $R^{\tilde{\mathfrak{H}}}_{\tilde{\mathfrak{H}}}(X,Y)$ of $\tilde{\mathfrak{H}}$. Let $\{X_3, X_4\}$ be a local frame for $\mathcal{V}$. Choose $\{X_1, X_2\}$ so that

$$g(X_1,X_3) = g(X_2,X_4) = 1 \quad \text{and} \quad g(X_1,X_4) = g(X_2,X_3) = 0\,. \tag{15.2.c}$$

We note that $\{X_1, X_2\}$ is not uniquely defined by these relations as we can add an arbitrary element of $\mathcal{V}$ to either X_1 or X_2 and preserve Equation (15.2.c). However, $\{\tilde{\pi}X_1, \tilde{\pi}X_2\}$ is uniquely defined by Equation (15.2.c). And, in particular, if we take $X_3 = \partial_{x_3}$ and $X_4 = \partial_{x_4}$, then we may take $X_1 = \partial_{x^1}$ and $X_2 = \partial_{x^2}$.

Definition 15.16 We use Lemma 15.15 to introduce some additional quantities.

1. Since $\rho(X,Y) = 0$ if either X or Y belongs to $\mathcal{V}$, ρ descends to a map from $\tilde{\mathfrak{H}} \oplus \tilde{\mathfrak{H}}$ to $\mathbb{R}$ that we shall denote by $\rho^{\tilde{\mathfrak{H}}} \in S^2(\tilde{\mathfrak{H}}^*)$. Let $\tilde{\pi} : T^*M \to M$. Because $\tilde{\pi}_*(\mathcal{V}) = 0$, $\tilde{\pi}_*$ induces a map from $\tilde{\mathfrak{H}}$ to TM. If $\Gamma_{11}{}^2 = 0$, if $2\Gamma_{11}{}^2 = \Gamma_{12}{}^2$, and if $\Phi_{11} = 0$, then $\rho^{\tilde{\mathfrak{H}}} = 2\tilde{\pi}^*\rho_s^\nabla$.
2. Let $\Omega(X,Y) = \mathrm{Tr}\{R^{\tilde{\mathfrak{H}}}_{\tilde{\mathfrak{H}}}(X,Y)\}$. Then $\Omega(X,Y) = 0$ if either X or Y belongs to $\mathcal{V}$ so Ω descends to an alternating bilinear map from $\tilde{\mathfrak{H}} \oplus \tilde{\mathfrak{H}}$ to $\mathbb{R}$ that we denote by $\Omega^{\tilde{\mathfrak{H}}} \in \Lambda^2(\tilde{\mathfrak{H}}^*)$. We have $\Omega^{\tilde{\mathfrak{H}}} = -2\tilde{\pi}^*\rho_a^\nabla$.
3. As $\mathcal{V}$ is parallel, ${}^g\nabla R(X,Y;Z)$ maps $\mathcal{V}$ to $\mathcal{V}$. Consequently, ${}^g\nabla R(X,Y;Z)$ extends to an endomorphism $({}^g\nabla R)^{\tilde{\mathfrak{H}}}(X,Y;Z)$ of $\tilde{\mathfrak{H}}$. A direct computation shows that $\mathrm{Tr}\{({}^g\nabla R)^{\tilde{\mathfrak{H}}}(X,Y;Z)\} = 0$ if X, Y, or Z belongs to $\mathcal{V}$. We may therefore regard $\mathrm{Tr}\{({}^g\nabla R)^{\tilde{\mathfrak{H}}}(X,Y;Z)\} \in \Lambda^2(\tilde{\mathfrak{H}}) \otimes \tilde{\mathfrak{H}}^*$. Assuming that $\Omega^{\tilde{\mathfrak{H}}} \neq 0$, we may decompose $\mathrm{Tr}\{({}^g\nabla R)^{\tilde{\mathfrak{H}}}\} = \omega^{\tilde{\mathfrak{H}}} \otimes \Omega^{\tilde{\mathfrak{H}}}$ for $\omega^{\tilde{\mathfrak{H}}} \in \tilde{\mathfrak{H}}^*$. Moreover, one has $d\omega^{\tilde{\mathfrak{H}}} = \Omega^{\tilde{\mathfrak{H}}}$.
4. Suppose that we are at a point of $\mathcal{N}$ where $\rho^{\tilde{\mathfrak{H}}}$ defines a non-degenerate symmetric bilinear form on $\tilde{\mathfrak{H}}$. We may then define

$$\beta_1 := \|\Omega^{\tilde{\mathfrak{H}}}\|^2_{\rho^{\tilde{\mathfrak{H}}}} = \frac{(R_{121}{}^1 + R_{122}{}^2)^2}{\rho_{11}\rho_{22} - \rho_{12}\rho_{12}}\,.$$

If we also assume that $\Omega^{\tilde{\mathfrak{H}}} \neq 0$ (i.e., $\rho_a^\nabla \neq 0$) or, equivalently, that $\beta_1 \neq 0$, then $\omega^{\tilde{\mathfrak{H}}}$ is well defined and we may set $\beta_2 := \|\omega^{\tilde{\mathfrak{H}}}\|^2_{\rho^{\tilde{\mathfrak{H}}}}$. We have that $\omega_1^{\tilde{\mathfrak{H}}} = \frac{R_{121}{}^1{}_{;1} + R_{122}{}^2{}_{;1}}{R_{121}{}^1 + R_{122}{}^2}$ and that

$\omega_2^{\tilde{\mathfrak{H}}} = \frac{R_{121}{}^1{}_{;2}+R_{122}{}^2{}_{;2}}{R_{121}{}^1+R_{122}{}^2}$. Consequently,

$$\beta_2 := \frac{\rho_{22}^{\tilde{\mathfrak{H}}}\omega_1^{\tilde{\mathfrak{H}}}\omega_1^{\tilde{\mathfrak{H}}} + \rho_{11}^{\tilde{\mathfrak{H}}}\omega_2^{\tilde{\mathfrak{H}}}\omega_2^{\tilde{\mathfrak{H}}} - 2\rho_{12}^{\tilde{\mathfrak{H}}}\omega_1^{\tilde{\mathfrak{H}}}\omega_2^{\tilde{\mathfrak{H}}}}{\rho_{11}^{\tilde{\mathfrak{H}}}\rho_{22}^{\tilde{\mathfrak{H}}} - \rho_{12}^{\tilde{\mathfrak{H}}}\rho_{12}^{\tilde{\mathfrak{H}}}}.$$

15.2.3 BACH FLAT VSI MANIFOLDS. It is obvious from the discussion given in Section 15.2.2 that β_1 and β_2 are isometry invariants of $\mathcal{N}$ where defined. Generically, β_1 and β_2 are very complicated expressions which involve non-trivial dependence on the fiber variables and which involve the endomorphism Φ. These invariants are uninteresting if $\Omega^{\tilde{\mathfrak{H}}} = 0$ or, equivalently in the setting where $g = g_{\nabla,\Phi,T}$ if ρ^∇ is symmetric because $\Omega^{\tilde{\mathfrak{H}}} = -2\pi^*\rho_a^\nabla$. Thus the Type $\mathcal{A}$ geometries do not lead to interesting examples in this setting. Let $\Gamma_{ij}{}^k = \frac{1}{x^1}A_{ij}{}^k$ define a Type $\mathcal{B}$ geometry and let $\mathcal{N} := (T^*M, g_{\nabla,\Phi,T})$. By Equation (14.1.a), the Ricci tensor is symmetric if and only if $A_{12}{}^1 + A_{22}{}^2 = 0$. Thus we shall assume that $A_{12}{}^1 + A_{22}{}^2 \neq 0$. It is not difficult to show that up to the action of the $ax+b$ group, if T is nilpotent, invariant, and non-trivial that either $T = \partial_{x^1} \otimes dx^2$ or $T = \partial_{x^2} \otimes dx^1$. We present the following examples to illustrate the nature of the invariants β_1 and β_2. We impose the relations of Theorem 15.4 to ensure the example are Bach flat in Case 1 and Case 2; we interchange the roles of x^1 and x^2 in Case 3 and Case 4 in Theorem 15.4.

Case 1. Let $\mathcal{M}$ be a Type $\mathcal{B}$ geometry with $\Gamma_{ij}{}^k = \frac{1}{x^1}A_{ij}{}^k$. Suppose that

$$A_{11}{}^2 = 0, \quad A_{11}{}^1 = 1, \quad T = \partial_{x^1} \otimes dx^2, \quad \text{and } \rho^{\tilde{\mathfrak{H}}} \text{ is non-degenerate}.$$

By Theorem 15.4, $\mathcal{N} = (T^*M, g_{\nabla,\Phi,T})$ is Bach flat. We have $\beta_1 = (A_{12}{}^1 + A_{22}{}^2)^2\Delta^{-1}$ where

$$\begin{aligned}\Delta \;=\;& 2(2-A_{12}{}^2)A_{12}{}^2(x^1)^2\Phi_{11} - 4(2-A_{12}{}^2)^2A_{12}{}^2x^1x_3\\ &-(4A_{12}{}^2+1)(A_{12}{}^1)^2 + 4(A_{12}{}^2-2)A_{22}{}^1(A_{12}{}^2)^2\\ &-(A_{22}{}^2)^2 + 2(1-2(A_{12}{}^2-1)A_{12}{}^2)A_{12}{}^1A_{22}{}^2.\end{aligned}$$

It now follows that $\beta_1 = 0$ if and only if the Ricci tensor ρ^∇ of $\mathcal{M}$ is symmetric. Moreover β_1 is a non-zero constant if and only if either $A_{12}{}^2 = 0$, in which case $\beta_1 = -\frac{(A_{12}{}^1+A_{22}{}^2)^2}{(A_{12}{}^1-A_{22}{}^2)^2}$, or $A_{12}{}^2 = 2$, and then $\beta_1 = -\frac{(A_{12}{}^1+A_{22}{}^2)^2}{(3A_{12}{}^1+A_{22}{}^2)^2}$. Further, if β_1 is non-zero, then β_2 is generically non-constant since

$$\begin{aligned}\beta_2 \;=\;& \{(A_{12}{}^2+3)^2(x^1)^2\Phi_{11} + 2(A_{12}{}^2-2)(A_{12}{}^2+3)^2x^1x_3\\ &-2(A_{12}{}^2+3)^2A_{12}{}^2A_{22}{}^1 - 2((A_{12}{}^2-1)A_{12}{}^2+3)(A_{22}{}^2)^2\\ &-2((4A_{12}{}^2+9)A_{12}{}^2+6)(A_{12}{}^1)^2\\ &-2((3A_{12}{}^2-4)A_{12}{}^2-9)A_{12}{}^1A_{22}{}^2\}\Delta^{-1}.\end{aligned}$$

Case 2. Let $\mathcal{M}$ be a Type $\mathcal{B}$ geometry with $\Gamma_{ij}{}^k = \frac{1}{x^1} A_{ij}{}^k$. Suppose

$$T = \partial_{x^1} \otimes dx^2, \quad A_{11}{}^2 = 0, \quad A_{12}{}^2 = A_{11}{}^1, \quad \text{and } \rho^{\tilde{\mathfrak{H}}} \text{ is non-degenerate}.$$

By Theorem 15.4, $\mathcal{N} = (T^*M, g_{\nabla,\Phi,T})$ is Bach flat. We have $\beta_1 = (A_{12}{}^1 + A_{22}{}^2)^2 \Delta^{-1}$ where

$$\begin{aligned} \Delta &= 2A_{11}{}^1(x^1)^2\Phi_{11} - 4(A_{11}{}^1)^2 x^1 x_3 - (A_{22}{}^2)^2 \\ &\quad -(4(A_{11}{}^1)^2 + 1)(A_{12}{}^1)^2 - 4A_{11}{}^1 A_{22}{}^1 + 2A_{12}{}^1 A_{22}{}^2. \end{aligned}$$

Therefore $\beta_1 = 0$ if and only if the Ricci tensor of $\mathcal{M}$ is symmetric. Moreover, one has that β_1 is a non-zero constant if and only if $A_{11}{}^1 = 0$, in which case $\beta_1 = -\frac{(A_{12}{}^1 + A_{22}{}^2)^2}{(A_{12}{}^1 - A_{22}{}^2)^2}$. Furthermore, if $\beta_1 \neq 0$, then

$$\begin{aligned} \beta_2 &= \{4(A_{11}{}^1 + 1)^2(x^1)^2\Phi_{11} - 8(A_{11}{}^1 + 1)^2 A_{11}{}^1 x^1 x_3 \\ &\quad -2(A_{11}{}^1 + 2)(A_{22}{}^2)^2 - 8(A_{11}{}^1 + 1)^2 A_{22}{}^1 \\ &\quad -2(A_{11}{}^1(8A_{11}{}^1 + 9) + 2)(A_{12}{}^1)^2 + 4(3A_{11}{}^1 + 2)A_{12}{}^1 A_{22}{}^2\}\Delta^{-1}. \end{aligned}$$

Case 3. Let $\mathcal{M}$ be a Type $\mathcal{B}$ geometry with $\Gamma_{ij}{}^k = \frac{1}{x^1} A_{ij}{}^k$. Suppose

$$T = \partial_{x^2} \otimes dx^1, \quad A_{22}{}^1 = 0, \quad A_{22}{}^2 = 0, \quad \text{and } \rho^{\tilde{\mathfrak{H}}} \text{ is non-degenerate}.$$

We interchange the roles of x^1 and x^2 in Theorem 15.4 to see that $\mathcal{N} = (T^*M, g_{\nabla,\Phi,T})$ is Bach flat. We have $\beta_1 = (A_{12}{}^1)^2\Delta^{-1}$ where

$$\begin{aligned} \Delta &= (A_{12}{}^1)^2\{-2(x^1)^2\Phi_{22} - 4A_{12}{}^1 x^1 x_4 \\ &\quad -4A_{11}{}^1 A_{12}{}^2 + 4A_{11}{}^2 A_{12}{}^1 - 1\}. \end{aligned}$$

The invariant β_1 is never constant in this case. Moreover, if $\beta_1 \neq 0$, then

$$\begin{aligned} \beta_2 &= (A_{12}{}^1)^2\{(x^1)^2\Phi_{22} + 2A_{12}{}^1 x^1 x_4 - 12A_{12}{}^2 - 2A_{11}{}^2 A_{12}{}^1 \\ &\quad -4 - 2(A_{11}{}^1)^2 - 8(A_{12}{}^2)^2 - 6(A_{12}{}^2 + 1)A_{11}{}^1\}\Delta^{-1}. \end{aligned}$$

It follows that β_2 is constant if and only if $2A_{11}{}^1 + 4A_{12}{}^2 + 3 = 0$, in which case $\beta_2 = -\frac{1}{2}$.

Case 4. Let $\mathcal{M}$ be a Type $\mathcal{B}$ geometry with $\Gamma_{ij}{}^k = \frac{1}{x^1} A_{ij}{}^k$. Suppose

$$T = \partial_{x^2} \otimes dx^1, \quad A_{22}{}^1 = 0, \quad A_{22}{}^2 = A_{12}{}^1, \quad \text{and } A_{12}{}^1 A_{12}{}^2 \neq 0.$$

We interchange the roles of x^1 and x^2 in Theorem 15.4 to see that $\mathcal{N} = (T^*M, g_{\nabla,\Phi,T})$ is Bach flat. The condition $A_{12}{}^1 A_{12}{}^2 \neq 0$ implies $\rho^{\tilde{\mathfrak{H}}}$ is non-degenerate. We then have

$$\begin{aligned} \beta_1 &= -(A_{12}{}^2)^{-2}, \\ \beta_2 &= -\left((x^1)^2\Phi_{22} - 2A_{12}{}^1 x^1 x_4 - 4(A_{12}{}^2)^2 - 2A_{12}{}^2\right)(A_{12}{}^2)^{-2}. \end{aligned}$$

In contrast with the previous cases, β_1 is constant while β_2 is never constant.

15.3 SIGNATURE (2, 2) BACH FLAT MANIFOLDS

The material in this section arises from work of Calviño-Louzao et al. [14]. Since we are interested in constructing Bach flat manifolds which are not half-conformally flat, we shall take T nilpotent in considering the metrics $g_{\nabla,\Phi,T}$. In Theorem 15.4, given T nilpotent, we showed how to construct an affine connection on the underlying manifold M so that $g_{\nabla,\Phi,T}$ was Bach flat. It follows from Theorem 15.14 that these geometries are VSI and we constructed scalar invariants which were not of Weyl type. In Section 15.2, we computed these invariants for various Type $\mathcal{B}$ geometries where T was nilpotent and where by Theorem 15.4, the resulting structures were Bach flat. In this section, we reverse the process and consider a perhaps more natural problem. We assume the affine structure $\mathcal{M}$ is given and look for non-trivial nilpotent operators so that $g_{\nabla,\Phi,T}$ is Bach flat. Since we are regarding the underlying structure Γ as fixed and looking for T, we cannot use Lemma 15.2 to renormalize the coordinate system. Our analysis is local. Since either $T^1{}_2(P) \neq 0$ or $T^2{}_1(P) \neq 0$, we assume for the sake of definiteness that $T^1{}_2(P) \neq 0$ and expand T near P in the form

$$T = \alpha(x^1, x^2) \begin{pmatrix} \xi(x^1, x^2) & 1 \\ -\xi^2(x^1, x^2) & -\xi(x^1, x^2) \end{pmatrix}. \tag{15.3.a}$$

Definition 15.17 We introduce the following operators:

$$\begin{aligned}
\mathcal{P}_1(\xi) &:= -\xi^{(1,0)} + \xi\,\xi^{(0,1)} + \Gamma_{22}{}^1\xi^3 - (2\Gamma_{12}{}^1 - \Gamma_{22}{}^2)\xi^2 + (\Gamma_{11}{}^1 - 2\Gamma_{12}{}^2)\xi + \Gamma_{11}{}^2, \\
\mathcal{P}_2(\xi, \alpha) &:= \alpha\alpha^{(2,0)} + \xi^2\alpha\alpha^{(0,2)} - 2\xi\alpha\alpha^{(1,1)} + (\alpha^{(1,0)})^2 + \xi^2(\alpha^{(0,1)})^2 - 2\xi\alpha^{(1,0)}\alpha^{(0,1)} \\
&\quad -\alpha\alpha^{(1,0)}\left(2\xi^{(0,1)} - 5\Gamma_{22}{}^1\xi^2 + 2(4\Gamma_{12}{}^1 - \Gamma_{22}{}^2)\xi - 3\Gamma_{11}{}^1 + 2\Gamma_{12}{}^2\right) \\
&\quad +\alpha\alpha^{(0,1)}\left(2\xi\xi^{(0,1)} - 6\Gamma_{22}{}^1\xi^3 + (10\Gamma_{12}{}^1 - 3\Gamma_{22}{}^2)\xi^2 - 4(\Gamma_{11}{}^1 - \Gamma_{12}{}^2)\xi - \Gamma_{11}{}^2\right) \\
&\quad +6\xi^4\alpha^2(\Gamma_{22}{}^1)^2 - 2\xi^3\alpha^2\left((\Gamma_{22}{}^1)^{(0,1)} + 9\Gamma_{12}{}^1\Gamma_{22}{}^1 - 3\Gamma_{22}{}^1\Gamma_{22}{}^2\right) \\
&\quad -\xi^2\alpha^2\left(4\Gamma_{22}{}^1\xi^{(0,1)} - 3(\Gamma_{12}{}^1)^{(0,1)} - 2(\Gamma_{22}{}^1)^{(1,0)} + (\Gamma_{22}{}^2)^{(0,1)}\right. \\
&\qquad\qquad \left. - 12(\Gamma_{12}{}^1)^2 - (\Gamma_{22}{}^2)^2 - 7\Gamma_{11}{}^1\Gamma_{22}{}^1 + 7\Gamma_{12}{}^1\Gamma_{22}{}^2 + 9\Gamma_{12}{}^2\Gamma_{22}{}^1\right) \\
&\quad +\xi\alpha^2\left(2(3\Gamma_{12}{}^1 - \Gamma_{22}{}^2)\xi^{(0,1)} - (\Gamma_{11}{}^1)^{(0,1)} - 3(\Gamma_{12}{}^1)^{(1,0)} + (\Gamma_{12}{}^2)^{(0,1)}\right. \\
&\qquad\qquad \left. +(\Gamma_{22}{}^2)^{(1,0)} - 2(\Gamma_{11}{}^1 - \Gamma_{12}{}^2)(4\Gamma_{12}{}^1 - \Gamma_{22}{}^2) + 4\Gamma_{11}{}^2\Gamma_{22}{}^1\right) \\
&\quad -\alpha^2\left(2(\Gamma_{11}{}^1 - \Gamma_{12}{}^2)\xi^{(0,1)} - (\Gamma_{11}{}^1)^{(1,0)} + (\Gamma_{12}{}^2)^{(1,0)}\right. \\
&\qquad\qquad \left. -(\Gamma_{11}{}^1)^2 + \Gamma_{11}{}^1\Gamma_{12}{}^2 + 3\Gamma_{11}{}^2\Gamma_{12}{}^1 - \Gamma_{11}{}^2\Gamma_{22}{}^2\right)
\end{aligned}$$

Theorem 15.18 *Let (M, ∇) be an affine surface. Let T have the form given in Equation (15.3.a) and let Φ be arbitrary. The modified Riemannian extension $(T^*M, g_{\nabla,\Phi,T})$ of Equation (15.2.a) is Bach flat if and only if α and ξ are solutions to the partial differential equations $\mathcal{P}_1(\xi) = 0$ and $\mathcal{P}_2(\xi, \alpha) = 0$.*

Remark 15.19 Suppose $\mathcal{M}$ is real analytic. The operator $\mathcal{P}_1(\xi)$ of Definition 15.17 takes the form: $\mathcal{P}_1(\xi) = -\xi^{(1,0)} + \xi\xi^{(0,1)} + f(\xi, \Gamma)$. Given a real analytic function $\xi_0(x^2)$, the Cauchy–Kovalevski Theorem shows that there is a unique local solution to the equation $\mathcal{P}_1(\xi) = 0$ with $\xi(0, x^2) = \xi_0(x^2)$. Once ξ is determined, the operator $\mathcal{P}_2(\xi, \alpha)$ of Definition 15.17 takes the form $\mathcal{P}_2(\xi, \alpha) = \alpha\alpha^{(2,0)} - 2\xi\alpha\alpha^{(1,1)} + \xi^2\alpha\alpha^{(0,2)} + F(\alpha, d\alpha; \Gamma, d\Gamma; \xi, d\xi)$. Given real analytic functions $\alpha_0(x^2)$ and $\alpha_1(x^2)$, there exists a unique local solution to the equation $\mathcal{P}_2(\xi, \alpha) = 0$ with $\alpha(0, x^2) = \alpha_0(x^2)$ and $\alpha^{(1,0)}(0, x^2) = \alpha_1(x^2)$. Thus given ∇, there are many nilpotent T so that $\mathcal{N}$ is Bach flat in this setting; the auxiliary tensor Φ plays no role in the analysis.

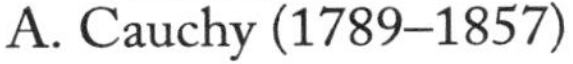

A. Cauchy (1789–1857) S. Kovalevskaya (1850–1891)

Proof. We suppose T is a nilpotent tensor field of Type $(1, 1)$. We then have that $\operatorname{Tr}\{T\} = 0$ and $\det(T) = 0$. If we assume that $T^1{}_2(P) \neq 0$, then T has the form given in Equation (15.3.a). A direct computation shows $\mathfrak{B}(\partial_{x^k}, \partial_{y_j}) = 0$ and $\mathfrak{B}(\partial_{y_i}, \partial_{y_j}) = 0$, and thus only $\mathfrak{B}_{11}$, $\mathfrak{B}_{12}$, and $\mathfrak{B}_{22}$, where $\mathfrak{B}_{ij} = \mathfrak{B}(\partial_{x^i}, \partial_{x^j})$, are relevant. We observe that

$$\begin{aligned}&\text{Coefficient}[\mathfrak{B}_{11}, \alpha^{(2,0)}] = -4\alpha\xi^2, \quad \text{Coefficient}[\mathfrak{B}_{12}, \alpha^{(2,0)}] = -4\alpha\xi,\\ &\text{Coefficient}[\mathfrak{B}_{22}, \alpha^{(2,0)}] = -4\alpha\,.\end{aligned}$$

We therefore define $\mathfrak{Q}_1 := \mathfrak{B}_{11} - \mathfrak{B}_{12}\xi$, $\mathfrak{Q}_2 := \mathfrak{B}_{11} - \mathfrak{B}_{22}\xi^2$, and $\mathfrak{Q}_3 := 2\mathfrak{Q}_1 - \mathfrak{Q}_2$. We may then express $\mathfrak{Q}_3 = -4\alpha^2(\mathcal{P}_1)^2$ and thus the vanishing of $\mathfrak{Q}_3$ is equivalent to the vanishing of $\mathcal{P}_1$. We set $\mathcal{P}_1 = 0$ and express $\xi^{(1,0)} = F_{(1,0)}(\xi, \Gamma, \xi^{(0,1)})$. Differentiating this relation permits us to express

$$\xi^{(1,1)} = F_{(1,1)}(\xi, \Gamma, d\Gamma, \xi^{(0,1)}, \xi^{(0,2)}) \quad \text{and} \quad \xi^{(2,0)} = F_{(2,0)}(\xi, \Gamma, d\Gamma, \xi^{(0,1)}, \xi^{(0,2)})\,.$$

Substituting these relations then yields $\mathfrak{Q}_1 = 0$ and $\mathfrak{Q}_2 = 0$. Thus only $\mathfrak{B}_{11}$ plays a role. Substituting these relations permits us to complete the proof by expressing $\mathfrak{B}_{11} = -4\xi^2\mathcal{P}_2$, $\mathfrak{B}_{12} = -4\xi\mathcal{P}_2$, and $\mathfrak{B}_{22} = -4\mathcal{P}_2$. □

Bibliography

[1] Z. Afifi, Riemann extensions of affine connected spaces, *Quart. J. Math., Oxford Ser. (2)* **5** (1954), 312–320. DOI: 10.1093/qmath/5.1.312. 119

[2] T. Arias-Marco and O. Kowalski, Classification of locally homogeneous affine connections with arbitrary torsion on 2-dimensional manifolds, *Monatsh. Math.* **153** (2008), no. 1, 1–18. DOI: 10.1007/s00605-007-0494-0. 2, 27, 38, 39

[3] H. W. Brinkmann, Riemann spaces conformal to Einstein spaces, *Math. Ann.* **91** (1924), no. 3-4, 269–278. DOI: 10.1007/bf01556083. 123

[4] S. Bromberg and A. Medina, A note on the completeness of homogeneous quadratic vector fields on the plane, *Qual. Theory Dyn. Syst.* **6** (2005), no. 2, 181–185. DOI: 10.1007/bf02972671. 68

[5] M. Brozos-Vázquez and E. García-Río, Four-dimensional neutral signature self-dual gradient Ricci solitons, *Indiana Univ. Math. J.* **65** (2016), no. 6, 1921–1943. DOI: 10.1512/iumj.2016.65.5938. 13

[6] M. Brozos-Vázquez, E. García-Río, and P. Gilkey, Homogeneous affine surfaces: moduli spaces, *J. Math. Anal. Appl.* **444** (2016), no. 2, 1155–1184. DOI: 10.1016/j.jmaa.2016.07.005. 43, 59, 74

[7] M. Brozos-Vázquez, E. García-Río, and P. Gilkey, Homogeneous affine surfaces: affine Killing vector fields and gradient Ricci solitons, *J. Math. Soc. Japan* **70** (2018), no. 1, 25–70. DOI: 10.2969/jmsj/07017479. 43, 47, 56, 94, 99

[8] M. Brozos-Vázquez, E. García-Río, P. Gilkey, S. Nikčević, and R. Vázquez-Lorenzo, *The geometry of Walker manifolds*. Synthesis Lectures on Mathematics and Statistics **5**, Morgan & Claypool Publishers, Williston, VT, 2009. DOI: 10.2200/s00197ed1v01y200906mas005. 118

[9] M. Brozos-Vázquez, E. García-Río, P. Gilkey, and X. Valle-Regueiro, Solutions to the affine quasi-Einstein equation for homogeneous surfaces. https://arxiv.org/abs/1802.07953. 43, 80, 82

[10] M. Brozos-Vázquez, E. García-Río, P. Gilkey, and X. Valle-Regueiro, The affine quasi-Einstein equation for homogeneous surfaces, *Manuscripta Math.* **157** (2018), no. 1-2, 279–294. DOI: 10.1007/s00229-017-0987-7.

[11] M. Brozos-Vázquez, E. García-Río, P. Gilkey, and X. Valle-Regueiro, Half conformally flat generalized quasi-Einstein manifolds of metric signature $(2,2)$, *Int. J. Math.* **29** (2018), no. 1, 1850002, 25 pp. DOI: 10.1142/s0129167x18500027. 124

[12] M. Brozos-Vázquez, E. García-Río, P. Gilkey, and X. Valle-Regueiro, A natural linear equation in affine geometry: the affine quasi-Einstein equation, *Proc. Amer. Math. Soc.* **146** (2018), no. 8, 3485–3497. DOI: 10.1090/proc/14090. 15, 26, 43

[13] E. Calviño-Louzao, E. García-Río, P. Gilkey, I. Gutiérrez-Rodríguez, and R. Vázquez-Lorenzo, Affine surfaces which are Kähler, para-Kähler, or nilpotent Kähler, *Results Math.* **73** (2018), no. 4, Art. 135, 24 pp. DOI: 10.1007/s00025-018-0895-5. 19, 102

[14] E. Calviño-Louzao, E. García-Río, P. Gilkey, I. Gutiérrez-Rodríguez, and R. Vázquez-Lorenzo, Constructing Bach flat manifolds of signature (2,2) using the modified Riemannian extension, *J. Math. Phys.* **60** (2019), no. 1, 013511, 14 pp. DOI: 10.1063/1.5080319. 117, 134

[15] H.-D. Cao and Q. Chen, On locally conformally flat gradient steady Ricci solitons, *Trans. Amer. Math. Soc.* **364** (2012), no. 5, 2377–2391.
DOI: 10.1090/s0002-9947-2011-05446-2. 123

[16] F. Catoni, R. Cannata, V. Catoni, and P. Zampetti, Lorentz surfaces with constant curvature and their physical interpretation, *Nuovo Cimento Soc. Ital. Fis. B* **120** (2005), no. 1, 37–51. DOI: 10.1393/ncb/i2004-10129-3. 111

[17] M. Christ, Some nonanalytic-hypoelliptic sums of squares of vector fields, *Bull. Amer. Math. Soc. (N.S.)* **26** (1992), no. 1, 137–140.
DOI: 10.1090/s0273-0979-1992-00258-6. 14

[18] A. Coley, S. Hervik, D. McNutt, N. Musoke, and D. Brooks, Neutral signature Walker-VSI metrics, *Class. Quantum Grav.* **31** (2014), 035015 (14pp). DOI: 10.1088/0264-9381/31/3/035015. 125

[19] V. Cortés, C. Mayer, T. Mohaupt, and F. Saueressig, Special geometry of Euclidean supersymmetry. I. Vector multiplets, *J. High Energy Phys.* (2004), no. 3, 028, 73 pp.
DOI: 10.1088/1126-6708/2004/03/028. 20

[20] V. Cruceanu, P. Fortuny, and P. M. Gadea, A survey on paracomplex geometry, *Rocky Mountain J. Math.* **26** (1996), no. 1, 83–115. DOI: 10.1216/rmjm/1181072105. 111

[21] D. D'Ascanio, P. Gilkey, and P. Pisani, Geodesic completeness for type $\mathcal{A}$ surfaces, *Differential Geom. Appl.* **54** (2017), part A, 31–43. DOI: 10.1016/j.difgeo.2016.12.008. 65

[22] D. D'Ascanio, P. Gilkey, and P. Pisani, The geometry of locally symmetric affine surfaces. *Vietnam J. Math* **47** (2019), 5–21. DOI: 10.1007/s10013-018-0280-4. 29, 112, 113

[23] D. D'Ascanio, P. Gilkey, and P. Pisani, Affine symmetric spaces with torsion, *in preparation*. 2, 9, 10, 104

[24] D. D'Ascanio, P. Gilkey, and P. Pisani, Affine Killing vector fields on homogeneous surfaces with torsion, *in preparation*. 39

[25] A. Derdzinski, Connections with skew-symmetric Ricci tensor on surfaces, *Results Math.* **52** (2008), no. 3-4, 223–245. DOI: 10.1007/s00025-008-0307-3. 16

[26] A. Derdzinski and W. Roter, Walker's theorem without coordinates, *J. Math. Phys.* **47** (2006), no. 6, 062504, 8 pp. DOI: 10.1063/1.2209167. 119

[27] S. Dumitrescu, Locally homogeneous rigid geometric structures on surfaces, *Geom. Dedicata* **160** (2012), 71–90. DOI: 10.1007/s10711-011-9670-4.

[28] S. Dumitrescu and A. Guillot, Quasihomogeneous analytic affine connections on surfaces, *J. Topol. Anal.* **5** (2013), no. 4, 491–532. DOI: 10.1142/s1793525313500222.

[29] L. P. Eisenhart, *Non-Riemannian geometry*. American Mathematical Society Colloquium Publications **8**, American Mathematical Society, Providence, RI, 1990. DOI: 10.1090/coll/008. 11

[30] P. Gilkey, *Invariance theory, the heat equation, and the Atiyah-Singer index theorem*. Second edition. Studies in Advanced Mathematics, CRC Press, Boca Raton, FL, 1995. 14

[31] P. Gilkey, The moduli space of type $\mathcal{A}$ surfaces with torsion and non-singular symmetric Ricci tensor, *J. Geom. Phys.* **110** (2016), 69–77. DOI: 10.1016/j.geomphys.2016.07.012. 2

[32] P. Gilkey, Moduli spaces of type $\mathcal{B}$ surfaces with torsion, *J. Geom.* **108** (2017), no. 2, 637–653. DOI: 10.1007/s00022-016-0364-9. 2

[33] P. Gilkey and S. Nikčević, Complete k-curvature homogeneous pseudo-Riemannian manifolds, *Ann. Global Anal. Geom.* **27** (2005), no. 1, 87–100. DOI: 10.1007/s10455-005-5217-y. 125, 128

[34] P. Gilkey, J. H. Park, and X. Valle-Regueiro, Affine Killing complete and geodesically complete homogeneous affine surfaces, *J. Math. Anal. Appl.* **474** (2019), no. 1, 179–193. DOI: 10.1016/j.jmaa.2019.01.038. 65

[35] P. Gilkey and X. Valle-Regueiro, Applications of PDEs to the study of affine surface geometry, *Matematicki Vesnik* **71** (2019), 45–62. 24, 43, 47, 59

[36] A. R. Gover and P. Nurowski, Obstructions to conformally Einstein metrics in n dimensions, *J. Geom. Phys.* **56** (2006), no. 3, 450–484. DOI: 10.1016/j.geomphys.2005.03.001. 123

[37] A. Guillot and A. Sánchez Godinez, A classification of locally homogeneous affine connections on compact surfaces, *Ann. Global Anal. Geom.* **46** (2014), no. 4, 335–349. DOI: 10.1007/s10455-014-9426-0. 28

[38] O. Kobayashi, A differential equation arising from scalar curvature function, *J. Math. Soc. Japan* **34** (1982), no. 4, 665–675. DOI: 10.2969/jmsj/03440665. 123

[39] S. Kobayashi and K. Nomizu, *Foundations of differential geometry. Vol. I and Vol. II.* Wiley Classics Library. A Wiley-Interscience Publication. John Wiley & Sons, Inc., New York, 1996. 2, 3, 11, 15

[40] O. Kobayashi and M. Obata, Certain mathematical problems on static models in general relativity. *Proc. of the 1980 Beijing Symposium on Differential Geometry and Differential Equations, Vol. 1, 2, 3*, (Beijing, 1980), 1333–1343, Science Press Beijing, Beijing, 1982. 123

[41] S. S. Koh, On affine symmetric spaces, *Trans. Amer. Math. Soc.* **119** (1965), 291–309. DOI: 10.1090/s0002-9947-1965-0184170-2. 9

[42] O. Kowalski, B. Opozda, and Z. Vlášek, A classification of locally homogeneous affine connections with skew-symmetric Ricci tensor on 2-dimensional manifolds, *Monatsh. Math.* **130** (2000), no. 2, 109–125. DOI: 10.1007/s006050070041. 28

[43] O. Kowalski, B. Opozda, and Z. Vlášek, On locally nonhomogeneous pseudo-Riemannian manifolds with locally homogeneous Levi-Civita connections, *Int. J. Math.* **14** (2003), no. 5, 559–572. DOI: 10.1142/s0129167x03001971. 29, 39

[44] O. Kowalski, B. Opozda, and Z. Vlášek, A classification of locally homogeneous connections on 2-dimensional manifolds via group-theoretical approach, *Cent. Eur. J. Math.* **2** (2004), no. 1, 87–102. DOI: 10.2478/bf02475953. 27

[45] O. Munteanu and N. Sesum, On gradient Ricci solitons, *J. Geom. Anal.* **23** (2013), no. 2, 539–561. DOI: 10.1007/s12220-011-9252-6. 123

[46] A. Newlander and L. Nirenberg, Complex analytic coordinates in almost complex manifolds, *Ann. of Math. (2)* **65** (1957), 391–404. DOI: 10.2307/1970051. 20

[47] K. Nomizu, Invariant affine connections on homogeneous spaces, *Amer. J. Math.* **76** (1954), 33–65. DOI: 10.2307/2372398. 9

[48] K. Nomizu, On local and global existence of Killing vector fields, *Ann. of Math. (2)* **72** (1960), 105–120. DOI: 10.2307/1970148. 53

[49] P. J. Olver, *Equivalence, Invariants, and Symmetry.* Cambridge University Press, Cambridge, 1995. DOI: 10.1017/cbo9780511609565. 28

[50] B. Opozda, Locally symmetric connections on surfaces, *Results Math.* **20** (1991), no. 3-4, 725–743. DOI: 10.1007/bf03323207. 9, 65

[51] B. Opozda, A class of projectively flat surfaces, *Math. Z.* **219** (1995), no. 1, 77–92. DOI: 10.1007/bf02572351. 10

[52] B. Opozda, Affine versions of Singer's theorem on locally homogeneous spaces, *Ann. Global Anal. Geom.* **15** (1997), no. 2, 187–199. DOI: 10.1023/A:1006585424144. 9

[53] B. Opozda, A classification of locally homogeneous connections on 2-dimensional manifolds, *Differential Geom. Appl.* **21** (2004), no. 2, 173–198. DOI: 10.1016/j.difgeo.2004.03.005. xiii, 27, 38

[54] B. Opozda, Locally homogeneous affine connections on compact surfaces, *Proc. Amer. Math. Soc.* **132** (2004), no. 9, 2713–2721. DOI: 10.1090/S0002-9939-04-07402-7. 28

[55] J. Patera, R. T. Sharp, P. Winternitz, and H. Zassenhaus, Invariants of real low dimension Lie algebras, *J. Math. Phys.* **17** (1976), no. 6, 986–994. DOI: 10.1063/1.522992. 34, 56

[56] E. M. Patterson and A. G. Walker, Riemann extensions, *Quart. J. Math., Oxford Ser. (2)* **3** (1952), 19–28. DOI: 10.1093/qmath/3.1.19. 119

[57] V. Pecastaing, On two theorems about local automorphisms of geometric structures, *Ann. Inst. Fourier (Grenoble)* **66** (2016), no 1, 175–208. DOI: 10.5802/aif.3009. 9

[58] C. Steglich, Invariants of conformal and projective structures, *Results Math.* **27** (1995), no. 1-2, 188–193. DOI: 10.1007/bf03322280. 11

[59] F. Trèves, Analytic-hypoelliptic partial differential equations of principal type, *Comm. Pure Appl. Math.* **24** (1971), 537–570. DOI: 10.1002/cpa.3160240407. 14

[60] A. Vanžurová, On metrizability of locally homogeneous affine 2-dimensional manifolds, *Arch. Math. (Brno)* **49** (2013), no. 5, 347–357. DOI: 10.5817/am2013-5-347. 39

[61] A. G. Walker, Canonical form for a Riemannian space with a parallel field of null planes, *Quart. J. Math., Oxford Ser. (2)* **1** (1950), 69–79. DOI: 10.1093/qmath/1.1.69. 118, 119

[62] Y.-C. Wong, Two dimensional linear connexions with zero torsion and recurrent curvature, *Monatsh. Math.* **68** (1964), 175–184. DOI: 10.1007/bf01307120. 15, 16, 17

Authors' Biographies

ESTEBAN CALVIÑO-LOUZAO[1] is a member of the research group in Riemannian Geometry at the Department of Geometry and Topology of the University of Santiago de Compostela (Spain). He received his Ph.D. in 2011 from the University of Santiago under the direction of E. García-Río and R. Vázquez-Lorenzo. His research specialty is Riemannian and pseudo-Riemannian geometry. He has published more than 20 research articles and books.

EDUARDO GARCÍA-RÍO[2] is a Professor of Mathematics at the University of Santiago de Compostela (Spain). He is a member of the editorial board of Differential Geometry and its Applications and The Journal of Geometric Analysis and leads the research group in Riemannian Geometry at the Department of Geometry and Topology of the University of Santiago de Compostela (Spain). He received his Ph.D. in 1992 from the University of Santiago under the direction of A. Bonome and L. Hervella. His research specialty is Differential Geometry. He has published more than 120 research articles and books.

[1] Dir. Xeral de Educación, Formación Profesional e Innovación Educativa, San Caetano, s/n, 15781 Santiago de Compostela, Spain.
email: `estebcl@edu.xunta.es`
[2] Department of Mathematics, Faculty of Mathematics, University of Santiago de Compostela, 15782 Santiago de Compostela, Spain
email: `eduardo.garcia.rio@usc.es`

PETER B. GILKEY[3] is a Professor of Mathematics and a member of the Institute of Theoretical Science at the University of Oregon. He is a fellow of the American Mathematical Society and is a member of the editorial board of Results in Mathematics, Differential Geometry and its Applications, and The Journal of Geometric Analysis. He received his Ph.D. in 1972 from Harvard University under the direction of L. Nirenberg. His research specialties are Differential Geometry, Elliptic Partial Differential Equations, and Algebraic topology. He has published more than 275 research articles and books.

JEONGHYEONG PARK[4] is a Professor of Mathematics at Sungkyunkan University and is an associate member of the KIAS (Korea). She received her Ph.D. in 1990 from Kanazawa University in Japan under the direction of H. Kitahara. Her research specialties are spectral geometry of Riemannian submersion and geometric structures on manifolds like eta-Einstein manifolds and H-contact manifolds. She organized the geometry section of AMC 2013 (The Asian Mathematical Conference 2013), the ICM 2014 satellite conference on Geometric analysis, and geometric structures on manifolds (2016). She has published more than 90 research papers and books.

[3]Mathematics Department, University of Oregon, Eugene OR 97403 U.S.
email: `gilkey@uoregon.edu`
[4]Mathematics Department, Sungkyunkwan University, Suwon, 16419, Korea
email: `parkj@skku.edu`

RAMÓN VÁZQUEZ-LORENZO[5] is a member of the research group in Riemannian Geometry at the Department of Geometry and Topology of the University of Santiago de Compostela (Spain). He is a member of the Spanish Research Network on Relativity and Gravitation. He received his Ph.D. in 1997 from the University of Santiago de Compostela under the direction of E. García-Río and R. Castro. His research focuses mainly on Differential Geometry, with special emphasis on the study of the curvature and the algebraic properties of curvature operators in the Lorentzian and in the higher signature settings. He has published more than 60 research articles and books.

[5]Department of Geometry and Topology, Faculty of Mathematics, University of Santiago de Compostela, 15782 Santiago de Compostela, Spain.
email: ravazlor@edu.xunta.es

Index

◎编辑手记

世界著名数学家 A. Weil 曾说过：

当一个数学分支不再引起除去其专家以外的任何人的兴趣时，这个分支就快要僵死了，只有把它重新栽入生气勃勃的科学土壤之中才能挽救它.

微分几何就是这样一个分支，如果说是陈省身当年靠一已之力使其重回巅峰，那么 Einstein 提出的广义相对论，特别是 Yang-Mills 提出的规范场更是激发了它的生机与活力. 本书是一部多卷本的英文微分几何学术专著，中文书名可译为《微分几何的各个方面，第四卷》.

本书的作者有五位：

埃斯特万·卡尔维尼奥-洛扎(Esteban Calviño-Louzao)，**爱德华多·加西亚-里奥**(Eduardo García-Río)　两个人都是西班牙圣地亚哥德孔波斯特拉大学的教授.

彼得·吉尔基(Peter Gilkey)　他是俄勒冈大学数学教授，也是理论科学研究所的成员，他是美国数学学会会员，也是《数学、微分几何与应用》和《几何分析》的编委会成员. 1972 年，在尼伦伯格的指导下，他从哈佛大学获得博士学位，他的研究方向是微分几何、椭圆型偏微分方程和代数拓扑学，他发表了 250 多篇研究论文和著作.

朴正阳(JeongHyeong Park)　她是韩国成均馆大学的数学教授，也是韩国高等科学院的准会员. 1990年，她在北原的指导下从日本金泽大学获得博士学位,她的研究方向是Riemann浸没的谱几何和流形上的几何结构,如eta-Einstein流形和H-contact流形.她组织了AMC 2013(2013年亚洲数学大会)几何部分和ICM 2014(2014年跨学科建模竞赛)几何分析卫星会议,她发表了70多篇研究论文和著作.

拉蒙·瓦茨克斯-洛伦佐(Ramón Vázquez-Lorenzo) 他是西班牙圣地亚哥德孔波斯特拉大学几何和拓扑学系Riemann几何研究小组的成员,也是西班牙相对论和引力研究网络的成员,他于1997年在圣地亚哥德孔波斯特拉大学获得博士学位.他的研究主要集中在微分几何上,特别关注代数性质的研究,他发表了50多篇研究论文和著作.

第四卷延续了前三卷中的讨论.虽然本书是针对一年级研究生的,但它也可以作为仿射微分几何研究人员的基本参考资料,对仿射微分几何感兴趣的本科生也可以使用它.我们主要关注局部齐次的仿射曲面的研究.我们讨论了仿射梯度Ricci孤子、仿射Killing矢量场和测地完备性.Opozda对局部齐次的仿射曲面几何进行了分类,我们按照她的分类进行阐述.直到同构为止,存在两个2维的单连通Lie群.平移群$\mathbb{R}^2$是Abel群,且$ax+b$群是非Abel群.第一章介绍了基础内容.第二章研究了A型曲面.这些是$\mathbf{R}^2$上的左不变仿射几何图形.将对应于特征值$\mu=-1$的准Einstein方程的解空间与每个A型曲面相关联,结果证明它是一种非常强大的方法,并且在我们的研究中起着核心作用,因为它将解析不变量与曲面的基础几何形状联系起来.第三章研究了B型曲面,这些是$ax+b$群的左不变仿射几何.这些几何结构形成了一个非常丰富的族,但人们对它的了解还很有限.唯一剩下的齐次几何是球体S^2.第四章介绍了仿射曲面与具有修正Riemann扩展的中性符号度量的余切丛的几何性质之间的关系.

本书的目录为:

本书中的许多专题,国内都有相应研究,如仿射 Killing p—形式.南平师专的欧阳立、张兰生两位教授早在 1998 年就定义了在仿射联络空间的仿射 Killing p—形式,把仿射 Killing 矢量场的一些结果推广到仿射 Killing p—形式,并研究了几个定理[①].

1. 引言

设 M 是一个 n 维 Riemann 流形,在 M 里,S. Bochner 推广了 Killing 矢量场并引入 Killing p—形式的概念. T. Kashimada 和 S. Tachibana 在 M 里引入了共形 Killing p—形式的概念,并把共形 Killing 矢量场的一些结果推广到共形 Killing p—形式中,S. Tachibana 在 M 里引入了射影 Killing p—形式,并把射影 Killing 矢量场的一些结果推广到射影 Killing p—形式中.他们将推广仿射 Killing 矢量场为仿射 Killing p—形式,把仿射 Killing 矢量场的一些结论推广到仿射 Killing p—形式中去,即有以下三个定理.

定理 1 在 Riemann 流形 M 中,任何 Killing p—形式都是仿射 Killing p—形式.

定理 2 在紧致可定向 Riemann 流形 M 里,仿射 Killing p—形式是 Killing p—形式.

定理 3 在 Riemann 流形 M 里,对于任意给定一点 $q\in M$ 和任意给定的关于指标 $j,i_1,i_2,\cdots,i_p$,称实数 $a_{i_1\cdots i_p}$ 和 $a_{j_1\cdots j_p}$ 在 q 的某个邻域里存在仿射 Killing p—形式 ω,使得 $(\omega_{i_1\cdots i_p})_q=a_{i_1\cdots i_p}$ 和 $(\nabla_j\omega_{i_1\cdots i_p})_q=a_{j_1\cdots j_p}$ 的充要条件是 M 为平坦 Riemann 流形.

2. 定义和定理的证明

我们知道,若 ω 是仿射 Killing 矢量场,则它满足

① 摘编自《南平师专学报(自然科学版)》,1998 年第 2 期.

$$\nabla_j \nabla_k \omega_i = R^l_{kij}\omega_l$$

反之亦然,现证以下命题:

命题 $\nabla_j \nabla_k \omega_i = R^l_{ijk}\omega_l + \nabla_i \nabla_j \omega_k = 0$ 与 $\nabla_j \nabla_k \omega_i = R^l_{kij}\omega_l$ 等价.

证 若

$$\nabla_j \nabla_k \omega_i + R^l_{ijk}\omega_l + \nabla_i \nabla_j \omega_k = 0 \tag{2.1}$$

作 $j \to k \to i \to j$ 的置换并相加,得到

$$\nabla_j \nabla_k \omega_i + \nabla_k \nabla_i \omega_j + \nabla_i \nabla_j \omega_k = 0 \tag{2.2}$$

但

$$\nabla_i \nabla_j \omega_k - \nabla_j \nabla_i \omega_k = R^l_{ijk}\omega_l \tag{2.3}$$

将(2.3)代入(2.2),得

$$\nabla_j \nabla_k \omega_i + R^l_{ikj}\omega_l = 0 \tag{2.4}$$

反之,若(2.4)成立,则(2.3),得到

$$\nabla_j \nabla_k \omega_i = R^l_{kji}\omega_l + \nabla_k \nabla_j \omega_i = -\nabla_i \nabla_j \omega_k - R^l_{ijk}\omega_l$$

上式即为(2.1),证明完毕.

于是,有以下的推广:

(1) 定义.

设 ω 是 Riemann 流形 M 里的 p—形式,如果 ω 满足

$$\nabla_k \nabla_j \omega_{i(p)} = -\frac{1}{p}\left[\sum_{a=1}^{p} R^l_{i_a kj}\omega_{\substack{i(l)\\ i(a)}} + \sum_{\substack{a<b\\ a=1}}^{p} R^l_{i_a i_b k}\omega_{\substack{(i,l)\\ (i_a)(i_b)}} + \sum_{a=1}^{p} \nabla_{i_a} \nabla_k \omega_{\substack{j(j)\\ (i_a)}}\right] \tag{2.5}$$

这里以 $\omega_{i(p)}$ 记 $\omega_{i_1 i_2 \cdots i_p}$;以 $\omega_{i(l)}$ 记 $\omega_{i_1 i_2 \cdots l \cdots i_p}$;以 $\omega_{i(j,k)}$ 记 $\omega_{i_1 \cdots j \cdots k \cdots i_p}$ $\cdots$(以下相同),则称 ω 为仿射 Killing p—形式.

注记:可假设 ω 为仿射联络空间里的 p—形式.

由定义可得以下结论

$$\triangle \omega = (p+1)\delta \nabla \omega + \mathrm{d}\delta\omega \tag{2.6}$$

$$\sum_{a=1}^{p}(\nabla_{i_a} \nabla_k \omega_{i(j)} - \nabla_{i_a} \nabla_j \omega_{i(k)}) = (p+1)\sum_{a=1}^{p} R^l_{kji_a}\omega_{i(l)} + \sum_{a<b}^{1\cdots p} R^l_{i_a i_b j}\omega_{\substack{i(k,l)\\ (i_a)(i_b)}} - R^l_{i_a i_b k}\omega_{\substack{i(j,l)\\ (i_a)(i_b)}} \tag{2.7}$$

这里 $\triangle$ 是拉氏算子,$\triangle = \delta\mathrm{d} + \mathrm{d}\delta$,$\delta$ 是上微分算子,而 d 是外微分算子.

(2) 定理的证明.

定理 1 的证明:因为 Killing p—形式 ω 满足

$$\nabla_k \omega_{i(p)} = -\frac{1}{p}\sum_{a=1}^{p} \nabla_{i_a}\omega_{i(k)}$$

所以

$$\nabla_j \nabla_k \omega_{i(p)} = -\frac{1}{p}\sum_{a=1}^{p} \nabla_j \nabla_{i_a}\omega_{i(k)}$$

$$=-\frac{1}{p}\Big[\sum_{a=1}^{p}(\nabla_j\nabla_{i_a}\omega_{i(k)}-\nabla_{i_a}\nabla_j\omega_{i(k)})+\sum_{a=1}^{p}\nabla_{i_a}\nabla_j\omega_{i(k)}\Big]$$

$$=-\frac{1}{p}\Big[\sum_{a=1}^{p}R^l_{i_ajk}\omega_{i(l)}+\sum_{a<b}^{p}(R^l_{i_ai_bj}\omega_{i(k,l)}+\sum_{a=1}^{p}\nabla_{i_a}\nabla_j\omega_{i(k)})\Big]$$

定理 2 的证明:因为

$$(\mathrm{d}\delta\omega)_{i(p)}=-\sum_{a=1}^{p}g^{jk}\nabla_{i_a}\nabla_j\omega_{i(k)}$$

$$=-\frac{1}{p}\Big[\sum_{a=1}^{p}R^l_{i_a}\omega_{i(l)}+\sum_{a<b}^{p}(R^{jk}_{i_ai_b}\omega_{i(j,k)}-g^{jk}\sum_{a=1}^{p}\nabla_k\nabla_{i_a}\omega_{i(j)}\Big]$$

$$=-\frac{1}{p}\Big[\sum_{a=1}^{p}R^l_{i_a}\omega_{i(l)}+\sum_{a<b}^{p}(R^{jk}_{i_ai_b}\omega_{i(j,k)}-$$

$$g^{jk}\sum_{a=1}^{p}(\nabla_k\nabla_{i_a}-\nabla_{i_a}\nabla_k)\omega_{i(j)}-$$

$$g^{jk}\sum_{a=1}^{p}\nabla_{i_a}\nabla_k\omega_{i(j)}\Big]$$

$$=-\frac{1}{p}(\mathrm{d}\delta\omega)_{i(p)}$$

所以

$$\mathrm{d}\delta\omega=0 \tag{2.8}$$

又因为 M 为紧致可走向的,所以

$$\delta\omega=0 \tag{2.9}$$

由(2.6)和(2.9),知 ω 为 Killing p— 形式.

定理 3 的证明:我们令

$$\omega_{ji(p)}=\nabla_j\omega_{i(p)}$$

则得下列偏微分方程组

$$\begin{cases}\nabla_j\omega_{i(p)}=\omega_{ji(p)} & (2.10)\\ \nabla_k\omega_{ji(p)}=-\frac{1}{p}\Big(\sum_{a=1}^{p}R^l_{i_akj}\omega_{i(l)}+\sum_{a<b}^{p}R^l_{i_ai_bk}\omega_{i(j,l)}+\sum_{a=1}^{p}\nabla_{i_a}\omega_{ki(j)}\Big) & (2.11)\end{cases}$$

(2.10) 的可积条件是

$$\nabla_k\nabla_j\omega_{i(p)}-\nabla_j\nabla_k\omega_{i(p)}=\sum_{a=1}^{p}R^l_{jki_a}\omega_{i(l)} \tag{2.12}$$

由(2.10)和(2.11),上式可改写为

$$(p+1)\sum_{a=1}^{p}R^l_{kji_a}\omega_{i(l)}=\sum_{a<b}^{p}(R^l_{i_ai_bk}\omega_{i(j,l)}-R^l_{i_ai_bj}\omega_{i(k,l)})+\sum_{a=1}^{p}\nabla_{i_a}(\omega_{ki(j)}-\omega_{ji(k)})$$

根据(2.7)和(2.10),上述恒等式成立.

(2.11) 的可积条件是

$$\nabla_h \nabla_k \omega_{ji(p)} - \nabla_k \nabla_h \omega_{ji(p)} = R^l_{khj} \omega_{li(p)} + \sum_{a=1}^{p} R^l_{khi_a} \omega_{i(j,l)} \tag{2.13}$$

根据(2.10) 和(2.11),(2.13) 可写为

$$\begin{aligned} R^l_{hkj} w_{li(p)} + \sum_{a=1}^{p} R^l_{hki_a} w_{ji(l)} = \frac{1}{p}\Big[& \sum_{a=1}^{p} (\nabla_h R^l_{i_a kj} w_{i(l)}) - \nabla_k R^l_{i_a hj} w_{i(l)} + \\ & (R^l_{i_a hj} w_{ki(l)} - R^l_{i_a kj} w_{hki(l)}) + \\ & (\nabla_h \nabla_{i_a} w_{ki(j)} - \nabla_k \nabla_{i_a} w_{hi(j)}) + \\ & \sum_{a<b}^{p} (\nabla_h R^l_{i_a i_b k} - \nabla_k R^l_{i_a i_b h}) w_{i(j,l)} + \\ & \sum_{a<b}^{p} (R^l_{i_a i_b k} w_{hi(j,l)} - R^l_{i_a i_b h} w_{ki(j,l)}) \Big] \end{aligned}$$

但是

$$\begin{aligned} \sum_{a=1}^{p} (\nabla_h \nabla_{i_a} w_{ki(j)} - \nabla_k \nabla_{i_a} w_{hi(j)}) = & \sum_{a=1}^{p} [(\nabla_h \nabla_{i_a} w_{ki(j)} - \nabla_{i_a} \nabla_h w_{ki(j)}) + \\ & (\nabla_{i_a} \nabla_k w_{hi(j)} - \nabla_k \nabla_{i_a} w_{hi(j)}) + \\ & (\nabla_{i_a} \nabla_h w_{ki(j)} - \nabla_{i_a} \nabla_k w_{hi(j)})] \\ = & \sum_{a=1}^{p} (R^l_{i_a hk} w_{li(j)} + R^l_{i_a hj} w_{ki(j)} + R^l_{ki_a h} w_{li(j)} + \\ & R^l_{ki_a j} w_{hi(l)} + R^l_{khj} w_{i_a j(l)}) + \\ & \sum_{a<b}^{p} (R^l_{i_a i_b h} w_{ki(j,l)} - R^l_{i_a i_b k} w_{hi(j,l)} + \\ & R^l_{khi_b} w_{i_a i(j,l)} - R^l_{khi_a} w_{i_b i(j,l)}) + \\ & \sum_{a=1}^{p} \nabla_{i_a} R^l_{khj} w_{i(l)} + \sum_{a<b}^{p} (\nabla_{i_a} R^l_{khi_b} - \nabla_{i_b} R^l_{khi_a}) w_{i(j,l)} \end{aligned}$$

于是,我们得到

$$\begin{aligned} & PR^l_{khj} w_{li(p)} + \sum_{a=1}^{p} (pR^l_{khi_a} w_{ji(l)} + R^l_{khi_a} w_{li(j)} + R^l_{j(l)} w_{i_a j(l)}) + \\ & \sum_{a<b}^{p} (R^l_{khi_b} w_{i_a j(l)} - R^l_{khi_a} w_{i_b i(j,l)}) + \\ & \sum_{a<b}^{p} (\nabla_h R^l_{i_a i_b k} - \nabla_k R^l_{i_a i_b h} + \nabla_{i_a} R^l_{khi_b} - \nabla_{i_b} R^l_{khi_a}) w_{i(j,l)} = 0 \end{aligned} \tag{2.14}$$

由(2.14),我们得到

$$\Big[pR^a_{khj} \delta^{a_1 \cdots a_p}_{i_1 \cdots i_p} + \sum_{s=1}^{p} \Big(pR^{a_s}_{khi_s} \delta^{aa_1 \cdots \tilde{a}_s \cdots a_p}_{ji_1 \cdots \tilde{i}_s \cdots i_p} + R^a_{khi_s} \delta^{a_1 \cdots a_s \cdots a_p}_{i_1 \cdots j \cdots i_p} + R^a_{khj} \delta^{aa_1 \cdots \tilde{a}_s \cdots a_p}_{i_s i_1 \cdots \tilde{i}_s \cdots i_p} \Big) +$$

$$\sum_{s<t}^{p}(R^{a_t}_{khi_t}\delta^{aa_1\cdots a_s\cdots\widetilde{a}_t\cdots a_p}_{i_si_1\cdots j\cdots\widetilde{i}_t\cdots i_p}-R^{a_t}_{khi_s}\delta^{aa_1\cdots a_s\cdots\widetilde{a}_t\cdots a_p}_{i_ti_1\cdots j\cdots\widetilde{i}_t\cdots i_p})\Big]w_{aa_1\cdots a_p}+$$

$$\sum_{s<t}^{p}(\nabla_h R^{a_t}_{i_si_tk}-\nabla_k R^{a_t}_{i_si_th}-\nabla_{i_s}R^{a_t}_{hki_t}-\nabla_{i_t}R^{a_t}_{khi_s})\delta^{a_1\cdots a_s\cdots\widetilde{a}_t\cdots a_p}_{i_1\cdots j\cdots\widetilde{i}_t\cdots i_p}w_{a\cdots a_p}=0 \qquad (2.15)$$

这里

$$\delta^{a_1\cdots\widetilde{a}_s\cdots a_p}_{i_1\cdots\widetilde{i}_s\cdots i_p}=\delta^{a_1}_{i_1}\delta^{a_2}_{i_2}\cdots\delta^{a_{s-1}}_{i_{s-1}}\delta^{a_{s+1}}_{i_{s+1}}\cdots\delta^{a_p}_{i_p}$$

对于任意一点 $q\in M$，存在仿射 Killing p－形式 w 使得 $(w_{i(p)})_q=a_{i(p)}$ 和 $\nabla_j w_{i(p)}=a_{ji(p)}$，这里 $a_{i(p)}$ 和 $a_{ji(p)}$ 关于指标 $j,i_1,\cdots,i_p$ 是反对称的，因 $w_{i(p)}$ 和 $w_{ji(p)}$ 满足(2.15)，故

$$\sum_{\sigma\in\pi_p}\operatorname{sgn}(\sigma)\theta_1^{a_{\sigma(1)\cdots\sigma(p)}}=0 \qquad (2.16)$$

和

$$\sum_{\sigma\in\pi_p}\operatorname{sgn}(\sigma)\theta_2^{aa_{\sigma(1)\cdots\sigma(p)}}=0 \qquad (2.17)$$

$$\operatorname{sgn}(\sigma)=\begin{cases}1,\text{当 }\sigma\text{ 是偶置换}\\-1,\text{当 }\sigma\text{ 是奇置换}\end{cases}$$

这里 π_p 是关于 p 的量换解，且

$$\theta_1^{a_1\cdots a_p}=\sum_{s<t}^{p}(\nabla_h R^{a_t}_{i_si_tk}-\nabla_k R^{a_t}_{i_si_th}-\nabla_{i_s}R^{a_t}_{khi_t})\delta^{a_1\cdots a_s\cdots\widetilde{a}_t\cdots a_p}_{i_1\cdots j\cdots\widetilde{i}_t\cdots i_p}$$

$$\theta_2^{aa_1\cdots a_p}=pR^{a}_{khj}\delta^{a_1\cdots a_p}_{i_1\cdots i_p}+\sum_{s=1}^{p}(pR^{a_s}_{khi_s}\delta^{aa_1\cdots\widetilde{a}_s\cdots a_p}_{ji_1\cdots\widetilde{i}_s\cdots i_p}+R^{a}_{khi_s}\delta^{a_1\cdots a_s\cdots a_p}_{i_1\cdots j\cdots i_p}+R^{a_s}_{khj}\delta^{aa_1\cdots\widetilde{a}_s\cdots a_p}_{i_si_1\cdots\widetilde{i}_s\cdots i_p})+$$

$$\sum_{s<t}^{p}(R^{a_t}_{khi_1}\delta^{aa_1\cdots a_s\cdots\widetilde{a}_t\cdots a_p}_{i_si_1\cdots j\cdots\widetilde{i}_t\cdots i_p}-R^{a_t}_{khi_s}\delta^{aa_1\cdots a_s\cdots\widetilde{a}_t\cdots a_p}_{i_si_1\cdots j\cdots\widetilde{i}_t\cdots i_p})$$

在(2.17)中，令 $i_1=a_1,i_2=a_2,\cdots,i_p=a_p$ 且分别关于 $a_1,a_2,\cdots,a_p$ 作和，得到

$$(pn^p+b_{p-1}n^{p-1}+b_{p-2}n^{p-2}+\cdots+b_0)R^{n}_{khj}=0$$

这里 $b_i(i=0,1,\cdots,p-1)$ 是整数，故得

$$R^{a}_{khj}=0 \qquad (2.18)$$

因此 M 是平坦流形.

反之，如果 M 是平坦的，那么有(2.18)，于是满足(2.14)，故(2.10)或(2.11)是完全可积的，证毕.

四川师范大学川北教育学院数学系的张三华教授2000年发表了《容有 Killing 矢量场的

Riemann 流形的超曲面》[①].

1. 预备知识

设(M^{n+1},g)是$n+1$维具有正定度量的C^∞级 Riemann 流形，$\boldsymbol{X}$是M^{n+1}上的矢量场. 在局部坐标系(u,x^A)下，$\boldsymbol{X}=X^A\dfrac{\partial}{\partial x^A}$，$X_A=X^B g_{BA}$，这里和下文中的$A,B,C,\cdots=1,2,\cdots,n+1$. 若$\boldsymbol{X}$满足$\mathscr{L}_{\boldsymbol{X}}g=0$，则称$\boldsymbol{X}$为 Killing 矢量场，局部地有

$$\mathscr{L}_{\boldsymbol{X}}g_{BC}=\nabla_B X_C+\nabla_C X_B=0 \tag{1.1}$$

这里，∇_A是关于度量g的共变微分算子，$\mathscr{L}_{\boldsymbol{X}}$是关于$\boldsymbol{X}$的 Lie 导数算子. 显然 Killing 矢量场一定是仿射 Killing 矢量场，即有

$$\nabla_A\nabla_B X_C+R_{DABC}X^D=0 \tag{1.2}$$

其中，R_{DABC}是M^{n+1}的 Riemann 曲率张量的分量.

再设$(\overline{M}^n,\overline{g})$是等距地浸入在$M^{n+1}$中的超曲面，其浸入映射为$x^A=x^A(u^a)$，这里$\{u^a\}$是$\overline{M}^n$的局部坐标，令$B_a^A=\dfrac{\partial x^A}{\partial u^a}$，则

$$\overline{g}_{ab}=g_{AB}B_a^A B_b^B \tag{1.3}$$

这里和下文中的$a,b,\cdots=1,2,\cdots,n$.

沿超曲面$\overline{M}^n$，矢量场$\boldsymbol{X}$有分解式：$\boldsymbol{X}=B\overline{\boldsymbol{X}}+\alpha\boldsymbol{N}$，局部地

$$X^A=B_a^A\overline{X}^a+\alpha N^A \tag{1.4}$$

其中，$\boldsymbol{N}=N^A\dfrac{\partial}{\partial x^A}$是$\overline{M}^n$在$M^{n+1}$中的单位法矢量场，$\overline{\boldsymbol{X}}=\overline{X}^a\dfrac{\partial}{\partial u^a}$是$\overline{\boldsymbol{X}}$在$\overline{M}^n$上的诱导矢量场，$\alpha=g(\boldsymbol{X},\boldsymbol{N})$是$\boldsymbol{X}$在$\boldsymbol{N}$上的投影，$|\alpha|$是$\boldsymbol{X}$的法部的长.

超曲面$\overline{M}^n$在M^{n+1}中的 Gauss 公式和 Weingarten 公式分别为

$$\overline{\nabla}_b B_a^A=h_{ba}N^A \tag{1.5}$$

$$\overline{\nabla}_b N^A=-h_b^a B_a^A \tag{1.6}$$

Codazzi 方程为

$$R_{CBAD}B_c^C B_b^B B_a^A N^D=\overline{\nabla}_c h_{ab}-\overline{\nabla}_b h_{ac} \tag{1.7}$$

这里，$\overline{\nabla}_b$是广义共变微分算子，h_{ab}是$\overline{M}^n$的第二基本张量的分量，$h_b^a=h_{bc}\overline{g}^{ca}$. 由(1.4)有

$$B_c^C B_b^B\nabla_B X_C=\overline{\nabla}_b\overline{X}_c-\alpha h_{bc} \tag{1.8}$$

---

① 摘编自《四川师范大学学报(自然科学版)》，1999 年，第 22 卷第 6 期.

$$N^C B_b^B \nabla_B X_C = h_{ab} X^a + \overline{\nabla}_b \alpha \tag{1.9}$$

容有 Killing 矢量场 $\boldsymbol{X}$ 的 Riemann 流形的超曲面 $\overline{M}^n$ 上的诱导矢量场 $\overline{\boldsymbol{X}}$ 有

$$\overline{\nabla}_a \overline{\nabla}_b \overline{X}_c = -\overline{R}_{dabc} \overline{X}_d + \overline{\nabla}_a (\alpha h_{bc}) + \overline{\nabla}_b (\alpha h_{ac}) - \overline{\nabla}_c (\alpha h_{ab}) \tag{1.10}$$

$$\overline{\nabla}_b \overline{X}_c + \overline{\nabla}_c \overline{X}_b = 2\alpha h_{bc} \tag{1.11}$$

$\overline{R}_{dabc}$ 是 $\overline{M}^n$ 的 Riemann 流形的曲率张量的分量.

在紧致可定向的 Riemann 流形 M 里,对于任意矢量场 $\boldsymbol{X}$,有 Green 定理成立

$$\int_M (\operatorname{div} \boldsymbol{X}) * 1 = 0 \tag{1.12}$$

这里,$\operatorname{div} \boldsymbol{X} = \nabla_A X^A$ 是矢量场 X^A 的散度,$*1$ 表示 Riemann 流形 M 的体积元素.

在紧致可定向的 Riemann 流形 M 里,对于任意的数量函数 f,有

$$\int_M \triangle f * 1 = 0 (\triangle \text{ 是 Laplace 算子}) \tag{1.13}$$

作者给出了容有 Killing 矢量场的 Riemann 流形 M^{n+1} 的紧致可定向的超曲面 $\overline{M}^n$ 上的几个积分公式,并给出这些公式的一些应用.

2. 超曲面的积分公式

定理 1 若 $\boldsymbol{X}$ 是 Riemann 流形 M^{n+1} 的 Killing 矢量场,则在紧致可定向的超曲面 $\overline{M}^n$ 上有积分公式

$$\int_{\overline{M}} (\alpha \boldsymbol{h}) * 1 = 0 \tag{2.1}$$

其中,$\boldsymbol{h} = \overline{g}^{ab} h^{ab}$.

证明 用 $\overline{g}^{bc}$ 与(1.11)缩并得

$$\overline{\nabla}_b \overline{X}^b = \alpha \boldsymbol{h} \tag{2.2}$$

由(1.12)有(2.1)成立.

定理 2 若 $\boldsymbol{X}$ 是 Riemann 流形 M^{n+1} 的 Killing 矢量场,则在紧致可定向的超曲面 $\overline{M}^n$ 上有积分公式

$$\int_{\overline{M}} [\alpha \sigma^2 + \overline{\boldsymbol{g}}(\delta \boldsymbol{H}, \overline{\boldsymbol{X}})] * 1 = 0 \tag{2.3}$$

其中,$\sigma^2 = h^{bd} h_{bd}$ 为第二基本张量的长度的平方,$\boldsymbol{H}$ 由 $h(\overline{\boldsymbol{X}}, \overline{\boldsymbol{Y}}) = \overline{\boldsymbol{g}}(\boldsymbol{H}\overline{\boldsymbol{X}}, \overline{\boldsymbol{Y}})$ 定义,$\delta \boldsymbol{H} : \overline{g}^{bd} \overline{\nabla}_d h_{ab}$,有 $\overline{\boldsymbol{g}}(\delta \boldsymbol{H}, \overline{\boldsymbol{X}}) : \overline{X}^a \overline{\nabla}_d h_a^d$.

证明 因为

$$\overline{\nabla}_d (h_a^d \overline{X}^a) = \overline{X}^a \overline{\nabla}_d h_a^d + h^{bd} \overline{\nabla}_d \overline{X}_b \tag{2.4}$$

把(1.8)代入(2.4),并用(1.1)化简有

$$\overline{\nabla}_d(h_a^d\overline{X}^a)=\overline{X}^a\overline{\nabla}_d h_a^d+\alpha h^{bd}h_{bd} \tag{2.5}$$

因而(2.3)成立.

定理 3 若 $\boldsymbol{X}$ 是 M^{n+1} 的 Killing 矢量场,则在紧致可定向的超曲面 $\overline{M}^n$ 上有积分公式

$$\int_{\overline{M}}[\boldsymbol{R}(\boldsymbol{X},\boldsymbol{N})+\alpha\sigma^2+\overline{\boldsymbol{g}}(\delta\boldsymbol{H},\overline{\boldsymbol{X}})]*1=0 \tag{2.6}$$

证明 用 $\overline{\nabla}$ 运算(1.9),并利用(1.5)(1.6)有

$$\begin{aligned}\overline{\nabla}_d\overline{\nabla}_b\alpha=&-h_d^cB_c^CB_b^B\nabla_BX_C+h_{db}N^CN^B\nabla_BX_C+\\&N^CB_b^BB_d^D\nabla_D\nabla_BX_C-\overline{\nabla}_d(h_{ab}\overline{X}^a)\end{aligned} \tag{2.7}$$

用 $\overline{g}^{db}$ 与(2.7)缩并,并用(1.1)化简有

$$\Delta\alpha=\overline{g}^{db}\overline{\nabla}_d\overline{\nabla}_b\alpha=\overline{g}^{db}B_b^BB_d^DN^C\nabla_D\nabla_BX_C-\overline{\nabla}_d(h_a^d\overline{X}^a) \tag{2.8}$$

$$\overline{g}^{db}B_d^BB_d^D=g^{DB}-N^DN^B \tag{2.9}$$

把(2.9)代入(2.8),并利用(1.1)化简有

$$\Delta\alpha=g^{DB}N^C\nabla_D\nabla_BX_C-\overline{\nabla}_d(h_a^d\overline{X}^a) \tag{2.10}$$

把(1.2)(2.5)代入(2.10),有

$$\Delta\alpha=-N^CX^AR_{AC}-\alpha h^{bd}h_{bd}-\overline{X}^a\overline{\nabla}_dh_a^d \tag{2.11}$$

因而得(2.6).

定理 4 若 $\boldsymbol{X}$ 是 M^{n+1} 的 Killing 矢量场,则在紧致可定向的超曲面上有积分公式

$$\int_{\overline{M}}[\alpha\boldsymbol{R}(\boldsymbol{N},\boldsymbol{N})+\alpha\sigma^2+\mathscr{L}_{\overline{\boldsymbol{X}}}\boldsymbol{h}]*1=0 \tag{2.12}$$

其中,$\mathscr{L}_{\boldsymbol{X}}$ 是关于 $\boldsymbol{X}$ 的 Lie 导数算子.

证明 由(1.7)有

$$R_{CD}B_c^CN^D=\overline{\nabla}_c\boldsymbol{h}-\overline{\nabla}_bh_c^b \tag{2.13}$$

把(1.4)代入(2.11),并利用(2.13)有

$$\Delta\alpha=-\overline{X}^a\overline{\nabla}_a\boldsymbol{h}-\alpha R_{AC}N^CN^A-\alpha h^{bd}h_{bd} \tag{2.14}$$

因而有(2.12).

定理 5 若 $\boldsymbol{X}$ 是 M^{n+1} 的 Killing 矢量场,则在紧致可定向的超曲面 $\overline{M}^n$ 上有积分公式

$$\begin{aligned}\int_{\overline{M}}[&\overline{\boldsymbol{R}}(\overline{\boldsymbol{X}},\overline{\boldsymbol{X}})-|\overline{\nabla}\,\overline{\boldsymbol{X}}|^2-2\alpha\overline{\boldsymbol{g}}(\delta\boldsymbol{H},\overline{\boldsymbol{X}})-\\&2\boldsymbol{h}(\overline{\boldsymbol{D}}\alpha,\overline{\boldsymbol{X}})+\alpha\mathscr{L}_{\overline{\boldsymbol{X}}}\boldsymbol{h}+\boldsymbol{h}\mathscr{L}_{\overline{\boldsymbol{X}}}\alpha]*1=0\end{aligned} \tag{2.15}$$

其中,$\overline{\boldsymbol{D}}\alpha:\overline{g}^{ac}\overline{\nabla}_c\alpha$,$\boldsymbol{h}(\overline{\boldsymbol{D}}\alpha,\overline{\boldsymbol{X}}):\overline{X}^ah_a^d\overline{\nabla}_d\alpha$.

证明 $\Delta|\overline{\boldsymbol{X}}^2| = \overline{g}^{dc}\overline{\nabla}_d\overline{\nabla}_c(\overline{X}^a\overline{X}_a) =$

$$2(\overline{\nabla}^c\overline{X}^a)(\overline{\nabla}_c\overline{X}_a) + 2\overline{X}^a\overline{g}^{dc}\overline{\nabla}_d\overline{\nabla}_c\overline{X}_a \tag{2.16}$$

把(1.10)代入(2.16),有

$$\Delta|\overline{\boldsymbol{X}}|^2 = 2(\overline{\nabla}^c\overline{X}^a)(\overline{\nabla}_c\overline{X}_a) - 2\overline{X}^a\overline{X}^b\overline{R}_{ab} + 4\overline{X}^a h_a^d \overline{\nabla}_d\alpha + 4\alpha\overline{X}^a\overline{\nabla}_d h_a^d - 2\alpha\overline{X}^a\overline{\nabla}_a\boldsymbol{h} - 2\boldsymbol{h}\overline{X}^a\overline{\nabla}_a\alpha \tag{2.17}$$

从而有(2.15).

定理 6 若 $\boldsymbol{X}$ 是 M^{n+1} 的 Killing 矢量场,则在紧致可定向的超曲面 $\overline{M}^n$ 上有积分公式

$$\int_{\overline{M}}[\overline{\boldsymbol{R}}(\overline{\boldsymbol{X}},\overline{\boldsymbol{X}}) - |\overline{\nabla}\,\overline{\boldsymbol{X}}|^2 + \frac{1}{2}|\mathscr{L}_{\overline{\boldsymbol{X}}}\overline{\boldsymbol{g}}|^2 + \alpha\mathscr{L}_{\overline{\boldsymbol{X}}}\boldsymbol{h} + \boldsymbol{h}\mathscr{L}_{\overline{\boldsymbol{X}}}^{\alpha}] * 1 = 0 \tag{2.18}$$

证明 $\overline{\nabla}_b[\overline{X}^a(\overline{\nabla}_a\overline{X}^b)] = (\overline{\nabla}_b\overline{X}_c)(\overline{\nabla}^c\overline{X}^b) +$

$$\overline{X}^a\overline{g}^{bc}\overline{\nabla}_b\overline{\nabla}_a\overline{X}_c \tag{2.19}$$

由 Yano 有

$$(\overline{\nabla}_b\overline{X}_c)(\overline{\nabla}^c\overline{X}^b) = \frac{1}{2}|\mathscr{L}_{\overline{\boldsymbol{X}}}\overline{\boldsymbol{g}}|^2 - (\overline{\nabla}_b\overline{X}_c)(\overline{\nabla}^b\overline{X}^c) \tag{2.20}$$

把(2.20)(1.10)代入(2.19),有

$$\overline{\nabla}_b[\overline{X}^a(\overline{\nabla}_a\overline{X}^b)] = \overline{X}^a\overline{X}^b\overline{R}_{ab} - (\overline{\nabla}^b\overline{X}^c)(\overline{\nabla}_b\overline{X}_c) + \frac{1}{2}|\mathscr{L}_{\overline{X}}\overline{g}|^2 + \alpha\overline{X}^a\overline{\nabla}_a\boldsymbol{h} + \boldsymbol{h}\overline{X}^a\overline{\nabla}_a\alpha \tag{2.21}$$

故有(2.18).

定理 7 若 $\boldsymbol{X}$ 是 M^{n+1} 的 Killing 矢量场,则在紧致可定向的超曲面 $\overline{M}^n$ 上有积分公式

$$\int_{\overline{M}}[(\alpha\boldsymbol{h})^2 + \alpha\mathscr{L}_{\overline{\boldsymbol{X}}}\boldsymbol{h} + \boldsymbol{h}\mathscr{L}_{\overline{\boldsymbol{X}}}^{\alpha}] * 1 = 0 \tag{2.22}$$

证明

$$\overline{\nabla}_b[\overline{X}^b(\overline{\nabla}_a\overline{X}^a)] = (\overline{\nabla}_a\overline{X}^a)^2 + \overline{X}^b\overline{\nabla}_b\overline{\nabla}_a\overline{X}^a \tag{2.23}$$

把(2.2)代入(2.23),有

$$\overline{\nabla}_b[\overline{X}^b(\overline{\nabla}_a\overline{X}^a)] = (\alpha\boldsymbol{h})^2 + \boldsymbol{h}\overline{X}^b\overline{\nabla}_b\alpha + \alpha\overline{X}^b\overline{\nabla}_b\boldsymbol{h} \tag{2.24}$$

所以有(2.22).

3. 积分公式的应用

推论 1 ~ 8 是在以下假定下作出的:

(i) $\boldsymbol{X}$ 是 Riemann 流形 M^{n+1} 的 Killing 矢量场.

(ii)$\overline{M}^n$ 是 M^{n+1} 的紧致可定向超曲面.

(iii) 沿 $\overline{M}^n$ 有 $\boldsymbol{X}=\boldsymbol{B}\overline{\boldsymbol{X}}+\alpha\boldsymbol{N}$.

因此,由上述定理可得下面的推论(为了简化叙述,下面推论中省略了这些大前提).

推论 1　若 $\boldsymbol{X}$ 与 $\boldsymbol{N}$ 的夹角不大于直角,且平均曲率非负,则 $\boldsymbol{X}$ 切于 $\overline{M}^n$ 或 $\overline{M}^n$ 是极小曲面.

推论 2　若第二基本张量平行,且 $\boldsymbol{X}$ 与 $\boldsymbol{N}$ 的夹角不大于 $\frac{\pi}{2}$,则 $\boldsymbol{X}$ 切于 $\overline{M}^n$ 或 $\overline{M}^n$ 是全测地曲面.

推论 3　若 $\overline{M}^n$ 的平均曲率是常数,且 $\boldsymbol{R}(\boldsymbol{N},\boldsymbol{N})$ 非负定,α 为非负数,则 $\boldsymbol{X}$ 切于 $\overline{M}^n$ 或 $\overline{M}^n$ 是全测地曲面,且 $\overline{M}^n$ 沿 $\boldsymbol{N}$ 方向的 Ricc 曲率为零.

推论 4　若 $\overline{\boldsymbol{R}}(\overline{\boldsymbol{X}},\overline{\boldsymbol{X}})$ 非正定,且 $\boldsymbol{X}$ 的法部有常长、第二基本张量平行,则 $\overline{M}^n$ 沿 $\overline{\boldsymbol{X}}$ 方向的 Ricc 曲率为零且 $\overline{\boldsymbol{X}}$ 是 $\overline{M}^n$ 的平行矢量场.

由(1.11)(2.20),有

$$|\mathscr{L}_{\overline{X}}\overline{\boldsymbol{g}}|^2=4\alpha^2\sigma^2 \tag{3.1}$$

把(3.1)代入(2.18),有

$$\int_{\overline{M}}[\overline{\boldsymbol{R}}(\overline{\boldsymbol{X}},\overline{\boldsymbol{X}})-|\overline{\nabla}\,\overline{\boldsymbol{X}}|^2+2\alpha^2\sigma^2+\alpha\mathscr{L}_{\overline{X}}\boldsymbol{h}+\boldsymbol{h}\mathscr{L}_{\overline{X}}\alpha]*1=0 \tag{3.2}$$

推论 5　若 $\overline{\boldsymbol{R}}(\overline{\boldsymbol{X}},\overline{\boldsymbol{X}})$ 非负定,且 $\boldsymbol{X}$ 的法部有常长,$\overline{M}^n$ 的平均曲率为常数,$\overline{\boldsymbol{X}}$ 是平行矢量场,则 $\overline{M}^n$ 沿 $\overline{\boldsymbol{X}}$ 方向的 Ricc 曲率为零,且 $\boldsymbol{X}$ 切于 $\overline{M}^n$ 或 $\overline{M}^n$ 是全测地曲面.

由(2.18)～(2.15),有

$$\int_{\overline{M}}[\frac{1}{2}|\mathscr{L}_{\overline{X}}\overline{\boldsymbol{g}}|^2+2\alpha\overline{\boldsymbol{g}}(\delta\boldsymbol{H},\overline{\boldsymbol{X}})+2\boldsymbol{h}(\overline{\boldsymbol{D}}\alpha,\overline{\boldsymbol{X}})]*1=0$$

推论 6　若 $\boldsymbol{X}$ 的法部有常长,且第二基本张量平行,则 $\overline{\boldsymbol{X}}$ 是 $\overline{M}^n$ 上的 Killing 矢量场.

推论 7　若 $\overline{M}^n$ 的平均曲率为常数,且 $\boldsymbol{X}$ 的法部有常长,则 $\boldsymbol{X}$ 切于 $\overline{M}^n$ 或 $\overline{M}^n$ 是极小曲面.

由(3.2)～(2.22),有

$$\int_{\overline{M}}[\overline{\boldsymbol{R}}(\overline{\boldsymbol{X}},\overline{\boldsymbol{X}})-|\overline{\nabla}\,\overline{\boldsymbol{X}}|^2+2\alpha^2\sigma^2-(2\boldsymbol{h})^2]*1=0$$

推论 8　若 $\boldsymbol{X}$ 切于 $\overline{M}^n$ 且 $\overline{\boldsymbol{R}}(\overline{\boldsymbol{X}},\overline{\boldsymbol{X}})$ 非正定,则 $\overline{M}^n$ 沿 $\overline{\boldsymbol{X}}$ 方向的 Ricc 曲率为零,且 $\overline{\boldsymbol{X}}$ 是 $\overline{M}^n$ 的平行矢量场.

再来介绍一下本书中论及的 Ricci 孤子. 2009 年武汉大学的陈立博士在陈文艺教授的指

导下完成了博士论文《有关 Ricci 孤子的几何与分析问题》，在文中的第一章就是“Ricci 孤子简介”，不妨摘录一段：

1. Ricci 流和 Ricci 孤子

自从 Riemann 几何从 19 世纪引入，曲率就被用来研究流形的拓扑性质. 这一领域著名的古典结果是 Gauss-Bonnet 定理：如果 M 是紧致无边的定向曲面，那么有公式成立

$$\frac{1}{2\pi}\int_M K\,\mathrm{d}A=\chi(M)$$

这个公式联系了曲率积分和 Euler 示性数这个拓扑不变量. 现在，因为与一个空间相容的度量有很多，所以在流形上引进度量增加它的复杂性. 我们试着通过考虑如下问题来了解这个复杂性.

(1) 是否可以在 M 上找到最好的度量，即曲率尽可能简单？

(2) M 的拓扑怎样影响 M 上度量的曲率，怎样得到最好的度量？

例如，如果 M 是一个二维环面，那么 Gauss-Bonnet 公式表明 M 上不能有严格正或严格负曲率的度量. 另外，单值化原理说任意曲面(紧或非紧) 上的任意完备度量能通过乘一个光滑函数变成一个非常好的度量，即常曲率中的一种. 有时如果流形有一个很好的度量，我们能从度量知道它的拓扑. 例如，如果 M 是一个具有常负曲率完备度量的单连通曲面，那么 M 一定是上半平面. 即使 M 不是单连通的，我们把常负曲率度量提升到它的万有覆盖 $\widetilde{M}$，$\widetilde{M}$ 因此也是上半平面. 而且这个覆盖的覆盖变换一定同构于 $\widetilde{M}$，因此它们一定在 $SL(2;R)$ 中. 所以，任何双曲曲面 M 一定是上半平面经 $SL(2;R)$ 的子群作用形成的商空间. 在这个意义下，我们说上半平面是双曲曲面的一个模型. 在高一维的情形，Thurston 猜想 3 维流形的拓扑能通过一般化上面的讨论而得到分类. 简而言之，我们希望表明每一个紧 3 维流形能切成一片片的，而每一片具有一个很好的度量以致它的万有覆盖是 8 个模型之一. Thurston 猜想本质上是利用对几何很好的理解来研究拓扑问题.

Ricci 流是 Riemann 度量的一种演化方式，是获得好的度量的方法之一. 它的演化由偏微分方程

$$\frac{\partial g}{\partial t}=-2\mathrm{Ric}(g)$$

确定，这就是著名的 Ricci 流方程. 一般地，我们从 M 上满足某个十分一般条件 C 的度量 g_0 开始，证明随 Ricci 的演化，度量 g_t 收敛于一个有更多限制的条件 C_0 的好度量. 例如：

(1) 如果 M 是紧致单连通 3 维流形，g_0 具有正 Ricci 曲率，那么 g_t 收敛于常正截面曲率的度量. 因此，M 一定是 $\mathbf{S}^3$ 的一个商空间.

(2) 如果 M 是紧致定向曲面，那么 g_t 收敛于和 M 的拓扑相对应的常曲率度量，而不管初始度量 g_0 如何.

(3) 如果 M 是紧致的 4 维流形，且当 g_0 的曲率张量视为 M 上全体 2 形式组成的空间的 2 次形式是半正定的，那么 g_t 收敛于 $\mathbf{S}^4$，$\mathbf{CP}^2$，$\mathbf{S}^3\times\mathbf{R}$，$\mathbf{S}^2\times\mathbf{S}^2$，$\mathbf{S}^2\times\mathbf{R}^2$ 上的标准度量之一. 因此，M 一定是这些空间之一的商空间.

(4) 如果 M 是紧致复曲面，g_0 是具有非负全纯的双截面曲率的 Kähler 度量，那么当 t 充分大时，g_t 仍旧是 Kähler 度量而且具有正的全纯的双截面曲率. 利用代数几何的讨论，我们知道 M 双全纯同构于 $\mathbf{CP}^2$.

严格地说，上面关于紧流形的结果是应用于归一化的 Ricci 流得到的，即

$$\frac{\partial g}{\partial t}=-2\mathrm{Ric}(g)+\frac{2}{n}r(g)g$$

其中 $r(g)$ 等于数量曲率的积分除以流形的体积. 归一化是不合理的操作以至当度量改变时，流形的体积仍旧是常数.（没有归一化，曲率为常数 1 的球 $\mathbf{S}^n$ 在时间为 $\frac{1}{2(n-1)}$ 时将会变成一个点. 另外，对于 Einstein 度量，上面方程的右边为 0，归一化后的 Ricci 流保持度量不变.）因为上面两个流在伸缩和时间从新参数化意义下是等价的. 所以，我们通常处理非归一化流，但利用归一化流证明收敛性结果.

然而，在 3 维紧流形上给定任意初始度量，标准化流在有限时间内可能产生奇点. 关于 Ricci 流最重要的猜想是 3 维流形的拓扑影响它的奇点的产生. 具体来说，就是 Thurston 的几何度量化猜想，已被 Perelamn 证明. 因此，理解 Ricci 流的奇点非常重要.

我们习惯称 Ricci 流方程的对称演化解为孤子，也叫作自相似解. 为了解释沿着对称性演化，考虑普通的一维热方程 $f_t=f_{xx}$. 不难看出，下列矢量场是方程对称性的无穷小生成元，也即每一个产生一个单参数变换群把解变成解

$$X_1=\frac{\partial}{\partial x}\quad 平移空间$$

$$X_2=\frac{\partial}{\partial t}\quad 时间空间$$

$$X_3=f\frac{\partial}{\partial f}\quad f\ 的伸缩$$

$$X_4=x\frac{\partial}{\partial x}+2t\frac{\partial}{\partial t}\quad x\ 和\ t\ 方向的伸缩$$

对于基本解 $f(x)=\frac{1}{\sqrt{t}}\mathrm{e}^{-\frac{x^2}{4t}}$，我们看到对于任意 $\lambda>0$

$$f(\lambda x,\lambda t)=\lambda^{-2}f(x,t)$$

这个解是沿着矢量场 $X=X_4-2X_3$ 演化的孤子(事实上,X 和 $f(x,t)$ 的图像相切).

流形 M 上的 Ricci 流 $\frac{\partial g}{\partial t}=-2\mathrm{Ric}(g)$ 是一个张量方程,所以 Ricci 流在 M 的微分同胚下保持对称. 我们定义 Ricci 孤子为 Ricci 流 $(M,g(t))$,$0\leqslant t<T\leqslant\infty$,它在 t 时刻与初始度量 $g(0)$ 在微分同胚映射 ϕ_t 下的拉回相差一个常数 $\sigma(t)$,这里 ϕ_t 是由矢量场 X 产生的单参数微分同胚群. 也就是说,在 Ricci 孤子上所有的 Riemann 流形 $(M,g(t))$ 在相差一个随 t 变化的收缩因子的意义下同构. 一种产生 Ricci 孤子的方法如下:假定在流形 M 上我们有一个矢量场 X,常数 λ 和初始度量 $g(0)$ 使得

$$-\mathrm{Ric}(g(0))=\frac{1}{2}\mathscr{L}_Xg(0)-\lambda g(0) \tag{1.1}$$

如果 $\lambda\leqslant 0$,$T=\infty$;如果 $\lambda>0$,$T=(2\lambda)^{-1}$. 于是,对于任意 $t\in[0,T)$,我们定义函数

$$\sigma(t)=1-2\lambda t$$

和矢量场

$$Y_t(x)=\frac{X(x)}{\sigma(t)}$$

于是,我们定义 ϕ_t 为时间依赖矢量场 Y_t 产生的单参数微分同胚群.

命题 1 流 $(M,g(t))$ 是孤子,其中 $0\leqslant t<T$,$g(t)=\sigma(t)\phi_t^*g(0)$.

证明 我们只需检查一下 Ricci 流方程即可. 我们有

$$\begin{aligned}\frac{\partial g(t)}{\partial t}&=\sigma'(t)\phi_t^*g(0)+\sigma(t)\phi_t^*\mathscr{L}_{Y(t)}g(0)\\&=\phi_t^*(-2\lambda+\mathscr{L}_{X(t)})g(0)\\&=\phi_t^*(-2\mathrm{Ric}(g(0)))=-2\mathrm{Ric}(\phi_t^*(g(0)))\end{aligned}$$

因为对于任意 α,有 $\mathrm{Ric}(g)=\mathrm{Ric}(\alpha g)$,所以有

$$\frac{\partial g}{\partial t}=-2\mathrm{Ric}(g(t))$$

定义 1 如果 λ 分别是负数、零或正数,那么流 $g(t)$ 称为扩张的、平稳的或收缩的孤子. 反之,对于某个 $\sigma(t)$,ϕ_t 和 $g(0)$ 给定 Ricci 流 $g(t)=\sigma(t)\phi_t^*g(0)$,我们在 $t=0$ 时微分 $g(t)$ 可得 $g(0)$ 是方程(1.1)的解.

因此,我们有时简单的称:

定义 2 n 维完备 Riemann 流形 (M,g) 叫作 Ricci 孤子,如果对于某个光滑函数 f 和常数 λ

$$\mathrm{Ric}+\frac{1}{2}\mathscr{L}_Xg=\lambda g$$

其中 $\mathscr{L}_X$ 是 X 方向的 Lie 导数，Ric 是 Ricci 张量．如果 λ 分别是负数、零或正数，那么我们称 (M,g) 为扩张的、平稳的或收缩的．

定义 3　Ricci 孤子的矢量场如果能写成某个函数 $f:M\to R$ 的梯度，那么我们称该孤子为梯度 Ricci 孤子．

这时候我们计算可得 $\mathscr{L}_X g=2D^2 f$，其中 $D^2 f$ 是 f 的 Hessian 矩阵．于是我们得到孤子方程

$$\mathrm{Ric}+D^2 f=\lambda g \tag{1.2}$$

2. Ricci 孤子的意义

如果我们把 Ricci 流看成度量空间上某个矢量场产生的流，孤子可以看成这个流的吸引子．事实上，在二维球面，流收敛到一个孤子．考虑这种情形，对给定紧流形上的孤子的度量分类变得十分重要．对非紧流形上的孤子感兴趣是由于如下原因．我们试着通过定义一列重新伸缩后的解来理解 Ricci 流奇性的形成．假设 p_k,T_k 是一系列点，在这些时间点 $g(t)$ 的度量的曲率的范数取得极大值，而且当 $k\to\infty$ 时，$T_k\to\infty$．对于 k，重新伸缩后曲率的范数在点 p_k 有界

$$g_k(t)=\lambda_k g\left(T_k+\frac{t}{\lambda_k}\right)$$

重新伸缩和平移时间使得 $g_k(t)$ 是 Ricci 流的解，而且在零时间处曲率有界．在重新伸缩变换下，曲率最大范数和时间乘积的有界性保持不变．如果它有界，我们说奇性“快速形成”；如果它无界，我们说奇性“慢速形成”．在后一种情形，如果度量 $g_k(0)$ 有一致有界曲率，且当 $k\to\infty$ 收敛某个度量，这个度量是对于任意时间都是 Ricci 流的解．我们确信他们收敛于一个完备 Ricci 孤子．如果这是对的，我们可以用孤子来理解 Ricci 流的奇性．

3. Ricci 孤子的几个重要例子

(1) 爱因斯坦流形．

爱因斯坦流形 (M,g) 满足方程

$$\mathrm{Ric}=\lambda g$$

显然，它是梯度 Ricci 孤子在 f 等于常数时的特例．

(2) Hamilton 的雪茄孤子．

设 $M=\mathbf{R}^2$，$g=\rho^2(\mathrm{d}x^2+\mathrm{d}y^2)$．于是，我们得到 Gauss 曲率的表达式

$$K=-\frac{1}{\rho 2}\Delta\ln\rho$$

其中 $\Delta=\frac{\partial^2}{\partial x^2}+\frac{\partial^2}{\partial y^2}$，Ricci 曲率能由 Gauss 曲率写成 $\mathrm{Ric}(g)=Kg$．如果我们设 $\rho^2=$

$\frac{1}{1+x^2+y^2}$，可得到 $K=\frac{2}{1+x^2+y^2}$，也就是

$$\operatorname{Ric}(g)=\frac{2}{1+x^2+y^2}g$$

同时，我们定义 X 为径向矢量场，$Y=-2\left(x\frac{\partial}{\partial x}+y\frac{\partial}{\partial y}\right)$，于是可计算得到

$$\mathscr{L}_X g=-\frac{4}{1+x^2+y^2}g$$

所以，$(\mathbf{R}^2,g)$ 是平稳$(\lambda=0)$ 孤子.

我们也可以把 g 写成测地坐标形式

$$g=\mathrm{d}s^2+\tanh^2 s\mathrm{d}\theta^2$$

其中 s 是到原点的测地距离，θ 是极角. 这显示 M 在无穷远处像一个圆柱面，因此像一个雪茄. 在测地坐标系，曲率可写成

$$K=\frac{2}{\cosh^2 s}$$

因为 X 径向，所以雪茄也是梯度 Ricci 孤子. 实际上，我们可以取 $f=-2\ln\cosh s$.

(3)Bryant 孤子.

在 $\mathbf{R}^3$ 上，Bryant 发现了一个相似的旋转对称平稳梯度 Ricci 孤子. 与雪茄孤子在无穷远处像一个圆柱不同，Bryant 孤子无穷远处像一个抛物面. 它有正的截面曲率.

(4)Gauss 孤子.

这是最简单的一个梯度 Ricci 孤子. 设 $M=\mathbf{R}^n$，g 为平坦度量以及 $f=\frac{1}{2}|x|^2$.

于是，我们知 M 是收缩$(\lambda=1)$ 孤子.

4. 梯度 Ricci 孤子的基本方程

首先，我们回顾一些有关 Riemann 几何的著名结果.

(1)Schur 引理，当 $n>2$ 时

$$2\operatorname{div}\operatorname{Ric}=\mathrm{d}R$$

(2) 某个形式的协变微分交换公式

$$\nabla_{ij}^2\omega_k-\nabla_{ji}^2\omega_k=R_{ijks}\omega^s$$

(3)Riemann 曲率张量的分解

$$R_{ijkl}=\frac{1}{n-2}(R_{ik}g_{jl}+R_{jl}g_{ik}-R_{il}g_{jk}-R_{jk}g_{il}-\frac{R}{(n-1)(n-2)}(g_{ik}g_{jl}-g_{il}g_{jk}))+W_{ijkl}\tag{1.3}$$

(4) 事实上,当 $n \leqslant 3$ 时 Weyl 张量为零.

下面我们得到一些有关梯度 Ricci 孤子的基本公式.

命题 1 设(M,g)是梯度 Ricci 孤子,则下列公式成立

$$R + \Delta f = n\lambda \tag{1.4}$$

$$\nabla_i R = 2R_{ij} \nabla^j f \tag{1.5}$$

$$\nabla_j R_{ik} - \nabla_i R_{jk} = R_{ijks} \nabla^s f \tag{1.6}$$

$$R + |\nabla f|^2 - 2\lambda f = C(C\text{ 是常数}) \tag{1.7}$$

$$\Delta R = \langle \nabla R, \nabla f \rangle + 2\lambda R - 2|\mathrm{Ric}|^2 \tag{1.8}$$

$$\Delta R_{ij} = \langle \nabla R_{ij}, \nabla f \rangle + 2\lambda R_{ij} - 2R_{ikjs}R^{ks} \tag{1.9}$$

证明 方程(1.4):我们缩并孤子方程(1.2) 即可得.

方程(1.5):对 Ricci 张量求散度,并利用孤子方程(1.2) 可得

$$\begin{aligned}
\mathrm{div}\ \mathrm{Ric}_i &= g^{ik} \nabla_k R_{ij} \\
&= -g^{jk} \nabla_k \nabla_i \nabla_j f \\
&= -g^{ik} \nabla_i \nabla_k \nabla_j f - g^{jk} R_{kijs} \nabla^s f \\
&= -\nabla_i \Delta f - R_{is} \nabla^s f
\end{aligned}$$

计算中我们使用了 1 — 形式的协变微分交换公式. 再利用孤子方程(1.2) 和 Schur 引理,我们得到

$$\frac{1}{2} \nabla_i R = -\nabla_i (n\lambda - R) - R_{is} \nabla_s f = \nabla_i R - R_{is} \nabla^s f$$

因此,我们得到方程(1.5).

方程(1.6):通过和前面一个相似的计算可得.

方程(1.7):微分左边并利用方程(1.5) 和孤子方程(1.2) 可得.

方程(1.8):一旦获得方程(1.9),我们可以用度量 g 缩并它而获得方程(1.8).

方程(1.9):我们利用方程(1.6) 和第二 Bianchi 恒等式,可得

$$\begin{aligned}
\Delta R_{ik} = \nabla^j \nabla_j R_{ik} &= \nabla^j \nabla_i R_{ik} + \nabla^j R_{ijks} \nabla^s f + R_{ijks} \nabla^j \nabla^s f \\
&= \nabla_i \nabla^j R_{jk} + R^j_{ijs} R R^s_k + R^j_{iks} R^s_j + \\
&\quad \nabla_k R^j_{sij} \nabla^f - \nabla_s R^j_{kij} \nabla^s f + R_{ijks} \nabla^j \nabla^s f \\
&= \nabla_i \nabla^j + R_{is} R^s_k - \\
&\quad \nabla_k R_{si} \nabla^s f + \nabla_s R_{ki} \nabla^s f + R_{ijks} \nabla^j \nabla^s f \\
&= \frac{1}{2} \nabla_i \nabla_k R + R_{is} R^s_k + R^j_{iks} - \\
&\quad \nabla_k R_{si} \nabla^s f + \langle \nabla R_{ik}, \nabla f \rangle - R_{ijks} R^{js} + \lambda R_{ik} \\
&= \langle \nabla R_{ik}, \nabla f \rangle + \lambda R_{ik} - 2R_{ijks} R^{js} + R_{is} R^s_k +
\end{aligned}$$

$$\frac{1}{2}\nabla_k\nabla_i R-\nabla_k R_{is}\nabla^s f$$

其中,我们利用了 Schur 引理,利用孤子方程(1.2)替代 $\nabla^j\nabla^s f$,并重新安排了最后一行的某些项.

由微分方程(1.5),我们得到

$$0=\frac{1}{2}\nabla_k(\nabla_i R-2R_{is}\nabla^s f)$$

因此

$$\frac{1}{2}\nabla_k\nabla_i R-\nabla_k R_{is}\nabla^s f=R_{is}\nabla^s\nabla_k f$$

因此,我们得到

$$\begin{aligned}\Delta R_{ik}&=\langle\nabla R_{ik},\nabla f\rangle+\lambda R_{ik}-2R_{ijks}R^{js}+R_{is}R^s_k+R_{is}\nabla^s\nabla_k f\\&=\langle\nabla R_{ik},\nabla f\rangle+\lambda R_{ik}-2R_{ijks}R^{js}+R_{is}R^s_k+\lambda R_{is}-R_{is}R^s_k\\&=\langle\nabla R_{ik},\nabla f\rangle+2\lambda R_{ik}-2R_{ijks}R^{js}\end{aligned}$$

本来在工作室成立之初,我们雄心万丈要做中国的"斯普林格",怎奈一晃近二十年过去,岁月蹉跎,在现实中逐渐清醒,求存为上.古人云:"志不在温饱",但对笔者这般普通之人,似乎要去掉那个"不"字了.

刘培杰

2024 年 10 月 9 日

于哈工大

刘培杰数学工作室
已出版(即将出版)图书目录——原版影印

书　　名	出版时间	定　价	编号
数学物理大百科全书.第1卷(英文)	2016—01	418.00	508
数学物理大百科全书.第2卷(英文)	2016—01	408.00	509
数学物理大百科全书.第3卷(英文)	2016—01	396.00	510
数学物理大百科全书.第4卷(英文)	2016—01	408.00	511
数学物理大百科全书.第5卷(英文)	2016—01	368.00	512
zeta函数,q-zeta函数,相伴级数与积分(英文)	2015—08	88.00	513
微分形式:理论与练习(英文)	2015—08	58.00	514
离散与微分包含的逼近和优化(英文)	2015—08	58.00	515
艾伦·图灵:他的工作与影响(英文)	2016—01	98.00	560
测度理论概率导论,第2版(英文)	2016—01	88.00	561
带有潜在故障恢复系统的半马尔柯夫模型控制(英文)	2016—01	98.00	562
数学分析原理(英文)	2016—01	88.00	563
随机偏微分方程的有效动力学(英文)	2016—01	88.00	564
图的谱半径(英文)	2016—01	58.00	565
量子机器学习中数据挖掘的量子计算方法(英文)	2016—01	98.00	566
量子物理的非常规方法(英文)	2016—01	118.00	567
运输过程的统一非局部理论:广义波尔兹曼物理动力学,第2版(英文)	2016—01	198.00	568
量子力学与经典力学之间的联系在原子、分子及电动力学系统建模中的应用(英文)	2016—01	58.00	569
算术域(英文)	2018—01	158.00	821
高等数学竞赛:1962—1991年的米洛克斯·史怀哲竞赛(英文)	2018—01	128.00	822
用数学奥林匹克精神解决数论问题(英文)	2018—01	108.00	823
代数几何(德文)	2018—04	68.00	824
丢番图逼近论(英文)	2018—01	78.00	825
代数几何学基础教程(英文)	2018—01	98.00	826
解析数论入门课程(英文)	2018—01	78.00	827
数论中的丢番图问题(英文)	2018—01	78.00	829
数论(梦幻之旅):第五届中日数论研讨会演讲集(英文)	2018—01	68.00	830
数论新应用(英文)	2018—01	68.00	831
数论(英文)	2018—01	78.00	832

刘培杰数学工作室
已出版(即将出版)图书目录——原版影印

书　　名	出版时间	定　价	编号
湍流十讲(英文)	2018－04	108.00	886
无穷维李代数:第3版(英文)	2018－04	98.00	887
等值、不变量和对称性(英文)	2018－04	78.00	888
解析数论(英文)	2018－09	78.00	889
《数学原理》的演化:伯特兰・罗素撰写第二版时的手稿与笔记(英文)	2018－04	108.00	890
哈密尔顿数学论文集(第4卷):几何学、分析学、天文学、概率和有限差分等(英文)	2019－05	108.00	891
偏微分方程全局吸引子的特性(英文)	2018－09	108.00	979
整函数与下调和函数(英文)	2018－09	118.00	980
幂等分析(英文)	2018－09	118.00	981
李群,离散子群与不变量理论(英文)	2018－09	108.00	982
动力系统与统计力学(英文)	2018－09	118.00	983
表示论与动力系统(英文)	2018－09	118.00	984
分析学练习.第1部分(英文)	2021－01	88.00	1247
分析学练习.第2部分,非线性分析(英文)	2021－01	88.00	1248
初级统计学:循序渐进的方法:第10版(英文)	2019－05	68.00	1067
工程师与科学家微分方程用书:第4版(英文)	2019－07	58.00	1068
大学代数与三角学(英文)	2019－06	78.00	1069
培养数学能力的途径(英文)	2019－07	38.00	1070
工程师与科学家统计学:第4版(英文)	2019－06	58.00	1071
贸易与经济中的应用统计学:第6版(英文)	2019－06	58.00	1072
傅立叶级数和边值问题:第8版(英文)	2019－05	48.00	1073
通往天文学的途径:第5版(英文)	2019－05	58.00	1074
拉马努金笔记.第1卷(英文)	2019－06	165.00	1078
拉马努金笔记.第2卷(英文)	2019－06	165.00	1079
拉马努金笔记.第3卷(英文)	2019－06	165.00	1080
拉马努金笔记.第4卷(英文)	2019－06	165.00	1081
拉马努金笔记.第5卷(英文)	2019－06	165.00	1082
拉马努金遗失笔记.第1卷(英文)	2019－06	109.00	1083
拉马努金遗失笔记.第2卷(英文)	2019－06	109.00	1084
拉马努金遗失笔记.第3卷(英文)	2019－06	109.00	1085
拉马努金遗失笔记.第4卷(英文)	2019－06	109.00	1086
数论:1976年纽约洛克菲勒大学数论会议记录(英文)	2020－06	68.00	1145
数论:卡本代尔1979:1979年在南伊利诺伊卡本代尔大学举行的数论会议记录(英文)	2020－06	78.00	1146
数论:诺德韦克豪特1983:1983年在诺德韦克豪特举行的Journees Arithmetiques数论大会会议记录(英文)	2020－06	68.00	1147
数论:1985－1988年在纽约城市大学研究生院和大学中心举办的研讨会(英文)	2020－06	68.00	1148

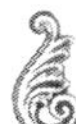

刘培杰数学工作室
已出版(即将出版)图书目录——原版影印

书名	出版时间	定价	编号
数论:1987年在乌尔姆举行的Journees Arithmetiques数论大会会议记录(英文)	2020—06	68.00	1149
数论:马德拉斯1987:1987年在马德拉斯安娜大学举行的国际拉马努金百年纪念大会会议记录(英文)	2020—06	68.00	1150
解析数论:1988年在东京举行的日法研讨会会议记录(英文)	2020—06	68.00	1151
解析数论:2002年在意大利切特拉罗举行的C. I. M. E. 暑期班演讲集(英文)	2020—06	68.00	1152
量子世界中的蝴蝶:最迷人的量子分形故事(英文)	2020—06	118.00	1157
走进量子力学(英文)	2020—06	118.00	1158
计算物理学概论(英文)	2020—06	48.00	1159
物质,空间和时间的理论:量子理论(英文)	2020—10	48.00	1160
物质,空间和时间的理论:经典理论(英文)	2020—10	48.00	1161
量子场理论:解释世界的神秘背景(英文)	2020—07	38.00	1162
计算物理学概论(英文)	2020—06	48.00	1163
行星状星云(英文)	2020—10	38.00	1164
基本宇宙学:从亚里士多德的宇宙到大爆炸(英文)	2020—08	58.00	1165
数学磁流体力学(英文)	2020—07	58.00	1166
计算科学:第1卷,计算的科学(日文)	2020—07	88.00	1167
计算科学:第2卷,计算与宇宙(日文)	2020—07	88.00	1168
计算科学:第3卷,计算与物质(日文)	2020—07	88.00	1169
计算科学:第4卷,计算与生命(日文)	2020—07	88.00	1170
计算科学:第5卷,计算与地球环境(日文)	2020—07	88.00	1171
计算科学:第6卷,计算与社会(日文)	2020—07	88.00	1172
计算科学.别卷,超级计算机(日文)	2020—07	88.00	1173
多复变函数论(日文)	2022—06	78.00	1518
复变函数入门(日文)	2022—06	78.00	1523
代数与数论:综合方法(英文)	2020—10	78.00	1185
复分析:现代函数理论第一课(英文)	2020—07	58.00	1186
斐波那契数列和卡特兰数:导论(英文)	2020—10	68.00	1187
组合推理:计数艺术介绍(英文)	2020—07	88.00	1188
二次互反律的傅里叶分析证明(英文)	2020—07	48.00	1189
旋瓦兹分布的希尔伯特变换与应用(英文)	2020—07	58.00	1190
泛函分析:巴拿赫空间理论入门(英文)	2020—07	48.00	1191
卡塔兰数入门(英文)	2019—05	68.00	1060
测度与积分(英文)	2019—04	68.00	1059
组合学手册.第一卷(英文)	2020—06	128.00	1153
—代数、局部紧群和巴拿赫—代数丛的表示.第一卷,群和代数的基本表示理论(英文)	2020—05	148.00	1154
电磁理论(英文)	2020—08	48.00	1193
连续介质力学中的非线性问题(英文)	2020—09	78.00	1195
多变量数学入门(英文)	2021—05	68.00	1317
偏微分方程入门(英文)	2021—05	88.00	1318
若尔当典范性:理论与实践(英文)	2021—07	68.00	1366
伽罗瓦理论.第4版(英文)	2021—08	88.00	1408
R统计学概论	2023—03	88.00	1614
基于不确定静态和动态问题解的仿射算术(英文)	2023—03	38.00	1618

刘培杰数学工作室

已出版(即将出版)图书目录——原版影印

书　　名	出版时间	定　价	编号
典型群,错排与素数(英文)	2020－11	58.00	1204
李代数的表示:通过 gln 进行介绍(英文)	2020－10	38.00	1205
实分析演讲集(英文)	2020－10	38.00	1206
现代分析及其应用的课程(英文)	2020－10	58.00	1207
运动中的抛射物数学(英文)	2020－10	38.00	1208
2－纽结与它们的群(英文)	2020－10	38.00	1209
概率,策略和选择:博弈与选举中的数学(英文)	2020－11	58.00	1210
分析学引论(英文)	2020－11	58.00	1211
量子群:通往流代数的路径(英文)	2020－11	38.00	1212
集合论入门(英文)	2020－10	48.00	1213
酉反射群(英文)	2020－11	58.00	1214
探索数学:吸引人的证明方式(英文)	2020－11	58.00	1215
微分拓扑短期课程(英文)	2020－10	48.00	1216
抽象凸分析(英文)	2020－11	68.00	1222
费马大定理笔记(英文)	2021－03	48.00	1223
高斯与雅可比和(英文)	2021－03	78.00	1224
π 与算术几何平均:关于解析数论和计算复杂性的研究(英文)	2021－01	58.00	1225
复分析入门(英文)	2021－03	48.00	1226
爱德华·卢卡斯与素性测定(英文)	2021－03	78.00	1227
通往凸分析及其应用的简单路径(英文)	2021－01	68.00	1229
微分几何的各个方面.第一卷(英文)	2021－01	58.00	1230
微分几何的各个方面.第二卷(英文)	2020－12	58.00	1231
微分几何的各个方面.第三卷(英文)	2020－12	58.00	1232
沃克流形几何学(英文)	2020－11	58.00	1233
彷射和韦尔几何应用(英文)	2020－12	58.00	1234
双曲几何学的旋转向量空间方法(英文)	2021－02	58.00	1235
积分:分析学的关键(英文)	2020－12	48.00	1236
为有天分的新生准备的分析学基础教材(英文)	2020－11	48.00	1237
数学不等式.第一卷.对称多项式不等式(英文)	2021－03	108.00	1273
数学不等式.第二卷.对称有理不等式与对称无理不等式(英文)	2021－03	108.00	1274
数学不等式.第三卷.循环不等式与非循环不等式(英文)	2021－03	108.00	1275
数学不等式.第四卷.Jensen 不等式的扩展与加细(英文)	2021－03	108.00	1276
数学不等式.第五卷.创建不等式与解不等式的其他方法(英文)	2021－04	108.00	1277

刘培杰数学工作室
已出版(即将出版)图书目录——原版影印

书　　名	出版时间	定　价	编号
冯・诺依曼代数中的谱位移函数:半有限冯・诺依曼代数中的谱位移函数与谱流(英文)	2021－06	98.00	1308
链接结构:关于嵌入完全图的直线中链接单形的组合结构(英文)	2021－05	58.00	1309
代数几何方法.第1卷(英文)	2021－06	68.00	1310
代数几何方法.第2卷(英文)	2021－06	68.00	1311
代数几何方法.第3卷(英文)	2021－06	58.00	1312

书　　名	出版时间	定　价	编号
代数、生物信息和机器人技术的算法问题.第四卷,独立恒等式系统(俄文)	2020－08	118.00	1199
代数、生物信息和机器人技术的算法问题.第五卷,相对覆盖性和独立可拆分恒等式系统(俄文)	2020－08	118.00	1200
代数、生物信息和机器人技术的算法问题.第六卷,恒等式和准恒等式的相等 问题、可推导性和可实现性(俄文)	2020－08	128.00	1201
分数阶微积分的应用:非局部动态过程,分数阶导热系数(俄文)	2021－01	68.00	1241
泛函分析问题与练习:第2版(俄文)	2021－01	98.00	1242
集合论、数学逻辑和算法论问题:第5版(俄文)	2021－01	98.00	1243
微分几何和拓扑短期课程(俄文)	2021－01	98.00	1244
素数规律(俄文)	2021－01	88.00	1245
无穷边值问题解的递减:无界域中的拟线性椭圆和抛物方程(俄文)	2021－01	48.00	1246
微分几何讲义(俄文)	2020－12	98.00	1253
二次型和矩阵(俄文)	2021－01	98.00	1255
积分和级数.第2卷,特殊函数(俄文)	2021－01	168.00	1258
积分和级数.第3卷,特殊函数补充:第2版(俄文)	2021－01	178.00	1264
几何图上的微分方程(俄文)	2021－01	138.00	1259
数论教程:第2版(俄文)	2021－01	98.00	1260
非阿基米德分析及其应用(俄文)	2021－03	98.00	1261
古典群和量子群的压缩(俄文)	2021－03	98.00	1263
数学分析习题集.第3卷,多元函数:第3版(俄文)	2021－03	98.00	1266
数学习题:乌拉尔国立大学数学力学系大学生奥林匹克(俄文)	2021－03	98.00	1267
柯西定理和微分方程的特解(俄文)	2021－03	98.00	1268
组合极值问题及其应用:第3版(俄文)	2021－03	98.00	1269
数学词典(俄文)	2021－01	98.00	1271
确定性混沌分析模型(俄文)	2021－06	168.00	1307
精选初等数学习题和定理.立体几何.第3版(俄文)	2021－03	68.00	1316
微分几何习题:第3版(俄文)	2021－05	98.00	1336
精选初等数学习题和定理.平面几何.第4版(俄文)	2021－05	68.00	1335
曲面理论在欧氏空间 En 中的直接表示(俄文)	2022－01	68.00	1444
维纳－霍普夫离散算子和托普利兹算子:某些可数赋范空间中的诺特性和可逆性(俄文)	2022－03	108.00	1496
Maple中的数论:数论中的计算机计算(俄文)	2022－03	88.00	1497
贝尔曼和克努特问题及其概括:加法运算的复杂性(俄文)	2022－03	138.00	1498

刘培杰数学工作室

已出版(即将出版)图书目录——原版影印

书　　名	出版时间	定　价	编号
复分析:共形映射(俄文)	2022—07	48.00	1542
微积分代数样条和多项式及其在数值方法中的应用(俄文)	2022—08	128.00	1543
蒙特卡罗方法中的随机过程和场模型:算法和应用(俄文)	2022—08	88.00	1544
线性椭圆型方程组:论二阶椭圆型方程的迪利克雷问题(俄文)	2022—08	98.00	1561
动态系统解的增长特性:估值、稳定性、应用(俄文)	2022—08	118.00	1565
群的自由积分解:建立和应用(俄文)	2022—08	78.00	1570
混合方程和偏差自变数方程问题:解的存在和唯一性(俄文)	2023—01	78.00	1582
拟度量空间分析:存在和逼近定理(俄文)	2023—01	108.00	1583
二维和三维流形上函数的拓扑性质:函数的拓扑分类(俄文)	2023—03	68.00	1584
齐次马尔科夫过程建模的矩阵方法:此类方法能够用于不同目上的的复杂系统研究、设计和完善(俄文)	2023—03	68.00	1594
周期函数的近似方法和特性:特殊课程(俄文)	2023—04	158.00	1622
扩散方程解的矩函数:变分法(俄文)	2023—03	58.00	1623
多赋范空间和广义函数:理论及应用(俄文)	2023—03	98.00	1632
分析中的多值映射:部分应用(俄文)	2023—06	98.00	1634
数学物理问题(俄文)	2023—03	78.00	1636
函数的幂级数与三角级数分解(俄文)	2024—01	58.00	1695
星体理论的数学基础:原子三元组(俄文)	2024—01	98.00	1696
素数规律:专著(俄文)	2024—01	118.00	1697
狭义相对论与广义相对论:时空与引力导论(英文)	2021—07	88.00	1319
束流物理学和粒子加速器的实践介绍:第2版(英文)	2021—07	88.00	1320
凝聚态物理中的拓扑和微分几何简介(英文)	2021—05	88.00	1321
混沌映射:动力学、分形学和快速涨落(英文)	2021—05	128.00	1322
广义相对论:黑洞、引力波和宇宙学介绍(英文)	2021—06	68.00	1323
现代分析电磁均质化(英文)	2021—06	68.00	1324
为科学家提供的基本流体动力学(英文)	2021—06	88.00	1325
视觉天文学:理解夜空的指南(英文)	2021—06	68.00	1326
物理学中的计算方法(英文)	2021—06	68.00	1327
单星的结构与演化:导论(英文)	2021—06	108.00	1328
超越居里:1903年至1963年物理界四位女性及其著名发现(英文)	2021—06	68.00	1329
范德瓦尔斯流体热力学的进展(英文)	2021—06	68.00	1330
先进的托卡马克稳定性理论(英文)	2021—06	88.00	1331
经典场论导论:基本相互作用的过程(英文)	2021—07	88.00	1332
光致电离量子动力学方法原理(英文)	2021—07	108.00	1333
经典域论和应力:能量张量(英文)	2021—05	88.00	1334
非线性太赫兹光谱的概念与应用(英文)	2021—06	68.00	1337
电磁学中的无穷空间并矢格林函数(英文)	2021—06	88.00	1338
物理科学基础数学.第1卷,齐次边值问题、傅里叶方法和特殊函数(英文)	2021—07	108.00	1339
离散量子力学(英文)	2021—07	68.00	1340
核磁共振的物理学和数学(英文)	2021—07	108.00	1341
分子水平的静电学(英文)	2021—08	68.00	1342
非线性波:理论、计算机模拟、实验(英文)	2021—06	108.00	1343
石墨烯光学:经典问题的电解解决方案(英文)	2021—06	68.00	1344
超材料多元宇宙(英文)	2021—07	68.00	1345
银河系外的天体物理学(英文)	2021—07	68.00	1346
原子物理学(英文)	2021—07	68.00	1347
将光打结:将拓扑学应用于光学(英文)	2021—07	68.00	1348
电磁学:问题与解法(英文)	2021—07	88.00	1364
海浪的原理:介绍量子力学的技巧与应用(英文)	2021—07	108.00	1365

刘培杰数学工作室
已出版(即将出版)图书目录——原版影印

书　名	出版时间	定　价	编号
多孔介质中的流体:输运与相变(英文)	2021—07	68.00	1372
洛伦兹群的物理学(英文)	2021—08	68.00	1373
物理导论的数学方法和解决方法手册(英文)	2021—08	68.00	1374
非线性波数学物理学入门(英文)	2021—08	88.00	1376
波:基本原理和动力学(英文)	2021—07	68.00	1377
光电子量子计量学.第1卷,基础(英文)	2021—07	88.00	1383
光电子量子计量学.第2卷,应用与进展(英文)	2021—07	68.00	1384
复杂流的格子玻尔兹曼建模的工程应用(英文)	2021—08	68.00	1393
电偶极矩挑战(英文)	2021—08	108.00	1394
电动力学:问题与解法(英文)	2021—09	68.00	1395
自由电子激光的经典理论(英文)	2021—08	68.00	1397
曼哈顿计划——核武器物理学简介(英文)	2021—09	68.00	1401
粒子物理学(英文)	2021—09	68.00	1402
引力场中的量子信息(英文)	2021—09	128.00	1403
器件物理学的基本经典力学(英文)	2021—09	68.00	1404
等离子体物理及其空间应用导论.第1卷,基本原理和初步过程(英文)	2021—09	68.00	1405
磁约束聚变等离子体物理:理想MHD理论(英文)	2023—03	68.00	1613
相对论量子场论.第1卷,典范形式体系(英文)	2023—03	38.00	1615
相对论量子场论.第2卷,路径积分形式(英文)	2023—06	38.00	1616
相对论量子场论.第3卷,量子场论的应用(英文)	2023—06	38.00	1617
涌现的物理学(英文)	2023—05	58.00	1619
量子化旋涡:一本拓扑激发手册(英文)	2023—04	68.00	1620
非线性动力学:实践的介绍性调查(英文)	2023—05	68.00	1621
静电加速器:一个多功能工具(英文)	2023—06	58.00	1625
相对论多体理论与统计力学(英文)	2023—06	58.00	1626
经典力学.第1卷,工具与向量(英文)	2023—04	38.00	1627
经典力学.第2卷,运动学和匀加速运动(英文)	2023—04	58.00	1628
经典力学.第3卷,牛顿定律和匀速圆周运动(英文)	2023—04	58.00	1629
经典力学.第4卷,万有引力定律(英文)	2023—04	38.00	1630
经典力学.第5卷,守恒定律与旋转运动(英文)	2023—04	38.00	1631
对称问题:纳维尔—斯托克斯问题(英文)	2023—04	38.00	1638
摄影的物理和艺术.第1卷,几何与光的本质(英文)	2023—04	78.00	1639
摄影的物理和艺术.第2卷,能量与色彩(英文)	2023—04	78.00	1640
摄影的物理和艺术.第3卷,探测器与数码的意义(英文)	2023—04	78.00	1641
拓扑与超弦理论焦点问题(英文)	2021—07	58.00	1349
应用数学:理论、方法与实践(英文)	2021—07	78.00	1350
非线性特征值问题:牛顿型方法与非线性瑞利函数(英文)	2021—07	58.00	1351
广义膨胀和齐性:利用齐性构造齐次系统的李雅普诺夫函数和控制律(英文)	2021—06	48.00	1352
解析数论焦点问题(英文)	2021—07	58.00	1353
随机微分方程:动态系统方法(英文)	2021—07	58.00	1354
经典力学与微分几何(英文)	2021—07	58.00	1355
负定相交形式流形上的瞬子模空间几何(英文)	2021—07	68.00	1356
广义卡塔兰轨道分析:广义卡塔兰轨道计算数字的方法(英文)	2021—07	48.00	1367
洛伦兹方法的变分:二维与三维洛伦兹方法(英文)	2021—08	38.00	1378
几何、分析和数论精编(英文)	2021—08	68.00	1380
从一个新角度看数论:通过遗传方法引入现实的概念(英文)	2021—07	58.00	1387
动力系统:短期课程(英文)	2021—08	68.00	1382
几何路径:理论与实践(英文)	2021—08	48.00	1385

刘培杰数学工作室
已出版(即将出版)图书目录——原版影印

书　名	出版时间	定　价	编号
论天体力学中某些问题的不可积性(英文)	2021—07	88.00	1396
广义斐波那契数列及其性质(英文)	2021—08	38.00	1386
对称函数和麦克唐纳多项式:余代数结构与 Kawanaka 恒等式(英文)	2021—09	38.00	1400
杰弗里·英格拉姆·泰勒科学论文集:第 1 卷.固体力学(英文)	2021—05	78.00	1360
杰弗里·英格拉姆·泰勒科学论文集:第 2 卷.气象学、海洋学和湍流(英文)	2021—05	68.00	1361
杰弗里·英格拉姆·泰勒科学论文集:第 3 卷.空气动力学以及落弹数和爆炸的力学(英文)	2021—05	68.00	1362
杰弗里·英格拉姆·泰勒科学论文集:第 4 卷.有关流体力学(英文)	2021—05	58.00	1363
非局域泛函演化方程:积分与分数阶(英文)	2021—08	48.00	1390
理论工作者的高等微分几何:纤维丛、射流流形和拉格朗日理论(英文)	2021—08	68.00	1391
半线性退化椭圆微分方程:局部定理与整体定理(英文)	2021—07	48.00	1392
非交换几何、规范理论和重整化:一般简介与非交换量子场论的重整化(英文)	2021—09	78.00	1406
数论论文集:拉普拉斯变换和带有数论系数的幂级数(俄文)	2021—09	48.00	1407
挠理论专题:相对极大值,单射与扩充模(英文)	2021—09	88.00	1410
强正则图与欧几里得若尔当代数:非通常关系中的启示(英文)	2021—10	48.00	1411
拉格朗日几何和哈密顿几何:力学的应用(英文)	2021—10	48.00	1412
时滞微分方程与差分方程的振动理论:二阶与三阶(英文)	2021—10	98.00	1417
卷积结构与几何函数理论:用以研究特定几何函数理论方向的分数阶微积分算子与卷积结构(英文)	2021—10	48.00	1418
经典数学物理的历史发展(英文)	2021—10	78.00	1419
扩展线性丢番图问题(英文)	2021—10	38.00	1420
一类混沌动力系统的分歧分析与控制:分歧分析与控制(英文)	2021—11	38.00	1421
伽利略空间和伪伽利略空间中一些特殊曲线的几何性质(英文)	2022—01	68.00	1422
一阶偏微分方程:哈密尔顿—雅可比理论(英文)	2021—11	48.00	1424
各向异性黎曼多面体的反问题:分段光滑的各向异性黎曼多面体反边界谱问题:唯一性(英文)	2021—11	38.00	1425
项目反应理论手册.第一卷,模型(英文)	2021—11	138.00	1431
项目反应理论手册.第二卷,统计工具(英文)	2021—11	118.00	1432
项目反应理论手册.第三卷,应用(英文)	2021—11	138.00	1433
二次无理数:经典数论入门(英文)	2022—05	138.00	1434

刘培杰数学工作室
已出版(即将出版)图书目录——原版影印

书　　名	出版时间	定　价	编号
数,形与对称性:数论,几何和群论导论(英文)	2022—05	128.00	1435
有限域手册(英文)	2021—11	178.00	1436
计算数论(英文)	2021—11	148.00	1437
拟群与其表示简介(英文)	2021—11	88.00	1438
数论与密码学导论:第二版(英文)	2022—01	148.00	1423
几何分析中的柯西变换与黎兹变换:解析调和容量和李普希兹调和容量、变化和振荡以及一致可求长性(英文)	2021—12	38.00	1465
近似不动点定理及其应用(英文)	2022—05	28.00	1466
局部域的相关内容解析:对局部域的扩展及其伽罗瓦群的研究(英文)	2022—01	38.00	1467
反问题的二进制恢复方法(英文)	2022—03	28.00	1468
对几何函数中某些类的各个方面的研究:复变量理论(英文)	2022—01	38.00	1469
覆盖、对应和非交换几何(英文)	2022—01	28.00	1470
最优控制理论中的随机线性调节器问题:随机最优线性调节器问题(英文)	2022—01	38.00	1473
正交分解法:涡流流体动力学应用的正交分解法(英文)	2022—01	38.00	1475
芬斯勒几何的某些问题(英文)	2022—03	38.00	1476
受限三体问题(英文)	2022—05	38.00	1477
利用马利亚万微积分进行 Greeks 的计算:连续过程、跳跃过程中的马利亚万微积分和金融领域中的 Greeks(英文)	2022—05	48.00	1478
经典分析和泛函分析的应用:分析学的应用(英文)	2022—03	38.00	1479
特殊芬斯勒空间的探究(英文)	2022—03	48.00	1480
某些图形的施泰纳距离的细谷多项式:细谷多项式与图的维纳指数(英文)	2022—05	38.00	1481
图论问题的遗传算法:在新鲜与模糊的环境中(英文)	2022—05	48.00	1482
多项式映射的渐近簇(英文)	2022—05	38.00	1483
一维系统中的混沌:符号动力学,映射序列,一致收敛和沙可夫斯基定理(英文)	2022—05	38.00	1509
多维边界层流动与传热分析:粘性流体流动的数学建模与分析(英文)	2022—05	38.00	1510
演绎理论物理学的原理:一种基于量子力学波函数的逐次置信估计的一般理论的提议(英文)	2022—05	38.00	1511
R^2 和 R^3 中的仿射弹性曲线:概念和方法(英文)	2022—08	38.00	1512
算术数列中除数函数的分布:基本内容、调查、方法、第二矩、新结果(英文)	2022—05	28.00	1513
抛物型狄拉克算子和薛定谔方程:不定常薛定谔方程的抛物型狄拉克算子及其应用(英文)	2022—07	28.00	1514
黎曼-希尔伯特问题与量子场论:可积重正化、戴森-施温格方程(英文)	2022—08	38.00	1515
代数结构和几何结构的形变理论(英文)	2022—08	48.00	1516
概率结构和模糊结构上的不动点:概率结构和直觉模糊度量空间的不动点定理(英文)	2022—08	38.00	1517

刘培杰数学工作室
已出版(即将出版)图书目录——原版影印

书　　名	出版时间	定　价	编号
反若尔当对:简单反若尔当对的自同构(英文)	2022—07	28.00	1533
对某些黎曼—芬斯勒空间变换的研究:芬斯勒几何中的某些变换(英文)	2022—07	38.00	1534
内诣零流形映射的尼尔森数的阿诺索夫关系(英文)	2023—01	38.00	1535
与广义积分变换有关的分数次演算:对分数次演算的研究(英文)	2023—01	48.00	1536
强子的芬斯勒几何和吕拉几何(宇宙学方面):强子结构的芬斯勒几何和吕拉几何(拓扑缺陷)(英文)	2022—08	38.00	1537
一种基于混沌的非线性最优化问题:作业调度问题(英文)	2023—03	38.00	1538
广义概率论发展前景:关于趣味数学与置信函数实际应用的一些原创观点(英文)	2023—03	48.00	1539
纽结与物理学:第二版(英文)	2022—09	118.00	1547
正交多项式和q—级数的前沿(英文)	2022—09	98.00	1548
算子理论问题集(英文)	2022—09	108.00	1549
抽象代数:群、环与域的应用导论:第二版(英文)	2023—01	98.00	1550
菲尔兹奖得主演讲集:第三版(英文)	2023—01	138.00	1551
多元实函数教程(英文)	2022—09	118.00	1552
球面空间形式群的几何学:第二版(英文)	2022—09	98.00	1566
对称群的表示论(英文)	2023—01	98.00	1585
纽结理论:第二版(英文)	2023—01	88.00	1586
拟群理论的基础与应用(英文)	2023—01	88.00	1587
组合学:第二版(英文)	2023—01	98.00	1588
加性组合学:研究问题手册(英文)	2023—01	68.00	1589
扭曲、平铺与镶嵌:几何折纸中的数学方法(英文)	2023—01	98.00	1590
离散与计算几何手册:第三版(英文)	2023—01	248.00	1591
离散与组合数学手册:第二版(英文)	2023—01	248.00	1592
分析学教程.第1卷,一元实变量函数的微积分分析学介绍(英文)	2023—01	118.00	1595
分析学教程.第2卷,多元函数的微分和积分,向量微积分(英文)	2023—01	118.00	1596
分析学教程.第3卷,测度与积分理论,复变量的复值函数(英文)	2023—01	118.00	1597
分析学教程.第4卷,傅里叶分析,常微分方程,变分法(英文)	2023—01	118.00	1598

刘培杰数学工作室
已出版(即将出版)图书目录——原版影印

书　　名	出版时间	定　价	编号
共形映射及其应用手册(英文)	2024－01	158.00	1674
广义三角函数与双曲函数(英文)	2024－01	78.00	1675
振动与波:概论:第二版(英文)	2024－01	88.00	1676
几何约束系统原理手册(英文)	2024－01	120.00	1677
微分方程与包含的拓扑方法(英文)	2024－01	98.00	1678
数学分析中的前沿话题(英文)	2024－01	198.00	1679
流体力学建模:不稳定性与湍流(英文)	2024－03	88.00	1680
动力系统:理论与应用(英文)	2024－03	108.00	1711
空间统计学理论:概述(英文)	2024－03	68.00	1712
梅林变换手册(英文)	2024－03	128.00	1713
非线性系统及其绝妙的数学结构.第1卷(英文)	2024－03	88.00	1714
非线性系统及其绝妙的数学结构.第2卷(英文)	2024－03	108.00	1715
Chip-firing 中的数学(英文)	2024－04	88.00	1716
阿贝尔群的可确定性:问题、研究、概述(俄文)	2024－05	716.00(全7册)	1727
素数规律:专著(俄文)	2024－05	716.00(全7册)	1728
函数的幂级数与三角级数分解(俄文)	2024－05	716.00(全7册)	1729
星体理论的数学基础:原子三元组(俄文)	2024－05	716.00(全7册)	1730
技术问题中的数学物理微分方程(俄文)	2024－05	716.00(全7册)	1731
概率论边界问题:随机过程边界穿越问题(俄文)	2024－05	716.00(全7册)	1732
代数和幂等配置的正交分解:不可交换组合(俄文)	2024－05	716.00(全7册)	1733
数学物理精选专题讲座:李理论的进一步应用	2024－10	252.00(全4册)	1775
工程师和科学家应用数学概论:第二版	2024－10	252.00(全4册)	1775
高等微积分快速入门	2024－10	252.00(全4册)	1775
微分几何的各个方面.第四卷	2024－10	252.00(全4册)	1775
具有连续变量的量子信息形式主义概论	2024－10	378.00(全6册)	1776
拓扑绝缘体	2024－10	378.00(全6册)	1776
论全息度量原则:从大学物理到黑洞热力学	2024－10	378.00(全6册)	1776
量化测量:无所不在的数字	2024－10	378.00(全6册)	1776
21 世纪的彗星:体验下一颗伟大彗星的个人指南	2024－10	378.00(全6册)	1776
激光及其在玻色－爱因斯坦凝聚态观测中的应用	2024－10	378.00(全6册)	1776

联系地址:哈尔滨市南岗区复华四道街10号　哈尔滨工业大学出版社刘培杰数学工作室
邮　　编:150006
联系电话:0451－86281378　　13904613167
E-mail:lpj1378@163.com